Global Tectonics

THE LATE PHILIP KEAREY
Formerly of the Department of Geology
University of Bristol
UK

KEITH A. KLEPEIS
Department of Geology
University of Vermont
Burlington, Vermont, USA

FREDERICK J. VINE
School of Environmental Sciences
University of East Anglia
Norwich, UK

THIRD EDITION

WILEY-BLACKWELL

A John Wiley & Sons, Ltd., Publication

This edition first published 2009, © 2009 by Philip Kearey, Keith A. Klepeis, Frederick J. Vine

Blackwell Publishing was acquired by John Wiley & Sons in February 2007. Blackwell's publishing program has been merged with Wiley's global Scientific, Technical and Medical business to form Wiley-Blackwell.

Registered office: John Wiley & Sons Ltd, The Atrium, Southern Gate, Chichester, West Sussex, PO19 8SQ, UK

Editorial offices: 9600 Garsington Road, Oxford, OX4 2DQ, UK
      The Atrium, Southern Gate, Chichester, West Sussex, PO19 8SQ, UK
      111 River Street, Hoboken, NJ 07030-5774, USA

For details of our global editorial offices, for customer services and for information about how to apply for permission to reuse the copyright material in this book please see our website at www.wiley.com/wiley-blackwell

Library of Congress Cataloguing-in-Publication Data

Kearey, P.
    Global tectonics. – 3rd ed. / Philip Kearey, Keith A. Klepeis, Frederick J. Vine
      p. cm.
    Includes bibliographical references and index.
    ISBN 978-1-4051-0777-8 (pbk. : alk. paper)   1. Plate tectonics–Textbooks.   I. Klepeis, Keith A.   II. Vine,
F. J.   III. Title.

    QE511.4.K43 2008
    551.1'36–dc22

                                        2007020963

A catalogue record for this book is available from the British Library.

Set in 9.5 on 11.5 pt Dante by SNP Best-set Typesetters Ltd., Hong Kong
Printed and bound in the United States of America by Sheridan Books, Inc.

8 2015

Contents

8   Continental transforms and strike-slip faults   210

9   Subduction zones   249

10   Orogenic belts   286

11 Precambrian tectonics and the supercontinent cycle

12 The mechanism of plate tectonics

13 Implications of plate tectonics

Color plates appear between pages 244 and 245

A companion resources website for this book is available at www.blackwellpublishing.com/kearey

Preface

As is well known, the study of tectonics, the branch of geology dealing with large-scale Earth structures and their deformation, experienced a major breakthrough in the 1960s with the formulation of plate tectonics. The simultaneous confirmation of sea floor spreading and continental drift, together with the recognition of transform faults and subduction zones, derived from the interpretation of new and improved data from the fields of marine geology and geophysics, and earthquake seismology. By 1970 the essentials of plate tectonics – the extent of plates, the nature of the plate boundaries, and the geometry and kinematics of their relative and finite motions – were well documented.

As further details emerged, it soon became apparent that plates and plate boundaries are well-defined in oceanic areas, where the plates are young, relatively thin, but rigid, and structurally rather uniform, but that this is not true for continental areas. Where plates have continental crust embedded in them they are generally thicker, older and structurally more complex than oceanic plates. Moreover the continental crust itself is relatively weak and deforms more readily by fracture and even by flow. Thus the nature of continental tectonics is more complex than a simple application of plate tectonic theory would predict and it has taken much longer to document and interpret. An important element in this has been the advent of Global Positioning data that have revealed details of the deformation field in complex areas.

The other major aspect of plate tectonics in which progress initially was slow is the driving mechanism for plate motions. Significant progress here had to await the development of new seismologic techniques and advances in laboratory and computer modeling of convection in the Earth's mantle.

Since 1990, when the first edition of Global Tectonics appeared, there have been many developments in our understanding of Earth structure and its formation, particularly in relation to continental tectonics and mantle convection. As a consequence, approximately two-thirds of the figures and two-thirds of the text in this third edition are new. The structure of the book is largely unchanged. The order in which data and ideas are presented is in part historical, which may be of some interest in itself, but it has the advantage of moving from simple to more complex concepts, from the recent to the distant past, and from the oceanic to the continental realms.

Thus one moves from consideration of the fundamentals of plate tectonics, which are best illustrated with reference to the ocean basins, to continental tectonics, culminating in Precambrian tectonics, and a discussion of the possible nature of the implied convection in the mantle.

The book is aimed at senior undergraduate students in the geological sciences and postgraduate students and other geoscientists who wish to gain an insight into the subject. We assume a basic knowledge of geology, and that for a full description of geophysical and geochemical methodology it will be necessary to refer to other texts. We have attempted to provide insights into the trends of modern research and the problems still outstanding, and have supplied a comprehensive list of references so that the reader can follow up any item of particular interest. We have included a list of questions for the use of tutors in assessing the achievement of their students in courses based on the book. These are mainly designed to probe the students' integrative powers, but we hope that in their answers students will make use of the references given in the text and material on relevant websites listed on the book's website at: http://www.blackwellpublishing.com/kearey

The initial impact of the plate tectonic concept, in the fields of marine geology and geophysics and seismology, was quickly followed by the realization of its relevance to igneous and metamorphic petrology, paleontology, sedimentary and economic geology, and all branches of goescience. More recently its potential relevance to the Earth system as a whole has been recognized. In the past, processes associated with plate tectonics may have produced changes in seawater and atmospheric chemistry, in sea level and ocean currents, and in the Earth's climate. These ideas are briefly reviewed in an extended final chapter on the implications of plate tectonics. This extension of the relevance of plate tectonics to the atmosphere and oceans, to the evolution of life, and possibly even the origin of life on Earth is particularly gratifying in that it emphasizes the way in which the biosphere, atmosphere, hydrosphere, and solid Earth are interrelated in a single, dynamic Earth system.

K.A. KLEPEIS
F.J. VINE

A companion resources website for this book is available at http://www.blackwellpublishing.com/kearey

Acknowledgments

The first two editions of Global Tectonics were largely written by Phil Kearey. Tragically Phil died, suddenly, in 2003 at the age of 55, just after starting work on a third edition. We are indebted to his wife, Jane, for encouraging us to complete a third edition. Phil had a particular gift for writing succinct and accessible accounts of often difficult concepts, which generations of students have been thankful for. We are very conscious of the fact that our best efforts to emulate his style have often fallen short.

We thank Cynthia Ebinger, John Hopper, John Oldow, and Peter Cawood for providing thoughtful reviews of the original manuscript. Ian Bastow, José Cembrano, Ron Clowes, Barry Doolan, Mian Liu, Phil Hammer, and Brendan Meade provided helpful comments on specific aspects of some chapters. KAK wishes to thank Gabriela Mora-Klepeis for her excellent research assistance and Pam and Dave Miller for their support.

K.A.K.
F.J.V.

Era	Period	Epoch		*Ma
Cenozoic	Neogene	Pleistocene		1.81
		Pliocene	Late	3.60
			Early	5.33
		Miocene	Late	11.61
			Middle	15.97
			Early	23.03
	Paleogene	Oligocene	Late	28.4
			Early	33.9
		Eocene	Late	37.2
			Middle	48.6
			Early	55.8
		Paleocene	Late	58.7
			Middle	61.7
			Early	65.5
Mesozoic	Cretaceous	Late		99.6
		Early		145.5
	Jurassic	Late		161.2
		Middle		175.6
		Early		199.6
	Triassic	Late		228.0
		Middle		245.0
		Early		251.0

Continued

Era	Period	Epoch	*Ma
Paleozoic	Permian	Late	260.4
		Middle	270.6
		Early	299.0
	Carboniferous	Pennsylvanian	318.1
		Mississippian	359.2
	Devonian	Late	385.3
		Middle	397.5
		Early	416.0
	Silurian	Late	422.9
		Early	443.7
	Ordovician	Late	460.9
		Middle	471.8
		Early	488.3
	Cambrian	Late	501.0
		Middle	513.0
		Early	542.0
	(Eon)	**(Era)**	
Precambrian	Proterozoic	Late	1000
		Middle	1600
		Early	2500
	Archean	Late	2800
		Middle	3200
		Early	3600
		Eoarchean	~4600

*Age, in millions of years (Ma), based on the timescale of Gradstein *et al.* (2004)

1 | Historical perspective

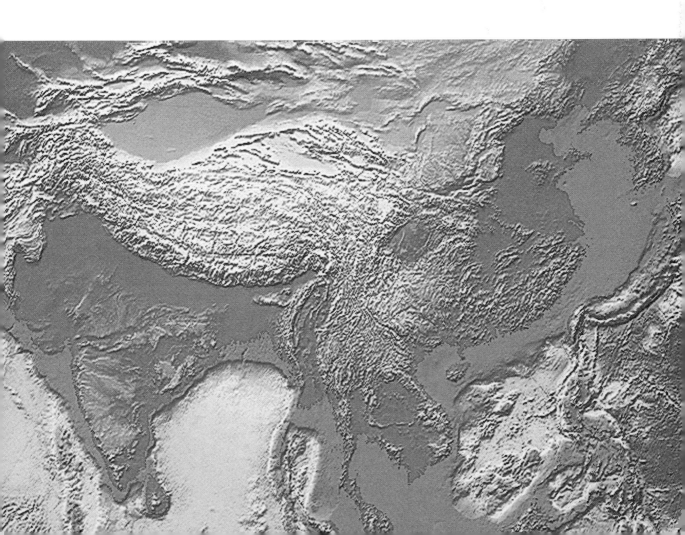

1.1 CONTINENTAL DRIFT

Although the theory of the new global tectonics, or plate tectonics, has largely been developed since 1967, the history of ideas concerning a mobilist view of the Earth extends back considerably longer (Rupke, 1970; Hallam, 1973a; Vine, 1977; Frankel, 1988). Ever since the coastlines of the continents around the Atlantic Ocean were first charted, people have been intrigued by the similarity of the coastlines of the Americas and of Europe and Africa. Possibly the first to note the similarity and suggest an ancient separation was Abraham Ortelius in 1596 (Romm, 1994). In 1620, Francis Bacon, in his *Novum Organum*, commented on the similar form of the west coasts of Africa and South America: that is, the *Atlantic* coast of Africa and the *Pacific* coast of South America. He also noted the similar configurations of the New and Old World, "both of which are broad and extended towards the north, narrow and pointed towards the south." Perhaps because of these observations, for there appear to be no others, Bacon is often erroneously credited with having been first to notice the similarity or "fit" of the Atlantic coastlines of South America and Africa and even with having suggested that they were once together and had drifted apart. In 1668, François Placet, a French prior, related the separation of the Americas to the Flood of Noah. Noting from the Bible that prior to the flood the Earth was one and undivided, he postulated that the Americas were formed by the conjunction of floating islands or separated from Europe and Africa by the destruction of an intervening landmass, "Atlantis." One must remember, of course, that during the 17th and 18th centuries geology, like most sciences, was carried out by clerics and theologians who felt that their observations, such as the occurrence of marine fossils and water-lain sediments on high land, were explicable in terms of the Flood and other biblical catastrophes.

Another person to note the fit of the Atlantic coastlines of South America and Africa and to suggest that they might once have been side by side was Theodor Christoph Lilienthal, Professor of Theology at Königsberg in Germany. In a work dated 1756 he too related their separation to biblical catastrophism, drawing on the text, "in the days of Peleg, the earth was divided." In papers dated 1801 and 1845, the German explorer

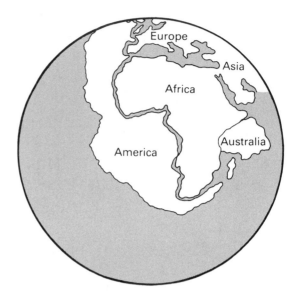

Figure 1.1 *Snider's reconstruction of the continents (Snider, 1858).*

Alexander von Humbolt noted the geometric and geologic similarities of the opposing shores of the Atlantic, but he too speculated that the Atlantic was formed by a catastrophic event, this time "a flow of eddying waters . . . directed first towards the north-east, then towards the north-west, and back again to the north-east . . . What we call the Atlantic Ocean is nothing else than a valley scooped out by the sea." In 1858 an American, Antonio Snider, made the same observations but postulated "drift" and related it to "multiple catastrophism" – the Flood being the last major catastrophe. Thus Snider suggested drift *sensu stricto*, and he even went so far as to suggest a pre-drift reconstruction (Fig. 1.1).

The 19th century saw the gradual replacement of the concept of catastrophism by that of "uniformitarianism" or "actualism" as propounded by the British geologists James Hutton and Charles Lyell. Hutton wrote "No powers are to be employed that are not natural to the globe, no action to be admitted of except those of which we know the principle, and no extraordinary events to be alleged in order to explain a common appearance." This is usually stated in Archibald Geikie's paraphrase of Hutton's words, "the present is the key to the past," that is, the slow processes going on at and beneath the Earth's surface today have been going on throughout geologic time and have shaped the surface record. Despite this change in the basis of geologic

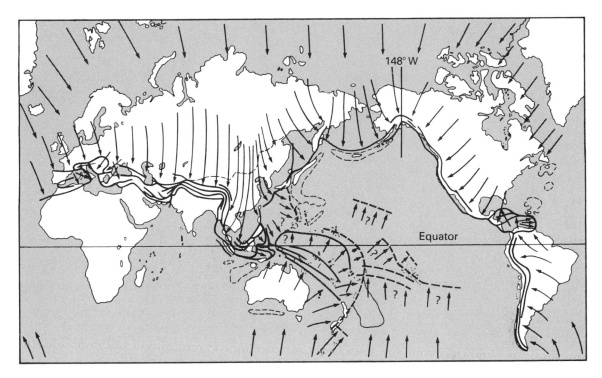

Figure 1.2 *Taylor's mechanism for the formation of Cenozoic mountain belts by continental drift (after Taylor, 1910).*

thought, the proponents of continental drift still resorted to catastrophic events to explain the separation of the continents. Thus, George Darwin in 1879 and Oswald Fisher in 1882 associated drift with the origin of the Moon out of the Pacific. This idea persisted well into the 20th century, and probably accounts in part for the reluctance of most Earth scientists to consider the concept of continental drift seriously during the first half of the 20th century (Rupke, 1970).

A uniformitarian concept of drift was first suggested by F.B. Taylor, an American physicist, in 1910, and Alfred Wegener, a German meteorologist, in 1912. For the first time it was considered that drift is taking place today and has taken place at least throughout the past 100–200 Ma of Earth history. In this way drift was invoked to account for the geometric and geologic similarities of the trailing edges of the continents around the Atlantic and Indian oceans and the formation of the young fold mountain systems at their leading edges. Taylor, in particular, invoked drift to explain the distribution of the young fold mountain belts and "the origin of the Earth's plan" (Taylor, 1910) (Fig. 1.2 and Plate 1.1 between pp. 244 and 245).

The pioneer of the theory of continental drift is generally recognized as Alfred Wegener, who as well as being a meteorologist was an astronomer, geophysicist, and amateur balloonist (Hallam, 1975), and he devoted much of his life to its development. Wegener detailed much of the older, pre-drift, geologic data and maintained that the continuity of the older structures, formations and fossil faunas and floras across present continental shorelines was more readily understood on a pre-drift reconstruction. Even today, these points are the major features of the geologic record from the continents which favor the hypothesis of continental drift. New information, which Wegener brought to his thesis, was the presence of a widespread glaciation in Permo-Carboniferous times which had affected most of the southern continents while northern Europe and Greenland had experienced tropical conditions. Wegener postulated that at this time the continents were joined into a single landmass, with the present southern continents centered on the pole and the northern continents straddling the equator (Fig. 1.3). Wegener termed this continental assembly Pangea (literally "all the Earth") although we currently prefer to think in terms of A. du

(a)

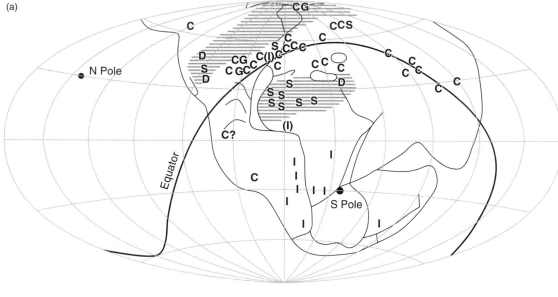

Carboniferous

(b)

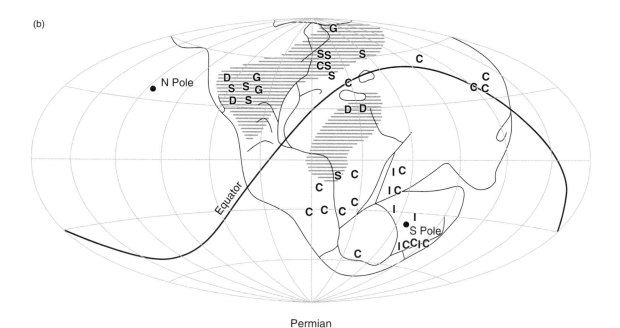

Permian

Figure 1.3 *Wegener's reconstruction of the continents (Pangea), with paleoclimatic indicators, and paleopoles and equator for (a) Carboniferous and (b) Permian time. I, ice; C, coal; S, salt; G, gypsum; D, desert sandstone; hatched areas, arid zones (modified from Wegener, 1929, reproduced from Hallam, 1973a, p. 19, by permission of Oxford University Press).*

Toit's idea of it being made up of two supercontinents (du Toit, 1937) (Fig. 11.27). The more northerly of these is termed Laurasia (from a combination of Laurentia, a region of Canada, and Asia), and consisted of North America, Greenland, Europe, and Asia. The southerly supercontinent is termed Gondwana (literally "land of the Gonds" after an ancient tribe of northern India), and consisted of South America, Antarctica, Africa, Madagascar, India, and Australasia. Separating the two supercontinents to the east was a former "Mediterranean" sea termed the paleo-Tethys Ocean (after the Greek goddess of the sea), while surrounding Pangea was the proto-Pacific Ocean or Panthalassa (literally "all-ocean").

Wegener propounded his new thesis in a book *Die Entstehung der Kontinente and Ozeane* (*The Origin of Continents and Oceans*), of which four editions appeared in the period 1915–29. Much of the ensuing academic discussion was based on the English translation of the 1922 edition which appeared in 1924, consideration of the earlier work having been delayed by World War I. Many Earth scientists of this time found his new ideas difficult to encompass, as acceptance of his work necessitated a rejection of the existing scientific orthodoxy, which was based on a static Earth model. Wegener based his theory on data drawn from several different disciplines, in many of which he was not an expert. The majority of Earth scientists found fault in detail and so tended to reject his work *in toto*. Perhaps Wegener did himself a disservice in the eclecticism of his approach. Several of his arguments were incorrect: for example, his estimate of the rate of drift between Europe and Greenland using geodetic techniques was in error by an order of magnitude. Most important, from the point of view of his critics, was the lack of a reasonable mechanism for continental movements. Wegener had suggested that continental drift occurred in response to the centripetal force experienced by the high-standing continents because of the Earth's rotation. Simple calculations showed the forces exerted by this mechanism to be much too small. Although in the later editions of his book this approach was dropped, the objections of the majority of the scientific community had become established. Du Toit, however, recognized the good geologic arguments for the joining of the southern continents and A. Holmes, in the period 1927–29, developed a new theory of the mechanism of continental movement (Holmes, 1928). He proposed that continents were moved by convection currents powered by the heat of radioactive decay (Fig. 1.4). Although differing consider-ably from the present concepts of convection and ocean floor creation, Holmes laid the foundation from which modern ideas developed.

Between the World Wars two schools of thought developed – the drifters and the nondrifters, the latter far outnumbering the former. Each ridiculed the other's ideas. The nondrifters emphasized the lack of a plausible mechanism, as we have already noted, both convection and Earth expansion being considered unlikely. The nondrifters had difficulty in explaining the present separation of faunal provinces, for example, which could be much more readily explained if the continents were formerly together, and their attempts to explain these apparent faunal links or migrations also came in for some ridicule. They had to invoke various improbable means such as island stepping-stones, isthmian links, or rafting. It is interesting to note that at this time many southern hemisphere geologists, such as du Toit, Lester King, and S.W. Carey, were advocates of drift, perhaps because the geologic record from the southern continents and India favors their assembly into a single supercontinent (Gondwana) prior to 200 Ma ago.

Very little was written about continental drift between the initial criticisms of Wegener's book and about 1960. In the 1950s, employing methodology suggested by P.M.S. Blackett, the paleomagnetic method was developed (Section 3.6), and S.K. Runcorn and his co-workers demonstrated that relative movements had occurred between North America and Europe. The work was extended by K.M. Creer into South America and by E. Irving into Australia. Paleomagnetic results became more widely accepted when the technique of magnetic cleaning was developed in which primary magnetization could be isolated. Coupled with dating by faunal or newly developed radiometric methods, the paleomagnetic data for Mesozoic to Recent times showed that there had been significant differences, beyond the scope of error, in the motions between various continents.

An important consideration in the development of ideas relating to continental drift was that prior to World War II geologists had, necessarily, only studied the land areas. Their findings had revealed that the continental crust preserves a whole spectrum of Earth history, ranging back to nearly 4000 Ma before the present, and probably to within a few hundred million years of the age of the Earth and the solar system itself. Their studies also revealed the importance of vertical movements of the continental crust in that the record was one of repeated uplift and erosion, subsidence, and

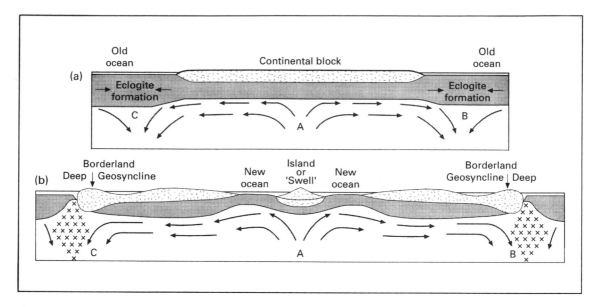

Figure 1.4 *The concept of convection as suggested by Holmes (1928), when it was believed that the oceanic crust was a thick continuation of the continental "basaltic layer". (a) Currents ascending at A spread laterally, place a continent under tension and split it, providing the obstruction of the old ocean floor can be overcome. This is accomplished by the formation of eclogite at B and C, where sub-continental currents meet sub-oceanic currents and turn downwards. The high density of the eclogite causes it to sink and make room for the continents to advance. (b) The foundering of eclogite at B and C contributes to the main convective circulation. The eclogite melts at depth to form basaltic magma, which rises in ascending currents at A, heals the gaps in the disrupted continent and forms new ocean floor. Local swells, such as Iceland, would be formed from old sial left behind. Smaller current systems, initiated by the buoyancy of the basaltic magma, ascend beneath the continents and feed flood basalts or, beneath "old" (Pacific) ocean floor, feed the outpourings responsible for volcanic islands and seamounts (redrawn from Holmes, 1928).*

sedimentation. But as J. Tuzo Wilson, a Canadian geophysicist, said, this is like looking at the deck of a ship to see if it is moving.

1.2 SEA FLOOR SPREADING AND THE BIRTH OF PLATE TECTONICS

If there is a possibility that the continental areas have been rifted and drifted apart and together, then presumably there should be some record of this within the ocean basins. However, it is only since World War II and notably since 1960 that sufficient data have been obtained from the 60% of the Earth's surface covered by deep water for an understanding of the origin and history of the ocean basins to have emerged. It transpires that, in contrast to the continents, the oceanic areas are very young geologically (probably no greater than 200 Ma in age) and that horizontal, or lateral, movements have been all-important during their history of formation.

In 1961, following intensive surveying of the sea floor during post-war years, R.S. Dietz proposed the mechanism of "sea floor spreading" to explain continental drift. Although Dietz coined the term "sea floor spreading," the concept was conceived a year or two earlier by H.H. Hess. He suggested that continents move in response to the growth of ocean basins between them, and that oceanic crust is created from the Earth's mantle at the crest of the mid-ocean ridge system, a

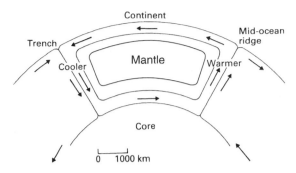

Figure 1.5 *The concept of sea floor spreading (after Hess, 1962).*

volcanic submarine swell or rise which occupies a median position in many of the world's oceans (Fig. 1.5). Oceanic crust is much thinner than continental crust, having a mean thickness of about 7 km, compared with the average continental thickness of about 40 km; is chemically different, and is structurally far less complex. The lateral motion of the oceanic crust was believed to be driven by convection currents in the upper mantle in the fashion of a conveyer belt. In order to keep the surface area of the Earth constant, it was further proposed that the oceanic crust is thrust back down into the mantle and resorbed at oceanic trenches. These are vast bathymetric depressions, situated at certain ocean margins and associated with intense volcanic and earthquake activity. Within this framework the continents are quite passive elements – rafts of less dense material which are drifted apart and together by ephemeral ocean floors. The continents themselves are a scum of generally much older material that was derived or separated from the Earth's interior either at a very early stage in its history or, at least in part, steadily throughout geologic time. Instead of blocks of crust, we now think in terms of "plates" of comparatively rigid upper mantle and crust, perhaps 50–100 km thick and which we term lithosphere (a term originally coined by R.A. Daly many years ago and meaning "rock layer"). Lithospheric plates can have both continental and oceanic crust embedded in them.

The theory of sea floor spreading was confirmed in the period 1963–66 following the suggestion of F.J. Vine and D.H. Matthews that the magnetic lineations of the sea floor might be explained in terms of sea floor spreading and reversals of the Earth's magnetic field (Section 4.1). On this model the conveyor belt of oceanic crust is viewed as a tape recorder which registers the history of reversals of the Earth's magnetic field.

A further precursor to the development of the theory of plate tectonics came with the recognition, by J.T. Wilson in 1965, of a new class of faults termed transform faults, which connect linear belts of tectonic activity (Section 4.2). The Earth was then viewed as a mosaic of six major and several smaller plates in relative motion. The theory was put on a stringent geometric basis by the work of D.P. McKenzie, R.L. Parker, and W.J. Morgan in the period 1967–68 (Chapter 5), and confirmed by earthquake seismology through the work of B. Isacks, J. Oliver, and L.R. Sykes.

The theory has been considerably amplified by intensive studies of the geologic and geophysical processes affecting plate margins. Probably the aspect about which there is currently the most contention is the nature of the mechanism that causes plate motions (Chapter 12).

Although the basic theory of plate tectonics is well established, understanding is by no means complete. Investigating the implications of plate tectonics will fully occupy Earth scientists for many decades to come.

1.3 GEOSYNCLINAL THEORY

Prior to the acceptance of plate tectonics, the static model of the Earth encompassed the formation of tectonically active belts, which formed essentially by vertical movements, on the site of geosynclines. A review of the development of the geosyncline hypothesis and its explanation in terms of plate tectonics is provided by Mitchell & Reading (1986).

Geosynclinal theory envisaged elongate, geographically fixed belts of deep subsidence and thick sediments as the precursors of mountain ranges in which the strata were exposed by folding and uplift of the geosynclinal sediments (Dickinson, 1971). A plethora of specific nomenclature evolved to describe the lithological associations of the sedimentary fill and the relative locations of the geosynclines.

The greatest failing of geosynclinal theory was that tectonic features were classified without there being an understanding of their origin. Geosynclinal nomenclature consequently represented an impediment to the recognition of a common causal mechanism. The relation of sedimentation to the mobilistic mechanism of plate tectonics (Mitchell & Reading, 1969) allowed the recognition of two specific environments in which geosynclines formed, namely rifted, or trailing, continental margins and active, or leading, continental margins landward of the deep oceanic trenches. The latter are now known as subduction zones (Chapter 9). Although some workers retain geosynclinal terminology to describe sedimentary associations (e.g. the terms eugeosyncline and miogeosyncline for sediments with and without volcanic members, respectively), this usage is not recommended, and the term geosyncline must be recognized as no longer relevant to plate tectonic processes.

1.4 IMPACT OF PLATE TECTONICS

Plate tectonics is of very great significance as it represents the first theory that provides a unified explanation of the Earth's major surface features. As such it has enabled an unprecedented linking of many different aspects of geology, which had previously been considered independent and unrelated. A deeper understanding of geology has ensued from the interpretation of many branches of geology within the basic framework provided by plate tectonics. Thus, for example, explanations can be provided for the past distributions of flora and fauna, the spatial relationships of volcanic rock suites at plate margins, the distribution in space and time of the conditions of different metamorphic facies, the scheme of deformation in mountain belts, or orogens, and the association of different types of economic deposit.

Recognition of the dynamic nature of the apparently solid Earth has led to the realization that plate tectonic processes may have had a major impact on other aspects of the Earth system in the past. Changes in volcanic activity in general, and at mid-ocean ridges in particular, would have changed the chemistry of the atmosphere and of seawater. Changes in the net accretion rate at mid-ocean ridges could explain major

changes in sea level in the past, and the changing configuration of the continents, and the uplift of mountain belts would have affected both oceanic and atmospheric circulation. The nature and implications of these changes, in particular for the Earth's climate, are explored in Chapter 13.

Clearly some of these implications were documented by Wegener, notably in relation to the distribution of fauna and flora in the past, and regional paleoclimates. Now, however, it is realized that plate tectonic processes impact on the physics and chemistry of the atmosphere and oceans, and on life on Earth, in many more ways, thus linking processes in the atmosphere, oceans, and solid Earth in one dynamic global system.

The fact that plate tectonics is so successful in unifying so many aspects of Earth science should not be taken to indicate that it is completely understood. Indeed, it is the critical testing of the implications of plate tectonic theory that has led to modifications and extrapolations, for example in the consideration of the relevance of plate tectonic processes in continental areas (Section 2.10.5) and the more distant geologic past (Chapter 11). It is to be hoped that plate tectonic theory will be employed cautiously and critically.

FURTHER READING

Hallam, A. (1973) *A Revolution in the Earth Sciences: from continental drift to plate tectonics.* Oxford University Press, Oxford, UK.

LeGrand, H.E. (1988) *Drifting Continents and Shifting Theories.* Cambridge University Press, Cambridge, UK.

Marvin, U.B. (1973) *Continental Drift: the evolution of a concept.* Smithsonian Institution, Washington, DC.

Oreskes, N. (1999) *The Rejection of Continental Drift: theory and method in American Earth Science.* Oxford University Press, New York.

Oreskes, N. (ed.) (2001) *Plate Tectonics: an insider's history of the modern theory of the Earth.* Westview Press, Boulder.

Stewart, J.A. (1990) *Drifting Continents and Colliding Paradigms: perspectives on the geoscience revolution.* Indiana University Press, Bloomington, IN.

2 The interior of the Earth

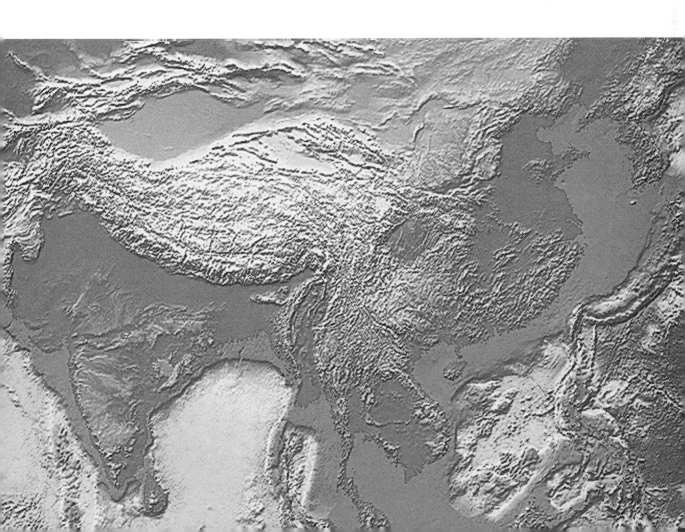

2.1 EARTHQUAKE SEISMOLOGY

2.1.1 Introduction

Much of our knowledge of the internal constitution of the Earth has come from the study of the seismic waves generated by earthquakes. These waves follow various paths through the interior of the Earth, and by measuring their travel times to different locations around the globe it is possible to determine its large-scale layering. It is also possible to make inferences about the physical properties of these layers from a consideration of the velocities with which they transmit the seismic waves.

2.1.2 Earthquake descriptors

Earthquakes are normally assumed to originate from a single point known as the *focus* or *hypocenter* (Fig. 2.1), which is invariably within about 700 km of the surface. In reality, however, most earthquakes are generated by movement along a fault plane, so the focal region may extend for several kilometers. The point on the Earth's surface vertically above the focus is the *epicenter*. The angle subtended at the center of the Earth by the epicenter and the point at which the seismic waves are detected is known as the *epicentral angle* Δ. The *magnitude* of an earthquake is a measure of its energy release on a logarithmic scale; a change in magnitude of one

on the *Richter* scale implies a 30-fold increase in energy release (Stein & Wysession, 2003).

2.1.3 Seismic waves

The strain energy released by an earthquake is transmitted through the Earth by several types of seismic wave (Fig. 2.2), which propagate by elastic deformation of the rock through which they travel. Waves penetrating the interior of the Earth are known as *body waves*, and consist of two types corresponding to the two possible ways of deforming a solid medium. *P waves*, also known as *longitudinal* or *compressional* waves, correspond to elastic deformation by compression/dilation. They cause the particles of the transmitting rock to oscillate in the direction of travel of the wave so that the disturbance proceeds as a series of compressions and rarefactions. The velocity of a P wave V_p is given by:

$$V_p = \sqrt{\frac{k + \frac{4}{3}\mu}{\rho}}$$

where k is the bulk modulus, μ the shear modulus (rigidity), and ρ the density of the transmitting medium. *S waves*, also known as *shear* or *transverse* waves, correspond to elastic deformation of the transmitting medium by shearing and cause the particles of the rock

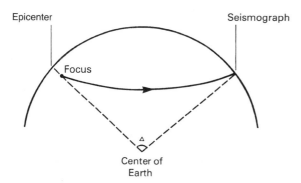

Figure 2.1 *Illustration of epicentral angle Δ.*

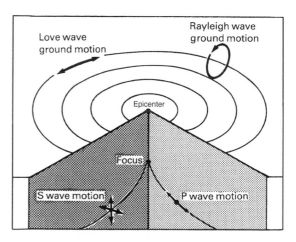

Figure 2.2 *Focus and epicenter of an earthquake and the seismic waves originating from it (after Davies, 1968, with permission from Iliffe Industrial Publications Ltd).*

to oscillate at right angles to the direction of propagation. The velocity of an S wave V_s is given by:

$$V_s = \sqrt{\frac{\mu}{\rho}}$$

Because the rigidity of a fluid is zero, S waves cannot be transmitted by such a medium.

A consequence of the velocity equations for P and S waves is that the P velocity is about 1.7 times greater than the S velocity in the same medium. Consequently, for an identical travel path, P waves arrive before S waves. This was recognized early in the history of seismology, and is reflected in the names of the body waves (P is derived from primus and S from secundus). The passage of body waves through the Earth conforms to the laws of geometric optics in that they can be both refracted and reflected at velocity discontinuities.

Seismic waves whose travel paths are restricted to the vicinity of a free surface, such as the Earth's surface, are known as *surface waves*. *Rayleigh waves* cause the particles of the transmitting medium to describe an ellipse in a vertical plane containing the direction of propagation. They can be transmitted in the surface of a uniform half space or a medium in which velocity changes with depth. *Love waves* are transmitted whenever the S wave velocity of the surface layer is lower than that of the underlying layer. Love waves are essentially horizontally polarized shear waves, and propagate by multiple reflection within this low velocity layer, which acts as a wave guide.

Surface waves travel at lower velocities than body waves in the same medium. Unlike body waves, surface waves are dispersive, that is, their different wavelength components travel at different velocities. Dispersion arises because of the velocity stratification of the Earth's interior, longer wavelengths penetrating to greater depths and hence sampling higher velocities. As a result, surface wave dispersion studies provide an important method of determining the velocity structure and seismic attenuation characteristics of the upper 600 km of the Earth.

2.1.4 Earthquake location

Earthquakes are detected by seismographs, instruments that respond to very small ground displacements, veloc-

ities, or accelerations associated with the passage of seismic waves. Since 1961 there has been an extensive and standardized global network of seismograph stations to monitor earthquake activity. The original World-Wide Standardized Seismograph Network (WWSSN), based on analogue instruments, has gradually been superseded since 1986 by the Global (Digital) Seismograph Network (GSN). By 2004 there were 136 well-distributed GSN stations worldwide, including one on the sea floor between Hawaii and California. It is hoped that this will be the first of several in oceanic areas devoid of oceanic islands for land-based stations. Digital equipment greatly facilitates processing of the data and also has the advantage that it records over a much greater dynamic range and frequency bandwidth than the earlier paper and optical recording. This is achieved by a combination of high frequency, low gain and very broadband seismometers (Butler *et al.*, 2004). Most countries have at least one GSN station and many countries also have national seismometer arrays. Together these stations not only provide the raw data for all global and regional seismological studies but also serve an important function in relation to monitoring the nuclear test ban treaty, and volcano and tsunami warning systems.

Earthquakes occurring at large, or *teleseismic*, distances from a seismograph are located by the identification of various *phases*, or seismic arrivals, on the seismograph records. Since, for example, the direct P and S waves travel at different velocities, the time separation between the arrival of the P phase and the S phase becomes progressively longer as the length of the travel path increases. By making use of a standard model for the velocity stratification of the Earth, and employing many seismic phases corresponding to different travel paths along which the seismic waves are refracted or reflected at velocity discontinuities, it is possible to translate the differences in their travel times into the distance of the earthquake from the observatory. Triangulation using distances computed in this way from many observatories then allows the location of the epicenter to be determined.

The focal depths of teleseismic events are determined by measuring the arrival time difference between the direct phase P and the phase pP (Båth, 1979). The pP phase is a short path multiple event which follows a similar path to P after first undergoing a reflection at the surface of the Earth above the focus, and so the P–pP time difference is a measure of focal depth. This method is least accurate for foci at depths of less than

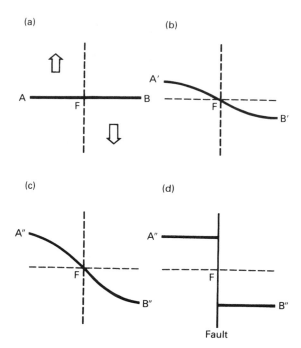

(a)   (b)   (c)   (d)

Fault

Figure 2.3 *Elastic rebound mechanism of earthquake generation.*

100 km as the P–pP time separation becomes very small. The focal depths of local earthquakes can be determined if a network of seismographs exists in the vicinity of the epicenter. In this case the focal depth is determined by triangulation in the vertical plane, using the P–S time difference to calculate the distance to the focus.

2.1.5 Mechanism of earthquakes

Most earthquakes are believed to occur according to the *elastic rebound theory*, which was developed after the San Francisco earthquake of 1906. In this theory an earthquake represents a sudden release of strain energy that has built up over a period of time.

In Fig. 2.3a a block of rock traversed by a pre-existing fracture (or fault) is being strained in such a way as eventually to cause relative motion along the plane of the fault. The line AB is a marker indicating the state of strain of the system, and the broken line the location of the fault. Relatively small amounts of strain can be

accommodated by the rock (Fig. 2.3b). Eventually, however, the strain reaches the level at which it exceeds the frictional and cementing forces opposing movement along the fault plane (Fig. 2.3c). At this point fault movement occurs instantaneously (Fig. 2.3d). The 1906 San Francisco earthquake resulted from a displacement of 6.8 m along the San Andreas Fault. In this model, faulting reduces the strain in the system virtually to zero, but if the shearing forces persist, strain would again build up to the point at which fault movement occurs. The elastic rebound theory consequently implies that earthquake activity represents a stepwise response to persistent strain.

2.1.6 Focal mechanism solutions of earthquakes

The seismic waves generated by earthquakes, when recorded at seismograph stations around the world, can be used to determine the nature of the faulting associated with the earthquake, to infer the orientation of the fault plane and to gain information on the state of stress of the lithosphere. The result of such an analysis is referred to as a *focal mechanism solution* or *fault plane solution*. The technique represents a very powerful method of analyzing movements of the lithosphere, in particular those associated with plate tectonics. Information is available on a global scale as most earthquakes with a magnitude in excess of 5.5 can provide solutions, and it is not necessary to have recorders in the immediate vicinity of the earthquake, so that data are provided from regions that may be inaccessible for direct study.

According to the elastic rebound theory, the strain energy released by an earthquake is transmitted by the seismic waves that radiate from the focus. Consider the fault plane shown in Fig. 2.4 and the plane orthogonal to it, the *auxiliary plane*. The first seismic waves to arrive at recorders around the earthquake are P waves, which cause compression/dilation of the rocks through which they travel. The shaded quadrants, defined by the fault and auxiliary planes, are compressed by movement along the fault and so the first motion of the P wave arriving in these quadrants corresponds to a compression. Conversely, the unshaded quadrants are stretched or dilated by the fault movement. The first motion of the P waves in these quadrants is thus dilational. The region around the earthquake is therefore divided into four quadrants on the basis of the P wave first motions,

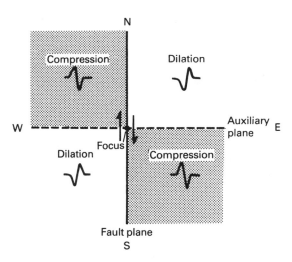

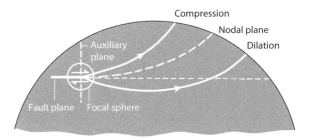

Figure 2.5 *Distribution of compressional and dilational first arrivals from an earthquake on the surface of a spherical Earth in which seismic velocity increases with depth.*

Figure 2.4 *Quadrantal distribution of compressional and dilational P wave first motions about an earthquake.*

defined by the fault plane and the auxiliary plane. No P waves propagate along these planes as movement of the fault imparts only shearing motions in their directions; they are consequently known as *nodal planes*.

Simplistically, then, a focal mechanism solution could be obtained by recording an earthquake at a number of seismographs distributed around its epicenter, determining the nature of the first motions of the P waves, and then selecting the two orthogonal planes which best divide compressional from dilational first arrivals, that is, the nodal planes. In practice, however, the technique is complicated by the spheroidal shape of the Earth and the progressive increase of seismic velocity with depth that causes the seismic waves to follow curved travel paths between the focus and recorders. Consider Fig. 2.5. The dotted line represents the continuation of the fault plane, and its intersection with the Earth's surface would represent the line separating compressional and dilational first motions if the waves generated by the earthquake followed straight-line paths. The actual travel paths, however, are curved and the surface intersection of the dashed line, corresponding to the path that would have been followed by a wave leaving the focus in the direction of the fault plane, represents the actual nodal plane.

It is clear then, that simple mapping of compressional and dilational first motions on the Earth's surface cannot readily provide the focal mechanism solution. However, the complications can be overcome

by considering the directions in which the seismic waves left the focal region, as it is apparent that compressions and dilations are restricted to certain angular ranges.

A focal mechanism solution is obtained firstly by determining the location of the focus by the method outlined in Section 2.1.4. Then, for each station recording the earthquake, a model for the velocity structure of the Earth is used to compute the travel path of the seismic wave from the focus to the station, and hence to calculate the direction in which the wave left the focal region. These directions are then plotted, using an appropriate symbol for compressional or dilational first motion, on an equal area projection of the lower half of the *focal sphere*, that is, an imaginary sphere of small but arbitrary radius centered on the focus (Fig. 2.5). An equal area net, which facilitates such a plot, is illustrated in Fig. 2.6. The scale around the circumference of such a net refers to the azimuth, or horizontal component of direction, while dips are plotted on the radial scale from 0° at the perimeter to 90° at the center. Planes through the focus are represented on such plots by great circles with a curvature appropriate to their dip; hence a diameter represents a vertical plane.

Let us assume that, for a particular earthquake, the fault motion is strike-slip along a near vertical fault plane. This plane and the auxiliary plane plot as orthogonal great circles on the projection of the focal sphere, as shown on Fig. 2.7. The lineation defined by the intersection of these planes is almost vertical, so it is apparent that the direction of movement along the fault is orthogonal to this intersection, that is, near horizontal. The two shaded and two unshaded regions of the projection defined by the nodal planes now correspond to the directions in which compressional and dilational

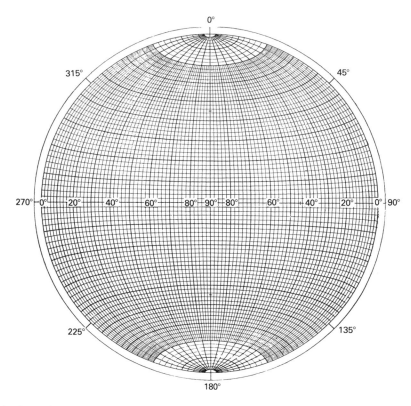

Figure 2.6 *Lambert equal area net.*

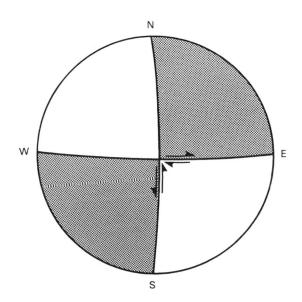

Figure 2.7 *Ambiguity in the focal mechanism solution of a strike-slip fault. Regions of compressional first motions are shaded.*

first motions, respectively, left the focal region. A focal mechanism solution is thus obtained by plotting all the observational data on the projection of the focal sphere and then fitting a pair of orthogonal planes which best divide the area of the projection into zones of compressional and dilational first motions. The more stations recording the earthquake, the more closely defined will be the nodal planes.

2.1.7 Ambiguity in focal mechanism solutions

It is apparent from Fig. 2.7 that the same distribution of compressional and dilational quadrants would be obtained if either nodal plane represented the actual fault plane. Thus, the same pattern of first motions would be obtained for sinistral motion along a north–south plane as for dextral motion along an east–west plane.

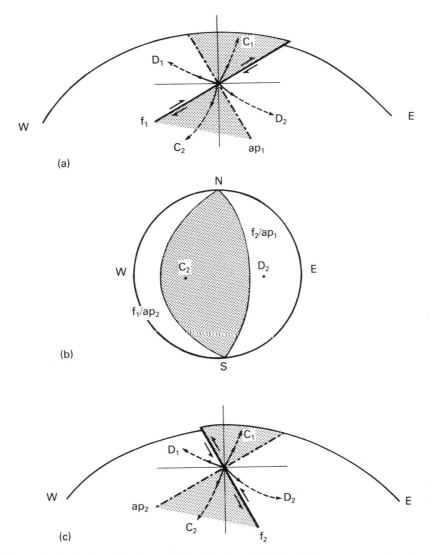

Figure 2.8 *Ambiguity in the focal mechanism solution of a thrust fault. Shaded areas represent regions of compressional first motions (C), unshaded areas represent regions of dilational first motions (D), f refers to a fault plane, ap to an auxiliary plane. Changing the nature of the nodal planes as in (a) and (c) does not alter the pattern of first motions shown in (b), the projection of the lower hemisphere of the focal sphere.*

In Fig. 2.8a an earthquake has occurred as a result of faulting along a westerly dipping thrust plane f_1. f_1 and its associated auxiliary plane ap_1 divide the region around the focus into quadrants which experience either compression or dilation as a result of the fault movement. The directions in which compressional first motions C_1 and C_2 and dilational first motions D_1 and D_2 leave the focus are shown, and C_2 and D_2 are plotted on the projection of the focal sphere in Fig. 2.8b, on

which the two nodal planes are also shown. Because Fig. 2.8a is a vertical section, the first motions indicated plot along an east–west azimuth. Arrivals at stations at other azimuths would occupy other locations within the projection space. Consider now Fig. 2.8c, in which plane ap_1 becomes the fault plane f_2 and f_1 the auxiliary plane ap_2. By considering the movement along the thrust plane it is obvious that the same regions around the fault are compressed or dilated, so that an identical

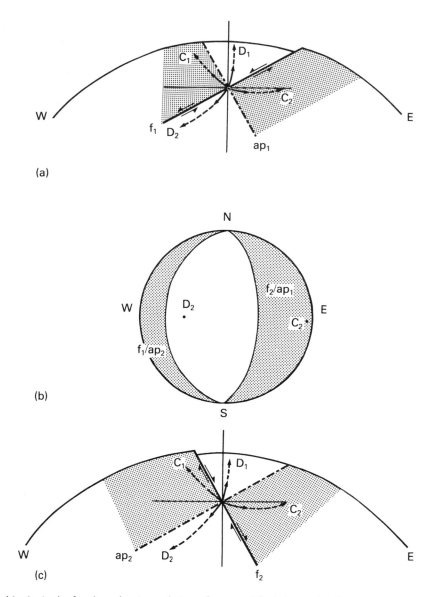

Figure 2.9 *Ambiguity in the focal mechanism solution of a normal fault. Legend as for Fig. 2.8.*

focal sphere projection is obtained. Similar results are obtained when the faulting is normal (Fig. 2.9). In theory the fault plane can be distinguished by making use of Anderson's simple theory of faulting (Section 2.10.2) which predicts that normal faults have dips of more than 45° and thrusts less than 45°. Thus, f_1 is the fault plane in Fig. 2.8 and f_2 the fault plane in Fig. 2.9.

It is apparent that the different types of faulting can be identified in a focal mechanism solution by the distinctive pattern of compressional and dilational regions on the resulting focal sphere. Indeed, it is also possible to differentiate earthquakes that have originated by a combination of fault types, such as dip-slip accompanied by some strike-slip movement. The precision with which the directions of the nodal planes can be determined is

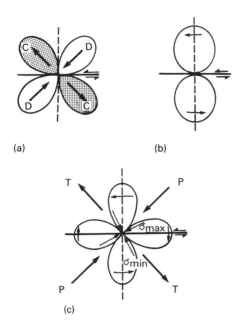

(a)   (b)

(c)

Figure 2.10 *(a) P wave radiation pattern for a type I and type II earthquake source mechanism; (b) S wave radiation pattern from a type I source (single couple); (c) S wave radiation pattern from a type II source (double couple).*

dependent upon the number and distribution of stations recording arrivals from the event. It is not possible, however, to distinguish the fault and auxiliary planes.

At one time it was believed that distinction between the nodal planes could be made on the basis of the pattern of S wave arrivals. P waves radiate into all four quadrants of the source region as shown in Fig. 2.10a. However, for this simple model, which is known as a type I, or single-couple source, S waves, whose corresponding ground motion is shearing, should be restricted to the region of the auxiliary plane (Fig. 2.10b). Recording of the S wave radiation pattern should then make it possible to determine the actual fault plane. It was found, however, that instead of this simple pattern, most earthquakes produce S wave radiation along the direction of both nodal planes (Fig. 2.10c). This observation initially cast into doubt the validity of the elastic rebound theory. It is now realized, however, that faulting occurs at an angle, typically rather less than 45% to the maximum compressive stress, σ_1, and the bisectors of the dilational and compressional quadrants, termed P and T, respectively, approximate to the directions of maximum and minimum principal compressive stress,

thus giving an indication of the stress field giving rise to the earthquake (Fig. 2.10c) (Section 2.10.2).

This type II, or double-couple source mechanism gives rise to a four-lobed S wave radiation pattern (Fig. 2.10c) which cannot be used to resolve the ambiguity of a focal mechanism solution. Generally, the only constraint on the identity of the fault plane comes from a consideration of the local geology in the region of the earthquake.

2.1.8 Seismic tomography

Tomography is a technique whereby three-dimensional images are derived from the processing of the integrated properties of the medium that rays encounter along their paths through it. Tomography is perhaps best known in its medical applications, in which images of specific plane sections of the body are obtained using X-rays. Seismic tomography refers to the derivation of the three-dimensional velocity structure of the Earth from seismic waves. It is considerably more complex than medical tomography in that the natural sources of seismic waves (earthquakes) are of uncertain location, the propagation paths of the waves are unknown, and the receivers (seismographs) are of restricted distribution. These difficulties can be overcome, however, and since the late 1970s seismic tomography has provided important new information on Earth structure. The method was first described by Aki *et al.* (1977) and has been reviewed by Dziewonski & Anderson (1984), Thurber & Aki (1987), and Romanowicz (2003).

Seismic tomography makes use of the accurately recorded travel times of seismic waves from geographically distributed earthquakes at a distributed suite of seismograph stations. The many different travel paths from earthquakes to receivers cross each other many times. If there are any regions of anomalous seismic velocity in the space traversed by the rays, the travel times of the waves crossing this region are affected. The simultaneous interpretation of travel time anomalies for the many criss-crossing paths then allows the anomalous regions to be delineated, providing a three-dimensional model of the velocity space.

Both body waves and surface waves (Section 2.1.3) can be used in tomography analysis. With body waves, the actual travel times of P or S phases are utilized. The procedure with surface waves is more complex, however, as they are dispersive; that is, their velocity depends upon their wavelength. The depth of penetration of surface waves is also wavelength-dependent, with the longer wavelengths reaching greater depths. Since seismic velocity generally increases with depth, the longer wavelengths travel more rapidly. Thus, when surface waves are utilized, it is necessary to measure the phase or group velocities of their different component wavelengths. Because of their low frequency, surface waves provide less resolution than body waves. However, they sample the Earth in a different fashion and, since either Rayleigh or Love waves (Section 2.1.3) may be used, additional constraints on shear velocity and its anisotropy are provided.

The normal procedure in seismic tomography is to assume an initial "one-dimensional" model of the velocity space in which the velocity is radially symmetrical. The travel time of a body wave from earthquake to seismograph is then equal to the sum of the travel times through the individual elements of the model. Any lateral velocity variations within the model are then reflected in variations in arrival times with respect to the mean arrival time of undisturbed events. Similarly, the dispersion of surface waves across a heterogeneous model differs from the mean dispersion through a radially symmetrical model. The method makes use of a simplifying assumption based on Fermat's Principle, which assumes that the ray paths for a radially symmetrical and laterally variable velocity model are identical if the heterogeneities are small and that the differences in travel times are caused solely by heterogeneity in the velocity structure of the travel path. This obviates the necessity of computing the new travel path implied by refractions at the velocity perturbations.

There are two main approaches to seismic tomography depending upon how the velocity heterogeneity of the model is represented. *Local methods* make use of body waves and subdivide the model space into a series of discrete elements so that it has the form of a three-dimensional ensemble of blocks. A set of linear equations is then derived which link the anomalies in arrival times to velocity variations over the different travel paths. A solution of the equations can then be obtained, commonly using matrix inversion techniques, to obtain the velocity anomaly in each block. *Global methods* express the velocity variations of the model in terms of some linear combination of continuous basic functions, such as spherical harmonic functions.

Local methods can make use of either teleseismic or local events. In the teleseismic method (Fig. 2.11) a large set of distant seismic events is recorded at a

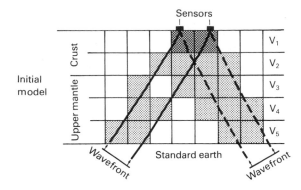

Figure 2.11 *Geometry of the teleseismic inversion method. Velocity anomalies within the compartments are derived from relative arrival time anomalies of teleseismic events (redrawn from Aki et al., 1977, by permission of the American Geophysical Union. Copyright © 1977 American Geophysical Union).*

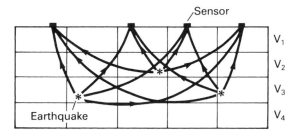

Figure 2.12 *Geometry of the local inversion method.*

network of seismographs over the volume of interest. Because of their long travel path, the incident wave fronts can be considered planar. It is assumed that deviations from expected arrival times are caused by velocity variations beneath the network. In practice, deviations from the mean travel times are computed to compensate for any extraneous effects experienced by the waves outside the volume of interest. Inversion of the series of equations of relative travel time through the volume then provides the relative velocity perturbations in each block of the model. The method can be extended by the use of a worldwide distribution of recorded teleseismic events to model the whole mantle. In the local method the seismic sources are located within the volume of interest (Fig. 2.12). In this case the location and time of the earthquakes must be accurately known, and ray-tracing methods used to construct the travel paths of the rays. The inversion

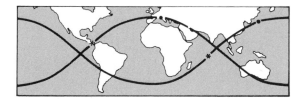

Figure 2.13 *Great circle paths from two earthquakes (stars) to recording stations (dots) (after Thurber & Aki, 1987).*

procedure is then similar to that for teleseisms. One of the uses of the resulting three-dimensional velocity distributions is to improve focal depth determinations.

Global methods commonly make use of both surface and body waves with long travel paths. If the Earth were spherically symmetrical, these surface waves would follow great circle routes. However, again making use of Fermat's Principle, it is assumed that ray paths in a heterogeneous Earth are similarly great circles, with anomalous travel times resulting from the heterogeneity. In the single-station configuration, the surface wave dispersion is measured for the rays traveling directly from earthquake to receiver. Information from only moderate-size events can be utilized, but the source parameters have to be well known. The great circle method uses multiple circuit waves, that is, waves that have traveled directly from source to receiver and have then circumnavigated the Earth to be recorded again (Fig. 2.13). Here the differential dispersion between the first and second passes is measured, eliminating any undesirable source effects. This method is appropriate to global modeling, but can only use those large magnitude events that give observable multiple circuits.

2.2 VELOCITY STRUCTURE OF THE EARTH

Knowledge of the internal layering of the Earth has been largely derived using the techniques of earthquake seismology. The shallower layers have been studied using local arrays of recorders, while the deeper layers have been investigated using global networks to detect seismic signals that have traversed the interior of the Earth.

The continental crust was discovered by Andrija Mohorovičić from studies of the seismic waves generated by the Croatia earthquake of 1909 (Fig. 2.14). Within a range of about 200 km from the epicenter, the first seismic arrivals were P waves that traveled directly from the focus to the recorders with a velocity of $5.6\,\mathrm{km\,s^{-1}}$. This seismic phase was termed P_g. At greater ranges, however, P waves with the much higher velocity of $7.9\,\mathrm{km\,s^{-1}}$ became the first arrivals, termed the P_n phase. These data were interpreted by the standard techniques of refraction seismology, with P_n representing seismic waves that had been critically refracted at a velocity discontinuity at a depth of some 54 km. This discontinuity was subsequently named the Mohorovičić discontinuity, or Moho, and it marks the boundary between the crust and mantle. Subsequent work has demonstrated that the Moho is universally present beneath continents and marks an abrupt increase in seismic velocity to about $8\,\mathrm{km\,s^{-1}}$. Its geometry and reflective character are highly diverse and may include one or more sub-horizontal or dipping reflectors (Cook, 2002). Continental crust is, on average, some 40 km thick, but thins to less than 20 km beneath some tectonically active rifts (e.g. Sections 7.3, 7.8.1) and thickens to up to 80 km beneath young orogenic belts (e.g. Sections 10.2.4, 10.4.5) (Christensen & Mooney, 1995; Mooney et al., 1998).

A discontinuity within the continental crust was discovered by Conrad in 1925, using similar methods. As well as the phases P_g and P_n he noted the presence of an additional phase $P^\star$ (Fig. 2.15) which he interpreted as the critically refracted arrival from an interface where the velocity increased from about 5.6 to $6.3\,\mathrm{km\,s^{-1}}$. This interface was subsequently named the Conrad discontinuity. Conrad's model was readily adopted by early petrologists who believed that two layers were necessarily present in the continental crust. The upper layer, rich in silicon and aluminum, was called the SIAL and was believed to be the source of granitic magmas, while the lower, silicon- and magnesium-rich layer or SIMA was believed to be the source of basaltic magmas. It is now known, however, that the upper crust has a composition more mafic than granite (Section 2.4.1), and that the majority of basaltic magmas originate in the mantle. Consequently, the petrological necessity of a two-layered crust no

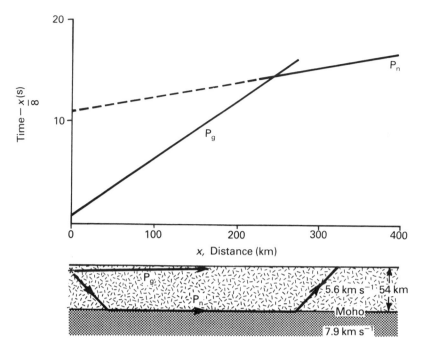

Figure 2.14 *Reduced time–distance relationship for direct waves (P_g) and waves critically refracted at the Moho (P_n) from an earthquake source.*

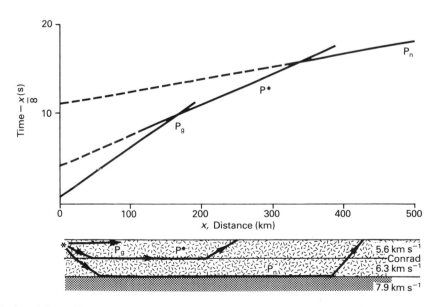

Figure 2.15 *Reduced time–distance relationship for direct waves (P_g), waves critically refracted at the Conrad discontinuity ($P*$) and waves critically refracted at the Moho (P_n) from an earthquake source.*

longer exists and, where applicable, it is preferable to use the terms upper and lower crust. Unlike the Moho, the Conrad discontinuity is not always present within the continental crust, although the seismic velocity generally increases with depth.

In some regions the velocity structure of continental crust suggests a natural division into three layers. The velocity range of the middle crustal layer generally is taken to be 6.4–6.7 km s^{-1}. The typical velocity range of the lower crust, where a middle crust is present, is 6.8–7.7 km s^{-1} (Mooney *et al.*, 1998). Examples of the velocity structure of continental crust in a tectonically active rift, a rifted margin, and a young orogenic belt are shown in Figs 7.5, 7.32a, and 10.7, respectively.

The oceanic crust has principally been studied by explosion seismology. The Moho is always present and the thickness of much of the oceanic crust is remarkably constant at about 7 km irrespective of the depth of water above it. The internal layering of oceanic crust and its constancy over very wide areas will be discussed later (Section 2.4.4).

In studying the deeper layering of the Earth, seismic waves with much longer travel paths are employed. The velocity structure has been built up by recording the travel times of body waves over the full range of possible epicentral angles. By assuming that the Earth is radially symmetrical, it is possible to invert the travel time data to provide a model of the velocity structure. A modern determination of the velocity–depth curve (Kennett *et al.*, 1995) for both P and S waves is shown in Fig. 2.16.

Velocities increase abruptly at the Moho in both continental and oceanic environments. A low velocity zone (LVZ) is present between about 100 and 300 km depth, although the depth to the upper boundary is very variable (Section 2.12). The LVZ appears to be universally present for S waves, but may be absent in certain regions for P waves, especially beneath ancient shield areas. Between 410 and 660 km velocity increases rapidly in a stepwise fashion within the mantle transition zone that separates the upper mantle from the lower mantle. Each velocity increment probably corresponds to a mineral phase change to a denser form at depth (Section 2.8.5). Both P and S velocities increase progressively in the lower mantle.

The Gutenberg discontinuity marks the core–mantle boundary at a depth of 2891 km, at which the velocity of P waves decreases abruptly. S waves are not transmitted through the outer core, which is consequently

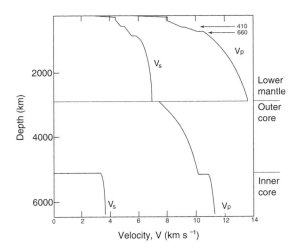

Figure 2.16 *Seismic wave velocities as a function of depth in the Earth showing the major discontinuities. AK 135 Earth model specified by Kennett et al., 1995 (after Helffrich & Wood, 2001, with permission from* Nature **412**, *501–7. Copyright © 2001 Macmillan Publishers Ltd.).*

believed to be in a fluid state. The geomagnetic field (Section 3.6.4) is believed to originate by the circulation of a good electrical conductor in this region. At a depth of 5150 km the P velocity increases abruptly and S waves are once again transmitted. This inner core is thus believed to be solid as a result of the enormous confining pressure. There appears to be no transition zone between inner and outer core, as was originally believed.

2.3 COMPOSITION OF THE EARTH

All bodies in the solar system are believed to have been formed by the condensation and accretion of the primitive interstellar material that made up the solar nebula. The composition of the Sun is the same as the average composition of this material. Gravitational energy was released during accretion, and together with the radioactive decay of short-lived radioactive nuclides eventually led to heating of the proto-Earth so that it differentiated into a radially symmetric body made up of a series of shells whose density increased towards its center. The differentiation prevents any estimate being made of the overall composition of the Earth by direct sampling. However, it is believed that meteorites are representatives of material within the solar nebula and that estimates of the Earth's composition can be made from them. The presence of metallic and silicate phases in meteorites is taken to indicate that the Earth consists of an iron/nickel core surrounded by a lower density silicate mantle and crust.

Seismic data, combined with knowledge of the mass and moment of inertia of the Earth, have revealed that the mean atomic weight of the Earth is about 27, with a contribution of 22.4 from the mantle and crust and 47.0 from the core. No single type of meteorite possesses an atomic weight of 27, the various types of chondrite being somewhat lower and iron meteorites considerably higher. However, it is possible to mix the proportions of different meteorite compositions in such a way as to give both the correct atomic weight and core/mantle ratio. Three such models are given in Table 2.1.

It is apparent that at least 90% of the Earth is made up of iron, silicon, magnesium, and oxygen, with the

bulk of the remainder comprising calcium, aluminum, nickel, sodium, and possibly sulfur.

2.4 THE CRUST

2.4.1 The continental crust

Only the uppermost part of the crust is available for direct sampling at the surface or from boreholes. At greater depths within the crust, virtually all information about its composition and structure is indirect. Geologic studies of high grade metamorphic rocks that once resided at depths of 20–50 km and have been brought to the surface by subsequent tectonic activity provide some useful information (Miller & Paterson, 2001a; Clarke *et al.*, 2005). Foreign rock fragments, or *xenoliths*, that are carried from great depths to the Earth's surface by fast-rising magmas (Rudnick, 1992) also provide samples of deep crustal material. In addition, much information about the crust has been derived from knowledge of the variation of seismic velocities with depth and how these correspond to experimental determinations of velocities measured over ranges of temperature and pressure consistent with crustal conditions. Pressure increases with depth at a rate of about $30\,MPa\,km^{-1}$, mainly due to the lithostatic confining pressure of the overlying rocks, but also, in some regions, with a contribution from tectonic forces. Temperature increases at an average rate of about $25°C\,km^{-1}$, but decreases to about half this value at the Moho because of the presence of radioactive heat sources within the crust (Section 2.13). Collectively, the observations from both geologic and geophysical studies show that the continental crust is vertically stratified in terms of its chemical composition (Rudnick & Gao, 2003).

The variation of seismic velocities with depth (Section 2.2) results from a number of factors. The increase of pressure with depth causes a rapid increase in incompressibility, rigidity, and density over the topmost 5 km as pores and fractures are closed. Thereafter the increase of these parameters with pressure is balanced by the decrease resulting from thermal expansion with increasing temperature so that there is little further change in velocity with depth. Velocities change with chemical composition, and also with changes in mineralogy resulting from phase changes. Abrupt velocity discontinuities are usually caused by

Table 2.1 *Estimates of the bulk composition of the Earth and Moon (in weight percent) (from Condie, 1982a).*

	Earth			Moon
	1	*2*	*3*	*4*
Fe	34.6	29.3	29.9	9.3
O	29.5	30.7	30.9	42.0
Si	15.2	14.7	17.4	19.6
Mg	12.7	15.8	15.9	18.7
Ca	1.1	1.5	1.9	4.3
Al	1.1	1.3	1.4	4.2
Ni	2.4	1.7	1.7	0.6
Na	0.6	0.3	0.9	0.07
S	1.9	4.7	–	0.3

1: 32.4% iron meteorite (with 5.3% FeS) and 67.6% oxide portion of bronzite chondrites.
2: 40% type I carbonaceous chondrite, 50% ordinary chondrite, and 10% iron meteorite (containing 15% sulfur).
3: Nonvolatile portion of type I carbonaceous chondrites with FeO/FeO + MgO of 0.12 and sufficient SiO_2 reduced to Si to yield a metal/silicate ratio of 32/68.
4: Based on Ca, Al, Ti = 5×type I carbonaceous chondrites, FeO = 12% to accommodate lunar density, and Si/Mg = chondritic ratio.

changes in chemical composition, while more gradational velocity boundaries are normally associated with phase changes that occur over a discrete vertical interval.

Models for the bulk chemical composition of the continental crust vary widely because of the difficulty of making such estimates. McLennan & Taylor (1996) pointed out that the flow of heat from the continental crust (Section 2.13) provides a constraint on the abundance of the heat producing elements, K, Th, and U, within it, and hence on the silica content of the crust. On this basis they argue that on average the continental crust has an andesitic or granodioritic composition with K_2O no more than 1.5% by weight. This is less silicic than most previous estimates. The abundance of the heat producing elements, and other "incompatible" elements, in the continental crust is of great importance because the degree to which they are enriched in the crust reflects the extent to which they are depleted in the mantle.

2.4.2 Upper continental crust

Past theories of crustal construction suggested that the upper continental crust was made up of rocks of granitic composition. That this is not the case is evident from the widespread occurrence of large negative gravity anomalies over granite plutons. These anomalies demonstrate that the density of the plutons (about $2.67\,Mg\,m^{-3}$) is some 0.10–$0.15\,Mg\,m^{-3}$ lower than the average value of the upper crust. The mean composition of the upper crust can be estimated, albeit with some uncertainty due to biasing, by determining the mean composition of a large number of samples collected worldwide and from analyses of sedimentary rocks that have sampled the crust naturally by the process of erosion (Taylor & Scott, 1985; Gao et al., 1998). This composition corresponds to a rock type between granodiorite and diorite, and is characterized by a relatively high concentration of the heat-producing elements.

2.4.3 Middle and lower continental crust

For a 40 km thick average global continental crust (Christensen & Mooney, 1995; Mooney et al., 1998), the middle crust is some 11 km thick and ranges in depth from 12 km, at the top, to 23 km at the bottom (Rudnick & Fountain, 1995; Gao et al., 1998). The average lower crust thus begins at 23 km depth and is 17 km thick. However, the depth and thickness of both middle and lower crust vary considerably from setting to setting. In tectonically active rifts and rifted margins, the middle and lower crust generally are thin. The lower crust in these settings can range from negligible to more than 10 km thick (Figs 7.5, 7.32a). In Mesozoic–Cenozoic orogenic belts where the crust is much thicker, the lower crust may be up to 25 km thick (Rudnick & Fountain, 1995).

The velocity range of the lower crust (6.8–$7.7\,km\,s^{-1}$, Section 2.2) cannot be explained by a simple increase of seismic velocity with depth. Consequently, either the chemical composition must be more mafic, or denser, high-pressure phases are present. Information derived from geologic studies supports this conclusion, indicating that continental crust becomes denser and more mafic with depth. In addition, the results from these studies show that the concentration of heat-producing elements decreases rapidly from the surface downwards. This decrease is due, in part, to an increase in metamorphic grade but is also due to increasing proportions of mafic lithologies.

In areas of thin continental crust, such as in rifts and at rifted margins, the middle and lower crust may be composed of low- and moderate-grade metamorphic rocks. In regions of very thick crust, such as orogenic belts, the middle and lower crust typically are composed of high-grade metamorphic mineral assemblages. The middle crust in general may contain more evolved and less mafic compositions compared to the lower crust. Metasedimentary rocks may be present in both layers. If the lower crust is dry, its composition could correspond to a high-pressure form of granulite ranging in composition from granodiorite to diorite (Christensen & Fountain, 1975; Smithson & Brown, 1977), and containing abundant plagioclase and pyroxene minerals. In the overthickened roots of orogens, parts of the lower crust may record the transition to the eclogite facies, where plagioclase is unstable and mafic rocks transform into very dense, garnet-, pyroxene-bearing assemblages (Section 9.9). If the lower crust is wet, basaltic rocks would occur in the form of amphibolite. If mixed with more silicic material, this would have a seismic velocity in the correct range. Studies of exposed sections of ancient lower crust suggest that both dry and wet rock types typically are present (Oliver, 1982; Baldwin et al., 2003).

Another indicator of lower crust composition is the elastic deformation parameter Poisson's ratio, which can be expressed in terms of the ratio of P and S wave velocities for a particular medium. This parameter varies systematically with rock composition, from approximately 0.20 to 0.35. Lower values are characteristic of rocks with high silica content, and high values with mafic rocks and relatively low silica content. For example, beneath the Main Ethiopian Rift in East Africa (Fig. 7.2) Poisson's ratios vary from 0.27 to 0.35 (Dugda *et al.*, 2005). By contrast, crust located outside the rift is characterized by varying from 0.23 to 0.28. The higher ratios beneath the rift are attributed to the intrusion and extensive modification of the lower crust by mafic magma (Fig. 7.5).

Undoubtedly, the lower crust is compositionally more complex than suggested by these simple geophysical models. Studies of deep crustal xenoliths and crustal contaminated magmas indicate that there are significant regional variations in its composition, age, and thermal history. Deep seismic reflection investigations (Jackson, H.R., 2002; van der Velden *et al.*, 2004) and geologic studies of ancient exposures (Karlstrom & Williams, 1998; Miller & Paterson, 2001a; Klepeis *et al.*, 2004) also have shown that this compositional complexity is matched by a very heterogeneous structure. This heterogeneity reflects a wide range of processes that create and modify the lower crust. These processes include the emplacement and crystallization of magma derived from the mantle, the generation and extraction of crustal melts, metamorphism, erosion, tectonic burial, and many other types of tectonic reworking (Sections 9.8, 9.9).

2.4.4 The oceanic crust

The oceanic crust (Francheteau, 1983) is in isostatic equilibrium with the continental crust according to the Airy mechanism (Section 2.11.2), and is consequently much thinner. Seismic refraction studies have confirmed this and show that oceanic crust is typically 6–7 km thick beneath an average water depth of 4.5 km. Thicker oceanic crust occurs where the magma supply rate is anomalously high due to higher than normal temperatures in the upper mantle. Conversely, thinner than normal crust forms where upper mantle temperatures are anomalously low, typically because of a very low rate of formation (Section 6.10).

Table 2.2 *Oceanic crustal structure (after Bott, 1982).*

	P velocity (km s^{-1})	Average thickness (km)
Water	1.5	4.5
Layer 1	1.6–2.5	0.4
Layer 2	3.4–6.2	1.4
Layer 3	6.4–7.0	5.0
	Moho	
Upper mantle	7.4–8.6	

The earliest refraction surveys produced time–distance data of relatively low accuracy that, on simple inversion using plane-layered models, indicated the presence of three principal layers. The velocities and thicknesses of these layers are shown in Table 2.2. More recent refraction studies, employing much more sophisticated equipment and interpretational procedures (Kennett B.L.N., 1977), have shown that further subdivision of the main layers is possible (Harrison & Bonatti, 1981) and that, rather than a structure in which velocities increase downwards in discrete jumps, there appears to be a progressive velocity increase with depth (Kennett & Orcutt, 1976; Spudich & Orcutt, 1980). Figure 2.17 compares the velocity structure of the oceanic crust as determined by early and more recent investigations.

2.4.5 Oceanic layer 1

Layer 1 has been extensively sampled by coring and drilling. Seabed surface materials comprise unconsolidated deposits including terrigenous sediments carried into the deep oceans by turbidity currents, and pelagic deposits such as brown zeolite clays, calcareous and silicic oozes, and manganese nodules. These deep-sea sediments are frequently redistributed by bottom currents or contour currents, which are largely controlled by thermal and haline anomalies within the oceans. The dense, cold saline water produced at the poles sinks and underflows towards equatorial regions, and is deflected by the Coriolis force. The resulting currents give rise to sedimentary deposits that are termed *contourites* (Stow & Lovell, 1979).

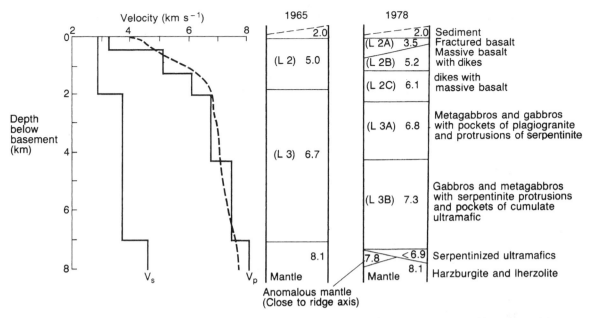

Figure 2.17 P and S wave velocity structure of the oceanic crust and its interpretation in terms of layered models proposed in 1965 and 1978. Numbers refer to velocities in km s⁻¹. Dashed curve refers to gradational increase in velocity with depth deduced from more sophisticated inversion techniques (after Spudich & Orcutt, 1980 and Harrison & Bonatti, 1981).

Layer 1 is on average 0.4 km thick. It progressively thickens away from the ocean ridges, where it is thin or absent. There is, however, a systematic difference in the sediment thicknesses of the Pacific and Atlantic/Indian oceans. The former is rimmed by trenches, that trap sediments of continental origin, and the latter are not, allowing greater terrestrial input. The interface between layer 1 and layer 2 is considerably more rugged than the seabed, because of the volcanic and faulted nature of layer 2. Within layer 1 are a number of horizons that show up as prominent reflectors on seismic reflection records. Edgar (1974) has described the acoustic stratigraphy in the North Atlantic, where up to four supra-basement reflectors are found (Fig. 2.18). Horizon A corresponds to an Eocene chert, although deep sea drilling indicates that it maintains its reflective character even when little or no chert is present. In such locations it may correspond to an early Cenozoic hiatus beneath the chert. Horizon A* occurs beneath A, and represents the interface between Late Cretaceous/Paleogene metal-rich clays and underlying euxinic black clays. Horizon B represents the base of the black clays, where they overlie a Late Jurassic/Lower Cretaceous limestone. Horizon B may represent a sedimentary horizon,

although it has also been identified as basalt similar to that at the top of layer 2.

Reflectors similar to A and B have been identified in the Pacific and Caribbean, where they are termed A', B' and A", B", respectively.

2.4.6 Oceanic layer 2

Layer 2 is variable in its thickness, in the range 1.0–2.5 km. Its seismic velocity is similarly variable in the range 3.4–6.2 km s⁻¹. This range is attributable to either consolidated sediments or extrusive igneous material. Direct sampling and dredging of the sediment-free crests of ocean ridges, and the necessity of a highly magnetic lithology at this level (Section 4.2), overwhelmingly prove an igneous origin. The basalts recovered are olivine tholeiites containing calcic plagioclase, and are poor in potassium, sodium, and the incompatible elements (Sun et al., 1979). They exhibit very little areal variation in major element composition, with the exception of locations close to oceanic islands (Section 5.4).

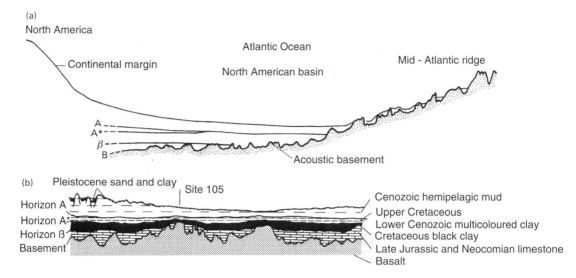

Figure 2.18 *(a) Major seismic reflectors in the western Atlantic Ocean. (b) Corresponding lithologies determined by deep sea drilling (after Edgar, 1974, Fig. 1. Copyright © 1974, with kind permission of Springer Science and Business Media).*

Three subdivisions of layer 2 have been recognized. Sublayer 2A is only present on ocean ridges near eruptive centers in areas affected by hydrothermal circulation of sea water, and ranges in thickness from zero to 1 km. Its porous, rubbly nature, as indicated by a P wave velocity of $3.6\,km\,s^{-1}$, permits such circulation. The very low velocities ($2.1\,km\,s^{-1}$) of the top of very young layer 2 located on the Mid-Atlantic Ridge (Purdy, 1987) probably indicate a porosity of 30–50%, and the much higher velocities of older layer 2 imply that the porosity must be reduced quite rapidly after its formation. Sublayer 2B forms the normal acoustic basement of layer 1 when sublayer 2A is not developed. Its higher velocity of 4.8–$5.5\,km\,s^{-1}$ suggests a lower porosity. With time layer 2A may be converted to layer 2B by the infilling of pores by secondary minerals such as calcite, quartz, and zeolites. Sublayer 2C is about 1 km thick, where detected, and its velocity range of 5.8–$6.2\,km\,s^{-1}$ may indicate a high proportion of intrusive, mafic rocks. This layer grades downwards into layer 3.

The DSDP/ODP drill hole 504B, that drilled through the top 1800 m of igneous basement in 6 Ma old crust on the Costa Rica Rift, in the eastern central Pacific, encountered pillow lavas and dikes throughout. It revealed that, at least for this location, the layer 2/3 seismic boundary lies within a dike complex and is associated with gradual changes in porosity and alteration (Detrick *et al.*, 1994).

2.4.7 Oceanic layer 3

Layer 3 is the main component of the oceanic crust and represents its plutonic foundation (Fox & Stroup, 1981). Some workers have subdivided it into sublayer 3A, with a velocity range of 6.5–$6.8\,km\,s^{-1}$, and a higher velocity lower sublayer 3B (7.0–$7.7\,km\,s^{-1}$) (Christensen & Salisbury, 1972), although the majority of seismic data can be explained in terms of a layer with a slight positive velocity gradient (Spudich & Orcutt, 1980).

Hess (1962) suggested that layer 3 was formed from upper mantle material whose olivine had reacted with water to varying degrees to produce serpentinized peridotite, and, indeed, 20–60% serpentinization can explain the observed range of P wave velocities. However for oceanic crust of normal thickness (6–7 km) this notion can now be discounted, as the value of Poisson's ratio for layer 3A, which can be estimated directly from a knowledge of both P and S wave velocities, is much lower than would be expected for serpentinized peridotite. In fact, Poisson's ratio for layer 3A is more in accord with a gabbroic composition, which also provides seismic velocities in the observed range. It is possible, however, that all or at least part of layer 3B, where recognized, consists of serpentinized ultramafic material.

The concept of a predominantly gabbroic layer 3 is in accord with models suggested for the origin of oceanic lithosphere (Section 6.10). These propose that layer 3 forms by the crystallization of a magma chamber or magma chambers, with an upper layer, possibly corresponding to sublayer 3A, of isotropic gabbro and a lower layer, possibly corresponding to 3B, consisting of cumulate gabbro and ultramafic rocks formed by crystal settling. This layering has been confirmed by direct observation and sampling by submersible on the Vema Fracture Zone in the North Atlantic (Auzende *et al.*, 1989).

2.5 OPHIOLITES

The study of oceanic lithosphere has been aided by investigations of characteristic rock sequences on land known as ophiolites (literally "snake rock", referring to the similarity of the color and texture to snakeskin; see Nicolas, 1989, for a full treatment of this topic). Ophiolites usually occur in collisional orogens (Section 10.4), and their association of deep-sea sediments, basalts, gabbros, and ultramafic rocks suggests that they originated as oceanic lithosphere and were subsequently thrust up into their continental setting by a process known as *obduction* (Dewey, 1976; Ben-Avraham *et al.*, 1982; Section 10.6.3). The complete ophiolite sequence (Gass, 1980) is shown in Table 2.3. The analogy of ophiolites with oceanic lithosphere is supported by the gross similarity in chemistry (although there is considerable difference in detail), metamorphic grades corresponding to temperature gradients existing under spreading centers, the presence of similar ore minerals, and the observation that the sediments were formed in deep water (Moores, 1982). Salisbury & Christensen (1978) have compared the velocity structure of the oceanic lithosphere with seismic velocities measured in samples from the Bay of Islands ophiolite complex in Newfoundland, and concluded that the determined velocity stratigraphies are identical. Figure 2.19 shows the correlation between the oceanic lithosphere and three well-studied ophiolite bodies.

At one time it seemed that investigations of the petrology and structure of the oceanic lithosphere could conveniently be accomplished by the study of

Table 2.3 *Correlation of ophiolite stratigraphy with the oceanic lithosphere (after Gass, 1980 with permission from the Ministry of Agriculture and Natural Resources, Cyprus).*

Complete ophiolite sequence	Oceanic correlation
Sediments	Layer 1
Mafic volcanics, commonly pillowed, merging into Mafic sheeted dike complex }	Layer 2

High level intrusives
    Trondhjemites
    Gabbros }
	Layer 3

Layered cumulates
    Olivine gabbros
    Pyroxenites
    Peridotites }
	— Moho —

Harzburgite, commonly serpentinized ⊥ lherzolite, dunite, chromitite	Upper mantle

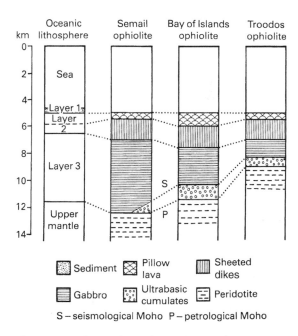

Figure 2.19 *Comparison of oceanic crustal structure with ophiolite complexes (after Mason, 1985, with permission from Blackwell Publishing).*

ophiolite sequences on land. However, this simple analogy has been challenged, and it has been suggested that ophiolites do not represent typical oceanic lithosphere, and were not emplaced exclusively during continental collision (Mason, 1985).

Dating of events indicates that obduction of many ophiolites occurred very soon after their creation. Continental collision, however, normally occurs a long time after the formation of a mid-ocean ridge, so that the age of the sea floor obducted should be considerably greater than that of the collisional orogeny. Ophiolites consequently represent lithosphere that was obducted while young and hot. Geochemical evidence (Pearce, 1980; Elthon, 1991) has suggested that the original sites of ophiolites were backarc basins (Section 9.10; Cawood & Suhr, 1992), Red Sea-type ocean basins, or the forearc region of subduction zones (Flower & Dilek, 2003). The latter setting seems at first to be an unlikely one. However, the petrology and geochemistry of the igneous basement of forearcs, which is very distinctive, is very comparable to that of many ophiolites. Formation in a forearc setting could also explain the short time interval between formation and emplacement, and the evidence for the "hot" emplacement of many ophiolites. A backarc or forearc origin is also supported by the detailed geochemistry of the lavas of most ophiolites, which indicates that they are derived from melts that formed above subduction zones.

There have been many different mechanisms proposed for ophiolite obduction, none of which can satisfactorily explain all cases. It must thus be recognized that there may be several operative mechanisms and that, although certainly formed by some type of accretionary process, ophiolite sequences may differ significantly, notably in terms of their detailed geochemistry, from lithosphere created at mid-ocean ridge crests in the major ocean basins.

Although many ophiolites are highly altered and tectonized, because of the way in which they are uplifted and emplaced in the upper crust, there are definite indications that there is more than one type of ophiolite. Some have the complete suite of units listed in Table 2.3 and illustrated in Fig. 2.19, others consist solely of deep-sea sediments, pillow lavas, and serpentinized peridotite, with or without minor amounts of gabbro. If present these gabbros often occur as intrusions within the serpentinized peridotite. These latter types are remarkably similar to the inferred nature of the thin oceanic crust that forms where magma supply rates are low. This type of crust is thought to form when the rate

of formation of the crust is very low (Section 6.10), in the vicinity of transform faults at low accretion rates (Section 6.7), and in the initial stages of ocean crust formation at nonvolcanic passive continental margins (Section 7.7.2). It seems probable that Hess (1962), in suggesting that layer 3 of the oceanic crust is serpentinized mantle, was in part influenced by his experience and knowledge of ophiolites of this type in the Appalachian and Alpine mountain belts.

2.6 METAMORPHISM OF OCEANIC CRUST

Many of the rocks sampled from the ocean basins show evidence of metamorphism, including abundant greenschist facies assemblages and alkali metasomatism: In close proximity to such rocks, however, are found completely unaltered species.

It is probable that this metamorphism is accomplished by the hydrothermal circulation of seawater within the oceanic crust. There is much evidence for the existence of such circulation, such as the presence of metalliferous deposits which probably formed by the leaching and concentration of minerals by seawater, observations of active hydrothermal vents on ocean ridges (Section 6.5), and the observed metamorphism within ophiolite sequences.

Hydrothermal circulation takes place by convective flow, probably through the whole of the oceanic crust (Fyfe & Lonsdale, 1981), and is of great significance. It influences models of heat production, as it has been estimated that approximately 25% of the heat escaping from the Earth's surface is vented at the mid-ocean ridges. The circulation must modify the chemistry of the ocean crust, and consequently will affect the chemical relationship of lithosphere and asthenosphere over geologic time because of the recycling of lithosphere that occurs at subduction zones. It is also responsible for the formation of certain economically important ore deposits, particularly massive sulfides.

These hydrothermal processes are most conveniently studied in the metamorphic assemblages of ophiolite complexes, and the model described below has been derived by Elthon (1981).

Hydrothermal metamorphism of pillow lavas and other extrusives gives rise to low-temperature (<230°C)

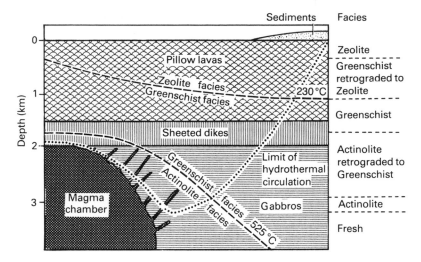

Figure 2.20 *Schematic model for hydrothermal metamorphism of the oceanic crust at a spreading center (redrawn from Elthon, 1981).*

and greenschist facies assemblages (Fig. 2.20). The distribution of alteration is highly irregular, and is controlled by the localized fissuring of the extrusive rocks. Higher temperature metamorphism is widespread within the sheeted dike complex, producing assemblages typical of the actinolite facies, although pockets of unaltered rock do occur. The highest metamorphic temperatures are achieved at the base of the sheeted dike complex and the upper part of the gabbroic section. Rarely, retrograde rocks of the greenschist facies occur at this level. Alteration decreases to only about 10% within the top kilometer of the gabbroic section and thereafter metamorphism is restricted to the locality of fissures and dikes, although metamorphism does not completely terminate at depth. According to this model, seawater circulation occurs extensively in the upper 3 km of the crust, producing the metamorphic assemblages and cooling the crust. High-temperature metamorphism only occurs near the spreading center. At depth the circulation becomes diminished as secondary minerals are deposited within the flow channels.

As the ridge spreads continuously, oceanic lithosphere is moved laterally from the heat source and undergoes retrograde metamorphism. This depends upon an adequate water supply, as water distribution is the major control of metamorphic grade. The absence of sufficient water allows the preservation of relict high temperature assemblages. The heterogeneous nature of the distribution of metamorphic facies is consequently explained by a similarly heterogeneous distribution of circulating fluids rather than extreme temperature variations. As indicated in Sections 2.4.7 and 2.5, parts of the oceanic crust consist of serpentinite, that is, hydrated ultramafic rock. The ultramafic rock may be formed by magmatic differentiation within the gabbro layer, or derived directly from the mantle.

2.7 DIFFERENCES BETWEEN CONTINENTAL AND OCEANIC CRUST

On the basis of information presented in this and following chapters, the major differences between continental and oceanic crust can be summarized as follows:

1 *Layering*. The large-scale layering of the continental crust is ill defined and highly variable, reflecting a complex geologic history. In places there is a broad subdivision by the

Conrad discontinuity, but this is not globally developed. By contrast, the layering of the majority of oceanic crust is well defined into three distinct layers. However, the nature of these layers, in particular layers 2 and 3, may change quite markedly with depth.

2 *Thickness.* The thickness of continental crust averages 40 km but is quite variable, thinning to only a few kilometers beneath rifts and thickening to up to 80 km beneath young mountain belts. Most oceanic crust has a remarkably constant thickness of about 7 km, although layer 1, the sedimentary layer, increases in thickness towards ocean margins that are not characterized by ocean trenches. Differences in the thickness and the creep strength (Section 2.10.4) of continental crust make the lower crust of continental regions much more likely to deform pervasively than in the lower layers of oceanic crust (Section 2.10.5).

3 *Age.* Continental crust is as at least as old as 4.0 Ga, the age of the oldest rocks yet discovered (Section 11.1). On a very broad scale the oldest crust consists of Precambrian cratons or shield areas that are surrounded by younger orogenic belts, both active and inactive. Oceanic crust, however, is nowhere older than 180 Ma, and progressively increases in age outwards from oceanic ridges (Section 4.1). Oceans are consequently viewed as essentially transient features of the Earth's surface. About 50% of the surface area of the present day ocean floor has been created during the last 65 Ma, implying that 30% of the solid Earth's surface has been created during the most recent 1.5% of geologic time.

4 *Tectonic activity.* Continental crust may be extensively folded and faulted and preserves evidence of being subjected to multiple tectonic events. Oceanic crust, however, appears to be much more stable and has suffered relatively little deformation except at plate margins.

5 *Igneous activity.* There are very few active volcanoes on the great majority of the continental crust. The only major locations of activity are mountain belts of Andean type (Section 9.8). The activity within the oceans is very much greater. Ocean ridges and island arcs are the location of the Earth's most active areas of volcanic and plutonic activity. Oceanic islands are a third distinct, but less prolific oceanic setting for igneous activity.

2.8 THE MANTLE

2.8.1 Introduction

The mantle constitutes the largest internal subdivision of the Earth by both mass and volume, and extends from the Moho, at a mean depth of about 21 km, to the core–mantle boundary at a depth of 2891 km. On a gross scale it is believed to be chemically homogeneous, apart from the abundances of minor and trace elements, and formed of silicate minerals. The mineralogy and structure of the silicates change with depth and give rise to a transition zone between 410 and 660 km depth, which separates the upper and lower mantle.

Mantle materials are only rarely brought to the surface, in ophiolite complexes (Section 2.5), in kimberlite pipes (Section 13.2.2), and as xenoliths in alkali basalts. Consequently, most of our information about the mantle is indirect and based on the variation of seismic velocities with depth combined with studies of mineral behavior at high temperatures and pressure, and in shock-wave experiments. Geochemical studies of meteorites and ultramafic rocks are also utilized in making predictions about the mantle.

2.8.2 Seismic structure of the mantle

The uppermost part of the mantle constitutes a high velocity lid typically 80–160 km thick in which seismic velocities remain constant at a figure in excess of 7.9 km s^{-1} or increase slightly with depth. This part of the mantle makes up the lower portion of the lithosphere (Section 2.12). Beneath the lithosphere lies a *low velocity zone* extending to a depth of approximately 300 km. This appears to be present beneath most regions of the Earth with the exception of the mantle beneath cratonic areas. From the base of this zone seismic velocities increase slowly until a major discontinuity is reached at a depth of 410 km, marking the upper region of the *transition zone*. There is a further velocity discontinuity at a depth of 660 km, the base of the transition zone.

Within the lower mantle velocities increase slowly with depth until the basal 200–300 km where gradients decrease and low velocities are present. This lowermost layer, at the core–mantle boundary, is known as *Layer D″* (Section 12.8.4) (Knittle & Jeanloz, 1991). Seismic studies have detected strong lateral heterogeneities and the presence of thin (5–50 km thick) *ultra-low velocity zones* at the base of Layer D″ (Garnero et al., 1998).

2.8.3 Mantle composition

The fact that much of the oceanic crust is made up of material of a basaltic composition derived from the upper mantle suggests that the upper mantle is composed of either peridotite or eclogite (Harrison & Bonatti, 1981). The main difference between these two rock types is that peridotite contains abundant olivine and less than 15% garnet, whereas eclogite contains little or no olivine and at least 30% garnet. Both possess a seismic velocity that corresponds to the observed upper mantle value of about 8 km s^{-1}.

Several lines of evidence now suggest very strongly that the upper mantle is peridotitic. Beneath the ocean basins the P_n velocity is frequently anisotropic, with velocities over 15% higher perpendicular to ocean ridges. This can be explained by the preferred orientation of olivine crystals, whose long [100] axes are believed to lie in this direction. None of the common minerals of eclogite exhibit the necessary crystal elongation. A peridotitic composition is also indicated by estimates of Poisson's ratio from P and S velocities, and the presence of peridotites in the basal sections of ophiolite sequences and as nodules in alkali basalts. The density of eclogites is also too high to explain the Moho topography of isostatically compensated crustal structures.

The bulk composition of the mantle can be estimated in several ways: by using the compositions of various ultramafic rock types, from geochemical computations, from various meteorite mixtures, and by using data from experimental studies. It is necessary to distinguish between undepleted mantle and depleted mantle which has undergone partial melting so that many of the elements which do not easily substitute within mantle minerals have been removed and combined into the crust. The latter, so called "incompatible" elements, include the heat producing elements K, Th, and U. It is clear from the composition of mid-ocean ridge basalts (MORB), however, that the mantle from which they are derived by partial fusion is relatively depleted in these elements. So much so that, if the whole mantle had this composition, it would only account for a small fraction of the heat flow at the Earth's surface emanating from the mantle (Hofmann, 1997). This, and other lines of geochemical evidence, have led geochemists to conclude that all or most of the lower mantle must be more enriched in incompatible elements than the upper mantle and that it is typically not involved in producing melts that reach the surface. However, seismological evidence relating to the fate of subducted oceanic lithosphere (Sections 9.4, 12.8.2) and the lateral heterogeneity of Layer D″ suggests mantle wide convection and hence mixing (Section 12.9). Helffrich & Wood (2001) consider that the various lines of geochemical evidence can be reconciled with whole mantle convection if various small- and large-scale heterogeneities in the lower mantle revealed by seismological studies are remnants of subducted oceanic and continental crust. They estimate that these remnants make up about 16% and 0.3% respectively of the mantle volume.

Although estimates of bulk mantle composition vary in detail, it is generally agreed that at least 90% of the mantle by mass can be represented in terms of the oxides FeO, MgO, and SiO$_2$, and a further 5–10% is made up of CaO, Al$_2$O$_3$, and Na$_2$O.

2.8.4 The mantle low velocity zone

The low velocity zone (Fig. 2.16) is characterized by low seismic velocities, high seismic attenuation, and a high electrical conductivity. The seismic effects are more pronounced for S waves than for P waves. The low seismic velocities could arise from a number of different mechanisms, including an anomalously high temperature, a phase change, a compositional change, the presence of open cracks or fissures, and partial melting. All but the latter appear to be unlikely, and it is generally accepted that the lower seismic velocities arise because of the presence of molten material. That melting is likely to occur in this region is supported by the fact that it is at this level that mantle material most closely approaches its melting point (Section 2.12, Fig. 2.36).

Only a very small amount of melt is required to lower the seismic velocity of the mantle to the observed values and to provide the observed attenuation properties. A liquid fraction of less than 1% would, if distributed along a network of fissures at grain boundaries, produce these effects (O'Connell & Budiansky, 1977). The melt may also be responsible for the high electrical conductivity of this zone. For the partial melting to occur, it is probable that a small quantity of water is required to lower the silicate melting point, and that this is supplied from the breakdown of hydrous mantle phases. The base of the low velocity zone and even its existence may be controlled by the availability of water in the upper mantle (Hirth & Kohlstedt, 2003).

The mantle low velocity zone is of major importance to plate tectonics as it represents a low viscosity layer along which relative movements of the lithosphere and asthenosphere can be accommodated.

2.8.5 The mantle transition zone

There are two major velocity discontinuities in the mantle at depths of 410 km and 660 km. The former marks the top of the transition zone and the latter its base. The discontinuities are rarely sharp and occur over a finite range in depth, so it is generally believed that they represent phase changes rather than changes in chemistry. Although these discontinuities could be due to changes in the chemical composition of the mantle at these depths, pressure induced phase changes are considered to be the more likely explanation. High-

pressure studies have shown that olivine, the dominant mineral in mantle peridotite, undergoes transformations to the spinel structure at the pressure/temperature conditions at 410 km depth and then to perovskite plus magnesiowüstite at 660 km (Table 2.4) (Helffrich & Wood, 2001). Within subducting lithosphere, where the temperature at these depths is colder than in normal mantle, the depths at which these discontinuities occur are displaced exactly as predicted by thermal modeling and high-pressure experiments (Section 9.5). This lends excellent support to the hypothesis that the upper and lower bounds of the transition zone are defined by phase transformations. The other components of mantle peridotite, pyroxene and garnet, also undergo phase changes in this depth range but they are gradual and do not produce discontinuities in the variation of seismic velocity with depth. Pyroxene transforms into the garnet structure at pressures corresponding to 350–500 km depth; at about 580 km depth Ca-perovskite begins to exsolve from the garnet, and at 660–750 km the remaining garnet dissolves in the perovskite phase derived from the transformation of olivine. Thus the lower mantle mostly consists of phases with perovskite structure.

2.8.6 The lower mantle

The lower mantle represents approximately 70% of the mass of the solid Earth and almost 50% of the mass of the entire Earth (Schubert et al., 2001). The generally smooth increase in seismic wave velocities with depth in most of this layer led to the assumption that it is relatively homogeneous in its mineralogy, having mostly a perovskite structure. However, more detailed seismo-

Table 2.4 *Phase transformations of olivine that are thought to define the upper mantle transition zone (after Helffrich & Wood, 2001).*

Depth	Pressure	
410 km	13–14 GPa	$(Mg,Fe)_2SiO_4 = (Mg,Fe)_2SiO_4$ Olivine Wadsleyite (β-spinel structure)
520 km	18 GPa	$(Mg,Fe)_2SiO_4 = (Mg,Fe)_2SiO_4$ Wadsleyite Ringwoodite (γ-spinel structure)
660 km	23 GPa	$(Mg,Fe)_2SiO_4 = (Mg,Fe)SiO_3 + (Mg,Fe)O$ Ringwoodite Perovskite Magnesiowüstite

logical studies have revealed that the lower mantle has thermal and/or compositional heterogeneity, probably as a result of the penetration of subducted oceanic lithosphere through the 660 km discontinuity (Section 2.8.3).

The lowest 200–300 km of the mantle, Layer D″ (Section 12.8.4), is often characterized by a decrease in seismic velocity, which is probably related to an increased temperature gradient above the mantle-core boundary. This lower layer shows large lateral changes in seismic velocity, indicating it is very heterogeneous. Ultra-low velocity zones, which show a 10% or greater reduction in both P and S wave velocities relative to the surrounding mantle, have been interpreted to reflect the presence of partially molten material (Williams & Garnero, 1996). These zones are laterally very heterogeneous and quite thin (5–40 km vertical thickness). Laboratory experiments suggest that the liquid iron of the core reacts with mantle silicates in Layer D″, with the production of metallic alloys and nonmetallic silicates from perovskite. Layer D″ thus is important because it governs core–mantle interactions and also may be the source of deep mantle plumes (Sections 12.8.4, 12.10).

2.9 THE CORE

The core, a spheroid with a mean radius of 3480 km, occurs at a depth of 2891 km and occupies the center of the Earth. The core–mantle boundary (Gutenberg discontinuity) generates strong seismic reflections and thus probably represents a compositional interface.

The outer core, at a depth of 2891–5150 km, does not transmit S waves and so must be fluid. This is confirmed by the generation of the geomagnetic field in this region by dynamic processes and by the long period variations observed in the geomagnetic field (Section 3.6.4). The convective motions responsible for the geomagnetic field involve velocities of ~10^4 m a^{-1}, five orders of magnitude greater than convection in the mantle. A fluid state is also indicated by the response of the Earth to the gravitational attraction of the Sun and Moon.

The boundary between the outer core and inner core at 5150 km depth is sharp, and not represented by any form of transition zone. The inner core is believed to be solid for several reasons. Certain oscillations of the Earth, produced by very large earthquakes, can only be explained by a solid inner core. A seismic phase has been recognized that travels to and from the inner core as a P wave, but traverses the inner core as an S wave. The amplitude of a phase reflected off the inner core also suggests that it must have a finite rigidity and thus be a solid.

Shock wave experiments have shown that the major constituents of both the inner and outer core must comprise elements of an atomic number greater than 23, such as iron, nickel, vanadium, or cobalt. Of these elements, only iron is present in sufficient abundance in the solar system to form the major part of the core. Again, by considering solar system abundances, it appears that the core should contain about 4% nickel. This iron–nickel mixture provides a composition for the outer core that is 8–15% too dense and it must therefore contain a small quantity of some lighter element or elements. The inner core, however, has a seismic velocity and density consistent with a composition of pure iron.

There are several candidates for the light elements present in the outer core, which include silicon, sulfur, oxygen, and potassium (Brett, 1976). Silicon requires an over-complex model for the formation of the Earth and sulfur conflicts with the idea that the interior of the Earth is highly depleted in volatile elements. Oxygen appears to be the most likely light element as FeO is probably sufficiently soluble in iron. The presence of potassium is speculative, but is interesting in that it would provide a heat source in the core that would be active over the whole of the Earth's history. It would also help to explain an apparent potassium deficiency in the Earth compared to meteorites.

2.10 RHEOLOGY OF THE CRUST AND MANTLE

2.10.1 Introduction

Rheology is the study of deformation and the flow of materials under the influence of an applied stress (Ranalli, 1995). Where temperature, pressure, and the magnitudes of the applied stresses are relatively low, rocks tend to break along discrete surfaces to form fractures and faults. Where these factors are relatively high rocks tend to deform by ductile flow. Measures of strain are used to quantify the deformation.

Stress (σ) is defined as the force exerted per unit area of a surface, and is measured in Pascals (Pa). Any stress acting upon a surface can be expressed in terms of a normal stress perpendicular to the surface and two components of shear stress in the plane of the surface. The state of stress within a medium is conveniently specified by the magnitudes and directions of three *principal stresses* that act on three planes in the medium along which the shear stress is zero. The principal stresses are mutually orthogonal and are termed σ_1, σ_2, and σ_3, referring to the maximum, intermediate and minimum principal stresses, respectively. In the geosciences, compressive stresses are expressed as positive and tensile stresses negative. The magnitude of the difference between the maximum and minimum principal stresses is called the *differential stress*. *Deviatoric stress* represents the departure of a stress field from symmetry. The value of the differential stress and the characteristics of deviatoric stress both influence the extent and type of distortion experienced by a body.

Strain (ε) is defined as any change in the size or shape of a material. Strains are usually expressed as ratios that describe changes in the configuration of a solid, such as the change in the length of a line divided by its original length. *Elastic* materials follow Hooke's law where strain is proportional to stress and the strain is reversible until a critical stress, known as the *elastic limit*, is reached. This behavior typically occurs at low stress levels and high strain rates. Beyond the elastic limit, which is a function of temperature and pressure, rocks deform by either brittle fracturing or by ductile flow. The *yield stress* (or yield strength) is the value of the differential stress above the elastic limit at which deformation becomes permanent. *Plastic* materials display continuous, irreversible deformation without fracturing.

The length of time over which stress is applied also is important in the deformation of Earth materials (Park, 1983). Rock rheology in the short term (seconds or days) is different from that of the same material stressed over durations of months or years. This difference arises because rocks exhibit higher strength at high strain rates than at low strain rates. For example, when a block of pitch is struck with a hammer, that is, subjected to rapid "instantaneous" strain, it shatters. However, when left for a period of months, pitch deforms slowly by flowing. This slow long-term flow of materials under constant stress is known as *creep*. On time scales of thousands of years, information about the strength and rheology of the lithosphere mainly comes from observations of isostasy and lithospheric flexure (Section 2.11.4). On time scales of millions of years, Earth rheology generally is studied using a continuum mechanics approach, which describes the macroscopic relationships between stress and strain, and their time derivatives. Alternatively, the long-term rheology of the Earth may be studied using a microphysical approach, where the results of laboratory experiments and observations of microstructures are used to constrain the behavior of rocks. Both of these latter approaches have generated very useful results (e.g. Sections 7.6.6, 8.6.2, 10.2.5).

2.10.2 Brittle deformation

Brittle fracture is believed to be caused by progressive failure along a network of micro- and meso-scale cracks. The cracks weaken rock by producing local high concentrations of tensile stress near their tips. The crack orientations relative to the applied stress determine the location and magnitude of local stress maxima. Fracturing occurs where the local stress maxima exceed the strength of the rock.

This theory, known as the Griffith theory of fracture, works well under conditions of applied tensile stress or where one of the principal stresses is compressional. When the magnitude of the tensile stress exceeds the tensile strength of the material, cracks orthogonal to this stress fail first and an extension fracture occurs. Below a depth of a few hundred meters, where all principal stresses are usually compressional, the behavior of cracks is more complex. Cracks close under compression and are probably completely closed at depths of >5 km due to increasing overburden pressure. This implies that the compressive strength of a material is much greater than the tensile strength. For example, the compressive strength of granite at atmospheric pressure is 140 MPa, and its tensile strength only about 4 MPa.

Where all cracks are closed, fracturing depends upon the inherent strength of the material and the magnitude of the differential stress (Section 2.10.1). Experiments show that shear fractures, or faults, preferentially form at angles of <45° on either side of the maximum principal compressive stress when a critical shear stress on the planes is exceeded. This critical shear stress (σ_s^*) depends upon the normal stress (σ_n) on planes of potential failure and the coefficient of internal friction (μ) on those planes, which resists relative motion across them.

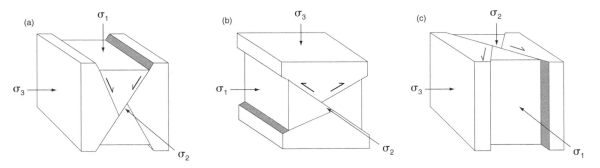

Figure 2.21 *Three classes of fault determined by the orientation of the principal stresses: (a) normal fault; (b) thrust fault; (c) strike-slip fault (after Angelier, 1994, with permission from Pergamon Press. Copyright Elsevier 1994).*

This relationship, called the Mohr–Coulomb fracture criterion, is described by the following linear equation:

$$|\sigma_s^*| = c + \mu\sigma_n$$

The cohesion (c) describes the resistance of the material to shear fracture on a plane of zero normal stress. Byerlee (1978) showed that many rock types have nearly the same coefficient of friction, within the range 0.6–0.8. The form of the equation, which is written using the absolute value of the critical shear stress, allows a pair of fractures to form that is symmetric about the axis of maximum principal compressive stress. Pore fluid pressure enhances fracturing by reducing the frictional coefficient and counteracting the normal stresses (σ_n) across the fault. The effect of pore fluid pressure explains faulting at depth, which would otherwise appear to require very high shear stresses because of the high normal stresses.

Under this compressional closed crack regime, the type of faulting which results, according to the theory of Anderson (1951), depends upon which of the principal stresses is vertical (Fig. 2.21). Normal, strike-slip, and thrust faults occur depending on whether σ_1, σ_2 or σ_3, respectively, is vertical. This theory is conceptually useful. However, it does not explain the occurrence of some faults, such as low-angle normal faults (Section 7.3), which display dips of $\leq 30°$, flat thrust faults, or faults that develop in previously fractured, anisotropic rock.

The strength of rock increases with the pressure of the surrounding rock, termed the *confining pressure*, but decreases with temperature. In the uppermost 10–15 km of the crust the former effect is dominant and rock strength tends to increase with depth. Confining pressure increases with depth at a rate of about $33\,\text{MPa}\,\text{km}^{-1}$

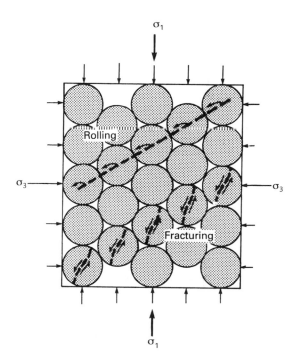

Figure 2.22 *Deformation of a brittle solid by cataclastic flow (redrawn from Ashby & Verrall, 1977, with permission from the Royal Society of London).*

depending on the density of the overlying rocks. Below 10–15 km the effect of temperature takes over, and rocks may progressively weaken downwards. However, this simple relationship can be complicated by local variations in temperature, fluid content, rock composition, and pre-existing weaknesses.

The deformation of brittle solids can take the form of *cataclasis* (Fig. 2.22) (Ashby & Verrall, 1977). This results from repeated shear fracturing, which acts to reduce the grain size of the rock, and by the sliding or rolling of grains over each other.

2.10.3 Ductile deformation

The mechanisms of ductile flow in crystalline solids have been deduced from studies of metals, which have the advantage that they flow easily at low temperatures and pressures. In general, where the temperature of a material is less than about half its melting temperature (T_m in Kelvin), materials react to low stresses by flowing slowly, or *creeping*, in the solid state. At high temperatures and pressures, the strength and flow of silicate minerals that characterize the crust (Tullis, 2002) and mantle (Li *et al.*, 2004) have been studied using experimental apparatus.

There are several types of ductile flow that may occur in the crust and mantle (Ashby & Verrall, 1977). All are dependent upon the ambient temperature and, less markedly, pressure. Increased temperature acts to lower the apparent viscosity and increase the strain rate, while increased pressure produces a more sluggish flow. In general, for ductile flow, the differential stress ($\Delta\sigma$) and the strain rate ($\delta\varepsilon/\delta\tau$) are related through a flow law of the form:

$$\Delta\sigma = [(\delta\varepsilon/\delta\tau)/A]^{1/n} \exp[E/nRT],$$

where E is the activation energy of the assumed creep process, T is temperature, R is the universal gas constant, n is an integer, and A is an experimentally determined constant.

Plastic flow occurs when the yield strength of the material is exceeded. Movement takes place by the gliding motions of large numbers of defects in the crystal lattices of minerals. Slip within a crystal lattice occurs as the individual bonds of neighboring atoms break and reform across glide planes (Fig. 2.23). This process results in linear defects, called *dislocations*, that separate slipped from unslipped parts of the crystal. The yield strength of materials deforming in this way is controlled by the magnitude of the stresses required to overcome the resistance of the crystal framework to the movement of the dislocations. The strain produced tends to be limited by the density of dislocations. The higher the density, the more difficult it is for dislocations to move in a process known as strain- or work-hardening.

Power-law creep (also known as *dislocation creep*) takes place at temperatures in excess of $0.55\,T_m$. In this form of creep the strain rate is proportional to the nth power of the stress, where $n \geq 3$. Power-law creep is similar to plastic flow, where deformation takes place by *dislocation glide*. However, in addition, the diffusion of atoms and of sites unoccupied by atoms called vacancies is permitted by the higher temperatures (Fig. 2.24). This diffusive process, termed *dislocation climb*, allows barriers to dislocation movement to be removed as they form. As a result work-hardening does not occur and steady state creep is facilitated. This balance results in *dynamic recrystallization* whereby new crystal grains form from old grains. Because of the higher temperature the yield strength is lower than for plastic flow, and strain results from lower stresses. Power-law creep is believed to be an important form of deformation in the upper mantle where it governs convective flow (Weertman, 1978). Newman & White (1997) suggest that the rheology of continental lithosphere is controlled by power-law creep with a stress exponent of three.

Diffusion creep dominates as temperatures exceed $0.85\,T_m$, and results from the migration of individual atoms and vacancies in a stress gradient (Fig. 2.25). Where the migration occurs through a crystal lattice it is known as *Nabarro–Herring creep*. Where it occurs along crystal boundaries it is known as *Coble creep*. In both forms of creep the strain rate ($\delta\varepsilon/\delta\tau$) is proportional to the differential stress ($\Delta\sigma$) with the constant of proportionality being the dynamic viscosity (η). This relationship is given by:

$$\Delta\sigma = 2\eta(\delta\varepsilon/\delta\tau)$$

The viscosity increases as the square of the grain radius so that a reduction in grain size is expected to result in rheological weakening. Diffusion creep is believed to occur in the asthenosphere (Section 2.12) and in the lower mantle (Section 2.10.6).

Superplastic creep has been observed in metals and may also occur in some rocks. This type of creep results from the coherent sliding of crystals along grain boundaries where the movement occurs without opening up gaps between grains. The sliding may be accommodated by both diffusion and dislocation mechanisms. Superplastic creep is characterized by a power-law rheology with a stress exponent of one or two and is associated with high strain rates. Some studies (e.g. Karato, 1998) have inferred that superplastic creep contributes

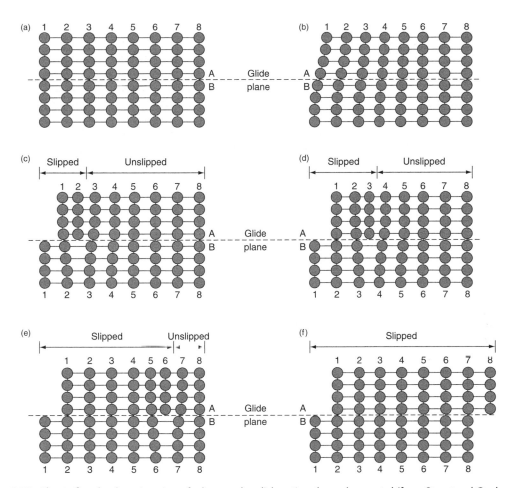

Figure 2.23 *Plastic flow by the migration of a linear edge dislocation through a crystal (from Structural Geology by Robert J. Twiss and Eldridge M. Moores. © 1992 by W.H. Freeman and Company. Used with permission).*

to deformation in the lower mantle, although this interpretation is controversial.

2.10.4 Lithospheric strength profiles

In most quantitative treatments of deformation at large scales, the lithosphere is assumed to consist of multiple layers characterized by different rheologies (e.g. Section 7.6.6). The rheologic behavior of each layer depends on the level of the differential stress ($\Delta\sigma$) and the lesser of the calculated brittle and ductile yield stresses (Section 2.10.1). The overall strength of the

lithosphere and its constituent layers can be estimated by integrating yield stress with respect to depth. This integrated strength is highly sensitive to the geothermal gradient as well as to the composition and thickness of each layer, and to the presence or absence of fluids.

The results of deformation experiments and evidence of compositional variations with depth (Section 2.4) have led investigators to propose that the lithosphere is characterized by a "jelly sandwich" type rheological layering (Ranalli & Murphy, 1987), where strong layers separate one or more weak layers. For example, Brace & Kohlstedt (1980) investigated the limits of lithospheric strength based on measurements on quartz and olivine, which are primary constituents of the

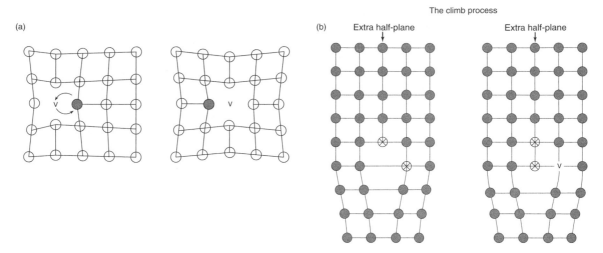

Figure 2.24 *(a) The diffusion of a vacancy (v) through a crystal; (b) the downward climb of an edge dislocation as adjacent atoms (crossed) exchange bonds leaving behind a vacancy that moves by diffusion (from Structural Geology by Robert J. Twiss and Eldridge M. Moores. © 1992 by W.H. Freeman and Company. Used with permission).*

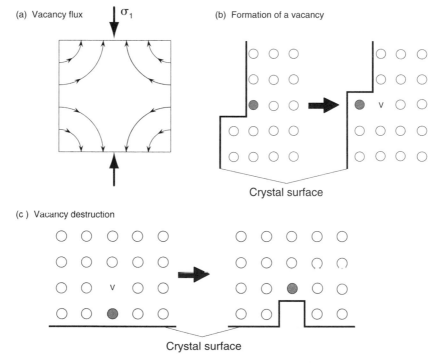

Figure 2.25 *Nabarro–Herring creep: (a) vacancies diffuse toward surfaces of high normal stress; (b) creation of a vacancy (v) at a surface of minimum compressive stress; (c) destruction of a vacancy at a surface of maximum compressive stress (from Structural Geology by Robert J. Twiss and Eldridge M. Moores. © 1992 by W.H. Freeman and Company. Used with permission). Solid lines in b and c mark crystal surface, solid circle marks the ion whose position changes during the creation of a vacancy.*

continental crust and upper mantle, respectively. The results of these and other measurements (e.g. Ranalli & Murphy, 1987; Mackwell *et al.*, 1998) suggest that within the oceanic lithosphere the upper brittle crust gives way to a region of high strength at a depth of 20–60 km, depending on the temperature gradient (Fig. 2.26a). Below this depth the strength gradually decreases and grades into that of the asthenosphere. Continental crust, however, is much thicker than oceanic crust, and at the temperatures of 400–700°C experienced in its lower layers the minerals are much weaker than the olivine found at these depths in the oceanic lithosphere. Whereas the oceanic lithosphere behaves as a single rigid plate because of its high strength, the continental lithosphere does not (Sections 2.10.5, 8.5) and typically is characterized by one or more layers of weakness at deep levels (Fig. 2.26b,c).

Figure 2.26c,d shows two other experimentally determined strength curves for continental lithosphere that illustrate the potential effects of water on the strength of various layers. These curves were calculated using rheologies for diabase and other crustal and mantle rocks, a strain rate of $(\delta\varepsilon/\delta\tau) = 10^{-17}\,s^{-1}$, a typical thermal gradient for continental crust with a surface heat flow of 60 mW m^{-2}, and a crustal thickness of 40 km (Mackwell *et al.*, 1998). The upper crust (0–15 km depth) is represented by wet quartz and Byerlee's (1978) frictional strength law (Section 2.10.2), and the middle crust (15–30 km depth) by wet quartz and power-law creep (Section 2.10.3). These and other postulated strength profiles commonly are used in thermomechanical models of continental deformation (Sections 7.6.6, 8.6.2, 10.2.5). However, it is important to keep in mind that the use of any one profile in a particular setting involves considerable uncertainty and is the subject of much debate (Jackson, J., 2002; Afonso & Ranalli, 2004; Handy & Brun, 2004). In settings where ambient conditions appear to change frequently, such as within orogens and magmatic arcs, several curves may be necessary to describe variations in rock strength with depth for different time periods.

2.10.5 Measuring continental deformation

Zones of continental deformation commonly are wider and more diffuse than zones of deformation affecting oceanic lithosphere. This characteristic results from the thickness, composition, and pressure–temperature profile of continental crust, which makes ductile flow in its lower parts more likely than it is in oceanic regions. The width and diffusivity of these zones make some of the concepts of plate tectonics, such as the rigid motion of plates along narrow boundaries, difficult to apply to the continents. Consequently, the analysis of continental deformation commonly requires a framework that is different to that used to study deformation in oceanic lithosphere (e.g. Section 8.5).

At the scale of large tectonic features such as wide intracontinental rifts (Section 7.3), continental transforms (Section 8.5), and orogenic belts (Section 10.4.3), deformation may be described by a regional horizontal velocity field rather than by the relative motion of rigid blocks (e.g. Fig. 8.18b). Methods of estimating the regional velocity field of deforming regions usually involves combining information from Global Positioning System (GPS) satellite measurements (Clarke *et al.*, 1998), fault slip rates (England & Molnar, 1997), and seismicity (Jackson *et al.*, 1992). One of the challenges of this approach is the short, decade-scale time intervals over which GPS data are collected. These short intervals typically include relatively few major earthquakes. Consequently, the measured surface motions mostly reflect nonpermanent, elastic strains that accumulate between major seismic events (i.e. interseismic) rather than the permanent strains that occur during ruptures (Bos & Spakman, 2005; Meade & Hager, 2005). This characteristic results in a regional velocity field that rarely shows the discontinuities associated with slip on major faults. Instead the displacements on faults are described as continuous functions and the velocity field is taken to represent the average deformation over a given region (Jackson, 2004). Nevertheless, regional velocity fields have proven to be a remarkably useful way of describing continental deformation. The methods commonly used to process and interpret them are discussed further in Sections 5.3 and 8.5.

Synthetic Aperture Radar (SAR) also is used to measure ground displacements, including those associated with volcanic and earthquake activity (Massonnet & Feigl, 1998). The technique involves using SAR data to measure small changes in surface elevations from satellites that fly over the same area at least twice, called repeat-pass Interferometric SAR, or InSAR. GPS data and strain meters provide more accurate and frequent observations of deformation in specific areas, but InSAR is especially good at revealing the spatial complexity of displacements that occur in tectonically active areas. In

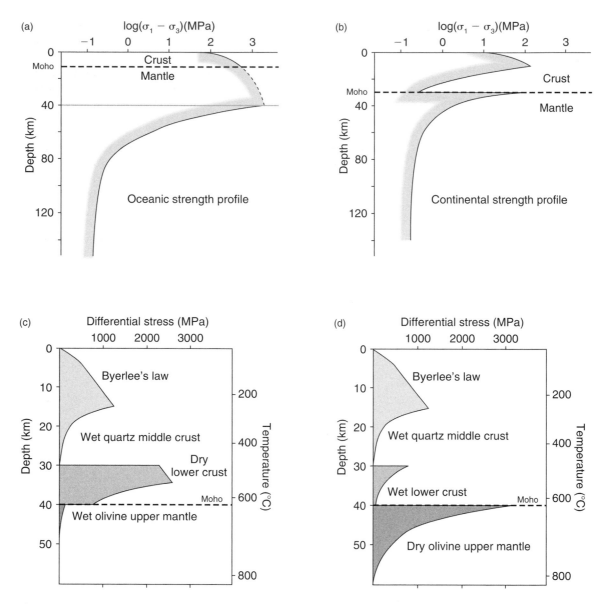

Figure 2.26 *Schematic strength profiles through (a) oceanic and (b) continental lithosphere (after Ranalli, 1995, fig. 12.2. Copyright © 1995, with kind permission of Springer Science and Business Media). Profile in (a) shows a 10-km-thick mafic crust and a 75-km-thick lithosphere. Profile in (b) shows a 30-km-thick unlayered crust and a thin, 50-km-thick lithosphere. Profiles in (c) and (d) incorporate a wet middle crust and show a dry lower crust and a wet upper mantle, and a wet lower crust and dry upper mantle, respectively (modified from Mackwell et al., 1998, by permission of the American Geophysical Union. Copyright © 1998 American Geophysical Union) See text for explanation.*

the central Andes and Kamchatka InSAR measurements have been used to evaluate volcanic hazards and the movement of magma in volcanic arcs (Pritchard & Simons, 2004). In southeast Iran, InSAR data have been used to determine the deformation field and source parameters of a magnitude $M_w = 6.5$ earthquake that affected the city of Bam in 2003 (Wang *et al.*, 2004). The combined use of GPS and InSAR data have revealed the vertical displacements associated with a part of the San Andreas Fault system near San Francisco (Fig. 8.7b).

2.10.6 Deformation in the mantle

Measurements of seismic anisotropy (Section 2.1.8) and the results of mineral physics experiments have been used to infer creep mechanisms and flow patterns in the mantle (Karato, 1998; Park & Levin, 2002; Bystricky, 2003). The deformation of mantle minerals, including olivine, by dislocation creep results in either a preferred orientation of crystal lattices or a preferred orientation of mineral shapes. This alignment affects how fast seismic waves propagate in different directions. Measurements of this directionality and other properties potentially allow investigators to image areas of the mantle that are deforming by dislocation creep (Section 2.10.3) and to determine whether the flow is mostly vertical or mostly horizontal. However, these interpretations are complicated by factors such as temperature, grain size, the presence of water and partial melt, and the amount of strain (Hirth & Kohlstedt, 2003; Faul *et al.*, 2004).

Most authors view power-law (or dislocation) creep as the dominant deformation mechanism in the upper mantle. Experiments on olivine, structural evidence in mantle-derived nodules, and the presence of seismic anisotropy suggest that power-law creep occurs to a depth of at least 200 km. These results contrast with many studies of post-glacial isostatic rebound (Section 2.11.5), which tend to favor a diffusion creep mechanism for flow in the upper mantle. Karato & Wu (1993) resolved this apparent discrepancy by suggesting that a transition from power-law creep to diffusion creep occurs with depth in the upper mantle. Diffusion creep may become increasingly prominent with depth as pressure and temperature increase and stress differences decrease. A source of potential uncertainty in studies of mantle rheology

using glacial rebound is the role of *transient creep*, where the strain rate varies with time under constant stress. Because the total strains associated with rebound are quite small ($\leq 10^{-3}$) compared to the large strains associated with mantle convection, transient creep may be important during post-glacial isostatic rebound (Ranalli, 2001).

In contrast to the upper mantle, much of the lower mantle is seismically isotropic, suggesting that diffusion creep is the dominant mechanism associated with mantle flow at great depths (Karato *et al.*, 1995). Unlike dislocation creep, diffusion creep (and also superplastic creep) result in an isotropic crystal structure in lower mantle minerals, such as perovskite and magnesiowüstite. Large uncertainties about lower mantle rheology exist because lower mantle materials are difficult to reproduce in the laboratory. Nevertheless, advances in high-pressure experimentation have allowed investigators to measure some of the physical properties of lower mantle minerals. Some measurements suggest that lower mantle rheology strongly depends on the occurrence and geometry of minor, very weak phases, such as magnesium oxide (Yamazaki & Karato, 2001). Murakami *et al.* (2004) demonstrated that at pressure and temperature conditions corresponding to those near the core–mantle boundary, $MgSiO_3$ perovskite transforms to a high-pressure form that may influence the seismic characteristics of the mantle below the D'' discontinuity (Section 12.8.4).

Unlike most of the lower mantle, observations at the base of the mesosphere, in the D'' layer (Section 2.8.5), indicate the presence of seismic anisotropy (Panning & Romanowicz, 2004). The dominance of V_{SH} polarization over V_{SV} in shear waves implies large-scale horizontal flow, possibly analogous to that found in the upper 200 km of the mantle. The origin of the anisotropy, whether it is due to the alignment of crystal lattices or to the preferred orientation of mineral shapes, is uncertain. However, these observations suggest that D'' is a mechanical boundary layer for mantle convection. Exceptions to the pattern of horizontal flow at the base of the lower mantle are equally interesting. Two exceptions occur at the bottom of extensive low velocity regions in the lower mantle beneath the central Pacific and southern Africa (Section 12.8.2) where anisotropy measurements indicate the onset of vertical upwelling (Panning & Romanowicz, 2004).

Another zone of seismic anisotropy and horizontal flow similar to that in the D'' layer also may occur at the top of the lower mantle or mesosphere (Karato, 1998). However, this latter interpretation is highly controversial and awaits testing by continued investigation. If such a zone of horizontal flow does exist then convection in the mantle probably occurs in layers and does not involve the whole mantle (Section 12.5.3).

2.11 ISOSTASY

2.11.1 Introduction

The phenomenon of isostasy concerns the response of the outer shell of the Earth to the imposition and removal of large loads. This layer, although relatively strong, is unable to support the large stresses generated by, for example, the positive weight of a mountain range or the relative lack of weight of an ocean basin. For such features to exist on the Earth's surface, some form of compensating mechanism is required to avoid the large stresses that would otherwise be generated.

Isostasy was first recognized in the 18th century when a party of French geodesists were measuring the length of a degree of latitude in Ecuador in an attempt to determine if the shape of the Earth corresponds to an oblate or a prolate ellipsoid. Plumb lines were used as a vertical reference in the surveying and it was recognized that a correction would have to be applied for the horizontal deflection caused by the gravitational attraction of the Andes. When this correction, based on the mass of the Andes above sea level, was applied, however, it was found that the actual vertical deflection was less than predicted (Fig. 2.27). This phenomenon was attributed to the existence of a negative mass anomaly beneath the Andes that compensates, that is to say, supports, the positive mass of the mountains. In the 19th century similar observations were made in the vicinity of the Himalaya and it was recognized that the compensation of surface loading at depth is a widespread phenomenon.

The presence of subsurface compensation is confirmed by the variation in the Earth's gravitational field over broad regions. Bouguer anomalies (Kearey *et al.,* 2002) are generally negative over elevated continental areas and positive over ocean basins (Fig. 2.28). These observations confirm that the positive topography of

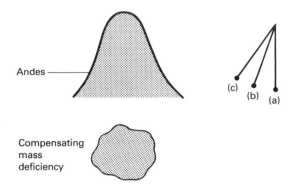

Figure 2.27 *Horizontal gravitational attraction of the mass of the Andes above sea level would cause the deflection (c) of a plumb bob from the vertical (a). The observed deflection (b) is smaller, indicating the presence of a compensating mass deficiency beneath the Andes (angles of deflection and mass distribution are schematic only).*

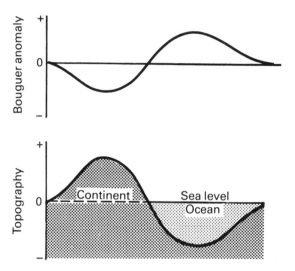

Figure 2.28 *Inverse correlation of Bouguer anomalies with topography indicating its isostatic compensation.*

continents and negative topography of oceans is compensated by regions at depth with density contrasts which are, respectively, negative and positive and whose mass anomaly approximates that of the surface features.

The principle of isostasy is that beneath a certain depth, known as the depth of compensation, the pressures generated by all overlying materials are everywhere equal; that is, the weights of vertical columns of unit cross-section, although internally variable, are identical at the depth of compensation if the region is in isostatic equilibrium.

Two hypotheses regarding the geometric form of local isostatic compensation were proposed in 1855 by Airy and Pratt.

2.11.2 Airy's hypothesis

Airy's hypothesis assumes that the outermost shell of the Earth is of a constant density and overlies a higher density layer. Surface topography is compensated by varying the thickness of the outer shell in such a way that its buoyancy balances the surface load. A simple analogy would be blocks of ice of varying thickness floating in water, with the thickest showing the greatest elevation above the surface. Thus mountain ranges would be underlain by a thick root, and ocean basins by a thinned outer layer or antiroot (Fig. 2.29a). The base of the outer shell is consequently an exaggerated mirror image of the surface topography. Consider the columns of unit cross-section beneath a mountain range and a region of zero elevation shown in Fig. 2.29a. Equating their weights gives:

$$g[h\rho_c + T_A\rho_c + r\rho_c + D_A\rho_m] = g[T_A\rho_c + r\rho_m + D_A\rho_m]$$

where g is the acceleration due to gravity.

Rearranging this equation gives the condition for isostatic equilibrium:

$$r = \frac{h\rho_c}{(\rho_m - \rho_c)}$$

A similar computation provides the condition for compensation of an ocean basin:

$$a = \frac{z(\rho_c - \rho_w)}{(\rho_m - \rho_c)}$$

If one substitutes appropriate densities for the crust, mantle, and sea water in these equations they predict that the relief on the Moho should be approximately seven times the relief at the Earth's surface.

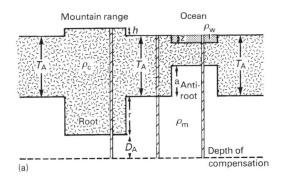

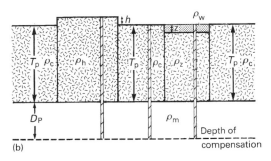

Figure 2.29 (a) Airy mechanism of isostatic compensation. h, height of mountain above sea level; z, depth of water of density ρ_w; T_A, normal thickness of crust of density ρ_c; r, thickness of root; a, thickness of antiroot; D_A, depth of compensation below root; ρ_m, density of mantle. (b) The Pratt mechanism of isostatic compensation. Legend as for (a) except T_p, normal thickness of crust; ρ_h, density of crust beneath mountain; ρ_z, density of crust beneath ocean; D_p, depth of compensation below T_p.

2.11.3 Pratt's hypothesis

Pratt's hypothesis assumes a constant depth to the base of the outermost shell of the Earth, whose density varies according to the surface topography. Thus, mountain ranges would be underlain by relatively low density material and ocean basins by relatively high density material (Fig. 2.29b). Equating the weights of columns of unit cross-section beneath a mountain range and a region of zero elevation gives:

$$g(T_p + h)\rho_h = gT_p\rho_c$$

which on rearrangement provides the condition for isostatic equilibrium of the mountain range:

$$\rho_h = \frac{T_p \rho_c}{(T_p + h)}$$

A similar computation for an ocean basin gives:

$$\rho_z = \frac{(T_p \rho_c - z\rho_w)}{(T_p - z)}$$

In these early models of isostasy it was assumed that the outer shell of the Earth, whose topography is compensated, corresponded to the crust. Certainly the large density contrast existing across the Moho plays a major part in the compensation. It is now believed, however, that the compensated layer is rather thicker and includes part of the upper mantle. This strong outer layer of the Earth is known as the lithosphere (Section 2.12). The lithosphere is underlain by a much weaker layer known as the asthenosphere which deforms by flow, and which can thus be displaced by vertical movements of the lithosphere. The density contrast across the lithosphere-asthenosphere boundary is, however, very small.

Both the Airy and Pratt hypotheses are essentially applications of Archimedes' Principle whereby adjacent blocks attain isostatic equilibrium through their buoyancy in the fluid substratum. They assume that adjacent blocks are decoupled by fault planes and achieve equilibrium by rising or subsiding independently. However, these models of *local* compensation imply unreasonable mechanical properties for the crust and upper mantle (Banks *et al.*, 1977), because they predict that independent movement would take place even for very small loads. The lithosphere is demonstrably not as weak as this implies, as large gravity anomalies exist over igneous intrusions with ages in excess of 100 Ma. The lithosphere must therefore be able to support stress differences of up to 20–30 MPa for considerable periods of time without the necessity of local compensation.

2.11.4 Flexure of the lithosphere

More realistic models of isostasy involve *regional* compensation. A common approach is to make the analogy

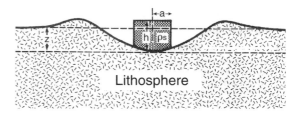

Figure 2.30 *Flexural downbending of the lithosphere as a result of a two-dimensional load of half-width a, height h, and density ρ_s.*

between the lithosphere and the behavior of an elastic sheet under load. Figure 2.30 illustrates the elastic response to loading; the region beneath the load subsides over a relatively wide area by displacing asthenospheric material, and is complemented by the development of peripheral bulges. Over long periods of time, however, the lithosphere may act in a viscoelastic manner and undergo some permanent deformation by creep (Section 2.10.3).

For example, the vertical displacement z of the oceanic lithosphere under loading can be calculated by modeling it as an elastic sheet by solving the fourth order differential equation:

$$D\frac{d^4 z}{dx^4} + (\rho_m - \rho_w)zg = P(x)$$

where $P(x)$ is the load as a function of horizontal distance x, g the acceleration due to gravity, and ρ_m, ρ_w the densities of asthenosphere and sea water, respectively. D is a parameter termed the flexural rigidity, which is defined by:

$$D = ET_c^3 / 12(1 - \sigma^2)$$

where E is Young's modulus, σ Poisson's ratio, and T_e the thickness of the elastic layer of the lithosphere.

The specific relationship between the displacement z and load for the two-dimensional load of half-width a, height h, and density ρ_s shown in Fig. 2.30 is:

$$z_{max} = h(\rho_s - \rho_w)(1 - e^{-\lambda a} \cos \lambda a)/(\rho_m - \rho_s)$$

where

$$\lambda = \sqrt[4]{(\rho_m - \rho_w)g / 4D}$$

and ρ_w, ρ_m the densities of water and the mantle, respectively.

Note that as the elastic layer becomes more rigid, D approaches infinity, λ approaches zero, and the depression due to loading becomes small. Conversely, as the layer becomes weaker, D approaches zero, λ approaches infinity, and the depression approaches $h(\rho_s - \rho_w)/(\rho_m - \rho_s)$ (Watts & Ryan, 1976). This is equivalent to Airy-type isostatic equilibrium and indicates that for this mechanism to operate the elastic layer and fluid substrate must both be very weak.

It can be shown that, for oceanic lithosphere away from mid-ocean ridges, loads with a half-width of less than about 50 km are supported by the finite strength of the lithosphere. Loads with half-widths in excess of about 500 km are in approximate isostatic equilibrium. Figure 2.31 illustrates the equilibrium attained by the oceanic lithosphere when loaded by a seamount (Watts et al., 1975). Thus, as a result of its flexural rigidity, the lithosphere has sufficient internal strength to support relatively small loads without sub-surface compensation. Such loads include small topographic features and variations in crustal density due, for example, to small granitic or mafic bodies within the crust. This more realistic model of isostatic compensation, that takes into account the flexural rigidity of the lithosphere, is referred to as *flexural isostasy* (Watts, 2001).

2.11.5 Isostatic rebound

The equilibrium flexural response of the lithosphere to loading is independent of the precise mechanical properties of the underlying asthenosphere as long as it facilitates flow. However, the reattainment of equilibrium after removal of the load, a phenomenon known as *isostatic rebound*, is controlled by the viscosity of the asthenosphere. Measurement of the rates of isostatic rebound provides a means of estimating the viscosity of the upper mantle. Fennoscandia represents an example of this type of study as precise leveling surveys undertaken since the late 19th century have shown that this region is undergoing uplift following the melting of the Pleistocene ice sheet (Fig. 2.32). The maximum uplift rates occur around the Gulf of Bothnia, where the land is rising at a rate of over 10 mm a^{-1}. Twenty thousand years ago the land surface was covered by an ice sheet about 2.5 km thick (Fig. 2.32a). The lithosphere accommodated this load by flexing (Fig. 2.32b), resulting in a subsidence of 600–700 m and a lateral displacement of asthenospheric material. This stage currently pertains in Greenland and Antarctica where, in Greenland, the land surface is depressed by as much as 250 m below sea level by the weight of ice. Melting of the ice was complete about 10,000 years ago (Fig. 2.32c), and since this time the lithosphere has been returning to its original position and the land rising in order to regain isostatic equilibrium. A similar situation pertains in northern Canada where the land surface around Hudson Bay is rising subsequent to the removal of an icecap. The rate of isostatic rebound provides an estimate for the viscosity of the upper mantle of 10^{21} Pa s (Pascal seconds), and measurements based on world-wide modeling of post-glacial recovery and its associated oceanic loading suggest that this figure generally applies throughout the upper mantle as a whole (Peltier & Andrews, 1976). Compared to the viscosity of water (10^{-3} Pa s) or a lava flow (4×10^3 Pa s), the viscosity of the sub-lithospheric mantle is extremely high and its fluid behavior is only apparent in processes with a large

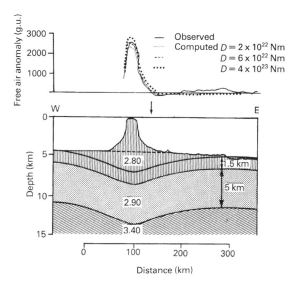

Figure 2.31 *Interpretation of the free air anomaly of the Great Meteor Seamount, northeast Atlantic Ocean, in terms of flexural downbending of the crust. A model with the flexural rigidity (D) of 6×10^{22} N m appears best to simulate the observed anomaly. Densities in Mg m^{-3}. Arrow marks the position 30°N, 28°W (redrawn from Watts et al., 1975, by permission of the American Geophysical Union. Copyright © 1975 American Geophysical Union).*

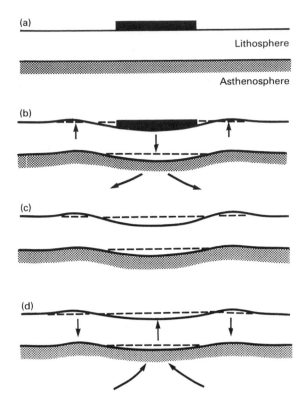

Figure 2.32 *Theory of isostatic rebound. (a) The load of an icecap on the lithosphere causes downbending accompanied by the elevation of the peripheral lithosphere and lateral flow in the asthenosphere (b). When the icecap melts (c), isostatic equilibrium is regained by reversed flow in the asthenosphere, sinking of the peripheral bulges and elevation of the central region (d).*

time constant. Knowledge of the viscosity of the mantle, however, provides an important control on the nature of mantle convection, as will be discussed in Section 12.5.2.

2.11.6 Tests of isostasy

The state of isostatic compensation of a region can be assessed by making use of gravity anomalies. The *isostatic anomaly*, IA, is defined as the Bouguer anomaly minus the gravity anomaly of the subsurface compensation. Consider a broad, flat plateau of elevation h compensated by a root of thickness r. The terrain correction

of such a feature is small in the central part of the plateau so that here the Bouguer anomaly, *BA*, is related to the free-air anomaly, *FAA* by the relationship:

$$BA = FAA - BC$$

where *BC* is the Bouguer correction, equal to $2\pi G\rho_c h$, where ρ_c is the density of the compensated layer. For such an Airy compensation:

$$IA = BA - A_{root}$$

where A_{root} is the gravity anomaly of the compensating root. Since the root is broad compared to its thickness, its anomaly may be approximated by that of an infinite slab, that is $2\pi G(\rho_c - \rho_m)r$, where ρ_m is the density of the substrate. Combining the above two equations:

$$IA = FAA - 2\pi G\rho_c h - 2\pi G(\rho_c - \rho_m)r$$

From the Airy criterion for isostatic equilibrium:

$$r = h\rho_c / (\rho_m - \rho_c)$$

Substitution of this condition into the equation reveals that the isostatic anomaly is equal to the free-air anomaly over a broad flat feature, and this represents a simple method for assessing the state of isostatic equilibrium. Figure 2.33 shows free-air, Bouguer and isostatic anomalies over a broad flat feature with varying degrees of compensation. Although instructive in illustrating the similarity of free-air and isostatic anomalies, and the very different nature of the Bouguer anomaly, this simple Airy isostatic anomaly calculation is clearly unsatisfactory in not taking into account topography and regional compensation due to flexure of the lithosphere.

To test isostasy over topographic features of irregular form more accurate computation of isostatic anomalies is required. This procedure involves calculating the shape of the compensation required by a given hypothesis of isostasy, computing its gravity anomaly, and then subtracting this from the observed Bouguer anomaly to provide the isostatic anomaly. The technique of computing the gravity anomaly from a hypothetical model is known as *forward modeling*.

Gravity anomalies can thus be used to determine if a surface feature is isostatically compensated at depth. They cannot, however, reveal the form of compensa-

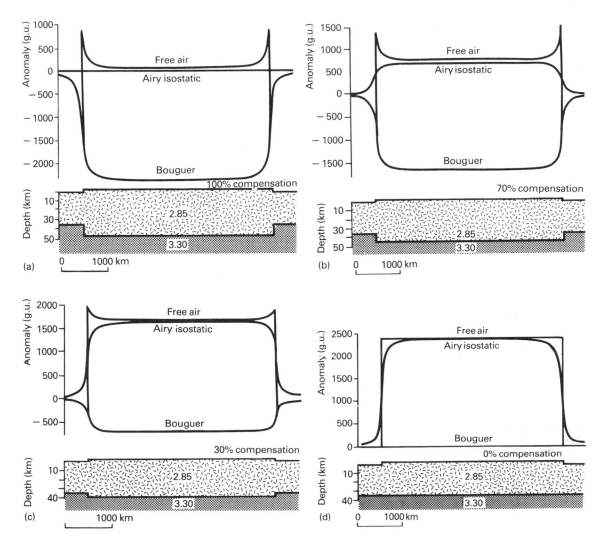

Figure 2.33 *Free air, Bouguer and Airy isostatic anomalies over an idealized mountain range (a) in perfect isostatic equilibrium, (b) with 70% isostatic compensation, (c) with 30% isostatic compensation, (d) uncompensated. Densities in Mg m⁻³.*

tion and indicate which type of mechanism is in operation. This is because the compensation occurs at a relatively deep level and the differences in the anomalies produced by a root/antiroot (according to the Airy hypothesis) or by different density units (according to the Pratt hypothesis) would be very small. Moreover, the gravity anomalies over most regions contain short wavelength components resulting from localized, uncompensated geologic structures that obscure the differences in the regional field arising from the different forms of compensation.

A more sophisticated test of isostasy involves the spectral analysis of the topography and gravity anomalies of the region being studied (Watts, 2001). The relationship between gravity and topography changes with wavelength. Moreover, the way in which it changes varies for different isostatic models. Thus by determining the frequency content of the gravity and topographic data it is possible to determine the type of compensation pertaining in the area. The technique also yields an estimate of T_e, the elastic thickness of the lithosphere (Sections 2.11.4, 2.12).

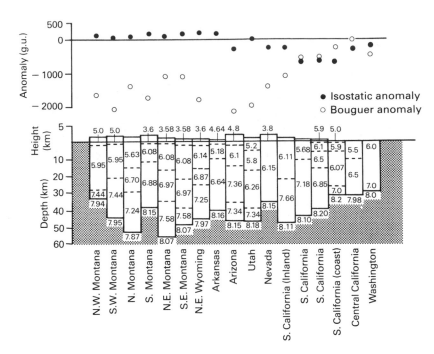

Figure 2.34 *Bouguer and isostatic gravity anomalies and their relation to seismic velocity sections from the western USA. Velocities in km s⁻¹ (redrawn from Garland, 1979).*

Information on the geometric form of isostatic compensation can also be gained by a combined analysis of gravity and seismic refraction data, as the latter technique can provide a reasonably detailed picture of the sub-surface structure of the region under consideration. Such studies have demonstrated that the broad isostatic equilibrium of continents and oceans is mainly accomplished by variations in crustal thickness according to the Airy hypothesis. Figure 2.34 shows seismic velocity sections from the western USA in which surface topography is largely compensated by Moho topography, although in several locations density variations in the upper mantle must be invoked to explain the isostatic compensation. A cross-section of the western USA (Fig. 2.35) reveals, however, that crustal thickness is not necessarily related to topographic elevation as the Great Plains, which reach a mean height of 1 km, are underlain by crust 45–50 km thick and the Basin and Range Province, at an average of 1.2 km above mean sea level, is underlain by a crustal thickness averaging 25–30 km (Section 7.3). Clearly, the Basin and Range Province must be partially compensated by a Pratt-type mechanism resulting from the presence of low density material in the upper mantle. Similarly, ocean ridges (Section

6.2) owe their elevation to a region of low density material in the upper mantle rather than to a thickened crust.

There are regions of the Earth's surface that do not conform to the concepts of isostasy discussed here. The hypotheses discussed above all assume that the support of surface features is achieved by their attaining hydrostatic equilibrium with the substrate. In certain areas, however, in particular convergent plate margins, surface features are supported dynamically by horizontal stresses. Such features provide the largest isostatic anomalies observed on the Earth's surface.

2.12 LITHOSPHERE AND ASTHENOSPHERE

It has long been recognized that for large-scale structures to attain isostatic equilibrium, the outermost shell of the Earth must be underlain by a weak layer

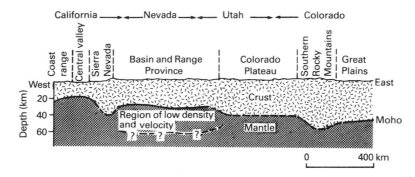

Figure 2.35 *Section from San Francisco, California to Lamar, Colorado based on seismic refraction data (redrawn from Pakiser, 1963, by permission of the American Geophysical Union. Copyright © 1963 American Geophysical Union).*

that deforms by flow. This concept has assumed fundamental importance since it was realized that the subdivisions of the Earth controlling plate tectonic movements must be based on rheology, rather than composition.

The lithosphere is defined as the strong, outermost layer of the Earth that deforms in an essentially elastic manner. It is made up of the crust and uppermost mantle. The lithosphere is underlain by the asthenosphere, which is a much weaker layer and reacts to stress in a fluid manner. The lithosphere is divided into plates, of which the crustal component can be oceanic and/or continental, and the relative movements of plates take place upon the asthenosphere.

However, having made these relatively simple definitions, examination of the several properties that might be expected to characterize these layers reveals that they lead to different ideas of their thickness. The properties considered are thermal, seismic, elastic, seismogenic, and temporal.

Temperature is believed to be the main phenomenon that controls the strength of subsurface material. Hydrostatic pressure increases with depth in an almost linear manner, and so the melting point of rocks also increases with depth. Melting will occur when the temperature curve intersects the melting curve (solidus) for the material present at depth (Fig. 2.36). The asthenosphere is believed to represent the location in the mantle where the melting point is most closely approached. This layer is certainly not completely molten, as it transmits S waves, but it is possible that a small amount of melt is present. The depth at which the asthenosphere occurs depends upon the geothermal gradient and the melting temperature of the mantle materials (Le Pichon

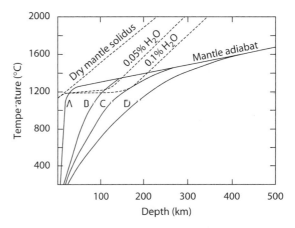

Figure 2.36 *Variation of temperature with depth beneath continental and oceanic regions. A, ocean ridge; B, ocean basin; C, continental platform; D, Archean Shield (redrawn from Condie, 2005b, with permission from Elsevier Academic Press).*

et al., 1973). Beneath ocean ridges, where temperature gradients are high, the asthenosphere must occur at shallow depth. Indeed, since it is actually created in the crestal region (Section 6.10), the lithosphere there is particularly thin. The gradient decreases towards the deep ocean basins, and the lithosphere thickens in this direction, the increase correlating with the depth of water as the lithosphere subsides as a result of contraction on cooling (Section 6.4). The mean lithosphere thickness on this basis beneath oceans is probably 60–70 km. Beneath continents a substantial portion of the observed heat flow is produced within the crust (Section

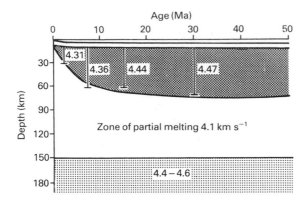

Figure 2.37 *Shear wave model of the thickening of oceanic lithosphere with age. Velocities in km s⁻¹ (redrawn from Forsyth, 1975, with permission from Blackwell Publishing). The 150 km transition may be somewhat deeper.*

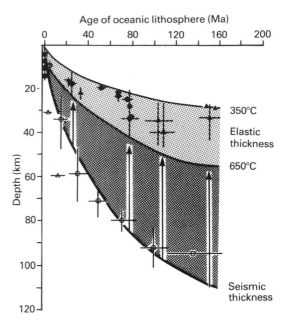

Figure 2.38 *Comparison of short-term "seismic" thickness and long-term "elastic" thickness for oceanic lithosphere of different ages (redrawn from Watts et al., 1980, by permission of the American Geophysical Union. Copyright © 1980 American Geophysical Union).*

2.13), so the temperature gradient in the sub-crustal lithosphere must be considerably lower than in oceanic areas. It is probable that the mantle solidus is not approached until a significantly greater depth, so that the continental lithosphere has a thickness of 100–250 km, being at a maximum beneath cratonic areas (Section 11.3.1).

The depth of the Low Velocity Zone (LVZ) for seismic waves (Section 2.2) agrees quite well with the temperature model of lithosphere and asthenosphere. Beneath oceanic lithosphere, for example, it progressively increases away from the crests of mid-ocean ridges, reaching a depth of approximately 80 km beneath crust 80 Ma in age (Forsyth, 1975) (Fig.2.37). Beneath continents it occurs at greater depths consistent with the lower geothermal gradients (Fig. 2.36). Within the LVZ attenuation of seismic energy, particularly shear wave energy, is very high. Both the low seismic velocities and high attenuation are consistent with the presence of a relatively weak layer at this level. As would be expected for a temperature-controlled boundary, the lithosphere–asthenosphere interface is not sharply defined, and occupies a zone several kilometers thick.

When the Earth's surface is loaded, the lithosphere reacts by downward flexure (Section 2.11.4). Examples include the loading of continental areas by ice sheets or large glacial lakes, the loading of oceanic lithosphere by seamounts, and the loading of the margins of both, at the ocean–continent transition, by large river deltas. The amount of flexure depends on the magnitude of the load and the flexural rigidity of the lithosphere. The latter, in turn, is dependent on the effective elastic thickness of the lithosphere, T_e (Section 2.11.4). Thus, if the magnitude of the load can be calculated and the amount of flexure determined, T_e may be deduced. However as indicated above (Section 2.11.6), T_e may be determined more generally from the spectral analysis of gravity and topographic data. Results obtained by applying this technique to oceanic areas are very consistent. They reveal that the elastic thickness of oceanic lithosphere is invariably less than 40 km and decreases systematically towards oceanic ridges (Watts, 2001) (Fig. 2.38). By contrast, the results obtained for continental areas vary from 5 to 110 km, the highest values being obtained for the oldest areas – the Precambrian cratons. However, McKenzie (2003) maintains that if there are sub-surface density contrasts that have no topographic expression, so-called *buried or hidden* loads, the technique yields an overestimate of the elastic thickness. Such loads are thought to be more common in continental areas, particularly in the cratons, because of their thick and rigid lithosphere. In oceanic areas loads are typically superimposed on the crust and expressed in the topography. McKenzie (2003) goes so far as to suggest that, if one makes allowance for buried loads, the elastic thickness of the lithosphere is probably less than 25 km in both oceanic and continental areas. By contrast, Perez-Gussinge & Watts (2005) maintain that T_e is greater than 60 km for continental lithosphere greater than 1.5 Ga in age and less than 30 km for continental areas less than 1.5 Ga in age. They suggest that this is a result of the change in thickness, geothermal gradient, and composition of continental lithosphere with time due to a decrease in mantle temperatures and volatile content (Section 11.3.3). Under tectonically active areas, such as the Basin and Range Province, the elastic thickness may be as small as 4 km (Bechtel *et al.*, 1990). Such very thin elastic thicknesses are undoubtedly due to very high geothermal gradients.

Yet another aspect of the lithosphere is the maximum depth to which the foci of earthquakes occur within it. This so-called *seismogenic thickness* is typically less than 25 km, that is, similar to or somewhat less than the elastic thickness in most areas (Watts & Burov, 2003). On the face of it this appears to lend support to the conclusion of McKenzie (2003) that the spectral analysis of topography and gravity anomalies systematically overestimates T_e, particularly in Precambrian shield areas because of the subdued topography and the presence of buried loads. However, there are alternative explanations that invoke the role of the ductile layer in the lower continental crust in decoupling the elastic upper layer from the lower lithosphere, the role of increased overburden pressure in inhibiting frictional sliding, and the fact that there is some evidence for earthquakes and faulting in the lower crust and upper mantle. It is thought that the latter may occur in the relatively rare instances where the lower crust and/or upper mantle are hydrated (Watts & Burov, 2003).

Thus, the concept of the lithosphere as a layer of uniformly high strength is seen to be over-simplistic when the rheological layering is considered. The upper 20–40 km of the lithosphere are brittle and respond to stress below the yield point by elastic deformation accompanied by transient creep. Beneath the brittle zone is a layer that deforms by plastic flow above a yield point of about 100 MPa. The lowest part, which is continuous with the asthenosphere, deforms by power-law creep and is defined as the region where the temperature increases with depth from 0.55 T_m to 0.85 T_m. The lithosphere is best thought of as a viscoelastic rather than an elastic layer (Walcott, 1970) for, as Walcott

demonstrated, the type of deformation experienced depends upon the duration of the applied loads. Over periods of a few thousand years, most of the region exhibiting power-law creep does not deform significantly and consequently is included within the elastic lithosphere. Long term loading, however, occurring over periods of a few million years, permits power-law deformation to occur so that this region then belongs to the asthenosphere.

The lithosphere can, therefore, be defined in a number of different ways that provide different estimates of its thickness. This must be borne in mind throughout any consideration of plate tectonic processes.

The asthenosphere is believed to extend to a depth of about 700 km. The properties of the underlying region are only poorly known. Seismic waves that cross this region do not suffer great attenuation (Section 9.4), and so it is generally accepted that this is a layer of higher strength, termed the *mesosphere*. The compositional and rheological layering of the Earth are compared in Fig. 2.39.

2.13 TERRESTRIAL HEAT FLOW

The study of thermal processes within the Earth is somewhat speculative because the interpretation of the distribution of heat sources and the mechanisms of heat transfer are based on measurements made at or near the surface. Such a study is important, however, as the process of heat escape from the Earth's interior is the direct or indirect cause of most tectonic and igneous activity.

The vast majority of the heat affecting the Earth's surface comes from the Sun, which accounts for some 99.98% of the Earth's surface energy budget. Most of this thermal energy, however, is reradiated into space, while the rest penetrates only a few hundred meters below the surface. Solar energy consequently has a negligible effect on thermal processes occurring in the interior of the Earth. The geothermal energy loss from heat sources within the Earth constitutes about 0.022% of its surface energy budget. Other sources of energy include the energy generated by the gradual deceleration of the

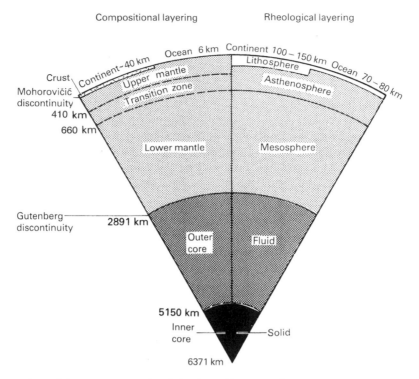

Figure 2.39 *Comparison of the compositional and rheological layering of the Earth.*

Earth's rotation and the energy released by earthquakes, but these make up only about 0.002% of the energy budget. It is thus apparent that geothermal energy is the major source of the energy which drives the Earth's internal processes.

It is believed that the geothermal energy is derived in part from the energy given off during the radioactive decay of long-lived isotopes, in particular K^{40}, U^{235}, U^{238}, and Th^{232}, and also from the heat released during the early stages of the formation of the Earth. These isotopes would account for the present geothermal loss if present in proportions similar to those of chondritic meteorites. Radioactive decay is exponential, so that during the early history of the Earth the concentration of radioactive isotopes would have been significantly higher than at present and the thermal energy available to power its internal processes would have been much greater (Section 12.2). Currently accepted models for the formation of the Earth require an early phase of melting and differentiation of its originally homogeneous structure. This melting is believed to have been powered in part by thermal energy provided by the decay of short-lived radioactive isotopes such as Al^{26},

Fe^{60}, and Cl^{36}. The differentiation of the Earth would also have contributed energy to the Earth arising from the loss in gravitational potential energy as the dense iron-nickel core segregated to a lower energy state at the center of the Earth.

The heat flow through a unit area of the Earth's surface, H, is given by:

$$H = K\frac{\delta T}{\delta z}$$

where $\delta T/\delta z$ is the thermal gradient perpendicular to the surface and K the thermal conductivity of the medium through which the heat is flowing. The units of H are $\mathrm{mW\,m^{-2}}$.

On land, heat flow measurements are normally made in boreholes. Mercury maximum thermometers or thermistor probes are used to determine the vertical temperature gradient. Thermal conductivity is measured on samples of the core using a technique similar to the Lee's disc method. Although appearing relatively simple, accurate heat flow measurements on land are difficult to accomplish. The drilling of a borehole necessitates the use of fluid lubricants that disturb the thermal regime of the borehole so that it has to be left for several months to allow the disturbance to dissipate. Porous strata have to be avoided as pore water acts as a heat sink and distorts the normal thermal gradients. Consequently, it is rarely possible to utilize boreholes sunk for the purposes of hydrocarbon or hydrogeologic exploration. In many areas readings may only be undertaken at depths below about 200 m so as to avoid the transient thermal effects of glaciations.

Heat flow measurements are considerably easier to accomplish at sea. The bottom temperatures in the oceans remain essentially constant and so no complications arise because of transient thermal perturbations. A temperature probe is dropped into the upper soft sediment layer of the seabed and, after a few minutes' stabilization, the temperature gradient is measured by a series of thermistor probes. A corer associated with the probe collects a sediment sample for thermal conductivity measurements; alternatively, the role of one of the thermistors can be changed to provide a source of heat. The change in the temperature of this probe with time depends on the rate at which heat is conducted away from it, and this enables a direct, *in situ* measurement of the thermal conductivity of the sediment to be made.

A large proportion of geothermal energy escapes from the surface by conduction through the solid Earth. In the region of the oceanic ridge system, however, the circulation of seawater plays a major role in transporting heat to the surface and about 25% of the geothermal energy flux at the Earth's surface is lost in this way.

The pattern of heat flow provinces on the Earth's surface broadly correlates with major physiographic and geologic subdivisions. On continents the magnitude of heat flow generally decreases from the time of the last major tectonic event (Sclater *et al.*, 1980). Heat flow values are thus low over the Precambrian shields and much higher over regions affected by Cenozoic orogenesis. Within the oceans the heat flow decreases with the age of the lithosphere (Section 6.5), with high values over the oceanic ridge system and active marginal seas and low values over the deep ocean basins and inactive marginal seas.

The average heat flux in continental areas is $65\,\mathrm{mW\,m^{-2}}$, and in oceanic areas $101\,\mathrm{mW\,m^{-2}}$, of which about 30% is contributed by hydrothermal activity at the mid-oceanic ridge system (Pollack *et al.*, 1993). As 60% of the Earth's surface is underlain by oceanic crust, about 70% of the geothermal energy is lost through oceanic crust, and 30% through continental crust.

FURTHER READING

Anderson, D.L. (2007) *New Theory of the Earth*, 2nd edn. Cambridge University Press, Cambridge, UK.

Bott, M.H.P. (1982) *The Interior of the Earth, its Structure, Constitution and Evolution*, 2nd edn. Edward Arnold, London.

Condie, K.C. (2005) *Earth as an Evolving Planetary System*. Elsevier, Amsterdam.

Fowler, C.M.R. (2005) *The Solid Earth: an introduction to global geophysics*, 2nd edn. Cambridge University Press, Cambridge.

Jacobs, J.A. (1991) *The Deep Interior of the Earth*. Chapman & Hall, London.

Nicolas, A. (1989) *Structure of Ophiolites and Dynamics of Oceanic Lithosphere*. Kluwer Academic Publishers, Dordrecht.

Park, R.G. (1988) *Geological Structures and Moving Plates*. Blackie, London and Glasgow.

Ranalli, G. (1995) *Rheology of the Earth*, 2nd edn. Chapman & Hall, London.

Stein, S. & Wysession, M. (2003) *An Introduction to Seismology, Earthquakes, and Earth Structure*. Blackwell Publishing, Oxford.

Twiss, R.J. & Moores, E.M. (2006) *Structural Geology*, 2nd edn. W.H. Freeman, New York.

3 | Continental drift

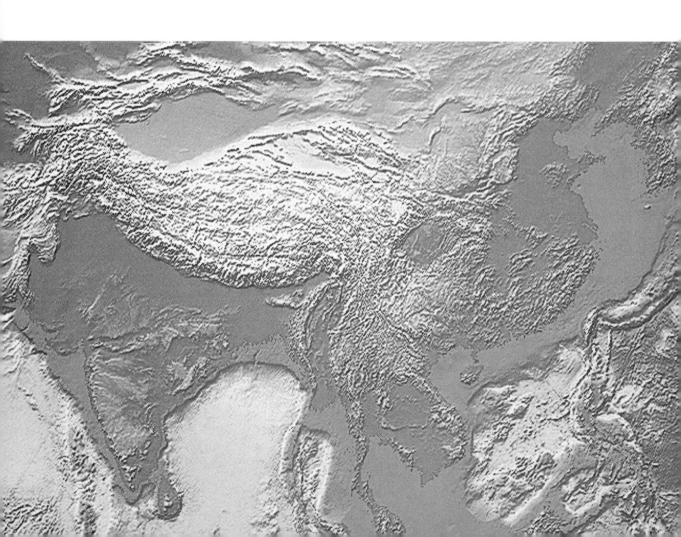

3.1 INTRODUCTION

As early as the 16th century it had been noted that the western and eastern coastlines of the Atlantic Ocean appeared to fit together like the pieces of a jigsaw puzzle (Section 1.1). The significance of this observation was not fully realized, however, until the 19th century, when the geometric fit of continental outlines was invoked as a major item of evidence in constructing the hypothesis of continental drift. The case for the hypothesis was further strengthened by the correspondence of geologic features across the juxtaposed coastlines. Application of the technique of paleomagnetism in the 1950s and 1960s provided the first quantitative evidence that continents had moved at least in a north–south direction during geologic time. Moreover, it was demonstrated that the continents had undergone relative motions, and this confirmed that continental drift had actually occurred.

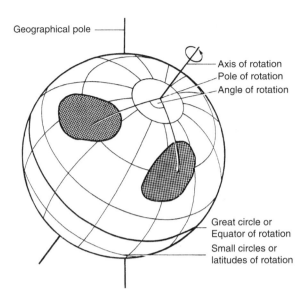

Figure 3.1 *Euler's theorem. Diagram illustrating how the motion of a continent on the Earth can be described by an angle of rotation about a pole of rotation.*

3.2 CONTINENTAL RECONSTRUCTIONS

3.2.1 Euler's theorem

In order to perform accurate continental reconstructions across closed oceans it is necessary to be able to describe mathematically the operation involved in making the geometric fit. This is accomplished according to a theorem of Euler, which states that the movement of a portion of a sphere across its surface is uniquely defined by a single angular rotation about a pole of rotation (Fig. 3.1). The pole of rotation, and its antipodal point on the opposite diameter of the sphere, are the only two points which remain in a fixed position relative to the moving portion. Consequently, the movement of a continent across the surface of the Earth to its pre-drift position can be described by its pole and angle of rotation.

3.2.2 Geometric reconstructions of continents

Although approximate reconstructions can be performed manually by moving models of continents across an accurately constructed globe (Carey, 1958), the most rigorous reconstructions are performed mathematically by computer, as in this way it is possible to minimize the degree of misfit between the juxtaposed continental margins.

The technique generally adopted in computer-based continental fitting is to assume a series of poles of rotation for each pair of continents arranged in a grid of latitude and longitude positions. For each pole position the angle of rotation is determined that brings the continental margins together with the smallest proportion of gaps and overlaps. The fit is not made on the coastlines, as continental crust extends beneath the surrounding shelf seas out to the continental slope. Consequently, the true junction between continental and oceanic lithosphere is taken to be at some isobath marking the midpoint of the continental slope, for example the 1000 m contour. Having determined the angle of rotation, the goodness of fit is quantified by some criterion based on the degree of mismatch. This goodness of fit is generally known as the objective function. Values of the objective function are entered on the grid of pole positions and contoured. The location of the minimum objective function revealed by this procedure then provides the pole of rotation for which the continental edges fit most exactly.

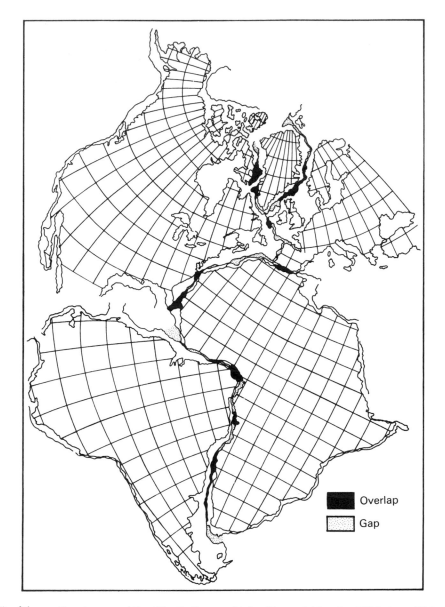

Figure 3.2 *Fit of the continents around the Atlantic Ocean, obtained by matching the 500 fathom (920 m) isobath (redrawn from Bullard* et al., *1965, with permission from the Royal Society of London).*

3.2.3 The reconstruction of continents around the Atlantic

The first mathematical reassembly of continents based solely on geometric criteria was performed by Bullard *et al.* (1965), who fitted together the continents on either side of the Atlantic (Fig. 3.2). This was accomplished by sequentially fitting pairs of continents after determining their best fitting poles of rotation by the procedure described in Section 3.2.2. The only rotation involving parts of the same landmass is that of the Iberian peninsular with respect to the rest of Europe. This is justified because of the known presence of oceanic lithosphere in the Bay of Biscay which is closed by this rotation. Geologic evidence (Section 3.3) and information provided by magnetic lineations in the Atlantic (Section 4.1.7) indicate that the reconstruction represents the continental configuration during late Triassic/early Jurassic times approximately 200 Ma ago.

Examination of Fig. 3.2 reveals a number of overlaps of geologic significance, some of which may be related to the process of stretching and thinning during the formation of rifted continental margins (Section 7.7). Iceland is absent because it is of Cenozoic age and its construction during the opening of the Atlantic postdates the reconstruction. The Bahama Platform appears to overlap the African continental margin and mainland. It is probable, however, that the platform represents an accumulation of sediment capped by coral on oceanic crust that formed after the Americas separated (Dietz & Holden, 1970). Similarly, the Niger Delta of Africa appears to form an overlap when in fact it also developed in part on oceanic crust formed after rifting.

A major criticism of the reconstruction is the overlap of Central America on to South America and absence of the Caribbean Sea. This must be viewed, however, in the light of our knowledge of the history of the opening of the Atlantic based, for the most part, on its pattern of magnetic lineations. Geologic and geometric considerations suggest that the Paleozoic crustal blocks which underlie Central America were originally situated within the region now occupied by the Gulf of México, an area existing within the reconstruction (Fig. 3.3). The North Atlantic started to open about 180 Ma ago, and the South Atlantic somewhat later, about 130 Ma ago. The poles of rotation of the North and South Atlantic were sufficiently different that the opening created the space between North and South America now occupied by the Caribbean. This also allowed a clockwise rotation of the Central American blocks out of the Gulf of México to their present locations. About 80 Ma ago the poles of rotation of the North and South Atlantic changed to an almost identical location in the region of the present north pole so that from this time the whole Atlantic Ocean effectively opened as a single unit.

3.2.4 The reconstruction of Gondwana

Geometric evidence alone has also been used in the reconstruction of the southern continents that make up

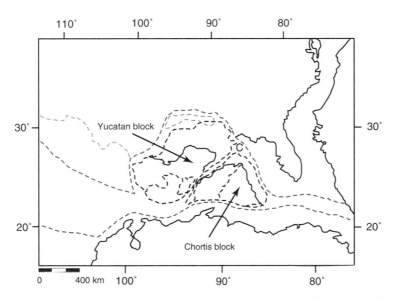

Figure 3.3 *Reconstruction of the Central American region within the Bullard* et al. *fit of the continents around the Atlantic (Fig. 3.2). C, location of pre-Mesozoic portions of Cuba (redrawn from White, 1980, with permission from* Nature ***283**, 823–6. Copyright 1980 Macmillan Publishers Ltd).*

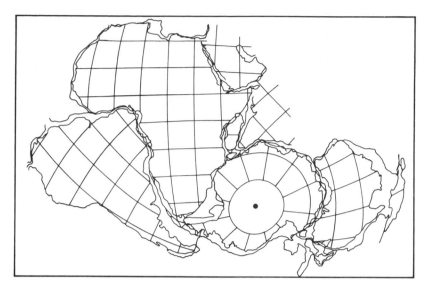

Figure 3.4 *Fit of the southern continents and India (redrawn from Smith & Hallam, 1970, with permission from* Nature *225, 139–44. Copyright 1970 Macmillan Publishers Ltd).*

Gondwana. The first such reconstruction was performed by Smith & Hallam (1970) and is illustrated in Fig. 3.4. The shapes of the continental edges of the east coast of Africa, Madagascar, India, Australia, and Antarctica are not quite so well suited to fitting as the circum-Atlantic continents. However this reconstruction has been confirmed by subsequent analysis of the record of magnetic lineations in the Indian Ocean (Section 4.1.7).

3.3 GEOLOGIC EVIDENCE FOR CONTINENTAL DRIFT

The continental reconstructions discussed in Sections 3.2.3 and 3.2.4 are based solely on the geometric fit of continental shelf edges. If they represent the true ancient configurations of continents it should be possible to trace continuous geologic features from one continent to another across the fits. The matching of features requires the rifting of the supercontinent across the general trend of geologic features. This does not always occur as the location of the rift is often controlled by the geology of the supercontinent, and takes place along lines of weakness that may run parallel to the geologic grain. However, there remain many geologic features that can be correlated across juxtaposed continental margins, some of which are listed below.

1 *Fold belts.* The continuity of the Appalachian fold belt of eastern North America with the Caledonian fold belt of northern Europe, illustrated in Fig. 3.5, is a particularly well-studied example (Dewey, 1969). Within the sedimentary deposits associated with fold belts there is often further evidence for continental drift. The grain size, composition, and age distribution of detrital zircon minerals in the sediments can be used to determine the nature and direction of their source. The source of sediments in the Caledonides of northern Europe lies to the west in a location now occupied by the Atlantic, indicating that, in the past, this location must have been occupied by continental crust (Rainbird *et al.*, 2001; Cawood *et al.*, 2003).

2 *Age provinces.* The correlation of the patterns of ages across the southern Atlantic is shown in Fig. 3.6, which illustrates the matching of both

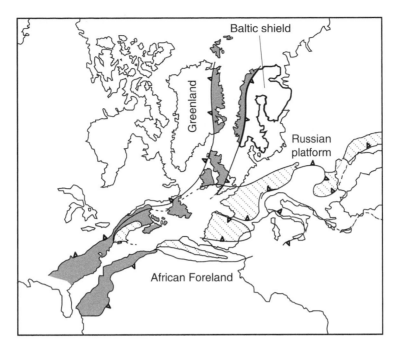

Figure 3.5 *The fit of the continents around the North Atlantic, after Bullard et al. (1965), and the trends of the Appalachian-Caledonian and Variscan (early and late Paleozoic) fold belts (dark and light shading respectively). The two phases of mountain building are superimposed in eastern North America (redraw from Hurley, 1968; the Confirmation of Continental Drift. Copyright © 1968 by Scientific American, Inc. All rights reserved.)*

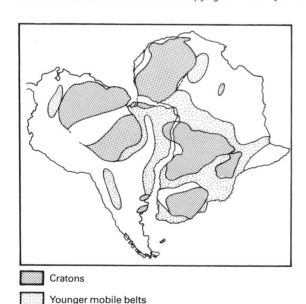

Cratons

Younger mobile belts

Figure 3.6 *Correlation of cratons and younger mobile belts across the closed southern Atlantic Ocean (redrawn from Hurley, 1968, the Confirmation of Continental Drift. Copyright © 1968 by Scientific American, Inc. All rights reserved.)*

Precambrian cratons and rocks of Paleozoic age (Hallam, 1975).

3 *Igneous provinces.* Distinctive igneous rocks can be traced between continents as shown in Fig. 3.7. This applies both to extrusive and intrusive rocks, such as the belt of Mesozoic dolerite, which extends through southern Africa, Antarctica, and Tasmania, and the approximately linear trend of Precambrian anorthosites (Section 11.4.1) through Africa, Madagascar, and India (Smith & Hallam, 1970).

4 *Stratigraphic sections.* Distinctive stratigraphic sequences can also be correlated between adjacent continents. Figure 3.8 shows stratigraphic sections of the Gondwana succession, a terrestrial sequence of sediments of late Paleozoic age (Hurley, 1968). Marker beds of tillite and coal, and sediments containing *Glossopteris* and *Gangamopteris* flora (Section 3.5) can be correlated through South America, South Africa, Antarctica, India, and Australia.

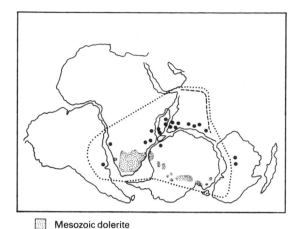

☐ Mesozoic dolerite

······· Limit of Permo - Carboniferous glaciation

• Precambrian anorthosite

Figure 3.7 *Correlation of Permo-Carboniferous glacial deposits, Mesozoic dolerites, and Precambrian anorthosites between the reconstructed continents of Gondwana (after Smith & Hallam, 1970, with permission from* Nature *225, 139–44. Copyright 1970 Macmillan Publishers Ltd).*

5 *Metallogenic provinces.* Regions containing manganese, iron ore, gold, and tin can be matched across adjacent coastlines on such reconstructions (Evans, 1987).

3.4 PALEOCLIMATOLOGY

The distribution of climatic regions on the Earth is controlled by a complex interaction of many phenomena, including solar flux (i.e. latitude), wind directions, ocean currents, elevation, and topographic barriers (Sections 13.1.2, 13.1.3). The majority of these phenomena are only poorly known in the geologic record. On a broad scale, however, latitude is the major controlling factor of climate and, ignoring small microclimatic regions dependent on rare combinations of other phenomena, it appears likely that the study of climatic indicators in ancient rocks can be used to infer, in a general sense, their ancient latitude. Consequently,

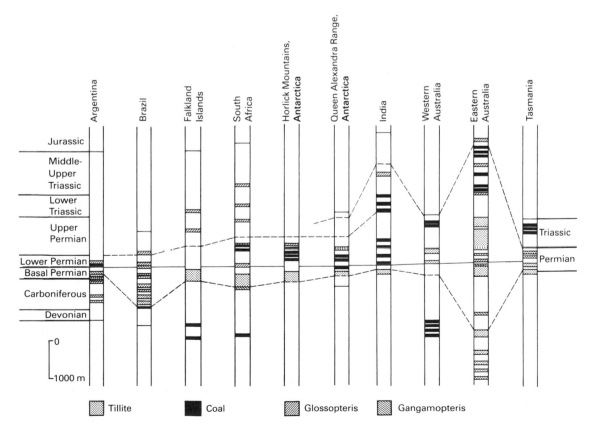

Figure 3.8 *Correlation of stratigraphy between Gondwana continents (redrawn from Hurley, 1968, the Confirmation of Continental Drift. Copyright © 1968 by Scientific American, Inc. All rights reserved.)*

paleoclimatology, the study of past climates (Frakes, 1979), may be used to demonstrate that continents have drifted at least in a north–south sense. It must be realized, however, that the Earth is presently in an interglacial period, and so parallels between modern and ancient climates may not be completely justified. The important paleolatitude indicators are listed below.

1 *Carbonates and reef deposits.* These deposits are restricted to warm water and occur within 30° of the equator at the present day where temperatures fall in the narrow range 25–30°C.

2 *Evaporites.* Evaporites are formed under hot arid conditions in regions where evaporation exceeds seawater influx and/or precipitation, and are usually found in basins bordering a sea with limited or intermittent connection to the ocean proper (Section 13.2.4). At the present day they do not form near the equator, but rather in the arid subtropical high pressure zones between about 10° and 50° where the required conditions prevail, and it is believed that fossil evaporites formed in a similar latitudinal range (Windley, 1984).

3 *Red-beds.* These include arkoses, sandstones, shales, and conglomerates that contain hematite. They form under oxidizing conditions where there is an adequate supply of iron. A hot climate is required for the dehydration of limonite into hematite, and at present they are restricted to latitudes of less than 30°.

4 *Coal.* Coal is formed by the accumulation and degradation of vegetation where the rate of accumulation exceeds that of removal and decay. This occurs either in tropical rain forests, where growth rates are very high, or in temperate forests where growth is slower but decay is inhibited by cold winters. Thus, coals may form in high or low latitudes, each type having a distinctive flora. In Wegener's compilation of paleoclimatic data for the Carboniferous and Permian (Fig. 1.3), the Carboniferous coals are predominantly of the low latitude type, whereas the Permian coals of Gondwana are of the high latitude type. Younger coals were typically formed at high latitudes.

5 *Phosphorites.* At the present day phosphorites form within 45° of the equator along the western margins of continents where upwellings of cold, nutrient-rich, deep water occur, or in arid zones at low latitudes along east-west seaways.

6 *Bauxite and laterite.* These aluminum and iron oxides only form in a strongly oxidizing environment. It is believed that they only originate under the conditions of tropical or subtropical weathering.

7 *Desert deposits.* Care must be employed in using any of these deposits because desert conditions can prevail in both warm and cold environments. However, the dune bedding of desert sandstones can be used to infer the ancient direction of the prevailing winds. Comparison of these with the direction of the modern wind systems found at their present latitudes can indicate if the continent has undergone any rotation.

8 *Glacial deposits.* Glaciers and icecaps, excluding those of limited size formed in mountain ranges, are limited to regions within about 30° of the poles at the present day.

The results of applying these paleoclimatic techniques strongly indicate that continents have changed their latitudinal position throughout geologic time. For example, during the Permian and Carboniferous the Gondwana continents were experiencing an extensive glaciation (Martin, 1981) and must have been situated near the south pole (Fig. 3.9). At the same time in Europe and the eastern USA, coal and extensive reef deposits were forming, which subsequently gave way to hot deserts with evaporite deposits. The northern continents were thus experiencing a tropical climate in equatorial latitudes (see also Fig. 1.3).

3.5 PALEONTOLOGIC EVIDENCE FOR CONTINENTAL DRIFT

Continental drift has affected the distribution of ancient animals and plants (Briggs, 1987) by creating barriers to

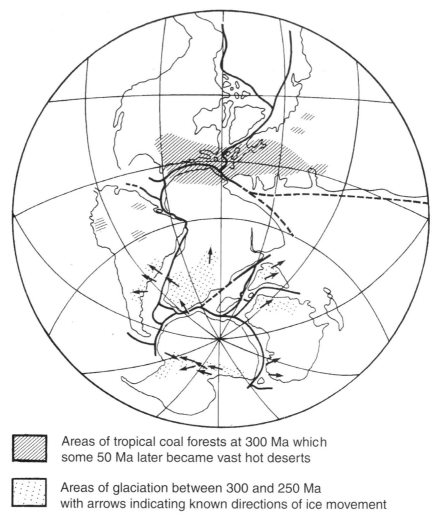

Areas of tropical coal forests at 300 Ma which some 50 Ma later became vast hot deserts

Areas of glaciation between 300 and 250 Ma with arrows indicating known directions of ice movement

Figure 3.9 *Use of paleoclimatic data to control and confirm continental reconstructions (redrawn from Tarling & Tarling, 1971).*

their dispersal (Hallam, 1972). An obvious example of this would be the growth of an ocean between two fragments of a supercontinent which prevented migration between them by terrestrial life-forms. The past distribution of tetrapods implies that there must have been easy communication between all parts of Gondwana and Laurasia. Remains of the early Permian reptile *Mesosaurus* are found in Brazil and southern Africa. Although adapted to swimming, it is believed that *Mesosaurus* was incapable of travelling large distances and could not have crossed the 5000 km of ocean now present between these two localities.

Oceans can also represent dispersal barriers to certain animals which are adapted to live in relatively shallow marine environments. The widespread dispersal of marine invertebrates can only occur in their larval stages when they form part of the plankton (Hallam, 1973b). For most species the larval stage is too short-lived to exist for the duration of the crossing of a large ocean. Consequently, ancient faunal province boundaries frequently correlate with sutures, which represent the join lines between ancient continents brought into juxtaposition by the consumption of an intervening ocean. The distribution of Cambrian trilobites strongly suggests that in Lower Paleozoic times there existed several continents separated by major ocean basins. The similarity between ammonite species now found in India, Madagascar, and Africa indicates that only shallow seas could have existed between these regions in Jurassic times.

Paleobotany similarly reveals the pattern of continental fragmentation. Before break-up, all the Gondwana continents supported, in Permo-Carboniferous times, the distinctive *Glossopteris* and *Gangamopteris* floras (Hurley, 1968; Plumstead, 1973) (Fig. 3.8), which are believed to be cold climate forms. At the same time a varied tropical flora existed in Laurasia (Fig. 3.10). After fragmentation, however, the flora of the individual continents diversified and followed separate paths of evolution.

A less obvious form of dispersal barrier is climate, as the latitudinal motions of continents can create climatic conditions unsuitable for certain organisms.

Indeed, relative continental movements can modify the pattern of ocean currents, mean annual temperature, the nature of seasonal fluctuations, and many other factors (Valentine & Moores, 1972) (Section 13.1.2). Also, plate tectonic processes can give rise to changes in topography, which modify the habitats available for colonization (Section 13.1.3).

The diversity of species is also controlled by continental drift. Diversity increases towards the equator so that the diversity at the equator is about ten times that at the poles. Consequently, drifting in a north–south direction would be expected to control the diversity on a continent. Diversity also increases with continental fragmentation (Kurtén, 1969). For example, 20 orders of reptiles existed in Paleozoic times on Pangea, but with its fragmentation in Mesozoic times 30 orders of mammals developed on the various continents. Each continental fragment becomes a nucleus for the adaptive radiation of the

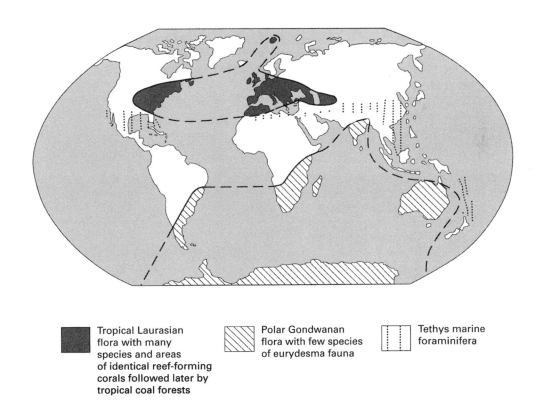

Tropical Laurasian flora with many species and areas of identical reef-forming corals followed later by tropical coal forests

Polar Gondwanan flora with few species of eurydesma fauna

Tethys marine foraminifera

Figure 3.10 *Present distributions of Pangean flora and fauna (redrawn from Tarling & Tarling, 1971).*

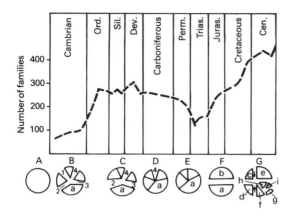

Figure 3.11 *Correlation of invertebrate diversity with time and continental distribution. A, earlier Pangea; B, fragmentation of earlier Pangea producing oceans preceding Caledonian (1), Appalachian (2), Variscan (3), and Uralian (4) orogenies; C, suturing during Caledonian and Acadian orogenies; D, suturing during Appalachian and Variscan orogenies; E, suturing of Urals and reassembly of Pangea; F, opening of Tethys Ocean; G, fragmentation of Pangea. a, Gondwana; b, Laurasia; c, North America; d, South America; e, Eurasia; f, Africa; g, Antarctica; h, India; i, Australia (after Valentine & Moores, 1970, with permission from* Nature **228**, *657–9. Copyright 1970 Macmillan Publishers Ltd).*

species as a result of genetic isolation and the morphological divergence of separate faunas. Consequently, more species evolve as different types occupy similar ecological niches. Figure 3.11, from Valentine & Moores (1970), compares the variation in the number of fossil invertebrate families existing in the Phanerozoic with the degree of continental fragmentation as represented by topological models. The correlation between number of species and fragmentation is readily apparent. An example of such divergence is the evolution of anteating mammals. As the result of evolutionary divergence this specialized mode of behavior is followed by different orders on separated continents: the antbears (*Edentata*) of South America, the pangolins (*Pholidota*) of northeast Africa and southeast Asia, the aardvarks (*Tubulidentata*) of central and southern Africa, and the spiny anteaters (*Monotremata*) of Australia.

Continental suturing leads to the homogenization of faunas by cross-migration (Hallam, 1972) and the extinction of any less well-adapted groups which face stronger competition. Conversely, continental rifting leads to the isolation of faunas which then follow their own distinct evolutionary development. For example, marsupial mammals probably reached Australia from South America in the Upper Cretaceous along an Antarctic migration route (Hallam, 1981) before the Late Cretaceous marine transgression removed the land connection between South America and Antarctica and closed the route for the later evolving placental mammals. Sea floor spreading then ensured the isolation of Australia when the sea level dropped, and the marsupials evolved unchallenged until the Neogene when the collision of Asia and New Guinea allowed the colonization of placental mammals from Asia.

3.6 PALEOMAGNETISM

3.6.1 Introduction

The science of paleomagnetism is concerned with studies of the fossil magnetism that is retained in certain rocks. If this magnetism originated at the time the rock was formed, measurement of its direction can be used to determine the latitude at which the rock was created. If this latitude differs from the present latitude at which the rock is found, very strong evidence has been furnished that it has moved over the surface of the Earth. Moreover, if it can be shown that the pattern of movement differs from that of rocks of the same age on a different continent, relative movement must have occurred between them. In this way, paleomagnetic measurements demonstrated that continental drift has taken place, and provided the first quantitative estimates of relative continental movements. For fuller accounts of the paleomagnetic method, see Tarling (1983) and McElhinny & McFadden (2000).

3.6.2 Rock magnetism

Paleomagnetic techniques make use of the phenomenon that certain minerals are capable of retaining a record of the past direction of the Earth's magnetic field. These minerals are all paramagnetic, that is, they contain atoms which possess an odd number of electrons. Magnetic fields are generated by the spin and orbital motions of the electrons. In shells with paired electrons, their magnetic fields essentially cancel each other. The unpaired electrons present in paramagnetic substances cause the atoms to act as small magnets or dipoles.

When a paramagnetic substance is placed in a weak external magnetic field, such as the Earth's field, the atomic dipoles rotate so as to become parallel to the external field direction. This induced magnetization is lost when the substance is removed from the field as the dipoles return to their original orientations.

Certain paramagnetic substances which contain a large number of unpaired electrons are termed ferromagnetic. The magnetic structure of these substances tends to devolve into a number of magnetic domains, within which the atoms are coupled by the interaction of the magnetic fields of the unpaired electrons. This interaction is only possible at temperatures below the Curie temperature, as above this temperature the energy level is such as to prohibit interatomic magnetic bonding and the substance then behaves in an ordinary paramagnetic manner.

Within each domain the internal alignment of linked atomic dipoles causes the domain to possess a net magnetic direction. When placed in a magnetic field the domains whose magnetic directions are in the same sense as the external field grow in size at the expense of domains aligned in other directions. After removal from the external field a preferred direction resulting from the growth and shrinkage of the domains is retained so that the substance exhibits an overall magnetic directionality. This retained magnetization is known as permanent or remanent magnetism.

3.6.3 Natural remanent magnetization

Rocks can acquire a natural remanent magnetization (NRM) in several ways. If the NRM forms at the same time as the rock it is referred to as primary; if acquired during the subsequent history of the rock it is termed secondary.

The primary remanence of igneous rocks is known as *thermoremanent magnetization* (TRM). It is acquired as the rock cools from its molten state to below the Curie temperature, which is realized after solidification. At this stage its ferromagnetic minerals pick up a magnetism in the same sense as the geomagnetic field at that time, which is retained during its subsequent history.

The primary remanence in clastic sedimentary rocks is known as *detrital remanent magnetization* (DRM). As the sedimentary particles settle through the water column, any ferromagnetic minerals present align in the direction of the geomagnetic field. On reaching bottom the particles flatten out, and if of elongate form preserve the azimuth of the geomagnetic field but not its inclination (Fig. 3.12). After burial, when the sediment is in a wet slurry state, the magnetic particles realign with the geomagnetic field as a result of microseismic activity, and this orientation is retained as the rock consolidates.

Secondary NRM is acquired during the subsequent history of the rock according to various possible mechanisms. *Chemical remanent magnetization* (CRM) is acquired when ferromagnetic minerals are formed as a result of a chemical reaction, such as oxidation. When of a sufficient size for the formation of one or more domains, the grains become magnetized in the direction of the geomagnetic field at the time of reaction. *Isothermal remanent magnetization* (IRM) occurs in rocks which have been subjected to strong magnetic fields, as in the case of a lightning strike. *Viscous remanent magnetization* (VRM) may arise when a rock remains in a relatively weak magnetic field over a long period of time as the magnetic domains relax and acquire the external field direction.

Some CRM may be acquired soon after formation, for example during diagenesis, or during a metamorphic event of known age, and hence preserve useful paleomagnetic information.

CRM, TRM, and DRM tend to be "hard," and remain stable over long periods of time, whereas certain secondary components of NRM, notably VRMs, tend to be "soft" and lost relatively easily. It is thus possible to destroy the "soft" components and isolate the "hard" components by the technique of *magnetic cleaning*. This involves monitoring the orientation and strength of the magnetization of a rock sample as it is subjected either to an alternating field of increasing intensity or to increasing temperature. Having isolated the primary remanent magnetization, its strength and direction are measured with either a spinner magnetometer or superconducting magne-

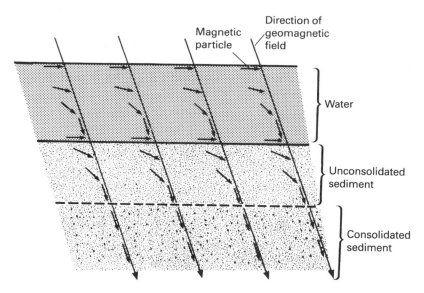

Figure 3.12 *Development of detrital remanent magnetization.*

tometer. The latter instrument is extremely sensitive and capable of measuring NRM orientations of rocks with a very low concentration of ferromagnetic minerals.

3.6.4 The past and present geomagnetic field

The magnetic field of the Earth approximates the field that would be expected from a large bar magnet embedded within it inclined at an angle of about 11° to the spin axis. The actual cause of the geomagnetic field is certainly not by such a magnetostatic process, as the magnet would have to possess an unrealistically large magnetization and would lie in a region where the temperatures would be greatly in excess of the Curie temperature.

The geomagnetic field is believed to originate from a dynamic process, involving the convective circulation of electrical charge in the fluid outer core, known as magnetohydrodynamics (Section 4.1.3). However, it is convenient to retain the dipole model as simple calculations can then be made to predict the geomagnetic field at any point on the Earth.

The geomagnetic field undergoes progressive changes with time, resulting from variations in the convective circulation pattern in the core, known as *secular variation*. One manifestation of this phenomenon is that the direction of the magnetic field at a particular geographic location rotates irregularly about the direction implied by an axial dipole model with a periodicity of a few thousand years. In a paleomagnetic study the effects of secular variation can be removed by collecting samples from a site which span a stratigraphic interval of many thousands of years. Averaging the data from these specimens should then remove secular variation so that for the purposes of paleomagnetic analysis the geomagnetic field in the past may be considered to originate from a dipole aligned along the Earth's axis of rotation.

Paleomagnetic measurements provide the intensity, azimuth and inclination of the primary remanent magnetization, which reflect the geomagnetic parameters at the time and place at which the rock was formed. By assuming the axial geocentric dipole model for the geomagnetic field, discussed above, the inclination I can be used to determine the paleolatitude ϕ at which the rock formed according to the relationship $2 \tan \phi = \tan I$. With a knowledge of the paleolatitude and the azimuth of the primary remanent magnetization, that is, the ancient north direction, the apparent location of the paleopole can be computed. Such computations, combined with age determinations of the samples by radiometric or biostratigraphic methods, make possible the calculation of the apparent location of the north magnetic pole at a particular time for the continent from which the samples were collected. Paleomagnetic analyses of samples of a wide age range can then be used to trace how the apparent pole position has moved over the Earth's surface.

It is important to recognize that remanent magnetization directions cannot provide an estimate of paleolongitude, as the assumed dipole field is axisymmetric. There is a consequent uncertainty in the ancient location of any sampling site, which could have been situated anywhere along a small circle, defined by the paleolatitude, centered on the pole position.

If a paleomagnetic study provides a magnetic pole position different from the present pole, it implies either that the magnetic pole has moved throughout geologic time, that is, the magnetic pole has wandered relative to the rotational pole, or if the poles have remained stationary that the sampling site has moved, that is, continental drift has occurred. It appears that wandering of the magnetic pole away from the geographic pole is unlikely because all theoretical models for the generation of the field predict a dominant dipole component paralleling the Earth's rotational axis (Section 4.1.3). Consequently, paleomagnetic studies can be used to provide a quantitative measure of continental drift.

An early discovery of paleomagnetic work was that in any one study about half of the samples analysed provided a primary remanent magnetization direction in a sense 180° different from the remainder. Although the possibility of self-reversal of rock magnetism remains, it is believed to be a rare phenomenon, and so these data are taken to reflect changes in the polarity of the geomagnetic field. The field can remain normal for perhaps a million years and then, over an interval of a few thousand years, the north magnetic pole becomes the south magnetic pole and a period of reversed polarity obtains. Polarity reversals are random, but obviously affect all regions of the Earth synchronously so that, coupled with radiometric or paleontologic dating, it is possible to construct a polarity timescale. This subject will be considered further in Chapter 4.

3.6.5 Apparent polar wander curves

Paleomagnetic data can be displayed in two ways. One way is to image what is believed to be the true situation, that is, plot the continent in a succession of positions according to the ages of the sampling sites (Fig. 3.13a). This form of display requires the assumption of the paleolongitudes of the sites. The other way is to regard the continent as remaining at a fixed position and plot the apparent positions of the poles for various times to provide an *apparent polar wander* (APW) path (Fig. 3.13b). As discussed above, this representation does not reflect real events, but it overcomes the lack of control of paleolongitude and facilitates the display of information from different regions on the same diagram.

The observation that the apparent position of the pole differed for rocks of different ages from the same continent demonstrated that continents had moved over the surface of the Earth. Moreover, the fact that APW paths were different for different continents demonstrated unequivocally that relative movements of the continents had taken place, that is, continental drift had occurred. Paleomagnetic studies thus confirmed and provided the first quantitative measurements of continental drift. Figure 3.14a illustrates the APW paths for North America and Europe from the Ordovician to the Jurassic. Figure 3.14b shows the result of rotating Europe and its APW path, according to the rotation parameters of Bullard *et al.* (1965), to close up the Atlantic Ocean. The APW paths for Europe and North America then correspond very closely from the time the continents were brought together at the end of the Caledonian orogeny, approximately 400 Ma ago, until the opening of the Atlantic.

APW paths can be used to interpret motions, collisions, and disruptions of continents (Piper, 1987), and are especially useful for pre-Mesozoic continents whose movements cannot be traced by the pattern of magnetic lineations in their surrounding ocean basins (Section 4.1.6). Figure 3.15 represents the full Wilson cycle (Section 7.9) of the opening and closure of an ocean basin between two continents. Before rifting, the two segments A and B of the initial continent have similar APW paths. They are unlikely to be identical as it is improbable that the initial rift

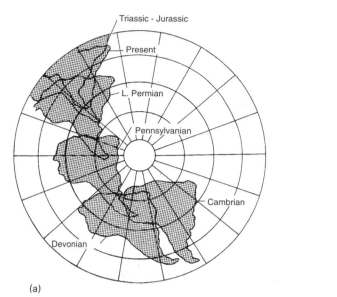
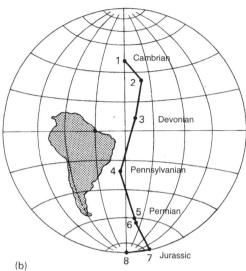

Figure 3.13 *Two methods of displaying paleomagnetic data: (a) assuming fixed magnetic poles and applying latitudinal shifts to the continent; (b) assuming a fixed continent and plotting a polar wander path. Subsequent work has modified the detail of the movements shown. Note that the south pole has been plotted (redrawn from Creer, 1965, with permission from the Royal Society of London).*

and final suture would coincide. After rifting the two segments describe diverging APW paths until the hairpin at time 8 signals a change in direction of motion to one of convergence. After suturing at time 12 the two segments follow a common polar track.

The southern continents, plus India, are thought to have formed a single continent, Gondwana, from late Pre-Cambrian to mid-Jurassic time. During this period, of approximately 400 Ma, they should have the same polar wander path when reassembled. Figure 3.16 illustrates a modern polar wander path for Gondwana (Torsvik & Van der Voo, 2002). The track of the path relative to South America can be compared with the very early path given by Creer (1965) (Fig. 3.13b). The seemingly greater detail of the path shown in Fig. 3.16 may however be unwarranted. There is considerable disagreement over the details of the APW path for Gondwana, presumably because of the paucity of sufficient reliable data (Smith, 1999; McElhinny & McFadden, 2000). Interestingly the path favored by Smith (1999), based on a detailed analysis

of paleomagnetic and paleoclimatic data, is very comparable to that of Creer (1965). All APW paths for Gondwana have the south pole during Carboniferous times in the vicinity of southeast Africa, as did Wegener (Fig. 1.3), and the Ordovician pole position in northwest Africa, where there is evidence for a minor glaciation in the Saharan region at this time (Eyles, 1993).

3.6.6 Paleogeographic reconstructions based on paleomagnetism

Reconstructions of the relative positions of the main continental areas at various times in the past 200 Ma are best achieved using the very detailed information on the evolution of the present ocean basins provided by the linear oceanic magnetic anomalies

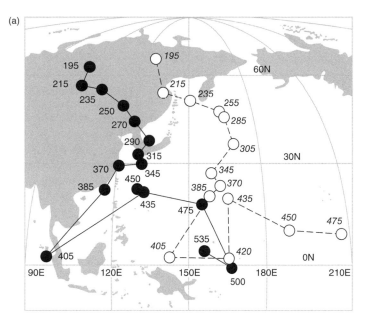

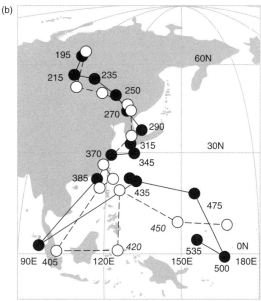

Figure 3.14 *Apparent polar wander paths for North America (solid circles and solid line) and Europe (open circles and dashed line) (a) with North America and Europe in their present positions, and (b) after closing the Atlantic ocean. Ages for each mean pole position are given in Ma with those for Europe in italics (redrawn from McElhinny & McFadden, 2000, with permission from Academic Press. Copyright Elsevier 2000).*

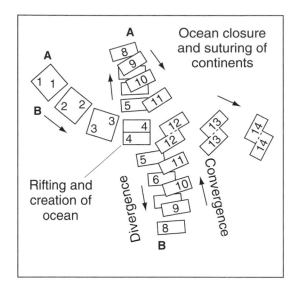
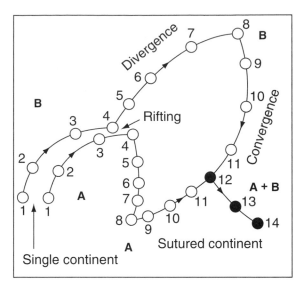

Figure 3.15 *Paleomagnetic signature of plate divergence and convergence (redrawn from Irving et al., 1974, by permission of the American Geophysical Union. Copyright © 1974 American Geoplysical Union).*

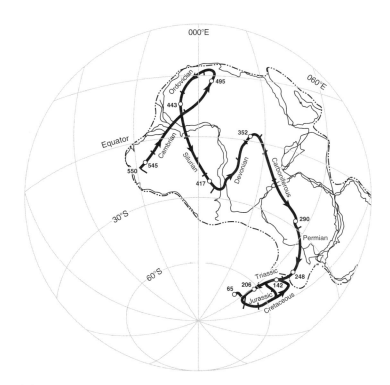

Figure 3.16 *APW path for Gondwana, based on the reconstruction of Lottes & Rowley (1990) (redrawn from Torsvik & Van der Voo, 2002, with permission from Blackwell Publishing).*

(Section 4.1.7). In order to position the continents in their correct paleolatitudes however, paleomagnetic results must be combined with these reconstructions to identify the positions of the paleopoles and paleo-equator. The sequence of paleogeographic maps in Chapter 13 (Figs 13.2–13.7) was obtained in this way. For any time prior to 200 Ma the constraints provided by the oceanic data are no longer available and reconstructions are based on paleomagnetic results and geologic correlations. Examples of these pre-Mesozoic reconstructions will be discussed in Chapter 11.

FURTHER READING

Frakes, L.A. (1979) *Climates Throughout Geologic Time*. Elsevier, New York.

McElhinny, M.W. & McFadden, P.L. (2000) *Paleomagnetism: continents and oceans*. Academic Press, San Diego.

Tarling, D.H. & Runcorn, S.K. (eds) (1973) *Implications of Continental Drift to the Earth Sciences*, vols 1 & 2. Academic Press, London.

Tarling, D.H. & Tarling, M.P. (1971) *Continental Drift: a study of the Earth's moving surface*. Bell, London.

4 | Sea floor spreading and transform faults

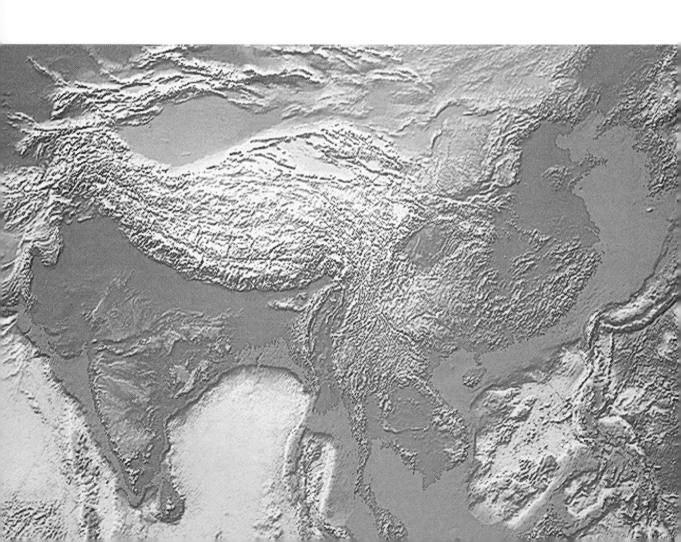

4.1 SEA FLOOR SPREADING

4.1.1 Introduction

By the late 1950s much evidence for continental drift had been assembled, but the theory was not generally accepted. Up to this time, work had concentrated upon determining the pre-drift configurations of the continents and assessing their geologic consequences. The paths by which the continents had attained their present positions had not been determined. In order to study the kinematics of continental drift it was necessary to study the regions now separating once juxtaposed continents. Consequently, at this time interest moved from the continents to the intervening ocean basins.

Any kind of direct observation of the sea floor, such as drilling, dredging, or submersible operations, is time consuming, expensive, and provides only a low density of data. Much of the information available over oceanic areas has therefore been provided by geophysical surveys undertaken from ships or aircraft. One such method involves measuring variations in the strength of the Earth's magnetic field. This is accomplished using either fluxgate, proton precession, or optical absorption magnetometers, which require little in the way of orientation so that the sensing element can be towed behind the ship or aircraft at a sufficient distance to minimize their magnetic effects. In this way total field values are obtained which are accurate to ± 1 nanotesla (nT) or about 1 part in 50,000. Magnetometers provide a virtually continuous record of the strength of the geomagnetic field along their travel paths. These absolute values are subsequently corrected for the externally induced magnetic field variations which give rise to a diurnal effect, and the regional magnetic field arising from that part of the magnetic field generated in the Earth's core. In theory the resulting magnetic anomalies should then be due solely to contrasts in the magnetic properties of the underlying rocks. The anomalies originate from the generally small proportion of ferromagnetic minerals (Section 3.6.2) contained within the rocks, of which the most common is magnetite. In general, ultramafic and mafic rocks contain a high proportion of magnetite and thus give rise to large magnetic anomalies. Metamorphic rocks are moderately

magnetic and acid igneous and sedimentary rocks are usually only weakly magnetic. A full account of the magnetic surveying method is given in Kearey *et al.* (2002).

On land, magnetic anomalies reflect the variable geology of the upper continental crust. The oceanic crust, however, is known to be laterally uniform (Section 2.4.4) and so unless the magnetic properties are heterogeneous it would be expected that marine magnetic anomalies would reflect this compositional uniformity.

4.1.2 Marine magnetic anomalies

Magnetic surveying is easily accomplished, and measurements have been carried out from survey vessels since the mid 1950s both on specific surveys and routinely on passage to the locations of other oceanographic investigations.

A most significant magnetic anomaly map (Fig. 4.1) was constructed after detailed surveys off the western seaboard of North America (Mason & Raff, 1961; Raff & Mason, 1961). The magnetic field was shown to be anything but uniform, and revealed an unexpected pattern of stripes defined by steep gradients separating linear regions of high amplitude positive and negative anomalies. These magnetic lineations are remarkably persistent, and can be traced for many hundreds of kilometers. Their continuity, however, is interrupted at major oceanic fracture zones, where the individual anomalies are offset laterally by distances of up to 1100 km.

Subsequent surveys have shown that magnetic lineations are present in virtually all oceanic areas. They are generally 10–20 km wide and characterized by a peak-to-peak amplitude of 500–1000 nT. They run parallel to the crests of the mid-ocean ridge system (Chapter 6), and are symmetrical about the ridge axes (Fig. 4.2).

The source of these linear magnetic anomalies cannot be oceanic layer 1, which is made up of non-magnetic sediments. They cannot originate at a depth corresponding to layer 3 as sources solely within this layer would be too deep to generate the steep anomaly gradients. The source of the anomalies must therefore be, at least in part, in oceanic layer 2. This conclusion is consistent with the basaltic composition of layer 2 determined by dredging and drilling (Section 2.4.6), since basalt is known to contain a relatively high proportion of magnetic minerals. The magnetic

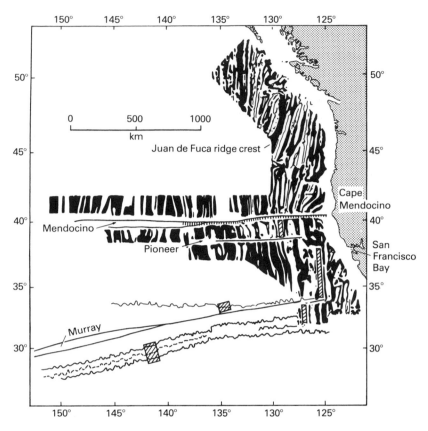

Figure 4.1 *Magnetic anomaly lineations in the northeastern Pacific Ocean. Positive anomalies in black; also shown are the oceanic fracture zones at which the lineations are offset (after Menard, 1964, with permission from the estate of the late Professor H. William Menard).*

lineations therefore confirm that layer 2 is everywhere composed of this rock type.

If magnetic lineations are generated by a layer of homogeneous composition, how do the magnetic contrasts originate that are responsible for the juxtaposition of large positive and negative magnetic anomalies? The shape of a magnetic anomaly is determined by both the geometric form of the source and the orientation of its magnetization vector. Oceanic layer 2 maintains a relatively constant depth and thickness. Any anomalies arising because of rugged topography on the top of the layer would attenuate too rapidly to account for the amplitude of the anomalies observed on the surface 3–7 km above the seabed. Consequently, the lineations must arise because adjacent blocks of layer 2 are magnetized in different directions. Figure 4.3 shows an interpretation of magnetic anomalies observed over the Juan de Fuca Ridge in the northeastern Pacific. Layer 2

has been divided into a series of blocks running parallel to the ridge crest which have been assigned magnetization vectors which are either in the direction of the ambient geomagnetic field or in the reversed direction. The interpretation shows that the observed anomalies are simulated by a model in which the intensities of the magnetization vary, and that relatively high values of some $10 \, A \, m^{-1}$ are required to produce the necessary contrasts.

4.1.3 Geomagnetic reversals

The possibility that the geomagnetic field reverses polarity was first suggested during the early part of the 20th century, when it was noted that reversed magnetizations were present in some rock samples, and that the low amplitudes of magnetic anomalies observed over

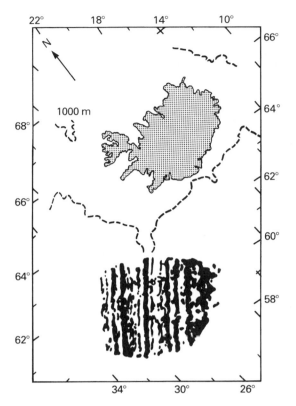

Figure 4.2 *Magnetic lineations either side of the mid-Atlantic ridge south of Iceland. Positive anomalies in black (after Heirtzler* et al., *1966, in* Deep Sea Research **13***, 428, with permission from Pergamon Press. Copyright Elsevier 1966).*

certain volcanic sequences were explicable in terms of a reversed magnetization vector.

By the early 1960s the concept of geomagnetic field reversals was being revived, both because of the large number of paleomagnetic measurements revealing reversed magnetization, and the demonstration that self-reversal, whereby a reversed magnetization can originate from interaction with normally magnetized material, was a very rare phenomenon. By the mid 1960s, following the work of Cox *et al.* (1964, 1967) on lava flows erupted within the past few million years, the concept was widely accepted. More recently, paleomagnetic studies of rapidly deposited sediments, lava flow sequences, and slowly cooled igneous intrusions have shown that a magnetic reversal occurs over a time interval of about 5000 years. It is accompanied by a reduction in field intensity to about 25% of its normal value which commences some time before the reversal and

continues for some time afterwards, with a total duration of about 10,000 years.

There is no general theory for the origin of the geomagnetic field. However, it is recognized that the main part originates within the Earth, and must be caused by dynamic processes. A magnetostatic origin appears impossible as no known material is sufficiently magnetic to give rise to the magnitude of the field observed at the surface, and subsurface temperatures would be well in excess of the Curie point, even given that its dependence upon pressure is largely unknown. The temporal variation of the internally generated field would also be inexplicable with such a model.

The geomagnetic field is believed to originate by magnetohydrodynamic processes within the fluid (outer) part of the Earth's core, magnetohydrodynamics being that branch of physics concerned with the interaction of fluid motions, electric currents, and magnetic fields. Indeed, this process is also believed to be responsible for the magnetic fields of other planets and certain stars. The process requires the celestial body to be rotating and to be partly or completely composed of a mobile fluid which is a good electrical conductor. The turbulent or convecting fluid constitutes a dynamo, because if it moves in a pre-existing magnetic field it generates an electric current which has a magnetic field associated with it. When the magnetic field is supplied solely by the electric currents, the dynamo is said to be "self-excited." Once "excited," the dynamo becomes self-perpetuating as long as there is a primary energy source to maintain the convection currents. The process is complex, and analytic solutions are only available for the very simplest configurations, which cannot approach the true configuration in the core. The field is thought to be maintained by convection in the outer core, which is thermally or gravitationally driven, either by heat sources in the core, such as potassium (^{40}K), or the latent heat and light constituents released during solidification of the inner core due to the slow cooling of the Earth (Section 2.9) (Merrill *et al.*, 1996).

A mathematical formulation of the geodynamo has not been possible because of the complexity of the physical processes occurring in the Earth's fluid core. Consequently theoreticians have had to resort to numerical modeling. Initially these simulations were severely limited by the computing power available for the very large number of numerical integrations involved. Large-scale, realistic simulations of the dynamo model had to await the advent of the so-called

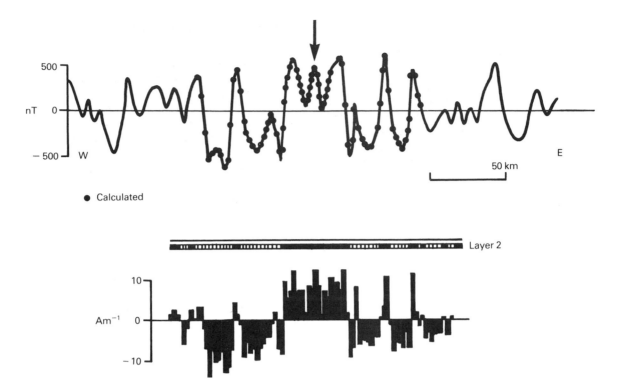

Figure 4.3 *Interpretation of a magnetic anomaly profile across the Juan de Fuca ridge, northeastern Pacific Ocean, in terms of normal and reversed magnetizations of two-dimensional rectangular blocks of oceanic layer 2. The arrow marks the ridge crest (redrawn from Bott, 1967, with permission Blackwell Publishing).*

supercomputers in the 1990s. The first results of numerical integrations of full three-dimensional, nonlinear, geodynamo models were published in 1995 (e.g. Glatzmaier & Roberts, 1995). These, and other comparable simulations through to 2000, were reviewed by Kono & Roberts (2002). The models simulate many of the features of the Earth's field, such as secular variation and a dominant axial dipole component, and in some cases magnetic reversals. Some of the latter are very similar in duration and characteristics to those deduced from paleomagnetic studies (Coe *et al.*, 2000).

The rates at which geomagnetic reversals have occurred in the geologic past is highly variable (see Figs 4.4, 4.13). There has been a gradual increase in the rate of reversals during the Cenozoic, following a period during the Cretaceous when the field was of constant normal polarity for 35 Ma. Paleomagnetic studies reveal a similar prolonged period of reverse polarity in the Late Carboniferous and Permian (McElhinny & McFadden, 2000). This seems to imply that the geodynamo can exist

in either of two states: one which generates a field of constant polarity for tens of millions of years, and one during which the field reverses in polarity at least once every million years. This is surprising in that convective overturn in the core is thought to be on a timescale of hundreds of years. It is difficult to imagine processes or conditions in the core that could account for two different states, which, once attained, persist for tens of millions of years. This timescale is characteristic of convection in the mantle. Changes in the pattern of convection in the mantle could produce changes in the physical conditions at the core–mantle boundary on the appropriate timescale. Small changes in seismic velocities in the mantle, revealed by seismic tomography, are interpreted in terms of temperature variations associated with convection, although they could in part be due to chemical inhomogeneity (Section 12.8.2). This raises the possibility that the heat flux at the core–mantle boundary is nonuniform, and changes significantly over periods of 10–100 Ma. The low viscosity and relatively rapid overturn in the outer core will ensure that the temperature at the core–mantle boundary is essentially uniform. The inferred temperature differences in the lower mantle, however, will give rise to a nonuniform distribution of heat flux at the core–mantle boundary. Anomalously cold material near the boundary will steepen the temperature gradient and increase the heat flow, whereas hotter material will decrease the gradient and heat flow. The new advances in computer simulations of the geodynamo make it possible to explore this possibility. The initial results of such computations (Glatzmaier *et al.*, 1999) are very interesting and encouraging in that different heat flow distributions do produce significant changes in the reversal frequency and might well explain the variations observed in Fig. 4.4.

The results obtained from numerical simulations of the geodynamo since the mid 1990s represent remarkable breakthroughs in our modeling and understanding of the possible origin of the Earth's magnetic field. However one has to bear in mind that, although the physical formulation of these models is thought to be complete, the parameters assumed are not in the range appropriate for the Earth. This is because the computing power available is still not adequate to cope with the spatial and temporal resolution that would be required in the integrations.

4.1.4 Sea floor spreading

In the early 1960s, Dietz (1961) and Hess (1962) had proposed that continental drift might be accomplished by a process that Dietz termed sea floor spreading (Section 1.2). It was suggested that new oceanic lithosphere is created by the upwelling and partial melting of material from the asthenosphere at the ocean ridges. As the ocean gradually grows wider with the progressive creation of lithosphere, the continents marginal to the ocean are moved apart. The drift between North America and Europe, for example, would have been accomplished by the gradual growth of the Atlantic Ocean over the past 180 Ma. Since the Earth is not increasing in surface area by any significant amount (Section 12.3), the increase in size of those oceans growing by sea floor spreading would be balanced by the destruction of lithosphere at the same rate in another, shrinking, ocean by subduction at deep sea trenches situated around its margins.

The driving mechanism of these movements was believed to be convection currents in the sub-lithospheric mantle (Fig. 1.5). These were thought to form cells in which mantle ascended beneath ocean ridges, bringing hot material to the surface and giving rise to new lithosphere. The flow then moved horizontally

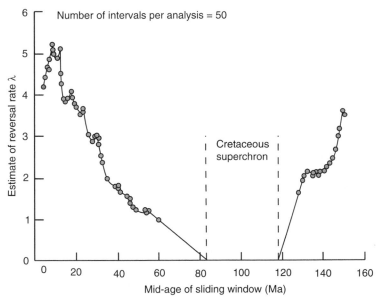

Figure 4.4 *Estimated frequency of geomagnetic reversals over the past 160 Ma (redrawn from Merrill* et al., *1996, with permission from Academic Press. Copyright Elsevier 1996).*

away from the ridge, driving the lithosphere laterally in the same direction by viscous drag on its base, and finally descended back into the deep mantle at the ocean trenches, assisting the subduction of the lithosphere. This possible mechanism will be discussed more fully in Section 12.7.

4.1.5 The Vine–Matthews hypothesis

It is perhaps surprising to note that magnetic maps of the oceans showing magnetic lineations (Section 4.2) were available for several years before the true significance of the lineations was realized. The hypothesis of Vine & Matthews (1963) was of elegant simplicity and combined the notion of sea floor spreading (Section 4.1.4) with the phenomenon of geomagnetic field reversals (Section 4.1.3).

The Vine–Matthews hypothesis explains the formation of magnetic lineations in the following way. New oceanic crust is created by the solidification of magma injected and extruded at the crest of an ocean ridge (Fig. 4.5). On further cooling, the temperature passes through the Curie point below which ferromagnetic behavior becomes possible (Section 3.6.2). The solidified magma then acquires a magnetization with the same orientation as the ambient geomagnetic field. The process of lithosphere formation is continuous, and proceeds symmetrically as previously formed lithosphere on either side of the ridge moves aside. But, if the geomagnetic field reverses polarity as the new lith-

osphere forms, the crust on either side of the ridge would consist of a series of blocks running parallel to the crest, which possess remanent magnetizations that are either normal or reversed with respect to the geomagnetic field. A ridge crest can thus be viewed as a twin-headed tape recorder in which the reversal history of the Earth's magnetic field is registered within oceanic crust (Vine, 1966).

The intensity of remanent magnetization in oceanic basalts is significantly larger than the induced magnetization. Since the shape of a magnetic anomaly is governed by the orientation of its total magnetization vector, that is, the resultant of the remanent and induced components, the shapes of magnetic lineations are effectively controlled by the primary remanent direction. Consequently, blocks of normally magnetized crust formed at high northern latitudes possess a magnetization vector that dips steeply to the north, and the vector of reversely magnetized material is inclined steeply upwards towards the south. The magnetic profile observed over this portion of crust will be characterized by positive anomalies over normally magnetized blocks and negative anomalies over reversely magnetized blocks. A similar situation pertains in high southern latitudes. Crust magnetized at low latitudes also generates positive and negative anomalies in this way, but because of the relatively shallow inclination of the magnetization vector the anomaly over any particular block is markedly dipolar, with both positive and negative components. This obscures the symmetry of the anomaly about the ridge crest, as individual blocks are no longer associated with a single positive or negative anomaly. However, at the magnetic equator, where

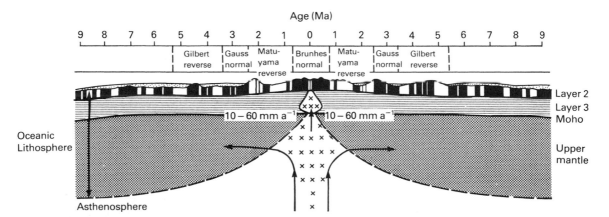

Figure 4.5 *Sea floor spreading and the generation of magnetic lineations by the Vine-Matthews hypothesis (redrawn from Bott, 1982, by permission of Edward Arnold (Publishers) Ltd).*

the field is horizontal, negative anomalies coincide with normally magnetized blocks and positive anomalies with reversely magnetized blocks, precisely the reverse situation to that at high latitudes. In addition, the amplitude of the anomaly decreases from the poles to the equator as the geomagnetic field strength, and hence the magnitude of the remanence, decreases in this direction. Figure 4.6 illustrates how the shape and amplitude of the magnetic anomalies over an ocean ridge striking east–west vary with latitude.

The orientation of the ridge also affects anomaly shape and amplitude, because only that component of the magnetization vector lying in the vertical plane through the magnetic profile affects the magnetic anomaly. This component is at a maximum when the ridge is east-west and the profile north-south, and at a minimum for ridges oriented north-south. The variation in amplitude and shape of the magnetic anomalies with orientation for a ridge of fixed latitude is shown in Fig. 4.7. In general, the amplitude of magnetic anomalies decreases as the latitude decreases and as the strike of the ridge progresses from east-west to north-south. The symmetry of the anomalies is most apparent for ridges at high magnetic latitudes (e.g. greater than 64°,

which is equivalent to geographic latitudes greater than 45°), north-south trending ridges at all latitudes and east-west trending ridges at the magnetic equator.

4.1.6 Magnetostratigraphy

Once the geomagnetic reversal timescale has been calibrated, oceanic magnetic anomalies may be used to date oceanic lithosphere. The method has been progressively refined so that it is now possible to deduce ages back to mid-Jurassic times with an accuracy of a few million years.

The Vine–Matthews hypothesis explains the sequence of magnetic anomalies away from ocean ridges in terms of normal and reversed magnetizations of the oceanic crust acquired during polarity reversals of the geomagnetic field. Verification of the hypothesis was provided by the consistency of the implied reversal sequence with that observed independently on land. Cox *et al.* (1967) had measured the remanent magnetization of lavas from a series of land sites. The lavas were dated by a newly refined potassium-argon method, which allowed the construction of a reversal timescale back to 4.5 Ma. The timescale could not be extended to earlier ages, as the errors involved in K-Ar dating become too large. Similarly, polarity events of less than 50,000 years duration could not be resolved. The timescale to 5 Ma before present, as later refined by Cande & Kent (1992), is given in Fig. 4.8. In magnetostratigraphic terminology, polarity chrons are

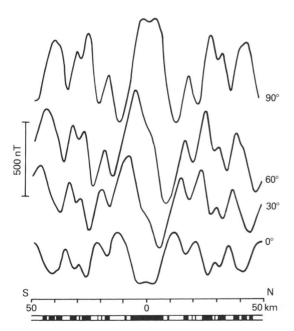

Figure 4.6 *Variation of the magnetic anomaly pattern with geomagnetic latitude. All profiles are north-south. Angles refer to magnetic inclination. No vertical exaggeration.*

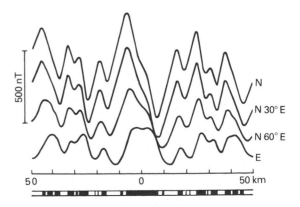

Figure 4.7 *Variation of the magnetic anomaly pattern with the direction of the profile at a fixed latitude. Magnetic inclination is 45° in all cases. No vertical exaggeration.*

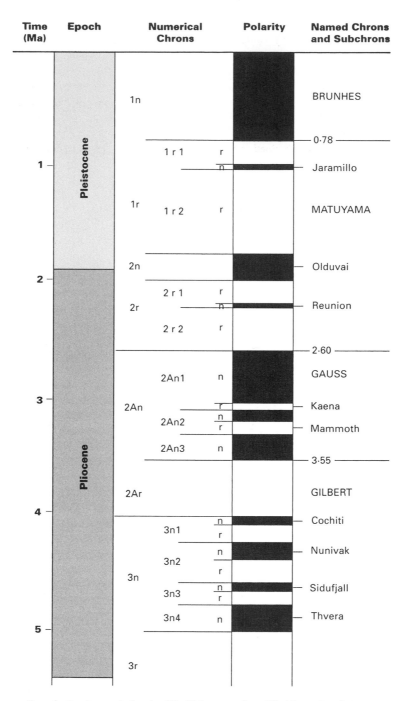

Figure 4.8 *Geomagnetic polarity timescale for the Plio-Pleistocene (modified from Cande & Kent, 1992, by permission of the American Geophysical Union. Copyright © 1992 American Geophysical Union). Numerical chrons are based on the numbered sequence of marine magnetic anomalies.*

defined with durations of the order of 10^6 years. Chrons may be dominantly of reversed or normal polarity, or contain mixed events.

Further verification of the geomagnetic reversal timescale was provided by paleomagnetic investigations of deep sea cores (Opdyke *et al.*, 1966). Unlike lava flows, these provide a continuous record, and permit accurate stratigraphic dating from their microfauna. This method is most conveniently applied to cores obtained in high magnetic latitudes where the geomagnetic inclination is high, because the cores are taken vertically and are not oriented azimuthally. Excellent correlation was found between these results and those from the lava sequences, and confirmed that at least 11 geomagnetic field reversals had occurred over the last 3.5 Ma. Subse-

quent work on other cores extended the reversal history back to 20 Ma (Opdyke *et al.*, 1974).

Pitman & Heirtzler (1966) and Vine (1966) used the radiometrically dated reversal timescale to compute the magnetic profiles that would be expected close to the crestal regions of mid-ocean ridges. By varying the spreading rate it was possible to obtain very close simulations of all observed anomaly sequences (Fig. 4.9), and consequently to determine the spreading rates. A compilation of such rates is shown in Table 4.1. Extensions of this work show that the same sequence of magnetic anomalies, resulting from spreading and reversals of the Earth's magnetic field, can be observed over many ridge flanks (e.g. Fig. 4.10). Later work has shown that similar linear magnetic anomalies are developed

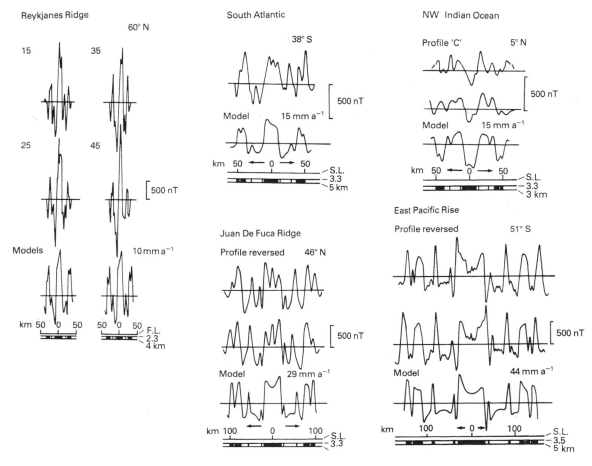

Figure 4.9 *Magnetic anomaly profiles and models of several spreading centers in terms of the reversal timescale (redrawn from Vine, 1966, Science **154**, 1405–15, with permission from the AAAS).*

Table 4.1 *Spreading rates at mid-ocean ridges ("spreading rate" is defined as the accretion rate per ridge flank).*

Ridge	Latitude	Observed rate (mm a^{-1})	Predicted rate (mm a^{-1})
Juan de Fuca	46.0°N	29	†
Gulf of California	23.4°N	25	24.7
Cocos –			
Pacific	17.2°N	37	39.4
Pacific	3.1°N	67	65.4
Galapagos	2.3°N	22	22.0
Galapagos	3.3°N	34	34.6
Nazca –			
Pacific	12.6°S	75	74.2
Chile Rise	43.4°S	31	30.2
Pacific –			
Antarctic	35.6°S	50	49.5
Antarctic	51.0°S	44	44.6
Antarctic	65.3°S	26	29.0
North Atlantic	86.5°N	6	5.7
North Atlantic	60.2°N	9.5	9.2
North Atlantic	42.7°N	11.5	11.9
Central Atlantic	35.0°N	10.5	11.0
Central Atlantic	23.0°N	12.5	12.6
Cayman	18.0°N	7.5	5.9
South Atlantic	38.5°S	18	17.6
Antarctic –			
South America	55.3°S	10	9.3
Africa –			
Antarctic	44.2°S	8	7.4
Northwest Indian Ocean	4.2°N	14	14.6
Northwest Indian Ocean	12.0°S	18.5	17.9
Northwest Indian Ocean	24.5°S	25	24.5
Southeast Indian Ocean	25.8°S	28	28.8
Southeast Indian Ocean	50.0°S	38	37.3
Southeast Indian Ocean	62.4°S	34.5	33.7
Gulf of Aden	12.1°N	8	8.6
Gulf of Aden	14.6°N	12	12.1
Red Sea	18.0°N	10	8.2

Based on data from DeMets *et al* (1990) and Vine (1966).
† Not available because Farallon plate is omitted from the model.

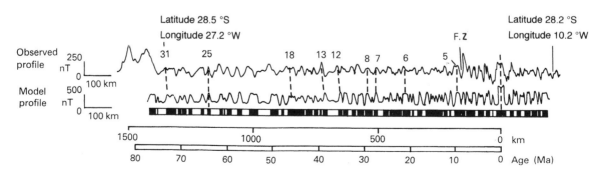

Figure 4.10 *Magnetic anomaly profile and model over the southern Mid-Atlantic Ridge (redrawn from Heirtzler* et al., *1968, by permission of the American Geophysical Union. Copyright © 1968 American Geophysical Union).*

over oceanic crust dating back to the Jurassic. Although there is no oceanic crust older than this, paleomagnetic investigations on land have shown that geomagnetic reversals have occurred at least back to 2.1 Ga.

That spreading rates have varied with time is apparent from an examination of magnetic profiles from different oceans. Examples are given in Fig. 4.11 in which the spreading rate in the South Atlantic is assumed to be constant and the distances to various magnetic anomalies from ridge crests in other oceans are plotted against the distance to the same anomaly in the South Atlantic. Inflection points in the curves for the other oceans indicate when the spreading rates changed there if the implicit assumption that the spreading rate has remained constant in the South Atlantic is correct. However, spreading rates may have changed with time in all oceans.

The first long-term geomagnetic timescale was constructed by Heirtzler et al. (1968). Again they made the assumption that spreading in the South Atlantic had remained constant at the same rate as had been deduced for the last 4 Ma. A model of normal and reversely magnetized blocks was constructed which simulated the observed anomaly pattern, and the distance axis transformed into a geomagnetic timescale of reversals extending back in time nearly 80 Ma. Prominent anomalies corresponding to periods of normal polarity were numbered from 1 to 32 with increasing time (Fig. 4.10).

Leg 3 of the Deep Sea Drilling Program (DSDP), in 1968, was specifically designed to test the hypothesis of sea floor spreading and the assumption of a constant rate of spreading in the South Atlantic (Maxwell et al., 1970). A series of holes was drilled in the South Atlantic along a traverse at right angles to the Mid-Atlantic Ridge (Fig. 4.12a). The age of the oceanic crust would ideally have been determined by radiometric dating of the layer 2 basalts that were penetrated in each hole. However the basalts were too weathered for this to be possible, and so their ages were determined, albeit slightly underestimated, by paleontologic dating of the basal sediments of layer 1. In Fig. 4.12b oldest sediment age is plotted against distance from the ridge axis, and it is readily apparent that there is a remarkable linear relationship, with crustal age increasing with distance from the ridge. The predicted ages imply a half spreading rate in this region of 20 mm a^{-1}, as predicted, and hence agree well with the age of the ocean floor and the reversal timescale proposed by Heirtzler et al. (1968) (Fig. 4.10).

A thorough review of the calibration of this polarity timescale was carried out by Cande & Kent (1992, 1995). It drew on oceanic magnetic anomaly data, magnetostratigraphic studies of sedimentary sequences on land and at sea, and radiometric dating of nine specific stratigraphic horizons. From this they concluded that sea floor spreading in the South Atlantic had been continuous, with some variation about an essentially constant rate, and that it was still appropriate to use the South Atlantic as a standard against which the spreading history in the other ocean basins could be compared. The revised timescale for the past 80 Ma suggested by Cande & Kent (1995) is illustrated in Fig. 4.13.

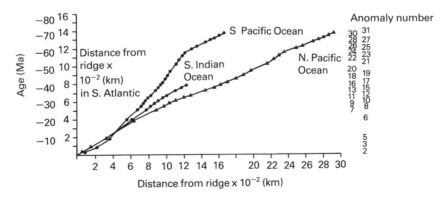

Figure 4.11 *Relationship between the distance to a given anomaly in the South Atlantic and the distance to the same anomaly in the South Indian, North Pacific and South Pacific Oceans. Numbers on the right refer to magnetic anomaly numbers (redrawn from Heirtzler et al., 1968, by permission of the American Geophysical Union. Copyright © 1968 American Geophysical Union).*

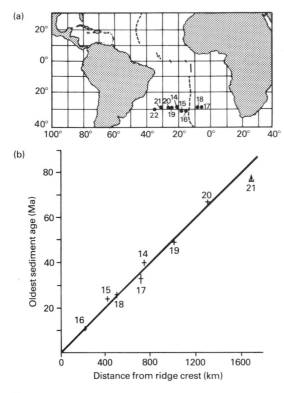

Figure 4.12 *(a) Location map of drilling sites on Leg 3 of the DSDP in the South Atlantic. (b) Relationship between greatest sediment age and distance from the Mid-Atlantic Ridge crest (after Maxwell* et al., *1970, Science* **168**, *1047–59, with permission from the AAAS).*

The discovery of Larson & Pitman (1972) of older magnetic anomalies in three regions of the western Pacific allowed the Heirtzler geomagnetic timescale to be extended back to 160 Ma. Lineations of similar pattern were also found in the Atlantic. The timescale was extended by assuming a constant spreading rate in the Pacific, calibrated by DSDP sites in the Pacific and Atlantic. The longer periods of *reversed* polarity in this sequence are numbered M0 to M28 (M representing Mesozoic). It appears that spreading in the major ocean basins has been continuous as all polarity events are present, although the rate of spreading has varied.

The version of the reversal timescale to 160 Ma shown in Fig. 4.13 combines the timescale of Cande & Kent (1995), for the Late Cretaceous and Cenozoic (anomalies 1–34), with that of Kent & Gradstein (1986) for the Early Cretaceous and Late Jurassic (anomalies M0–M28).

4.1.7 Dating the ocean floor

The use of the geomagnetic timescale to date the oceanic lithosphere is based on the identification of characteristic patterns of magnetic anomaly lineations and their relation to the dated reversal chronology. Particularly conspicuous markers which are widely used are anomalies 5, 12–13, 21–26, and 31–32. Also of interest is the prolonged period of normal polarity in the Cretaceous. This period corresponds to magnetic quiet zones within the oceans where there are no linear magnetic anomalies. In many instances, however, the recognition of particular anomalies is not possible, and the usual approach is to construct the anomaly pattern expected for relevant parts of the timescale and to compare it with the observed sequence.

Once the reversal chronology has been established, lineations of known age can be identified on magnetic maps and transformed into isochrons so that the sea floor can be subdivided into age provinces (Scotese *et al.*, 1988). Summaries of the isochrons derived from the linear oceanic magnetic anomalies are also provided by Cande *et al.* (1989) and Müller R.D. *et al.* (1997). (Plate 4.1 between pp. 244 and 245). Lineations of the same age on either side of a mid-ocean ridge can be fitted together by employing techniques similar to those used for continental margins (Section 3.2.2). In this way reconstructions of plate configurations can be made for different times, and the whole evolution of the present day ocean basins determined (Scotese *et al.*, 1988). Figure 4.14 shows this method applied to the Mesozoic and Cenozoic history of the North Atlantic. Examples of areas with more complex spreading histories, involving extinct ridges and ridge jumps, include the Indian Ocean (Norton & Sclater, 1979) and the Greenland–Iceland–Scotland region (Nunns, 1983).

4.2 TRANSFORM FAULTS

4.2.1 Introduction

The theory of sea floor spreading proposes that oceanic lithosphere is created at mid-ocean ridges and is

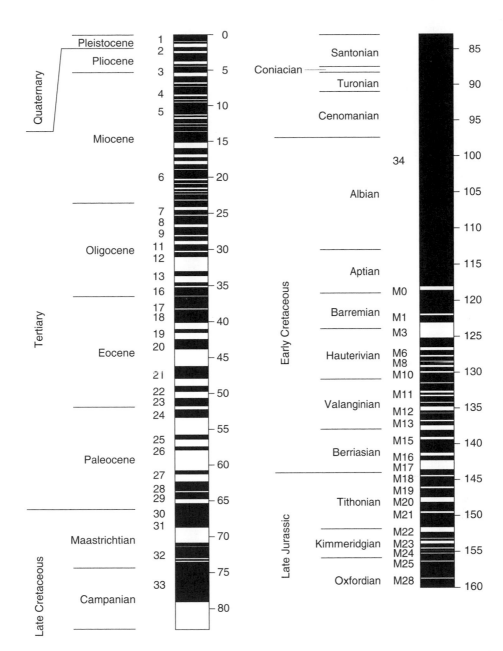

Figure 4.13 *A geomagnetic polarity timescale for the past 160 Ma together with oceanic magnetic anomaly numbers (after McElhinny & McFadden, 2000, with permission from Academic Press. Copyright Elsevier 2000).*

balanced by the complementary destruction of oceanic lithosphere at subduction zones. While this theory neatly explains the geometry of lithospheric behavior in two dimensions, a problem arises when the third dimension is considered, namely where do ridges and trenches terminate horizontally? This problem was addressed by Wilson (1965), who proposed that the ends of these features were linked by a new class of faults which he called transform faults. At these faults there is neither creation nor destruction of lithosphere,

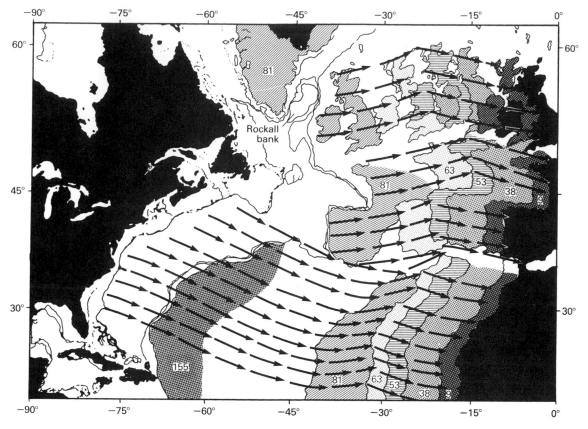

Figure 4.14 *Relative positions of Europe and Africa with respect to North America illustrating their separation during the Mesozoic and Cenozoic. Ages of reconstruction shown in millions of years (redrawn from Pitman & Talwani, 1972, with permission from the Geological Society of America).*

but rather the motion is strike-slip, with adjacent lithosphere in tangential motion.

The existence of large lateral relative movements of the lithosphere was first suggested from marine magnetic anomalies in the northeastern Pacific (Fig. 4.1), which were found to be offset along fracture zones. Combined left lateral offsets along the Mendocino and Pioneer faults amount to 1450 km, while the right lateral offset across the Murray Fault is 600 km in the west and only 150 km in the east (Vacquier, 1965).

However, in interpreting these fracture zones as large scale strike-slip faults, a major problem arises in that there is no obvious way in which the faults terminate, as it is certain that they do not circumnavigate the Earth to join up with themselves. Wilson (1965) proposed that the faults terminate at the ends of ridges or trenches, which they commonly meet at right angles.

Wilson termed this new class of faults transform faults, because the lateral displacement across the fault is taken up by transforming it into either the formation of new lithosphere at a terminated ocean ridge segment or lithosphere subduction at a trench. Figure 4.15 shows the plan view of an ocean ridge crest that has been displaced by transcurrent and transform faulting. The transcurrent, or strike-slip, fault (Fig. 4.16b) causes a sinistral offset along a vertical plane which must stretch to infinity beyond the ridge crests. The transform fault (Fig. 4.15a), however, is only active between the offset ridge crests, and the relative movement of the lithosphere on either side of it is dextral. Transform faults differ from other types of fault in that they imply, indeed derive from, the fact that the area of the faulted medium, in this case lithosphere, is not conserved at ridges and trenches.

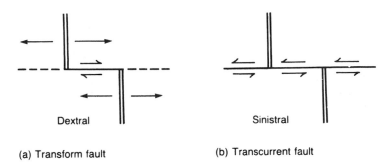

(a) Transform fault   (b) Transcurrent fault

Figure 4.15 *Comparison of transform and transcurrent faults.*

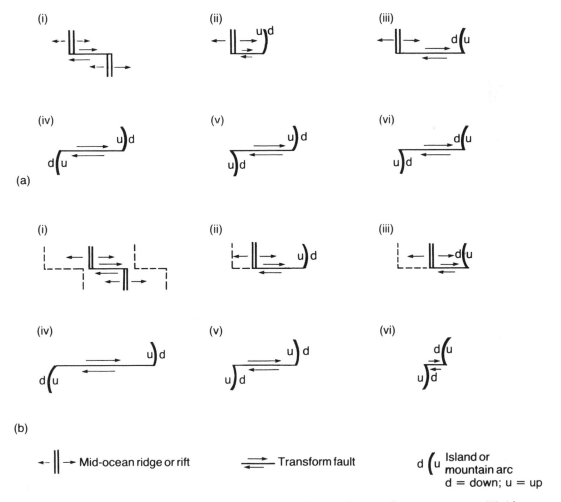

Figure 4.16 *(a) Six possible types of dextral transform fault: (i) ridge to ridge; (ii) ridge to concave arc; (iii) ridge to convex arc; (iv) concave arc to concave arc; (v) concave arc to convex arc; (vi) convex arc to convex arc. (b) Appearance of the dextral transform faults after a period of time (redrawn from Wilson, 1965, with permission from* Nature ***207**, 334–47. Copyright © 1965 Macmillan Publishers Ltd).*

Wilson (1965) defined six classes of transform fault that depend upon the types of nonconservative features they join (Fig. 4.16). These may be an ocean ridge, the overriding plate at a trench or the underthrusting plate at a trench. Figure 4.16a shows the six possible kinds of dextral transform fault; a further six based on sinistral movement are also possible. Figure 4.16b shows how the transform faults would develop with time. Cases (i) and (v) will remain unchanged, cases (ii) and (iv) will grow, and cases (iii) and (vi) will diminish in length with the passage of time.

4.2.2 Ridge–ridge transform faults

Sykes (1967) determined focal mechanism solutions for earthquakes occurring in the vicinity of the fracture zones that offset the Mid-Atlantic Ridge to the left at equatorial latitudes (Fig. 4.17). Events along the ridge axis are consistent with normal faulting along north–south planes. Events along the fracture zones are much more common and the energy release is about a hundred times greater than along the ridge crest. Between the offset ridge segments events are of strike-slip type with one nodal plane consistent with dextral transform motion. Events along the fracture zone beyond the ridge extremities are rare. These results provided striking confirmation of the transform fault concept and further, independent, confirmation of the hypothesis of sea floor spreading.

Before the recognition of transform faulting, the parallel fracture zones which appear to displace the crest of the Mid-Atlantic Ridge in equatorial latitudes between Africa and South America were believed to represent sinistral transcurrent faults that displaced an originally straight crest (Fig. 4.15b). However their

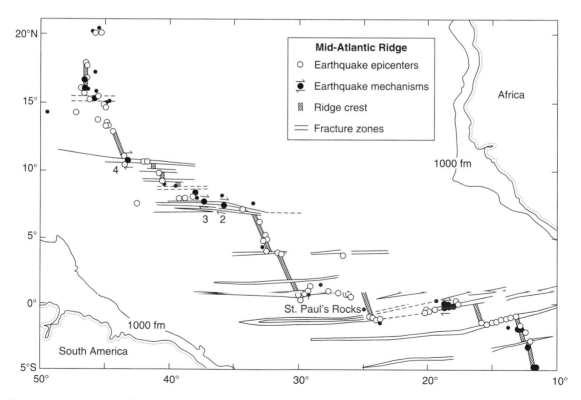

Figure 4.17 *Epicenters of earthquakes that occurred on the Mid-Atlantic ridge in the equatorial Atlantic between 1955 and 1965. The arrows beside four of the earthquakes indicate the sense of shear and the strike of the fault plane inferred from focal mechanism solutions (modified from Sykes, 1967, by permission of the American Geophysical Union. Copyright © 1967 American Geophysical Union).*

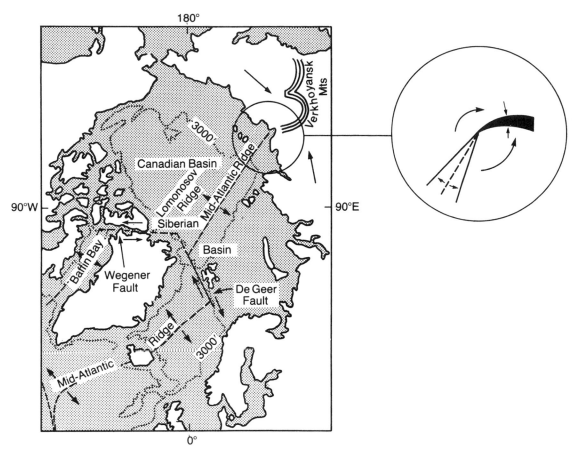

Figure 4.18 *Northern termination of the Mid-Atlantic Ridge (redrawn from Wilson, 1965, with permission from* Nature *207, 334–47. Copyright © 1965 Macmillan Publishers Ltd).*

re-interpretation, as ridge–ridge transform faults (Fig. 4.15a), implies that the offsets on them do not change with time. Thus the geometry, or locus, of the step-like ridge crest-transform fault sequence in the equatorial Atlantic has remained essentially unchanged throughout the opening of the South Atlantic. As a result the locus parallels the continental shelf edges of South America and Africa and reflects the geometry of the original rifting of the Gondwana supercontinent in this area.

Wilson (1965) also suggested examples of transform faults in the extreme North Atlantic area (Fig. 4.18). In early Paleogene times the Mid-Atlantic Ridge bifurcated to the south of Greenland. The western branch, which is now inactive, passed through Baffin Bay and terminated against the Wegener Fault, an extinct, sinistral ridge–ridge transform fault. The active eastern branch

passes through Iceland, and terminates southwest of Spitsbergen at the De Geer Fault. This dextral ridge-ridge transform fault connects to the Gakkel Ridge in the Arctic Basin. Wilson predicted that this is a very slow spreading ridge that is transformed into the Verkhoyansk Mountains of Siberia by rotation about a fulcrum near the New Siberian Islands (Fig. 4.18 inset).

4.2.3 Ridge jumps and transform fault offsets

The different offsets observed across the Murray Fracture Zone from magnetic lineations (Fig. 4.1) are thought to be due to a change in location of the ridge crest to the south of the fracture zone about 40 Ma ago. The change in offset of anomalies of the same age implies a "ridge jump" of approximately 500 km to the east (Harrison & Sclater, 1972). Similar but better documented ridge jumps, which also greatly reduce the offset of crust of the same age on either side of a fracture zone, occur in the extreme south of the Atlantic Ocean (Barker, 1979). Here, to the south of the Falkland–Agulhas Fracture Zone, the ridge crest has jumped westward on three occasions since the opening of the South Atlantic, that is at 98, 63 and 59 Ma. In so doing it has reduced the original offset of 1400 km to approximately 200 km. Other ridge jumps, producing major changes in ridge crest geometry within the past 10 Ma, have occurred to the north of Iceland (Vogt *et al.*, 1970) and along the crest of the East Pacific Rise in the east central Pacific (Herron, 1972).

In general, however, ridge jumps are relatively rare, as evidenced by the median position of oceanic ridges between separated continents. The geometry of ridge crest segments and transform faults is more likely to be modified, less dramatically, by ridge propagation (Section 6.11) and changes in the spreading direction (Section 5.9).

FURTHER READING

Cox, A. & Hart, R.B. (1986) *Plate Tectonics. How it works.* Blackwell Scientific Publications, Oxford.

Jacobs, J.A. (1994) *Reversals of the Earth's Magnetic Field.* Cambridge University Press, Cambridge.

Jones, E.J.W. (1999) *Marine Geophysics.* Wiley, Chichester, England.

Merrill, R.T., McElhinny, M.W. & McFadden, P.L. (1996) *The Magnetic Field of the Earth: paleomagnetism, the core and the deep mantle.* Academic Press, San Diego.

Opdyke, N.D. & Channel, J.E.T. (1996) *Magnetic Stratigraphy.* Academic Press, San Diego.

5 | The framework of plate tectonics

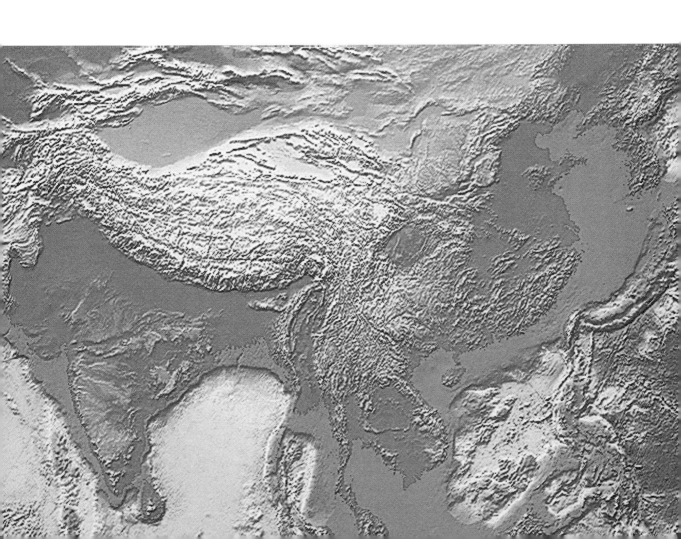

5.1 PLATES AND PLATE MARGINS

The combination of the concept of transform faults with the hypothesis of sea floor spreading led to the construction of the theory of plate tectonics. In this theory the lithosphere is divided into an interlocking network of blocks termed plates. The boundaries of plates can take three forms (Isacks *et al.*, 1968).

1 *Ocean ridges* (accretive or constructive plate margins) mark boundaries where plates are diverging. Magma and depleted mantle upwell between the separating plates, giving rise to new oceanic lithosphere. The divergent motion of the plates is frequently perpendicular to the strike of the boundary, although this is not always the case and is not a geometric necessity. In the Pacific it appears to be an intrinsic characteristic of spreading whenever a steady direction has been established for some time (Menard & Atwater, 1968).

2 *Trenches* (destructive plate margins) mark boundaries where two plates are converging by the mechanism of the oceanic lithosphere of one of the plates being thrust under the other, eventually to become resorbed into the sub-lithospheric mantle. Since the Earth is not expanding significantly (Section 12.3), the rate of lithospheric destruction at trenches must be virtually the same as the rate of creation at ocean ridges. Also included in this category are Himalayan-type orogens caused by the collision of two continental plates (Section 10.1), where continued compressional deformation may be occurring. The direction of motion of the underthrusting plate need not be at right angles to the trench, that is, oblique subduction can occur.

3 *Transform faults* (conservative plate margins) are marked by tangential motions, in which adjacent plates in relative motion undergo neither destruction nor construction. The relative motion is usually parallel to the fault. There are, however, transform faults that possess a sinuous trace, and on the bends of these faults relatively small regions of extension and compression are created (Section 8.2). For

the time being such structural elements are ignored.

Within the basic theory of plate tectonics plates are considered to be internally rigid, and to act as extremely efficient stress guides. A stress applied to one margin of a plate is transmitted to its opposite margin with no deformation of the plate interior. Deformation, then, only takes place at plate margins. This behavior is rather surprising when it is appreciated that plates are typically only about 100 km thick but may be many thousands of kilometers in width. When plate behavior is examined in more detail, however, it is recognized that there are many locations where intra-plate deformation occurs (Gordon & Stein, 1992; Gordon, 1998, 2000), especially within the continental crust (Section 2.10.5). Zones of extension within continental rifts may be many hundreds of kilometers wide (Section 7.3). Continental transforms are more complex than oceanic varieties (Section 8.1). Orogenic belts are characterized by extensive thrust faulting, movements along large strike-slip fault zones, and extensional deformation that occur deep within continental interiors (Section 10.4.3). Within oceanic areas there also are regions of crustal extension and accretion in the backarc basins that are located on the landward sides of many destructive plate margins (Section 9.10).

Plates are mechanically decoupled from each other, although plate margins are in intimate contact. A block diagram illustrating schematically the different types of plate boundaries is presented in Fig. 5.1.

5.2 DISTRIBUTION OF EARTHQUAKES

Plate tectonic theory predicts that the majority of the Earth's tectonic activity takes place at the margins of plates. It follows, then, that the location of earthquake epicenters can be used to define plate boundaries. Figure 5.2 shows the global distribution of the epicenters of large magnitude earthquakes for the period 1961–67 (Barazangi & Dorman, 1969). Although in terms of most geologic processes this represents only a very short period of observation, the relatively rapid motions experienced by plates generate very large numbers of earthquakes over a short interval of time.

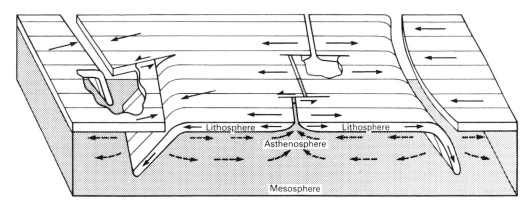

Figure 5.1 *Block diagram summarizing the principal features of plate tectonics. Arrows on lithosphere represent relative motions. Arrows in asthenosphere may represent complementary flow in the mantle (redrawn from Isacks et al., 1968, by permission of the American Geophysical Union. Copyright © 1968 American Geophysical Union).*

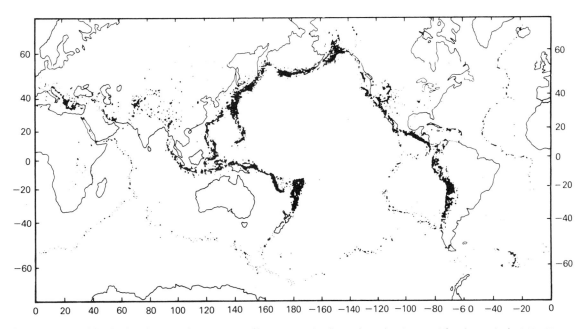

Figure 5.2 *Worldwide distribution of epicenters of large magnitude earthquakes ($m_b > 4$) for the period 1961–67 (after Barazangi & Dorman, 1969, with permission from the Seismological Society of America).*

The significance of 1961, as the start of this time window, is that prior to the setting up of the World Wide Standardized Seismograph Network in 1961 (Section 2.1.4), epicentral locations, particularly in oceanic areas, were very poorly determined. For a more detailed discussion of earthquake distribution see Engdahl *et al.* (1998).

Earthquakes are classified according to their focal depths: 0–70 km shallow focus, 70–300 km intermediate focus, greater than 300 km deep focus.

An important belt of shallow focus earthquakes follows the crest of the ocean ridge system (Fig. 5.2), where focal mechanism solutions indicate tensional events associated with plate accretion and strike-slip events where the ridges are offset by transform faults (Section 4.2.1). On land, shallow focus tensional events are also associated with rifts, including the Basin and Range Province of the western USA (Section 7.3), the East African Rift system (Section 7.2), and the Baikal Rift system.

All intermediate and deep events are associated with destructive plate margins. The northern, eastern and western Pacific Ocean is ringed by a belt of earthquakes which lie on planes, in places offset by transform faults, dipping at an angle of about 45° beneath the neighboring plates. These planes of earthquake foci, known as Benioff (or Benioff Wadati) zones, are typically associated with volcanic activity at the surface. The deepest events recorded lie at a depth of about 670 km. Collisional mountain belts such as the Alpine-Himalayan chain are similarly characterized by intermediate and deep focus earthquakes although, since there is no longer a Benioff zone present in such regions, the seismic activity occurs within a relatively broad belt (Fig. 10.17). Careful examination of epicenter locations has revealed, however, that some of the shallow events lie on arcuate strike-slip fault zones associated with the collisional event.

The intra-plate areas are relatively aseismic on this timescale, although occasionally large magnitude earthquakes do occur. Although insignificant in their release of seismic energy, intra-plate earthquakes are important as they can indicate the nature and direction of stress within plates (Section 12.7).

5.3 RELATIVE PLATE MOTIONS

The present day motion of plates can now be measured using the techniques of space geodesy (Section 5.8). However, these techniques were only developed in the 1980s, and, ideally, measurements are required over a period of 10–20 years (Gordon & Stein, 1992). Prior to this relative plate motions, averaged over the past few million years, were determined using geologic and geophysical data.

The motion of plates over the Earth's surface can be described by making use of Euler's theorem (Section 3.2.1), which says that the relative motion between two plates is uniquely defined by an angular separation about a pole of relative motion known as an Euler pole. The pole and its antipole are the two unique points on the surface of the Earth that do not move relative to either of the two plates. An important aspect of relative plate motion is that the pole of any two plates tends to remain fixed relative to them for long periods of time. Plate velocities are similarly constant for periods of several million years (Wilson, 1993).

There are three methods by which the pole of relative motion for two plates can be determined. The first, and most accurate, is based on the fact that for true tangential motion to occur during the relative movement of two plates, the transform faults along their common boundary must follow the traces of small circles centered upon the pole of relative motion (McKenzie & Parker, 1967; Morgan, 1968). The pole of rotation of two plates can thus be determined by constructing great circles at right angles to the trends to transform faults affecting their common margin and noting their point of intersection. The most convenient type of plate margin to which to apply this technique is the accretive type (Fig. 5.3), as ocean ridges

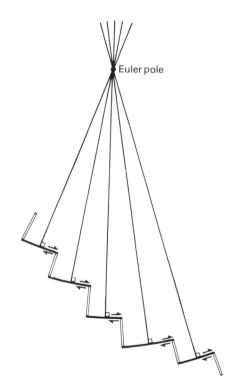

Figure 5.3 *Determination of the Euler pole for a spreading ridge from its offsetting transform faults that describe small circles with respect to the pole.*

are frequently offset laterally by transform faults (Section 4.2.1). Because of inaccuracies involved in mapping oceanic fracture zones, the great circles rarely intersect at a single point. Consequently, statistical methods are applied which are able to predict a circle within which it is most probable that the relative rotation pole lies.

A second method is based on the variation of spreading rate with angular distance from the pole of rotation. Spreading rates are determined from magnetic lineations (Section 4.1.6) by identifying anomalies of the same age (usually number 3 or less so that the movement represents a geologically *instantaneous* rotation) on either side of an ocean ridge and measuring the distance between them. The velocity of spreading is at a maximum at the equator corresponding to the Euler pole and thence decreases according to the cosine of the Euler pole's latitude (Fig. 5.4). The determination of the spreading rate at a number of points along the ridge then allows the pole of relative rotation to be found.

The final, and least reliable, method of determining the directions of relative motion between two plates makes use of focal mechanism solutions of earthquakes (Section 2.1.6) on their common margins. If the inclination and direction of slip along the fault plane are known, then the horizontal component of the slip vector is the direction of relative motion. The data are less accurate than the other two methods described above because, except in very well determined cases, the nodal planes could be drawn in a range of possible orientations and the detailed geometry of fault systems at plate boundaries is often more complex than implied here (Section 8.2 and below).

Divergent plate boundaries can be studied using spreading rates and transform faults. Convergent boundaries, however, present more of a problem, and it is often necessary to use indirect means to determine relative velocities. This is possible by making use of information from adjoining plates and treating the rotations between plate pairs as vectors (Morgan, 1968). Thus, if the relative movements between plates A and B and between plates B and C are known, the relative movement between plates A and C can be found by vector algebra.

This approach can be extended so that relative motions can be determined for any number of interlocking plates. Indeed, the method can be applied to the complete mosaic of plates that make up the Earth's surface, provided that there are sufficient divergent

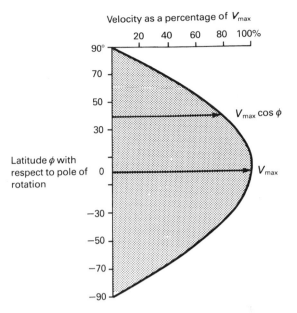

Figure 5.4 *Variation of spreading rate with latitudinal distance from the Euler pole of rotation.*

plate margins to be able to compute relative velocities at convergent margins.

The first study of this type was undertaken by Le Pichon (1968). He made use of globally distributed estimates of relative plate velocities derived from transform faults and spreading rates, but not of information obtained from focal mechanism solutions. Le Pichon used a subdivision of the Earth's surface based on only six large plates: the Eurasian, African, Indo-Australian, American, Pacific, and Antarctic plates. In spite of this simplification his model provided estimates of spreading rates that agreed well with those derived from magnetic anomalies (Section 4.1.6).

Subsequently, more detailed analyses of global plate motions were performed by Chase (1978), Minster & Jordan (1978), and DeMets *et al.* (1990). These studies recognized a number of additional plate boundaries and hence additional plates. The latter included the Caribbean and Philippine Sea plates, the Arabian plate, the Cocos and Nazca plates of the east Central Pacific, and the small Juan de Fuca plate, east of the Juan de Fuca ridge, off western North America (Fig. 5.5). The American plate was divided into two, the North American and South American plates, and the Indo-Australian plate similarly, into the Indian and Australian plates. The new boundaries identified within the American and

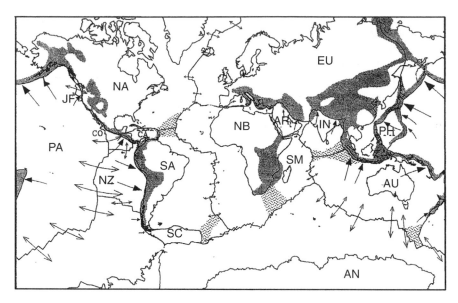

Figure 5.5 *Map showing the relative motion between the major plates, and regions of diffuse deformation within plates (shaded areas). Solid arrowheads indicate plate convergence, with the arrow on the underthrusting plate; open arrowheads indicate plate divergence at mid ocean ridges. The length of the arrows represents the amount of plate accretion or subduction that would occur if the plates were to maintain their present relative velocities for 25 Ma. Note that, because of the Mercator projection, arrows at high latitudes are disproportionately long compared to those at low latitudes. AN, Antarctica; AR, Arabia; AU, Australia; CA, Caribbean; CO, Cocos; EU, Eurasia; IN, India; JF, Juan de Fuca; NA, North America; NB, Nubia; NZ, Nazca; PA, Pacific; PH, Philippine; SA, South America; SC, Scotia Sea; SM, Somalia (modified from Gordon, 1995, by permission of the American Geophysical Union. Copyright © 1995 American Geophysical Union).*

Indo-Australian plates are rather indistinct and characterized by diffuse zones of deformation and seismicity (Gordon, 2000) (Fig. 5.5). Thus, the analysis of DeMets *et al.* (1990) involved 14 plates. Other plates have been recognized, but the relative movement across one or more of their boundaries is difficult to quantify. Examples include the Scotia Sea plate, and the diffuse boundary through the African plate, associated with the East African Rift system, that divides the African plate into the Nubian and Somali plates (Fig. 5.5). The only well-defined plate boundaries invariably omitted from these analyses are the spreading ridges in certain backarc basins (Section 9.10), for example, those in the east Scotia Sea, the east Philippine Sea and the South Fiji basin.

These analyses of relative plate motions all used large datasets of relative motion vectors derived from transform faults, spreading rates and focal mechanism solutions; that of DeMets *et al.* (1990) employing a dataset three times larger than those used in the earlier

models. In all cases so many data were available that the problem became over-determined, and in inverting the data set to provide the global distribution of plate motions, they used a technique whereby the sum of the squares of residual motions was minimized. Errors in determining spreading rates were generally less than $3 \, mm \, a^{-1}$, in transform fault orientation between $3°$ and $10°$, and in earthquake slip vector direction no more than $15°$.

Figure 5.5 illustrates the directions and rates of spreading and subduction predicted by the model of DeMets *et al.* (1990), at specific points on the respective plate boundaries. The rates have been corrected for a subsequent revision of the geomagnetic reversal times-cale (DeMets *et al.*, 1994). In Table 4.1, predicted rates of spreading, at various points on the mid-ocean ridge system, are compared with observed rates derived from the magnetic anomalies observed over these ridge crests. Along the length of the East Pacific Rise accretion rates per ridge flank vary from 25 to $75 \, mm \, a^{-1}$. By contrast, subduction rates around the margins of the Pacific are typically between 60 and 95 mm a^{-1}. Thus the oceanic plates of the Pacific are steadily reducing in size as they are being consumed at subduction zones at a higher rate than they are being created at the East Pacific Rise. By contrast, plates containing parts of the Atlantic and Indian oceans are increasing in size. A corollary of this is that the Mid-Atlantic Ridge and Carlsberg Ridge of the northwestern Indian Ocean must be moving apart. This has important implications for the nature of the driving mechanism of plate tectonics discussed in Chapter 12. Not all ocean ridges spread in a direction perpendicular to the strike of their magnetic lineations. It may be significant that the major obliquities of this type are found in the more slowly spreading areas, in particular the North Atlantic, Gulf of Aden, Red Sea, and southwestern Indian Ocean (Plate 4.1 between pp. 244 and 245).

In contrast to accretionary plate margins, where the spreading boundary is typically perpendicular to the direction of relative motion, convergent margins are not constrained in this way and the relative motion vector typically makes an oblique angle with the plate boundary. Extreme examples, with very high obliquity, occur at the western end of the Aleutian arc and the northern end of the Indonesian arc (Fig. 5.5). In subduction zones therefore, in addition to the component of motion perpendicular to the plate boundary, that produces underthrusting, there will be a component of relative motion parallel to the plate boundary. This "trench parallel" component often gives rise to strike-slip faulting within the overriding plate immediately landward of the forearc region. As a consequence, focal mechanism solutions, for earthquakes occurring on the interface between the two plates beneath the forearc region, do not yield the true direction of motion between the plates. They tend to underestimate the trench parallel component of motion because part of this is taken up by the strike-slip faulting (DeMets *et al.*, 1990). Classic examples of such trench parallel strike-slip faults include the Philippine Fault, the Median Tectonic Line of southwest Japan (Section 9.9), and the Atacama Fault and the Liquiñe–Ofqui Fault (Section 10.2.3) in Chile.

As indicated in Fig. 5.5, approximately 15% of the Earth's surface is covered by regions of deforming lithosphere; for example in the Alpine–Himalayan belt, southeast Asia, and western North America. Within these areas it is now possible to identify additional small plates, albeit often with diffuse boundaries, using GPS (Global Positioning System) data (Section 5.8). GPS measurements also make it possible to determine the motion of these plates relative to adjacent plates, whereas this is not possible using the techniques based on geologic and geophysical data described above. Most of the poorly defined zones of deformation surrounding these plates occur within continental lithosphere, reflecting the profound difference between oceanic and continental lithosphere and the ways in which they deform (Sections 2.10, 8.5.1).

5.4 ABSOLUTE PLATE MOTIONS

The relative motion between the major plates, averaged over the past few million years, can be determined with remarkable precision, as described in the preceding section. It would be of considerable interest, particularly in relation to the driving mechanism for plate motions, if the motion of plates, and indeed plate boundaries, across the face of the Earth could also be determined. If the motion of any one plate or plate boundary across the surface of the Earth is known, then the motion of all other plates and plate boundaries can be determined because the relative motions are known. In general, within the framework of plate tectonics, all plates and plate boundaries must move across the face of the Earth. If one or more plates and/or plate boundaries are stationary, then this is fortuitous. A particular point on a plate, or, less likely, on a plate boundary, will be stationary if the Euler vector of the motion of that plate or plate boundary passes through that point (Fig. 5.6).

The *absolute* motion of plates is much more difficult to define than the relative motion between plates at plate boundaries, not least because the whole solid Earth is in a dynamic state. It is generally agreed that absolute plate motions should specify the motion of the lithosphere relative to the lower mantle as this accounts for 70% of the mass of the solid Earth and deforms more slowly than the asthenosphere above and the outer core below. In theory if the lithosphere and asthenosphere were everywhere of the same thickness and effective viscosity, there would be no net torque on the plates and hence no net rotation of the

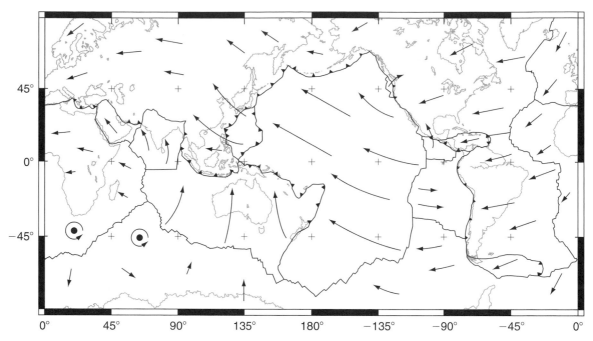

Figure 5.6 *The absolute velocities of plates, assuming the hotspot reference frame. The arrows indicate the displacement of points within the plates if the plates were to maintain their current angular velocities, relative to the hotspots, for 40 Ma. Filled circles indicate the pole (or antipole) of rotation for the plate if this occurs within the plate. The medium solid lines are approximate plate boundaries; where barbed, they indicate subduction zones with the barb on the overriding plate. Note that, because of the Mercator projection, arrows at high latitudes are disproportionately long compared to those at low latitudes (modified and redrawn from Gripp & Gordon, 2002 with permission from Blackwell Publishing).*

lithosphere relative to the Earth's deep interior. If plate velocities are specified in the no net rotation (NNR) reference frame, the integration of the vector product of the velocity and position vectors for the whole Earth's surface will equal zero. By convention, space geodesists specify absolute plate motions in terms of the NNR criterion (Prawirodirdjo & Bock, 2004).

An alternative model for the determination of absolute motions utilizes the information provided by volcanic *hotspots* on the Earth's surface. Wilson (1963) suggested that the volcanic ridges and chains of volcanoes associated with certain major centers of igneous activity such as Hawaii, Iceland, Tristan da Cunha in the South Atlantic, and Reunion Island in the Indian Ocean, might be the result of the passage of the Earth's crust over a hotspot in the mantle beneath. Morgan (1971) elaborated on this idea by suggesting that these hotspots are located over *plumes* of hot material rising from

the lower mantle, and hence provide a fixed reference frame with respect to the lower mantle. This hypothesis is considered further in the next section, and in Chapter 12. The hotspot model is attractive to many geologists and geophysicists in that the tracks of hotspots across the face of the Earth offer the possibility of determining the absolute motion of plates throughout the past 200 Ma (Morgan, 1981, 1983).

The model of Gripp & Gordon (2002) for the current absolute motion of plates, based on the trends and rates of propagation of active hotspot tracks, is illustrated in Fig. 5.6. It averages plate motions over the past 5.8 Ma, approximately twice the length of time over which relative velocities are averaged. Two propagation rates and 11 segment trends from four plates were used in deriving this model.

Several other frames of reference for absolute motions have been suggested, but not pursued. One of these proposed that the African plate has remained stationary during the past 25 Ma. Following a long period of quiescence, in terms of tectonic and volcanic activity, large parts of Africa have been subjected to uplift and/or igneous activity during the late Cenozoic. This was considered to be a result of the plate becoming stationary over hot spots in the upper mantle. Another proposal was that the Caribbean plate is likely to be stationary as it has subduction zones of opposite polarity along its eastern and western margins. Subducting plates would appear to extend through the asthenosphere and would be expected to inhibit lateral motion of the overlying plate boundary. Similar reasoning led Kaula (1975) to suggest a model in which the lateral motion of plate boundaries in general is minimized.

5.5 HOTSPOTS

The major part of the Earth's volcanic activity takes place at plate margins. However, a significant fraction occurs within the interiors of plates. In oceans the intra-plate volcanic activity gives rise to linear island and seamount chains such as the Hawaiian–Emperor and Line Islands chains in the Pacific (Fig. 5.7). Moreover, several of these Pacific island chains appear to be mutually parallel. Where the volcanic centers in the chains are closely spaced, aseismic ridges are constructed, such as the Ninety-East Ridge in the Indian Ocean, the Greenland–Scotland Ridge in the North Atlantic, and the Rio Grande and Walvis ridges in the South Atlantic. These island chains and ridges are associated with broad crustal swells which currently occupy about 10% of the surface of the Earth, making them a major cause of uplift of the Earth's surface (Crough, 1979).

The island chains are invariably younger than the ocean crust on which they stand. The lower parts of these volcanic edifices are believed to be formed predominantly of tholeiitic basalt, while the upper parts are alkali basalts (Karl et al., 1988) enriched in Na and K and, compared to mid-ocean ridge basalts, have higher concentrations of Fe, Ti, Ba, Zr, and rare earth elements (REE) (Bonatti et al., 1977). Their composition is compatible with the mixing of juvenile mantle material and depleted asthenosphere (Schilling et al., 1976) (Section 6.8). They are underlain by a thickened crust but thinned lithosphere, and represent a type of anom-

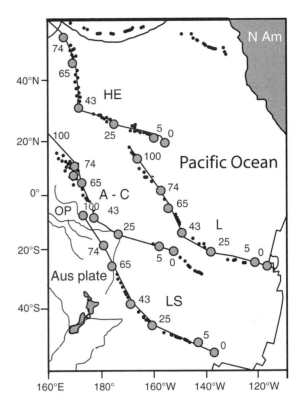

Figure 5.7 Hotspot tracks on the Pacific plate. HE, Hawaiian–Emperor chain; A-C, Austral-Cook islands; L, Line islands; LS, Louisville chain; OP, Ontong-Java Plateau. Numbers on chains indicate the predicted age of seamounts in Ma (redrawn from Gaina et al., 2000, by permission of the American Geophysical Union. Copyright © 2000 American Geophysical Union).

alous feature that will eventually become welded to a continental margin as a suspect terrane (Section 10.6.1).

An example of an oceanic island chain is the Hawaiian–Emperor chain in the north-central Pacific Ocean (Fig. 5.7). This chain is some 6000 km long and shows a trend from active volcanoes at Hawaii in the southeast to extinct, subsided guyots (flat-topped seamounts) in the northwest. Dating of the various parts of the chain confirmed this trend, and revealed that the change in direction of the chain occurred at 43 Ma (Clague & Dalrymple, 1989). The Hawaiian–Emperor chain parallels other chains on the Pacific Plate, along which volcanism has progressed at a similar rate (Fig. 5.7).

As indicated above, a possible explanation of the origin of island chains was proposed by Wilson (1963). It was suggested that the islands formed as the lithosphere passed over a hotspot. These hotspots are now thought to originate from mantle plumes rising from the lower mantle that thin the overlying lithosphere (Section 12.10). The volcanic rocks are then derived from pressure-release melting and differentiation within the plume. Such plumes represent material of low seismic velocity and can be detected by seismic tomography (Section 2.1.8; Montelli *et al.*, 2004a). Although the mantle plume mechanism has been widely adopted, some workers (e.g. Turcotte & Oxburgh, 1978; Pilger, 1982) have questioned the necessity for mantle hotspots and suggest that magmas simply flow to the surface from the asthenosphere through fractures in the lithosphere resulting from intra-plate tensional stresses. This mechanism obviates the problem of maintaining a mantle heat source for long periods. It does not,

however, explain why fractures in the same plate should trend in the same direction and develop at similar rates (Condie, 1982a).

Morgan (1971, 1972a) proposed that mantle plumes remain stationary with respect to each other and the lower mantle, and are of long duration. If so, the hotspots represent a fixed frame of reference by which absolute motions of plates can be determined (Section 5.4).

Between 40 and 50 present day hotspots have been suggested (Fig. 5.8) (Duncan & Richards, 1991; Courtillot *et al.*, 2003). It seems unlikely, however, that all of these centers of intra-plate volcanism, or enhanced igneous activity at or near ridge crests, are of the same type or origin. Many are short-lived, and consequently have no tracks reflecting the motion of the plate on which they occur. By contrast, others have persisted for tens of millions of years, in some cases over 100 million years, and can be traced back to a major episode of igneous activity giving rise to flood basalts on land or

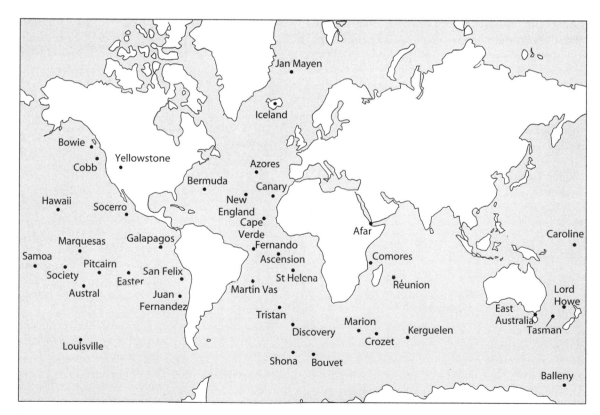

Figure 5.8 *World-wide distribution of hotspots (modified from Duncan & Richards, 1991, by permission of the American Geophysical Union. Copyright © 1991 American Geophysical Union).*

an oceanic plateau under the sea. These remarkable episodes of localized enhanced partial melting in the mantle punctuate the geologic record and are collectively termed Large Igneous Provinces (LIPs) (Section 7.4.1). It seems probable therefore that there are at least two types of hotspot and that those originating as LIPs are the most likely to be a result of plume heads rising from deep within the mantle, probably from the thermal boundary layer at the core–mantle boundary (Section 12.10).

Courtillot *et al.* (2003) proposed five criteria for distinguishing such *primary* hotspots (Section 12.10). They suggest that, on the basis of existing knowledge, only seven present day hotspots satisfy these criteria, although ultimately 10–12 may be recognized. The seven are Iceland, Tristan da Cunha, Afar, Reunion, Hawaii, Louisville, and Easter (Fig. 5.8). The first four of these hotspots are within the "continental hemisphere," which consists of the Indian and Atlantic Oceans and the continents that surround them. All four were initially LIPs characterized by continental flood basalts, and associ-

ated with the rifting of continental areas, followed by the initiation of sea floor spreading (Sections 7.7, 7.8). The Parana flood basalts of Uruguay and Brazil, and the Etendeka igneous province of Namibia, emplaced 130 Ma ago, were the first expression of the Tristan da Cunha hotspot, and precursors of the opening of the South Atlantic. The Deccan Traps of western India were extruded 65 Ma ago coinciding with the creation of a new spreading center in the northwest Indian Ocean. This hotspot would appear to be located at the present position of Reunion Island (Fig. 5.9). The first igneous activity associated with the Iceland hotspot would appear to have occurred 60 Ma ago giving rise to the North Atlantic igneous province of Greenland and northwest Scotland, and heralding the initiation of sea floor spreading in this area. The Afar hotspot first appeared approximately 40 Ma ago with the outpouring of flood basalts in the Ethiopian highlands, and igneous activity in the Yemen, precursors of rifting and spreading in the Red Sea and Gulf of Aden. The remaining three primary hotspots of Courtillot *et al.* (2003) occur

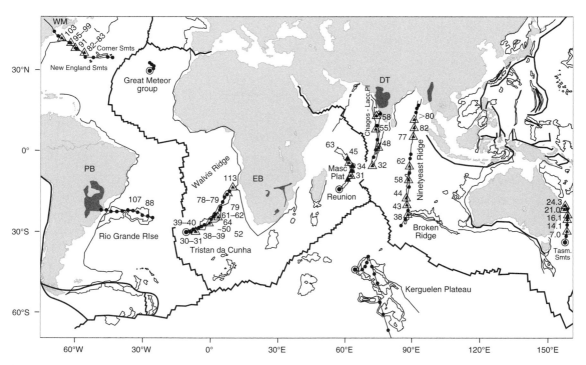

Figure 5.9 *Hotspot tracks in the Atlantic and Indian Oceans. Large filled circles are present day hotspots. Small filled circles define the modeled paths of hotspots at 5 Ma intervals. Triangles on hotspot tracks indicate radiometric ages. WM, White Mountains; PB, Parana flood basalts; EB, Etendeka flood basalts; DT, Deccan Traps (modified and redrawn from Müller et al., 1993, courtesy of the Geological Society of America).*

within the "oceanic hemisphere," i.e. the Pacific ocean, and have produced distinctive traces across the Pacific plate (Fig. 5.7). The Louisville Ridge originates at the Ontong Java Plateau of the western Pacific. This formed approximately 120 Ma ago and is the largest LIP in terms of the volume of mafic igneous material emplaced. The Hawaiian–Emperor seamount chain may well have had a similar origin but the earlier part of this track has been subducted, the oldest seamounts in the chain dating at approximately 80 Ma. The Easter Island–Line Islands track originated about 100 Ma ago, not as an LIP, but in an area with an unusually high density of submarine volcanoes known as the mid-Pacific mountains.

The relative positions of the continents around the Atlantic and Indian oceans, for the past 200 Ma, are well constrained by the detailed spreading history contained within these oceans (Section 4.1.7). If one or more hotspot tracks within this Indo-Atlantic hemisphere are used to determine the absolute motions of the relevant plates in the past, tracks for the remaining hotspots in this hemisphere can be predicted. Comparison of the observed and predicted tracks provides a test of the fixed hotspots hypothesis, and a measure of the relative motion between the hotspots. Such an analysis by Müller *et al.* (1993) suggests that the relative motion between hotspots in the Indo-Atlantic reference frame is less than 5 mm a^{-1}, i.e. an order of magnitude less than average plate velocities. A similar analysis for Pacific hotspots by Clouard & Bonneville (2001) yields a similar result for the Pacific reference frame. However, there are problems in linking together the two reference frames; in other words, in predicting Pacific hotspot traces using the Indo-Atlantic reference frame or vice-versa. This is because, for most of the Mesozoic and Cenozoic, the oceanic plates of the Pacific hemisphere are surrounded by outward dipping subduction zones, except in the south. This means that in order to determine the motion of the Pacific Ocean plates relative to the Indo-Atlantic hemisphere one must have a detailed knowledge of the nature and evolution of the plate boundaries around and within the Antarctic plate in the South Pacific area. Unfortunately there are still uncertainties about this, but an analysis based on the model of Cande *et al.* (1999) for the evolution of these boundaries suggests that the two reference frames or *domains* are not compatible, despite the compatibility of hotspot tracks within each domain (Fig. 5.10). The discrepancy is greatest before 40–50 Ma, when the relative motion between the two hotspot frames is approximately

50 mm a^{-1}. Intriguingly, this corresponds with a period of major reorganization of global plate motions (Rona and Richardson, 1978), the age of the major bend in the Hawaiian–Emperor seamount chain, and a period in which the rate of true polar wander (Section 5.6) was much greater than during the period 10–50 Ma ago, when it was virtually at a standstill (Besse & Courtillot, 2002).

If hotspots remain fixed, and provide a framework for absolute plate motions, then paleomagnetic studies should be able to provide a test of their unchanging latitude. Paleomagnetic data for the oceanic plates of the Pacific are sparse, and subject to greater uncertainties than those obtained for continental areas. Nevertheless preliminary results (Tarduno & Cottrell, 1997) suggest that the Hawaiian hotspot may have migrated south through as much as 15–20° of latitude during the period 80–43 Ma. Paleomagnetic results obtained from Ocean Drilling Program drill core, from which any latitudinal change in of the Reunion hotspot could be deduced (Vandamme & Courtillot, 1990), suggest that this hotspot may have moved northwards through approximately 5° of latitude between 65 and 43 Ma. These latitudinal shifts are compatible with the discrepancy between the two hotspot reference frames prior to 43 Ma ago, and support the assumptions regarding Cenozoic – late Mesozoic plate boundaries within and around the Antarctic plate. These results also imply that the major bend in the Hawaiian–Emperor seamount chain at approximately 43 Ma does not reflect a major change in the absolute motion of the Pacific plate, as originally thought, but can be accounted for almost entirely by the southward motion of the Hawaiian hotspot (Norton, 1995).

Predicted hotspot traces in the Atlantic and Indian Ocean (Müller *et al.*, 1993) are shown in Fig. 5.9, super-imposed on volcanic structures on the sea floor and on land. The correlation between the two is excellent. For example, the Reunion hotspot began beneath western India and was responsible for the Deccan Traps flood basalts: India's northwards motion was then recorded by the Maldive-Chagos Plateau and the Mascarene Plateau. The gap between these two features results from the passage of the mid-ocean ridge over the hot spot approximately 33 Ma ago. The hotspot is currently beneath a seamount 150 km west of the volcanically active island of Réunion.

It will be noted that Iceland has not been included in Fig. 5.9. If one assumes that this hotspot was initi-ated 60 Ma ago beneath East Greenland then its track

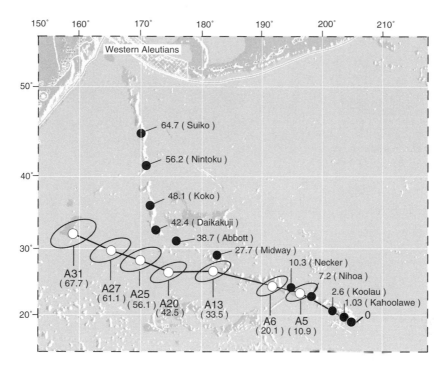

Figure 5.10 *Predicted Hawaiian hotspot track (solid line) from plate reconstructions assuming that the Indo-Atlantic hotspots are fixed. Ages in Ma (redrawn from Steinberger & O'Connell, 2000, by permission of the American Geophysical Union. Copyright © 2000 American Geophysical Union).*

implies that its position is not fixed relative to the other major hotspots in the Indo-Atlantic domain. However one can use the absolute motions derived from the other hotspots (Müller *et al.*, 1993) to predict the track of the Iceland hotspot on the assumption that it is fixed in relation to this frame of reference. Such an analysis has been conducted by Lawver & Müller (1994) with intriguing results (Fig. 5.11). The track can be projected back to 130 Ma, at which time the hotspot would have been beneath the northern margin of Ellesmere Island in the Canadian Arctic. Lawver & Müller (1994) suggest that such a track might explain the formation of the Mendeleyev and Alpha Ridges in the Canadian Basin of the Arctic Ocean and the mid-Cretaceous volcanic rocks of Axel Heiberg Island and northern Ellesmere Island. At 60 Ma the hotspot is predicted to have been beneath West Greenland where there are volcanics of this age, for example on Disko Island. At 40 Ma it would have been beneath East Greenland which may explain the anomalous post-drift uplift of this area. On this model the North Atlantic igneous province, initiated at

approximately 60 Ma, was a result of rifting of lithosphere that had already been thinned by its proximity to a hotspot, rather than the arrival of a plume head. In contrast to this interpretation, however, there is considerable doubt, on the basis of geochemical and geophysical data, that the Iceland hotspot is fed by a deep mantle plume (Section 12.10). The Iceland hotspot is therefore something of an enigma.

5.6 TRUE POLAR WANDER

In Section 3.6 it was demonstrated that paleomagnetic techniques can be used to construct apparent polar wandering paths which track the motions of plates with respect to the magnetic north pole and hence, using an axial geocentric dipole model, the spin axis of the Earth. In Section 5.5 it was suggested that hotspots are nearly

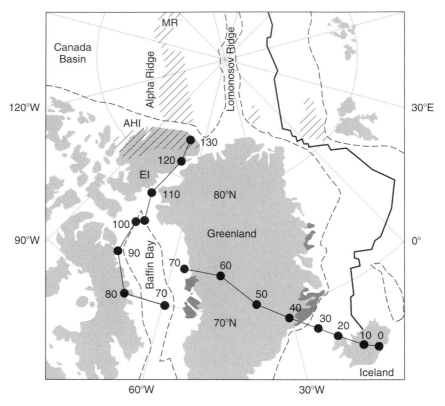

Figure 5.11 *Predicted hotspot track assuming that the Iceland hotspot is fixed relative to the other Indo-Atlantic hotspots of Fig. 5.9. Position of hotspot at 10 Ma intervals is indicated by solid dots. AHI, Axel Heiberg Island; EI, Ellesmere Island; MR, Mendeleyev Ridge. Dashed line, continent–ocean boundary based on bathymetry. Gap between 70 Ma positions results from sea floor created after the passage of the Labrador Sea Ridge over the hotspot at 70 Ma (modified and redrawn from Lawver & Müller, 1994, courtesy of the Geological Society of America).*

stationary in the mantle, and so their trajectories provide a record of the motions of plates with respect to the mantle. A combination of these two methods can be used to test if there has been any relative movement between the mantle and the Earth's spin axis. This phenomenon is known as true polar wander (TPW).

The method employed to investigate TPW is as follows. Paleomagnetic pole positions for the past 200 Ma are compiled for a number of continents that are separated by spreading oceans so that their relative motions can be reconstructed from magnetic lineation data (Section 4.1.7). The pole positions are then corrected for the rotations relative to a single continent (usually Africa) experienced as a result of sea floor spreading since the time for which they apply. In this way a composite or global apparent polar wander path

is obtained. This is then compared with the track of the axis of the hotspot reference frame as viewed from the fixed continent. The TPW path is then determined by calculating the angular rotation that shifts the global mean paleomagnetic pole of a certain age to the north pole, and then applying the same rotation to the hotspot pole of the same age (Courtillot & Besse, 1987).

The TPW path for the past 200 Ma, obtained by Besse & Courtillot (2002), is shown in Fig. 5.12. Their analysis utilizes paleomagnetic data from six continents, sea floor spreading data from the Atlantic and Indian Oceans, and the Indo-Atlantic hotspot reference frame of Müller *et al.* (1993) for the past 130 Ma, and of Morgan (1983) for the period from 130 to 200 Ma. They conclude that as much as 30° of true polar wander has occurred in the past 200 Ma, and that the movement of the pole

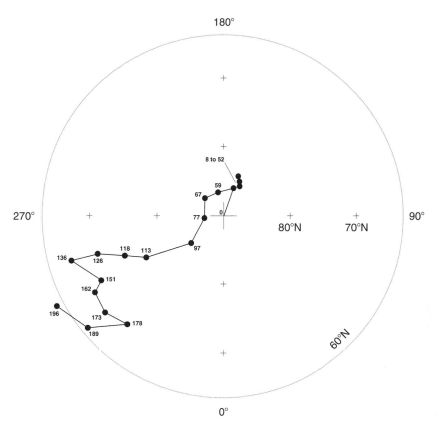

Figure 5.12 *True Polar Wander (TPW) path for the past 200 Ma. TPW is defined as the movement of the "geographic" pole of the Indo-Atlantic hotspot reference frame with respect to the magnetic pole defined by paleomagnetic data, the latter being equated to the Earth's rotational axis (redrawn from Besse & Courtillot, 2002, by permission of the American Geophysical Union. Copyright © 2002 American Geophysical Union).*

has been episodic. A period of relatively fast TPW, averaging 30 mm a^{-1}, separates periods of quasi-standstill between 10 and 50 Ma, and 130 and 160 Ma. During the past 5–10 Ma the rate has been high, of the order of 100 mm a^{-1}. This analysis does not include the oceanic plates of the Pacific hemisphere. This is because there are problems with the quality and quantity of data from the Pacific, and doubts about the fixity of the Pacific hotspots relative to the Indo-Atlantic hotspots (Section 5.5). Notwithstanding these problems, Besse & Courtillot (2002) carried out an analysis for the Pacific plate using nine paleomagnetic poles, between 26 and 126 Ma, derived from analyses of the pattern of the linear magnetic anomalies and the magnetic anomalies developed over seamounts (Petronotis & Gordon, 1999). They assumed the hotspot kinematic model for the Pacific

plate of Engebretson *et al.* (1985), and derived a TPW path for this period of time that is remarkably similar in length and direction to that of the path shown in Fig. 5.12, but offset from it in a way that is compatible with the southward motion of the Hawaiian hotspot discussed in Section 5.5. This, taken together with the similarities between the path shown in Fig. 5.12 and those derived in earlier analyses, based on smaller data sets (e.g. Livermore *et al.*, 1984; Besse & Courtillot, 1991; Prevot *et al.*, 2000), suggests a robust result. One must bear in mind however that these conclusions are only as good as the underlying assumptions: the axial dipole nature of the Earth's magnetic field, and hotspot tracks as indicators of the motion of plates with respect to the Earth's deep interior throughout the past 200 Ma.

The relative motion between the mantle and the rotation axis, as illustrated by the TPW path, may be interpreted as a shifting of the whole or part of the Earth in response to some form of internal mass redistribution that causes a change in the direction about which the moment of inertia of the mantle is a maximum (Andrews, 1985). For example, Anderson (1982) relates TPW to the development of elevations of the Earth's surface resulting from the insulating effect of supercontinents that prevents heat loss from the underlying mantle. It is possible that only the lithosphere or the mantle or both lithosphere and mantle together shift during polar wander. It is highly unlikely that the lithosphere and mantle are sufficiently decoupled to move independently, and so it appears probable that shifting of lithosphere and mantle as a single unit takes place during TPW. Indeed, if there is coupling between core and mantle, the whole Earth may be affected. Andrews's interpretation of TPW is supported by astronomical data which shows that during the 20th century the location of the Earth's rotational axis has moved at a rate similar to that computed from paleomagnetic and hotspot data, namely about $1°\,Ma^{-1}$. This suggests that at least part of the mass redistribution takes place in the mantle, as the continents do not move this rapidly. Sabadini & Yuen (1989) have shown that both viscosity and chemical stratification in the mantle are important in determining the rate of polar wander. Another mechanism proposed for driving TPW is the surface mass redistribution arising from major glaciations and deglaciations (Sabadini *et al.*, 1982). However, mantle flow is required to explain TPW during periods with no evidence of significant continental glaciation, and, indeed, may be responsible for the majority of TPW. It has also been suggested that TPW is excited by the mass redistributions associated with subduction zones (Section 12.9) (Spada *et al.*, 1992), mountain building, and erosion (Vermeersen & Vlaar, 1993).

5.7 CRETACEOUS SUPERPLUME

Certain hotspots, as described in Section 5.5, are thought to be the surface manifestation of plumes of hot material ascending from the deep mantle. These are of moderate size and can be considered to form part of the normal mantle convecting system. It has been proposed, however, that at least once during the history of the Earth there has been an episode of much more intense volcanic activity. The cause has been ascribed to a phenomena termed superplumes, large streams of overheated material rising buoyantly from the D″ layer at the base of the mantle (Section 2.8.6), that derived their heat from the core. These spread laterally at the base of the lithosphere to affect an area ten times larger than more normal plume activity.

Larson (1991a, 1991b, 1995) proposed that a superplume was responsible for the widespread volcanic and intrusive igneous activity that affected abnormally large amounts of ocean floor during the mid-Cretaceous. One manifestation of this activity was the creation of numerous seamounts and ocean plateaux in the western Pacific (Fig. 7.15) at a rate some five times greater during this period than at other times. Similarly there were extrusions of thick, areally extensive flood basalts on the continents, such as the Paraná Basalts of Brazil.

Phenomena attributed to the mid-Cretaceous superplume episode are illustrated in Fig. 5.13. At 120–125 Ma the rate of formation of oceanic crust doubled over a period of 5 Ma, decreased within the next 40–50 Ma, and returned to previous levels about 80 Ma ago (Fig. 5.13d). The additional production of crust required increased subduction rates, and it is significant that major batholiths of the Andes and the Sierra Nevada were emplaced at this time.

Coupled to the increased crust production, and caused by the consequent general rise in the level of the sea floor, was a worldwide increase in sea level to an elevation some 250 m higher than at the present day (Fig. 5.13b). At high latitudes the surface temperature of the Earth increased by about 10°C, as shown by oxygen isotope measurements made on benthic foraminifera from the North Pacific (Fig. 5.13a). This effect was probably caused by the release of large amounts of carbon dioxide during the volcanic eruptions, which created an enhanced *"greenhouse"* effect (Sections 13.1.1, 13.1.2). During the superplume episode the rates of carbon and carbonate sequestration in organisms increased due to the greater area of shallow seas and the increased temperature, which caused plankton to thrive. This is reflected in the presence of extensive black shale deposits at this time (Force, 1984) and in the estimated oil reserves of this period (Tissot, 1979;

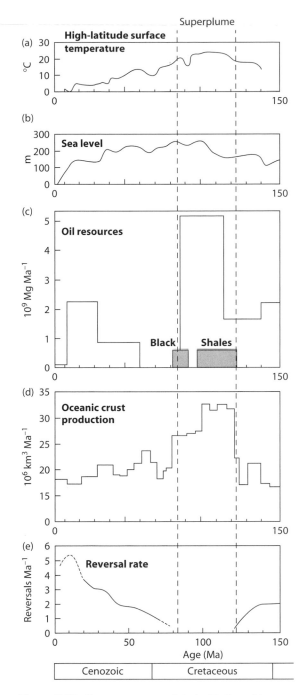

Figure 5.13 *Phenomena associated with the mid-Cretaceous superplume (after Larson, 1991a, 1991b, with permission from the Geological Society of America).*

Fig. 5.13c), which may constitute about 50% of the world's supply. Also of economic significance is the placement of a large percentage of the world's diamond supply at this time, probably as a result of the diamonds' having been translated to the surface by the rising plumes. During the plume episode the rate of geomagnetic reversals (Section 4.1.4) was very low (Fig. 5.13e), with the field remaining in normal polarity for some 35 Ma. This indicates that activity in the core, where the geomagnetic field originates (Section 3.6.4), was low, perhaps related to the transfer of considerable quantities of heat to the mantle.

Acceptance of a mid-Cretaceous superplume episode is not universal. For example, Anderson (1994) suggests that the phenomena of this period were caused by a general reorganization of plates on a global scale associated with the break-up of Pangea and reorganization of the Pacific plate. The mantle upwelling in the latter may then have been a passive reaction to plates being pulled apart by their attached slabs. The episode would thus be viewed as a period when mantle ascended passively as a result of changing plate motions.

5.8 DIRECT MEASUREMENT OF RELATIVE PLATE MOTIONS

It is now possible to measure the relative motion between plates using methods of space geodesy (Gordon & Stein, 1992). Before about 1980 the only methods available for this type of investigation were the standard terrestrial geodetic methods of baseline measurement using optical techniques or laser ranging instruments such as the geodolite (Thatcher, 1979). These methods are certainly sufficiently precise to measure relative plate motions of a few tens of millimeters a year. However, as noted in Section 5.3, in some regions the strain between plates is not all dissipated across a narrow plate boundary, but may extend into the adjacent plates for great distances, particularly in continental areas (Fig. 5.5). In order to study these large-scale problems it is necessary to be able to measure across very large distances to very great accuracy. Terrestrial methods are extremely time consuming on land, and impossible to use across major oceans. Since 1980, however, the measurement of very long baselines using extraterrestrial methods has become possible via the application of space technology.

Three independent methods of extraterrestrial surveying are available. These are very long baseline interferometry, satellite laser ranging, and satellite radio positioning. The most common and best known example of the latter method is the Global Positioning System (GPS).

The technique of very long baseline interferometry (VLBI) makes use of the radio signals from extragalactic radio sources or quasars (Niell *et al.*, 1979; Carter & Robertson, 1986; Clark *et al.*, 1987). The signal from a particular quasar is recorded simultaneously by two or more radio telescopes at the ends of baselines which may be up to 10,000 km long. Because of their different locations on the Earth's surface, the signals received at the telescopes are delayed by different times, the magnitude of the delays between two stations being proportional to the distance between them and the direction from which the signals are coming. Typically, during a 24-hour experiment, 10–15 quasars are each observed 5–15 times. This scheme provides estimates of baseline length that are accurate to about 20 mm (Lyzenga *et al.*, 1986). The usefulness of this system has been greatly enhanced by the development of mobile radio telescopes that frees the technique from the necessity of using fixed observatory installations.

The technique of satellite laser ranging (SLR) calculates the distance to an orbiting artificial satellite or a reflector on the Moon by measuring the two-way travel time of a pulse of laser light reflected from the satellite (Cohen & Smith, 1985). The travel time is subsequently converted to range using the speed of light. If two laser systems at different sites simultaneously track the same satellite, the relative location of the sites can be computed by using a dynamic model of satellite motion, and repeated measurements provide an accuracy of about 80 mm. Periodic repetition of the observations can then be used to observe relative plate motions (Christodoulidis *et al.*, 1985).

The technique of satellite radio positioning makes use of radio interferometry from the GPS satellites (Dixon, 1991). It is a three-dimensional method by which the relative positions of instruments at the ends of baselines are determined from the signals received at the instruments from several satellites. The simultaneous observation of multiple satellites makes extremely accurate measurements possible with small portable receivers. This is now the most efficient and accurate method of establishing geodetic control on both local and regional surveys (e.g. Sections 8.5.2, 10.4.3).

Gordon & Stein (1992) summarized the early determinations of relative plate motions by these methods. Generally, plate velocities averaged over a few years of observation agree remarkably well with those averaged over millions of years. The methods were first applied to the measurement of the rate of movement across the San Andreas Fault in California. Smith *et al.* (1985), using SLR, reported that a 900 km baseline that crossed the fault at an angle of 25° had been shortened at an average rate of 30 mm a^{-1}. Lyzenga *et al.* (1986) have used VLBI to measure the length of several baselines in the southwestern USA and have found that over a period of 4 years movement on the fault was 25 ± 4 mm a^{-1}. These direct measurements of the rate of displacement across the San Andreas Fault are lower than the 48–50 mm a^{-1} predicted from global models of plate movements (DeMets *et al.*, 1990). However, during the period of observation, no major earthquakes occurred. Over longer time intervals, the discrete jumps in fault movement associated with the elastic rebound mechanism of large earthquakes (Section 2.1.5) would contribute to the total displacement and provide a somewhat higher figure for the average rate of movement. Alternatively, motion between the Pacific and North American plates may be occurring along other major faults located adjacent to the San Andreas Fault (Fig. 8.1, Section 8.5.2).

Tapley *et al.* (1985), using SLR, measured changes in length of four baselines between Australia and the North American and Pacific plates, and found that the rates differ by no more than 3 mm a^{-1} from average rates over the last 2 Ma. Similarly Christodoulidis *et al.* (1985) and Carter & Robertson (1986) measured the relative motion between pairs of plates and found a strong correlation with the kinematic plate model of Minster & Jordan (1978). Herring *et al.* (1986) made VLBI measurements between various telescopes in the USA and Europe and determined that the present rate of movement across the Atlantic Ocean is 19 ± 10 mm a^{-1}. This agrees well with the rate of 23 mm a^{-1} averaged over the past 1 Ma.

Sella *et al.* (2002) provided a comprehensive review of the determinations of relative plate velocities, using the techniques of space geodesy, up to the year 2000. Most of the data summarized were obtained by the GPS method after 1992, when the system was upgraded and the accuracy greatly improved. They presented a model for recent relative plate velocities (REVEL-2000), based on this data, that involves 19 plates. The velocities obtained for numerous plate pairs within this model were then compared with those predicted by the "geologic" model for current plate motions (NUVEL-1A) that averages plate velocities over the past 3 Ma (DeMets *et al.*, 1990, 1994). The velocities for two-thirds of the plate pairs tested were in very close agreement. An example of the comparison between the two models, for the Australian–Antarctica boundary, is shown in Fig. 5.14. Some of the exceptions are thought to be due to inaccuracies in the NUVEL-1A model, for example the motion of the Caribbean plate relative to North and South America; others could well be due

to real changes in relative velocities over the past few million years. Examples of the latter include Arabia-Eurasia and India-Eurasia, which may well reflect long term deceleration associated with continental collision.

Most of the space geodetic data points in stable plate interiors confirm the rigidity of plates and hence the rigid plate assumption of plate tectonics. Of the major plates the only exception to this generalization is the Australian plate.

These techniques of direct measurement are clearly extremely important in that they provide estimates of relative plate movements that are independent of plate tectonic models. It is probable that their accuracy will continue to improve, and that observations will become more widely distributed over the globe. The determination of intra-plate deformation and its relationship to intra-plate stress fields, earthquakes, and magmatic activity should also become possible. Important new findings are anticipated over the next few decades.

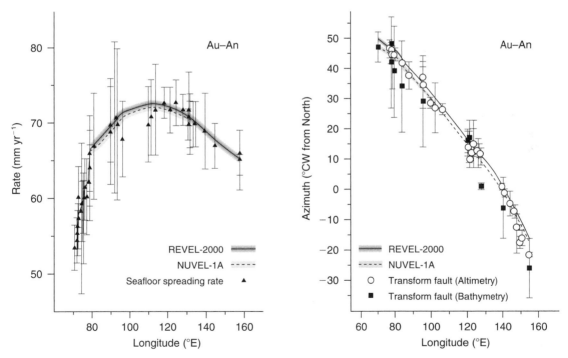

Figure 5.14 *Measured sea floor spreading rates and transform fault azimuths for the Australian–Antarctic plate boundary, compared to predicted rates and azimuths from REVEL 2000 and NUVEL-1A. Details of NUVEL-1A, measured spreading rates, and transform azimuths obtained from bathymetry, from DeMets* et al., *1990, 1994. Transform azimuths from altimetry from Spitzak & DeMets, 1996 (redrawn from Sella* et al., *2002, by permission of the American Geophysical Union. Copyright © 2002 American Geophysical Union).*

5.9 FINITE PLATE MOTIONS

The motions of the plates described in Section 5.3 are termed geologically *instantaneous* as they refer to movements averaged over a very short period of geologic time. Such rotations cannot, therefore, provide information on the paths followed by the plates in arriving at the point at which the instantaneous motion is measured. Although it is a basic tenet of plate tectonics that poles of rotation remain fixed for long periods of time, consideration of the relationships between plates forming an interlinked spherical shell reveals that this cannot be the case for all plates (McKenzie & Morgan, 1969).

Consider the three plates on a sphere A, B, and C shown in Fig. 5.15a. P_{BA}, P_{BC}, and P_{AC} represent Euler poles for pairs of plates that describe their instantaneous angular rotation. Let plate A be fixed. Clearly the poles P_{BA} and P_{AC} can remain fixed with respect to the relevant pairs of plates. Thus, for example, any transform faults developing along common plate margins would follow small circles centered on the poles. Consider now the relative movements between plates B and C. It is apparent that if A, P_{BA}, and P_{AC} remained fixed, the rotation vector of C relative to B ($_B\omega_C$) acts through P_{BC} and is given by the sum of the

vectors $_B\omega_A$ and $_A\omega_C$ that act about P_{BA} and P_{AC}, respectively. Thus, P_{BC} lies within the plane of P_{BA} and P_{AC} and is fixed relative to A. Such a point, however, does not remain stationary with respect to B and C. Consequently, relative motion between B and C must take place about a pole that constantly changes position relative to B and C (Fig. 5.15b). Transform faults developed on the B–C boundary will not then follow simple small circle routes.

Even when a moving pole is not a geometric necessity, it is not uncommon for Euler poles to jump to a new location (Cox & Hart, 1986). In Fig. 5.16 the pole of rotation of plates A and B was initially at P_1, and gave rise to a transform fault with a small circle of radius 30°. The new pole location is P_2, 60° to the north of P_1, so that the transform fault is now 90° from P_2, that is, on the equator of this pole. The occurrence of this pole jump is easily recognizable from the abrupt change in curvature of the transform fault.

Menard & Atwater (1968) have recognized five different phases of spreading in the northeastern Pacific. In Fig. 5.17 it is shown that the numerous large fracture zones of this region appear to lie on small circles centered on a pole at 79°N, 111°E. If the fracture zone patterns are analysed in more detail, however, it can be seen that the fracture zones in fact consist of five different segments with significantly different orientations that can be correlated between adjacent fracture zones. The apparent gross small circle form of the fractures only represents the third phase of movement.

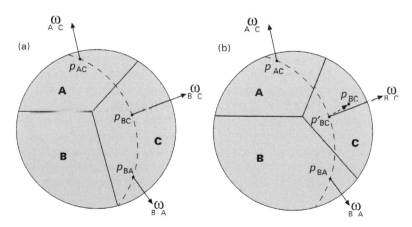

Figure 5.15 *The three plate problem. P_{AC}, P_{BC}, and P_{BA} refer to instantaneous Euler poles between plates A and C, B and C, B and A respectively, and $_A\omega$, $_B\omega_C$, and $_B\omega_A$ to their relative rotation vectors. In (b) P'_{BC} is the present location of P_{BC}. See text for explanation.*

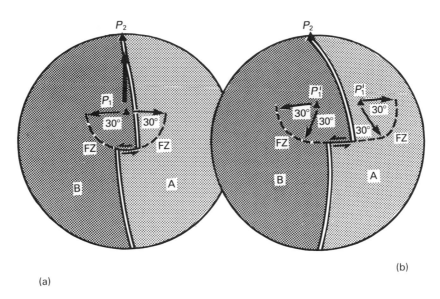

Figure 5.16 *(a) Rotation of plates A and B about pole P_1 produces arcuate fracture zones with a radius of curvature of 30°; (b) a jump of the pole of rotation to P_2 causes the fracture zones to assume a radius of curvature of 90°. P'_1 represents the positions of pole P_1 after rotation about pole P_2 (after Cox & Hart, 1986, with permission from Blackwell Publishing).*

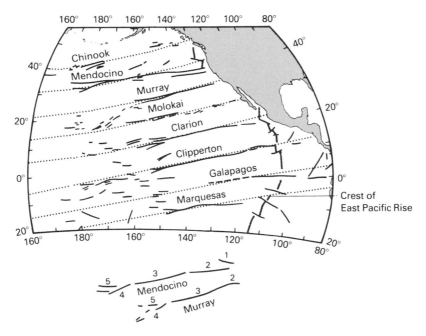

Figure 5.17 *Fracture zones in the northeastern Pacific showing trends corresponding to five possible spreading episodes, each with a new pole of rotation (redrawn from Menard & Atwater, 1968, with permission from Nature **219**, 463–7. Copyright © 1968 Macmillan Publishers Ltd).*

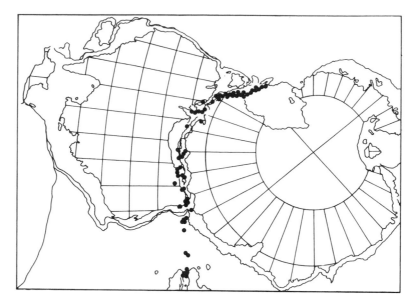

Fig: 5.18 *Earthquake epicenters superimposed on a reconstruction of Australia and Antarctica (redrawn from McKenzie & Sclater, 1971, with permission from Blackwell Publishing).*

It is thus apparent that in the northeastern Pacific sea floor spreading has taken place about a pole of rotation that was continually changing position by small discrete jumps. This progression has been analyzed and illustrated in greater detail by Engebretson *et al.* (1985).

Changes in the direction of relative motions of plates do not cause large-scale deformation of the plate boundaries but rather result in geometric adjustments of transform faults and ocean ridge crests. This may be a consequence of the lithosphere being thin at accretive margins and consequently of smaller mechanical strength (Le Pichon *et al.*, 1973). That the adjustments are only minor, however, is appreciated from continental reconstructions such as shown in Fig. 5.18, where the earthquake foci associated with present day activity are superimposed on the pre-drift reconstruction. The coincidence of shape of the initial rift and modern plate margins indicates that there has been little post-drift modification of the latter.

The past relative positions of plates can be determined by the fitting of lineaments that are known to have been juxtaposed originally. One approach is to fit former plate margins. Fossil accretive margins are usually readily identified from their symmetric magnetic lineations (Section 4.1.7), and fossil transform faults from the offsets they cause of the lineations. Ancient transform faults on continents are more difficult to identify, as their direction may be largely controlled by the pre-existing crustal geology. Their trace, however, normally approximately follows a small circle route, with any deviations from this marked by characteristic tectonic activity (Section 8.2). Ancient destructive margins can be recognized from their linear belts of calc-alkaline magmatism, granitic batholiths, paired metamorphic belts, and, possibly, ophiolite bodies (Sections 9.8, 9.9).

The features most commonly used for determining earlier continental configurations are continental margins and oceanic magnetic anomalies. The former are obviously used to study the form of pre-drift supercontinents (Section 3.2.2). Because magnetic anomalies can be reliably dated (Section 4.1.6), and individual anomalies identified on either side of their parental spreading ridge, the locus of any particular anomaly represents an isochron. Fitting together pairs of isochrons then allows reconstructions to be made of plates at any time during the history of their drift (Section 4.1.7). With the additional information provided by the orientation of fracture zones, instantaneous rates and poles of spreading can be determined for any time during the past 160 Ma or so; the period for which the necessary information, from oceanic magnetic anomalies and fracture zones, is available.

5.10 STABILITY OF TRIPLE JUNCTIONS

The stability of the boundaries between plates is dependent upon their relative velocity vectors. If a boundary is unstable it will exist only instantaneously and will immediately devolve into a stable configuration.

Figure 5.19a shows an unstable boundary between two plates where plate X is underthrusting plate Y at bc in a northeasterly direction and plate Y is underthrusting plate X at ab in a southwesterly direction. The boundary is unstable because a trench can only consume in one direction, so to accommodate these movements a dextral transform fault develops at b (Fig 5.19b). This sequence of events may have occurred in the develop-

ment of the Alpine Fault of New Zealand (Fig. 5.19c), which is a dextral transform fault linking the Tonga-Kermadec Trench, beneath which Pacific lithosphere is underthrusting in a southwesterly direction, to a trench to the south of New Zealand where the Tasman Sea is being consumed in a northeasterly direction (McKenzie & Morgan, 1969).

A more complex and potentially unstable situation arises when three plates come into contact at a triple junction. Quadruple junctions are always unstable, and immediately devolve into a pair of stable *triple junctions*, as will be shown later.

The Earth's surface is covered by more than two plates, therefore there must be points at which three plates come together to form triple junctions. In a similar fashion to a boundary between two plates, the stability of triple junctions depends upon the relative directions of the velocity vectors of the plates in contact. Figure 5.20 shows a triple junction between a ridge (R), trench (T), and transform fault (F). From this figure it can be appreciated that, in order to be stable, the triple junction must be capable of migrating up or

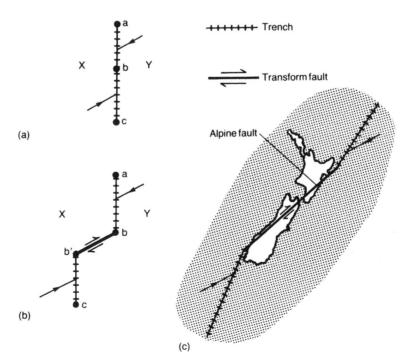

Figure 5.19 *(a,b) Evolution of a trench. (c) Alpine Fault of New Zealand (redrawn from McKenzie & Morgan, 1969, with permission from Nature* **224**, *125–33. Copyright © 1969 Macmillan Publishers Ltd).*

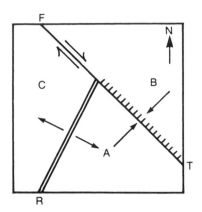

Figure 5.20 *Ridge (R)–trench (T)–transform fault (F)–triple junction between plates, A, B, and C.*

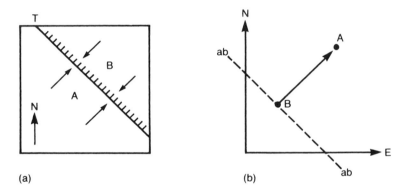

(a)   (b)

Figure 5.21 *(a) Trench (T) between plates A and B; (b) its representation in velocity space with the velocity line ab corresponding to its related triple junction.*

down the three boundaries between pairs of plates. It is easier to visualize the conditions for stability of the triple junction if each boundary is first considered individually.

Figure 5.21a shows the trench, at which plate A is underthrusting plate B in a northeasterly direction. Figure 5.21b shows the relative movement between A and B in velocity space (Cox & Hart, 1986), that is, on a figure in which the velocity of any single point is represented by its north and east components, and lines joining two points represent velocity vectors. Thus, the direction of line AB represents the direction of relative movement between A and B, and its length is proportional to the magnitude of their relative velocity. Line ab must represent the locus of a point that travels up and down the trench. This line, then, is the locus of a stable

triple junction. B must lie on ab because there is no motion of the overriding plate B with respect to the trench.

Now consider the transform boundary (Fig. 5.22a) between plates B and C, and its representation in velocity space (Fig. 5.22b). Again, line BC represents the relative velocity vector between the plates, but the locus of a point traveling up and down the fault, bc, is now in the same sense as vector BC, because the relative motion direction of B and C is along their boundary.

Finally, consider the ridge separating two plates A and C (Fig. 5.23a), and its representation in velocity space (Fig. 5.23b). The relative velocity vector AC is now orthogonal to the plate margin, and so the line ac now represents the locus of a point traveling along the ridge. The ridge crest must pass through the midpoint

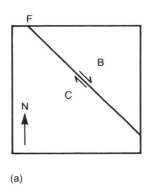
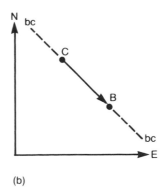

(a)   (b)

Figure 5.22 *(a) Transform fault (F) between plates B and C; (b) its representation in velocity space with the velocity line bc corresponding to its related triple junction.*

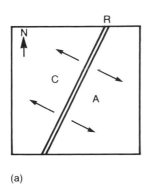
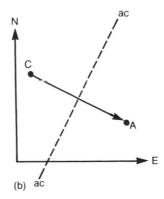

(a)   (b) ac

Figure 5.23 *(a) Ridge (R) between plates A and C; (b) its representation in velocity space with the velocity line ac corresponding to its related triple junction.*

of velocity vector CA if the accretion process is symmetric with plates A and C each moving at half the rate of accretion.

By combining the velocity space representations (Fig. 5.24), the stability of the triple junction can be determined from the relative positions of the velocity lines representing the boundaries. If they intersect at one point, it implies that a stable triple junction exists because that point has the property of being able to travel up and down all three plate margins. In the case of the RTF triple junction, it can be appreciated that a stable triple junction exists only if velocity line ac passes through B, or if ab is the same as bc, that is, the trench and transform fault have the same trend, as shown here. If the velocity lines do not all intersect at a single point the triple junction is unstable. The more general case

of an RTF triple junction, which is unstable, is shown in Fig. 5.25.

Figure 5.26 illustrates how an unstable triple junction can evolve into a stable system, and how this evolution can produce a change in direction of motion. The TTT triple junction shown in Fig. 5.26a is unstable, as the velocity lines representing the trenches do not intersect at a single point (Fig. 5.26b). In time the system evolves into a stable configuration (Fig. 5.26c) in which the new triple junction moves northwards along trench AB. The dashed lines show where plates B and C would have been if they had not been subducted. The point X (Fig. 5.26a,c) undergoes an abrupt change in relative motion as the triple junction passes. This apparent change in underthrusting direction can be distinguished from a global change as it occurs at

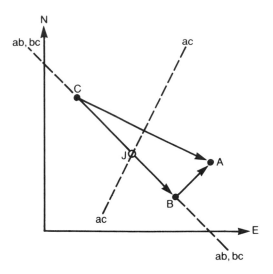

Figure 5.24 *Velocity space representation of the plate system shown in Fig. 5.20. Velocity lines ab, bc, and ac intersect at the single point J, which thus represents a stable triple junction.*

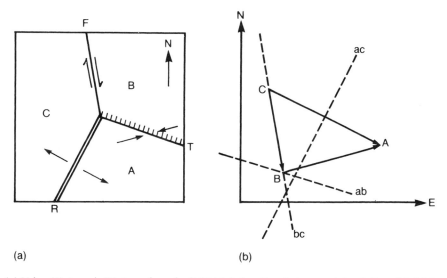

(a)

(b)

Figure 5.25 *(a) Ridge (R)–trench (T)–transform fault (F) triple junction between plates A, B, and C. (b) Its representation in velocity space. As the velocity lines ab, bc, and ac do not intersect at a single point, the triple junction must be unstable.*

different times and locations along the plate boundary. In order to be stable, the plate configuration shown in Fig. 5.26a must be as in Fig. 5.26d. When plotted in velocity space (Fig. 5.26e) the velocity lines then intersect at a single point.

McKenzie & Morgan (1969) have determined the geometry and stability of the 16 possible combina-

tions of trench, ridge, and transform fault (Fig. 5.27), taking into account the two possible polarities of trenches, but not transform faults. Of these, only the RRR triple junction is stable for any orientation of the ridges. This comes about because the associated velocity lines are the perpendicular bisectors of the triangle of velocity vectors, and these always intersect at a

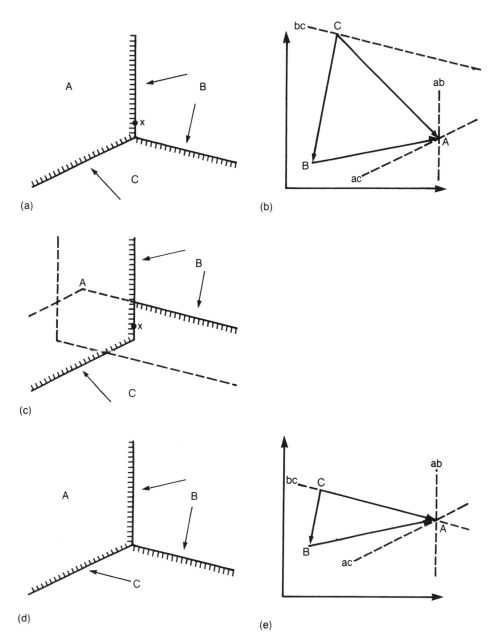

Figure 5.26 *(a) Triple junction between three trenches separating plates A, B, and C. (b) Its representation in velocity space, illustrating its instability. (c) The positions plates B and C would have reached if they had not been consumed are shown as dashed lines. (d) Stable configuration of a trench–trench–trench triple junction. (e) Its representation in velocity space. ((a) and (c) redrawn from McKenzie & Morgan, 1969, with permission from* Nature **224***, 125–33. Copyright © 1969 Macmillan Publishers Ltd).*

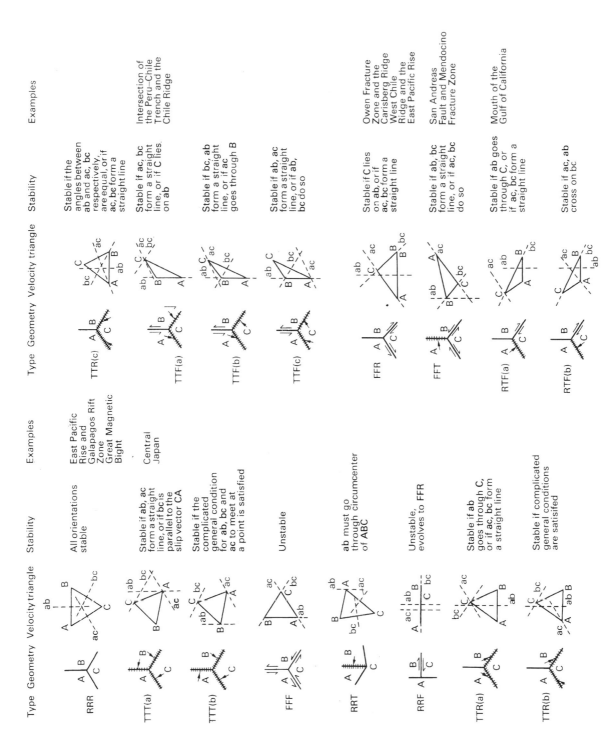

Figure 5.27 *Geometry and stability of all possible triple junctions (redrawn from McKenzie & Morgan, 1969, with permission from Nature **224**, 125–33. Copyright © 1969 Macmillan Publishers Ltd).*

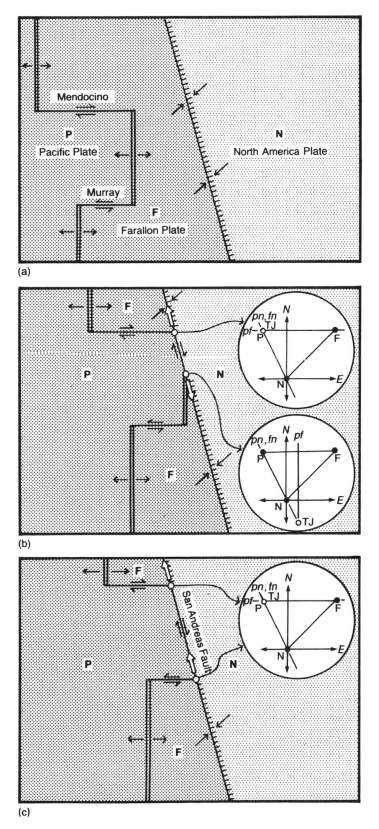

(a)

(b)

(c)

Figure 5.28 *Evolution of the San Andreas Fault (redrawn from Cox & Hart, 1986, with permission from Blackwell Publishing).*

single point (the circumcenter of the triangle). The FFF triple junction is never stable, as the velocity lines coincide with the vector triangle, and, of course, the sides of a triangle never meet in a single point. The other possible triple junctions are only stable for certain particular orientations of the juxtaposed plate margins.

5.11 PRESENT DAY TRIPLE JUNCTIONS

Only six types of triple junction are present during the current phase of plate tectonics. These are RRR (e.g. the junction of East Pacific Rise and Galapagos Rift Zone), TTT (central Japan), TTF (junction of Peru–Chile Trench and West Chile Rise), FFR (possibly at the junction of Owen Fracture Zone and Carlsberg ridge), FFT (junction of San Andreas Fault and Mendocino Fracture Zone), and RTF (mouth of Gulf of California).

The evolution of the San Andreas Fault illustrates the importance of the role of triple junctions. In Oligocene times (Fig. 5.28a), the East Pacific Rise separated the Pacific and Farallon plates. The transform faults associated with this ridge have been simplified,

and only the Mendocino and Murray fracture zones are shown. The Farallon Plate was being underthrust beneath the North American Plate, and, since the rate of consumption exceeded the rate of spreading at the East Pacific Rise, the ridge system moved towards the trench. The first point of the ridge to meet the trench was the eastern extremity of the Mendocino Fracture Zone. A quadruple junction existed momentarily at about 28 Ma, but this devolved immediately into two triple junctions (Fig. 5.28b). The more northerly was of FFT type, the more southerly of RTF type, and both were stable (insets on Fig. 5.28b). Because of the geometry of the system the northern triple junction moved north along the trench and the southern triple junction moved south. Thus the dextral San Andreas Fault formed in response to the migration of these triple junctions. The southerly migration of the southern triple junction ceased as the eastern extremity of the Murray Fracture Zone reached the trench (Fig. 5.28c). The triple junction changed to FFT type and began to move northwards. The Farallon Plate continued to be subducted to the north and south of the San Andreas Fault, until the geometry changed back to that shown in Fig. 5.28b when the East Pacific Rise to the south of the Murray Fracture Zone reached the trench. The triple junction then reverted to RTF type and changed to a southerly motion along the trench. This represents the situation at the present day at the mouth of the Gulf of California.

6 | Ocean ridges

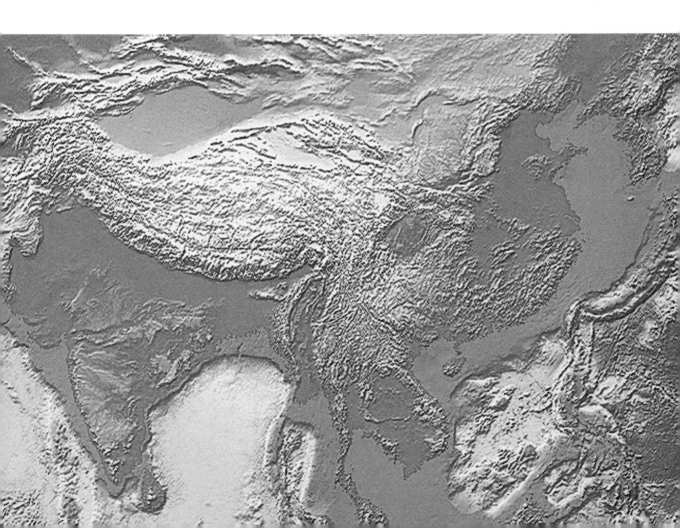

6.1 OCEAN RIDGE TOPOGRAPHY

Ocean ridges mark accretive, or constructive plate margins where new oceanic lithosphere is created. They represent the longest, linear uplifted features of the Earth's surface, and can be traced by a belt of shallow focus earthquakes that follows the crestal regions and transform faults between offset ridge crests (Fig. 5.2). The total length of the spreading margins on mid-ocean ridges is approximately 55,000 km. The total length of the active ridge–ridge transform faults is in excess of 30,000 km. The topographic expression of mid-ocean ridges is typically between 1000 and 4000 km in width. Their crests are commonly 2–3 km higher than neighboring ocean basins, and locally the topography can be quite rugged and runs parallel to the crests.

The gross morphology of ridges appears to be controlled by separation rate (Macdonald, 1982). Spreading rates at different points around the mid-ocean ridge system vary widely. In the Eurasian basin of the Arctic Ocean, and along the Southwest Indian Ocean Ridge, the full spreading rate (the accretion rate) is less than 20 mm a^{-1}. On the East Pacific Rise, between the Nazca and Pacific plates, the accretion rate ranges up to 150 mm a^{-1}. It is not surprising therefore that many of the essential characteristics of the ridges, such as topography, structure, and rock types, vary as a function of spreading rate. Very early on it was recognized that the gross topography of the East Pacific Rise, which is relatively smooth, even in the crestal region, contrasts with the rugged topography of the Mid-Atlantic Ridge, which typically has a median rift valley at its crest. This can now be seen to correlate with the systematically different spreading rates on the two ridges (Fig. 5.5), that is, fast and slow respectively. These two types of ridge crest are illustrated in Fig. 6.1, which is based on detailed bathymetric data obtained using deeply towed instrument packages. In each case, the axis of spreading is marked by a narrow zone of volcanic activity that is flanked by zones of fissuring. Away from this volcanic zone, the topography is controlled by vertical tectonics on normal faults. Beyond distances of 10–25 km from the axis, the lithosphere becomes stable and rigid. These stable regions bound the area where oceanic lithosphere is generated – an area known as the "crestal accretion zone" or "plate boundary zone".

The fault scarps on fast-spreading ridges are tens of meters in height, and an axial topographic high, up to 400 m in height and 1–2 km in width, commonly is present. Within this high a small linear depression, or graben, less than 100 m wide and up to 10 m deep is sometimes developed (Carbotte & Macdonald, 1994). The axial high may be continuous along the ridge crest for tens or even hundreds of kilometers. On slow-spreading ridges the median rift valley is typically 30–50 km wide and 500–2500 m deep, with an inner valley floor, up to 12 km in width, bounded by normal fault scarps approximately 100 m in height. Again there is often an axial topographic high, 1–5 km in width, with hundred of meters of relief, but extending for only tens

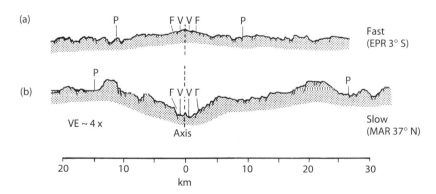

Fig. 6.1 *Bathymetric profiles of ocean ridges at fast and slow spreading rates. EPR, East Pacific Rise; MAR, Mid-Atlantic Ridge. Neovolcanic zone bracketed by Vs, zone of fissuring by Fs, extent of active faulting by Ps (redrawn with permission from MacDonald, 1982,* Annual Review of Earth and Planetary Sciences **10**. *Copyright © 1982 by Annual Reviews).*

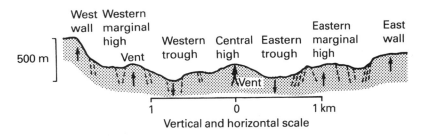

Fig. 6.2 *Diagrammatic cross-section of the inner rift valley of the Mid-Atlantic Ridge at 36°50′N in the FAMOUS area (redrawn from Ballard & van Andel, 1977, with permission from the Geological Society of America).*

of kilometers along the axis. At fast rates of spreading the high may arise from the buoyancy of hot rock at shallow depth, but on slowly spreading ridges it is clearly formed by the coalescence of small volcanoes 1–2 km in width, and hence is known as an axial volcanic ridge (Smith & Cann, 1993).

A detailed study of a median rift valley was made in the Atlantic Ocean between latitudes 36°30′ and 37°N, a region known as the FAMOUS (Franco American Mid-Ocean Undersea Study) area, using both surface craft and submersibles (Ballard & van Andel, 1977). The median rift in this area is some 30 km wide, bounded by flanks about 1300 m deep, and reaches depths between 2500 and 2800 m. In some areas the inner rift valley is 1–4 km wide and flanked by a series of fault-controlled terraces (Fig. 6.2). Elsewhere, however, the inner floor is wider with very narrow or no terraces developed. The normal faults that control the terracing and walls of the inner rift are probably the locations where crustal blocks are progressively raised, eventually to become the walls of the rift and thence ocean floor, as they are carried laterally away from the rift by sea floor spreading. Karson *et al.* (1987) described investigations of the Mid-Atlantic Ridge at 24°N using a submersible, deep-towed camera and side-scan sonar. Along a portion of the ridge some 80 km long they found considerable changes in the morphology, tectonic activity, and volcanism of the median valley. By incorporating data supplied by investigations of the Mid-Atlantic Ridge elsewhere, they concluded that the development of the style of the median valley may be a cyclic process between phases of tectonic extension and volcanic construction.

Bicknell *et al.* (1988) reported on a detailed survey of the East Pacific Rise at 19°30′S. They found that faulting is more prevalent than on slow-spreading ridges, and conclude that faulting accounts for the vast majority of the relief. They observed both inward and outward facing fault scarps that give rise to a horst and graben topography. This differs from slower spreading ridges, where the topography is formed by back-tilted, inward-facing normal faults. Active faulting is confined to the region within 8 km of the ridge axis, and is asymmetric with the greater intensity on the eastern flank. The half extension rate due to the faulting is 4.1 mm a^{-1}, compared to 1.6 mm a^{-1} observed on the Mid-Atlantic Ridge in the FAMOUS area.

Historically, for logistical reasons, the slowest spreading ridges, the Southwest Indian Ocean Ridge and the Gakkel Ridge of the Arctic Ocean, were the last to be studied in detail. In the Arctic the year-round ice cover necessitated the use of two research icebreakers (Michael *et al.*, 2003). The results of these studies led Dick *et al.* (2003) to suggest that there are three types of ridge as a function of spreading rate: fast, slow, and ultraslow (Fig. 6.3). Although the topography of the ultraslow Gakkel Ridge is analogous to that of slow-spreading ridges, typically with a well-developed median rift, the distinctive crustal thickness (Fig. 6.3), the lack of transform faults, and the petrology of this ridge set it apart as a separate class. Note that there are two additional categories of ridge with spreading rates between those of fast and slow, and slow and ultraslow, termed intermediate and very slow respectively. Intermediate spreading rate ridges may exhibit the characteristics of slow or fast-spreading ridges, and tend to alternate between the two with time. Similarly, a very slow-spreading ridge may exhibit the characteristics of a slow or ultraslow ridge. It is interesting to note that at the present day the East Pacific Rise is the only example of a fast-spreading ridge and the Gakkel Ridge of the Arctic is the only ultraslow-spreading ridge. Differences between the crustal structure and petrology of fast, slow and ultraslow ridges are discussed in Sections 6.6–6.9.

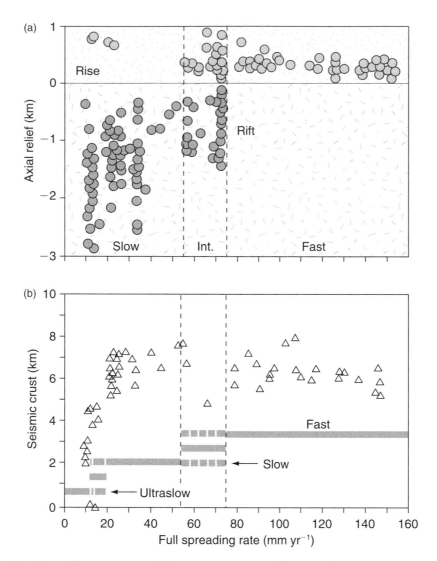

Fig. 6.3 *(a) Axial relief and (b) seismic crustal thickness as a function of full spreading rate at mid-ocean ridge crests. A ridge classification scheme is shown by the heavy black straight lines which indicate the spreading rate ranges for ultraslow, slow, fast and two intermediate classes (modified from Dick* et al., *2003, with permission from* Nature ***426**, 405–12. Copyright © 2003 Macmillan Publishers Ltd).*

6.2 BROAD STRUCTURE OF THE UPPER MANTLE BELOW RIDGES

Gravity measurements have shown that free air anomalies are broadly zero over ridges (Figs 6.4, 6.5), indicating that they are in a state of isostatic equilibrium (Section 2.11.6), although small-scale topographic features are uncompensated and cause positive and negative free air anomalies. The small, long wavelength, positive and negative free air anomalies over the crests and flanks, respectively, of ridges are a consequence of the compensation, with the positives being caused by the greater elevation of the ridge and the negatives from the compensating mass deficiency. The gravitational effects of the compensation dominate the gravity field away from the ridge crest, and indicate that the compensation is deep.

Seismic refraction experiments by Talwani *et al.* (1965) over the East Pacific Rise showed that the crust is slightly thinner than encountered in the main ocean basins, and that the upper mantle velocity beneath the crestal region is anomalously low (Fig. 6.4). Oceanic layer 1 rocks (Section 2.4.5) are only present within topographic depressions, but layers 2 and 3 appear to be continuous across the ridge except for a narrow region at the crest. A similar structure has been determined for the Mid-Atlantic Ridge (Fig. 6.5). The suggestion of this latter work that layer 3 is not continuous across the ridge was subsequently disproved (Whitmarsh, 1975; Fowler, 1976).

As the crust does not thicken beneath ridges, isostatic compensation must occur within the upper mantle by a Pratt-type mechanism (Section 2.11.3). Talwani *et al.* (1965) proposed that the anomalously low upper mantle velocities detected beneath ridges correspond to the tops of regions of low density. The densities were determined by making use of the Nafe–Drake relationship between P wave velocity and density (Nafe & Drake, 1963), and a series of models produced that satisfied both the seismic and gravity data. One of these is shown in Fig. 6.6, and indicates the presence beneath the ridge of a body with a density contrast of $-0.25\,Mg\,m^{-3}$ extending to a depth of some 30 km. This large density contrast is difficult to explain geologically. An alternative interpretation, constructed by Keen & Tramontini (1970), is shown in Fig. 6.7. A much lower,

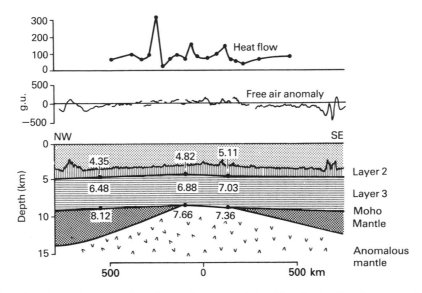

Fig. 6.4 *Heat flow, free air gravity anomaly and crustal structure defined by seismic refraction across the East Pacific Rise at 15–17°S. P wave velocities in km s^{-1} (redrawn from Talwani et al., 1965, by permission of the American Geophysical Union. Copyright © 1965 American Geophysical Union).*

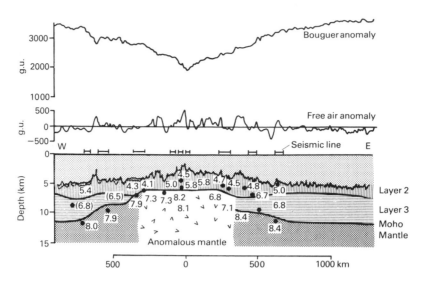

Fig. 6.5 *Gravity anomalies and crustal structure defined by seismic refraction across the Mid-Atlantic Ridge at about 31°N. Bouguer anomaly reduction density 2.60 Mg m^{-3}, P wave velocities in km s^{-1} (redrawn from Talwani et al., 1965, by permission of the American Geophysical Union. Copyright © 1965 American Geophysical Union).*

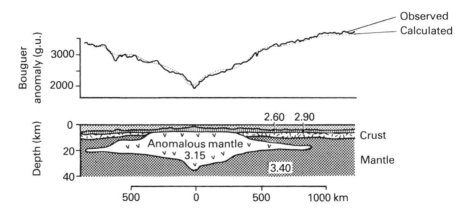

Fig. 6.6 *Possible model of the structure beneath the Mid-Atlantic Ridge from gravity modeling with seismic refraction control. Densities in Mg m^{-3} (redrawn from Talwani et al., 1965, by permission of the American Geophysical Union. Copyright © 1965 American Geophysical Union).*

more realistic density contrast of −0.04 Mg m^{-3} is employed, and the anomalous body is considerably larger, extending to a depth of 200 km. However, this model can also be criticized in that the densities employed are rather too high, and provide too low a density contrast, and the depth to the base of the anomalous mass is too great. A model that employs densities of 3.35 and 3.28 Mg m^{-3} for normal and anomalous

mantle, respectively, with the anomalous mass extending to a depth of 100 km, would be more in accord with geologic and geophysical data. Indeed, seismic tomography (Section 2.1.8) suggests that the low velocity region beneath ocean ridges extends to a depth of 100 km (Anderson *et al.*, 1992).

Given the ambiguity inherent in gravity modeling, the two interpretations shown probably represent end

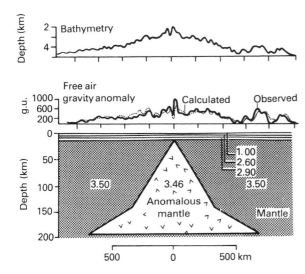

Fig. 6.7 *Alternative model of the structure beneath the Mid-Atlantic Ridge from gravity modeling. Profile at 46°N. Densities in Mg m⁻³ (redrawn from Keen & Tramontini, 1970, with permission from Blackwell Publishing).*

members of a suite of possible interpretations. They demonstrate without ambiguity, however, that ridges are underlain by large, low-density bodies in the upper mantle whose upper surfaces slope away from the ridge crests.

6.3 ORIGIN OF ANOMALOUS UPPER MANTLE BENEATH RIDGES

There are three possible sources of the low-density regions which underlie ocean ridges and support them isostatically (Bott, 1982): (i) thermal expansion of upper mantle material beneath the ridge crests, followed by contraction as sea floor spreading carries it laterally away from the source of heat, (ii) the presence of molten material within the anomalous mantle,

(iii) a temperature-dependent phase change. The high temperatures beneath ocean ridge crests might cause a transition to a mineralogy of lower density.

Suppose the average temperature to a depth of 100 km below the Moho is 500°C greater at the ridge crest than beneath the flanking regions, the average density to this depth is 3.3 Mg m⁻³ and the volume coefficient of thermal expansion is 3×10^{-5} per degree. In this case the average mantle density to a depth of 100 km would be 0.05 Mg m⁻³ less than that of the flanking ocean basins. If isostatic equilibrium were attained, this low-density region would support a ridge elevated 2.2 km above the flanking areas. If the degree of partial melting were 1%, the consequent decrease in density would be about 0.006 Mg m⁻³. Extended over a depth range of 100 km this density contrast would support a relative ridge elevation of 0.25 km. The aluminous minerals within the upper mantle that might transform to a lower density phase are also the minerals that enter the melt that forms beneath the ridge crest. They are absent therefore in the bulk of the mantle volume under consideration, which consists of depleted mantle; mantle from which the lowest melting point fraction has been removed. It is unlikely then that a phase change contributes significantly to the uplift.

Partial melting of the upper mantle clearly is a reality because of the magmatic activity at ridge crests, but its extent was a matter of conjecture. However, in the mid-1990s a very large-scale experiment, the Mantle Electromagnetic and Tomography (MELT) experiment, was carried out on the crest of the East Pacific Rise specifically to define the vertical and lateral extent of the region of partial melting beneath it (MELT seismic team, 1998). Fifty-one ocean bottom seismometers and 47 instruments that measure changes in the Earth's magnetic and electric fields were deployed across the ridge, between 15° and 18°S, in two linear arrays each approximately 800 km long. This location was chosen because it is in the middle of a long, straight section of the ridge between the Nazca and Pacific plates, and has one of the fastest spreading rates: 146 mm a⁻¹ at 17°S. The extent of any partial melt in the mantle should therefore be well developed in terms of low seismic velocities and high electrical conductivity. Seismic waves from regional and teleseismic earthquakes, and variations in the Earth's electric and magnetic fields, were recorded for a period of approximately 6 months. Analysis of the data revealed an asymmetric region of low seismic velocities extending to a depth of 100 km, with its shallowest point beneath the ridge crest, but extending to 350 km to the west and 150 km to the east of the ridge crest (Fig. 6.8). Both the velocity anomalies and electrical conductivity are consistent with 1–2% partial melting (Evans *et al.*, 1999). There is an indication of incipient melting to a depth of 180 km. The asymmetry of the region of partial melting is thought to be due to a combination of two effects. Within the hot spot framework the western flank of the ridge is moving at more than twice the rate of the eastern flank (Fig. 6.8). It is also close to the South Pacific superswell (Section 12.8.3). Enhanced upwelling and hence flow in the asthenosphere from the superswell and viscous drag beneath the fast moving Pacific plate are thought to produce higher rates of flow and hence higher temperatures beneath the western flank of the ridge. These elevated temperatures are reflected in shallower bathymetry (Section 6.4) and a higher density of seamount volcanism on the western flank compared to the eastern flank.

The width of the region of partial melt defined by the MELT experiment seems to be quite wide. One must recall however that the spreading rate at this point is very high, five times higher than that on much of the Mid-Atlantic Ridge. In fact the region of primary melt only underlies crust 2–3 Ma in age, whereas the anomalous uplift of ridges extends out to crust of 70–80 Ma in age. Partial melt in the upper mantle may therefore account for some of the uplift of ridge crests but cannot account for the uplift of ridge flanks.

6.4 DEPTH–AGE RELATIONSHIP OF OCEANIC LITHOSPHERE

The major factor contributing to the uplift of mid-ocean ridges is the expansion and contraction of the material of the upper mantle. As newly formed oceanic lithosphere moves away from a mid-ocean

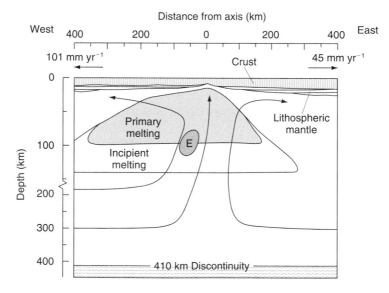

Fig. 6.8 *Schematic cross-section beneath the East Pacific Rise at 17°S illustrating the extent of partial melting in the mantle deduced from the results of the MELT experiment. Plate velocities are in the hot spot reference frame. The region labeled E (embedded heterogeneity) indicates enhanced melting due to anomalously enriched mantle or localized upwelling (modified from MELT seismic team, 1998,* Science **280**, *1215–18, with permission from the AAAS).*

ridge, it becomes removed from underlying heat sources and cools. This cooling has two effects. First, the lithosphere contracts and increases in density. Second, because the lithosphere–asthenosphere boundary is controlled by temperature (Section 2.12), the cooling causes the lithosphere to increase in thickness away from the mid-ocean ridge. This latter phenomenon has been confirmed by lithosphere thickness estimates derived from surface wave dispersion studies in the Pacific Ocean, which indicate that the thickness increases from only a few kilometers at the ridge crest to 30 km at 5 Ma age and 100 km at 50 Ma (Forsyth, 1977).

The cooling and contraction of the lithosphere cause a progressive increase in the depth to the top of the lithosphere away from the ridge (Sclater & Francheteau, 1970), accompanied by a decrease in heat flow. It follows that the width of a ridge depends upon the spreading rate, and so provides an explanation for the relative widths of the rapidly spreading East Pacific Rise and more slowly spreading Mid-Atlantic Ridge. Parsons & Sclater (1977) determined the nature of the age–depth relationships of oceanic lithosphere, and suggested that the depth d (meters) is related to age t (Ma) by:

$$d = 2500 + 350t^{1/2}$$

It was found, however, that this relationship only holds for oceanic lithosphere younger than 70 Ma. For older lithosphere the relationship indicates a more gradual increase of depth with age. In order to explain this, Parsons & McKenzie (1978) suggested a model in which the cooling layer comprises two units rather than the single unit implied by Parsons & Sclater (1977). In this model the upper unit, through which heat moves by conduction, is mechanically rigid, and the lower unit is a viscous thermal boundary layer. As the lithosphere travels away from a spreading center, both units thicken and provide the relationship – depth proportional to the square root of age – described above. However, the lower unit eventually thickens to the point at which it becomes unstable and starts to convect. This brings extra heat to the base of the upper layer and prevents it thickening at the same rate. They suggested that the age–depth relationship for oceanic lithosphere older than 70 Ma is then given by:

$$d = 6400 - 3200\exp(-t/62.8)$$

These two models, for the cooling and contraction of oceanic lithosphere with age, are referred to as the half space and plate models respectively. In the former the lithosphere cools indefinitely, whereas in the latter it ultimately attains an equilibrium situation determined by the temperature at the lithosphere–asthenosphere boundary and the depth at which this occurs as a result of convection in the asthenosphere. Clearly the main constraints on these models are the observed depth (corrected for sediment loading) and heat flux at the ocean floor as a function of age. Stein & Stein (1992), using a large global data set of depth and heat flow measurements, derived a model (GDH1 – global depth and heat flow model 1) that gave the best fit to the observations. Any such model must make assumptions about the depth to the ridge crest and the thermal expansion coefficient, the thermal conductivity, the specific heat, and the density of the lithosphere. However Stein & Stein (1992) showed that the crucial parameters in determining the best fit to the data are the limiting plate thickness and the temperature at the base of the lithospheric plate. In the GDH1 model these have the values 95 km and 1450°C respectively.

A comparison of the age–depth relationship predicted by the half space model, the Parsons, Sclater & McKenzie model and GDH1, is shown in Fig. 6.9a and the depth–age equations for GDH1 are:

$$d = 2600 + 365t^{1/2} \quad \text{for } t < 20 \text{ Ma}$$
$$\text{and } d = 5650 - 2473\exp(-t/36) \quad \text{for } t > 20 \text{ Ma}.$$

6.5 HEAT FLOW AND HYDROTHERMAL CIRCULATION

The half space model of lithospheric cooling with age predicts that the heat flux through the ocean floor on ridge flanks will vary in proportion to the inverse square root of its age, but across older ocean floor measured heat flow values vary more slowly than this, again favoring a plate model. The GDH1 model of Stein & Stein (1992) predicts the following values for heat flow, q (mWm^{-2}) as a function of age, t (Ma):

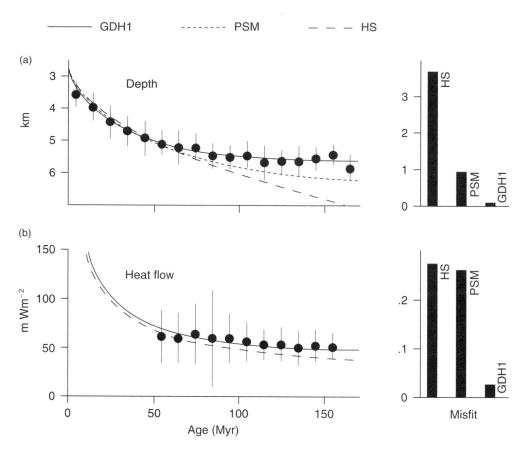

Fig. 6.9 *Observed depth and heat flow data for oceanic ridges plotted as a function of lithospheric age, and compared to the predictions of three thermal models: HS, half space model; PSM, model of Parsons, Sclater and McKenzie; GDH1, global depth and heat flow model of Stein and Stein (redrawn from Stein & Stein, 1996, by permission of American Geophysical Union. Copyright © 1996 American Geophysical Union).*

$$q = 510t^{-1/2} \quad \text{for } t \leq 55 \, \text{Ma}$$
$$\text{and } q = 48 + 96\exp(-t/36) \quad \text{for } t > 55 \, \text{Ma}.$$

The variation of heat flow with age predicted by all three thermal models is illustrated in Fig. 6.9b and compared to observed heat flow values. It will be noted that observed values for younger lithosphere have not been plotted. This is because there are large variations in the heat flux measured in young oceanic crust (Fig. 6.4). The values obtained are typically less than those predicted by the models and there is now thought to be good reason for this. In particular, there is a large scatter in heat flow magnitude near the crests of ocean ridges. Thermal lows tend to occur in flat-floored valleys and highs within areas of rugged topography (Lister, 1980).

Blanketing by sediment does not appear to be the cause of the low heat flow because the troughs are within the least sedimented areas of the ridge and also the youngest and therefore hottest. To explain these phenomena it was proposed that the pattern of heat flow is controlled by the circulation of seawater through the rocks of the oceanic crust.

Although the penetration of water through the hard rock of the sea floor at first seems unlikely, it has been shown that thermal contraction can induce sufficient permeability for efficient convective flow to exist. The cracks are predicted to advance rapidly and cool a large volume of rock in a relatively short time, so that intense localized sources of heat are produced at the surface. Active geothermal systems that are driven by water coming into contact with near-molten material are expected to be short-lived, but the relatively gentle circulation of cool water, driven by heat conducted from below, should persist for some time. However, as the oceanic crust moves away from the ridge crest, and subsides, it is blanketed by impermeable sediments, and the pores and cracks within it become clogged with minerals deposited from the circulating water. Ultimately heat flux through it is by conduction alone and hence normal heat flow measurements are obtained. This "sealing age" of oceanic crust would appear to be approximately 60 Ma.

Detailed heat flow surveys on the Galapagos Rift revealed that the pattern of large-scale zoning and the wide range of individual values are consistent with hydrothermal circulation (Williams *et al.*, 1974). Small-scale variations are believed to arise from variations in the near-surface permeability, while larger-scale variations are due to major convection patterns which exist in a permeable layer several kilometers thick which is influenced by topography, local venting, and recharge at basement outcrops. The penetration of this convection is not known, but it is possible that it is crust-wide. It is thought that hydrothermal circulation of seawater in the crust beneath ocean ridges transports about 25% of the global heat loss, and is clearly a major factor in the Earth's thermal budget (Section 2.13).

The prediction of hydrothermal circulation on mid-ocean ridges, to explain the heat flow values observed, was dramatically confirmed by detailed investigations at and near the sea floor at ridge crests, most notably by submersibles. Numerous hydrothermal vent fields have been discovered on both the East Pacific Rise and the Mid-Atlantic Ridge, many of them revealed by the associated exotic and previously unknown forms of life that survive without oxygen or light. The physical and chemical properties of the venting fluids and the remarkable microbial and macrofaunal communities associated with these vents, have been reviewed by Kelly *et al.* (2002). The temperature of the venting fluids can, exceptionally, be as high as 400°C. The chemistry of the hydrothermal springs on the East Pacific Rise and Mid-Atlantic Ridge is remarkably similar, in spite of the great difference in spreading rates, and suggests that they have equilibrated with a greenschist assemblage of minerals (Campbell *et al.*, 1988). Surprisingly perhaps, because of the cooler environment at the ridge crest, there are high levels of hydrothermal activity at certain locations on the very slow- and ultraslow-spreading Gakkel Ridge. This appears to result from the focusing of magmatic activity at these points, producing higher temperatures at shallow depths (Michael *et al.*, 2003).

Further evidence that hydrothermal circulation occurs comes from the presence of metalliferous deposits at ridge crests. The metals are those known to be hydrothermally mobile, and must have been leached from the oceanic crust by the ingress of seawater which permitted their extraction in a hot, acidic, sulfide-rich solution (Rona, 1984). On coming into contact with cold seawater on or just below the sea floor the solutions precipitate base metal sulfide deposits. The presence of such deposits is corroborated by studies of ophiolites (Section 13.2.2).

6.6 SEISMIC EVIDENCE FOR AN AXIAL MAGMA CHAMBER

Models for the formation of oceanic lithosphere normally require a magma chamber beneath the ridge axis from which magma erupts and intrudes to form the lava flows and dikes of layer 2. Solidification of magma within the chamber is thought to lead to the formation of most of oceanic layer 3 (Section 6.10). Evidence for the presence of such a magma chamber has been sought from detailed seismic surveys at ridge crests employing refraction, reflection, and tomographic techniques.

On the fast-spreading East Pacific Rise many of the surveys have been carried out in the area north of the Siquieros Fracture Zone between 8° and 13°N. The area centered on the ridge crest at 9°30'N has been particularly intensively studied (e.g. Herron *et al.*, 1980; Detrick *et al.*, 1987; Vera *et al.*, 1990). More recently additional experiments have been carried out at 14°15'S, on one of the fastest spreading sections of the ridge (Detrick *et al.*, 1993a; Kent *et al.*, 1994). All of these studies have revealed a region of low seismic velocities in the lower crust, 4–8 km wide, and evidence for the top of a magma chamber at varying depths, but typically 1–2 km below the sea floor. There is some indication that the depth to the magma chamber is systematically less at 14°S compared to 9°N on the East

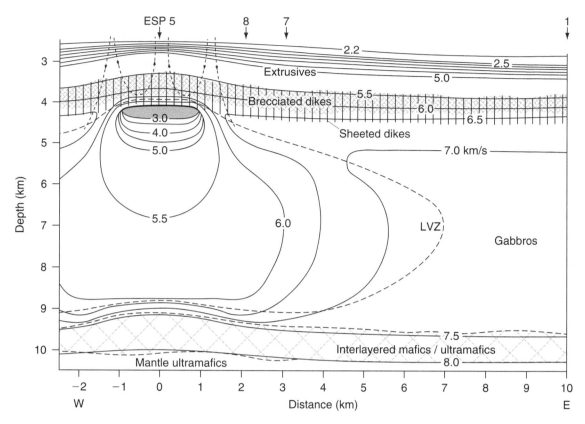

Fig. 6.10 *The variation of P wave velocity in the oceanic crust, at the crest of the East Pacific Rise at 9°30′N, deduced from expanded spread (ESP) and common depth point seismic profiling. Shaded area indicates a region with a high percentage of melt. An interpretation of the velocities in terms of rock units, and an indication of the extent of the zone of anomalously low seismic velocities (LVZ), are also shown (redrawn from Vera et al., 1990, by permission of the American Geophysical Union. Copyright © 1990 American Geophysical Union).*

Pacific Rise, suggesting an inverse correlation between magma chamber depth and spreading rate (Detrick *et al.*, 1993b). The interpretation of Vera *et al.* (1990) of results obtained at 9°30′N, using multi-channel, expanded spread reflection profiling, is shown in Fig. 6.10. They considered that only the volume in which the P-wave velocity is less than $3 \, \mathrm{km \, s^{-1}}$ can be regarded as a melt lens, and that the region in which the P-wave velocity is greater than $5 \, \mathrm{km \, s^{-1}}$, which includes much of the low velocity zone, behaves as a solid. Detrick *et al.* (1987) demonstrated that a strong reflector, thought to be associated with the top of the magma chamber, can be traced as a nearly continuous feature for tens of kilometers along the ridge axis. Much of the

more recent work, typically employing tomographic techniques (Section 2.1.8), suggests that the region in which there is a high melt fraction, probably no more than 30% crystals so that the shear wave velocity is zero, is remarkably small, perhaps no more than a few tens of meters thick, and less than 1 km wide (Kent *et al.*, 1990, 1994; Caress *et al.*, 1992; Detrick *et al.*, 1993a; Collier & Singh, 1997). Thus most of the low velocity zone beneath the ridge crest behaves as a solid and is interpreted as a region of anomalously hot rock.

In contrast to the picture that has emerged for the East Pacific Rise, most seismic studies of the slowly spreading Mid-Atlantic Ridge recognize a low velocity zone in the lower crust beneath the ridge crest but have not yielded any convincing evidence for a magma chamber or melt lens (Whitmarsh, 1975; Fowler, 1976; Purdy & Detrick, 1986; Detrick *et al.*, 1990). However, Calvert (1995), in reanalyzing the data of Detrick *et al.* (1990) acquired at 23°17′N, isolated reflections from a presumed magma chamber at a depth of 1.2 km and with a width of 4 km.

It seems unlikely therefore that steady state magma chambers exist beneath the axes of slowly spreading ridges. Transient magma chambers, however, related to influxes of magma from the mantle, may exist for short periods. In order to test this hypothesis a very detailed combined seismic and electromagnetic experiment was carried out across the Reykjanes Ridge south of Iceland (Sinha *et al.*, 1998). This study was deliberately centered on a magmatically active axial volcanic ridge (AVR) on the Reykjanes Ridge at 57°45′N, and did reveal a melt lens and crystal mush zone analogous to those imaged on the East Pacific Rise. In this instance the melt lens occurs at a depth 2.5 km beneath the sea floor. The results of this study provide strong support for the hypothesis that the process of crustal accretion on slow-spreading ridges is analogous to that at fast-spreading ridges but that the magma chambers involved are short-lived rather than steady state. Despite its proximity to the Iceland hot spot, the ridge crest south of 58°N on the Reykjanes Ridge has the characteristics of a typical slow-spreading ridge: a median valley, and normal crustal thickness and depth.

The logistically complicated seismic experiments required to test for the presence or absence of a melt lens have yet to be carried out on the very slow- and ultraslow-spreading Gakkel Ridge. It seems extremely unlikely that melt lenses exist beneath the amagmatic segments of this ridge, in that these consist of mantle peridotite with only a thin carapace of basalts, but possible that transient melt lenses occur beneath the magmatic segments and volcanic centers (Section 6.9). However, in 1999 seismological and ship-borne sonar observations recorded a long-lived magmatic-spreading event on the Gakkel Ridge that had characteristics more consistent with the magma being derived directly from mantle depths than from a crustal magma chamber (Tolstoy *et al.*, 2001).

Sinton & Detrick (1992), taking account of the seismic data available at that time and incorporating new ideas on magma chamber processes, proposed a model in which the magma chambers comprise narrow, hot, crystal-melt mush zones. In this model magma chambers are viewed as composite structures compris-

ing an outer transition zone made up of a hot, mostly solidified crust with small amounts of interstitial melts and an inner zone of crystal mush with sufficient melt for it to behave as a very viscous fluid. A melt lens only develops in fast-spreading ridges where there is a sufficiently high rate of magma supply for it to persist at the top of the mush zone (Fig. 6.11a). This lens may extend for tens of kilometers along the ridge crest, but is only 1–2 km wide and tens or hundreds of meters in thickness. Slow-spreading ridges are assumed to have an insufficient rate of magma supply for a melt lens to develop (Fig. 6.11b) and that eruptions only occur when there are periodic influxes of magma from the mantle. Such a model is consistent with the seismic data from ocean ridges and petrologic observations which require magma to have been modified by fractionation within the crust, which could not occur in a large, well-mixed chamber. It also explains why less fractionation occurs in the volcanic rocks of slow-spreading ridges. A problem with this model, however, is that it is not apparent how the layered gabbros of layer 3 might develop.

Subsequent work by Singh *et al.* (1998), involving further processing of the seismic reflection data obtained by Detrick *et al.* (1993a) near to 14°S on the East Pacific Rise, was specifically targeted at identifying any along-axis variations in the seismic properties and thickness of the melt lens. Their results suggest that only short, 2–4 km lengths of the melt lens contain pure melt capable of erupting to form the upper crust. The intervening sections of the melt lens, 15–20 km in length, are rich in crystal mush and are assumed to contribute to the formation of the lower crust. It seems probable that the pockets of pure melt are related to the most recent injections of magma from the mantle.

6.7 ALONG-AXIS SEGMENTATION OF OCEANIC RIDGES

Many early investigations of ocean ridges were essentially two-dimensional in that they were based on quite widely spaced profiles oriented perpendicular to their strike. More recently "swath"-mapping systems have

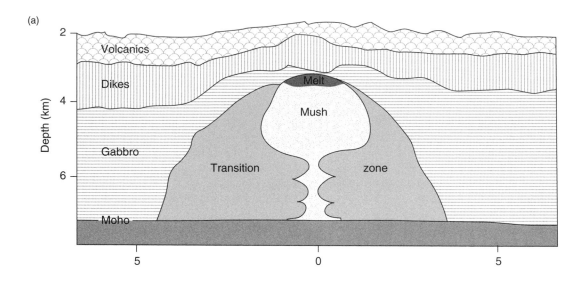

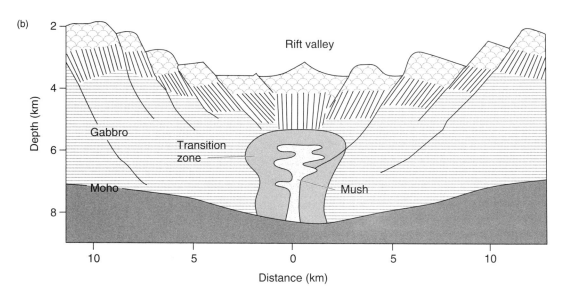

Fig. 6.11 *Interpretive models of magma chambers beneath a fast (a) and slow (b) spreading ridge (modified from Sinton & Detrick, 1992, by permission of the American Geophysical Union. Copyright © 1992 American Geophysical Union).*

been employed which provide complete areal coverage of oceanic features. These systems have been used to reveal variations in the structure of ocean ridges along strike. A review of these developments was provided by Macdonald *et al.* (1988).

Studies of the East Pacific Rise have shown that it is segmented along its strike by nontransform ridge axis discontinuities such as propagating rifts (Section 6.11) and overlapping spreading centers (OSC), which occur at local depth maxima, and by smooth variations in the depth of the ridge axis. These features may migrate up or down the ridge axis with time.

OSCs (MacDonald & Fox, 1983) are nonrigid discontinuities where the spreading center of a ridge is offset by a distance of 0.5–10 km, with the two ridge portions overlapping each other by about three times the offset. It has been proposed that OSCs originate on fast-spreading ridges where lateral offsets are less than 15 km, and true transform faults fail to develop because the lithosphere is too thin and weak. The OSC geometry is obviously unstable, and its development has been deduced from the behavior of slits in a solid wax film floating on molten wax, which appears to represent a reasonable analogue (Fig. 6.12a). Tension applied orthogonal to the slits (spreading centers) causes their lateral propagation (Fig. 6.12b) until they overlap (Fig. 6.12c), and the enclosed zone is subjected to shear and rotational deformation. The OSCs continue to advance until one tip links with the other OSC (Fig. 6.12d). A single spreading center then develops as one OSC becomes inactive and is moved away as spreading continues (Fig. 6.12e).

Fast-spreading ridges are segmented at several different scales (Fig. 6.13). First order segmentation is defined by fracture zones (Section 4.2) and propagating rifts (Section 6.11), which divide the ridge at intervals of 300–500 km by large axial depth anomalies. Second order segmentation at intervals of 50–300 km is caused by nonrigid transform faults (which affect crust that is still thin and hot) and large offset (3–10 km) OSCs that cause axial depth anomalies of hundreds of meters. Third order segmentation at intervals of 30–100 km is defined by small offset (0.5–3 km) OSCs, where depth anomalies are only a few tens of meters. Finally, fourth order segmentation at intervals of 10–50 km is caused by very small lateral offsets (<0.5 km) of the axial rift and small deviations from axial linearity of the ridge axis (DEVALS). These are rarely associated with depth anomalies and may be represented by gaps in the volcanic activity within the central rift or by geochemical variation. Clearly fourth order segmentation is on the same along-axis length scale as the intervals between pure melt pockets in the melt lens documented by Singh *et al.* (1998) (Section 6.6).

Third and fourth order segmentations appear to be short-lived, as their effects can only be traced for a few kilometers in the spreading direction. Second order segmentations, however, create off axis scars on the spreading crust consisting of cuspate ridges and elongate basins that cause differential relief of several hundred

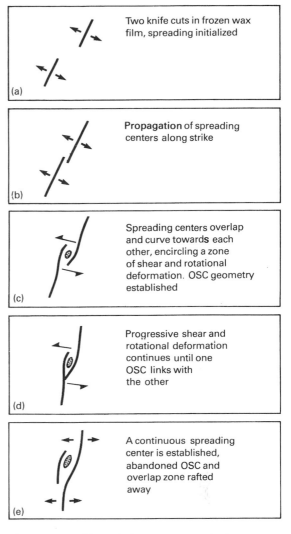

Fig. 6.12 *Possible evolutionary sequence in the development of an overlapping spreading center (redrawn from MacDonald & Fox, 1983, with permission from* Nature ***302**, 55–8. Copyright © 1983 Macmillan Publishers Ltd).*

meters. The scars do not follow small circle routes about the spreading pole, but form V-shaped wakes at 60–80° to the ridge. This indicates that the OSCs responsible for the segmentation migrate along the ridge at velocities of up to several hundred millimeters per year. Figure 6.14 summarizes the three general cases for the evolution of such ridge–axis discontinuities in terms of the movement of magma pulses.

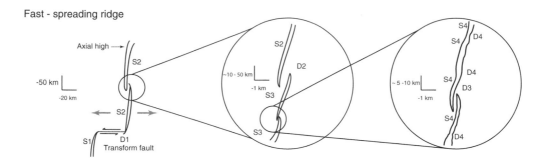

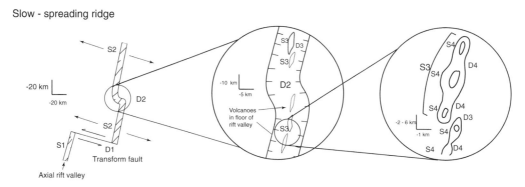

Fig. 6.13 *Summary of the hierarchy of segmentation on fast- and slow-spreading ridges. S_1, S_2, S_3, and S_4 – first to fourth order ridge segments. D_1, D_2, D_3, and D_4, – first to fourth order discontinuities (redrawn from Macdonald et al., 1991,* Science **253***, 986–94, with permission from the AAAS).*

The different scales and hence "orders" of ridge segmentation were first recognized on the fast-spreading East Pacific Rise. Segmentation also exists on the slow-spreading Mid-Atlantic Ridge but takes on somewhat different forms, presumably because the ridge crest is cooler and hence more brittle (Sempéré *et al.*, 1990; Gente *et al.*, 1995) (Fig. 6.13). First order segmentation is defined by transform faults, but overlapping spreading centers are absent and second order segments are bounded by oblique offsets of the ridge axis associated with deep depressions in the sea floor. Third and fourth order segmentation is in the form of geochemical variations and breaks in volcanic activity in the inner valley floor. The latter generate discrete linear volcanic ridges 2–20 km long and 1–4 km wide (Smith & Cann, 1993). Again first and second order segmentation is long-lived and third and fourth order segmentation is short-lived. Segmentation on the ultraslow-spreading Gakkel Ridge,

which does not even exhibit transform faulting, is in the form of volcanic and tectonic, or magmatic and amagmatic, segments (Michael *et al.*, 2003) (Section 6.9).

The first order segment boundaries, transform faults, are marked by pronounced bathymetric depressions (Section 6.12). They are often underlain by thinner crust than normal and anomalously low sub-Moho seismic velocities that may be due to partial serpentinization of the mantle as a result of seawater percolating down through the fractured crust. This thinning of the crust in the vicinity of fracture zones is particularly marked on the slow-spreading Mid-Atlantic Ridge (White *et al.*, 1984; Detrick *et al.*, 1993b). By contrast, the central portions of segments are elevated, have crust of normal thickness, and thinner lithosphere. This implies that the supply of magma from the mantle is focused at discrete points along the ridge axis at segment centers. These regions of thicker crust and enhanced

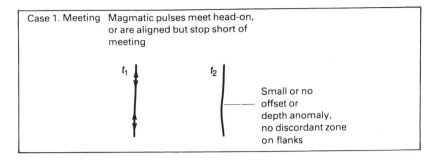

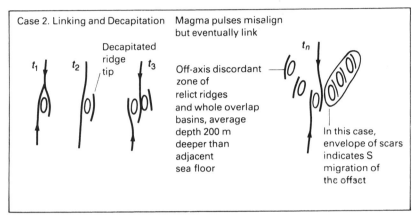

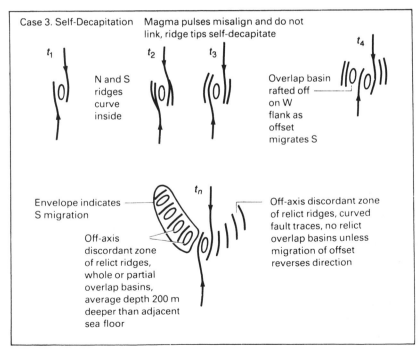

Fig. 6.14 *Three possible general cases for the evolution of ridge–axis discontinuities. Arrows along axis refer to direction of propagation of magmatic pulses. t_1, t_2, . . . , t_n refer to a time sequence. Cases 2 and 3 apply to second and third order discontinuities (after Macdonald* et al.*, 1988, with permission from* Nature ***335**, 217–25. Copyright © 1988 Macmillan Publishers Ltd).*

magma supply are characterized by negative mantle Bouguer anomalies (MBAs) and the areas of thinned crust between them by positive MBAs (Lin *et al.*, 1990). These latter areas include second order discontinuities in addition to transform faults. Evidence for the focusing of magma supply at segment centers is less obvious on the fast-spreading East Pacific Rise, but discontinuities and variations in the size and width of the magma chamber correlate with segmentation (Toomey *et al.*, 1990). These observations suggest that magma is emplaced at segment centers and migrates laterally along the ridge axis towards the segment ends. Increasingly it has been recognized that on cooler slow-spreading crust this can mean that segment ends are starved of magma and that parts of the crustal section consist of serpentinized mantle.

The extension of oceanic crust at ridge crests can occur either by the intrusion of magma or by extensional faulting. If ridge crests in the vicinity of transform faults are deprived of magma, amagmatic extension becomes more important. Perhaps the most spectacular expression of this is the occurrence of major low-angle detachment faults (Ranero & Reston, 1999; MacLeod *et al.*, 2002) (Section 7.3) on the inside corners of slow-spreading ridge–transform intersections that give rise to large corrugated and striated domes of serpentinized peridotite and gabbro (Plate 6.1 between pp. 244 and 245). These corrugated domes are exposed fault planes that deform the upper mantle and lower crust of oceanic lithosphere. The corrugations parallel the spreading direction and indicate the direction of motion on the fault. The offset on these faults is typically at least 10–15 km (Cann *et al.*, 1997). These exposures are thought to result from processes involving extension, detachment faulting, and crustal flexure that are similar to those that form metamorphic core complexes in zones of continental extension (Sections 7.3, 7.6.2, 7.7.3). For this reason the zones of exhumed peridotite at ridge–transform intersections are referred to as *oceanic core complexes*. Examples include the Atlantis Massif at the Mid-Atlantic Ridge–Atlantis transform intersection (Plate 6.1 between pp. 244 and 245) (Blackman *et al.*, 1998; Schroeder & John, 2004; Karson *et al.*, 2006) and along the Southeast Indian Ridge south of Australia (Baines *et al.*, 2003; Okino *et al.*, 2004).

Figure 6.15 illustrates the along axis variation of oceanic crust for slow- and fast-spreading ridges as envisioned by Cannat *et al.* (1995) and Sinton & Detrick (1992) respectively. Indeed Cannat *et al.* (1995) suggested that serpentinized rocks may be much more common in the oceanic crust than previously assumed, even in areas distant from fracture zones. They dredged in the region of the North Atlantic Ridge at 22–24°N over areas of positive gravity anomalies, indicative of relatively thin crust, and over areas with a normal gravity field. Over the former, which comprised some 23% of the area surveyed, they encountered serpentinite with very few of the basaltic rocks which normally characterize oceanic layer 2. They suggested that as magmatic centers grow, migrate along the ridge axis and decline, the normal oceanic crust would similarly migrate and would enclose those regions of serpentinitic crust that originate where magma was absent. This work is important as it implies that serpentinized peridotite is more common in slow-spreading oceans than previously recognized. There are wide ranging implications. Peridotite is much more reactive with seawater than basalt and on weathering would release magnesium, nickel, chromium and noble metals. Sepentinite also contains far more water than altered basalt, which could account for much of the water supplied to the mantle in subduction zones (Section 9.8), although at the present day the only examples of oceanic crust formed at slow-spreading rates entering subduction zones are the Caribbean and Scotia arcs.

Segmentation of ocean ridges appears to be controlled by the distribution of partial melts beneath them (Toomey *et al.*, 1990; Gente *et al.*, 1995; Singh *et al.*, 1998), which feed magma chambers at discrete locations along them and create local depth anomalies. The ridge model of Sinton & Detrick (1992), described above, precludes extensive mixing within the small axial magma chamber along the ridge, and could explain the observed geochemical segmentation. With time the magma may migrate away from its sources, creating a gradual increase in depth of the axis as the pressure within it gradually wanes. This phenomenon may explain the noncoincidence of magma chamber and rise culmination noted by Mutter *et al.* (1988). The brittle shell overlying the magma stretches and cracks and magma intrudes so that eruptions follow the path of magma migration. After eruption the removal of supporting magma gives rise to the formation of an axial summit graben. Evidence for the pulse-like, episodic spreading of ridges, in which sea floor spreading occurs by fracturing, dike injection and copious volcanism, has been provided by seismological studies and direct observation (e.g. Dziak & Fox, 1999;

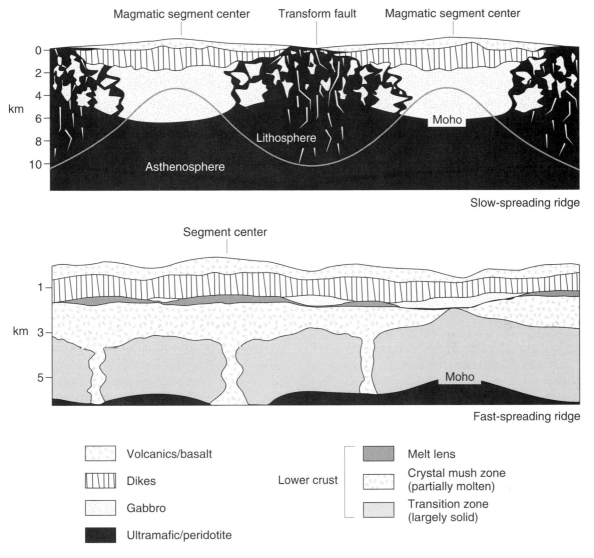

Fig. 6.15 *Along axis sections illustrating the variation in crustal structure between segment centers and segment ends on slow- and fast-spreading ridges, as envisioned by Cannat* et al. *(1995) and Sinton and Detrick (1992) respectively (redrawn from Cannat* et al., *1995, with permission from the Geological Society of America, and modified from Sinton and Detrick, 1992, by permission of the American Geophysical Union. Copyright © 1992 American Geophysical Union).*

Tolstoy *et al.*, 2001) and studies of ophiolites (Harper, 1978). Ridge axis discontinuities consequently occur where the magmatic pulses finally come to rest. The variable bathymetry and associated geophysical and geochemical differences imply that adjacent segments of ridge have distinct, different sources in the mantle. First to third order segmentation is caused by the variable depth associated with magma migration; fourth order effects are caused by the geochemical differences in magma supply.

6.8 PETROLOGY OF OCEAN RIDGES

Under normal conditions the peridotite of the upper mantle does not melt. However, the high heat flow at ocean ridges implies that the geothermal gradient crosses the peridotite solidus at a depth of about 50 km (Wyllie, 1981, 1988), giving rise to the parental magma of the oceanic crust (Fig. 2.36). A similarly high geothermal gradient is believed to exist beneath oceanic islands as oceanic lithosphere traverses a mantle plume or hotspot (Section 5.5), so generating basaltic rocks by a similar mechanism.

Mid-ocean ridge basalts (MORB) have the composition of olivine tholeiite (Kay et al., 1970), and exhibit only minor variation in major element composition caused by variable alumina and iron contents. They may contain phenocrysts of olivine or plagioclase or, rarely, clinopyroxene (Nisbet & Fowler, 1978). The simplest interpretation of the chemistry of oceanic basalts, suggested from experimental petrology, is that separation of the partial melt occurs at a depth of 15–25 km. However, a wide range of alternative interpretations exist. The analysis of trace elements reveals that much of the compositional variation in the basalts is explicable in terms of high-level fractionation. To explain the most extreme variations, however, it is necessary to invoke the mixing of batches of magma. The frequent presence of xenocrysts of deep-level origin indicates that the rocks only spend a very short time in a high-level magma chamber.

On a smaller scale, a detailed sampling of the East Pacific Rise by Langmuir et al. (1986) revealed a series of basalts that are diverse in their major and trace element chemistry. This compositional variation has been interpreted in terms of a series of magmatic injection centers along the crest of the ridge which correlate with bathymetric highs spaced about 50–150 km apart. Magma moves outwards from the injection points along the ridge so that the temperature of eruption decreases regularly from maxima at the bathymetric

highs, which correspond to the centers of segments (Section 6.7). Batiza et al. (1988) sampled along the axis of the southern Mid-Atlantic Ridge, and showed that there are regular patterns of chemical variation along it caused by differences in the depth and extent of partial melting and degree of fractionation. They conclude that these patterns imply the presence of a deep central magma supply, with limited melt migration along the axis and no large, well-mixed magma chamber in the crust.

Flower (1981) has shown that differences in the lithology and chemistry of basalts generated at mid-ocean ridges show a simple correlation with spreading rate. The differences are not related to processes in the upper mantle, as the primary melts appear to be identical. They are believed to reflect the fractionation environment after partial melting. Slow-spreading systems are characterized by a complex magma chamber in which there is widespread accumulation of calcic plagioclase, the presence of phenocryst-liquid reaction morphologies, and pyroxene-dominated fractionation extracts. These phenomena are consistent with fractionation at many different pressures in a chamber that appears to be transient. This conclusion is in accord with the pattern of rare earth elements in basalts sampled from the Mid-Atlantic Ridge (Langmuir et al., 1986). Although a homogeneous mantle source is suggested, the variations in rare earth chemistry apparent in samples from adjacent areas indicate a complex subsequent history of differentiation. Fast-spreading ridges, however, suggest low-pressure basalt fractionation trends to iron-rich compositions with little plagioclase accumulation or crystal–liquid interaction. This is consistent with the magma chamber being a stable and steady state feature.

Basalts from very slow- and ultraslow-spreading ridges have lower sodium and higher iron contents than typical MORB, reflecting a smaller degree of mantle melting and melting at greater depths. The geochemistry of the peridotites dredged from such ridges also indicates that the extent of mantle melting beneath the ridge is low. The great variation in the rate at which magma is supplied along the length of the Gakkel Ridge, and its lack of correlation with spreading rate, suggests that additional factors must be involved. Different thermal regimes or varying mantle composition along the length of the ridge, or lateral migration of melts in the upper mantle are some of the possibilities. Indeed, because of the smaller vertical extent of melting beneath such ridges (Section 6.9), small variations in mantle temperature and/or composition would lead to greater proportional changes in the volume of magma produced (Michael *et al.*, 2003).

6.9 SHALLOW STRUCTURE OF THE AXIAL REGION

As noted above (Section 6.1), normal oceanic crust, that is, not formed in the vicinity of hot spots or transform faults, has a remarkably uniform seismic thickness of $7 \pm 1\,km$ if generated at a full spreading rate in excess of $20\,mm\,a^{-1}$. For a homogeneous mantle this implies a comparable thermal gradient beneath all such ridge crests, and a similar degree of partial melting of the mantle, which produces the uniform thickness of mafic crust. The essential uniformity of the thermal regime beneath ridges is also implied by the lithospheric age versus depth relationship (Section 6.4). However, the rate at which magma is supplied to the crust will depend on the spreading rate. On fast-spreading ridges the magma supply rate is such that the whole crestal region at relatively shallow depth is kept hot and a steady state magma chamber exists. Indeed the crust above the magma chamber would be even hotter and weaker but for the cooling effect of hydrothermal circulation (Section 6.5). On slow-spreading ridges the lower rate of magma supply enables the crust to cool by conduction, as well as hydrothermal circulation, between injections of magma from the mantle. As a result the crust is cooler, and a steady state magma chamber cannot be maintained. At spreading rates of less than $20\,mm\,a^{-1}$ this conductive cooling between injections of magma extends into the mantle and inhibits melt generation. This reduces the magma supply, as well as the magma supply rate, and hence the thickness of mafic crust produced, as observed on the Southwest Indian Ocean and Gakkel ridges (Section 6.1). It also makes the existence of even transient magma chambers beneath such ridges rather unlikely except beneath the volcanic centers (Section 6.6).

The relatively smooth axial topography of fast-spreading ridges is characterized by an axial high, up to 400 m in height and 1–2 km wide, and fault scarps with a relief of tens of meters, the fault planes dipping either towards or away from the ridge axis. Active volcanism is largely confined to the axial high, and the smooth topography is thought to result from the high eruption rate and the low viscosity of the magma. The axial high appears to correspond to and to be supported by the buoyancy of the axial magma chamber beneath. Studies of major fault scarps and drill core from DSDP/ODP drill hole 504B, all in Pacific crust, reveal that at depth the lava flows dip towards the ridge axis at which they were erupted and that the dikes beneath them dip away from the ridge axis (Karson, 2002). This geometry indicates a very narrow and persistent zone of dike intrusion, and isostatic subsidence as the thickness of the lava flow unit increases away from the point of extrusion (Section 6.10). This relatively simple structure of the upper crust at the crests of fast-spreading ridges is illustrated in Fig. 6.16.

The shallow structure at the crests of slow-spreading ridges is fundamentally different to that on fast-spreading ridges (Smith & Cann, 1993). As a result of less frequent eruptions of magma and a cooler, more brittle upper crust, extension by normal faulting is more pronounced. The fault scarps have approximately 100 m of relief and the fault planes dip towards the ridge axis. Volcanism is essentially confined to the inner valley floor, and at any one time appears to be focused along specific axis-parallel fissures, forming axial volcanic ridges 1–5 km wide and tens of kilometers in length. As these ridges move off axis, as a result of further accretion, they may be cut by the faults that ultimately form the bounding scarps of the median valley. The spacing of these bounding faults appears to be about one-third to one-half of the width of the inner valley, that is, several kilometers. Within the inner valley floor the topography is fissured and cut by small throw normal faults, the density of these features giving an indication of its age. There is clear evidence of alternate phases of volcanic and tectonic (magmatic and amagmatic) extension of the crust, as one would expect if there are transient magma chambers beneath, which supply discrete packets of magma to the inner valley floor.

Very slow-spreading ridges are characterized by thin mafic crust and large regions of peridotite exposures where the mantle appears to have been emplaced directly to the sea floor. However there are also magmatic segments analogous to the second order

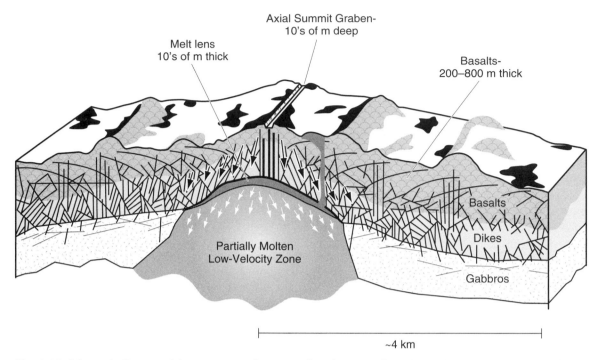

Melt lens
10's of m thick

Axial Summit Graben-
10's of m deep

Basalts-
200–800 m thick

Partially Molten
Low-Velocity Zone

Basalts

Dikes

Gabbros

~4 km

Fig. 6.16 *Schematic diagram of the upper crustal structure for a fast-spreading ridge (redrawn with permission from Karson, et al., 2002 by permission of the American Geophysical Union. Copyright © 2002, American Geophysical Union).*

segments on slow-spreading ridges. These have abundant volcanoes, typically in the form of axial volcanic ridges. These are 15–25 km long, and rise 400–1500 m from the axial valley floor. In the amagmatic sections the rift valley is often deeper than on slow-spreading ridges, up to 5000 m deep in places, and the rift valley walls have up to 2000 m of relief. On the Gakkel Ridge the western section is magmatic, the central section essentially amagmatic, less than 20% of the rift valley having a basaltic cover, and the ultraslow-spreading eastern section is very different again. It has six large volcanic centers on it that extend for 15–50 km along axis and are 50–160 km apart. These volcanic edifices are larger and more circular than those on other ridges. The amplitudes of the magnetic anomalies recorded between the volcanic centers suggest that the basaltic cover is thin in these tectonized zones. These marked along-axis contrasts in the extent of magma supply, which do not correlate with spreading rate, pose interesting questions regarding the generation and/or migration of melts beneath the ridge (Section 6.8) (Michael *et al.*, 2003).

6.10 ORIGIN OF THE OCEANIC CRUST

A widely accepted model of the petrologic processes occurring at ocean ridges was proposed by Cann (1970, 1974). In this model hot asthenospheric material ascends buoyantly (Nicolas *et al.*, 1994) sufficiently rapidly up a narrow zone to pass through the basalt melting curve and provides an interstitial melt of basaltic composition. The molten fraction increases in volume as the asthenosphere rises, and eventually departs the parental material to ascend independently and produce a magma chamber within the lower part of the oceanic crust at the level of layer 3. Part of this magma rises through fissures in the crust and erupts onto the ocean floor to produce pillowed lava flows. Beneath the flows is a zone of dikes formed by solidification of magma in the fissures that feed the flows. The lavas and dikes together make up layer 2 of the oceanic crust. Kidd (1977) modeled these processes of extrusion and intrusion and compared them with observations of ophiolite complexes. Layer 2C was found to consist entirely of sheeted dikes, which were intruded through zones less than 50 m wide. The dikes show some 10% more chilled margins on one side than the other, showing that approximately 10% of the dikes are cut by later dikes, such that the margins of the original dikes ended up on opposite sides of the ridge crest. The symmetry of sea floor spreading about the ridge axis is explained because dike intrusion will proceed preferentially into the hot central axis where existing dikes are weakest. It was suggested that the lavas extruded above the dikes cool rapidly in contact with sea water and flow less than 2 km before solidification. Lavas and dikes are predicted to rotate towards the ridge crest as they move away from the zone of extrusion as a result of isostatic adjustment (Fig. 6.17). They also undergo metamorphism near the ridge axis as they equilibrate at high temperatures in the presence of seawater.

This model for the origin of layer 2 has received striking confirmation from studies of sections through the upper crust revealed by major fault scarps and drill core from DSDP/ODP drill hole 504B, all in fast-spreading Pacific crust (Karson, 2002) (Section 6.9). Furthermore the model predicts that beneath the axial high the extrusive layer should be very thin and the dikes correspondingly closer to the sea floor (Fig. 6.17). This is confirmed by seismic studies that reveal a narrow central band of high seismic velocities beneath the axial high (Toomey *et al.*, 1990; Caress *et al.*, 1992) and a thin extrusive layer that thickens rapidly off axis within 1–2 km (Detrick *et al.*, 1993b; Kent *et al.*, 1994).

In the model of Cann (1974) the crust at lower levels develops from the crystallization of the axial magma chamber. The first minerals to crystallize in the magma chamber, olivine and chrome spinel, fall through the magma and form a basal layer of dunite with occasional accumulations of chromite. With further cooling pyroxene crystallizes and cumulate peridotitic layers (i.e. of olivine and pyroxene) are produced, giving way upwards to pyroxenites as the crystallization of pyroxene begins to dominate. Ultimately, plagioclase also crystallizes and layered olivine gabbros form. Much of the residual liquid, still volumetrically quite large, then solidifies over a very small temperature range to form an upper, "isotropic" gabbro. A small volatile-rich residuum of this differentiation process, consisting essentially of plagioclase and quartz, is the last fraction to crystallize, sometimes intruding upwards to form veins and small

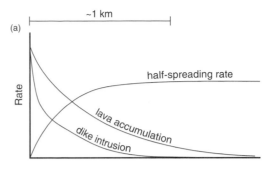

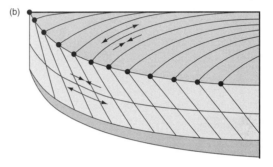

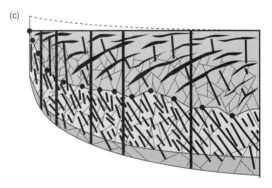

Fig. 6.17 *Geologic interpretation of the model of Kidd (1977) for the construction of Layer 2 at a fast-spreading ridge crest. Note the prediction of a rapid increase in the thickness of the extrusive layer away from the ridge axis and the presence of dikes at shallow depths near the ridge axis (redrawn with permission from Karson, et al., 2002, by permission of the American Geophysical Union. Copyright © 2002 American Geophysical Union).*

pockets of "plagiogranite" within the overlying sheeted dike complex. The abundance of volatiles, notably water, in the uppermost part of the magma chamber may be due, at least in part, to interaction with seawater percolating downwards and/or stoping of the

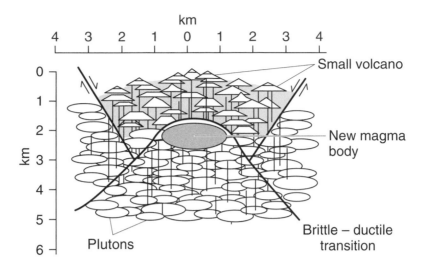

Fig. 6.18 *Model for the construction of oceanic crust at a slow-spreading ridge. Transient magma bodies rise to the brittle–ductile transition within the crust and shoulder aside and depress older plutons. Part of the magma body erupts through a fissure to produce a volcano or hummocky lava flow on the sea floor and the remainder solidifies to form part of the main crustal layer (redrawn from Smith & Cann, 1993, with permission from* Nature *365, 707–15. Copyright © 1993 Macmillan Publishers Ltd).*

overlying, hydrothermally altered dikes into the magma chamber.

The two gabbro units, isotropic and layered, are often correlated with seismic layers 3A and 3B, respectively (Section 2.4.7). The ultramafic cumulates, rich in olivine and pyroxene, would then account for the sub-Moho seismic velocities. Thus, the Moho occurs within the crystallized magma chamber at the base of the mafic section. Off axis, however, in a lower temperature environment, the uppermost ultramafics may become partially hydrated (i.e. serpentinized) and as a result acquire lower seismic velocities more characteristic of layer 3B. The seismic Moho would then occur at a somewhat greater depth, within the ultramafic section. As a result of this uncertainty in defining the seismic Moho, petrologists have tended to define the base of the crust as the base of the presumed magma chamber, that is, the dunite/chromitite horizon. Hence, this level is termed the "petrologic Moho".

The model of Cann (1974) and Kidd (1977) has met with considerable success in explaining the known structure and petrology of oceanic crust created at fast-spreading ridge crests, where there is a steady state magma chamber. At slow-spreading ridge crests, however, the zone of crustal accretion is wider and it seems probable that magma chambers are only tran-

sient. In this case the alternative model derived from early reinterpretations of ophiolites in terms of sea floor spreading may be more applicable. This invoked multiple small magma chambers within the main crustal layer in the light of the multiple intrusive relationships observed at all levels in the Troodos ophiolite of southern Cyprus (Moores & Vine, 1971) (Section 2.5). Smith & Cann (1993) favor such a model for the creation of oceanic crust at slow-spreading ridge crests (Fig. 6.18). However away from segment centers, and particularly in the vicinity of transform faults, the magma supply may be greatly reduced, and serpentinized mantle peridotite appears to be a common constituent of the thinned oceanic crust. This type of crust becomes even more common on very slow-spreading ridges and ultimately most of the crust is effectively exposed mantle with or without a thin carapace of basalts. On the ultra-slow Gakkel Ridge the crust is essentially serpentinized and highly tectonized mantle peridotite with volcanic centers at intervals of $100 \times 50 \, km$.

An alternative approach to understanding the accretionary processes at mid-ocean ridge crests is by way of thermal modeling (Sleep, 1975; Kusznir & Bott, 1976; Chen & Morgan, 1990). Chen & Morgan (1990) made significant improvements to such models by including the effects of hydrothermal circulation at ridge crests and the different rheological properties of the crust compared to the mantle, oceanic crust being more ductile at high temperatures than the mantle. As outlined in Section 6.9, the thermal regime beneath a ridge crest is influenced by the rate at which magma is supplied to the crust, which depends on the spreading rate. As a consequence the brittle–ductile transition (at approximately 750°C) occurs at a shallower depth in the crust at a fast-spreading ridge compared to a slow-spreading ridge that has a lower rate of magma supply. This in turn implies that at a fast-spreading ridge there is a much greater volume, and hence width, of ductile lower crust. This ductile crust effectively decouples the overlying brittle crust from the viscous drag of the convecting mantle beneath, and the tensile stresses pulling the plates apart are concentrated in a relatively thin and weak layer that extends by repeated tensile fracture in a very narrow zone at the ridge axis. On a slow-spreading ridge the brittle layer is thicker and the volume of ductile crust much smaller. As a result the tensile stresses are distributed over a larger area and there is more viscous drag on the brittle crust. In this situation the upper brittle layer deforms by steady state attenuation or "necking" in the form of a large number of normal faults creating a median valley.

Chen & Morgan (1990) demonstrated that for crust of normal thickness and appropriate model parameters the transition from smooth topography with a buoyant axial high to a median rift valley is quite abrupt, at a full spreading rate of approximately 70 mm a^{-1} as observed. The model also predicts that for thicker crust forming at a slow rate of spreading, as for example on the Reykjanes Ridge immediately south of Iceland, there will be a much larger volume of ductile crust, and smooth topography is developed rather than a rift valley. Conversely, where the crust is thin on a slow-spreading ridge, for example in the vicinity of fracture zones on the Mid-Atlantic Ridge, the median valley will be more pronounced than at a segment center. Such instances of thicker or thinner crust than normal are also likely to be areas of higher or lower than normal upper mantle temperatures respectively which will enhance the effect in each case. The model was extended by Morgan & Chen (1993) to incorporate a magma chamber as observed on the East Pacific Rise. This enhanced model predicts that a steady state magma chamber can only exist at spreading rates greater than 50 mm a^{-1} and that the depth to the top of the chamber will decrease as spreading rate increases, whilst retaining the essential features of the Chen & Morgan (1990) model.

In general therefore there is good agreement between the theoretical models for the creation of oceanic crust and observations made on *in situ* ocean floor and on ophiolites. Certain aspects however are still problematic. The evolution of a median valley as accretion occurs, that is, the way in which its flanks are uplifted and the normal faults ultimately reversed, is poorly understood. This is particularly true for the amagmatic segments of very slow- and ultraslow-spreading ridges where mantle material is emplaced directly to the sea floor. The details of the formation of the gabbroic layer 3, from a steady state or transient magma chamber, are also the subject of much debate.

6.11 PROPAGATING RIFTS AND MICROPLATES

The direction of spreading at an ocean ridge does not always remain constant over long periods of time, but may undergo several small changes. Menard & Atwater (1968) proposed that spreading in the northeastern Pacific had changed direction five times on the basis of changes in the orientation of major transform faults (Section 5.9) and magnetic anomaly patterns. Small changes in spreading direction have also been proposed as an explanation of the anomalous topography associated with oceanic fracture zones (Section 6.12).

Menard & Atwater (1968) made the assumption that the reorientation of a ridge would take place by smooth, continuous rotations of individual ridge segments until they became orthogonal to the new spreading direction (Fig. 6.19a). The ridge would then lie at an angle to the original magnetic anomaly pattern. Long portions of ridges affected in this way might be expected to devolve into shorter lengths, facilitating ridge rotation and creating new transform faults (Fig. 6.19b). The change in spreading direction is thus envisaged as a gradual, continuous rotation that produces a fan-like pattern of magnetic anomalies that vary in width according to position.

An alternative model of changes in spreading direction envisions the creation of a new spreading center

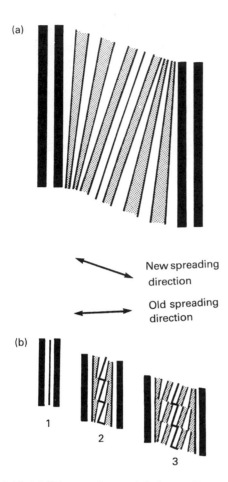

Fig. 6.19 (a) Ridge rotation model of spreading center adjustment; (b) evolution of a stepped ridge following rotation (modified from Hey et al., 1988, by permission of the American Geophysical Union. Copyright © 1988 American Geophysical Union).

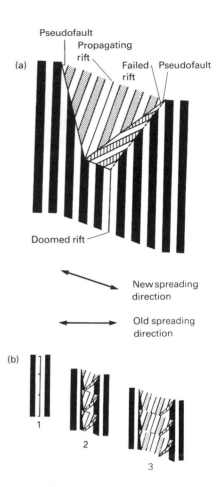

Fig. 6.20 (a) Ridge adjustment by rift propagation; (b) evolution of a stepped ridge following propagation (modified from Hey et al., 1988, by permission of the American Geophysical Union. Copyright © 1988 American Geophysical Union).

and its subsequent growth at the expense of the old ridge. This mechanism has been termed the *propagating rift* model (Hey, 1977; Hey et al., 1980). Thus the old, "doomed," rift is progressively replaced by a propagating spreading center orthogonal to the new spreading direction (Fig. 6.20a). Kleinrock & Hey (1989) have described the complex processes that take place at the tip of the propagating rift. The boundaries between lithosphere formed at old and new ridges are termed *pseudofaults*. Pseudofaults define a characteristic V-shaped wake pointing in the direction of propagation. Between the propagating and failing rifts, lithosphere is progressively transferred from one plate to the other,

giving rise to a sheared zone with a quite distinctive fabric. Therefore, abrupt changes in both the topographic and magnetic fabric of the sea floor occur at the pseudofaults and failed rift, and the new ridge propagates by the disruption of lithosphere formed by symmetric accretion at the old ridge. Figure 6.20b shows a possible way in which the propagating model could give rise to evenly spaced fracture zones. These new fracture zones are bounded by pseudofaults and/or failed rifts, because the fracture zones do not form until propagation is completed. They thus contrast with the ridge rotation model (Fig. 6.19b) which does not produce failed rifts and in which the fracture zones are areas of highly asymmetric sea floor spreading. The propagation model predicts abrupt boundaries between areas of uniform magnetic anomaly and bathymetric trends of different orientation. The rotation model predicts a continuous fanlike configuration of magnetic anomalies whose direction changes from the old to new spreading direction. Consequently, detailed bathymetric and magnetic surveys should be able to distinguish between the two models.

Hey *et al.* (1988) reported the results of a detailed investigation of the region where the direction of spreading of the Pacific–Farallon boundary changed direction at about 54 Ma, just north of the major bend of the Surveyor Fracture Zone, using side-scan sonar, magnetometry, and seismic reflection. They found that the change in direction of sea floor fabric revealed by sonar is abrupt, in accord with the propagating rift model. Similar conclusions were reached by Caress *et al.* (1988). Hey *et al.* (1980) described the results of a survey of an area west of the Galapagos Islands at 96°W. They concluded that here a new ridge is progressively breaking through the Cocos plate, and the magnetic data in particular (Fig. 6.21) provide convincing evidence that the ridge propagation mechanism is operative. This interpretation was confirmed by detailed mapping of the bathymetry in this area (Hey *et al.*, 1986). This clearly revealed the V-shaped pattern of the pseudofaults, the active and failed rifts, and the oblique tectonic fabric in the sheared zone of transferred lithosphere. The propagating rift model also elegantly explains the way in which the change in orientation of the Juan de Fuca Ridge (Fig. 4.1) has been achieved within the past 10 Ma (Wilson *et al.*, 1984).

Engeln *et al.* (1988) pointed out that the propagating rift model described above assumes that the newly formed rift immediately attains the full accretion rate between the two plates, thereby rendering the pre-existing rift redundant. However, if spreading on the new rift is initiated at a slow rate, and only gradually builds up to the full rate over a period of millions of years, the failing rift continues to spread, albeit at a slower and decreasing rate, in order to maintain the net accretion rate. In contrast to the original propagating rift model, in this model the two rifts overlap and the area of oceanic lithosphere between them increases with time. In addition, as a result of the gradients in spreading rate along each rift, the block of intervening lithosphere rotates. This rotation in turn produces compression in the oceanic lithosphere adjacent to the tip of the propagating rift and transtension (Section 8.2) in

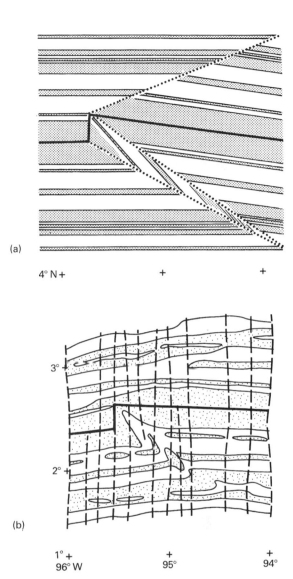

(a)

4° N +   +   +

(b)

1° +   +   +
96° W   95°   94°

Fig. 6.21 (a) Predicted magnetic lineation pattern resulting from ridge propagation; (b) observed magnetic anomalies near 96°W west of the Galapagos Islands (redrawn from Hey et al., 1980, by permission of the American Geophysical Union. Copyright © 1980 American Geophysical Union).

the region between the points where the propagating rift was initiated and the original rift started to fail. After a few million years this transpression gives rise to an additional propagating rift.

This second propagating rift model was put forward to explain the remarkable phenomenon of microplates

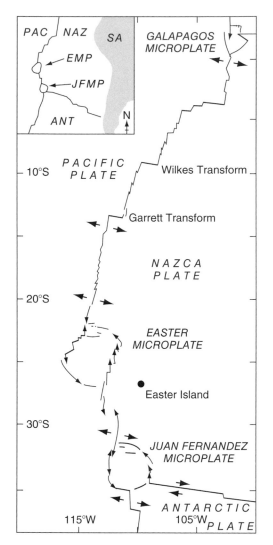

Fig. 6.22 *Map showing the location and extent of the Galapagos, Easter and Juan Fernandez microplates in the southeast Pacific Ocean. Arrows on ridge segments indicate active or previously active propagating rifts (modified from Bird* et al.*, 1998, by permission of the American Geophysical Union. Copyright © 1998 American Geophysical Union).*

in the southeast Pacific (Fig. 6.22). Detailed studies of the Easter and Juan Fernandez microplates show that their bathymetric fabrics and structural evolution are very similar, and fit well with the predictions of the model of Engeln *et al.* (1988) (Searle *et al.*, 1989; Rusby & Searle, 1995; Larson *et al.*, 1992; Bird *et al.*, 1998). The

tectonic elements of the Juan Fernandez microplate (Fig. 6.23) clearly show the characteristic pseudofaults of the original propagating rift to the east, and the subsequent propagating rift to the southwest of the microplate. Microplates are thought to exist for no more than 5–10 million years, by which time the initial rift succeeds in transferring the oceanic lithosphere of the microplate from one plate to another, in the case of the Juan Fernandez microplate, probably from the Nazca to the Antarctic plate (Bird *et al.*,1998). Tebbens *et al.* (1997) have documented an analogous example in the late Miocene when a newly formed rift, propagating northwards from the Valdivia Fracture Zone on the Chile Ridge, ultimately transferred lithosphere from the Nazca to the Antarctic plate. Brozena & White (1990) have reported ridge propagation from the South Atlantic, so this phenomenon appears to be independent of spreading rate.

The cause of the initiation of ridge propagation is unknown but several researchers have noted that propagating rifts tend to form in the vicinity of hot spots and on the hot spot side of the pre-existing ridge crest (e.g. Bird *et al.*, 1998; Brozena & White, 1990). An important corollary of the mere existence of propagating rifts is that the ridge-push force at spreading centers (Section 12.6) is not a primary driving mechanism as it appears to be quite easily overridden during ridge propagation.

6.12 OCEANIC FRACTURE ZONES

Transform faults in the oceans are well defined, in the absence of sedimentary cover, by fracture zones. These are long, linear, bathymetric depressions that normally follow arcs of small circles on the Earth's surface perpendicular to the offset ridge (Bonatti & Crane, 1984). The apparent relative simplicity of oceanic fracture zones is no doubt due in part to the fact that they are commonly studied from the sea surface several kilometers above the ocean floor. Direct observations of a fracture zone on the Mid-Atlantic Ridge (Choukroune *et al.*, 1978) have shown that it consists of a complex swarm of faults occupying a zone 300–1000 m in width. Searle (1983) suggests that these multi-fault zones are

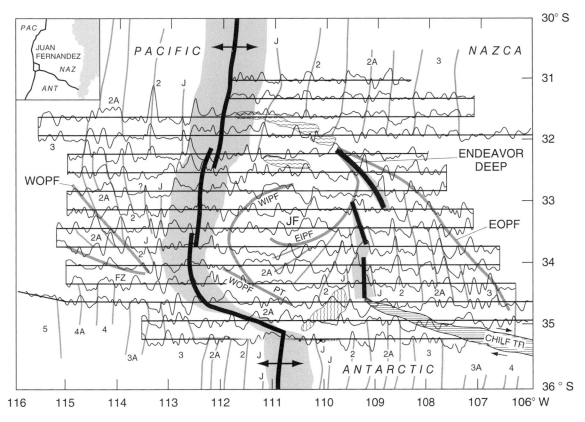

Fig. 6.23 *The tectonic elements of the Juan Fernandez microplate, together with magnetic anomalies numbered according to the timescale of Fig. 4.8. TR, transform; PT, paleo-transform; FZ, fracture zone; WIPF, WOPF, EIPF, EOPF: western/eastern, inner/outer pseudofaults (redrawn from Larson* et al., *1992, with permission from* Nature **356**, *571–6. Copyright © 1992 Macmillan Publishers Ltd).*

wider and more common on fast-spreading ridges such as the East Pacific Rise.

Fracture zones mark both the active transform segment and its fossilized trace. It has been suggested (Collette, 1979) that the fractures result from thermal contraction in the direction of the ridge axis. The internal stresses caused by contraction are much larger than the breaking strength of the rocks, and it is possible that fracture zones develop along the resulting lines of weakness.

Dredging of fracture zones has recovered both normal oceanic crustal rocks and rocks which show much greater metamorphism and shearing. Very commonly large blocks of serpentinite lie at the bases of the fracture zones. Bonatti & Honnorez (1976) and Fox et al. (1976) have examined specimens recovered

from the thick crustal sections exposed in the large equatorial Atlantic fracture zones, which were found to consist of ultramafic, gabbroic and basaltic rock types and their metamorphosed and tectonized equivalents. Serpentinite intrusion appears to be quite common within fracture zones, accompanied by alkali basalt volcanism, hydrothermal activity, and metallogenesis. Investigations of the Vema Fraction Zone (Auzende et al., 1989) have indicated a sequence similar to normal oceanic layering. St Peter and St Paul rocks, in the equatorial Atlantic, which lie on a ridge associated with the St Paul Fracture Zone, are composed of mantle peridotite.

In the North Atlantic, fracture zone crust is very heterogeneous in thickness and internal structure (Detrick et al., 1993b). It is often thin (<3 km) with low

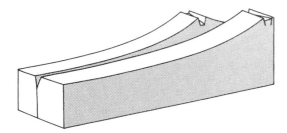

Fig. 6.24 *Differential topography resulting from transform faulting of a ridge axis.*

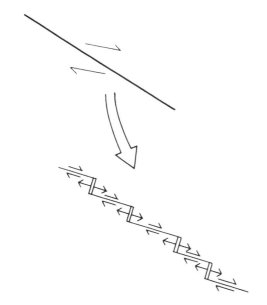

Fig. 6.26 *Development of a leaky transform fault because of a change in the pole of rotation.*

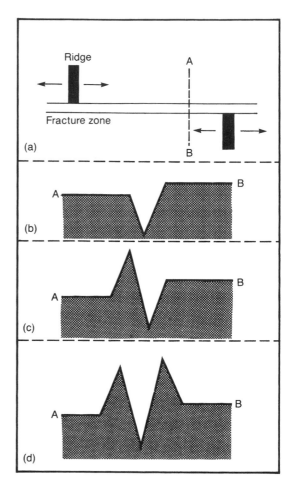

Fig. 6.25 *Different types of basement morphology across fracture zones (redrawn from Bonatti, 1978, with permission from Elsevier).*

seismic velocities, and layer 3 is absent. The crustal thinning may extend several tens of kilometers from the fracture zone. Geologically, this structure may represent a thin, intensely fractured and hydrothermally altered basaltic layer underlain by serpentinized ultra-mafic rocks. The apparent thickness variations may reflect different extents of serpentinization. The thin mafic crust is thought to be a result of reduced magma supply at ridge offsets as noted in Section 6.7.

Ocean fracture zones must bring oceanic crust of different ages into juxtaposition. The depth of the sea floor is dependent upon its age (Section 6.4), and so it would be expected that a scarp would develop across

the fracture zone from the younger, higher crust to the lower, older crust (Menard & Atwater, 1969; DeLong *et al.*, 1977) (Figs 6.24, 6.25b). The rate of subsidence of oceanic lithosphere is inversely dependent upon the square root of its age (DeLong *et al.*, 1977), so the higher, younger crust subsides more rapidly than the lower, older side. The combination of contraction in the vertical plane and horizontally perpendicular to the direction of the ridge axis would result in a small component of dip-slip motion along the fracture zone away from the active transform fault. DeLong *et al.* (1977) have suggested that this small amount of dip-slip motion could give rise to fracture zone seismicity and deformation of rocks within the floor and walls.

Transverse ridges are often found in association with major fracture zones and can provide vertical relief of over 6 km. These run parallel to the fractures (Bonatti, 1978) on one or both margins. They are frequently anomalous in that their elevation may be greater than that of the crest of the spreading ridge (Fig. 6.25c,d). Consequently, the age–depth relationship of normal oceanic lithosphere (Section 6.4) does not apply and depths differ from "normal" crust of the same age. The ridges do not originate from volcanic activity within the fracture zone, nor by hotspot activity (Section 5.5), but appear to result from the tectonic uplift of blocks of crust and upper mantle. Transverse ridges, therefore, cannot be explained by normal processes of lithosphere accretion. Bonatti (1978) considers that the most reasonable mechanism for this uplift is compressional and tensional horizontal stresses across the fracture zone that originate from small changes in the direction of spreading, so that transform movement is no longer exactly orthogonal to the ridge. Several small changes in spreading direction can give rise to episodic compression and extension affecting different parts of the fracture zone. This has caused, for example, the emergence of parts of transverse ridges as islands, such as St Peter and St Paul rocks, and their subsequent subsidence (Bonatti & Crane, 1984).

Lowrie *et al.* (1986) have noted that, in some fracture zones, the scarp height may be preserved even after 100 Ma. They accept that some parts of fracture zones are weak, characterized by active volcanism, and maintain the theoretical depths predicted for cooling lithosphere. Other parts, however, appear to be welded together and lock in their initial differential bathymetry. The differential cooling stresses would then cause flexure of the lithosphere on both sides of the fracture zone. Future work will reveal if there is any systematic pattern in the distribution of strong and weak portions of fracture zones.

There are certain oceanic transform faults in which the direction of the fault plane does not correspond exactly to the direction of spreading on either side so that there is a component of extension across the fault. When this occurs the fault may adjust its trajectory so as to become approximately parallel to the spreading direction by devolving into a series of fault segments joined by small lengths of spreading center (Fig. 6.26). A fault system in which new crust originates is termed a leaky transform fault (Thompson & Melson, 1972; Taylor *et al.*, 1994). An alternative mechanism for leaky transform fault development occurs when there is a small shift in the position of the pole of rotation about which the fault describes a small circle. The fault would then adjust to the new small circle direction by becoming leaky.

7 Continental rifts and rifted margins

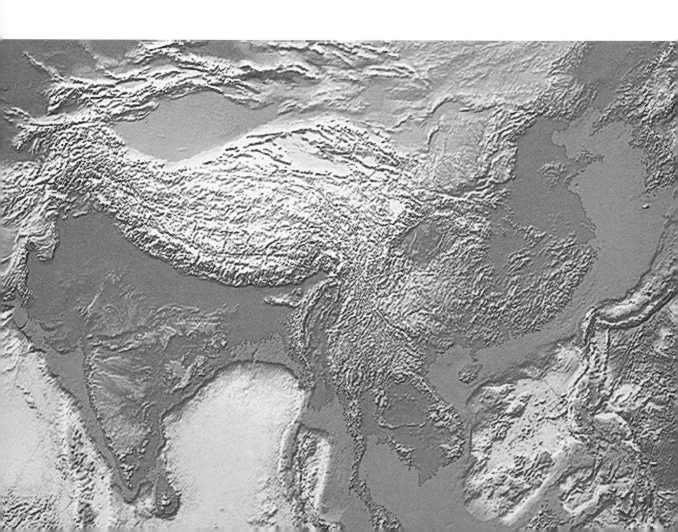

7.1 INTRODUCTION

Continental rifts are regions of extensional deformation where the entire thickness of the lithosphere has deformed under the influence of deviatoric tension. The term "rift" thus applies only to major lithospheric features and does not encompass the smaller-scale extensional structures that can form in association with virtually any type of deformation.

Rifts represent the initial stage of continental break-up where extension may lead to lithospheric rupture and the formation of a new ocean basin. If it succeeds to the point of rupture the continental rift eventually becomes inactive and a *passive* or rifted continental margin forms. These margins subside below sea level as a result of isostatic compensation of thinned continental crust and as the heat that was transferred to the plate from the asthenosphere during rifting dissipates. However, not all rifts succeed to the point where new ocean crust is generated. Failed rifts, or *aulacogens*, become inactive during some stage of their evolution. Examples of failed rifts include the Mesozoic Connecticut Valley in the northeastern United States and the North Sea Basin.

Studies of active rifts show that their internal structure, history, and dimensions are highly variable (Ruppel, 1995). Much of this variability can be explained by differences in the strength and rheology of the lithosphere (Section 2.10) at the time rifting initiates and by processes that influence these properties as rifting progresses (Section 7.6.1). Where the lithosphere is thick, cool, and strong, rifts tend to form narrow zones of localized strain less than 100 km wide (Section 7.2). The Baikal Rift, the East African Rift system, and the Rhine Graben are examples of this type of rift (Fig. 7.1). Where the lithosphere is thin, hot, and weak, rifts tend to form wide zones where strain is delocalized and distributed across zones several hundreds of kilometers wide (Section 7.3). Examples of this type of rift include the Basin and Range Province and the Aegean Sea. Both varieties of rift may be associated with volcanic activity (Section 7.4). Some rift segments, such as those in Kenya, Ethiopia, and Afar, are characterized by voluminous magmatism and the eruption of continental flood

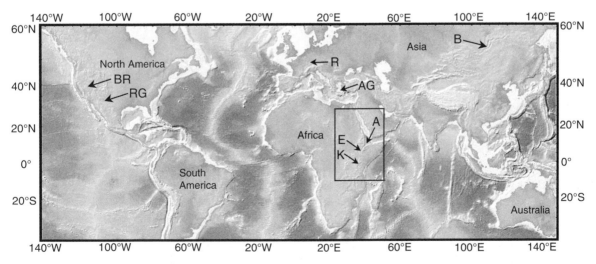

Figure 7.1 *Shaded relief map showing selected tectonically active rifts. Map constructed using digital seafloor topography of Smith & Sandwell, 1997, USGS Global 30 arc second elevation data (GTOPO30) for land areas (data available from USGS/EROS, Sioux Falls, SD, http://eros.usgs.gov/), and software provided by the Marine Geoscience Data System (http://www.marine-geo.org), Lamont-Doherty Earth Observatory, Columbia University. BR, Basin and Range; RG, Río Grande Rift; R, Rhine Graben; AG, Aegean Sea; B, Baikal Rift; E, Main Ethiopian Rift; A, Afar depression; K, Kenya Rift. Box shows location of Fig. 7.2.*

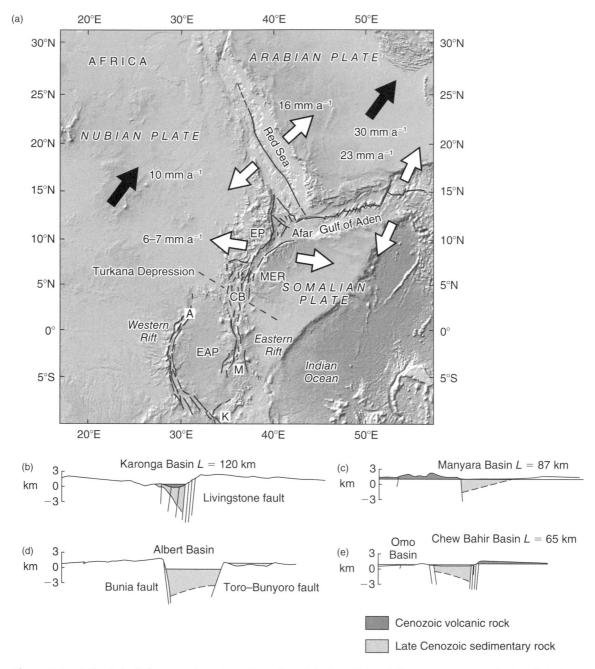

Figure 7.2 *(a) Shaded relief map and geodynamic setting of the East African Rift system constructed using digital topography data and software cited in Fig. 7.1. White arrows indicate relative plate velocities. Black arrows indicate absolute plate motion in a geodetic, no-net-rotation (NNR) framework (Section 5.4). (b–e) Cross-sections showing fault and half-graben morphology (after compilation of Ebinger* et al., *1999, with permission from the Royal Society of London). M, Manyara basin (from Foster* et al., *1997); K, Karonga basin (from van der Beek* et al., *1998); A, Albert basin (from Upcott* et al., *1996); CB, Chew Bahir basin (from Ebinger & Ibrahim, 1994); EAP, East African Plateau; EP, Ethiopian Plateau; MER, Main Ethiopian rift; L, the length of the border fault.*

basalts. Others, such as the Western branch of the East African Rift system (Fig. 7.2) and the Baikal Rift, are magma starved and characterized by very small volumes of volcanic rock.

In this chapter, several well-studied examples of rifts and rifted margins are used to illustrate how strain and magmatism are distributed as rifting proceeds to sea floor spreading. The examples also show how geoscientists combine different data types and use spatial and temporal variations in the patterns of rifting to piece together the tectonic evolution of these features.

7.2 GENERAL CHARACTERISTICS OF NARROW RIFTS

Some of the best-studied examples of tectonically active, narrow intracontinental rifts occur in East Africa (Fig. 7.2). Southwest of the Afar triple junction, the Nubian and Somalian plates are moving apart at a rate of approximately 6–7 mm a^{-1} (Fernandes et al., 2004). This divergent plate motion results in extensional deformation that is localized into a series of discrete rift segments of variable age, including the Western Rift, the Eastern Rift, the Main Ethiopian Rift, and the Afar Depression. These segments display characteristics that are common to rifts that form in relatively strong, cool continental lithosphere. Key features include:

1 *Asymmetric rift basins flanked by normal faults.* Continental rifts are associated with the formation of sedimentary basins that are bounded by normal faults. Most tectonically active rift basins show an asymmetric *half graben* morphology where the majority of the strain is accommodated along border faults that bound the deep side of the basins (Fig. 7.2b–e). The polarity of these half grabens may change along the strike of the rift axis, resulting in a segmentation of the rift valley (Fig. 7.3a). In plan view, the border faults typically are the longest faults within each individual basin. Slip on these faults combined

with flexural isostatic compensation of the lithosphere (Section 7.6.4) leads to uplift of the rift flanks, creating a characteristic asymmetric topographic profile. The lower relief side of the basin may be faulted and exhibits a monocline that dips toward the basin center. Deposition during slip on the bounding normal faults produces sedimentary and volcanic units that thicken towards the fault plane (Fig. 7.3b). The age of these *syn-rift* units, as well as units that pre-date rifting, provide control on the timing of normal faulting and volcanism. In plan view, displacements decrease toward the tips of border faults where they interact with other faults bounding adjacent basins. Within these *transfer zones* faults may accommodate differential horizontal (including strike-slip) and vertical displacements between adjacent basins.

2 *Shallow seismicity and regional tensional stresses.* Beneath the axis of most continental rifts earthquakes generally are confined to the uppermost 12–15 km of the crust, defining a *seismogenic layer* that is thin relative to other regions of the continents (Section 2.12). Away from the rift axis, earthquakes may occur to depths of 30 km or more. These patterns imply that rifting and thinning locally weaken the crust and affect its mechanical behavior (Section 7.6).

In Ethiopia, the record of seismicity from 1960 to 2005 (Fig. 7.4a) shows that the majority of large earthquakes occur between the Afar Depression and the Red Sea. Analyses of seismic moment release for this period shows that more than 50% of extension across the Main Ethiopian Rift is accommodated aseismically (Hofstetter & Beth, 2003). The earthquakes show combinations of normal, oblique and strike-slip motions. North of the Afar Depression, the horizontal component of most axes of minimum compressive stress strike to the north and northeast at high angles to the trend of the rift segments.

Keir *et al.* (2006) used nearly 2000 earthquakes to determine seismicity patterns within the northern Ethiopian Rift and its flanks

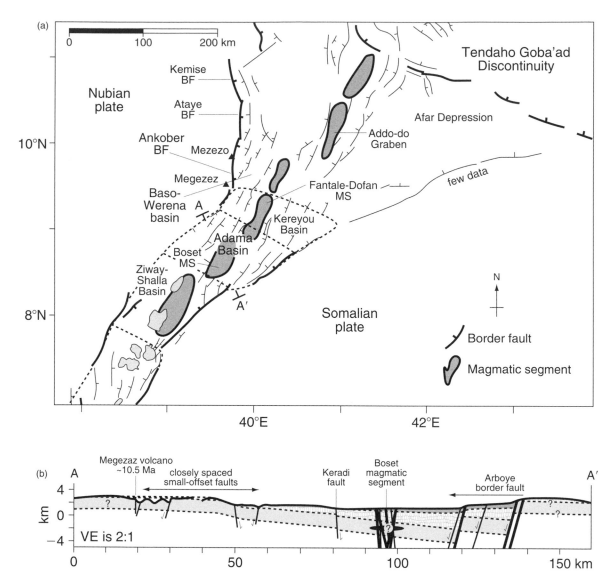

Figure 7.3 *(a) Major faults and segmentation pattern of the northern Main Ethiopian rift and (b) cross-section of Adama Rift Basin showing half graben morphology (images provided by C. Ebinger and modified from Wolfenden et al., 2004, with permission from Elsevier). MS, magmatic segment; BF, border fault. In (b) note the wedge-shaped geometry of the syn-rift Miocene and younger ignimbrite and volcanic units (vertical lined pattern and upper shaded layer). Pre-rift Oligocene flood basalts (lowest shaded layer) show uniform thickness.*

(Fig. 7.4b). Inside the rift, earthquake clusters parallel faults and volcanic centers in a series of 20 km wide, right-stepping zones of magmatism (Fig. 7.4c). Up to 80% of the total extensional strain is localized within these magmatic segments (Bilham *et al.*,

1999; Ebinger & Casey, 2001). The largest earthquakes typically occur along or near major border faults, although the seismicity data indicate that the border faults are mostly aseismic. Earthquakes are concentrated around volcanoes and fissures at depths of

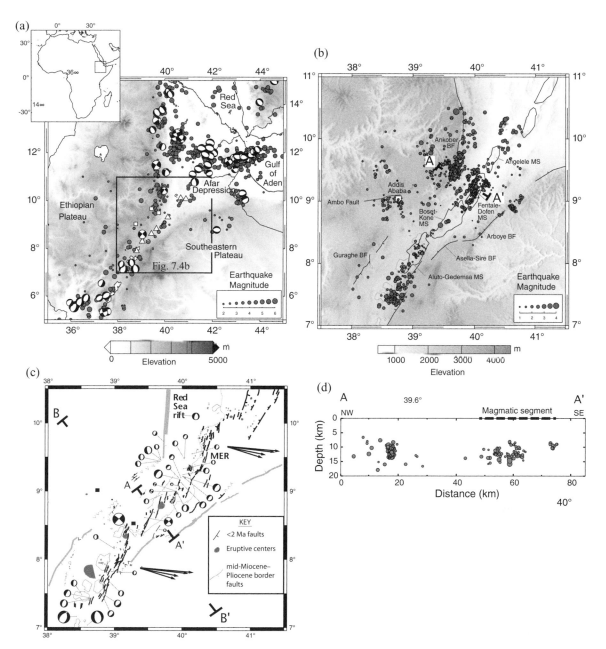

Figure 7.4 *(a) Seismicity and focal mechanisms of East Africa between 1960 and 2005. Late Cenozoic volcanoes shown by triangles. (b) Seismicity of rift segments in the northern Main Ethiopian Rift (MER) between October 2001 and January 2003. (c) Faults that cut <1.9 Ma lavas and late Cenozoic eruptive centers comprising magmatic segments (MS). Miocene border faults in rift basins also shown. Size of focal mechanism solutions indicates relative magnitude of earthquakes. Black arrows show approximate range of plate velocity vectors derived from geodetic data. (d) Earthquake depth distribution across profile A–A' shown in (c). Cross-section B–B' shown in Fig. 7.5 (images provided by D. Keir and modified from Keir et al., 2006, by permission of the American Geophysical Union. Copyright © 2006 American Geophysical Union).*

less than 14 km (Fig. 7.4d), probably reflecting magma movement in dikes. In the rift flanks, seismic activity may reflect flexure of the crust (Section 7.6.4) as well as movement along faults. The orientation of the minimum compressive stress determined from earthquake focal mechanisms is approximately horizontal, parallel to an azimuth of 103°. This stress direction, like that in Afar, is consistent with determinations of extension directions derived from tension fractures in young <7000 year old lavas, geodetic measurements, and global plate kinematic data (Fig. 7.4c).

3 *Local crustal thinning modified by magmatic activity.* Geophysical data indicate that continental rifts are characterized by thinning of the crust beneath the rift axis. Crustal thicknesses, like the fault geometries in rift basins, are variable and may be asymmetric. Thick crust may occur beneath the rift flanks as a result of magmatic intrusions indicating that crustal thinning is mostly a local phenomenon (Mackenzie et al., 2005; Tiberi et al., 2005). Variations in crustal thickness may also reflect inherited (pre-rift) structural differences.

Mackenzie et al. (2005) used the results of controlled-source seismic refraction and seismic reflection studies to determine the crustal velocity structure beneath the Adama Rift Basin in the northern part of the Main Ethiopian Rift (Fig. 7.5a). Their velocity model shows an asymmetric crustal structure with maximum thinning occurring slightly west of the rift valley. A thin low velocity layer (3.3 km s^{-1}) occurs within the rift valley and thickens eastward from 1 to 2.5 km. A 2–5-km-thick sequence of intermediate velocity (4.5–5.5 km s^{-1}) sedimentary and volcanic rock lies below the low velocity layer and extends along the length of the profile. Normal crustal velocities (P$_n$ = 6.0–6.8 km s^{-1}) occur to depths of 30–35 km except in a narrow 20–30 km wide region in the upper crust beneath the center of the rift valley where P$_n$ velocities are 5–10% higher (>6.5 km s^{-1}) than those outside the rift (Fig. 7.5a). These differences probably reflect the presence of mafic intrusions associated with magmatic centers. A nearly continuous intracrustal reflector at 20–25 km depth and Moho depths of 30 km show crustal thinning beneath the rift axis. The western flank of the rift is underlain by a ~45 km thick crust and displays a ~15 km thick high velocity (7.4 km s^{-1}) lower crustal layer. This layer is absent from the eastern side, where the crust is some 35 km thick. Mackenzie et al. (2005) interpreted the high velocity lower crustal layer beneath the western flank as underplated material associated with pre-rift Oligocene flood basalts and, possibly, more recent magmatic activity. Variations in intracrustal seismic reflectivity also suggest the presence of igneous intrusions directly below the rift valley (Fig. 7.5b).

Gravity data provide additional evidence that the crustal structure of rift zones is permanently modified by magmatism that occurs both prior to and during rifting. In Ethiopia and Kenya, two long-wavelength (>1000 km) negative Bouguer gravity anomalies coincide with two major ~2 km high topographic uplifts: the Ethiopian Plateau and the Kenya Dome, which forms part of the East African Plateau (Figs 7.2, 7.6a). The highest parts of the Ethiopian Plateau are more than 3 km high. This great height results from the eruption of a large volume of continental flood basalts (Section 7.4) between 45 and 22 Ma, with the majority of volcanism coinciding with the opening of the Red Sea and Gulf of Aden at ~30 Ma (Wolfenden et al., 2005). The negative gravity anomalies reflect the presence of anomalously low density upper mantle and elevated geotherms (Tessema & Antoine, 2004). In each zone, the rift valleys display short-wavelength positive Bouguer gravity anomalies (Fig. 7.6b) that reflect the presence of cooled, dense mafic intrusions (Tiberi et al., 2005).

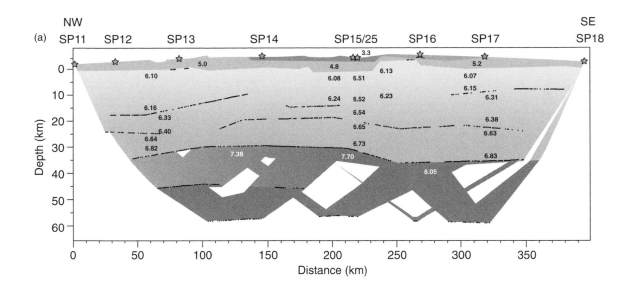

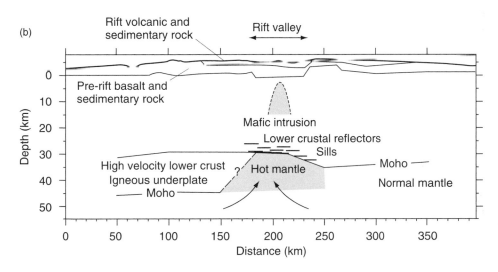

Figure 7.5 *(a) P-wave velocity model and (b) interpretation of the Main Ethiopian Rift (after Mackenzie et al., 2005, with permission from Blackwell Publishing). Location of profile (B–B′) shown in Fig. 7.4c.*

4 *High heat flow and low velocity, low density upper mantle.* Heat flow measurements averaging 70–90 mW m^{-2} and low seismic velocities in many rift basins suggest temperature gradients (50–100°C km^{-1}) that are higher than those in the adjacent rift flanks and nearby cratons. Where the asthenosphere is anomalously hot, such as in East Africa, domal uplifts and pervasive volcanism result. Nevertheless, there is a large degree of variability in temperature and volcanic activity among rifts. The Baikal Rift, for example, is much cooler. This rift displays low regional heat flow of 40–60 mW m^{-2} (Lysack, 1992) and lacks volcanic activity.

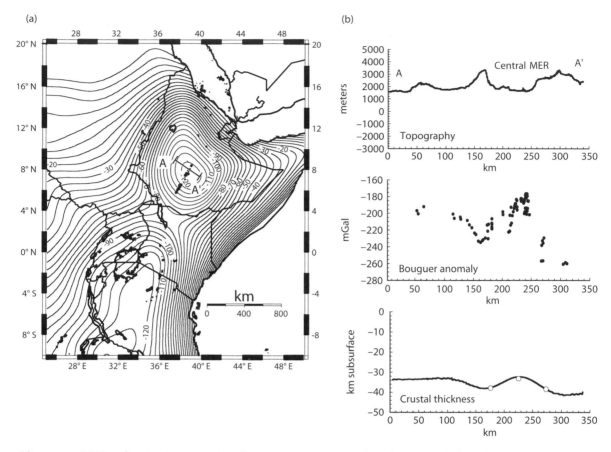

Figure 7.6 (a) Map showing long-wavelength Bouguer gravity anomalies after removal of the short wavelength (image provided by A. Tessema and modified from Tessema & Antoine, 2004, with permission from Elsevier). (b) Profiles (A–A') of topography, short-wavelength Bouguer gravity anomaly, and crustal thickness estimates of the central Main Ethiopian Rift (MER) (images provided by C. Tiberi and modified from Tiberi et al., 2005, with permission from Blackwell Publishing). Profile location shown in (a). Circles in crustal thickness profile indicate depths estimated from receiver function studies.

In East Africa, relatively slow P_n wave velocities of $7.7\,km\,s^{-1}$ in the upper mantle beneath the Adama Rift Basin in Ethiopia (Fig. 7.5a) suggest elevated temperatures (Mackenzie et al., 2005). Elsewhere upper mantle P_n wave velocities are in the range $8.0–8.1\,km\,s^{-1}$, which is expected for stable areas with normal heat flow. Tomographic inversion of P- and S-wave data (Fig. 7.7a–c) indicate that the low velocity zone below the rift is tabular, approximately $75\,km$ wide, and extends to depths of $200–250\,km$ (Bastow et al., 2005). The zone is segmented and offset away from the rift axis in the upper $100\,km$ but becomes

more central about the rift axis below this depth (Fig. 7.7c). In the more highly extended northern section of Ethiopian Rift (Fig. 7.7d), the low velocity anomaly broadens laterally below $100\,km$ and may be connected to deeper low velocity structures beneath the Afar Depression (Section 7.4.3). This broadening of the low velocity zone is consistent with the propagation of the Main Ethiopian Rift, during Pliocene–Recent times, toward the older spreading centers of the Red Sea and Gulf of Aden.

In addition to the high temperatures, the low velocity zones beneath rifts may also reflect the

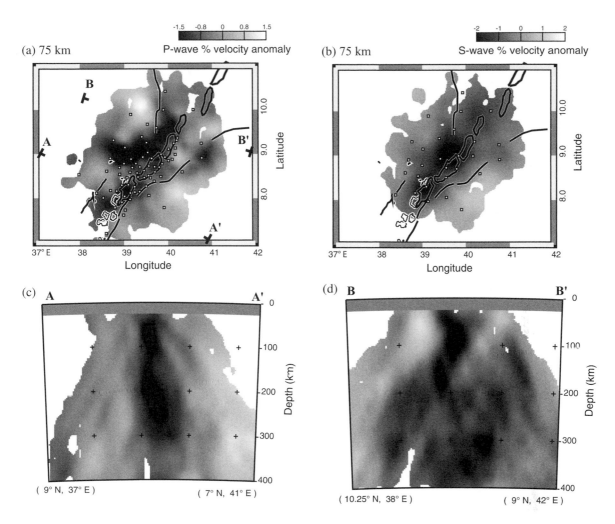

Figure 7.7 *Depth slices through (a) P-wave and (b) S-wave velocity models at 75 km depth in the Main Ethiopian Rift. (c,d) Vertical profiles through the P-wave velocity model (images provided by I. Bastow and modified from Bastow et al., 2005, with permission from Blackwell Publishing). Heavy black lines in (a) and (b) are Pleistocene magmatic segments and mid-Miocene border faults (cf. Fig. 7.3). The locations of stations contributing to the tomographic inversions are shown with white squares in (a) and (b). Profile locations shown in (a). Velocity scales in (c) and (d) are same as in (a).*

presence of partial melt. Observations of shear wave splitting and delay times of teleseismic waves traveling beneath the Kenya Rift (Ayele *et al.*, 2004) and northern Ethiopian Rift (Kendall *et al.*, 2005) suggest the alignment of partial melt in steep dikes within the upper 70–90 km of the lithosphere or the lattice preferred orientation of olivine in the asthenosphere as hot material flows laterally into the rift zone. These observations indicate that the upper mantle underlying rifts is characterized by low velocity, low density and anomalously high temperature material.

7.3 GENERAL CHARACTERISTICS OF WIDE RIFTS

One of the most commonly cited examples of a wide intracontinental rift is the Basin and Range Province of western North America (Fig. 7.1). In this region, large extensional strains have accumulated across a zone ranging in width from 500 to 800 km (Fig. 7.8). In the central part of the province, some 250–300 km of horizontal extension measured at the surface has occurred since ~16 Ma (Snow & Wernicke, 2000). In eastern Nevada and western Utah alone the amount of total horizontal surface extension is approximately 120–150 km (Wernicke, 1992). These values, and the width of the zone over which the deformation occurs, greatly

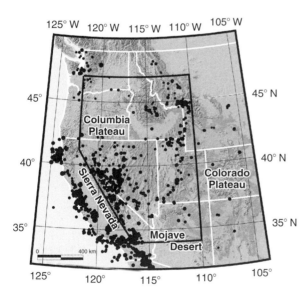

Figure 7.8 *Shaded relief map of the western United States showing topography and earthquakes with $M \geq 4.8$ in the northern and central sectors of the Basin and Range (image provided by A. Pancha and A. Barron and modified from Pancha et al., 2006, with permission from the Seismological Society of America). Circle radius is proportional to magnitude. The area outlined with a bold polygon encloses all major earthquakes that are associated with deformation of the Basin and Range.*

exceed those observed in narrow continental rifts (Section 7.2).

The Basin and Range example thus shows that continental lithosphere may be highly extended without rupturing to form a new ocean basin. This pattern is characteristic of rifts that form in relatively thin, hot, and weak continental lithosphere. Here, the key features that distinguish wide rifts from their narrow rift counterparts are illustrated using the Basin and Range and the Aegean Sea provinces as examples:

1 *Broadly distributed deformation.* The Basin and Range Province is bounded on the west by the greater San Andreas Fault system and Sierra Nevada–Great Valley microplate and on the east by the Colorado Plateau (Figs 7.8, 7.9). Both the Sierra and the Plateau record comparatively low heat flow values (40–60 mW m^{-2}) and virtually no Cenozoic extensional deformation (Sass *et al.*, 1994; Bennett *et al.*, 2003). In between these two rigid blocks Cenozoic deformation has resulted in a broad zone of linear, north-trending mountain ranges of approximately uniform size and spacing across thousands of square kilometers. The mountain ranges are about 15–20 km wide, spaced approximately 30 km apart, and are elevated ~1.5 km above the adjacent sedimentary basins. Most are delimited on one side by a major range-bounding normal fault. Some strike-slip faulting also is present. In the northern part of the province (latitude 40°N) roughly 20–25 basin-range pairs occur across 750 km.

The present day deformation field of the Basin and Range is revealed by patterns of seismicity (Figs 7.8, 7.10) and horizontal velocity estimates (Fig. 7.11) derived from continuous GPS data (Section 5.8) (Bennett *et al.*, 2003). The data show two prominent bands of high strain rate along the eastern side of the Sierra Nevada and the western side of the Colorado Plateau. These are the eastern California/central Nevada seismic belt and the Intermountain seismic belt, respectively (Fig. 7.9). Focal mechanisms (Fig. 7.10) indicate that the former accommodates both right lateral and normal displacements and the latter accommodates mostly normal motion.

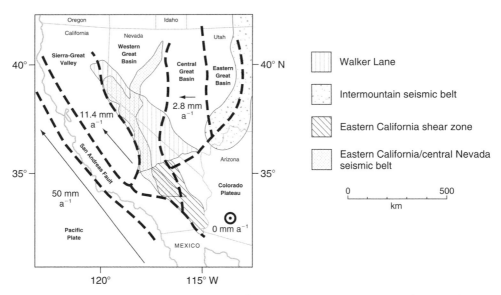

Figure 7.9 *Map showing the various tectonic provinces of the Basin and Range determined from geodetic and geologic data (modified from Bennett et al., 2003, by permission of the American Geophysical Union. Copyright © 2003 American Geophysical Union). Rates given are relative to the North American plate.*

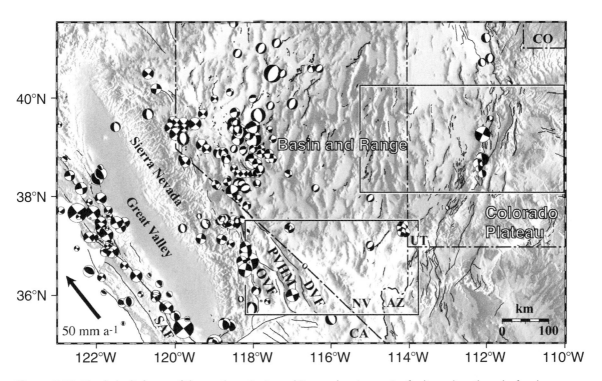

Figure 7.10 *Shaded relief map of the northern Basin and Range showing major faults and earthquake focal mechanisms (image provided by R. Bennett and modified from Bennett et al., 2003, and Shen-Tu et al., 1998, by permission of the American Geophysical Union. Copyright © 2003 and 1998 American Geophysical Union). SAF, San Andreas Fault. Northward translation of the Sierra Nevada–Great Valley microplate is accommodated by strike-slip motion on the Owens Valley (OVF), Panamint Valley–Hunter Mountain (PVHM), and Death Valley (DVF) fault zones. Black boxes show approximate area of Fig. 7.13 (lower box) and Fig. 7.14 (upper box).*

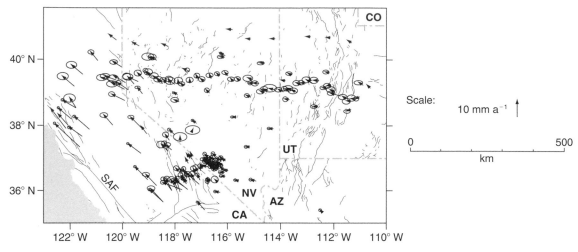

Figure 7.11 *GPS velocities of sites in the Sierra Nevada–Great Valley microplate, northern Basin and Range, and Colorado Plateau with respect to North America (image provided by R. Bennett and modified from Bennett et al., 2003, by permission of the American Geophysical Union. Copyright © 2003 American Geophysical Union). Error ellipses represent the 95% confidence level. Velocity estimates were derived from continuous GPS data from GPS networks in and around the northern Basin and Range. SAF, San Andreas Fault.*

In the intervening area, deformation is diffusely distributed and, in some places, absent from the current velocity field. Three sub-provinces, designated the eastern, central and western Great basins, show distinctive patterns of strain (Fig. 7.9). Relative motion between the central Great Basin and Colorado Plateau occurs at a rate of 2.8 mm a^{-1} and is partly accommodated by diffuse east–west extension across the eastern Great Basin. Relative motion between the Sierra Nevada–Great Valley and the central Great Basin occurs at a rate of 9.3 mm a^{-1} toward N37°W and is accommodated by diffuse deformation across the western Great Basin (Section 8.5.2). The central Great Basin records little current internal deformation. Similar patterns of distributed deformation punctuated by zones of high strain rate occur in the extensional provinces of central Greece and the Aegean Sea (Goldsworthy et al., 2002).

Two other zones of deformation in the Basin and Range have been defined on the basis of Middle Miocene–Recent geologic patterns. The Walker Lane (Fig. 7.9) displays mountain ranges of variable orientation and complex displacements involving normal faulting and both left lateral and right lateral strike-slip faulting. This belt overlaps with the Eastern California Shear Zone (Fig. 8.1 and Section 8.5.2) to the south. Hammond & Thatcher (2004) reasoned that the concentration of right lateral motion and extension within the western Basin and Range results from weak lithosphere in the Walker Lane. Linear gradients in gravitational potential energy and viscosity also may concentrate the deformation (Section 7.6.3). Together these data suggest that the broad region of the Basin and Range currently accommodates some 25% of the total strain budget between the Pacific and North American plates (Bennett et al., 1999). The data also indicate that, at least currently, deformation in the Basin and Range involves a heterogeneous combination of normal and strike-slip displacements.

The depth distribution of microearthquakes also shows that the Basin and Range Province is characterized by a seismogenic layer that is thin relative to other regions of the continent. Approximately 98% of events occur at depths less than 15 km for all of

Utah (1962–1999) and 17 km for Nevada (1990–1999) (Pancha *et al.*, 2006). This thickness of the seismogenic layer is similar to that displayed by most other rifts, including those in East Africa, except that in the Basin and Range it characterizes thousands of square kilometers of crust. The pattern implies that high geothermal gradients and crustal thinning have locally weakened a very large area.

Because deformation is distributed over such a broad region, most of the major faults in the Basin and Range have recurrence times of several thousand years (Dixon *et al.*, 2003). In the northern part of the province, several hundred faults show evidence of slip since 130 ka, yet contemporary seismicity and large historical earthquakes are clustered on only a few of them. This observation raises the possibility that a significant portion of strain is accommodated by aseismic displacements. Niemi *et al.* (2004) investigated this possibility by combining geologic data from major faults with geodetic data in the eastern Great Basin. The results suggest that both data types define a ~350 km wide belt of east–west extension over the past 130 ka. Reconciling deformation patterns measured over different timescales is a major area of research in this and most other zones of active continental tectonics.

2 *Heterogeneous crustal thinning in previously thickened crust.* Wide rifts form in regions where extension occurs in thick, weak continental crust. In the Basin and Range and the Aegean Sea the thick crust results from a history of convergence and crustal shortening that predates rifting. Virtually the entire western margin of North America was subjected to a series of compressional orogenies during Mesozoic times (Allmendinger, 1992). These events thickened sedimentary sequences that once formed part of a Paleozoic passive margin. The ancient margin is marked now by an elongate belt of shallow marine sediments of Paleozoic and Proterozoic age that thicken to the west across the eastern Great Basin and are deformed by thrust faults and folds of the Mesozoic Sevier thrust belt (Fig. 7.12). This deformation created a thick pile of weak sedimentary rocks that has contributed to a delocalization of strain (Section 7.6.1) during Cenozoic extension (Sonder & Jones, 1999). Some estimates place parts of the province at a pre-rift crustal thickness of 50 km, similar to that of the unextended Colorado Plateau (Parsons *et al.*, 1996). Others have placed it at more than 50 km (Coney & Harms, 1984). This pre-extensional history is one of the most important factors that has contributed to a heterogeneous style of extensional deformation in the Basin and Range.

The uniformity in size and spacing of normal faults in the Basin and Range, and the apparent uniform thickness of the seismogenic layer, at first suggests that strain and crustal thinning, on average, might also be uniformly distributed across the province. However, this assertion is in conflict with the results of geologic and geophysical surveys. Gilbert & Sheehan (2004) found Moho depths ranging from 30 to 40 km beneath the eastern Basin and Range (Plate 7.1a), with the thinnest crust occurring in northern Nevada and Utah (Plate 7.1b) and thicknesses of 40 km in southern Nevada (Plate 7.1c) (Plate 7.1a–c between pp. 244 and 245). Louie *et al.* (2004) also found significant variations in Moho depths with the thinnest areas showing depths of only 19–23 km beneath the Walker Lane and northwest Nevada. This southward thickening of the crust coincides with variations in the pre-Cenozoic architecture of the lithosphere, including differences in age and pre-extensional thickness. Similar nonuniform variations in crustal thickness occur beneath the Aegean Sea (Zhu *et al.*, 2006). These results illustrate that crustal thinning in wide rifts is nonuniform and, like narrow rifts, is strongly influenced by the pre-existing structure of the lithosphere.

The nonuniformity of crustal thinning in the Basin and Range is expressed in patterns of faulting within the upper crust. The Death Valley region of eastern California contains some of the youngest examples of large-magnitude extension in the world adjacent to

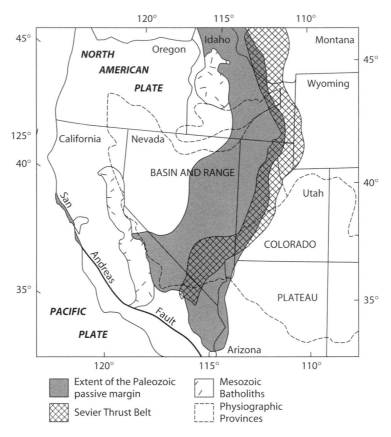

Figure 7.12 *Map of the western United States showing extent of Paleozoic passive margin sequences and the Sevier thrust belt (after Niemi et al., 2004, with permission from Blackwell Publishing).*

areas that record virtually no upper crustal strain. East–west extension beginning about ~16 Ma has resulted in ~250 km of extension between the Sierras and the Colorado Plateau (Wernicke & Snow, 1998). The intervening region responded to this divergence by developing a patchwork of relatively unextended crustal blocks separated by regions strongly deformed by extension, strike-slip faulting, and contraction (Fig. 7.13a). The heterogeneous distribution of extension is illustrated in Fig. 7.13b, which shows estimates of the thickness of the pre-Miocene upper crust that remains after extension assuming an original thickness of 15 km. In some areas, such as the Funeral and Black mountains, the upper crust has been dissected and pulled apart to such a

degree that pieces of the middle crust are exposed (Snow & Wernicke, 2000).

One of the most enigmatic characteristics of the Basin and Range Province involves local relationships between large-scale extension in the upper crust and the distribution of strain in the lower part of the crust. Some studies have shown that despite highly variable patterns of upper crustal strain, local crustal thickness appears to be surprisingly uniform (Gans, 1987; Hauser *et al.*, 1987; Jones & Phinney, 1998). This result implies that large strains have been compensated at depth by lateral flow in a weak lower crust, which acted to smooth out any Moho topography (Section 7.6.3). Park & Wernicke (2003) used magnetotelluric data to show that this lateral flow and flattening out of the Moho in the

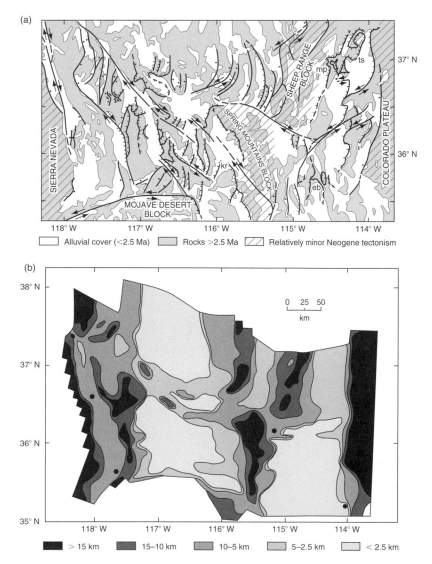

Figure 7.13 *Maps showing (a) major Cenozoic faults (heavy black lines) in the central Basin and Range Province and (b) the distribution of upper crustal thinning estimated by reconstructing Cenozoic extension using pre-extensional markers (images provided by B. Wernicke and modified from Snow & Wernicke, 2000. Copyright 2000 by* American Journal of Science. *Reproduced with permission of* American Journal of Science *in the format Textbook via Copyright Clearance Center). Symbols in (a) indicate strike-slip faults (arrows), high-angle normal faults (ball and bar symbols), low-angle normal faults (tick marks), and thrust faults (teeth). Large-magnitude detachment faults in metamorphic core complexes include the Eldorado-Black Mountains (eb), the Mormon Peak (mp), Tule Springs (ts), and the Kingston Range (kr) detachments. Contours in (b) represent the remaining thickness of a 15-km-thick pre-extensional Cenozoic upper crust, such that the lightly shaded areas represent the areas of greatest thinning. Black dots are points used in the reconstruction.*

Basin and Range probably occurred during the Miocene. By contrast, other regions, such as the Aegean Sea and the D'Entrecasteaux islands (Section 7.8.2), do not show this relationship, implying a more viscous lower crust that resists flow beneath highly extended areas.

3 *Thin mantle lithosphere and anomalously high heat flow.* Like most wide rifts, the Basin and Range is characterized by high surface heat flow, negative long-wavelength Bouguer gravity anomalies, and low crustal P_n and S_n velocities (Catchings & Mooney, 1991; Jones *et al.*, 1992; Zandt *et al.*, 1995; Chulick & Mooney, 2002). Regional topography in the Basin and Range also is unusually high with an average of 1.2 km above mean sea level. Low seismic velocities are discernible down to 300–400 km depth. Seismic tomographic models indicate that adiabatic mantle temperatures of 1300°C occur as shallow as 50 km under most of the province. For comparison, temperatures at 50–100 km in the cratonic mantle beneath the stable eastern part of North America are on average 500°C cooler than under the Basin and Range. All of these characteristics indicate a shallow asthenosphere and very thin, warm upper mantle (Goes & van der Lee, 2002).
  Temperatures at 110 km depth inferred from seismic velocity models suggest the presence of small melt and fluid pockets in the shallow mantle beneath the Basin and Range (Goes & van der Lee, 2002). Warm, low-density subsolidus mantle also may contribute to the high average elevation and large-scale variations in topography of the region. Other factors contributing to the high elevations probably include isostatic effects caused by previously thickened continental crust and magmatic intrusions. However, a lack of correlation between crustal thickness variations and surface topography indicates that simple Airy isostasy is not at play and the high elevations across the southwestern United States must involve a mantle component (Gilbert & Sheehan, 2004).
  Volcanic activity is abundant, including eruptions that occurred both before and during extension. This activity is compatible with evidence of high heat flow, elevated geotherms, and shallow asthenosphere. Pre-rift volcanism is mostly calc-alkaline in composition. Magmatism that accompanied extension is mostly basaltic. Basalts from Nevada have an isotopic signature suggesting that they were derived from sublithospheric mantle. This pattern matches evidence of mantle upwelling beneath the rift (Savage & Sheehan, 2000).

4 *Small- and large-magnitude normal faulting.* Large extensional strains and thinning of the crust in wide rifts is partly accommodated by slip on normal faults. Two contrasting patterns are evident. First, the deformation can involve distributed normal faulting where a large number of more or less regularly spaced normal faults each accommodate a relatively small amount (<10 km) of the total extension. Second, the strain may be highly localized onto a relatively small number of normal faults that accommodate large displacements of several tens of kilometers. Both patterns are common and may occur during different stages of rift evolution.
  Many of the range-bounding normal faults in the Basin and Range record relatively small offsets. These structures appear similar to those that characterize narrow rift segments. Asymmetric half graben and footwall uplifts are separated by a dominant normal fault that accommodates the majority of the strain. The morphology of these features is governed by the elastic properties of the lithosphere (Section 7.6.4) and the effects of syn-rift sedimentation and erosion. The asymmetry of the half graben and the dips of the range-bounding faults also commonly change in adjacent basin-range pairs. Many of the tectonically active faults maintain steep dips (>45°) that may penetrate through the upper crust. However, unlike the border faults of East Africa, some of the range-bounding faults of the Basin and Range exhibit geometries that involve low-angle extensional detachment faults. A few of these low-angle normal faults accommodate very large displacements and penetrate tens of

kilometers into the middle and, possibly, the lower crust.

Extensional detachment faults are low-angle ($<30°$), commonly domed fault surfaces of large areal extent that accommodate displacements of 10–50 km (Axen, 2004). The footwalls of these faults may expose a thick (0.1–3 km) ductile shear zone that initially formed in the middle or lower crust and later evolved into a frictional (brittle) slip surface as it was unroofed during the extension (Wernicke, 1981). In the Basin and Range, these features characterize regions that have been thinned to such an extent (100–400% extension) that the upper crust has been completely pulled apart and metamorphic rocks that once resided in the middle and lower crust have been exhumed. These domed regions of deeply denuded crust and detachment faulting are the hallmarks of the Cordilleran extensional *metamorphic core complexes* (Crittenden et al., 1980; Coney & Harms, 1984). Core complexes are relatively common in the Basin and Range (Figs 7.13, 7.14), although they are not unique to this province. Their ages are diverse with most forming during Late Oligocene–Middle Miocene time (Dickinson, 2002). Similar features occur in many other settings, including the southern Aegean Sea, in rifts that form above subduction zones, such as the D'Entrecasteaux Islands (Section 7.8.2), near oceanic spreading centers (Section 6.7), and in zones of extension within collisional orogens (Section 10.4.4).

Most authors view core complexes as characteristic of regions where weak crustal rheologies facilitate lateral flow in the deep crust and, in some cases, the mantle, causing upper crustal extension to localize into narrow zones (Sections 7.6.2, 7.6.5). Nevertheless, the mechanics of slip on low-angle normal faults is not well understood. Much of the uncertainty is centered on whether specific examples initially formed at low angles or were rotated from a steep orientation during deformation (Axen, 2004). The consensus is that both types probably occur (Section 7.8.2). Some low-angle,

large-offset normal faults may evolve from high-angle faults by flexural rotation (Section 7.6.4). As the hanging wall is removed by slip on the fault, the footwall is mechanically unloaded and results in isostatic uplift and doming (Buck, 1988; Wernicke & Axen, 1988). The doming can rotate the normal fault to gentler dips and lead to the formation of new high-angle faults.

The variety of Cenozoic fault patterns that typify the Basin and Range is illustrated in Fig. 7.14, which shows a segment of the eastern Great Basin in Utah and eastern Nevada (Niemi et al., 2004). The 350 km long Wasatch Fault Zone is composed of multiple segments with the largest displaying dips ranging from 35° to 70° to the west. Its subsurface geometry is not well constrained but it probably penetrates at least through the upper crust. The Sevier Desert Detachment Fault dips 12° to the west and can be traced continuously on seismic reflection profiles to a depth of at least 12–15 km (Fig. 7.14b). The range-bounding Spring Valley and Egan Range faults penetrate to at least 20 km depth and possibly through the entire 30 km thickness of the crust at angles of ~30°. The Snake Range Detachment also dips ~30° through most of the upper crust. Large-magnitude extension along the Snake Range (Miller et al., 1999) and Sevier Desert (Stockli et al., 2001) detachment faults began in Early Miocene time and Late Oligocene or Early Miocene time, respectively. In most areas, high-angle normal faults are superimposed on these older structures.

7.4 VOLCANIC ACTIVITY

7.4.1 Large igneous provinces

Many rifts and rifted margins (Section 7.7.1) are associated with the subaerial eruption of continental flood

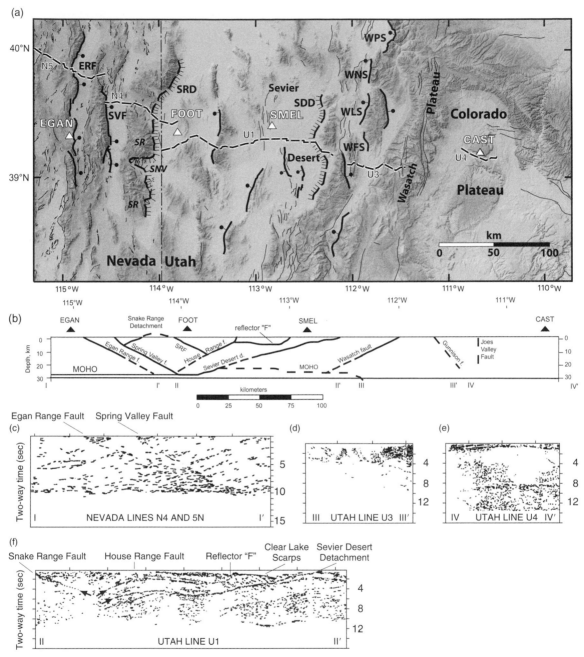

Figure 7.14 (a) Shaded relief map of a part of the eastern Basin and Range showing range-bounding faults and locations of seismic reflection profiles (black dashed lines) and GPS sites (white triangles) (image provided by N. Niemi and modified from Niemi et al., 2004, with permission from Blackwell Publishing). High-angle faults show ball and bar symbol in the hanging wall, low-angle faults show hachured pattern. Faults mentioned in the text include the Egan Range Fault (ERF), the Spring Valley Fault (SVF), the Sevier Desert Detachment (SDD), the Wasatch Fault Zone (WFS, WLS, WNS, WPS), and the Snake Range Detachment (SRD). Cross-section (b) constructed using seismic reflection data from (c) Hauser et al., 1987 and (d) Allmendinger et al., 1983 (with permission from the Geological Society of America). (e,f) Allmendinger et al., 1986 (redrawn from Allmendinger et al., 1986, by permission of the American Geophysical Union. Copyright © 1986 American Geophysical Union). SR, Snake Range Metamorphic Core Complex.

basalts. These eruptions represent one major subcategory of a broad group of rocks known as *Large Igneous Provinces* (LIPs).

Large Igneous Provinces are massive crustal emplacements of mostly mafic extrusive and intrusive rock that originated from processes different from normal sea floor spreading. LIPs may cover areas of up to several million km^2 and occur in a wide range of settings. Within oceanic plates, LIPs form oceanic plateaux such as Kerguelen and Ontong Java (Fig. 7.15). This latter example occupies an area two-thirds that of Australia. The Siberian and Columbia River basalts are examples that have erupted in the interior of continental plates. In East Africa, the Ethiopian and Kenyan flood basalts are associated with active continental rifting and the Deccan Traps in India and the Karoo basalts in southern Africa were emplaced near rifted continental margins. This diversity indicates that not all LIPs are associated with zones of extension. Within rifts their eruption can occur synchronously with rifting or million of years

prior to or after the onset of extension (Menzies *et al.*, 2002).

Estimation of the total volumes of lava in LIPs is complicated by erosion, dismemberment by sea floor spreading, and other tectonic processes that postdate their eruption. The Hawaiian Islands are well studied in this respect. Seismic data have shown that beneath the crust is a zone of rocks with a particularly high seismic velocity, which is probably derived from the same mantle source as the surface volcanic rocks. For Hawaii a basic relationship exists between velocity structure and the total volume of igneous rock (Coffin & Eldholm, 1994). This relationship has been applied to other LIPs to determine their volumes. For example, the Columbia River basalts are composed of 1.3 million km^3, whereas the Ontong Java Plateau is composed of at least 27 million km^3 of volcanic rock and possibly twice this amount (Section 5.5). These values are much higher than the continental flood basalts of East Africa. In Kenya the total volume of flood basalts has been

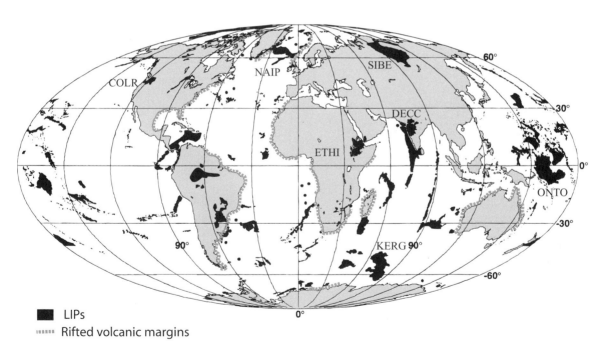

■ LIPs

▪▪▪▪▪ Rifted volcanic margins

Figure 7.15 *Map showing global distribution of Large Igneous Provinces (modified from Coffin & Edholm, 1994, by permission of the American Geophysical Union. Copyright © 1994 American Geophysical Union). Rifted volcanic margins are from Menzies* et al. *(2002). Labeled LIPs: Ethiopian flood basalts (ETHI); Deccan traps (DECC); Siberian basalts (SIBE); Kerguelen plateau (KERG); Columbia River basalts (COLR); Ontong Java (ONTO); North Atlantic igneous province (NAIP).*

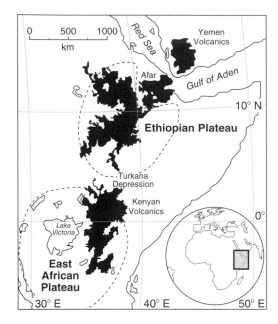

Figure 7.16 *Map showing the location of Cenozoic flood basalts of the Ethiopian Plateau and East African Plateau (Kenya Dome) (after Macdonald* et al., *2001, by permission of Oxford University Press).*

estimated at approximately 924,000 km³ (Latin *et al.*, 1993). In Ethiopia (Fig. 7.16), layers of basaltic and felsic rock reach thickness of >2 km with a total volume estimated at 350,000 km³ (Mohr & Zanettin, 1988). The eruption of such large volumes of mafic magma has severe environmental consequences, such as the formation of greenhouse gases, the generation of acid rain, and changes in sea level (Coffin & Eldholm, 1994; Ernst *et al.*, 2005). The eruptions also make significant contributions to crustal growth.

Some LIPs appear to form very quickly. For many continental flood volcanics, 70–80% of the basaltic rock erupted in less than 3 million years (Menzies *et al.*, 2002). Geochronologic studies have shown that the main flood event in Greenland (Tegner *et al.*, 1998), the Deccan Traps (Hofmann *et al.*, 2000), and the bulk of the Ethiopian Traps (Hofmann *et al.*, 1997) all erupted in less than one million years. Nevertheless, this latter example (Section 7.2) also shows that pulses of volcanism between 45 and 22 Ma contributed to the formation of the flood basalts in the Afar region. Submarine plateaux probably formed at similar rates, although less information is available from these types of LIPs. The North Atlantic Province and Ontong Java Plateau

formed in less than 3 Ma and the Kerguelen Plateau in 4.5 Ma. Most of the volcanic activity occurred in short, violent episodes separated by long periods of relative quiescence. Estimates of the average rates of formation, which include the periods of quiescence, are 12–18 km³ a⁻¹ for the Ontong Java Plateau and 2–8 km³ a⁻¹ for the Deccan Traps. Ontong Java's rate of emplacement may have exceeded the contemporaneous global production rate of the entire mid-ocean ridge system (Coffin & Eldholm, 1994).

The outpouring of large volumes of mafic magma in such short periods of time requires a mantle source. This characteristic has encouraged interpretations involving deep mantle plumes (Sections 5.5, 12.10), although the existence and importance of these features are debated widely (Anderson & Natland, 2005). Mantle plumes may form large oceanic plateaux and some continental flood basalts also may be attributed to them. Beneath the Ethiopian Plateau and the Kenya Dome (in the East African Plateau), extensive volcanism and topographic uplift appear to be the consequences of anomalously hot asthenosphere (Venkataraman *et al.*, 2004). The isotopic characteristics of the volcanic rock and the large volume of mafic lava erupted over a short period of time (Hofmann *et al.*, 1997; Ebinger & Sleep, 1998) suggest that a plume or plumes below the uplifts tap deep undegassed mantle sources (Marty *et al.*, 1996; Furman *et al.*, 2004). As the deep plumes ascend they undergo decompression melting with the amount of melt depending on the ambient pressure (Section 7.4.2). Consequently, less melting is expected under thick continental lithosphere than under thick oceanic lithosphere. Nevertheless, the sources of magma that generated many LIPs are not well understood and it is likely that no single model explains them all. Ernst *et al.* (2005) review the many aspects of LIP research and models of their formation, including links to ore deposits (Section 13.2.2).

7.4.2 Petrogenesis of rift rocks

The geochemistry of mafic volcanic rocks extruded at continental rifts provides information on the sources and mechanisms of magma generation during rifting. Rift basalts typically are enriched in the alkalis (Na_2O, K_2O, CaO), large ion lithophile elements (LILE) such as K, Ba, Rb, Sr, Pb^{2+} and the light rare earths, and volatiles, in particular CO_2 and the halogens. Tholeiitic

(a)

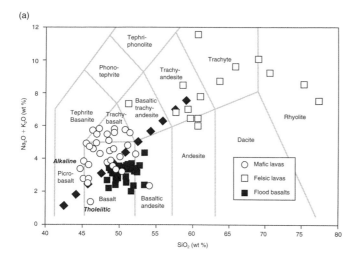

(b)

(c)

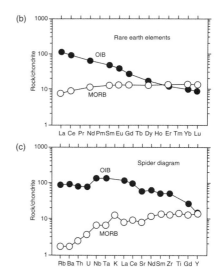

Figure 7.17 *(a) Total alkali-silica diagram showing the geochemical characteristics of lavas from Ethiopia (after Kieffer et al., 2004, by permission of Oxford University Press). Dashed line separates alkaline from tholeiitic basalts. Rare earth element (b) and spider diagram (c) showing a typical alkaline oceanic island basalt (OIB) and a typical tholeiitic mid-ocean ridge basalt (MORB)* (from Winter, John D., An Introduction to Igneous and Metamorphic Petrology, *1st edition © 2001, p. 195. Reprinted by permission of Pearson Education, Inc., Upper Saddle River, NJ).*

flood basalts also are common and may be associated with silicic lavas, including rhyolite. Observations in East Africa indicate that a continuum of mafic rocks generally occurs, including alkaline, ultra-alkaline, tholeiitic, felsic, and transitional compositions (Fig. 7.17a). This diversity reflects both the compositional heterogeneity of mantle source regions and processes that affect the genesis and evolution of mafic magma.

There are three ways in which the mantle may melt to produce basaltic liquids beneath rifts. First, melting may be accomplished by heating the mantle above the normal geotherm (Fig. 7.18a). Perturbations in the geotherm could be related to the vertical transfer of heat by deep mantle plumes. It is probable, for example, that the volcanism and topographic uplift associated with the Ethiopian and East African plateaux reflect anomalously hot mantle. Investigations of P_n wave attenuation beneath the Eastern branch of the East African Rift suggest sublithospheric temperatures that are significantly higher than those in the ambient mantle (Venkataraman *et al.*, 2004). A second mechanism for melting the mantle is to lower the

ambient pressure (Fig. 7.18b). The ascent of hot mantle during lithospheric stretching (Section 7.6.2) or the rise of a mantle plume causes a reduction in pressure that leads to decompression melting at a variety of depths, with the degree of melting depending on the rate of ascent, the geotherm, the composition of the mantle, and the availability of fluids. A third mechanism of melting involves the addition of volatiles, which has the effect of lowering the solidus temperature. All three of these mechanisms probably contribute to generation of basaltic melts beneath continental rifts.

Once formed, the composition of mafic magmas may be affected by *partial melting*. This process results in the separation of a liquid from a solid residue, which can produce a variety of melt compositions from a single mantle source. Primary mafic melts also tend to *fractionate*, whereby crystals are physically removed from melts over a wide range of crustal pressures, resulting in suites of compositionally distinctive rocks. Current models generally favor fractional crystallization of basaltic melts in shallow magma chambers as the dominant process that generates rhyolite.

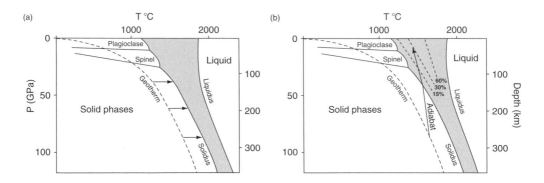

Figure 7.18 *(a) Melting by raising temperature. (b) Melting by decreasing pressure (from Winter, John D.,* An Introduction to Igneous and Metamorphic Petrology, *1st edition © 2001, p. 195. Reprinted by permission of Pearson Education, Inc., Upper Saddle River, NJ). In (b) melting occurs when the adiabat enters the shaded melting zone. Percentages of melting are shown.*

Compositional variability also reflects the *assimilation* of crustal components and *magma mixing*. The bimodal basalt-rhyolite eruptions are thought to reflect combinations of mantle and silica-rich crustal melts.

A comparison of trace element concentrations and isotopic characteristics indicates that basalts generated in continental rifts are broadly similar to those of oceanic islands (Section 5.5). Both rock types preserve evidence of a mantle source enriched in incompatible trace elements, including the LILE, and show relatively high radiogenic strontium ($^{87}Sr/^{86}Sr$) and low neodymium ($^{143}Nd/^{144}Nd$) ratios. These patterns are quite different to those displayed by mid-ocean ridge basalts, which are depleted in incompatible trace elements (Fig. 7.17b,c) and display low strontium and high neodymium ratios. Trace elements are considered *incompatible* if they are concentrated into melts relative to solid phases. Since it is not possible to explain these differences in terms of the conditions of magma genesis and evolution, the mantle from which these magmas are derived must be heterogeneous. In general, the asthenosphere is recognized as depleted in incompatible elements, but opinions diverge over whether the enriched sources originate above or below the asthenosphere. Undepleted mantle plumes offer one plausible source of enriched mantle material. Enrichment also may result from the trapping of primitive undepleted asthenosphere at the base of the lithosphere or the diffusion of LILE-rich volatiles from the asthenosphere or deeper mantle into the lithosphere.

On the basis of trace element concentrations and isotopic characteristics, Macdonald et al. (2001) inferred

that mafic magmas in the Eastern branch of the East African Rift system were derived from at least two mantle sources, one of sublithospheric origin similar to that which produces ocean island basalts and one within the subcontinental lithosphere. Contributions from the subcontinental mantle are indicated by xenoliths of lithospheric mantle preserved in lavas, distinctive rare earth element patterns, and by the mineralogy of basaltic rock. In southern Kenya, the presence of amphibole in some mafic lavas implies a magma source in the subcontinental lithosphere rather than the asthenosphere (le Roex *et al.*, 2001; Späth *et al.*, 2001). This conclusion is illustrated in Fig. 7.19 where the experimentally determined stability field of amphibole is shown together with a probable continental geotherm and adiabats corresponding to normal asthenospheric mantle and a 200°C hotter mantle plume. It is only in the comparatively cool lithospheric mantle that typical hydrous amphibole can exist. The additional requirement of garnet in the source, which is indicated by distinctive rare earth element patterns, constrains the depth of melting to 75–90 km. These and other studies show that the generation of lithospheric melts is common in rifts, especially during their early stages of development. They also indicate that the identification of melts derived from the subcontinental lithosphere provides a potentially useful tool for assessing changes in lithospheric thickness during rifting.

In addition to compositional variations related to source regions, many authors have inferred systematic relationships between basalt composition and the depth and amount of melting in the mantle beneath rifts

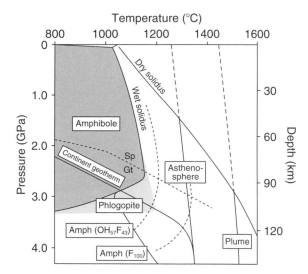

Figure 7.19 *Pressure–temperature diagram showing the stability field of amphibole (after le Roex et al., 2001, Fig. 10. Copyright © 2001, with kind permission of Springer Science and Business Media). Amphibole is stable in the subcontinental mantle but not under conditions characteristic of the asthenospheric mantle or a mantle plume. Gt, garnet; Sp, spinel.*

(Macdonald *et al.*, 2001; Späth *et al.*, 2001). Tholeiitic basalts originate from relatively large amounts of melting at shallow mantle depths of 50 km or less. Transitional basalts are produced by less melting at intermediate depths and highly alkaline magmas originate at even greater depths (100–200 km) by relatively small amounts of melting. These relationships, and the general evolution of mafic magmas toward mid-oceanic ridge compositions as rifting progresses to sea floor spreading, imply a decrease in the depth of melting and a coincident increase in the amount of melting with time. In support of this generalization, tomographic images from East Africa show the presence of small melt fractions in relatively thick mantle lithosphere below juvenile rift segments, such as those in northern Tanzania and Kenya (Green *et al.*, 1991; Birt *et al.*, 1997). Larger melt fractions occur at shallower depths beneath more mature rift segments, such as those in northern Ethiopia and the Afar Depression (Bastow *et al.*, 2005). However, as discussed below, compositional trends in basaltic lavas erupted at continental rifts may not follow a simple progression, especially prior to lithospheric rupture.

Although there may be broad trends of decreasing alkalinity with time, defining systematic compositional trends in basalts is often difficult to achieve at the local and regional scales. For example, attempts to document a systematic decrease in the degree of lithospheric contamination as rifting progresses have proven elusive. Such a decrease might be expected if, as the lithosphere thins and eventually ruptures, melts from the sublithospheric mantle begin to penetrate the surface without significant interaction with lithosphere-derived melts. However, studies in Kenya and Ethiopia show no systematic temporal or spatial patterns in the degree of lithospheric contamination in rift basalts (Macdonald *et al.*, 2001). This indicates that rift models involving the progressive evolution of alkaline magmas toward more tholeiitic magmas during the transition to sea floor spreading are too simplistic. Instead, the data suggest that the full compositional range of mafic melts can coexist in continental rifts and that magma genesis may involve multiple sources at any stage of the rifting process. Tholeiites, for example, commonly are present during all stages of rifting and can precede the generation of alkaline and transitional basalts.

7.4.3 Mantle upwelling beneath rifts

The three-dimensional velocity structure of the upper mantle beneath rifts can be ascertained using teleseismic travel-time delays and seismic tomography. Davis & Slack (2002) modeled these types of data from beneath the Kenya Dome using two Gaussian surfaces that separate undulating layers of different velocities (Plate 7.2 between pp. 244 and 245). An upper layer (mesh surface) peaks at the Moho beneath the rift valley and has a velocity contrast of −6.8% relative to 8 km s^{-1} mantle. A lower layer (grayscale surface) peaks at about 70 km depth and has a −11.5% contrast extending to a depth of about 170 km. This model, which is in good agreement with the results of seismic refraction studies, shows a domal upper mantle structure with sides that dip away from the center of the Kenya Rift. The authors suggested that this structure results from the separation of upwelling asthenosphere into currents that impinge on the base of the lithosphere and form a low velocity, low density zone of melting between 70 and 170 km depth.

Park & Nyblade (2006) used teleseismic P-wave travel times to image the upper mantle beneath the

Eastern branch of the East African Rift system to depths of 500 km. They found a steep-sided, west-dipping low velocity anomaly that is similar to the one modeled by Davis & Slack (2002) above 160 km depth. Below this depth, the anomaly broadens to the west indicating a westerly dip. Similar structures have been imaged below Tanzania (Ritsema *et al.*, 1998; Weeraratne *et al.*, 2003) and parts of Ethiopia (Benoit *et al.*, 2006). Bastow *et al.* (2005) found that a tabular (75 km wide) low velocity zone below southern Ethiopia broadens at depths of >100 km beneath the more highly extended northern section of the rift (Fig. 7.7c,d). The anomalies are most pronounced at ~150 km depth. These broad, dipping structures are difficult to reconcile with models of a simple plume with a well-defined head and tail. Instead they appear to be more consistent with either multiple plumes or tomographic models (Plate 7.3 between pp. 244 and 245) where the hot asthenosphere connects to a broad zone of anomalously hot mantle beneath southern Africa.

In the deep mantle below South Africa, Ritsema *et al.* (1999) imaged a broad (4000 by 2000 km^2 area) low velocity zone extending upward from the core–mantle boundary and showed that it may have physical links to the low velocity zones in the upper mantle beneath East Africa (Plate 7.3 between pp. 244 and 245). The tilt of the deep velocity anomaly shows that the upwelling is not vertical. Between 670 and 1000-km depth the anomaly weakens, suggesting that it may be obstructed. These observations support the idea that anomalously hot asthenosphere beneath Africa is related in some way to this broad deep zone of upwelling known as the African superswell (Section 12.8.3). Nevertheless, a consensus on the location, depth extent and continuity of hot mantle material below the East African Rift system has yet to be reached (*cf.* Montelli *et al.*, 2004a).

A comparison of the mantle structure beneath rifts in different settings indicates that the size and strength of mantle upwellings are highly variable. Achauer & Masson (2002) showed that in relatively cool rifts, such as the Baikal Rift and the southern Rhine Graben, low velocity zones are only weakly negative (−2.5% relative to normal mantle P-wave velocities) and occur mostly above depths of 160 km. In these relatively cool settings, the low velocity zones in the uppermost mantle show no continuation to deeper levels (>160 km) and no broadening of an upwelling asthenosphere with depth below the rift. In still other settings, such as the Rio Grande rift, low velocity zones in the upper mantle may form parts of small-scale convection cells where upwell-ing occurs beneath the rift and downwelling beneath its margins (Gao *et al.*, 2004).

7.5 RIFT INITIATION

Continental rifting requires the existence of a horizontal deviatoric tensional stress that is sufficient to break the lithosphere. The deviatoric tension may be caused by stresses arising from a combination of sources, including: (i) plate motions; (ii) thermal buoyancy forces due to asthenospheric upwellings; (iii) tractions at the base of the lithosphere produced by convecting asthenosphere; and/or (iv) buoyancy (gravitational) forces created by variations in crustal thickness (Huismans *et al.*, 2001). These stresses may be inherited from a previous tectonic regime or they may develop during extension. Full rupture of the lithosphere leading to the formation of a new ocean basin only occurs if the available stresses exceed the strength of the entire lithosphere. For this reason lithospheric strength is one of the most important parameters that governs the formation and evolution of continental rifts and rifted margins.

The horizontal force required to rupture the entire lithosphere can be estimated by integrating yield stress with respect to depth. The integrated yield stress, or lithospheric strength, is highly sensitive to the geothermal gradient as well as to crustal composition and crustal thickness (Section 2.10.4). A consideration of these factors suggests that a force of $3 \times 10^{13}\,N\,m^{-1}$ may be required to rupture lithosphere with a typical heat flow value of $50\,mW\,m^{-2}$ (Buck *et al.*, 1999). In areas where lithosphere exhibits twice the heat flow, such as in the Basin and Range Province, it may take less than $10^{12}\,N\,m^{-1}$ (Kusznir & Park, 1987; Buck *et al.*, 1999). Several authors have estimated that the tectonic forces available for rifting are in the range $3–5 \times 10^{12}\,N\,m^{-1}$ (Forsyth & Uyeda, 1975; Solomon *et al.*, 1975). If correct, then only initially thin lithosphere or lithosphere with heat flow values greater than $65–70\,mW\,m^{-2}$ is expected to undergo significant extension in the absence of any other weakening mechanism (Kusznir & Park, 1987). Elsewhere, magmatic intrusion or the addition of water may be required to sufficiently weaken the lithosphere to allow rifting to occur.

Another important factor that controls whether rifting occurs, is the mechanism that is available to

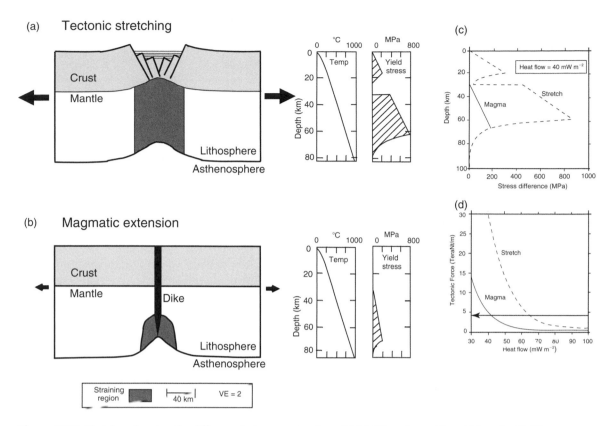

Figure 7.20 *Sketches showing the difference between extension of thick lithosphere without (a) and with (b) magmatic intrusion by diking. Temperature and yield stress curves for each case are show to the right of the sketches. VE, vertical exaggeration. (c) Example of yield stresses for strain rate $10^{-14}\,s^{-1}$ for 30-km-thick crust. Solid line, stress difference for magmatic rifting; dashed line, stress difference for lithospheric stretching. (d) Tectonic force for rifting with and without magma as a function of heat flow. The bold black line in (d) shows the estimated value of driving forces (from Buck, 2004. Copyright © 2004 from Columbia University Press. Reprinted with permission of the publisher).*

accommodate the extension. At any depth, deviatoric tension can cause yielding by faulting, ductile flow, or dike intrusion, depending on which of these processes requires the least amount of stress. For example, if a magma source is available, then the intrusion of basalt in the form of vertical dikes could permit the lithosphere to separate at much lower stress levels than is possible without the diking. This effect occurs because the yield stress that is required to allow basaltic dikes to accommodate extension mostly depends on the density difference between the lithosphere and the magma (Buck, 2004). By contrast, the yield stresses required to cause faulting or ductile flow depend upon many other factors that result in yield strengths that can be up to an order of magnitude greater than those required for

lithospheric separation by diking (Fig. 7.20). High temperatures (>700°C) at the Moho, such as those that can result from the thermal relaxation of previously thickened continental crust, also may contribute to the tectonic forces required for rift initiation. For high Moho temperatures gravitational forces become increasingly important contributors to the stresses driving rifting.

Finally, the location and distribution of strain at the start of rifting may be influenced by the presence of pre-existing weaknesses in the lithosphere. Contrasts in lithospheric thickness or in the strength and temperature of the lithosphere may localize strain or control the orientations of rifts. This latter effect is illustrated by the change in orientation of the Eastern branch of the East African Rift system where the rift axis meets the cool, thick lithospheric root of the Archean Tanzanian craton (Section 7.8.1). The Tanzanian example suggests that lateral heterogeneities at the lithosphere–asthenosphere boundary rather than shallow level structures in the crust are required to significantly alter rift geometry (Foster *et al.*, 1997).

7.6 STRAIN LOCALIZATION AND DELOCALIZATION PROCESSES

7.6.1 Introduction

The localization of strain into narrow zones during extension is achieved by processes that lead to a mechanical weakening of the lithosphere. Lithospheric weakening may be accomplished by the elevation of geotherms during lithospheric stretching, heating by intrusions, interactions between the lithosphere and the asthenosphere, and/or by various mechanisms that control the behavior of faults and shear zones during deformation. Working against these *strain softening* mechanisms are processes that promote the mechanical strengthening of the lithosphere. Lithospheric strengthening may be accomplished by the replacement of weak crust by strong upper mantle during crustal thinning and by the crustal thickness variations that result from extension. These and other *strain hardening* mechanisms promote the delocalization of strain during rifting. Competition among these mechanisms, and whether they result in a net weakening or a net strengthening of the lithosphere, controls the evolution of deformation patterns within rifts.

To determine how different combinations of lithospheric weakening and strengthening mechanisms control the response of the lithosphere to extension, geoscientists have developed physical models of rifting using different approaches. One approach, called *kinematic modeling*, involves using information on the geometry, displacements, and type of strain to make predictions about the evolution of rifts and rifted

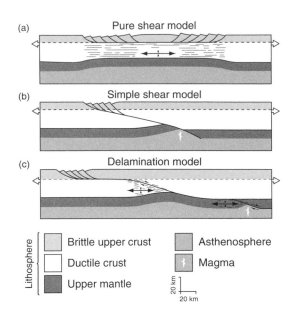

Figure 7.21 *Kinematic models of continental extension (after Lister et al., 1986, with permission from the Geological Society of America).*

margins. Figures 7.4c, 7.10, and 7.11 illustrate the data types that frequently are used to generate these types of models. Among the most common kinematic examples are the *pure shear* (McKenzie, 1978), the *simple shear* (Wernicke, 1985), and the *crustal delamination* (Lister *et al.*, 1986) models of extension (Fig. 7.21). The predictions from these models are tested with observations of subsidence and uplift histories within rifts and rifted margins, and with information on the displacement patterns recorded by faults and shear zones. This approach has been used successfully to explain differences in the geometry of faulting and the history of extension among some rifts and rifted margins. However, one major limitation of kinematic modeling is that it does not address the underlying causes of these differences. By contrast, *mechanical models* employ information about the net strength of the lithosphere and how it changes during rifting to test how different physical processes affect rift evolution. This latter approach permits inhomogeneous strains and a quantitative evaluation of how changes to lithospheric strength and rheology influence rift behavior. The main physical processes involved in rifting and their effects on the evolution of the lithosphere are discussed in this section.

7.6.2 Lithospheric stretching

During horizontal extension, lithospheric stretching results in a vertical thinning of the crust and an increase in the geothermal gradient within the zone of thinning (McKenzie, 1978). These two changes in the physical properties of the extending zone affect lithospheric strength in contrasting ways. Crustal thinning or *necking* tends to strengthen the lithosphere because weak crustal material is replaced by strong mantle lithosphere as the latter moves upward in order to conserve mass. The upward movement of the mantle also may result in increased heat flow within the rift. This process, called *heat advection*, results in higher heat flow in the rift because the geotherms become compressed rather than through any addition of heat. The compressed geotherms tend to result in a net weakening of the lithosphere, whose integrated strength is highly sensitive to temperature (Section 2.10). However, the weakening effect of advection is opposed by the diffusion of heat away from the zone of thinning as hot material comes into contact with cooler material. If the rate of heat advection is faster than the rate of thermal diffusion and cooling then isotherms at the base of the crust are compressed, the geotherm beneath the rift valley increases, and the integrated strength of the lithosphere decreases. If thermal diffusion is faster, isotherms and crustal temperatures move toward their pre-rift configuration and lithospheric weakening is inhibited.

England (1983) and Kusznir & Park (1987) showed that the integrated strength of the lithosphere in rifts, and competition between cooling and heat advection mechanisms, is strongly influenced by the rate of extension. Fast strain rates ($10^{-13}\,s^{-1}$ or $10^{-14}\,s^{-1}$) result in larger increases in geothermal gradients than slow rates ($10^{-16}\,s^{-1}$) for the same amount of stretching. This effect suggests that high strain rates tend to localize strain because inefficient cooling keeps the thinning zone weak, allowing deformation to focus into a narrow zone. By contrast, low strain rates tend to delocalize strain because efficient cooling strengthens the lithosphere and causes the deformation to migrate away from the center of the rift into areas that are more easily deformable. The amount of net lithospheric weakening or strengthening that results from any given amount of stretching also depends on the initial strength of the lithosphere and on the total amount of extension. The total amount of thinning during extension usually is described by the stretching factor (β), which is the ratio of the initial and final thickness of the crust (McKenzie, 1978).

The thermal and mechanical effects of lithospheric stretching at different strain rates are illustrated in Fig. 7.22, which shows the results of two numerical experiments conducted by van Wijk & Cloetingh (2002). In these models, the lithosphere is divided into an upper crust, a lower crust, and a mantle lithosphere that have been assigned different rheological properties (Fig. 7.22a). Figures 7.22b–d show the thermal evolution of the lithosphere for uniform extension at a rate of $16\,mm\,a^{-1}$. At this relatively fast rate, heating by thermal advection outpaces thermal diffusion, resulting in increased temperatures below the rift and strain localization in the zone of thinning. As the crust thins, narrow rift basins form and deepen. Changes in stretching factors for the crust (β) and mantle (δ) are shown in Fig. 7.22e,f. The total strength of the lithosphere (Fig. 7.22g), obtained by integrating the stress field over the thickness of the lithosphere, gradually decreases with time due to stretching and the strong temperature dependence of the chosen rheologies. Eventually, at very large strains, the thermal anomaly associated with rifting is expected to dissipate. These and many other models of rift evolution that are based on the principles of lithospheric stretching approximate the subsidence patterns measured in some rifts and at some rifted continental margins (van Wijk & Cloetingh, 2002; Kusznir et al., 2004) (Section 7.7.3).

The experiment shown in Fig. 7.22h–j shows the evolution of rift parameters during lithospheric stretching at the relatively slow rate of $6\,mm\,a^{-1}$. During the first 30 Ma, deformation localizes in the center of the rift where the lithosphere is initially weakened as isotherms and mantle material move upward. However, in contrast with the model shown in Fig. 7.22b–d, temperatures begin to decrease with time due to the efficiency of conductive cooling at slow strain rates. Mantle upwelling in the zone of initial thinning ceases and the lithosphere cools as temperatures on both sides of the central rift increase. At the same time, the locus of thinning shifts to both sides of the first rift basin, which does not thin further as stretching continues. The mantle thinning factor (Fig. 7.22l) illustrates this behavior. During the first 45 Ma, upwelling mantle causes δ to be larger in the central rift than its surroundings. After this time, δ decreases in the central rift as new upwelling zones develop on its sides. The total strength of the lithosphere (Fig. 7.22m) for this low strain rate model shows that the central rift is weakest until about 55 Ma.

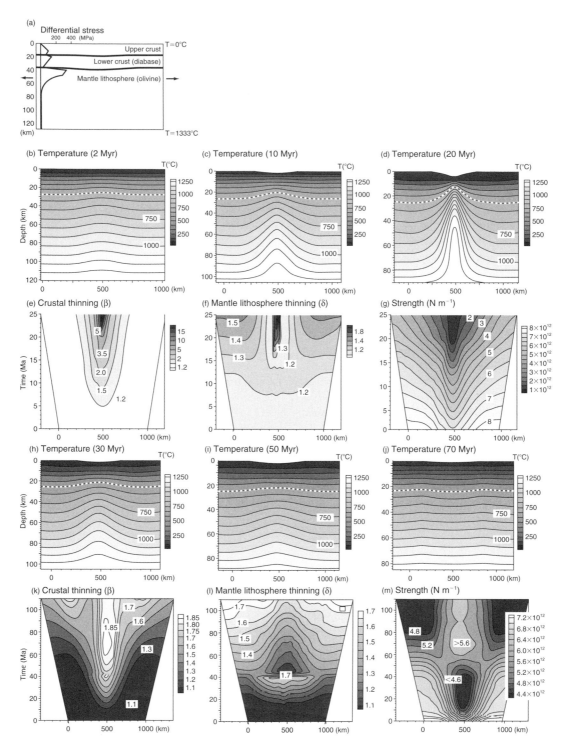

Figure 7.22 (a) Three-layer lithospheric model where the base of the lithosphere is defined by the 1300°C isotherm at 120 km. Differential stress curves show a strong upper crust and upper mantle and a lower crust that weakens with depth. Thermal evolution of the lithosphere (b–d) during stretching for a horizontal extensional velocity of 16 mm a⁻¹. Evolution of lithospheric strength (g) and of thinning factors for the crust (e) and mantle (f) for a velocity of 16 mm a⁻¹. Thermal evolution of the lithosphere (h–j) during stretching for a velocity of 6 mm a⁻¹. Evolution of lithospheric strength (m) and of thinning factors for the crust (k) and mantle (l) for a velocity of 6 mm a⁻¹ (image provided by J. van Wijk and modified from van Wijk & Cloetingh, 2002, with permission from Elsevier).

After this time the weakest areas are found on both sides of the central rift basin. This model shows how the strong dependence of lithospheric strength on temperature causes strain delocalization and the formation of wide rifts composed of multiple rift basins at slow strain rates. The model predicts that continental break-up will not occur for sufficiently slow rift velocities.

7.6.3 Buoyancy forces and lower crustal flow

In addition to crustal thinning and the compression of geotherms (Section 7.6.2), lithospheric stretching results in two types of buoyancy forces that influence strain localization during rifting. First, lateral variations in temperature, and therefore density, between areas inside and outside the rift create a *thermal buoyancy* force that adds to those promoting horizontal extension (Fig. 7.23). This positive reinforcement tends to enhance those aspects of lithospheric stretching (Section 7.6.2) that promote the localization of strain. Second, a *crustal buoyancy* force is generated by local (Airy) isostatic effects as the crust thins and high density material is brought to shallow levels beneath the rift (Fleitout & Froidevaux, 1982). Because the crust is less dense than the underlying mantle, crustal thinning lowers surface elevations in the center of the rift (Fig. 7.23). This subsidence places the rift into compression, which opposes the forces driving extension. The opposing force makes it more difficult to continue deforming in the same locality, resulting in a delocalization of strain as the deformation migrates into areas that are more easily deformable (Buck, 1991).

Several processes may either reduce or enhance the effects of crustal buoyancy forces during lithospheric stretching. Buck (1991) and Hopper & Buck (1996) showed that where the crust is initially thin and cool, and the mantle lithosphere is relatively thick, the overall strength (the effective viscosity) of the lithosphere remains relatively high under conditions of constant strain rate (Fig. 7.24a). In this case, the effects of crustal buoyancy forces are reduced and the thermal effects of lithospheric necking are enhanced. Narrow rifts result because the changes in yield strength and thermal buoyancy forces that accompany lithospheric stretching dominate the force balance, causing extensional strains to remain localized in the region of necking. By contrast, where the crust is initially thick and hot, and the mantle lithosphere is relatively thin, the overall strength of the lithosphere remains relatively low. In this case, crustal buoyancy forces dominate because the amount of possible weakening due to lithospheric necking is relatively small, resulting in strain delocalization and the formation of wide zones of rifting (Fig. 7.24b) as the necking region migrates to areas that require less force to deform. These models illustrate how crustal thickness and the thermal state of the lithosphere at the start of rifting greatly influence the style of extension.

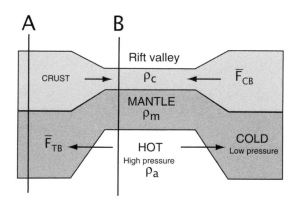

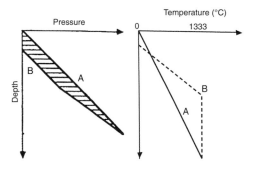

Figure 7.23 *Schematic diagram illustrating thermal and crustal buoyancy forces generated during rifting. A and B represent vertical profiles outside and inside the rift valley, respectively. Pressure and temperature as a function of depth for each profile are shown to the right of sketch (modified from Buck, 1991, by permission of the American Geophysical Union. Copyright © 1991 American Geophysical Union). Differences in profiles generate lateral buoyancy forces.*

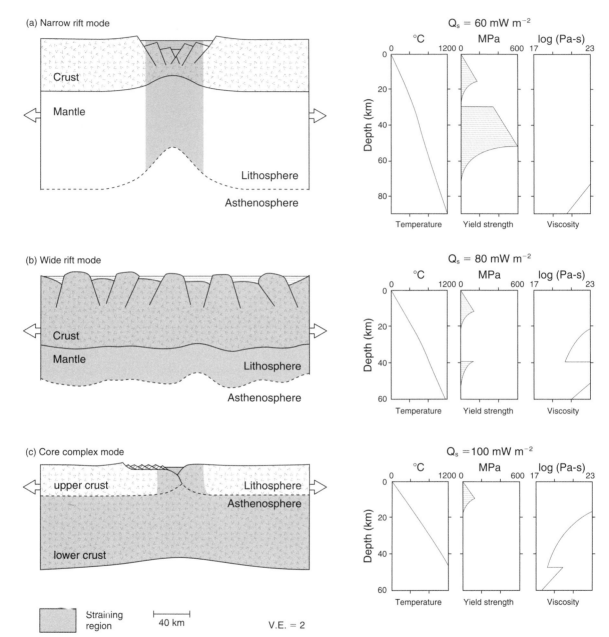

Figure 7.24 *Sketches of the lithosphere illustrating three modes of extension emphasizing the regions undergoing the greatest amount of extensional strain (modified from Buck, 1991, by permission of the American Geophysical Union. Copyright © 1991 American Geophysical Union). (a) Narrow mode, (b) wide mode, (c) core complex mode. Lithosphere is defined as areas with effective viscosities of $>10^{21}$ Pa s^{-1}. The plots to the right of each sketch show initial model geotherms, yield strengths (for a strain rate of 8×10^{-15} s^{-1}) and effective viscosities for a dry quartz crust overlying a dry olivine mantle. From top to bottom the crustal thicknesses are 30 km, 40 km, and 50 km. Q_s, initial surface heat flow. (c) shows layers labeled at two scales: the upper crust and lower crust labels on the left side of diagram show a weak, deforming lower crust (shaded); the lithosphere and asthenosphere labels on the right side of diagram show a scale emphasizing that the zone of crustal thinning (shaded column) is localized into a relatively narrow zone of weak lithosphere.*

Models of continental extension that emphasize crustal buoyancy forces incorporate the effects of ductile flow in the lower crust. Buck (1991) and Hopper & Buck (1996) showed that the pressure difference between areas inside and outside a rift could cause the lower crust to flow into the zone of thinning if the crust is thick and hot. Efficient lateral flow in a thick, hot, and weak lower crust works against crustal buoyancy forces by relieving the stresses that arise from variations in crustal thickness. This effect may explain why the present depth of the Moho in some parts of the Basin and Range Province, and therefore crustal thickness, remains fairly uniform despite the variable amounts of extension observed in the upper crust (Section 7.3). In cases where low yield strengths and flow in the lower crust alleviate the effects of crustal buoyancy, the zone of crustal thinning can remain fixed as high strains build up near the surface. Buck (1991) and Hopper & Buck (1996) defined this latter style of deformation as core complex-mode extension (Fig. 7.24c). Studies of flow patterns in ancient lower crust exposed in metamorphic core complexes (e.g. Klepeis *et al.*, 2007) support this view.

The relative magnitudes of the thermal and crustal buoyancy forces may be affected by two other parameters: strain rate and strain magnitude. Davis & Kusznir (2002) showed that the strain delocalizing effects of the crustal buoyancy force are important at low strain rates, when thermal diffusion is relatively efficient (e.g. Fig. 7.22h–j), and after long (>30 Myr) periods of time. In addition, thermal buoyancy forces may dominate over crustal buoyancy forces immediately after rifting when strain magnitudes are relatively low. This latter effect occurs because variations in crustal thicknesses are relatively small at low stretching (β) factors. This study, and the work of Buck (1991) and Hopper & Buck (1996), suggests that shifts in the mode of extension are expected as continental rifts evolve through time and the balance of thermal and crustal forces within the lithosphere changes.

7.6.4 Lithospheric flexure

Border faults that bound asymmetric rift basins with uplifted flanks are among the most common features in continental rifts (Fig. 7.25). Some aspects of this characteristic morphology can be explained by the elastic response of the lithosphere to regional loads caused by normal faulting.

Plate flexure (Section 2.11.4) describes how the lithosphere responds to long-term (>10^5 years) geologic loads. By comparing the flexure in the vicinity of

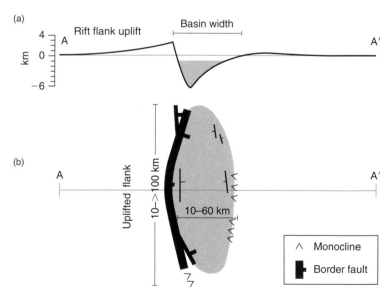

Figure 7.25 *Generalized form of an asymmetric rift basin showing border fault in (a) cross-section and (b) plan view (after Ebinger et al., 1999, with permission from the Royal Society of London). Line of section (A–A') shown in (b). Shading in (b) shows areas of depression.*

different types of load it has been possible to estimate the effective long-term elastic thickness (T_e) of continental lithosphere (Section 2.12) using forward models of topography and gravity anomaly profiles (Weissel & Karner, 1989; Petit & Ebinger, 2000). The value of T_e in many rifts, such as the Basin and Range, is low (4 km) due to the weakening effects of high geothermal gradients. However, in other rifts, including those in East Africa and in the Baikal Rift, the value of T_e exceeds 30 km in lithosphere that is relatively strong (Ebinger *et al.*, 1999). The physical meaning of T_e, and its relationship to the thickness (T_s) of the seismogenic layer, is the subject of much discussion. Rheological considerations based on data from experimental rock mechanics suggest that T_e reflects the integrated brittle, elastic, and ductile strength of the lithosphere. It, therefore, is expected to differ from the seismogenic layer thickness, which is indicative of the depth to which short term (periods of years) anelastic deformation occurs as unstable frictional sliding (Watts & Burov, 2003). For these reasons, T_e typically is larger than T_s in stable continental cratons and in many continental rifts.

The deflection of the crust by slip on normal faults generates several types of vertical loads. A mechanical unloading of the footwall occurs as crustal material in the overlying hanging wall is displaced downward and the crust is thinned. This process creates a buoyancy force that promotes surface uplift. Loading of the hanging wall may occur as sediment and volcanic material are deposited into the rift basin. These loads combine with those that are generated during lithospheric stretching (Section 7.6.2). Loads promoting surface uplift are generated by increases in the geothermal gradient beneath a rift, which leads to density contrasts. Loads promoting subsidence may be generated by the replacement of thinned crust by dense upper mantle and by conductive cooling of the lithosphere if thermal diffusion outpaces heating.

Weissel & Karner (1989) showed that flexural isostatic compensation (Section 2.11.4) following the mechanical unloading of the lithosphere by normal faulting and crustal thinning leads to uplift of the rift flanks. The width and height of the uplift depend upon the strength of the elastic lithosphere and, to a lesser extent, on the stretching factor (β) and the density of the basin infill. Other factors may moderate the degree and pattern of the uplift, including the effects of erosion, variations in depth of lithospheric necking (van der Beek & Cloetingh, 1992; van der Beek, 1997) and, possibly, small-scale convection in the underlying mantle

(Steckler, 1985). Ebinger *et al.* (1999) showed that increases in the both T_e and T_s in several rift basins in East Africa and elsewhere systematically correspond to increases in the length of border faults and rift basin width. As the border faults grow in size, small faults form to accommodate the monoclinal bending of the plate into the depression created by slip on the border fault (Fig. 7.25). The radius of curvature of this bend is a measure of flexural rigidity. Strong plates result in a narrow deformation zone with long, wide basins and long border faults that penetrate deeper into the crust. Weak plates result in a very broad zone of deformation with many short, narrow basins and border faults that do not penetrate very deeply. These studies suggest that the rheology and flexural rigidity of the upper part of the lithosphere control several primary features of rift structure and morphology, especially during the first few million years of rifting. They also suggest that the crust and upper mantle may retain considerable strength in extension (Petit & Ebinger, 2000).

Lithospheric flexure also plays an important role during the formation of large-magnitude normal faults (Section 7.3). Large displacements on both high- and low-angle fault surfaces cause isostatic uplift of the footwall as extension proceeds, resulting in dome-shaped fault surfaces (Buck *et al.*, 1988; Axen & Bartley, 1997; Lavier *et al.*, 1999; Lavier & Manatschal, 2006). Lavier & Manatschal (2006) showed that listric fault surfaces whose dip angle decreases with depth (i.e. concave upward faults) are unable to accommodate displacements large enough (>10 km) to unroof the deep crust. By contrast, low-angle normal faults whose dips increase with depth (i.e. concave downward faults) may unroof the deep crust efficiently and over short periods of time if faulting is accompanied by a thinning of the middle crust and by the formation of serpentinite in the lower crust and upper mantle. The thinning and serpentinization weaken the crust and minimize the force required to bend the lithosphere upward during faulting, allowing large magnitudes of slip.

7.6.5 Strain-induced weakening

Although differences in the effective elastic thickness and flexural strength of the lithosphere (Section 7.6.4) may explain variations in the length of border faults and the width of rift basins, they have been much less successful at explaining another major source of variability in rifts: the degree of strain localization in faults and shear zones. In some settings normal faulting is widely distributed across large areas where many faults accommodate a relatively small percentage of the total extension (Section 7.3). However, in other areas or at different times, extension may be highly localized on relatively few faults that accommodate a large percentage of the total extension. Two approaches have been used to explain the causes of this variability. The first incorporates the effects of a strain-induced weakening of rocks that occurs during the formation of faults and shear zones. A second approach, discussed in Section 7.6.6, shows how vertical contrasts in the rheology of crustal layers affect the localization and delocalization of strain during extension.

In order for a normal fault to continue to slip as the crust is extended it must remain weaker than the surrounding rock. As discussed in Section 7.6.4, the deflection of the crust by faulting changes the stress field surrounding the fault. Assuming elastic behavior, Forsyth (1992) showed that these changes depend on the dip of the fault, the amount of offset on the fault, and the inherent shear strength or *cohesion* of the faulted material. He argued that the changes in stresses by normal faulting increase the yield strength of the layer and inhibit continued slip on the fault. For example, slip on high-angle faults create surface topography more efficiently than low-angle faults, so more work is required for large amounts of slip on the former than on the latter. These processes cause an old fault to be replaced with a new one, leading to a delocalization of strain. Buck (1993) showed that if the crust is not elastic but can be described with a finite yield stress (elastic-plastic), then the amount of slip on an individual fault for a given cohesion depends on the thickness of the elastic-plastic layer. In this model the viscosity of the elastic-plastic layer is adjusted so that it adheres to the Mohr–Coulomb criterion for brittle deformation (Section 2.10.2). For a brittle layer thickness of >10 km and a reasonably low value of cohesion a fault may slip only a short distance (a maximum of several kilometers) before a new one replaces it. If the brittle layer is very thin, then the offset magnitude can increase because the increase in yield strength resulting from changes in the stress field due to slip is small.

Although layer thickness and its inherent shear strength play an important role in controlling fault patterns, a key process that causes strain localization and may lead to the formation of very large offset (tens of kilometers) faults is a reduction in the cohesion of the faulted material. During extension, cohesion can be reduced by a number of factors, including increased fluid pressure (Sibson, 1990), the formation of fault gouge, frictional heating (Montési & Zuber, 2002), mineral transformations (Bos & Spiers, 2002), and decreases in strain rate (Section 2.10). Lavier *et al.* (2000) used simple two-layer models to show that the formation of a large-offset normal fault depends on two parameters: the thickness of the brittle layer and the rate at which the cohesion of the layer is reduced during faulting (Plate 7.4a,b between pp. 244 and 245). The models include an upper layer of uniform thickness overlying a ductile layer having very little viscosity. In the ductile layer the yield stress is strain-rate- and temperature-dependent following dislocation creep flow laws (Section 2.10.3). In the upper layer brittle deformation is modeled using an elastic-plastic rheology. The results show that where the brittle layer is especially thick (>22 km) extension always leads to multiple normal faults (Plate 7.4c between pp. 244 and 245). In this case the width of the zone of faulting is equivalent to the thickness of the brittle layer. However, for small brittle layer thicknesses (<22 km), the fault pattern depends on how fast cohesion is reduced during deformation (Plate 7.4d,e between pp. 244 and 245). To obtain a single large-offset fault, the rate of weakening must be high enough to overcome the resistance to continued slip on the fault that results from flexural bending.

These studies provide some insight into how layer thickness and the loss of cohesion during faulting control the distribution of strain, its symmetry, and the formation of large-offset faults. However, at the scale of rifts, other processes also impact fault patterns. In ductile shear zones changes in mineral grain size may promote a switch from dislocation creep to grain-size-sensitive diffusion creep (Section 2.10.3), which can reduce the yield strengths of layers in the crust and mantle. In addition, the rate at which a viscous material flows has an important effect on the overall strength of the material. The faster it flows, the larger the stresses that are generated by the flow and the stronger the material becomes. This latter process may counter the effects of cohesion loss during faulting and could result in a net strengthening of the lithosphere by increasing the depth of the brittle–ductile transition (Section 2.10.4). At the scale of the lithosphere, it therefore becomes necessary to examine the interplay among the various weakening mechanisms in both brittle and ductile layers in order to reproduce deformation patterns in rifts.

Huismans & Beaumont (2003, 2007) extended the work of Lavier et al. (2000) by investigating the effects of strain-induced weakening in both brittle (frictional-plastic) and ductile (viscous) regimes on deformation patterns in rifts at the scale of the lithosphere and over time periods of millions of years. This study showed that strain softening in the crust and mantle can produce large-offset shear zones and controls the overall symmetry of the deformation. Figure 7.26a shows a simple three-layer lithosphere where brittle deformation is modeled by using a frictional-plastic rheology that, as in most physical experiments, is adjusted so that it adheres to the Mohr–Coulomb failure criterion. Ductile deformation is modeled using a thermally activated power law rheology. During each experiment, ambient conditions control whether the deformation is frictional-plastic (brittle) or viscous (ductile). Viscous flow occurs when the state of stress falls below the frictional-plastic yield point. Variations in the choice of crustal rheology also allow an investigation of cases where the crust is either coupled or decoupled to the mantle lithosphere. Coupled models involve deformation that is totally within the frictional-plastic regime. Decoupled models involve a moderately weak viscous lower crust. Strain-induced weakening is specified by linear changes in the effective angle of internal friction (Section 2.10.2) for frictional-plastic deformation and in the effective viscosity for viscous deformation. The deformation is seeded using a small plastic weak region.

A reference model (Fig. 7.26b,c) shows how a symmetric style of extensional deformation results when strain softening is absent. An early phase of deformation is controlled by two conjugate frictional-plastic shear zones (S1A/B) that are analogous to faults and two forced shear zones in the mantle (T1A/B). During a subsequent phase of deformation, second generation shear zones develop and strain in the mantle occurs as focused pure shear necking beneath the rift axis. Figures 7.26d and e show the results of another model where frictional-plastic (brittle) strain softening occurs and the resulting deformation is asymmetric. An initial stage is very similar to the early stages of the reference model, but at later times strain softening focuses deformation into one of the conjugate faults (S1B). The asymmetry is caused by a positive feedback between increasing strain and the strength reduction that results from a decreased angle of internal friction (Section 2.10.2). Large displacements on the S2A and T1B shear zones cut out a portion of the lower crust (LC) at point C (Fig. 7.26, insert) and begin to exhume the lower plate. By 40 Ma, a

symmetric necking of the lower lithosphere and continued motion on the asymmetric shear zones results in the vertical transport of point P until mantle lithosphere is exposed. The model shown in Fig. 7.26f and g combines both frictional-plastic and viscous weakening mechanisms. The early evolution is similar to that shown in Fig. 7.26d, except that S1B continues into the ductile mantle. The two softening mechanisms combine to make deformation asymmetric at all levels of the lithosphere where displacements are mostly focused onto one shear zone. These models show how a softening of the dominant rheology in either frictional-plastic or viscous layers influences deformation patterns in rifts through a positive feedback between weakening and increased strain.

The effect of strain-dependent weakening on fault asymmetry also is highly sensitive to rift velocity. This sensitivity is illustrated in the models shown in Fig. 7.27. The first model (Fig. 7.27a) is identical to that shown in Fig. 7.26d and e except that the velocity is decreased by a factor of five to $0.6\,\mathrm{mm\,a^{-1}}$. Reducing the velocity has the effect of maintaining the thickness of the frictional-plastic layer, which results in deformation that is more strongly controlled by the frictional regime than that shown in Fig. 7.26e. The overall geometry matches a lithospheric-scale simple shear model (cf. Fig. 7.21b) in which the lower plate has been progressively uplifted and exhumed beneath a through-going ductile shear zone that remains the single major weakness during rifting. By contrast, a velocity that is increased to $100\,\mathrm{mm\,a^{-1}}$ (Fig. 7.27b) results in deformation that is more strongly controlled by viscous flow at the base of the frictional layer than that in the model involving slow velocities. However, at high velocities the strain softening does not develop in part because of the high viscous stresses that result from high strain rates. The model shows no strong preference for strain localization on one of the frictional fault zones. The deformation remains symmetrical as the ductile mantle undergoes narrow pure shear necking. These results suggest that increasing or decreasing rift velocities can either promote or inhibit the formation of large asymmetric structures because varying the rate changes the dominant rheology of the deforming layers.

These experiments illustrate the sensitivity of deformation patterns to strain-induced weakening mechanisms during faulting and ductile flow. The results suggest that extension is most likely to be asymmetric in models that include frictional-plastic fault zone weakening mechanisms, a relatively strong lower crust, and slow rifting velocities. However, before attempting to

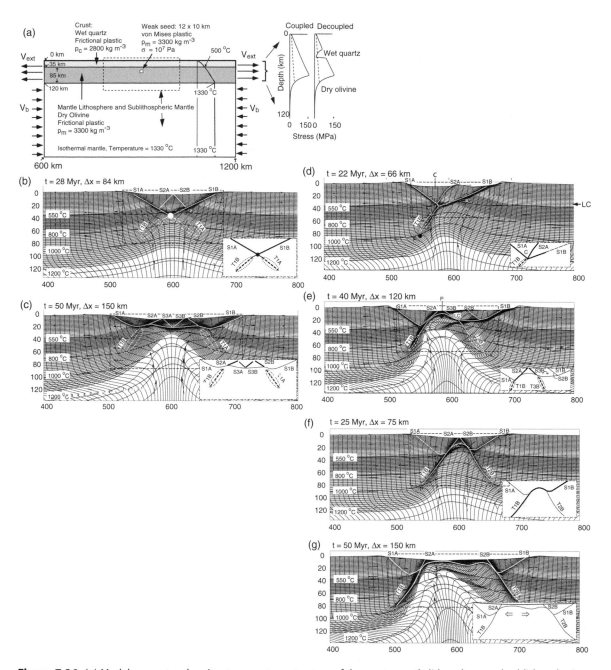

Figure 7.26 *(a) Model geometry showing temperature structure of the crust, mantle lithosphere and sublithospheric mantle (images provided by R. Huismans and modified from Huismans & Beaumont, 2003, by permission of the American Geophysical Union. Copyright © 2003 American Geophysical Union). Initial (solid lines) and strain softened (dashed lines) strength envelopes are shown for an imposed horizontal extensional velocity of $V_{ext} = 3 \, mm \, a^{-1}$, with V_b chosen to achieve mass balance. Decoupling between crust and mantle is modeled using a wet quartzite rheology for the lower crust. (b,c) Reference model of extension when strain softening is absent. Models of extension involving (d,e) frictional-plastic (brittle) strain softening and (f,g) both frictional-plastic and viscous weakening mechanisms. Models in (b–g) show a subdivision of the crust and mantle into an upper and lower crust, strong frictional upper mantle lithosphere, ductile lower lithosphere, and ductile sublithospheric mantle. Scaling of quartz viscosity makes the three upper layers frictional-plastic in all models shown. t, time elapsed in millions of years; Δx, amount of horizontal extension. Vertical and horizontal scales are in kilometers. $V_{ext} = 3 \, mm \, a^{-1}$ for every model.*

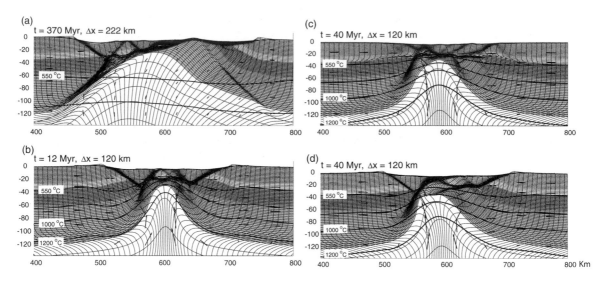

Figure 7.27 *Models of extension involving frictional-plastic (brittle) strain softening at (a) low extensional velocities ($V_{ext} = 0.6\,mm\,a^{-1}$) and (b) high extensional velocities ($V_{ext} = 100\,mm\,a^{-1}$). Models also show rift sensitivity to (c) a weak and (d) a strong middle and lower crust at $V_{ext} = 3\,mm\,a^{-1}$ (images provided by R. Huismans and modified from Huismans & Beaumont, 2007, with permission from the Geological Society of London). t, time elapsed in millions of years; Δx amount of horizontal extension. Vertical and horizontal scales are in kilometers.*

apply these results to specific natural settings, it is important to realize that the effects of strain-induced weakening can be suppressed by other mechanisms that affect the rheology of the lithosphere. For example, a comparison of two models, one incorporating a weak lower crust (Fig. 7.27c) and the other a strong lower crust (Fig. 7.27d), illustrates how a weak crust can diminish crustal asymmetry. This suppression occurs because conjugate frictional shears that develop during rifting sole out in the weak ductile lower crust where they propagate laterally beneath the rift flanks. As rifting progresses, viscous flow in a weak lower crust results in a nearly symmetric ductile necking of the lower lithosphere. These examples show that the degree of rift asymmetry depends not only on strain softening mechanisms and rifting velocities, but also on the strength of the lower crust.

7.6.6 Rheological stratification of the lithosphere

In most quantitative models of continental rifting, the lithosphere is assumed to consist of multiple layers that

are characterized by different rheologies. (Section 2.10.4). This vertical stratification agrees well with the results from both geophysical investigations of continental lithosphere and with the results of laboratory experiments that reveal the different behaviors of crust and mantle rocks over a range of physical conditions. In the upper part of the lithosphere strain is accommodated by faulting when stress exceeds the frictional resistance to motion on fault planes. In the ductile layers, strain is described using temperature-dependent power law rheologies that relate stress and strain-rate during flow (Section 2.10.3). Using these relationships, experimentally derived friction and flow laws for crustal and mantle rocks can be incorporated into models of rifting. This approach has allowed investigators to study the effects of a rheological stratification of the lithosphere on strain localization and delocalization processes during extension, including the development of large-offset normal faults (Sections 7.3, 7.6.4). The sensitivity of strain patterns to the choice of crustal rheology for different initial conditions are illustrated below using three different physical models of continental rifting.

Behn *et al.* (2002) explored how the choice of crustal rheology affects the distribution of strain within the lithosphere during extension using a simple two-layer model composed of an upper crustal layer and a lower mantle layer (Fig. 7.28a). These authors incorporated a strain-rate softening rheology to model brittle behavior and the development of fault-like shear zones. Ductile deformation was modeled using temperature-dependent flow laws that describe dislocation creep in the crust and mantle. Variations in the strength (effective viscosity) of the crust at any given temperature and strain rate are defined by material parameters that are derived from rock physics experiments. The use of several flow laws for rocks with different mineralogies and water contents allowed the authors to classify the rheologies as either weak, intermediate, or strong. Variations in crustal thickness and thermal structure were added to a series of models to examine the interplay among these parameters and the different rheologies. The results show that when crustal thickness is small, so that no ductile layer develops in the lower crust, deformation occurs mostly in the mantle and the width of the rift is controlled primarily by the vertical geothermal gradient (Fig. 7.28b,f). By contrast, when the crustal thickness is large the stress accumulation in the upper crust becomes much greater than the stress accumulation in the upper mantle (Fig. 7.28c,d). In these cases the deformation becomes crust-dominated and the width of the rift is a function of both crustal rheology and the vertical geothermal gradient (Fig. 7.28e,f).

Figure 7.28e illustrates the effects of the strong, intermediate and weak crustal rheologies on rift morphology (half-width). The models predict the same rift half-width for mantle-dominated deformation. However, the transition between mantle- and crust-dominated deformation begins at a slightly larger crustal thickness for the strong rheology than for the intermediate or weak rheologies. In addition, the strong crustal rheology results in a rift half-width for the crust-dominated regime that is ~1.5 times greater than the value predicted by the intermediate rheology and ~4 times greater than that predicted by the weak rheology. Figure 7.28f summarizes the combined effects of crustal thickness, crustal rheology, and a vertical geothermal gradient on rift half-width. These results illustrate that the evolution of strain patterns during lithospheric stretching is highly sensitive to the choice of crustal rheology, especially in situations where the crust is relatively thick.

A similar sensitivity to crustal rheology was observed by Wijns *et al.* (2005). These authors used a simple two-layer crustal model where a plastic yield law controlled brittle behavior below a certain temperature and the choice of temperature gradient controlled the transition from a brittle upper crust into a ductile lower crust.

This formulation and a 20-km-thick upper crust lying above a 40-km-thick lower crust allowed them to investigate how a mechanically stratified crust influenced fault spacing and the distribution of strain during extension. They found that the ratio of the integrated strength of the upper and lower crust governs the degree of strain localization on fault zones. When this ratio is small, such that the lower crust is relatively strong, extension results in widely distributed, densely spaced faults with a limited amount of slip on each fault. By contrast, a large strength ratio between the upper and lower crust, such that the lower crust is very weak, causes extension to localize onto relatively few faults that accommodate large displacements. In this latter case, the large-offset faults dissect the upper crust and exhume the lower crust, leading to the formation of metamorphic core complexes (Section 7.3). Wijns *et al.* (2005) also concluded that secondary factors, such as fault zone weakening and the relative thicknesses of the upper and lower crust (Section 7.6.5), determine the exact value of the critical ratio that controls the transition between localized and delocalized extension.

The results of Wijns *et al.* (2005), like those obtained by Behn *et al.* (2002), suggest that a weak lower crust promotes the localization of strain into narrow zones composed of relatively few faults. This localizing behavior reflects the ability of a weak lower crust to flow and transfer stress into the upper crust, which may control the number of fault zones that are allowed to develop. This interpretation is consistent with field studies of deformation and rheology contrasts in ancient lower crust exposed in metamorphic core complexes (e.g. Klepeis *et al.* 2007). It is also consistent with the results of Montési & Zuber (2003), who showed that for a brittle layer with strain localizing properties overlying a viscous layer, the viscosity of the ductile layer controls fault spacing. In addition, a weak lower crust allows fault blocks in the upper crust to rotate, which can facilitate the dissection and dismemberment of the upper crust by faulting.

Lastly, a third numerical model of rifting illustrates how the interplay among strain-induced weakening, layer thickness, and rheological contrasts can influence deformation patterns in a four-layer model of the lithosphere. Nagel & Buck (2004) constructed a model that consisted of a 12-km-thick brittle upper crust, a relatively strong 10-km-thick lower crust, a thin (3 km) weak mid-crustal layer, and a 45-km-thick upper mantle (Fig. 7.29a). The model incorporates temperature-dependent power law rheologies that determine viscous behavior in the crust and mantle. The mantle and upper and

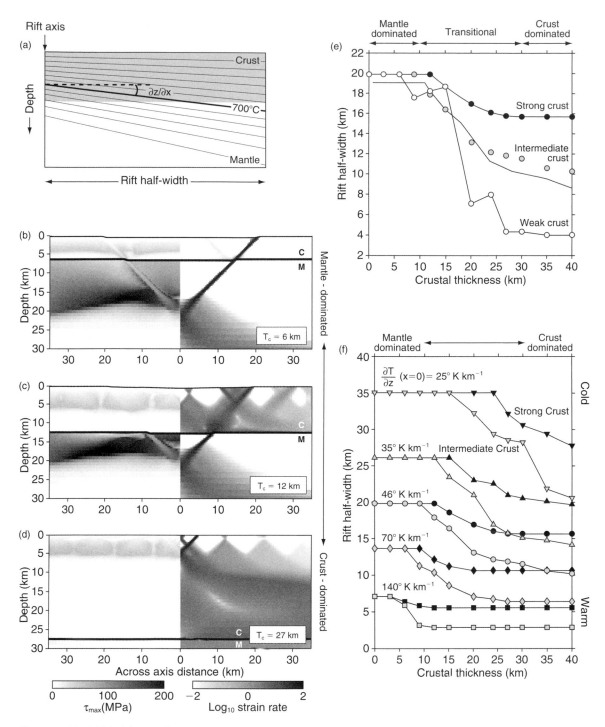

Figure 7.28 *(a) Model setup for numerical simulations of lithospheric stretching. The transition from mantle- to crust-dominated deformation is illustrated by (b), (c), and (d), which show the deformation grid after 1% total strain for a crustal thickness (T_c) of 6, 12 and 27 km, respectively. Grayscale indicates the magnitude of shear stress on left and normalized strain-rate on right. C and M mark the base of the crust and top of the mantle, respectively. (e) Effect of crustal thickness on predicted rift half-width. (f) Effect of vertical geothermal gradient on predicted rift half-width (images provided by M. Behn and modified from Behn* et al.*, 2002, with permission from Elsevier). Each point in (e) and (f) represents an experiment. Black, strong; gray, intermediate; and white, weak rheology.*

lower crust also follow the Mohr–Coulomb failure criterion and cohesion loss during faulting is included. The model also incorporates a predefined bell-shaped thermal perturbation at its center that serves to localize deformation at the beginning of extension. The horizontal thermal gradient created by this perturbation, and the predetermined vertical stratification, control the mechanical behavior of the lithosphere during rifting.

As extension begins, the upper mantle and lower crust undergo localized necking in the hot, weak center of the rift. Deformation in the upper crust begins as a single graben forms above the area of necking in the lower crust and mantle and subsequently evolves into an array of parallel inward dipping normal faults. The faults root down into the weak mid-crustal layer where distributed strain in the upper crust is transferred into the necking area in the strong lower parts of the model (Fig. 7.29b,c).

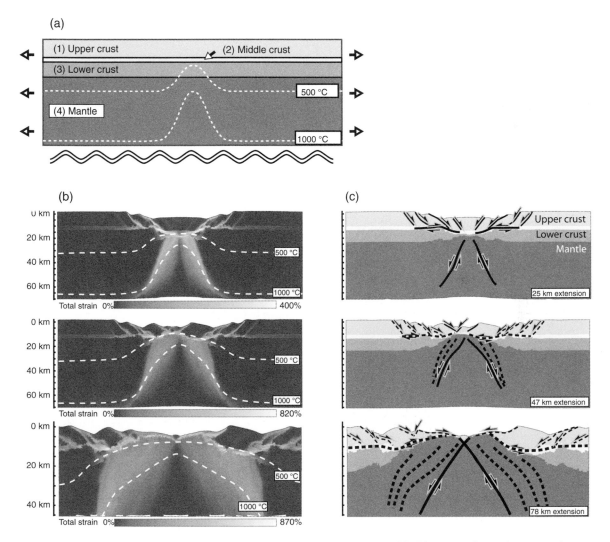

Figure 7.29 *Model of symmetric rifting (images provided by T. Nagel and modified from Nagel & Buck, 2004, with permission from the Geological Society of America). (a) Model setup. (b) Total strain and (c) distribution of upper, middle and lower crust and mantle after 25, 47 and 78 km of extension. Solid black lines, active zones of deformation; dashed lines, inactive zones; thin black lines, brittle faults; thick black lines, ductile shear zones.*

After ~25 km of extension, the lower crust pulls apart and displacements on the normal faults lead the collapse and dismemberment of the upper crust at the margins of the rift. Mantle material wells upward into the zone of thinning where the collapsing upper crust is placed in direct contact with mantle rocks. After 40 km of extension, the array of normal faults is abandoned and upper crustal deformation is concentrated in the center of the rift. Finally, after ~75 km, new ocean lithosphere is generated, leaving behind two tectonically quiet passive margins. This, and the other physical models described in this section, show how combinations of competing processes that either weaken or strengthen the crust can be used to explain much of the variability in deformation patterns observed in rifts.

7.6.7 Magma-assisted rifting

Most quantitative treatments of continental rifting focus on the effects of variations in lithospheric conditions. This emphasis reflects both the success of these models at explaining many aspects of rifting and the relative ease at which geoscientists can constrain the physical properties of the lithosphere compared to those of the asthenosphere. Nevertheless, it is evident that interactions between the asthenosphere and the lithosphere form crucial components of rift systems (Ebinger, 2005). One of the most important aspects of these interactions involves magmatism (Section 7.4), which weakens the lithosphere and causes strain localization.

Among its possible effects, mafic magmatism may allow rifting to initiate in regions of relatively cold or thick continental lithosphere (Section 7.5). In addition to its weakening effects, the availability of a significant source of basaltic magma influences the thickness, temperature, density, and composition of the lithosphere. The presence of hot, partially molten material beneath a rift valley produces density contrasts that result in thermal buoyancy forces (Section 7.6.3). As the two sides of the rift separate, magma also may accrete to the base of the crust where it increases in density as it cools and may lead to local crustal thickening (Section 7.2, Fig. 7.5). These processes can create bending forces within the lithosphere as the plate responds to the changing load, and affect the manner in which strain is accommodated during rifting. The changes may be recorded in patterns of uplift and subsidence across rifts and rifted margins.

Buck (2004) developed a simple two-dimensional thermal model to illustrate how rifting and magma intrusion can weaken the lithosphere and influence subsidence and uplift patterns. The emplacement of large quantities of basalt in a rift can accommodate extension without crustal thinning. This process has been observed in the mature rift segments of northern Ethiopia (Section 7.8.1) where strain accommodation by faulting has been greatly reduced as magmatism increased (Wolfenden *et al.*, 2005). If enough material intrudes, the crustal thickening that can result from magmatism can lessen the amount of subsidence in the rift and may even lead to regional uplift. This effect is illustrated in Fig. 7.30, which shows the average isostatic elevation through time for magma-assisted rifting compared to a typical subsidence curve for lithospheric stretching due to thermal relaxation (McKenzie, 1978). The uplift or subsidence result from changes in density related to the combined effects of crustal thinning, basalt intrusion and temperature differences integrated over a 100 km wide rift to a depth of 150 km. Buck (2004) suggested that this process might explain why some continental margins, such as those off the east coast of Canada (Royden & Keen, 1980), show less initial tectonic subsidence related to crustal thinning compared to the

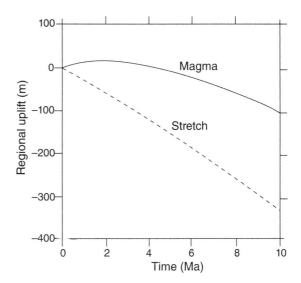

Figure 7.30 *Comparison of the predicted average regional isostatic elevation changes for magma-assisted rifting (solid line) and pure shear necking (dashed line) (from Buck, 2004. Copyright © 2004 from Columbia University Press. Reprinted with permission of the publishers).*

long-term (tens of millions of years) subsidence induced by cooling.

Two other problems of rift evolution that also might be resolved by incorporating the effects of magmatism and/or flow of the asthenosphere include the extra subsidence observed at some rifted margins and the lack of magma that characterize nonvolcanic margins (Buck, 2004). These effects are discussed in the context of the evolution of rifted continental margins in Section 7.7.3.

7.7 RIFTED CONTINENTAL MARGINS

7.7.1 Volcanic margins

Rifted volcanic margins are defined by the occurrence of the following three components: Large Igneous Provinces (Section 7.4.1) composed of thick flood basalts and silicic volcanic sequences, high velocity (V_p > 7 km s^{-1}) lower crust in the continent–ocean transition zone, and thick sequences of volcanic and sedimentary strata that give rise to *seaward-dipping reflectors* on seismic reflection profiles (Mutter *et al.*, 1982). The majority of rifted continental margins appear to be volcanic, with some notable exceptions represented by the margins of the Goban Spur, western Iberia, eastern China, South Australia, and the Newfoundland Basin–Labrador Sea. Relationships evident in the Red Sea and southern Greenland suggest that a continuum probably exists between volcanic and nonvolcanic margins.

The high velocity lower crust at volcanic margins occurs between stretched continental crust and normal thickness oceanic crust (Figs 7.31, 7.32). Although these layers have never been sampled directly, the high P_n wave velocities suggest that they are composed of thick accumulations of gabbro that intruded the lower crust during continental rifting. The intrusion of this material helps to dissipate the thermal anomaly in the mantle that is associated with continental rifting.

The Lofoten–Vesterålen continental margin off Norway (Figs 7.31, 7.32) illustrates the crustal structure of a volcanic margin that has experienced moderate extension (Tsikalas *et al.*, 2005). The ocean–continent

transition zone between the shelf edge and the Lofoten basin is 50–150 km wide, includes an abrupt lateral gradient in crustal thinning, and is covered by layers of volcanic material that display shallow seaward dipping reflectors (Fig. 7.32a). The 50–150 km width of this zone is typical of many rifted margins, although in some cases where there is extreme thinning the zone may be several hundred kilometers wide. Crustal relief in this region is related to faulted blocks that delineate uplifted highs. In the Lofoten example, the continent–ocean boundary occurs landward of magnetic anomaly 24B (53–56 Ma) and normal ocean crust occurs seaward of magnetic anomaly 23 (Fig. 7.31b). Crustal thinning is indicated by variations in Moho depth. The Moho reaches a maximum depth of 26 km beneath the continental shelf and 11–12 km beneath the Lofoten basin. Along profile A–A′ a region of 12–16 km thick crust within the ocean–continent transition zone coincides with a body in the lower crust characterized by a high lower crustal velocity (7.2 km s^{-1}) (Fig. 7.32a,c). This body thins to the north along the margin, where it eventually disappears, and thickens to the south, where at one point it has a thickness of 9 km (Fig. 7.31c). Oceanic layers display velocities of 4.5–5.2 km s^{-1}, sediments show velocities of ≤2.45 km s^{-1}. These seismic velocities combined with gravity models (Fig. 7.32b) provide information on the nature of the material within the margin (Fig. 7.32c).

In most volcanic margins the wedges of seaward-dipping reflectors occur above or seaward of the high velocity lower crust in the continent–ocean transition zone. Direct sampling of these sequences indicates that they are composed of a mixture of volcanic flows, volcaniclastic deposits, and nonvolcanic sedimentary rock that include both subaerial and submarine types of deposits. Planke *et al.* (2000) identified six units that are commonly associated with these features (Fig. 7.33): (i) an outer wedge of seaward-dipping reflectors; (ii) an outer high; (iii) an inner wedge of seaward-dipping reflectors; (iv) landward flows; (v) lava deltas; and (vi) inner flows. The wedge-like shape of the reflector packages is interpreted to reflect the infilling of rapidly subsiding basement rock. The outer reflectors tend to be smaller and weaker than the inner variety. The outer high is a mounded, commonly flat-topped feature that may be up to 1.5 km high and 15–20 km wide. In some places this may be a volcano or a pile of erupted basalt. Landward flows are subaerially erupted flood basalts that display little to no sediment layers between the flows. The inner flows are sheet-like bodies located

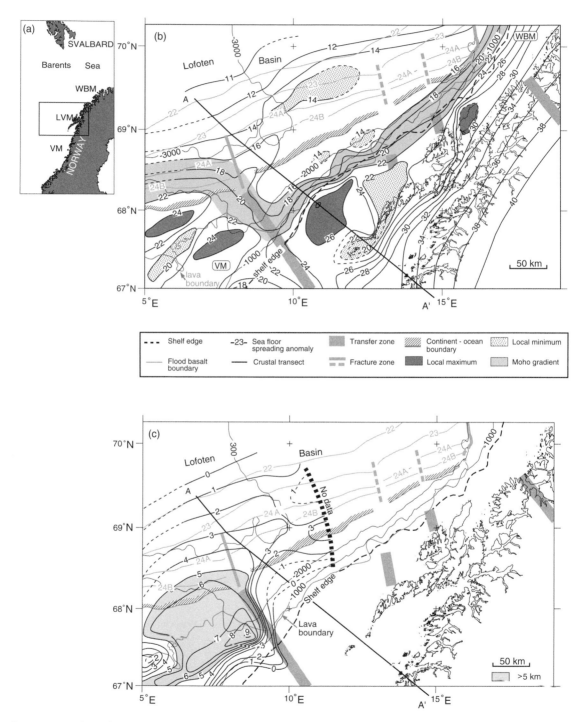

Figure 7.31 *The Lofoten–Vesterålen continental margin. Inset (a) shows Vøring (VM), Lofoten–Vesterålen (LVM), and Western Barents Sea (WBM) margins. (b) Map showing Moho depths with 2 km contour interval. (c) Thickness of high velocity lower crustal body with contour interval of 1 km (images provided by F. Tsikalas and modified from Tsikalas et al., 2005, with permission from Elsevier). A–A' indicates the location of the cross-sections shown in Fig. 7.32.*

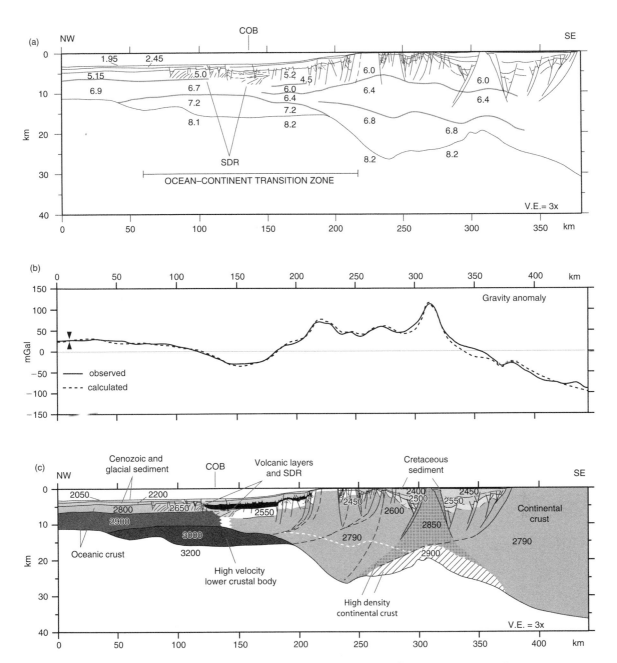

Figure 7.32 *(a) Seismic velocity structure along the southern Lofoten–Vesterålen margin. COB, continent–ocean boundary. (b,c) Gravity modeled transect and interpretation of the geology (images provided by F. Tsikalas and modified from Tsikalas et al., 2005, with permission from Elsevier). Densities in (c) are shown in kilograms per cubic meter. SDR, seaward dipping reflectors. For location of profile see Fig. 7.31.*

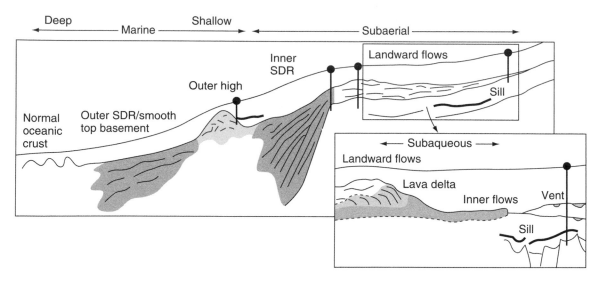

Figure 7.33 *Interpretation of the main seismic facies of extrusive units at volcanic margins (modified from Planke et al., 2000, by permission of the American Geophysical Union. Copyright © 2000 American Geophysical Union). Inset shows enlargement of a region of landward subaqueous flows where lava deltas and inner flow units commonly occur. Solid circles with vertical lines show locations of wells where drill holes have penetrated the various units. SDR, seaward dipping reflectors (shaded). Bold black lines, sills.*

landward and, typically, below the lava delta. Lava deltas form as flowing basalt spills outward in front of the growing flood basalts. The emplacement of these features is associated with the establishment of thicker than normal ocean crust within the continent to ocean transition zone (Planke *et al.*, 2000).

The conditions and processes that form volcanic rifted margins are the subject of much debate. In general, the formation of the thick igneous crust appears to require larger amounts of mantle melting compared to that which occurs at normal mid-ocean ridges. The origin of this enhanced igneous activity is uncertain but may be related to asthenospheric temperatures that are higher than those found at mid-ocean ridges or to unusually high rates of upwelling mantle material (Nielson & Hopper, 2002, 2004). Both of these mechanisms could occur in association with mantle plumes (Sections 5.5, 12.10), although this hypothesis requires rigorous testing.

7.7.2 Nonvolcanic margins

The occurrence of nonvolcanic margins (Fig. 7.34a) shows that extreme thinning and stretching of the crust

is not necessarily accompanied by large-scale volcanism and melting. Nonvolcanic margins lack the large volume of extrusive and intrusive material that characterizes their volcanic counterparts. Instead, the crust that characterizes this type of margin may include highly faulted and extended continental lithosphere, oceanic lithosphere formed by very slow sea floor spreading, or continental crust intruded by magmatic bodies (Sayers *et al.*, 2001). In addition, these margins may contain areas up to 100 km wide that are composed of exhumed, serpentinized upper mantle (Fig. 7.34b,c) (Pickup *et al.*, 1996; Whitmarsh *et al.*, 2001). Dipping reflectors in seismic profiles also occur within nonvolcanic margins. However, unlike in volcanic varieties, these reflectors may be preferentially tilted continentward and do not represent sequences of volcanic rock (Pickup *et al.*, 1996). Some of these *continentward-dipping* reflectors represent detachment faults (Section 7.3) that formed during rifting (Boillot & Froitzheim, 2001).

Two end-member types of nonvolcanic margins have been identified on the basis of relationships preserved in the North Atlantic region (Louden & Chian, 1999). The first case is derived from the southern Iberia Abyssal Plain, Galicia Bank, and the west Greenland margins. In these margins rifting of the continent

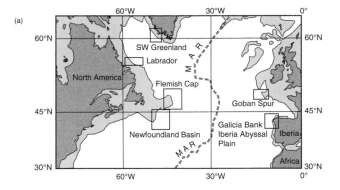

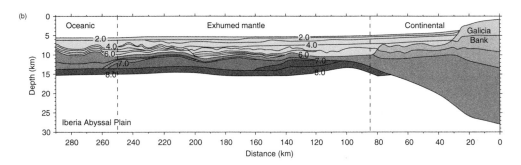

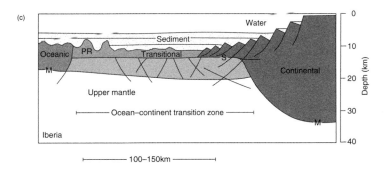

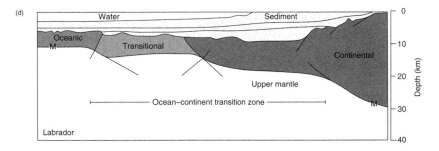

Figure 7.34 (a) Map of the North Atlantic showing location of selected nonvolcanic margins. MAR, Mid-Atlantic Ridge. (b) Velocity model of the West Iberia margin and the Iberia Abyssal Plain (image provided by T. Minshull and modified from Minshull, 2002 with permission from Royal Society of London). Data are from Dean et al. (2000). The dashed lines mark the approximate edges of the ocean–continent transition zone. Velocities in km s^{-1}. (c,d) Two end-member types of nonvolcanic margin (images provided by K. Louden and modified from Louden & Chian, 1999, with permission from the Royal Society of London). PR, peridotite ridge; S, reflections interpreted to represent a detachment fault or shear zone; M, Moho reflections.

produced a zone of extremely thin continental crust. This thin crust is characterized by tilted fault blocks that are underlain by a prominent subhorizontal reflector (S) that probably represents a serpentinized shear zone at the crust–mantle boundary (Fig. 7.34c) (Reston et al., 1996). The reflector occurs seaward of stretched continental basement and above a high velocity lower layer of serpentinized mantle. Below the reflector seismic velocities increase gradually with depth and approach normal mantle velocities at depths of 15–20 km. Seaward of the thinned continental crust and landward of the first oceanic crust, a transitional region is characterized by low basement velocities, little reflectivity, and a lower layer of serpentinized mantle showing velocities ($V_p >$ 7.0 km s^{-1}) that are similar to high velocity lower crust. Farther seaward, the basement is characterized by a complex series of peridotite ridges (PR), which contain sea floor spreading magnetic anomalies that approximately parallel the strike of the oceanic spreading center. Although this zone is composed mostly of serpentinized mantle, it may also contain minor intrusions. Thus, basement at these margins consists of faulted continental blocks, a smooth transitional region, and elevated highs. Moho reflections (M) are absent within the ocean–continent transition zone. Instead, this region displays landward and seaward dipping reflectors that extend to depths of 15–20 km.

In the second type of nonvolcanic margin (Fig. 7.34d), based primarily on the Labrador example, only one or two tilted fault blocks of upper continental crust are observed and the S-type horizontal reflection is absent. A zone of thinned mid-lower continental crust occurs beneath a thick sedimentary basin. A transitional region occurs farther seaward in a manner similar to the section shown in Fig. 7.34c. However, dipping reflections within the upper mantle are less prevalent. For Labrador, the region of extended lower continental crust is very wide with a thick sedimentary basin, while for Flemish Cap and the Newfoundland basin, the width of extended lower continental crust is narrow or absent. Moho reflections (M) indicate very thin (~5 km) oceanic crust.

7.7.3 The evolution of rifted margins

The evolution of rifted continental margins is governed by many of the same forces and processes that affect the formation of intracontinental rifts (Section 7.6). Thermal and crustal buoyancy forces, lithospheric flexure, rheological contrasts, and magmatism all may affect margin behavior during continental break-up, although the relative magnitudes and interactions among these factors differ from those of the pre-break-up rifting stage. Two sets of processes that are especially important during the transition from rifting to sea floor spreading include: (i) post-rift subsidence and stretching; and (ii) detachment faulting, mantle exhumation, and ocean crust formation at nonvolcanic margins.

Post-rift subsidence and stretching

As continental rifting progresses to sea floor spreading, the margins of the rift isostatically subside below sea level and eventually become tectonically inactive. This subsidence is governed in part by the mechanical effects of lithospheric stretching (Section 7.6.2) and by a gradual relaxation of the thermal anomaly associated with rifting. Theoretical considerations that incorporate these two effects for the case of uniform stretching predict that subsidence initially will be rapid as the crust is tectonically thinned and eventually slow as the effects of cooling dominate (McKenzie, 1978). However, the amount of subsidence also is influenced by the flexural response of the lithosphere to loads generated by sedimentation and volcanism and by changes in density as magmas intrude and melts crystallize and cool (Section 7.6.7). Subsidence models that include the effects of magmatism and loading predict significant departures from the theoretical thermal subsidence curves.

The amount of subsidence that occurs at rifted margins is related to the magnitude of the stretching factor (β). There are several different ways of estimating the value of this parameter, depending on the scale of observation (Davis & Kusznir, 2004). For the brittle upper crust, the amount of extension typically is derived from summations of the offsets on faults imaged in seismic reflection profiles that are oriented parallel to fault dips. Estimates of the combined upper crustal extension and lower crustal stretching are obtained from variations in crustal thickness measured using wide-angle seismic surveys, gravity studies, and seismic reflection data. This latter approach relies on the assumption that the variations are a consequence of crustal extension and thinning. At the scale of the entire lithosphere, stretching factors are obtained through considerations of the flexural isostatic response to

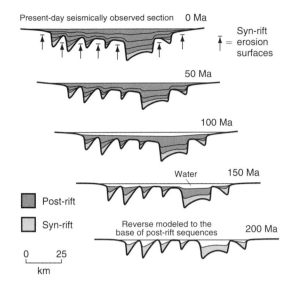

Figure 7.35 *Schematic diagram showing application of flexural backstripping and the modeling of post-rift subsidence to predict sequential restorations of stratigraphy and paleobathymetry. Restored sections are dependent on the β stretching factor used to define the magnitude of lithospheric extension and lithospheric flexural strength (after Kusznir* et al., *2004, with permission from Blackwell Publishing).*

loading (Section 7.6.4) and thermal subsidence. One of the most commonly used approaches to obtaining lithospheric-scale stretching factors employs a technique known as flexural backstripping.

Flexural backstripping involves reconstructing changes in the depth to basement in an extensional sedimentary basin by taking into account the isostatic effects of loading. The concept behind the method is to exploit the stratigraphic profile of the basin to determine the depth at which basement rock would be in the absence of loads produced by both water and all the overlying layers. This is accomplished by progressively removing, or *backstripping*, the loads produced by each layer and restoring the basement to its depth at the time each layer was deposited (Fig. 7.35). These results combined with knowledge of water depth theoretically allow determination of the stretching factor (β). Nevertheless, as discussed further below, relationships between stretching factor and subsidence curves may be complicated by interactions between the lithosphere and the sublithospheric mantle. In practice, flexural backstripping is carried out by assigning each layer

a specific density and elastic thickness (T_e) (Section 7.6.4) and then summing the effects of each layer for successive time intervals. Corrections due to sediment compaction, fluctuations in sea level, and estimates of water depth using fossils or other sedimentary indicators are then applied. This approach generally involves using information derived from post-rift sediments rather than syn-rift units because the latter violate assumptions of a closed system during extension (Kusznir *et al.*, 2004). The results usually show that the depth of rifted margins at successive time intervals depends upon both the magnitude of stretching factor (β) and the flexural strength of the lithosphere. Most applications indicate that the elastic thickness of the lithosphere increases as the thermal anomaly associated with rifting decays.

Investigations of lithospheric-scale stretching factors at both volcanic and nonvolcanic margins have revealed several characteristic relationships. Many margins show more subsidence after an initial tectonic phase due to stretching than is predicted by thermal subsidence curves for uniform stretching. Rifted margins off Norway (Roberts *et al.*, 1997), near northwest Australia (Driscoll & Karner, 1998), and in the Goban Spur and Galicia Bank (Davis & Kusznir, 2004) show significantly more subsidence than is predicted by the magnitude of extension indicated by upper crustal faulting. In addition, many margins show that the magnitude of lithospheric stretching increases with depth within ~150 km of the ocean–continent boundary (Kusznir *et al.*, 2004). Farther toward the continent, stretching and thinning estimates for the upper crust, whole crust, and lithosphere converge as the stretching factor (β) decreases. These observations provide important boundary conditions on the processes that control the transition from rifting to sea floor spreading. However, the causes of the extra subsidence and depth-dependent stretching are uncertain. One possibility is that the extra subsidence results from extra uplift during the initial stage of sea floor spreading, perhaps as a result of upwelling anomalously hot asthenosphere (Hopper *et al.*, 2003; Buck, 2004). Alternatively, greater stretching in the mantle lithosphere than in the crust, or within a zone of mantle lithosphere that is narrow than in the crust, also may result in extra uplift. Once these initial effects decay the ensuing thermal subsidence during cooling would be greater than models of uniform stretching would predict. These hypotheses, although seemingly plausible, require further testing.

Observations of the southeast Greenland volcanic margin support the idea that the flow of low-density mantle during the transition to sea floor spreading strongly influences subsidence and stretching patterns. Hopper *et al.* (2003) found distinctive changes in the morphology of basaltic layers in the crust that indicate significant vertical motions of the ridge system. At the start of spreading, the system was close to sea level for at least 1 Myr when spreading was subaerial. Later subsidence dropped the ridge to shallow water and then deeper water ranging between 900 and 1500 m depth. This history appears to reflect the dynamic support of the ridge system by upwelling of hot mantle material during the initiation of spreading. Exhaustion of this thermal anomaly then led to loss of dynamic support and rapid subsidence of the ridge system over a 2 Ma period. In addition, nearly double the volume of dikes and volcanic material occurred on the Greenland side of the margin compared to the conjugate Hatton Bank margin located south of Iceland on the other side of the North Atlantic ocean. These observations indicate that interactions between hot asthenosphere and the lithosphere continue to influence the tectonic development of rifted margins during the final stages of continental break-up when sea floor spreading centers are established.

The flow of low-density melt-depleted asthenosphere out from under a rift also may help explain the lack of magmatic activity observed at rifted nonvolcanic margins. The absence of large volumes of magma could be linked to the effects of prior melting episodes, convective cooling of hot asthenosphere, and/or the rate of mantle upwelling (Buck, 2004). As sublithospheric mantle wells up beneath a rift it melts and cools. This process could result in shallow mantle convection due to the presence of cool, dense melt-depleted material overlying hotter, less dense mantle. Cooling also restricts further melting by bringing the mantle below its solidus temperature (Section 7.4.2). If some of this previously cooled, melt-depleted asthenosphere is pulled up under the active part of the rift during the transition to sea floor spreading, its presence would suppress further melting, especially if the rate of rifting or sea floor spreading is slow. The slow rates may not allow the deep, undepleted asthenosphere to reach the shallow depths that generate large amounts of melting.

Magma accretion, mantle exhumation, and detachment faulting

The transition from rifting to sea floor spreading at nonvolcanic margins is marked by the exhumation of large sections of upper mantle. Seismic reflection data collected from the Flemish Cap off the Newfoundland margin provide insight into the mechanisms that lead to this exhumation and how they relate to the formation of ocean crust.

The Flemish Cap is an approximately circular shaped block of 30-km-thick continental crust that formed during Mesozoic rifting between Newfoundland and the Galicia Bank margin near Iberia (Fig. 7.36a). The two conjugate margins show a pronounced break-up asymmetry. Seismic images from the Galicia Bank show a transition zone composed of mechanically unroofed continental mantle (Fig. 7.36b) and a strong regional west-dipping S-type reflection (Fig. 7.36b, stages 1 & 2) (Section 7.7.2). The transition zone is several tens of kilometers wide off the Galicia Bank and widens to 130 km to the south off southern Iberia. The S-reflection is interpreted to represent a detachment fault between the lower crust and mantle that underlies a series of fault-bounded blocks. By contrast, the Newfoundland margin lacks a transition zone and shows no evidence of any S-type reflections or detachment faults (Hopper *et al.*, 2004). Instead, this latter margin shows an abrupt boundary between very thin continental crust and a zone of anomalously thin (3 to 4 km thick), highly tectonized oceanic crust (Fig. 7.36b, stages 3, 4, and 5). Seaward of this boundary the oceanic crust thins even further to <1.3 km and exhibits unusual very reflective layering (I).

The five stage model of Hopper *et al.* (2004) explains these structural differences and the evolution of the conjugate margins. In Fig. 7.36b, the top panel shows a reconstruction of the two margins emphasizing their asymmetry at final break-up when the continental crust was thinned to a thickness of only a few kilometers (stage 1). During break-up, displacement within an extensional detachment fault (labeled S in Fig. 7.36b) unroofed a peridotite ridge (PR) above a zone of weak serpentinized upper mantle. Break-up west of the ridge isolated it on the Galicia Bank margin when, during stage 2, mantle melts reached the surface and sea floor spreading was established. Limited magmatism produced the thinner than normal (3–4 km), highly tectonized ocean crust. During stage 3, a

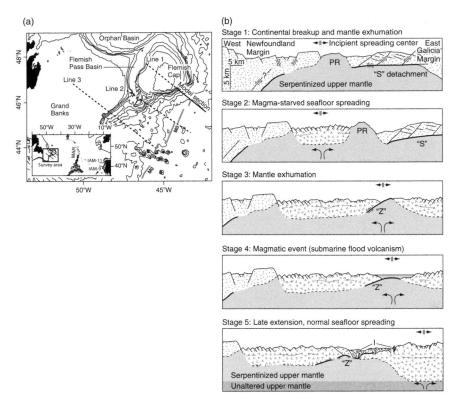

Figure 7.36 *(a) Location of seismic surveys of the Flemish Cap and (b) five stage model of nonvolcanic margins (after Hopper* et al.*, 2004, with permission from the Geological Society of America). MO in (a) is magnetic anomaly. Random-dash pattern, continental crust; v pattern, oceanic crust; light gray shading, serpentinized upper mantle; dark gray shading, unaltered upper mantle; thick lines, strong reflections; dashed lines, inferred crust–mantle boundary; dotted lines, oceanic layers; PR, peridotite ridge; S, reflections interpreted to represent a detachment fault; I, unusual very reflective oceanic crust; Z, reflections interpreted to represent a detachment fault buried by deep marine flood basalts.*

reduction in magma supply led to about 20 km of extension that was accommodated mostly by detachment faulting. The detachment faulting led to the exhumation of the mantle and formed an oceanic core complex that is similar to those found in slow-spreading environments at ridge–transform intersections (Section 6.7). Voluminous but localized magmatism during stage 4 resulted in a 1.5-km-thick layer of deep marine flood basalts that buried the detachment surface (reflection

Z). The intrusion of gabbroic material may have accompanied this volcanism. This magmatic activity marked the beginning of sea floor spreading that formed normal (6 km) thickness ocean crust (stage 5).

This example shows that, to a first order, the transition from rift to oceanic crust at nonvolcanic margins is fundamentally asymmetric and involves a period of magmatic starvation that leads to the exhumation of the mantle. This type of margin may typify slow-spreading systems (Section 6.6) where large fluctuations in melt supply occur in transient magma chambers during the early stages of sea floor spreading.

7.8 CASE STUDIES: THE TRANSITION FROM RIFT TO RIFTED MARGIN

7.8.1 The East African Rift system

The East African Rift system (Fig. 7.2) is composed of several discrete rift segments that record different stages in the transition from continental rift to rifted volcanic margin (Ebinger, 2005). The Eastern Rift between northern Tanzania and southern Kenya is an example of a youthful rift that initiated in thick, cold and strong continental lithosphere. Volcanism and sedimentation began by ~5 Ma with the largest fault escarpments forming by ~3 Ma. Strain and magmatism are localized within narrow asymmetric rift basins with no detectable deformation in the broad uplifted plateau adjacent to the rifts (Foster *et al.*, 1997). Earthquake hypocenters occur throughout the entire 35 km thickness of the crust, indicating that crustal heating is at a minimum (Foster & Jackson, 1998). The basins are shallow (~3 km deep) with 100-km-long border faults that accommodate small amounts of extension. The border faults have grown from short fault segments that propagated along their lengths to join with other nearby faults, creating linkages between adjacent basins (Foster *et al.*, 1997). Faults that were oriented unfavorably with respect to the opening direction were abandoned as strain progressively localized onto the border faults (Ebinger, 2005). Geophysical (Green *et al.*, 1991; Birt *et al.*, 1997) and geochemical (Chesley *et al.*, 1999) data show that the mantle lithosphere has been thinned to about 140 km. Elsewhere the lithosphere is at least 200 km and possibly 300–350 km thick (Ritsema *et al.*, 1998). These patterns conform to the predictions of lithospheric stretching models (Section 7.6.2, 7.6.3) in regions of relatively thick lithosphere. They also illustrate that the cross-sectional geometry and the along-axis segmentation in youthful rifts are controlled by the flexural strength of the lithosphere (Section 7.6.4).

The effects of pre-existing weaknesses on the geometry of rifting are also illustrated in the southern segment of the Eastern Rift in Tanzania. Border faults and half graben preferentially formed in a zone of weakness created by a contrast between thick, cool lithosphere of the Archean Tanzanian craton and thin, weak Proterozoic lithosphere located to the east (Foster *et al.*, 1997). From north to south, the axis of the rift diverges from a single ~50 km wide rift to a ~200 km wide zone composed of three narrow segments (Fig. 7.2b). This segmentation and a change in orientation of faults occurs where the rift encounters the Archean Tanzanian craton (Fig. 7.37), indicating that the thick lithosphere has deflected the orientation of the rift. These observations illustrate that lateral heterogeneities at the lithosphere–asthenosphere boundary exert a strong control on the initial location and distribution of strain at the start of rifting (Section 7.4).

An example of a rift that is slightly more evolved than the Tanzanian example occurs in central and northern Kenya where rifting began by 15 Ma. In this rift segment the crust has been thinned by up to 10 km and the thickness of the lithosphere has been reduced to about 90 km (Mechie *et al.*, 1997). A progressive shallowing of the Moho occurs between central and northern Kenya where the rift widens from ~100 km to ~175 km (Fig. 7.2b). In northern Kenya, crustal thickness is about 20 km and the total surface extension is about 35–40 km ($\beta = 1.55$–1.65) (Hendrie *et al.*, 1994). In the south, crustal thickness is 35 km with estimates of total extension ranging from 5 to 10 km (Strecker *et al.*, 1990; Green *et al.*, 1991). As the amount of crustal stretching increases, and the lithosphere–asthenosphere boundary rises beneath a rift, the amount of partial melting resulting from decompression melting also increases (Section 7.4.2). Young lavas exposed in central and northern Kenya indicate source regions that are shallower than those in Tanzania (Furman *et al.*, 2004). High velocity, high density material is present in the upper crust and at the base of the lower crust, suggesting the presence of cooled basaltic intrusions (Mechie *et al.*, 1997; Ibs-von Seht *et al.*, 2001). These relationships indicate that as a continental rift enters maturity magmatic activity increases and a significant component of the extension is accommodated by magmatic intrusion below the rift axis (Ebinger, 2005).

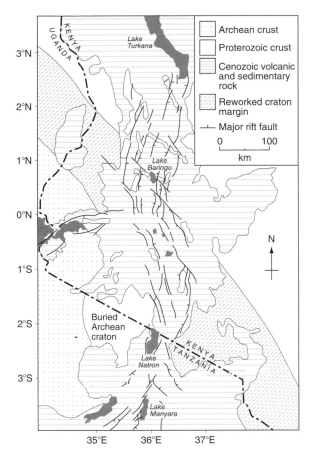

Figure 7.37 *Structural map of the Eastern branch of the East African Rift system in Kenya and northern Tanzania showing the deflection of faults at the boundary of the Archean Tanzanian craton (after Macdonald et al., 2001, with permission from the Journal of Petrology **42**, 877–900. Copyright © 2001 by permission of Oxford University Press, and Smith & Mosely, 1993, by permission of the American Geophysical Union. Copyright © 1993 American Geophysical Union).*

The increase in magmatic activity that accompanies a shallowing of the asthenosphere–lithosphere boundary beneath the Kenya Rift also results in increased crustal heating and contributes to a decrease in lithospheric strength (Section 7.6.7). This effect is indicated by a progressive decrease in the depth of earthquake hypocenters and in the depth of faulting from 35 km to 27 km (Ibs-von Seht et al., 2001). These patterns suggest a decrease in the effective elastic

thickness (T_e) of the lithosphere (Section 7.6.4) compared to the rift in northern Tanzania. Although both the mantle and crust have thinned, the thinning of mantle lithosphere outpaces crustal thinning. This asymmetry occurs because a sufficient amount of magma has accreted to the base of the crust, resulting in a degree of crustal thickening. It also results because the mantle lithosphere is locally weakened by interactions with hot magmatic fluids, which further localizes stretching.

Extension in the central and southern part of the Main Ethiopian Rift began between 18 and 15 Ma and, in the north, after 11 Ma (Wolfenden et al., 2004). The deformation resulted in the formation of a series of high-angle border faults that are marked by chains of volcanic centers (Fig. 7.38a). Since about 1.8 Ma the loci of magmatism and faulting have become progressively more localized, concentrating into ~20-km-wide, 60-km-long magmatic segments (Fig. 7.38b). This localization involved the formation of new, shorter and narrower rift segments that are superimposed on old long border faults in an old broad rift basin. This narrowing of the axis into short segments reflects a plate whose effective elastic thickness is less than it was when the long border faults formed (Ebinger et al., 1999). The extrusion of copious amounts of volcanic rock also has modified both the surface morphology of the rift and its internal structure. Relationships in this rift segment indicate that magma intrusion in the form of vertical dikes first becomes equally and then more important than faulting as rifting approaches sea floor spreading (Kendall et al., 2005). Repeated eruptions create thick piles that load the weakened plate causing older lava flows to bend down toward the rift axis. This process creates the seaward-dipping wedge of lavas (Section 7.7.1) that is typical of rifted volcanic margins (Section 7.7.1).

The rift segments in the Afar Depression illustrate that, as extension increases and the thickness of the lithosphere decreases, the asthenosphere rises and decompresses, and more melt is generated. Eventually all the border faults in the rift are abandoned as magmatism accommodates the extension (Fig. 7.38c). At this stage the rift functions as a slow-spreading mid-ocean ridge that is bordered on both sides by thinned continental lithosphere (Wolfenden et al., 2005). As the melt supply increases and/or strain rate increases, new oceanic lithosphere forms in the magmatic segments and the crust and mantle lithosphere subside below sea level. This transition has occurred in the Gulf of Aden

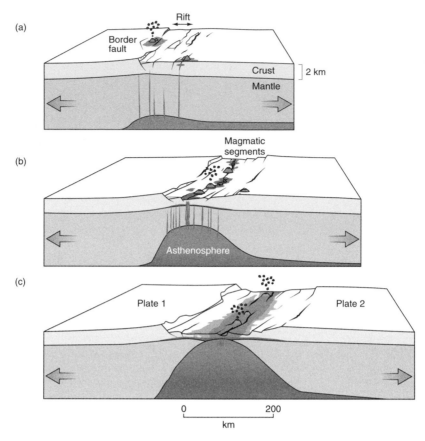

Figure 7.38 *Three-stage model for continental break-up leading to the formation of a volcanic passive margin (after Ebinger, 2005, with permission from Blackwell Publishing).*

(Fig. 7.2b) where conjugate rifted margins have formed recently. The margins on the western side of the Gulf are mostly buried by Oligocene-Miocene lavas from the Afar mantle plume. Those on the eastern side are starved of sediment and volcanic material and preserve 19–35 Ma structures that formed during oblique rifting and the transition to sea floor spreading (d'Acremont *et al.*, 2005). Seismic reflection studies of these latter margins indicate that the southern rifted margin is about twice as wide as the northern one and displays thicker post-rift deposits and greater amounts of subsidence. As rifting gave way to sea floor spreading in this area, deformation localized in a 40-km-wide transition zone where magma intruded into very thin continental crust and, possibly, in the case of the northern side, exhumed mantle. The different widths and structure of the two margins indicate that the transition to sea floor

spreading in the Gulf of Aden was an asymmetric process.

7.8.2 The Woodlark Rift

The Woodlark Basin and adjacent Papuan Peninsula (Fig. 7.39a) record a continuum of active extensional processes that vary laterally from continental rifting in the west to sea floor spreading in the east. This example provides an important record of how sea floor spreading segments develop spatially during continental break-up and the formation of nonvolcanic margins. It also illustrates the type of lithospheric conditions that promote the development of metamorphic core complexes during rifting. Continental rifting occurs presently in the Papuan Peninsula where core complexes

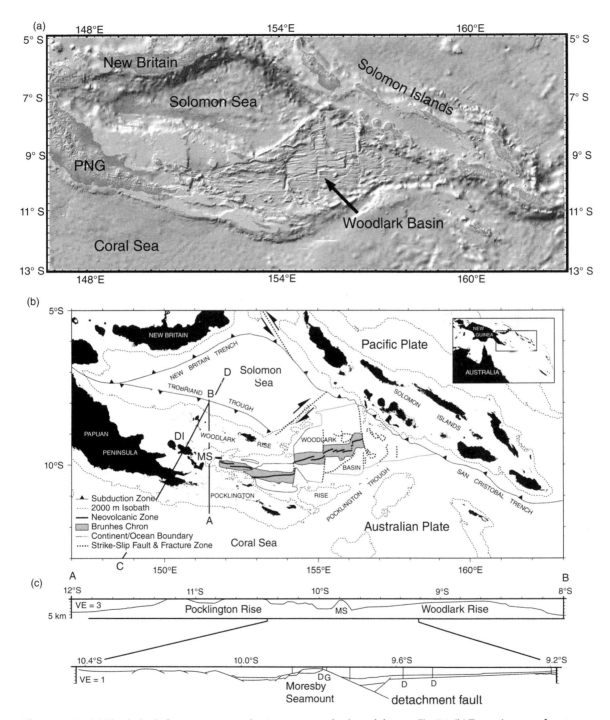

Figure 7.39 *(a) Shaded relief map constructed using same methods and data as Fig. 7.1. (b) Tectonic map of eastern Papua New Guinea (PNG) and the Solomon Islands showing present-day tectonic setting and (c) cross-section (A–B) of western Woodlark Rift showing topography and detachment fault (images in (b) and (c) provided by B. Taylor and modified from Taylor & Huchon, 2002, with permission from the Ocean drilling Program, Texas A & M University). DI, D'Entrecasteaux Islands; MS, Moresby Seamount; D, dolerite; G, gabbro. Line C–D indicates the line of the section shown in Fig. 7.40a.*

and both high-angle (≥45°) and low-angle (<30°) normal faults have formed in the D'Entrecasteaux islands since the Pliocene. Ocean crust in the easternmost and oldest part of the Woodlark Basin is now being consumed to the north beneath the Solomon Islands (Fig. 7.39b).

The pre-rift evolution of the Woodlark region involved subduction, arc volcanism, and arc-continent collision (Section 10.5) along a relic Paleogene convergent plate boundary that now coincides with the Pocklington Rise and southern margin of the Papuan Peninsula (Fig. 7.39b). As the Coral Sea opened from 62 to 56 Ma, fragments of continental crust rifted away from Australia and collided with a Paleogene volcanic arc during north-directed subduction along this plate boundary (Weissel & Watts, 1979). The Trobriand Trough, located to the north of the Woodlark Rise (Fig. 7.39b), is a Neogene subduction zone that accommodates south-directed motion of the Solomon sea floor. This region thus records a history of convergence and crustal thickening that pre-dates the onset of extension during the Pliocene.

Rift initiation in the Pliocene split the rheologically weak continental fragments and volcanic arc of the Woodlark and Pocklington rises. This weak zone lay between two regions of strong oceanic lithosphere in the Coral and Solomon seas and helped to localize strain during rifting (Taylor *et al.*, 1995). Rifting began more or less synchronously along 1000 km of the margin at ~6 Ma. However, strain localization and sea floor spreading developed in a time transgressive fashion from east to west within this large zone. Sea floor spreading began east of about 157° E longitude and was focused there up until ~3.6 Ma. At ~3.6 Ma a spreading ridge abruptly propagated ~300 km westward to ~154° E longitude. Seismic studies (Abers *et al.*, 2002; Ferris *et al.*, 2006) indicate that the crust thickens from <20 km beneath the D'Entrecasteaux islands to 30–35 km beneath the eastern Papuan Peninsula.

Rifting eventually led to the formation of nonvolcanic margins along the northern and southern boundaries of the Woodlark Basin. Currently, continental break-up is focused on an asymmetric rift basin bounded by a low-angle (27°) extensional detachment fault (Fig. 7.39c) that extends though the entire thickness (3–9 km) of the seismogenic layer north of the Moresby Seamount (Abers *et al.*, 1997). Abers & Roecker (1991) identified several possible earthquake events that may indicate active slip on this low-angle

detachment. By contrast, break-up at 2 Ma occurred along a symmetric rift basin bounded by high-angle normal faults. Extension and the slip on low-angle shear zones has resulted in the very rapid (>10 mm a^{-1}) exhumation of deep (up to 75 km) Pliocene plutonic and metamorphic rocks that formed during prior subduction (Baldwin *et al.*, 2004). These core complexes formed when thick upper crust was pulled apart by extension. This process was aided by the emplacement of dense ophiolitic material over less dense crust during Paleogene collision (Abers *et al.*, 2002). Focused extension locally raised temperatures in the lithosphere and allowed buoyant lower crust and mantle to flow beneath the core complexes (Fig. 7.40a). Presently, the Moho is elevated beneath the core complexes, indicating that the lower crust maintains some strength and has not yet flowed sufficiently to smooth out these variations.

The Woodlark Rift indicates that continental break-up occurs in a step-wise fashion by successive phases of rift localization, spreading center nucleation, spreading center propagation, and, finally, a jump to the next site of localized rifting (Taylor *et al.*, 1999). Extension within the rifted nonvolcanic margins continued for up to 1 Myr after sea floor spreading initiated. The transition from rifting to sea floor spreading occurred after a uniform degree of continental extension of 200 ± 40 km and some 130–300% strain (Taylor *et al.*, 1999). Spreading segments nucleated in rift basins that were separated from one another by accommodation zones (Fig. 7.40b). The initial spreading segments achieved much of their length at nucleation, and subsequently lengthened further as spreading propagated into rifting continental crust. Offset margins were controlled by the geometry and location of rheological weaknesses in the Papuan Peninsula. The spreading centers nucleated in orientations approximately orthogonal to the opening direction but, because the developing margins were oblique to this direction, nucleation jumps occurred in order to maintain the new spreading centers within rheologically weak zones. Transform faults, which cut across previous rift structures, link spreading segments that had nucleated in, and/or propagated into, offset continental rifts. This relationship indicates that transform faults do not evolve from transfer faults between rift basins. In addition, the Woodlark example shows how rheological weaknesses in the lithosphere continue to control how continents break-up during the final stages of the transition from rifting to sea floor spreading.

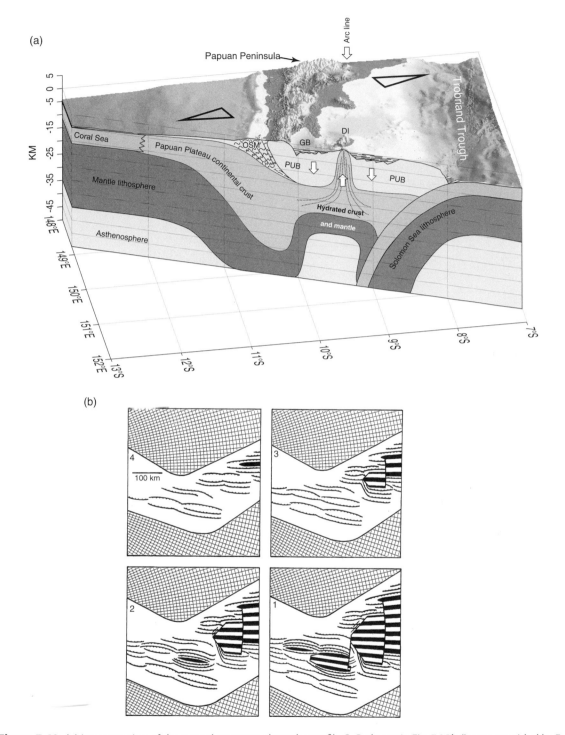

Figure 7.40 (a) Interpretation of the crustal structure along the profile C–D shown in Fig. 7.39b (image provided by F. Martínez and modified from Martínez et al., 2001, with permission from Nature **411**, 930–4. Copyright © 2001 Macmillan Publishers Ltd). DI, D'Entrecasteaux Islands; GB, Goodenough basin; PUB, Papuan ultramafic belt; OSM, Owen Stanley metamorphic belt. (b) Model of continental break-up and ocean formation derived from the Woodlark Basin and Papua New Guinea (modified from Taylor et al., 1999, by permission of the American Geophysical Union. Copyright © 1999 American Geophysical Union). White areas are continental lithosphere. Nonextending regions are represented by a pattern of small and great circles to the pole of opening. Black and white stripes are new oceanic lithosphere. Four stages are shown from 4 Ma to 1 Ma.

7.9 THE WILSON CYCLE

The transition from intracontinental rift to ocean basin has occurred repeatedly on Earth since at least the Late Archean (Section 11.3.5). The relatively young Mesozoic-Cenozoic age of the current ocean basins implies that there have been many cycles of ocean creation and destruction during the Earth's history. Very little remains of these ancient oceans, although their existence is implied by continental reconstructions (Figs 3.4, 3.5) and by fragments of ancient ocean crust that are preserved as ophiolite assemblages (Section 2.5) in orogenic belts

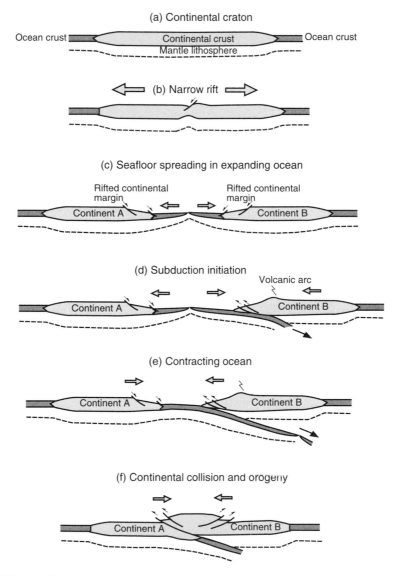

Figure 7.41 *The Wilson cycle showing: (a) continental craton; (b) formation of a narrow rift; (c) initiation of seafloor spreading and formation of rifted continental margins in an expanding ocean basin; (d) initiation of subduction; (e) a closing ocean basin; (f) continental collision and orogeny.*

(Section 10.6.1). This periodicity of ocean formation and closure is known as the *Wilson cycle*, named after J. Tuzo Wilson in recognition of his contributions to the theory of plate tectonics (Dewey & Burke, 1974).

Figure 7.41 shows a schematic illustration of the various stages in the Wilson cycle beginning with the initial break-up of a stable continental craton (Fig. 7.41a) and the thinning of continental lithosphere. Rifting (Fig. 7.41b) is followed by the development of a thinned, rifted continental margin and eventually gives way to sea floor spreading as the two continents separate across an expanding ocean (Fig. 7.41c). The termination of basin opening may occur in response to plate collisions, which could trigger subduction at one or more rifted margins (Fig. 7.41d). Basin closure also may compensate for oceanic lithosphere that is newly formed elsewhere. The contracting ocean is a consequence of subduction at one or both continental margins (Fig. 7.41e). This phase will continue until the two continents collide and the ocean basin closes completely (Fig. 7.41f). Continent–continent collision leads to the formation of a Himalaya-type orogen (Section 10.1) and the exhumation of deep crustal rocks. At this time subduction zones must initiate at other continental margins in order to maintain constant global surface area. The forces associated with these new subduction zones place the continent under tension and, if other conditions are extant (Section 7.5), the rifting process begins again. Present day analogues of the oceans shown in Fig. 7.41 are: Fig. 7.41c (expanding oceans) = the Gulf of Aden, Woodlark Rift, and the Atlantic Ocean; Fig. 7.41d,e (contracting oceans) = the Pacific Ocean. Chapters 9 and 10 provide discussions of the processes that operate during the destructive part of the Wilson cycle as ocean basins close and continents collide.

8 Continental transforms and strike-slip faults

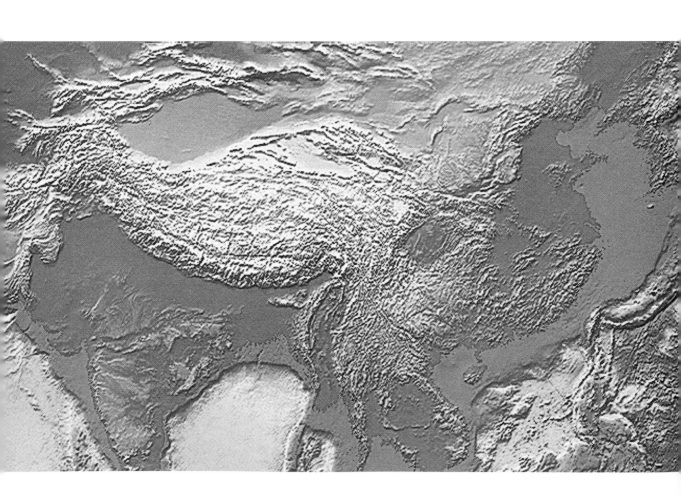

8.1 INTRODUCTION

Continental transforms, like their oceanic counterparts (Section 4.2.1), are conservative plate boundaries where lithosphere is neither created nor destroyed and strike-slip deformation results in lateral displacements across the fault zone. Strike-slip faults generally may occur at a variety of scales in virtually any tectonic setting. Only transform faults represent plate boundaries.

In contrast to oceanic fracture zones, which are characterized by a relatively simple linear trough (Section 6.12), continental transforms exhibit a structural complexity that reflects differences in the thickness, composition, and pressure–temperature profile of oceanic and continental lithosphere (Sections 2.7, 2.10.4). In the southwestern United States, for example, relative motion between the Pacific and North American plates is distributed across a zone that ranges from hundreds to a thousand kilometers wide (Fig. 8.1). Similarly, in New Zealand (Fig. 8.2), oblique convergence on the South Island has produced a >100-km-wide zone of deformation on the continental portion of the Pacific plate. These diffuse, commonly asymmetric patterns generally reflect lateral contrasts in lithospheric strength and areas where continental lithosphere is especially weak (Section 8.6.2). In areas where continental lithosphere is relatively cool and strong, transforms tend to display narrow zones of deformation. The Dead Sea Transform is an example of this latter type of system where deformation has localized into a zone that is only 20–40 km wide (Fig. 8.3).

In this chapter, the shallow (Section 8.2) and deep (Section 8.3) structure of continental transforms and major strike-slip faults is illustrated using examples from the southwestern U.S., New Zealand, the Middle East, and elsewhere. Other topics include the evolution of transform continental margins (Section 8.4), the use of velocity fields to describe crustal motion (Section 8.5), and the mechanisms that control the localization and delocalization of strain during strike-slip faulting (Section 8.6). This latter subject, and the overall strength of large strike-slip faults (Section 8.7), are especially important for explaining how continental transforms accomplish large magnitudes of slip.

8.2 FAULT STYLES AND PHYSIOGRAPHY

The following fault styles and physiographic features characterize the surface and upper crust of continental transforms and major continental strike-slip faults:

1 *Linear fault scarps and laterally offset surface features.* Large continental strike-slip faults typically display linear scarps and troughs that result from the differential erosion of juxtaposed material and the erosion of fault gouge (Allen, 1981). Surface features along active or recently active fault traces may be displaced laterally due to the strike-slip motion. The age and magnitude of these offsets provide an important means of determining slip rates. In New Zealand, for example, the Alpine Fault is marked by a nearly continuous, linear fault trace that extends across the South Island for a distance of ~850 km (Fig. 8.2). Glacial moraines, rivers, valleys, lake shores, and other topographic features are offset laterally across the fault (Fig. 8.4), suggesting late Pleistocene slip rates of 21–24 mm a^{-1} (Sutherland *et al.*, 2006). Vertical motion between parallel fault segments also is common and may create areas of localized uplift and subsidence that are expressed as pressure ridges and sag ponds, respectively (Sylvester, 1988).

2 *Step-overs, push-ups, and pull-apart basins.* Most large strike-slip faults are composed of multiple fault segments. Where one active segment terminates in proximity to another sub-parallel segment, motion is transferred across the intervening gap, resulting in zones of localized extension or contraction (Fig. 8.5a). In these *step-overs*, the initial geometry and sense of slip on the adjacent faults control whether the area separating them is extended or shortened (Dooley & McClay, 1997; McClay & Bonora, 2001). Normal faults and extensional troughs called *pull-apart basins* characterize step-overs where the intervening region is thrown into tension. Thrust faults, folds, and topographic uplifts known as *push-ups* form where the intervening region is compressed. In these

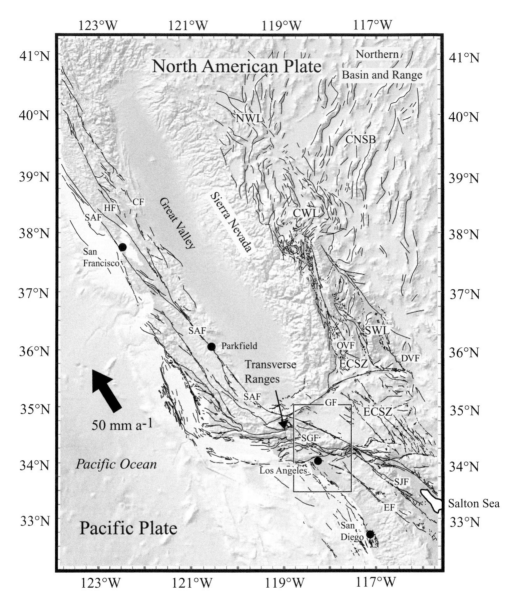

Figure 8.1 *Shaded relief map showing major faults and topographic features in California and western Nevada. Fault traces are from Jennings (1994) and Oldow (2003). SAF, San Andreas Fault; HF, Hayward Fault; CF, Calaveras Fault; GF, Garlock Fault; SGF, San Gabriel Fault; EF, Elsinore Fault; SJF, San Jacinto Fault; ECSZ, Eastern California Shear Zone; OVF, Owens Valley Fault; DVF, Death Valley Fault; CNSB, Central Nevada Seismic Belt. SWL, CWL, and NWL are the southern, central, and northern Walker Lane, respectively. Box shows location of Fig. 8.8a. Map was constructed using the same topographic data and methods as in Fig. 7.1.*

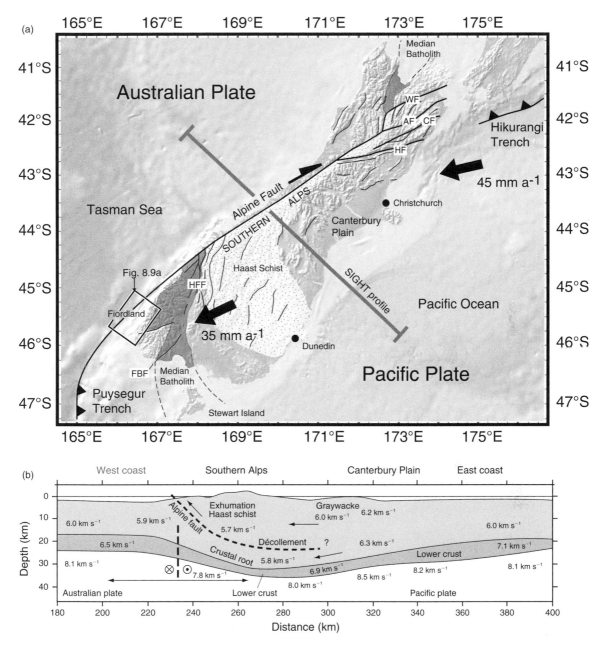

Figure 8.2 *(a) Shaded relief map showing major faults and tectonic features of the Australian-Pacific plate boundary on the South Island of New Zealand. Map was constructed using the same topographic data and methods as in Fig. 7.1. WF, Wairau Fault; AF, Awatere Fault; CF, Clarence Fault; HF, Hope Fault; HFF, Hollyford Fault; FBF, Fiordland Boundary Fault. (b) Seismic velocity profile constructed without vertical exaggeration (modified from Van Avendonk et al., 2004, by permission of the American Geophysical Union. Copyright © 2002 American Geophysical Union).*

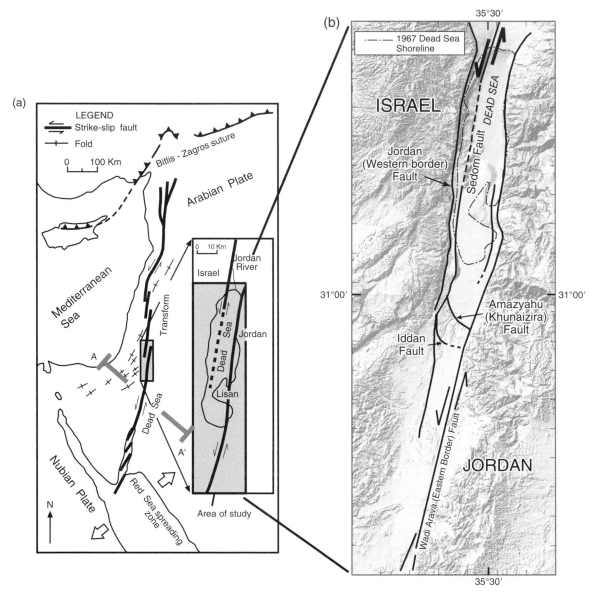

Figure 8.3 *(a) Tectonic map and (b) shaded relief map showing major fault segments of the Dead Sea Transform and pull-part basin (images provided by U. ten Brink and modified from Al-Zoubi & ten Brink, 2002, with permission from Elsevier). Digital topography in (b) is from Hall (1993), 1967 coastline of the Dead Sea, showing subsidence of the basin, is from Neev & Hall (1979). Profile A–A' is shown in Fig. 8.11. Folds reflect Mesozoic–Early Cenozoic shortening.*

settings, the combination of strike-slip motion and extension is known as *transtension*. The combination of strike-slip motion and contraction is known as *transpression*.

The El Salvador Fault Zone in Central America illustrates many of the physiographic and structural features that are common to extensional step-overs. In this region, oblique

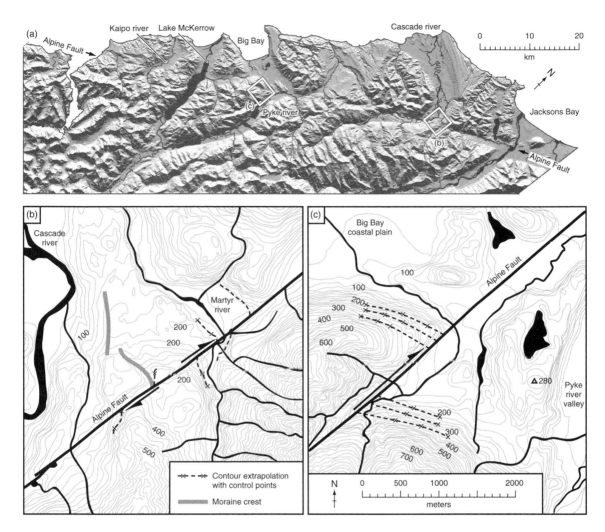

Figure 8.4 *(a) Shaded relief map and (b, c) 1:50,000 topographic maps showing linear scarp and offset surface features along a segment of the Alpine Fault on the South Island of New Zealand (images provided by R. Sutherland and modified from Sutherland et al., 2006, with permission from the Geological Society of America). Maps are derived from 1:50,000 NZMS 260 digital data. Curved bold black lines in (b) and (c) are rivers or creeks. Contour interval is 20 m.*

convergence between the Cocos and Caribbean plates (Fig. 8.6a) results in a component of dextral motion within a volcanic arc above the Middle America Trench (Martínez-Díaz *et al.*, 2004). The Río Lempa pull-apart basin is marked by several irregular depressions and oblique normal faults that have formed in an extensional

step-over between the San Vicente and Berlin fault segments (Fig. 8.6b) (Corti *et al.*, 2005). Late Pleistocene volcanic edifices, river terraces, and alluvial fans are offset across prominent fault scarps.

In northern California, dextral strike-slip faults in the San Francisco Bay area record crustal shortening and topographic uplift related to

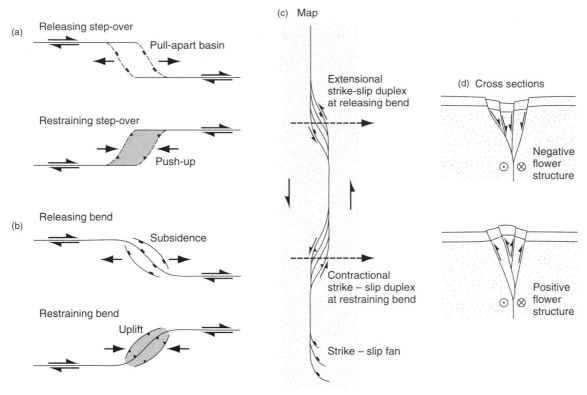

Figure 8.5 *Map views of (a) step-overs and (b) bends and associated structures (after McClay & Bonora, 2001,* Bull. Am. Assoc. Petroleum Geols. *AAPG © 2001, reprinted by permission of the AAPG whose permission is required for further use). (c) Map and (d) cross-sections of strike-slip duplexes, fans and flower structures developed at bends (after Woodcock & Rickards, 2003, with permission from Elsevier).*

a series of contractional step-overs. East of the bay, Mt. Diablo (Fig. 8.7a) marks the core of an anticlinorium that has formed between the Greenville and Concord faults (Unruh & Sawyer, 1997). The transfer of about 18 km of dextral strike-slip motion across this step-over during the late Cenozoic has resulted in a series of oblique anticlines, thrust faults, and surface uplifts that form a typical stepped, overlapping *en echelon* pattern. Mt. Diablo is the largest push-up in the region. Studies of deformed fluvial terraces suggest an uplift rate of 3 mm a^{-1} over the last 10,000 years, which is comparable to the rates of slip on the adjacent faults (Sawyer, 1999). Bürgmann *et al.* (2006) resolved the rates of vertical crustal motion associated with several

contractional step-overs near San Francisco Bay by combining GPS velocities with interferometric synthetic aperture radar (InSAR) data (Section 2.10.5) collected over an 8 year period. After filtering out seasonally varying ground motions, the InSAR residuals (Fig. 8.7b) showed that the highest uplift rates occur over the southern foothills of Mount Diablo. Other zones of rapid uplift occur in the Mission Hills step-over between the Hayward and Calaveras faults, and between faults in the Santa Cruz Mountains. In the former area, seismicity is consistent with the transfer of slip on the Calaveras Fault onto the northern Hayward Fault through the Mission Hills (Waldhauser & Ellsworth, 2002). The origin of other vertical

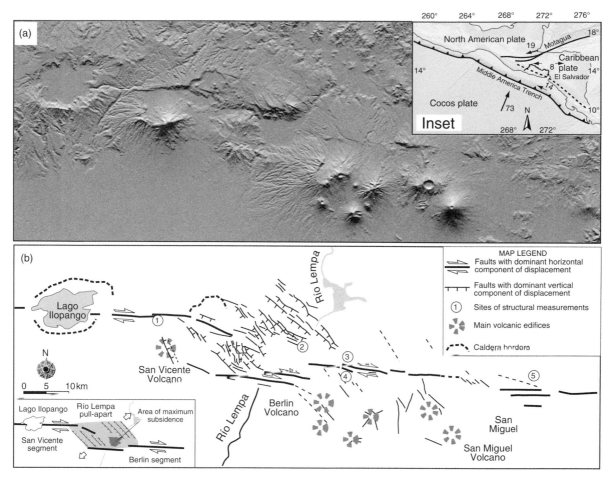

Figure 8.6 *(a) Shaded relief map and (b) interpretation of the Río Lempa pull-apart basin between the San Vicente and Berlin fault segments of the El Salvador Fault Zone (images provided by G. Corti and modified from Corti et al., 2005, with permission from the Geological Society of America). Digital elevation model is derived from SRTM data (http://srtm.usgs.gov/) and Landsat ETM 7 satellite images processed by the University of Maryland, Global Land Cover facility. Inset shows the plate tectonic setting of Central America and plate velocities (mm a^{-1}) after DeMets, 2001.*

movements is more uncertain and may reflect some nontectonic displacements, such as active landslides, subsidence and rebound over aquifers, and the settling of unconsolidated sediments along the bay margins. Nevertheless, the data reveal a pattern of highly localized vertical motion associated with regions of active strike-slip faulting.

3 *Releasing and restraining bends.* In zones where strike-slip faults are continuous, the strike of the faults may locally depart from a simple linear trend following a small circle on the Earth's surface. In these areas, the curvature of the fault plane creates zones of localized shortening and extension according to whether the two sides of the bend converge or diverge (Harding, 1974; Christie-Blick & Biddle, 1985) (Fig. 8.5b). These zones are similar to those that form in step-overs. Pull-apart basins, zones of subsidence and deposition, and normal faults characterize *releasing bends.* *Restraining bends* display thrust faults, folds, and push-ups.

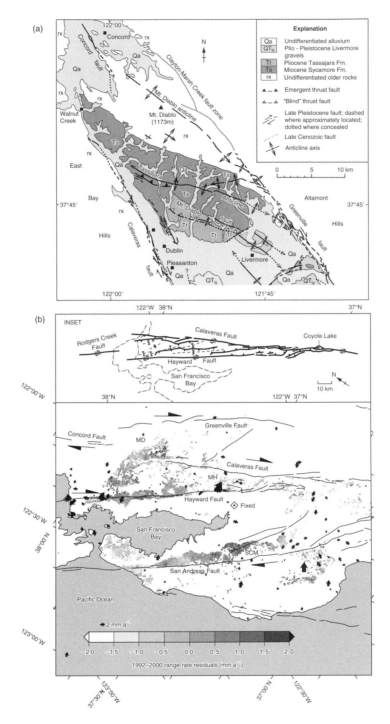

Figure 8.7 *(a) Geologic map of the Mount Diablo contractional step-over (after Wakabayashi et al., 2004, with permission from Elsevier). Geologic data are from Wagner et al. (1990) and Unruh & Sawyer (1997). (b) Fault map showing residual permanent scatterer InSAR rates (dots) after removing the contribution of tectonic horizontal motions and all points located on late Pleistocene substrate (image provided by R. Bürgmann and modified from Bürgmann et al., 2006, with permission from the Geological Society of America). Permanent scatterers are stable radar-bright points such as buildings, outcrops, utility poles etc. that are used to identify time-dependent surface motions. Modeled range rates include 115,487 permanent scatterers relative to point labeled FIXED for the years 1992–2000. Positive residuals correspond to uplift. MD, Mount Diablo; MH, Mission Hills; SCM, Santa Cruz Mountains. Black arrows show residual (observed minus modeled) GPS horizontal velocities, which provide a measure of how well the model fits the observations (Section 8.5.3). Inset in (b) shows the geometry of folds and thrust faults in a contractional step-over between the Hayward and Calaveras faults (after Aydin & Page, 1984, with permission from the Geological Society of America).*

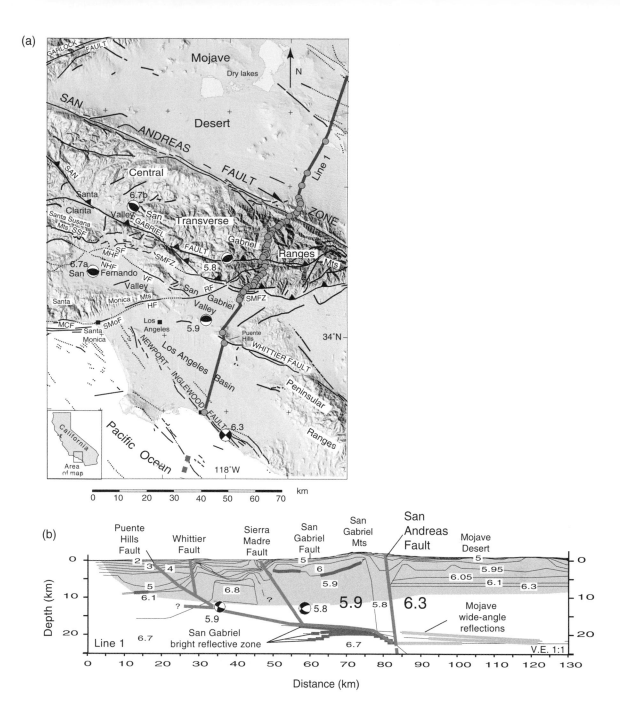

Figure 8.8 *(a) Shaded relief map and (b) seismic profile of the Central Transverse Ranges, southern California showing faults (thin black lines, dotted where buried), shotpoints (gray circles), seismographs (thick black line), and epicenters of earthquakes greater than M = 5.8 since 1933 (images provided by G. Fuis and modified from Fuis et al., 2003, with permission from the Geological Society of America). Focal mechanisms with attached magnitudes: 6.7a, Northridge (Hauksson et al., 1995); 6.7b, San Fernando (Heaton, 1982); 5.9, Whittier Narrows (Hauksson et al., 1988); 5.8, Sierra Madre (Hauksson, 1994); 6.3, Long Beach (Hauksson, 1987). HF, Hollywood Fault; MCF, Malibu Coast Fault; MHF, Mission Hills Fault; NHF, Northridge Hills Fault; RF, Raymond Fault; SF, San Fernando surface breaks; SSF, Santa Susana Fault; SmoF, Santa Monica Fault; SMFZ, Sierra Madre Fault Zone; VF, Verdugo Fault. In (b) gray area represents refraction coverage, thin black lines represent velocity contours or boundaries; contour interval 0.5 km s⁻¹ to 5.5 km s⁻¹ and arbitrary above 5.5 km s⁻¹. Large numbers on either side of the San Andreas Fault are average basement velocities.*

The Transverse Ranges in southern California (Figs 8.1, 8.8a) illustrate the characteristics of a large restraining bend in the San Andreas Fault. These ranges have been uplifted in response to a combination of dextral motion and compression across a portion of the fault that strikes more westerly than the general strike of the fault system. Seismic reflection profiles and information from wells indicate that thrust faults dip northward at $25–35°$ beneath the San Gabriel Mountains and intersect the near vertical ($83°$) San Andreas Fault at mid-crustal depths of $\sim21\,km$ (Fuis *et al.*, 2001, 2003). Earthquake focal mechanisms show thrust solutions on fault splays that branch upward off the dipping décollement surface (Fig. 8.8b). This combination of motion has resulted in a zone of transpression and topographic uplift commonly referred to as the *Big Bend*.

Examples of active releasing bends and strike-slip basins occur along the southernmost part of the Alpine Fault in southwest New Zealand. Near Fiordland, three semicontinuous fault segments accommodate dextral strike-slip motion between the Australian and Pacific plates. Along the Resolution segment of the plate boundary (Fig. 8.9a), geophysical surveys have revealed the presence of a Pleistocene pull-apart called the Dagg Basin (Barnes *et al.*, 2001, 2005). A seismic reflection profile across the northern part of the basin (Fig. 8.9b) shows that it is bounded on the northwest by a ridge above an active reverse fault. Inactive faults are buried beneath the ridge. At the center of the basin, upward splaying faults accommodate oblique extension, forming a graben. Some west-dipping splays (labeled IA in Fig. 8.9b) presently are inactive, although the deposition of wedge-shaped strata between the development of two unconformities (surfaces DB3 and DB4) indicates that they once were active simultaneously with the east-dipping splays. This geometry suggests that the pull-apart basin initially formed in an extensional step-over prior to unconformity DB3 (Barnes *et al.*, 2001, 2005). Another pull-apart, called the Five Fingers Basin, formed in a similar step-

over 10 km farther south (Fig. 8.9a). The current smooth shape of the releasing bends formed later, after unconformity DB3, as subsequent strike-slip motion formed faults that joined across the gap between the step-overs (Fig. 8.10a,b).

In contrast to the extension that characterizes the northern Dagg Basin, the southern end of the basin shows evidence of reverse faulting and uplift. A combination of shortening and strike-slip faulting associated with a restraining bend in this region formed the Dagg Ridge, which has been squeezed upward between the main trace of the Alpine Fault on the west and a curved oblique-slip fault beneath its eastern margin (Fig. 8.9c). South of the ridge, the Breaksea Basin preserves features that indicates it was once continuous with the Dagg Basin, suggesting that the reverse faulting occurred after the pull apart had formed (Barnes *et al.*, 2005). As the total plate motion and amount of slip increased, some faults were abandoned and others formed linkages that cut through the extensional basins, resulting in localized push-ups and ridges where they formed restraining bends (Fig. 8.10c). These relationships illustrate how large strike-slip faults typically evolve very rapidly and that localized strike-slip basins and uplifts develop along different parts of the fault zone (Fig. 8.10c) on timescales of tens to hundreds of thousands of years.

4 *Strike-slip duplexes, fans, and flower structures.* A *strike-slip duplex* is an imbricate array of two or more fault-bounded blocks and basins that occur between two or more large bounding faults (Woodcock & Fischer, 1986). These structures are analogous to the duplexes that form on the ramps of dip-slip faults but differ in that vertical movements are not constrained at the upper (ground) surface. The fault-bounded basins that characterize the duplex typically are lens-shaped. The individual blocks defined by the strike-slip faults are shortened and uplifted when the faults converge and stretched and downthrown where the faults diverge (Fig. 8.5c). This tendency for strike-slip faults to diverge and converge creates a characteristic

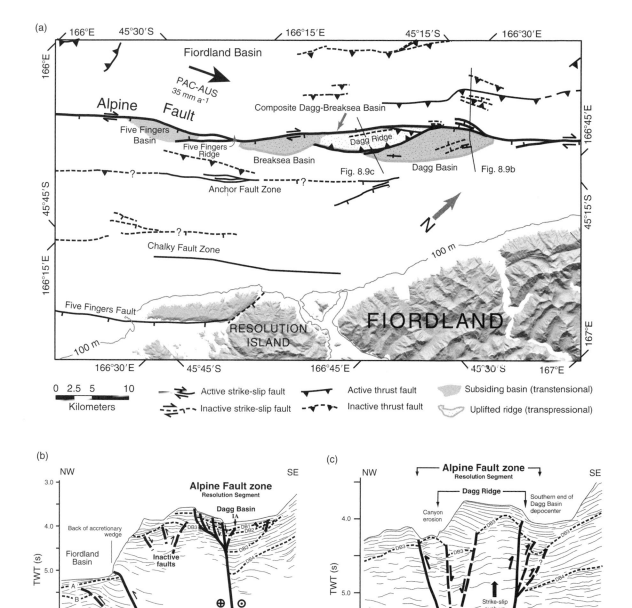

Figure 8.9 (a) Map of the south-central coast of Fiordland and (b,c) line drawings of seismic reflection profiles showing active faults, strike-slip basins, and other physiographic features along the southern segment of the Alpine Fault (images provided by P. Barnes and modified from Barnes et al., 2005, with permission from the Geological Society of America). Map location shown in Fig. 8.2a. Profile (b) shows the subsiding part of the Dagg Basin, profile (c) the uplifting part. Profile locations are shown in (a). Solid faults are active, dashed faults are inactive. Labeled reflections are unconformities.

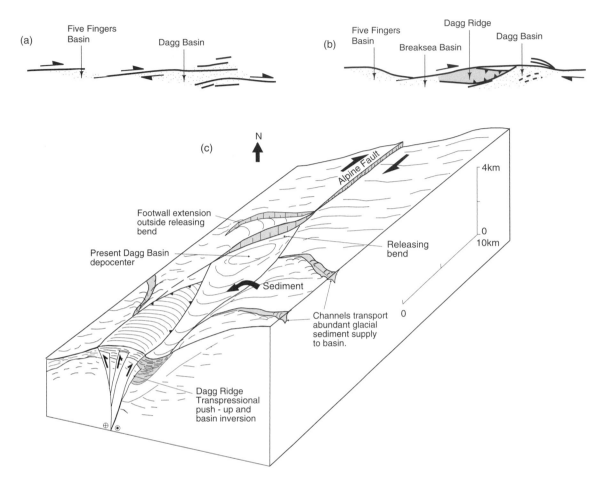

Figure 8.10 *Sketches showing the progressive evolution of the Resolution segment of the Alpine Fault near Fiordland at (a) ~1 Ma when a series of pull-apart basins formed between extensional step-overs and (b) presently when linkages among faults have cut through the Dagg Basin forming the Dagg Ridge (shaded) (after Barnes* et al., *2005, with permission from the Geological Society of America). (c) Schematic block diagram showing the three-dimensional geometry of adjacent releasing and restraining bends (image provided by P. Barnes and modified from Barnes* et al., *2001, with permission from Elsevier). Southern end of basin displays a positive flower structure, push-up, and transpression. Northern end of basin displays pull-apart basin, subsidence, and transtension.*

braided pattern in plan view. The faults that most closely follow the direction of plate movements predominate, grow longer, and assume near vertical dips. Other faults at an angle to the overall direction of movement may then rotate farther out of alignment and develop dips significantly less than vertical, so that the fault involves a component of dip-slip motion. If the fault's curvature carries it to a region of extension, a normal oblique-slip fault develops; if to a region of compression, a

reversed oblique-slip fault forms. Significant rotations about vertical or near vertical axes also commonly occur (Section 8.5). At the ends of large strike-slip faults, displacements may be dissipated along arrays of curved faults that link to the main fault forming *fans* or *horsetail splays* (Fig. 8.5c). These structures may record either contractional or extensional deformation according to the geometry of the curvature and the sense of motion on the main fault.

In profile, the various splays of a strike-slip fault zone may converge downward at depth to produce a characteristic geometry in profile known as a *flower structure* (Fig. 8.5d) (Harding, 1985; Christie-Blick & Biddle, 1985). Negative flower structures are those where the upward-branching faults display mostly normal offsets beneath a synform or surface depression (e.g. Fig. 8.9b). Positive flower structures are those where the upward-branching faults display mostly reverse offsets beneath an antiform or surface culmination. A positive flower structure is illustrated by the geometry of faults in southern Dagg Basin (Fig. 8.10).

5 *Strike-slip partitioning in transpression and transtension.* There are several ways in which displacements may be distributed between the boundaries of obliquely converging or diverging blocks and plates. One common way is by simultaneous motion on separate strike-slip and contractional or extensional structures. In this scenario, strike-slip faults accommodate the component of oblique convergence/divergence that parallels the plate boundary and the contractional or extensional structures accommodate the component oriented orthogonal to the plate boundary. Such systems, where strike-slip and dip-slip motion occur in different places and on separate structures, are *strike-slip partitioned.* Alternatively, both strike-slip and margin-perpendicular components of the deformation may occur either on the same structure, such as occurs presently on the central oblique-slip section of the Alpine Fault in New Zealand (Section 8.3.3), or both components may be distributed more or less uniformly across a zone. The relative contributions of strike-slip and margin-perpendicular deformation allow further classification into strike-slip-dominated and thrust- (or normal-) dominated systems. The southern segment of the Alpine Fault illustrates a strike-slip partitioned style of transpression. Near the Fiordland margin (Fig. 8.9a), the fault lies at a low angle (11–25°) to the azimuth of Pacific–Australian plate motion (Barnes *et al.*, 2005). This low angle results in almost pure strike-slip motion along the active trace of the Alpine Fault,

which in this area is nearly vertical. The contractional component of deformation that arises from the oblique plate convergence is accommodated by structures located both west and east of the Alpine Fault. On the western side, a 25-km-wide thrust wedge is composed of a series of active thrust, reverse, and oblique-slip faults that steepen downward toward the Alpine Fault. This fault segment illustrates how shortening occurs simultaneously with dextral strike-slip motion in different places along the plate boundary.

A similar strike-slip partitioned system occurs in the "Big Bend" region of southern California where thrust faults accommodate contraction simultaneously with dextral strike-slip motion. Within the San Gabriel Mountains (Fig. 8.8a), the San Andreas Fault lies at about 35° to the direction of relative motion between the Pacific and North American plates. This oblique angle results in a component of contraction that is accommodated by reverse faulting and folding within the mountains north of the Los Angeles Basin. The oblique angle also results in strike-slip motion, which is accommodated by a series of steep west–northwest-trending faults includes the San Andreas Fault itself (Fuis *et al.*, 2003).

An example of a very weakly or nonpartitioned style of transpressional deformation occurs along the central segment of the Alpine Fault on the South Island of New Zealand. Here, the Alpine Fault strikes to the northeast (55°) and dips moderately to the southeast (Fig. 8.2b, Section 8.3.3). Norris & Cooper (2001) showed that, unlike the Fiordland segment, slip on the central segment of the fault is oblique and approximately parallels the interplate vector (Section 8.3.3). At the eastern and western limits of deformation on the South Island, reverse faults approximately parallel the Alpine fault but are inferred to have relatively low rates and minor components of strike-slip motion (Norris & Cooper, 2001; Sutherland *et al.*, 2006). These characteristics indicate that the central segment of the Alpine Fault system is at best weakly partitioned and appears to be nonpartitioned in some areas.

8.3 THE DEEP STRUCTURE OF CONTINENTAL TRANSFORMS

8.3.1 The Dead Sea Transform

The Dead Sea Transform forms part of the Arabia–Nubia plate boundary between the Red Sea and the Bitlis suture zone in eastern Turkey (Fig. 8.3a). The southern part of this plate boundary provides an important example of a transtensional transform that has formed in relatively cool (45–53 mW m^{-2}), strong continental lithosphere (Eckstein & Simmons, 1978; Galanis et al., 1986).

Since its inception in Middle Miocene times, approximately 105 km of left lateral strike-slip motion and ~4 km of fault-perpendicular extension has occurred within the southern part of the plate boundary (Quennell, 1958; Garfunkel, 1981). The component of extension was initiated during the Pliocene (Shamir et al., 2005). Horizontal velocities derived from GPS data (Section 5.8) suggest that relative motion between the Arabian and Nubian plates is occurring at the relatively slow rate of 4.3 mm a^{-1} (Mahmoud et al., 2005). Most of this motion is accommodated by faults that form a series of en echelon step-overs within a narrow, 20- to 40-km-wide transform valley (Fig. 8.3a). Rhomb-shaped grabens, elongate pull-apart basins, and steep normal faults have formed where the fault segments step to the left. One of the largest of these extensional features is the Dead Sea Basin, which is ~135 km long, 10–20 km wide, and filled with at least 8.5 km of sediment (Fig. 8.3b).

Superficially the pull-apart basins and normal faults along the transtensional Dead Sea Transform resemble features that characterize narrow intracontinental rift basins (Section 7.2). Both types of basin typically are asymmetric, bounded by border faults, and display along-strike segmentations (Lazar et al., 2006). However, there are important differences between the two tectonic settings. Among the most significant of these is that, along transtensional transforms, the extension is confined mostly to the crust and displays minimal involvement of the upper mantle (Al-Zoubi & ten

Brink, 2002). Both gravity data (ten Brink et al., 1993) and wide-angle seismic reflection and refraction profiles (DESERT Group, 2004; Mechie et al., 2005) support this conclusion by indicating that the Moho is elevated only slightly (<2 km) under the Dead Sea Basin. These characteristics suggest that, although extension influences the surface morphology and shapes of extensional basins that form along transforms, it does not play a dominant role in shaping the deep structure of the fault system (Section 8.6.2) like it does in rift basins.

Seismic reflection and refraction data collected across the Arava Fault (Fig. 8.3a) reveal the deep structure of the Dead Sea Transform. Beneath the surface trace of the fault, the base of a 17- to 18-km-thick upper crust (seismic basement) is vertically offset by 3–5 km (DESERT Group, 2004; Mechie et al., 2005). The fault descends vertically into the lower crust where it broadens downward into a zone of ductile deformation (Fig. 8.11). The width of this lower crustal zone is constrained by a ~15-km-wide gap in a series of strong subhorizontal reflectors. These reflectors may represent either compositional contrasts related to lateral displacements within a narrow zone or the effects of localized horizontal flow (Al-Zoubi & ten Brink, 2002). Below the gap, the Moho displays a small amount of topography, suggesting that a narrow zone of deformation beneath the Arava Fault may extend into the mantle.

These physical characteristics provide important constraints on the dynamics of transform faults. The results from the DESERT geophysical survey (DESERT Group, 2004; Mechie et al., 2005) suggest that the ~105 km of left lateral displacement between the Arabian and Nubian plates (Fig. 8.12a) has resulted in a profile with a significantly different crustal structure east and west of the Arava Fault (Fig. 8.12b). The occurrence of extension and transtension between fault segments results in localized subsidence and crustal flexure west of the fault and a minor, similar deflection of the Moho (Fig. 8.12c). Erosion and sedimentation result in the present day structure of the plate boundary (Fig. 8.12d).

8.3.2 The San Andreas Fault

The San Andreas Fault formed in Oligocene times (Atwater, 1970, 1989) when the Pacific–Farallon spreading ridge collided with the western margin of North

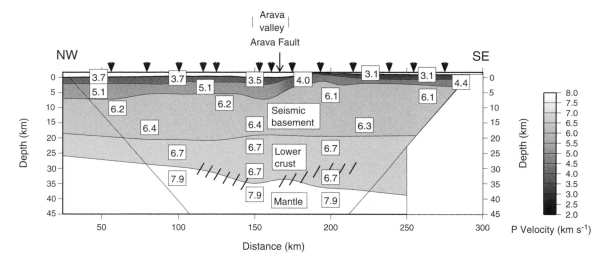

Figure 8.11 *P-wave velocity model of the crust and mantle below the Arava Fault within the southern segment of the Dead Sea Transform (image provided by M. Weber and modified from the DESERT Group, 2004, with permission from Blackwell Publishing). Profile location is shown in Fig. 8.3a. Vertical exaggeration is 2:1. Triangles indicate shot points along a wide angle seismic reflection and refraction survey used to obtain the velocities (km s⁻¹). Hatched area near crust–mantle boundary represents zone of strong lower crustal reflections. The boundaries and P-wave velocities located northwest of the fault are from Ginzburg et al. (1979a,b) and Makris et al. (1983). Those to the southeast of the fault are based on El-Isa et al. (1987a,b).*

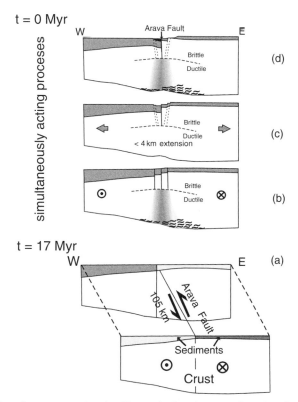

Figure 8.12 *Sketches showing the processes involved in producing the crustal section shown in Fig. 8.11 (image provided by J. Mechie and modified from the DESERT Group, 2004, with permission from Blackwell Publishing). (a) Crustal structure at ~17 Myr, before initiation of strike-slip motion. (b) Strike-slip displacement of ~105 km results in significantly different structure east and west of the Arava Fault. (c) Minor (~4 km) of extension results in subsidence and flexure of the western block and uplift of the eastern block. Moho shows a similar deflection. (d) Erosion and sedimentation produces present structure. Note that processes (b–d) act simultaneously.*

America (Fig. 5.28a, Section 5.11). The fault joins the Mendocino Triple Junction with the Gulf of California and is the only continuous structure within the plate boundary zone (Fig. 8.1). Displacement on the fault is dominantly strike-slip (Fig. 7.10), although in places it also is associated with localized transpression and transtension (Section 8.2). Heat flow measurements (Sass *et al.*, 1994), seismicity (Fig. 7.8), and seismic reflection and refraction surveys (Henstock & Levander, 2000; Godfrey *et al.*, 2002) indicate that the fault has formed in very heterogeneous lithosphere characterized by large lateral variations in thickness, strength, and thermal properties.

Most of the evidence from northern and central California suggests that the San Andreas Fault penetrates into the lower crust as a near vertical structure and may offset the Moho (Holbrook *et al.*, 1996; Hole *et al.*, 2000). From west to east, the top and bottom of a 5- to 6-km-thick lower crust drops by up to 4 km across the San Andreas Fault. The Moho is similarly offset, albeit by only ~2 km. Seismic velocities in the upper mantle show a small change across the profile, from 8.1 km s^{-1} beneath the Pacific to about 7.9 km s^{-1} beneath the Coast Ranges, suggesting that the latter are characterized by slightly lower densities and temperatures

that are higher by ~550 K between 30 and 50 km depth (Henstock & Levander, 2000).

Beneath the Transverse Ranges, a velocity model (Fig. 8.13), constructed using active-source seismic data, reveals the presence of an 8-km-thick crustal root centered beneath the surface trace of the San Andreas Fault (Godfrey *et al.*, 2002). The presence of this crustal root indicates that the transpression associated with the Big Bend in the San Andreas Fault (Section 8.2) affects the entire crust. The data also show an offset Moho. Estimates of the magnitude of the offset are variable, mostly because they depend on the specific velocities used. Published estimates show the Moho to be at least one and possibly several kilometers deeper on the northern side of the fault. A similar, small Moho disruption also occurs beneath the Eastern California Shear Zone (Zhu, 2000) (Fig. 8.1). These offsets suggest that a narrow zone of brittle and ductile deformation surrounding the southern segment of the San Andreas Fault also extends vertically through the entire crust.

In addition to an offset Moho, the velocity model shown in Fig. 8.13 indicates that relatively slow seismic velocities (6.3 km s^{-1}), which are consistent with weak, quartz-rich lithologies, characterize the middle and

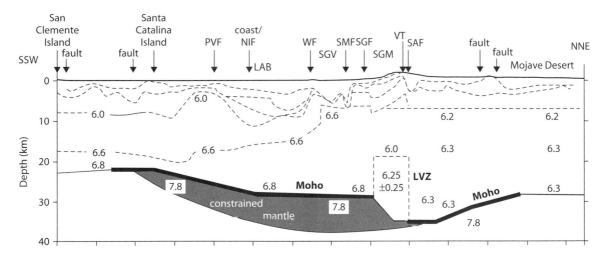

Figure 8.13 *Velocity model of the crust and mantle below the Los Angeles Basin (LAB) (modified from Godfrey* et al., *2002, by permission of the American Geophysical Union. Copyright © 2002 American Geophysical Union). Section parallels thick black profile line shown in Fig. 8.8a. Dashed lines are velocity contours. Numbers are velocities in km s^{-1}. Vertical exaggeration is 2:1. Upper 10 km is constrained by crustal refractions. The shaded region shows the part of the mantle constrained by P$_n$ wave data, thick black lines indicate constrained part of the Moho. LVZ, low velocity zone; NIF, Newport–Inglewood Fault; PVF, Palos Verde Fault; SAF, San Andreas Fault; SGF, San Gabriel Fault; SGM, San Gabriel Mountains; SGV, San Gabriel Valley; SMF, Sierra Madre Fault; VT, Vincent Thrust; WF, Whittier fault.*

lower crust beneath the Mojave Desert. By contrast, the lower crust located south of the San Andreas Fault is characterized by relatively fast velocities (6.6–6.8 km s^{-1}), suggesting that the region south of the fault is composed of strong feldspar- and/or olivine-rich rocks. This velocity structure is compatible with the idea that the weak crust north of the fault has flowed southward, creating the thick root beneath the Transverse Ranges (Fig. 8.14). In support of this hypothesis, several prominent bright spots beneath the San Gabriel Mountains, where reflector amplitudes are especially high (zones A and B in Fig. 8.14), suggest the presence of fractures and fluids that have penetrated along a thrust décollement

surface (Section 8.2) at the base of the brittle seismogenic zone (Fuis *et al.*, 2001). The pattern implies that the décollement is associated with a weak, ductilely flowing crust beneath the brittle upper crust.

The structure of the subcontinental mantle beneath the Transverse Ranges has been studied using the principles of seismic anisotropy (Section 2.1.8). In this region a near vertical, 60- to 80-km-wide, high velocity, high density body extends some 200 km downward into the upper mantle below the surface trace of the San Andreas Fault (Kohler, 1999). The significance of the anomaly is uncertain, but it may represent a zone of sinking material that helps to drive lower crustal flow

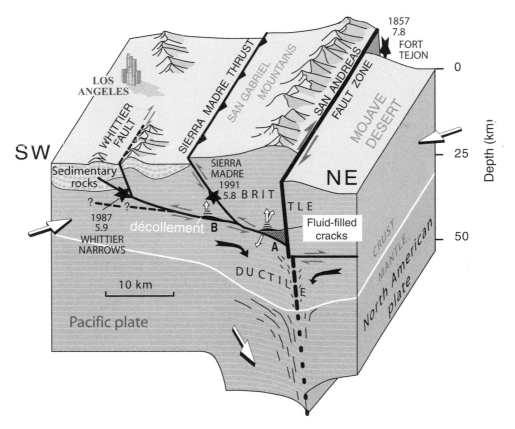

Figure 8.14 *Schematic block diagram showing the three-dimensional geometry of active faults of the Los Angeles region (image provided by G. Fuis and modified from Fuis* et al., *2001, with permission from the Geological Society of America). Moderate and large earthquakes are shown with black stars, dates, and magnitudes. Small white arrows show block motions in vicinities of bright reflective regions A and B. Large white arrows show relative convergence direction of Pacific and North American plates. Regions A and B are zones of cracks that transport fluids migrating up from depth. A décollement surface ascends from cracked region A at San Andreas Fault, above which brittle upper crust is imbricated along thrust and reverse faults and below which lower crust is flowing toward San Andreas Fault (black arrows), depressing the Moho. Mantle of Pacific plate sinks beneath the San Gabriel Mountains.*

and enhance crustal contraction beneath the Transverse Ranges (Godfrey *et al.*, 2002).

Measurements of shear wave (SKS) splitting have revealed an anisotropic upper mantle whose properties change with depth beneath the northern and central segments of the San Andreas Fault (Özalaybey & Savage, 1995; Hartog & Schwartz, 2001). Özalaybey & Savage (1995) interpreted these data in terms of two superimposed layers. The lower layer contains an east–west direction of fast polarization that may originate from asthenospheric flow caused by the migration of the Mendocino triple junction ~15 million years ago. Alternatively, the pattern may reflect a fossil anisotropy. The upper layer contains a fast polarization direction that parallels the trace of the San Andreas Fault and is well expressed on the northeast side of the San Andreas Fault where the lithosphere is relatively thin and hot. It is poorly developed on the southwest side where the lithosphere is relatively thick. The localization of this upper layer near the San Andreas Fault suggests that the anisotropy originates from deformation in a steep 50- to 100-km-wide mantle shear zone (Teyssier & Tikoff, 1998). Its thickness is not well constrained but it may reach 115–125 km thick and involve the asthenospheric mantle. The change in polarization direction with depth directly below the fault could result from either a change in the amount of strain due to right lateral shearing (Savage, 1999) or a change in strain direction (Hartog & Schwartz, 2001). Additional work is needed to establish the relationship between the postulated mantle shear zone and faulting in the upper crust.

8.3.3 The Alpine Fault

The Alpine Fault system in New Zealand (Fig. 8.2a) provides an example of a continental transform whose structure reflects a large component of fault-perpendicular shortening. Geophysical observations of the sea floor south of New Zealand suggest that contraction originated with changes in the relative motion between the Australian and Pacific plates between 11 and 6 Ma (Walcott, 1998; Cande & Stock, 2004). Prior to ~11 Ma, relative plate motion resulted in mostly strike-slip movement on the Alpine Fault with a small component of fault-perpendicular shortening. After ~11 Ma and again after ~6 Ma, changes in the relative motion between the Pacific and Australian plates resulted in an increased

component of compression across the pre-existing Alpine Fault, and led to increased shortening and rapid uplift of the Southern Alps (Norris *et al.*, 1990; Cande & Stock, 2004). The changes produced an oblique continent–continent collision on the South Island. In the central part of the island uplift rates range from 5 to 10 mm a^{-1} (Bull & Cooper, 1986) and are accompanied by high rates of erosion. Together with the crustal shortening, these processes have led to the exhumation of high grade schist that once resided at depths of 15–25 km (Little *et al.*, 2002; Koons *et al.*, 2003).

The Alpine Fault crosses the South Island between the Puysegur subduction zone in the south and the Hikurangi subduction zone in the north (Fig. 8.2a). During the late Cenozoic, the fault increasingly became the locus of slip between the Australian and Pacific plates. Geodetic measurements (Beavan *et al.*, 1999) and offset glacial deposits (Fig. 8.4) suggest that it has accommodated some 60–80% of relative plate motion since the late Pleistocene (Norris & Cooper, 2001; Sutherland *et al.*, 2006). The remaining motion is accommodated by slip on dipping thrust and oblique-slip faults in a >100-km-wide zone located mostly to the east of the fault (Fig. 8.2a). Geologic reconstructions of basement units suggest that a total of 850 ± 100 km of dextral movement has accumulated along the plate boundary since about 45 Ma (Sutherland, 1999). At least 460 km of this motion has been accommodated by the Alpine Fault (Wellman, 1953; Sutherland, 1999), as indicated by the dextral offset of the Median Batholith (Fig. 8.2a) and other Mesozoic and Paleozoic belts. About 100 km of shortening has occurred across the South Island since ~10 Ma (Walcott, 1998).

The subsurface structure of the Alpine Fault beneath the central South Island differs from that displayed by strike-slip-dominated transforms, such as the San Andreas and Dead Sea faults. Seismic imaging (Davey *et al.*, 1995) indicates that the central segment of the Alpine Fault dips southeastward at angles of 40–50° to a depth in excess of 25 km (Fig. 8.2b). Motion on the fault is in a direction that plunges approximately 22°, indicating that the fault in this region is an oblique thrust (Norris *et al.*, 1990). By contrast, motion on the Fiordland segment of the fault is almost purely strike-slip (Barnes *et al.*, 2005).

A 600-km-long seismic velocity profile, constructed as part of the **S**outh **I**sland **G**eophysical **T**ransect (SIGHT), has revealed the presence of a large crustal root beneath the Southern Alps (Fig. 8.2b). On the Pacific side, the Moho deepens from ~20 km beneath the Canterbury Plain to a maximum depth of 37 km below a point located 45 km southeast of the surface trace of the Alpine Fault. The root is asymmetric and mimics the tapered profile of the Southern Alps at the surface: Moho depths southeast of the fault decrease more gradually than those on its northwest side (Scherwath *et al.*, 2003; Henrys *et al.*, 2004). The root is composed mostly of thickened upper crust with seismic velocities ranging between 5.7 and 6.2 km s^{-1} (Scherwath *et al.*, 2003; Van Avendonk *et al.*, 2004). At large distances from the plate boundary, the upper crust shows a normal thickness of ~15 km. A thin (3–5 km) lower crust with a velocity range of 6.5–7.1 km s^{-1} occurs at the base of the root. A low velocity zone occurs in the middle and lower crust below the fault trace, most likely as a result of high fluid pressure (Section 8.6.3) (Stern *et al.*, 2001, 2002), and extends downward into the upper mantle.

Below the crustal root, teleseismic data show that deformation becomes progressively wider with depth. Measurements of P_n wave speeds (Scherwath *et al.*, 2002; Baldock & Stern, 2005) and shear wave (SKS) splitting (Klosko *et al.*, 1999; Duclos *et al.*, 2005) suggest the presence of a zone of distributed ductile deformation in the upper mantle beneath the Alpine Fault. Fast polarization directions generally are oriented subparallel to the fault strike (Fig. 8.15), suggesting flow parallel to the plate boundary. Baldock & Stern (2005) found evidence for two distinctive domains beneath the South Island: a 335-km-wide zone of mantle deformation in the south and a narrower, ~200-km-wide zone in the north (Fig. 8.15). These widths and the orientation of the mantle anisotropy are consistent with a model of transpression involving 800 ± 200 km of right lateral strike-slip displacement, which is close to that predicted by geologic reconstructions.

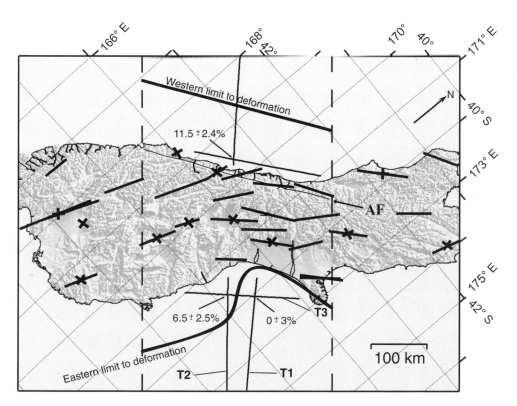

Figure 8.15 *Map showing the geometry of the SIGHT experiment and SKS measurements with an interpretation of mantle deformation below the Alpine Fault (AF) (image provided by T. Stern and modified from Baldock & Stern, 2005, with permission from the Geological Society of America). Three seismic transects (T1, T2, T3) are shown. Black bars indicate direction of maximum seismic velocity. Bar length is proportional to the amplitude of shear wave splitting determined from the SKS results of Klosko et al. (1999). P_n anisotropy measurement of 11.5 ± 2.4% is from Scherwath et al. (2002).*

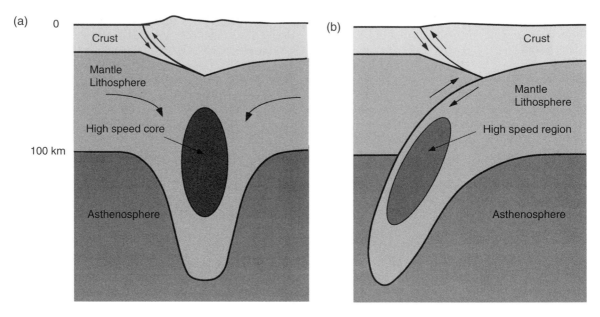

Figure 8.16 *Cartoons showing two possible modes of convergence in the mantle below the Alpine Fault (after Stern* et al.*, 2002). (a) Symmetric root formed by homogeneous shortening and thickening. (b) Westward underthrusting of Pacific mantle lithosphere beneath the Australian plate forming a zone of intracontinental subduction.*

The vertical thickness of the mantle root beneath the South Island is at least 100 km (Stern *et al.*, 2002). Earthquakes occur between 30 and 70 km depth (Kohler & Eberhart-Phillips, 2003). The root has a core of relatively cool, dense, high velocity mantle lithosphere that has been displaced into hotter, less dense, slower asthenosphere (Scherwath *et al.*, 2006). This excess mass in the mantle is required by observed gravity anomalies and provides sufficient force to maintain the crustal root, which is twice as thick as necessary to support the topography of the Southern Alps (Stern *et al.*, 2000). One possible interpretation of the root geometry is that it is symmetric and has formed in response to distributed deformation and a uniform thickening of the lithosphere (Fig. 8.16a). Alternatively, the mantle root may be asymmetric, requiring the deformation to be concentrated on a dipping thrust surface that results from intracontinental subduction (Fig. 8.16b). These and other processes that contribute to mantle root formation and its tectonic modification are key elements of studies in virtually all zones of continental deformation (e.g. Sections 7.5, 7.8.1, 10.2.5, 10.4.6), and are discussed in more detail in Section 11.3.3. Whichever of these hypotheses is correct, the anomaly suggests that

the low upper mantle temperatures beneath the Southern Alps are caused by cold downwelling beneath the collision zone. In addition, the teleseismic data indicate that the displacements associated with continental transforms can be accommodated by distributed deformation in the mantle without requiring discrete faulting. The great width of the deforming zone found in the New Zealand setting compared to other continental transforms may reflect the large component of convergence across the plate boundary (Stern *et al.*, 2002).

8.4 TRANSFORM CONTINENTAL MARGINS

Where a transform fault develops during continental rifting the continental margin is defined by the transform fault and is termed a *transform continental margin*. The history of such a margin, first considered by

Scrutton (1979), reflects its initial contact with its continental counterpart on the adjacent plate and subsequent contact with oceanic lithosphere and an ocean ridge as the separation proceeds. These margins differ from rifted or passive margins (Section 7.7) by a narrow (<30 km) continental shelf and a steep ocean–continent transition zone.

One of the best-studied transform margins is the Ivory Coast–Ghana margin in the north of the Gulf of Guinea. This margin formed during the Early Creta-

ceous opening of the South Atlantic, which was accompanied by transform motion within what is now the Romanche Fracture Zone (Fig. 8.17a) (Mascle & Blarez, 1987; Attoh et al., 2004). The margin has undergone little subsequent modification and so can be considered to represent a fossil transform margin.

The Ivory Coast–Ghana margin displays a triangular-shaped continental shelf, a steep (15°) continental slope, and a narrow (6–11 km) ocean–continent transition zone (Fig. 8.17b). Seismic reflection data provide

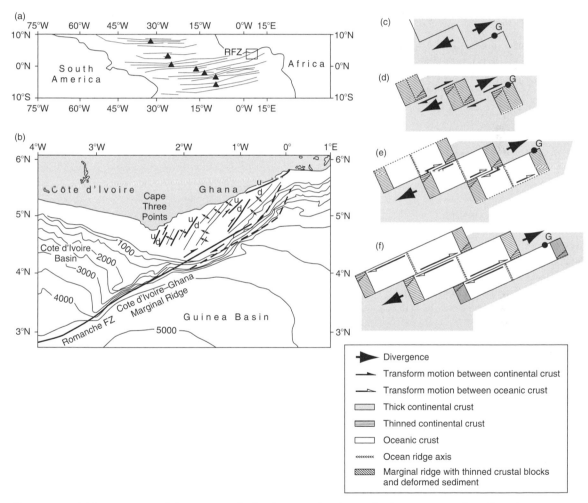

Figure 8.17 (a) Tectonic map of the equatorial Atlantic ocean showing major fracture zones that offset the Mid-Atlantic Ridge (triangles) and location of (b) the Ivory Coast–Ghana transform margin (modified from Edwards et al., 1997, by permission of the American Geophysical Union. Copyright © 1997 American Geophysical Union). RFZ, Romanche Fracture Zone. Faults and folds in (b) are modified from data presented by Attoh et al. (2004). u, up; d, down. (c–f) Simplified model of the formation of a transform continental margin (after Mascle & Blarez, 1987, with permission from Nature **326**, 378–81. Copyright © 1987 Macmillan Publishers Ltd). G, position of the Ghana transform margin.

evidence of folding and faulting associated with dextral motion within a 10- to 20-km-wide zone along the Côte d'Ivoire–Ghana marginal ridge (Edwards *et al.*, 1997; Attoh *et al.*, 2004). The folds display northeast-trending axes that are compatible with dextral motion. The faults record both strike-slip and dip-slip (south-side down) displacements that appear to reflect at least two episodes of strike-slip deformation (Attoh *et al.*, 2004). The first involved a combination of strike-slip motion and extension on northeast-trending faults, leading to the formation of pull-apart basins (Section 8.2). The second involved strike-slip motion and folding, possibly as a result of a change in the direction of motion in the transform.

On the basis of these and other observations, it has been possible to reconstruct the large-scale evolution of the Ivory Coast–Ghana margin. Four main phases are illustrated diagrammatically in Fig. 8.17c–f. In phase 1 (Fig. 8.17c) there is contact between two continents. Strike-slip motion results in brittle deformation of the upper crust and ductile deformation at depth (Section 2.10), giving rise to pull-apart basins and rotated crustal blocks (Section 8.5). In phase 2 (Fig. 8.17d), as rifting and crustal thinning accompany the formation of a divergent margin, the contact is between normal thickness continental lithosphere and thinner, stretched continental lithosphere. The newly created rift basin experiences rapid sedimentation from the adjacent continent and subsidence associated with the crustal thinning (Section 7.7.3). The sediments are folded and faulted by the transform motion and blocks of material are uplifted (Basile & Allemand, 2002), forming scarps and marginal ridges (see also Section 6.2). This tectonism is recorded in unconformities in the sedimentary sequence and other structures imaged in seismic reflection profiles (Attoh *et al.*, 2004). In phase 3 (Fig. 8.17e) new oceanic lithosphere emerges along a spreading center to establish an active ocean–continent transform. At this stage there is contact between the faulted continental margin and oceanic crust. The faulted margin passes adjacent to the hot oceanic crust of the spreading center and the thermal exchange it experiences results in heating and differential uplift within the faulted margin, especially near the continent–ocean boundary. Seismic data suggest magmatic underplating in the deep portions of the continental crust, where the magmatic features align with the transform faults (Mohriak & Rosendahl, 2003). In phase 4 (Fig. 8.17f) the transform is only active between blocks of oceanic crust and thus appears as a fracture zone (Section 6.12). The faulted margin is then in contact with cooling oceanic lithosphere and its subsidence evolves in a manner similar to other rifted passive margins (Section 7.7.3).

8.5 CONTINUOUS VERSUS DISCONTINUOUS DEFORMATION

8.5.1 Introduction

The distributed nature of deformation on the continents compared to most oceanic regions has led to the invention of a unique framework for describing continental deformation (Sections 2.10.5, 5.3). One of the most important aspects of developing this framework involves determining whether the motion is accommodated by the movement of many coherent blocks separated by discrete zones of deformation or by a more spatially continuous process. In some areas, the presence of large aseismic regions such as the Great Valley and Sierra Nevada in the southwestern USA (Figs 7.8, 7.10) imply that part of the continental lithosphere behaves rigidly. However, in other areas, such as the Walker Lane and Eastern California Shear Zone, seismicity reveals the presence of diffuse zones of deformation that are better approximated by a regional velocity field rather than by the relative motions of rigid blocks.

To distinguish between the possibilities, geoscientists use combinations of geologic, geodetic, and seismologic data to determine the degree to which deformation is continuous or discontinuous across a region (Thatcher, 2003; McCaffrey, 2005). Determining the characteristics of these regional velocity fields is important for developing accurate kinematic and rheological models of deforming continental lithosphere (Section 8.6), and for estimating where strain is accumulating most rapidly and, thus, where earthquakes are most likely to occur.

In models involving continuous velocity fields, even though the upper brittle crust is broken into faults, the faults are predicted to be relatively closely spaced, have small slip rates, and extend only through the elastic part of the crust. In this view, the velocity field commonly is assumed to represent the average deformation of the whole lithosphere, which consists of a thin layer (10–20 km) that deforms by faulting above a thick layer (80–100 km) that deforms by ductile creep (Jackson, 2004). In rigid block models, faults are predicted to be widely spaced, slip rapidly, and extend vertically through the entire lithosphere, terminating as large ductile shear zones in the upper mantle (McCaffrey, 2005). These latter properties suggest that deforming continental lithosphere exhibits a type of behavior that resembles plate tectonics. In both types of model, the deformation may be driven by a combination of forces, including those acting along the edges of crustal blocks, basal tractions due to the flow of the lower crust and upper mantle, and gravity.

Determining the degree to which continental deformation is continuous or discontinuous has proven difficult to achieve with certainty in many areas. One reason for the difficulty is that the short-term (decade-scale) surface velocity field measured with geodetic data usually appears continuous at large scales (kilometers to hundreds of kilometers) (Section 2.10.5). This characteristic results because geodetic positioning techniques provide velocity estimates at specific points in space, with the density of available points depending on the region and the scale of the investigation (Bos & Spakman, 2005). Most interpretation methods start with some interpolation of the geodetic data, with the final resolution depending on the number and distribution of the available stations (Jackson, 2004). In addition, available information on fault slip rates is incomplete and those that are calculated over short timescales may not be representative of long-term slip rates (Meade & Hager, 2005; McCaffrey, 2005). If accurate long-term slip rates on all crustal faults were available, then the problem could be solved directly. Furthermore, even though continental deformation is localized along faults over the long term, the steady-state motion of an elastic upper crust over the short-term contains little information about the rheology of the deforming material, so the importance of the faults in the overall mechanical behavior of the lithosphere is unclear. This latter uncertainty clouds the issue of whether deformation along continental transforms is driven mostly by edge forces, basal tractions, or gravitational forces (Savage, 2000; Zatman, 2000; Hetland & Hager, 2004) (Section 8.5.3).

Despite the difficulties involved in quantifying continental deformation, geoscientists have been able to show that elements of both continuous and discontinuous representations fit the observations in many areas. In this section, the results of geodetic measurements and velocity field modeling are discussed in the context of the San Andreas Fault system, whose structural diversity illustrates the variety of ways in which strain may be accommodated along and adjacent to continental transforms.

8.5.2 Relative plate motions and surface velocity fields

In the southwestern USA, relative motion between the Pacific and North American plates occurs at a rate of about 48–50 mm a^{-1} (DeMets & Dixon, 1999; Sella et al., 2002). Geodetic and seismologic data suggest that up to 70% of this motion presently may be accommodated by dextral slip on the San Andreas Fault (Argus & Gordon, 2001). Out of a total of approximately 1100–1500 km of strike-slip motion, since the Oligocene (Stock & Molnar, 1988), only 300 and 450 km of right lateral slip have accumulated along the southern and northern reaches of the San Andres Fault, respectively (Dillon & Ehlig, 1993; James et al., 1993). The remaining movements, therefore, must be accommodated elsewhere within the diffuse zone of deformation that stretches from the Coast of California to the Basin and Range (Fig. 8.1).

Along its ~1200 km length, the San Andreas Fault is divided into segments that exhibit different short-term mechanical behaviors. Some segments, such as those located north of Los Angeles and north of San Francisco, have ruptured recently, generating large historical earthquakes, and now exhibit little evidence of slip. These segments appear locked at depth and are now accumulating significant nonpermanent (elastic) strains near the surface, making them a major potential earthquake hazard (Section 2.1.5). Between the two locked segments is a 175-km-long fault segment in central California that is characterized by aseismic slip, shallow (<15 km depth) microearthquakes, and few large historical earthquakes. Along this segment, the aseismic slip reflects a relatively steady type of creep that results from frictional properties promoting stable sliding on the fault plane (Scholz, 1998). These different behaviors, and especially the occurrence of aseismic creep on or near faults at the surface, complicate the estimation of horizontal velocity fields (Section 8.5.3).

In northern California, earthquakes reveal the presence of a ~120-km-wide zone of faulting within the Coast Ranges between the Pacific plate on the west and the Great Valley–Sierra Nevada microplate on the east (Fig. 8.18a). In this region, horizontal velocities (Fig. 8.18b) show an approximately uniform distribution of right lateral motion toward N29°W (Savage et al., 2004a). This direction is close to the local strike (N34°W) of the San Andreas Fault and results in dominantly strike-slip motion along the major faults in the area. A velocity profile along a great circle passing through the Pacific–North America pole of rotation (Fig. 8.18c) illustrates this result by showing the components of motion that occur parallel to and perpendicular to the trace of the San Andreas Fault. Slip rates parallel to the fault are highest. Other faults display lower rates. In addition, the westward movement of the Great Valley–Sierra Nevada block relative to the Pacific plate (Dixon et al., 2000; Williams et al., 2006) and the slight obliquity between this motion and the trace of the San Andreas Fault (Prescott et al., 2001; Savage et al., 2004a) produces a small component of contraction across the Coast Ranges. This latter result is supported by both geologic data (Fig. 8.7b, inset) and by earthquake focal mechanisms that show thrust solutions west of the Great Valley (Fig. 7.10). By contrast, little deformation occurs across the Great Valley and within the Sierra Nevada, suggesting that these regions form a coherent, rigid block.

In southern California, the distribution of earthquakes indicates that relative plate motion is accommodated across a zone that is several hundreds of kilometers wide (Fig. 7.8). South of latitude 34°N, some motion is distributed between the San Jacinto and San Andreas faults (Fig. 8.1). Becker et al. (2005) estimated that the former accommodates some $15\,mm\,a^{-1}$ of slip and the latter $\sim23\,mm\,a^{-1}$. Within the Transverse Ranges, where crustal shortening and surface uplift accommodate a component of the motion (Fig. 8.8), slip on the San Andreas Fault appears to be significantly slower (Meade & Hager, 2005). Other displacements occur along major sinistral faults, such as the Garlock, Raymond Hill, and Cucamonga faults (Fig. 8.8a), and by the clockwise and anticlockwise rotation of crustal blocks about vertical axes (Savage et al., 2004b; Bos & Spakman, 2005).

Figure 8.19a shows an example of a velocity field for southern California in a local reference frame (Meade & Hager, 2005). Stations on the North American plate move toward the southeast at about half of the relative plate velocity, those on the Pacific plate move toward the northwest. The observed velocities vary smoothly across the San Andreas Fault. Two profiles (Fig. 8.19b) (gray shaded areas) show similar total velocity changes of $\sim42\,mm\,a^{-1}$ but distinctly different velocity gradients. Along the northern profile, the fault-parallel velocity drops by $\sim30\,mm\,a^{-1}$ across the San Andreas Fault and decreases slightly through the San Joaquin Valley before distributing some $12\,mm\,a^{-1}$ across the Eastern California Shear Zone. By contrast, the southern profile shows a total velocity drop across a distance that is about 50% that in the northern profile. This reflects the difference in the geometry of the fault system from north to south. The flat portion of the northern profile mirrors the relative stability of the ~200-km-wide Great Valley–Sierra Nevada microplate, with the deforming central segment of the San Andreas Fault to the west and Eastern California Shear Zone to the east. By contrast, the southern profile shows that the 40-km-wide zone between the San Andreas and San Jacinto faults accommodates approximately 80% of the relative plate motion.

North of the Garlock Fault, relative motion is deflected east of the Sierra Nevada by deformation in the southern Walker Lane (Figs 7.9, 8.1). This eastward deflection reflects an extensional step-over between the northwest-trending faults of the Eastern California Shear Zone and those located along the eastern margin of the Sierra Nevada (Oldow, 2003). North of the step-over, the zone of deformation broadens into the central and northern Walker Lane and central Nevada seismic belt (Fig. 8.20a). Earthquake focal mechanisms (Fig. 7.10) indicate that displacements in these latter belts involve both strike-slip and normal fault motion on variably oriented faults. Horizontal velocities increase from $2–3\,mm\,a^{-1}$ to $\sim14\,mm\,a^{-1}$ from the central Great Basin (Section 7.3) toward the Sierra Nevada (Oldow, 2003). Accompanying this rate increase, the directions of motion rotate clockwise from west-northwest to northwest (Fig. 8.20b), indicating an increase in a component of dextral strike-slip deformation from east to west. Oldow (2003) showed that two distinctive zones of transtension characterize this belt: one that is dominated by extension on the west (domain III) and another that is dominated by strike-slip motion on the east (domain II in Fig. 8.20b). Together with deformation in the central and eastern Basin and Range (Section 7.3), these belts accommodate up to 25% of the relative motion between the Pacific and North American plates (Bennett et al., 1999). This transfer of motion east of

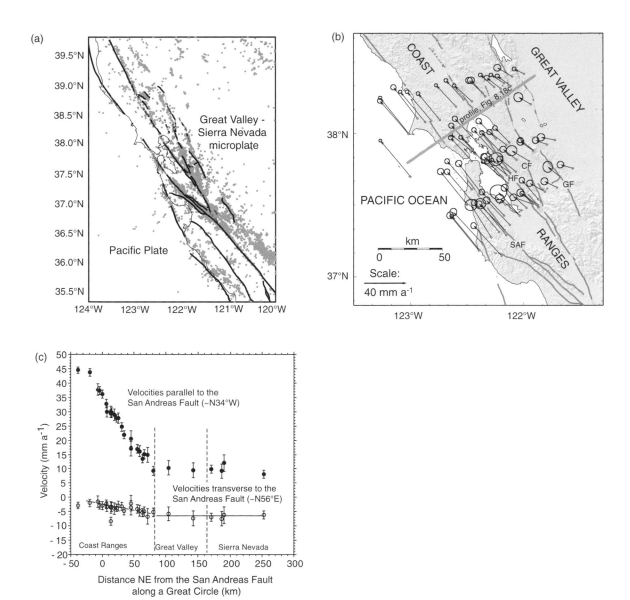

Figure 8.18 *(a) Earthquakes recorded by the Northern California Seismic Network between 1968 and 1999 (images provided by G. Bokelmann and modified from Bokelmann & Beroza, 2000, by permission of the American Geophysical Union. Copyright © 2000 American Geophysical Union). Over 58,000 seismic events show the seismogenic segments of the greater San Andreas Fault system. Map (b) and (c) profile showing velocities derived from GPS surveys in the San Francisco Bay area (images provided by J. Savage and modified from Savage et al., 2004a, by permission of the American Geophysical Union. Copyright © 2004 American Geophysical Union). Error ellipses at ends of velocity arrows in (b) define 95% confidence limits. SAF, San Andreas Fault; HF, Hayward Fault; CF, Calaveras Fault; GF, Greenville Fault. Velocity profile in (c) shows components of motion parallel and perpendicular to the San Andreas Fault. Profile passes through the Pacific–North American pole of rotation and the trajectory shown in (b). Error bars represent two standard deviations on either side of the plotted points.*

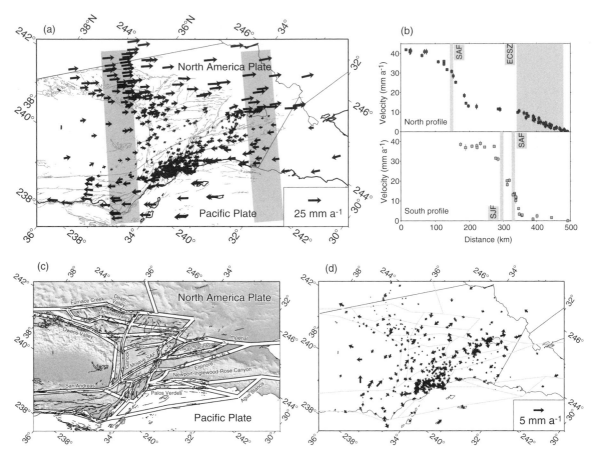

Figure 8.19 *Results from GPS measurements and block modeling of crustal motion in southern California (images provided by B. Meade and modified from Meade & Hager, 2005, by permission of the American Geophysical Union. Copyright © 2005 American Geophysical Union). (a) Velocities observed during periods between earthquakes (i.e. interseismic velocities), when strain accumulations are elastic and appreciable slip on faults is absent. Confidence ellipses have been removed to reduce clutter. The two shaded swaths show regions in which fault parallel velocities are drawn in two profiles (b). Vertical lines in profiles give uncertainties of one standard deviation. Gray shaded areas show locations of the San Andreas Fault (SAF), San Jacinto Fault (SJF), and the Eastern California Shear Zone (ECSZ). Differences in velocity gradients reflect fault spacing. (c) Block model boundaries (white zones) superimposed on a shaded relief map showing major fault traces. (d) Residual velocities. Gray lines show block boundaries. Note that velocity vectors are drawn at a scale that is five times larger than in part (a).*

the Sierra Nevada helps to explain the limited amount of slip that is observed on the San Andreas Fault.

8.5.3 Model sensitivities

An important means of evaluating a modeled velocity field involves comparing the short-term slip rates on major faults implied by the model with the average

long-term slip rates derived from geologic data. In some settings, these comparisons show that the continuum approach to estimating velocity fields explains most of the observed displacements. For example, Savage *et al.* (2004a) showed that a uniform velocity field involving distributed right lateral shear within a 120-km-wide zone in the Coast Ranges (Fig. 8.18) matches the vector sum of all the average slip rates determined independently for all major faults across the zone. Their approx-

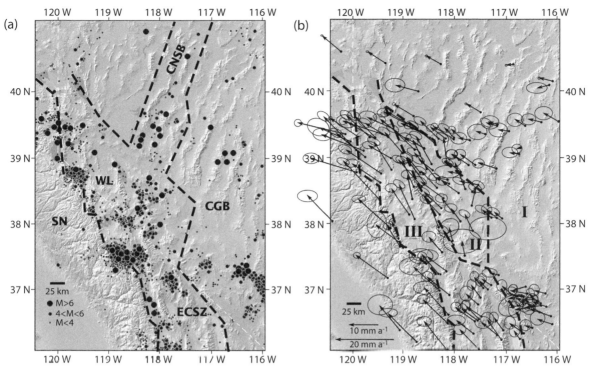

Figure 8.20 *Shaded relief map of the Sierra Nevada (SN) and central Great Basin (CGB) showing (a) seismicity and (b) GPS velocities in a fixed North America reference frame (images provided by J. Oldow and modified from Oldow, 2003, with permission from the Geological Society of America). Seismicity data include 1967–2000 events for M < 6 and 1850–2000 events for M > 6 from the United States Geologic Survey (USGS) National Earthquake Information Center (NEIC) and Rogers et al. (1991). CNSB, Central Nevada Seismic Belt; WL, Walker Lane; ECSZ, Eastern California Shear Zone. Ellipses in (b) represent 95% uncertainty limits. Tectonic domains (dashed lines) in (b) are: I, extension; II, strike-slip-dominated transtension; III, normal fault-dominated transtension.*

imation of a uniform strain rate across the Coast Ranges yielded a slip rate of ~39 mm a^{-1}, which is consistent with the average slip rates assigned to the main faults using offset geologic and cultural features over time periods ranging from hundreds to several tens of thousands of years. However, in other areas, such as southern and eastern California (Fig. 8.19a) where the fault geometry is very complex, there are large mismatches between the geodetic and geologic slip rates. These mismatches have prompted investigators to use alternatives to the continuum model approach to describe the surface deformation (McCaffrey, 2005; Meade & Hager, 2005; Bos & Spakman, 2005). One of the most useful of these alternative approaches employs block rotations.

Block models of continental deformation provide a framework for incorporating aspects of the long-term,

discontinuous deformation caused by faulting into estimates of the velocity field. In these models, calculated fault slip rates take into account the effects of both the rotation of fault-bounded blocks about vertical or inclined axes and the steady-state elastic accumulation of strain (i.e. creep) on or near faults. The blocks are defined as any number of closed polygons on the Earth's surface that cover the modeled region (Fig. 8.19c). In most applications, the block boundaries coincide with major faults; however, in some cases the choice is less clear. Each point inside the blocks is assumed to rotate with the same angular velocity (McCaffrey, 2005). The description of the motion is mathematically similar to methods of estimating the rotations of large tectonic plates (Section 5.3). However, a potential problem is that the use of short-term geodetic data results in elastic strain rates inside the blocks as well as along their boundaries, causing the surface velocities to deviate from the "rigid plate" requirement of plate tectonics.

In the case of southern California, the incorporation of block rotations and the small-scale displacement discontinuities associated with creep on and near major faults has provided a relatively good fit to the available geodetic data (Becker *et al.*, 2005; McCaffrey, 2005; Meade & Hager, 2005). A common way of evaluating the fit of the models involves the calculation of residual velocities, which represent the difference between the modeled and observed values. An example of one of these comparisons is shown in Fig. 8.19d. In this application, crustal blocks were chosen to minimize the residuals, while still conforming to known boundary conditions, such as the orientation of fault traces and the sense of slip on them. The comparison shows that despite the improvement over some continuous models there are still areas of mismatch. In the Eastern California Shear Zone, for example, Meade & Hager (2005) found that slip rates estimated using geodetic data and the results of block models are almost twice as fast as the $2\,mm\,a^{-1}$ geologic estimates (Beanland & Clark, 1994) for the past 10,000 years. A similar discrepancy occurs on the San Jacinto Fault. In addition, the modeled slip rates on the San Bernadino segment of the San Andreas Fault are much slower than geologically determined rates for the past 14,000 years. Finding ways to explain and minimize these mismatches remains an important area of research.

One possible explanation of why geodetic and geologic rates commonly mismatch lies with the mechanical behavior of large faults and the vertical extent of brittle faulting within the lithosphere. Because slip on a fault plane near the surface is controlled by its frictional properties (Section 2.1.5), there is a tendency for faults to become stuck or locked for certain periods of time (Section 8.5.2). This locking may result in elastic strain rates that are evident in short-term geodetic data but not in the long-term record of permanent displacements (McCaffrey, 2005). To address this problem, investigators utilize the concept of the *elastic locking depth* (Savage & Burford, 1973). This depth is defined as the level below which there is a transition from localized elastic strain accumulations on a fault plane to distributed aseismic flow. The value of the parameter is related directly to the mechanical strength of the fault and the geometry of deformation at the surface. Strong faults and wide zones of surface deformation correspond to deeper locking depths.

Published estimates of locking depths for the San Andreas Fault typically range from 0 to 25 km. However, locking depths are not known *a priori* and, therefore, must be inferred on the basis of seismicity, long-term geologic slip rates, deformation patterns at the surface, or inferences about the rheology of the lithosphere. Locking depths that fall significantly below the predicted depth of the brittle–ductile transition (8–15 km) for a typical geotherm, or below the seismogenic layer, usually require some sort of explanation. In some cases, slow slip rates on the faults have been used to infer relatively deep locking depths for some segments of the San Andreas Fault (Meade & Hager, 2005; Titus *et al.*, 2005). These and other studies illustrate how the choice of locking depth is directly related to inferences about slip rates on or near major faults.

Other reasons why geodetic and geologic slip rates commonly differ may include inherent biases during sampling or changes in the behavior of faults over time. This latter possibility is especially important when the effects of long-term, permanent strains are considered (Jackson, 2004). Meade & Hager (2005) concluded that the differences between their calculated slip rates and geologic slip rates on faults might be explained by the time-dependent behavior of the fault system. In this interpretation, the San Bernadino segment of the San Andreas Fault is less active now than it has been in the past. By contrast, the San Jacinto Fault and faults in the Eastern California Shear Zone are relatively more active now compared with geologic estimates, possibly due to the effects of earthquake clusters. This possibility highlights the importance of combining geologic, geodetic, and seismologic information to better understand the relationship between the short- and long-term (permanent) behaviors of faults.

By incorporating elements of permanent deformation into block rotation models, McCaffrey (2005) found that the largest blocks in the southwestern US, including the Sierra Nevada–Great Valley and the eastern Basin and Range Province, show approximately rigid behavior after all nonpermanent (elastic) strain has been removed from the data. Most of the blocks rotate about vertical axes at approximately the same rate as the Pacific plate (relative to North America), suggesting that, locally, rotation rates are communicated from block to block. This and several other properties of the model support a plate tectonic-style description of deformation in the western USA, where the rotating blocks behave like microplates. Nevertheless, the problem of determining the mechanisms of the deformation is far from resolved. Many other models have been proposed for this same region (e.g. Flesch *et al.*, 2000) that also fit the geodetic observations and most investigators agree that the deformation probably results from a combination of mechanisms rather than a single one.

Part of the problem of determining the specific mechanisms of continental deformation is that success in fitting geodetic observations neither proves any given model nor precludes other possibilities (McCaffrey, 2005). In addition, the results of mechanical modeling have shown that the steady-state motion of an elastic upper crust is insensitive to the properties of any flow field below it (Savage, 2000; Zatman, 2000; Hetland & Hager, 2004). This latter result means that short-term geodetic observations of deformation between large earthquakes (i.e. interseismic deformation) provide no diagnostic information about the long-term behavior of a viscous layer in the deep crust or mantle. One especially promising area of research suggests that transient deformation following large earthquakes offers the prospect of inferring the rheology of lower viscous layers (Hetland & Hager, 2004). However, presently, the specific mechanisms and the relative contribution of edge forces, basal tractions, and buoyancy forces to the deformation in most regions remain highly speculative, with the results of models depending strongly on the imposed boundary conditions.

8.6 STRAIN LOCALIZATION AND DELOCALIZATION MECHANISMS

8.6.1 Introduction

One of the most interesting questions about continental transforms and major strike-slip faults is how these structures accomplish the large displacements that are observed on them. To determine the mechanisms that allow these displacements to occur, geoscientists have developed mechanical models to investigate the processes that lead to a localization or delocalization of

strain during strike-slip faulting. As is the case in other tectonic settings (e.g. Section 7.6.1), competition among these processes, and whether they result in a net weakening or a net strengthening of the lithosphere, ultimately controls the large-scale patterns of the deformation.

8.6.2 Lithospheric heterogeneity

The distribution of strain within deforming continental lithosphere is strongly influenced by horizontal variations in temperature, strength, and thickness (e.g. Sections 2.10.4, 2.10.5). In New Zealand, for example, oblique convergence on the central part of the South Island has resulted in deformation that occurs almost entirely on the Pacific plate side, leaving the Australian plate relatively undisturbed (Fig. 8.2a). This asymmetry reflects the greater initial crustal thickness and weaker rheology of the Pacific plate compared to that of the Australian plate, causing the former to deform more easily (Gerbault *et al.*, 2002; Van Avendonk *et al.*, 2004).

To investigate the effects of initial variations in crustal thickness and lithospheric temperature on strike-slip deformation patterns, Sobolev *et al.* (2005) conducted numerical experiments of a simple transform fault (Fig. 8.21). In these models, the crust consists of two layers overlying mantle lithosphere. Velocities of $30 \, \text{mm a}^{-1}$ are applied to the sides of the lithosphere, forming a zone of left lateral strike-slip deformation. Although motion takes place in and out of the plane of observation, all other model parameters vary in only two dimensions. The rheological description of the crustal layers allows both brittle and ductile styles of deformation to develop, whichever is the most energetically efficient. Brittle deformation is approximated with a Mohr–Coulomb elastic-plastic rheology. Ductile flow employs a nonlinear, temperature-dependent, viscous-elastic rheology (see also Section 7.6.6). Both rheologies allow for heating during deformation as a result of friction or ductile flow.

In the first model, (Fig. 8.21a) the lower crust is thicker on the left than on the right and temperature is kept constant at the base of the lithosphere. The second (Fig. 8.21b) shows a constant crustal thickness and a thermal perturbation in the central part of the model. In the third model (Fig. 8.21c), both crustal thickness and temperature heterogeneities are present. This latter

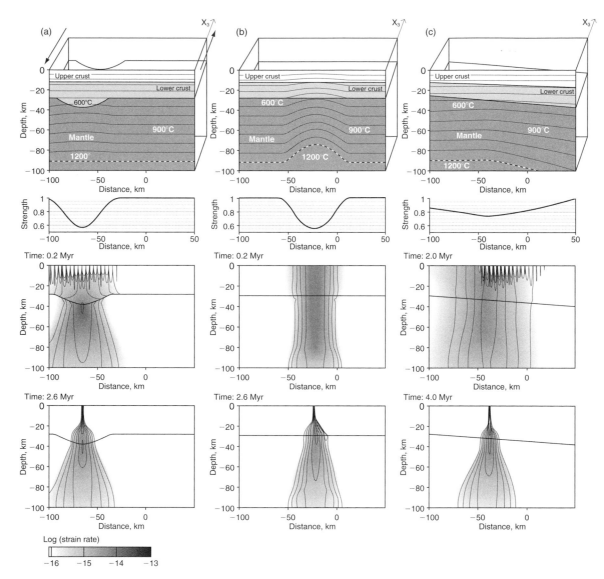

Figure 8.21 *Thermomechanical models of simple strike-slip faulting in a three-layer lithosphere that incorporate initial variations in (a) crustal thickness, (b) lithospheric temperature, and (c) both crustal thickness and lithospheric temperature (after Sobolev* et al.*, 2005, with permission from Elsevier). Top row of diagrams shows model setup geometry with corresponding lithospheric strength curves before deformation below them. Thin black lines are isotherms prior to deformation. Lower two rows of diagrams show snapshots of the distribution of strain rate demonstrating the strain localization process.*

model mimics the structure of a rifted continental margin (Section 7.7) where crustal thickness decreases linearly from right to left and high lithospheric temperatures occur at shallow depths below the thinnest crust. In all three models, multiple faults form in the brittle upper crust during the first 1–2 Ma. The number of active faults gradually decreases over time until a single fault dominates the upper crust at about 2 Ma. Over time a zone of high strain rate in the ductile lower crust and mantle lithosphere narrows and stabilizes.

These results show that, for each model, strain localizes where lithospheric strength is at a minimum, regardless of the cause of the weakening. They also show that crustal thickness and the initial thermal state of the lithosphere play key roles in localizing strike-slip deformation. These effects may explain why strike-slip faulting has localized in some areas of the southwestern USA (Figs 7.9, 8.1), such as the Eastern California Shear Zone and Walker Lane, leaving others, such as the Great Valley–Sierra Nevada and central Great Basin, virtually undeformed (Bennett et al., 2003). In this case, strain localization may be related to differences in heat flow between the western Basin and Range Province and the Sierra Nevada (Section 7.3). However, it has not been demonstrated whether the elevated heat flux is a cause or a product of strain localization. Alternatively, crustal thickness variations and horizontal gradients in gravitational potential energy and viscosity may concentrate the deformation (Section 7.6.3).

In addition to horizontal variations in strength, a vertical stratification of the lithosphere into weak and strong layers greatly influences how strain is accommodated during strike-slip deformation. To illustrate this effect, Sobolev et al. (2005) compared patterns of strain localization and delocalization in two models of pure strike-slip deformation that incorporate two different crustal rheologies. In the first model (Plate 8.1a between pp. 244 and 245), the crust is strong and modeled using laboratory data on hydrous quartz and plagioclase. Three layers correspond to a brittle upper crust, a brittle-ductile middle crust, and a mostly ductile lower crust. In the second model (Plate 8.1b between pp. 244 and 245), the effective viscosity of the crust at a fixed strain rate is reduced tenfold. The models also incorporate a reduction in crustal thickness from right (east) to left (west) in a manner similar to that observed in the Dead Sea Transform (Fig. 8.11). Lithospheric thickness is defined by the 1200 °C isotherm and increases to the east, simulating the presence of a thick continental shield on the right side of the model.

The results of these two experiments show that in both the strong crust and weak crust models, strain localizes into a sub-vertical, lithospheric-scale zone at the margin of the thick shield region, where the temperature-controlled lithospheric strength is at a minimum (Plate 8.1a,b between pp. 244 and 245). In the case of the strong crust, the zone of largest crustal deformation is located above a zone of mantle deformation and is mostly symmetric (Plate 8.1c between pp. 244 and 245). These characteristics result from the strong mechanical coupling between the crust and upper mantle layers. In the 15-km-thick brittle upper crust, shear strain localizes onto a single vertical fault. The deformation widens with depth into a zone of diffuse deformation in the middle crust and then focuses slightly in the uppermost part of the lower crust. In the model with weak crust, the lower crust is partially decoupled from both the upper mantle and the upper and middle crusts (Plate 8.1d between pp. 244 and 245). Consequently, the deformation is delocalized, asymmetric, and involves more upper crustal faults. The distribution of viscosity (Plate 8.1e,f between pp. 244 and 245) also illustrates the mechanical decoupling of layers in the weak crustal model. This decoupling results because the deforming lithosphere becomes very weak due to the dependency of the viscosity on strain rate and temperature, which increases due to strain-induced heating. In several variations of this model, which involve the addition of a minor component of transform-perpendicular extension, second order effects appear, such as a small deflection of the Moho, the development of deep sedimentary basins, and asymmetric topographic uplift (Sobolev et al., 2005). These latter features match observations in the Dead Sea Transform (Section 8.3.1).

These numerical models illustrate that the localization and delocalization of strain during strike-slip deformation is influenced by vertical contrasts in rheology as well as initial horizontal contrasts in crustal thickness and temperature. The width of the deforming zone is controlled mostly by strain-induced heating and the temperature- and strain-rate dependency of the viscosity of the rock layers. Lithospheric thickness appears to play a minor role in controlling fault zone width. The results also highlight how the interplay between forces applied to the edges of plates or blocks and the effects of ductile flow in the lower crust and mantle result in a vertical and horizontal partitioning of strain within the lithosphere.

8.6.3 Strain-softening feedbacks

Once strain starts to localize (Section 8.6.2), several mechanisms may enhance crustal weakening and reduce the amount of work required to continue the deformation. Two of the most influential of these strain-softening mechanisms involve increased pore fluid pressure, which results from crustal thickening, and the vertical advection of heat, which results from concentrated surface erosion and the exhumation of deep crustal rocks. These processes may cause strain to continue to localize as deformation progresses, resulting in a *positive feedback*. The transpressional plate boundary on the South Island of New Zealand illustrates how these strain-softening feedbacks allow a dipping fault plane to accommodate large amounts of strain.

One of the principal results of the SIGHT program (Section 8.3.3) is an image of a low velocity zone below the surface trace of the Alpine Fault (Fig. 8.2b). In addition to low seismic wave speeds, this zone includes an elongate region of very low (40 ohm-m) resistivity in the middle to lower crust that generally parallels the dip of the Alpine Fault (Fig. 8.22). Magnetotelluric soundings show that the region forms part of a U-shaped pattern of elevated conductivity that rises northwestward toward the trace of the Alpine Fault, attains a near-vertical orientation at ~10 km depth, and approaches the surface about 5–10 km southeast of the fault trace (Wannamaker et al., 2002). Stern et al. (2001) concluded that the low velocities and resistivities result from the release of fluids during deformation and prograde metamorphism in thickening continental crust (Koons et al., 1998). In support of this interpretation, areas of hydrothermal veining and gold mineralization of deep crustal origin coincide with the shallow continuation of the conductive zone (Wannamaker et al., 2002). Similar steeply dipping conductive features coincide with active strike-slip faults in other settings, including the San Andreas Fault (Unsworth & Bedrosian, 2004), the Eastern California Shear Zone, and the southern Walker Lane (Park & Wernicke, 2003). These observations suggest that elevated pore fluid

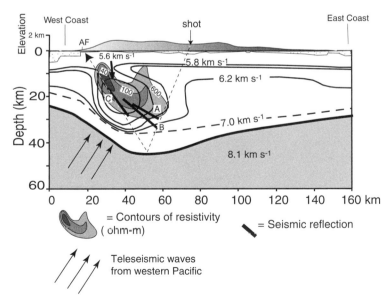

Figure 8.22 *Crustal structure below the Alpine Fault (AF) showing region of low P-wave velocities and low resistivity that satisfies wide-angle reflections and teleseismic delays (image provided by T. Stern and modified from Stern et al., 2002). Contours of wave speed shown by solid and dashed lines (km s⁻¹). Shading is resistivity ranging from 40 ohm-m for darkest zone to 600 ohm-m for lightest. Zones of strong crustal reflectivity (A, B, C) are from Stern et al. (2001). Dashed lines represent ray path for wide-angle reflections and P-wave delays.*

pressures characterize the Alpine Fault and other major strike-slip fault zones.

Laboratory experiments on the mechanics of faulting show that high fluid pressures in the crust result in a reduction in the magnitude of differential stress required to slip on a fault (Section 2.10.2). In New Zealand, this reduction in the crustal strength is implied by an unusually thin (8 km) seismogenic layer, which coincides with the top of the low velocity zone beneath the Alpine Fault (Leitner *et al.*, 2001; Stern *et al.*, 2001). These relationships suggest that high fluid pressures have reduced the amount of work required for deformation on the Alpine Fault, allowing large magnitudes of slip. As the convergent component of deformation on the South Island increased during the late Cenozoic, the magnitude of crustal thickening and fluid release also increased, resulting in a positive feedback that led to a further focusing of strain in the fault zone.

In addition to high fluid pressures, surface uplift and enhanced erosional activity may result in a strain-softening feedback. The removal of surface material due to

high rates of erosion unloads the lithosphere and causes the upward advection of heat as deep crustal rocks are exhumed (Koons, 1987; Batt & Braun, 1999; Willett, 1999). If the exhumation is faster than the rate at which the advected heat diffuses into the surrounding region, then the temperature of the shallow crust rises (Beaumont *et al.*, 1996). This thermal disturbance weakens the lithosphere because of the high sensitivity of rock strength to temperature (Section 2.10).

In the case of the Southern Alps, moisture-laden winds coming from the west have concentrated erosion on the western side of the mountains, resulting in rapid (5–10 mm a^{-1}) surface uplift, an asymmetric topographic profile, and the exhumation of deep crustal rocks on the southeast side of the mountain range (Fig. 8.23a). Thermochronologic data and exposures of metamorphic rock show an increase in the depth of exhumation toward the Alpine Fault from the southeast (Kamp *et al.*, 1992; Tippet & Kamp, 1993). Surface uplift and exhumation have progressively localized near the Alpine Fault since the Early Miocene, resulting in the exposure

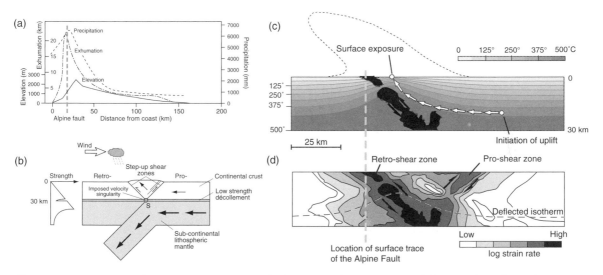

Figure 8.23 (a) Average elevation, precipitation and exhumation along a transect of the central Southern Alps (modified from Willett, 1999, by permission of the American Geophysical Union. Copyright © 1999 American Geophysical Union). Exhumation is from Tippett & Kamp (1993). Numerical model setup (b) and results (c) illustrating the thermal evolution of a 30-km-thick upper crust and the exhumation history (white arrows) of a particle passing through a convergent orogenic system modeled on the Southern Alps (modified from Batt & Braun, 1999, and Batt et al., 2004, by permission of Blackwell Publishing and the American Geophysical Union, respectively. Copyright © 2004 American Geophysical Union). Convergence involved a rate of 10 mm a^{-1} over 10 Ma. Horizontal and vertical scales are equal. Black region marks the peak strain rates and is interpreted to equate with the Alpine Fault for the Southern Alps. Dashed envelope above the model represents the approximate volume of eroded material lost from the system. (d) Strain rates after Batt & Braun (1999) showing pro- and retro-shear zones.

of rocks that once resided at mid-crustal depths (Batt *et al.*, 2004).

To investigate how erosion, exhumation, and heat advection cause these asymmetries and result in the localization of strain on a dipping fault plane, researchers have developed numerical experiments of plate convergence and transpression (Koons, 1987; Beaumont *et al.*, 1996; Batt & Braun, 1999; Willet, 1999). In most of these experiments, crustal deformation is driven by underthrusting the mantle lithosphere of one plate beneath an adjacent, stationary plate (Fig. 8.23b). As mantle lithosphere subducts, the crust accommodates the convergence by deforming. A doubly vergent accretionary wedge develops, whose geometry is determined by the internal strength of the crust and mantle, the coefficient of friction on the basal detachment (Dahlen & Barr, 1989), and patterns of erosion at the surface (Willett, 1992; Naylor *et al.*, 2005).

Figure 8.23c,d show the results of an experiment applied to the Southern Alps. In this case, the moving and stationary blocks represent the Pacific and Australian plates, respectively. Initial conditions include a 30-km-thick crust with a feldspar-dominated rheology and a fixed temperature of $500\,°C$ at its base (Batt & Braun, 1999; Batt *et al.*, 2004). Over a period of $10\,Ma$, two ductile shear zones form and define a doubly vergent wedge that becomes progressively more asymmetric through time (Fig. 8.23c,d). A retro-shear zone develops into a major, crustal-scale thrust. A pro-shear zone also forms but does not accumulate significant strain. Surface erosion and crustal exhumation are concentrated between the two shear zones, reaching maxima at the retro-shear zone. The effects of these processes are illustrated in Fig. 8.23c by the white arrows, which show the exhumation trajectory of a selected particle. The dashed envelope above the model represents the approximate volume of eroded material. As heat is advected upwards in response to the exhumation the mechanical behavior of the deforming region changes. The heat decreases the strength of the retro-shear zone, which brings hot material from the base of the crust to the surface, and weakens the fault. This preferential weakening of the retro-shear zone relative to the pro-shear zone increases the localization of strain on the former and enhances the asymmetry of the model (Fig. 8.23d).

The results of this experiment explain how erosion, exhumation, and thermal weakening result in a concentration of strain along a dipping thrust surface in the upper crust during continental collision. The model predictions match many of the patterns observed in the Southern Alps. Nevertheless, discrepancies also exist. For example, despite the thermal weakening and strain localization caused by exhumation and thermal advection, the retro-shear zone in Fig. 8.23b remains several kilometers thick and does not narrow toward the surface. Batt & Braun (1999) speculated that this lack of fit between the model and observations in New Zealand reflects the absence of strain-induced weakening, high fluid pressures, and other processes that affect strain localization (e.g. Section 7.6.1). Nevertheless, the model explains the prominence of the Alpine Fault as a discrete, dipping surface that accommodates large amounts of slip in the Southern Alps.

To determine whether positive strain-softening feedbacks allow the Alpine Fault to accommodate *oblique-slip* along a single dipping fault, Koons *et al.* (2003) developed a three-dimensional numerical description of transpression for two end-member cases. In both cases, a three-layered Pacific plate is dragged along its base toward an elastic block located on the left side of the model (Fig. 8.24a). The elastic block simulates the behavior of the strong, relatively rigid Australian plate; the crustal layers of the Pacific plate accommodate the majority of the strain. A pressure-dependent Mohr–Coulomb rheology simulates brittle behavior in a strong upper crust. Ductile deformation in a weak lower crust is described using a thermally activated plastic rheology. As in most other models of this type, a zone of basal shear separates the lower crust from Pacific mantle lithosphere. Oblique plate convergence results in velocities of $40\,mm\,a^{-1}$ parallel to and $10\,mm\,a^{-1}$ normal to a vertical plate boundary. Maintaining the western slope at a constant elevation simulates asymmetric erosion at the surface.

In the first experiment (Figs. 8.24a–f), the Pacific plate exhibits a horizontally layered crust. As deformation proceeds, two well-defined fault zones extend down from the plate boundary through the upper crust, forming a doubly vergent wedge. This wedge includes a vertical fault that accommodates lateral (strike-slip) movement and an east dipping convergent (thrust) fault along which deep crustal rocks are exhumed (Fig. 8.24f). In the second experiment (Figs. 8.24g–l), the Pacific plate exhibits a thermally perturbed crust in which advection of hot rock has weakened the upper crust and elevated the $350\,°C$ isotherm to within the upper $10\,km$ of the crust. In this model, strain is concentrated within the thermally perturbed region. Through the upper crust, the lateral and convergent components of strain occur along the same

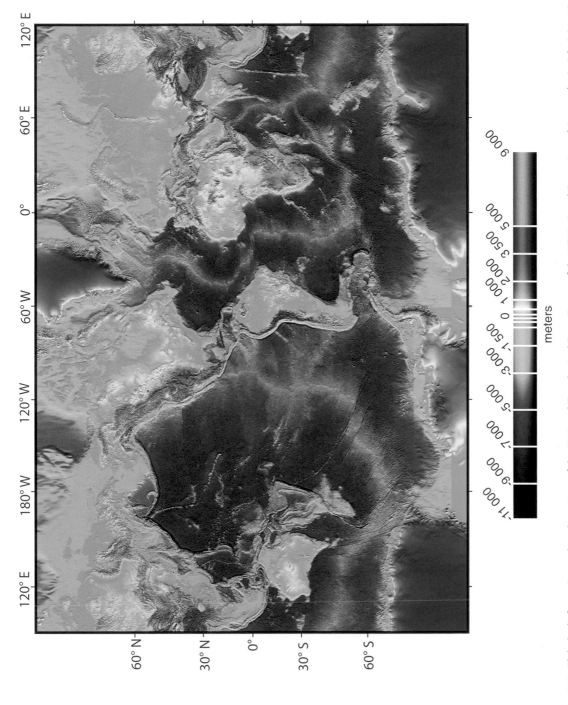

Plate 1.1 Global relief map (reproduced courtesy of the National Geophysical Data Center of the US National Oceanic and Atmospheric Administration).

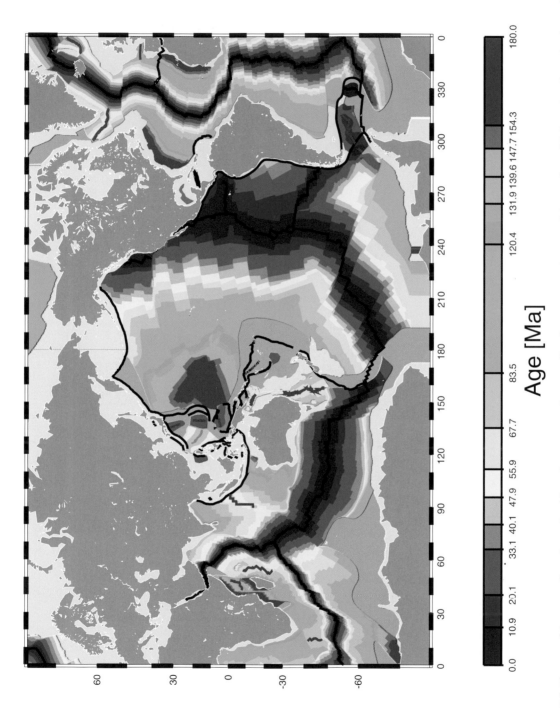

Plate 4.1 *The age of the ocean floor determined from linear magnetic anomalies, the geomagnetic reversal timescale, and Euler poles of rotation (reproduced courtesy of R.D. Müller).*

Age [Ma]

0.0 10.9 20.1 33.1 40.1 47.9 55.9 67.7 83.5 120.4 131.9 139.6 147.7 154.3 180.0

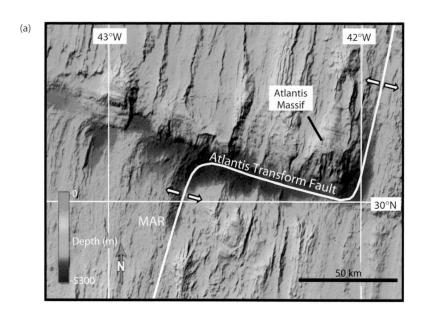

(a)

43°W 42°W

Atlantis Massif

Atlantis Transform Fault

MAR

30°N

Depth (m)

-0

-5300

N

50 km

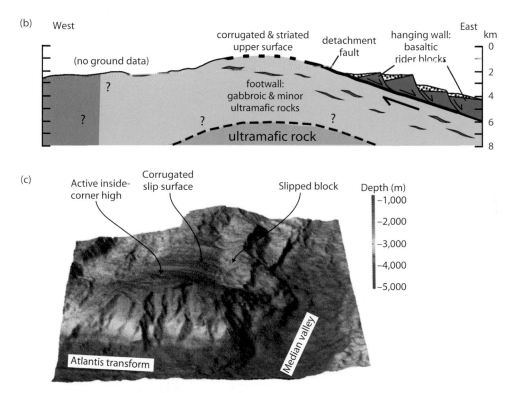

(b)

West   East   km

corrugated & striated   detachment   hanging wall:
upper surface   fault   basaltic
                                        rider blocks   0

(no ground data)   2

?   footwall: 4
                     gabbroic & minor
?   ultramafic rocks   ?   6
        ?   ?
                     ultramafic rock   8

(c)

Corrugated
slip surface

Active inside-   Slipped block   Depth (m)
corner high   -1,000

                                         -2,000

                                         -3,000

                                         -4,000

                                         -5,000

Atlantis transform   Median valley

Plate 6.1 *(a) Colored relief map showing bathymetry and plate boundary geometry of the Atlantis ridge–transform intersection and (b) schematic cross section across the central part of the Atlantis Massif oceanic core complex (modified from Karson et al., 2006, by permission of the American Geophysical Union. Copyright © 2006 American Geophysical Union). (c) Three-dimensional color relief map showing the morphology of the Atlantis Massif viewed from the south (after Cann et al., 1997, with permission from* Nature **385**, *329–32. Copyright © 1997 Macmillan Publishers Ltd). The corrugated surface in (c) curves from near horizontal to about 10° toward the spreading axis below a slipped block of basaltic rock.*

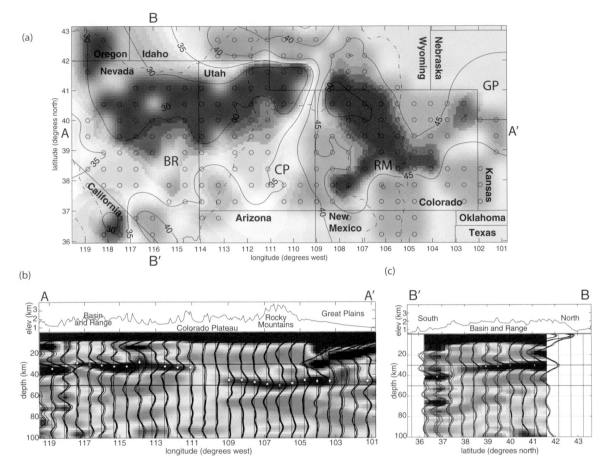

Plate 7.1 (a) Contour map and (b), (c) vertical profiles showing depth to the Moho beneath the northern Basin and Range (BR), Colorado Plateau (CP), the Rocky Mountains (RM), and the Great Plains (GP) (image provided by H. Gilbert and modified from Gilbert & Sheehan, 2004, with permission from the American Geophysical Union. Copyright © 2004 American Geophysical Union). Thin dashed-dotted lines mark the outlines of the four physiographic provinces. Circles in (a) show locations of points where crustal thicknesses have been determined by examining discontinuities in seismic shear wave speeds in the crust and mantle. White dots in (b) indicate the interpreted Moho depths found below each point (circle) shown in (a). Vertical wavy lines in (b) show the range of teleseismic receiver functions obtained in the study. Horizontal lines at 30 km and 50 km depth are shown for reference.

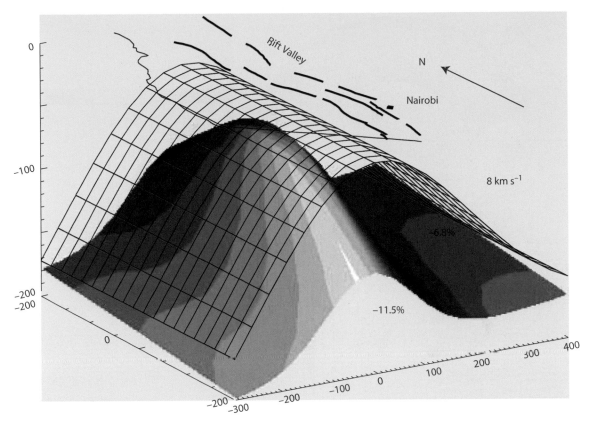

Plate 7.2 *Three dimensional velocity model of the upper mantle below the Kenya Dome derived from the tomographic inversion of teleseismic P-wave residuals (after Davis & Slack, 2002, with permission from the American Geophysical Union. Copyright © 2002 American Geophysical Union). Horizontal and vertical scales in kilometers. The upper layer (gridded) peaks at the Moho (35 km) and has a velocity contrast of −6.8% relative to 8 km s⁻¹ mantle. The lower layer (colored) forms a lobe and peaks at ~70 km, and has a −11.5% contrast extending to a depth of ~170 km.*

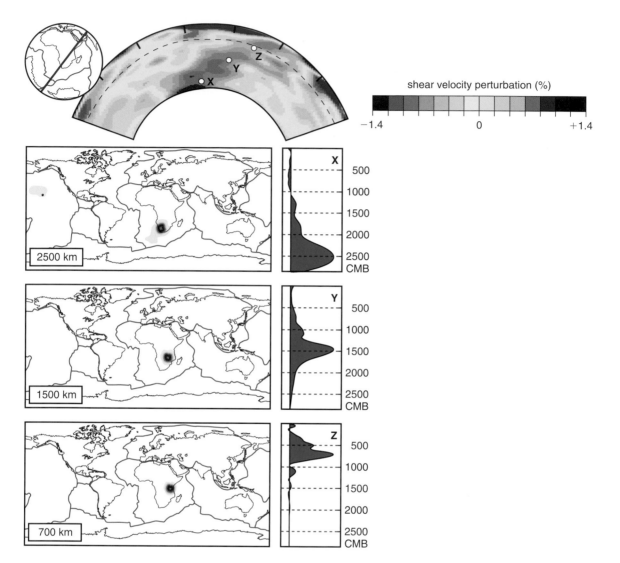

Plate 7.3 *Tomographic model of three-dimensional shear wave velocity variations for locations X, Y, and Z at 2500, 1500, and 700 km depth, respectively, in the mantle beneath Africa (after Ritsema et al., 1999, Science **286**, 1925–8, with permission from the AAAS). Vertical profile shown at the top with location map. High and low velocity regions are indicated by blue and red colors, respectively, with an intensity that is proportional to the percentage amplitude of the velocity perturbations compared with shear wave velocities from a reference model. Horizontal cross sections at the three depths are shown on the left. Lateral (radial) variability in the shapes of the anomalies at the three locations are shown on the right. CMB, core-mantle boundary.*

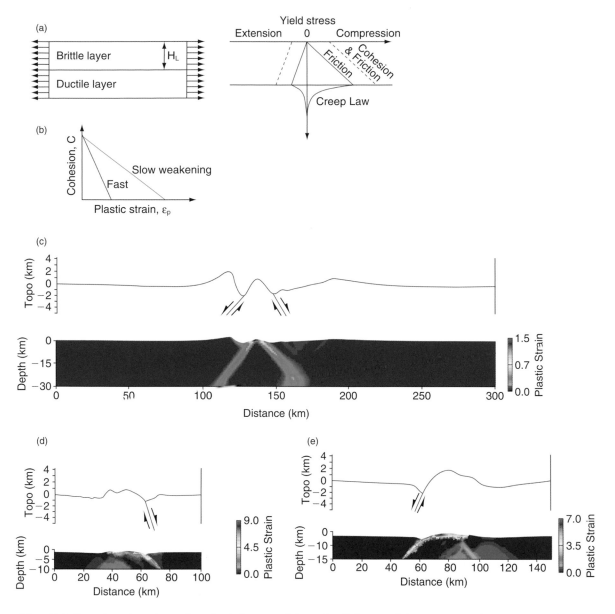

Plate 7.4 *Extension of an elasto-viscoplastic layer overlying a ductile layer with low viscosity (after Lavier & Buck, 2002, with permission from the American Geophysical Union. Copyright © 2002 American Geophysical Union). (a) Model setup. H$_L$, layer thickness. (b) The linear reduction of cohesion, C, with plastic strain. (c–e) Extension of a uniform brittle layer with a cohesion of 44 MPa and a thickness of 30 km, 10 km, and 15 km, respectively, after 25 km of extension. In each model the critical amount of offset that a fault must accumulate to lose all its cohesional strength is 1.5 km.*

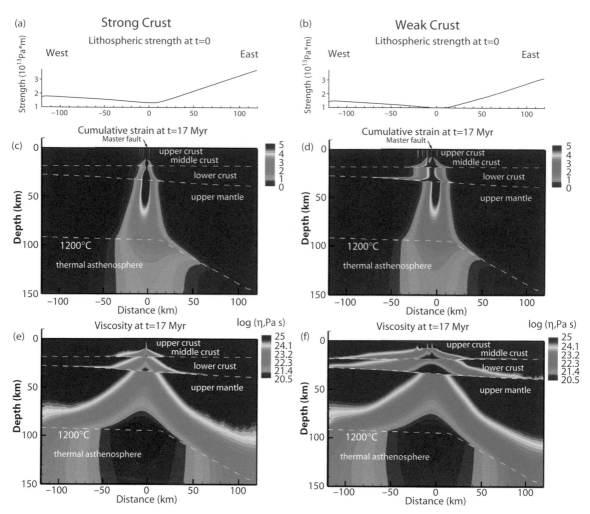

Plate 8.1 *Results of thermomechanical models involving pure strike-slip faulting and cold thick lithosphere, with strong crust and weak crust (after Sobolev et al., 2005, with permission from Elsevier). The upper plots (a,b) show the dependence of lithospheric strength prior to deformation (t = 0). The middle sections (c,d) show the distribution of cumulative finite strain at 17 Myr. Dashed white lines indicate major lithospheric boundaries. The lower sections (e,f) show the distribution of the viscosity at 17 Myr.*

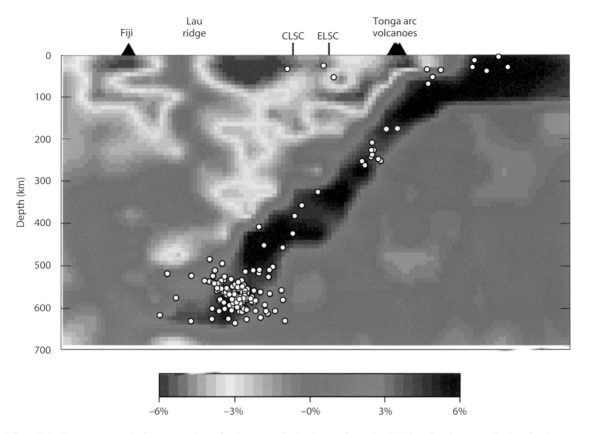

Plate 9.1 *East–west vertical cross section of a P-wave velocity image from 0 to 700 km depth across the Lau basin (after Zhao et al., 1997, Science **278**, 254–7 with permission from the AAAS). Colors indicate extent to which the velocities are less than (e.g. red) or greater than (e.g. blue) the mean P-wave velocity at that depth. Solid triangles denote active volcanoes. CLSC, Central Lau spreading center; ELSC, Eastern Lau spreading center. White circles, earthquakes.*

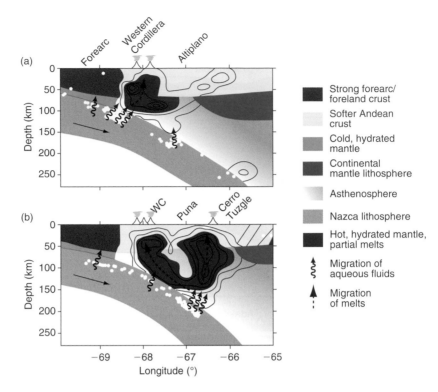

Plate 9.4 *(Top) Interpretative cross sections across the Andean continental arc at (a) 22.1°S and (b) 24.2°S latitude showing regions of low Q (Q ≤ 200) (redrawn from Schurr et al., 2003, with permission from Elsevier). Water is released from the oceanic lithosphere at the discrete earthquake clusters (white dots) causing partial melting in the overlying hot mantle wedge. Regions of low Q are interpreted to represent mantle and crustal rocks that contain significant amounts of partial melt.*

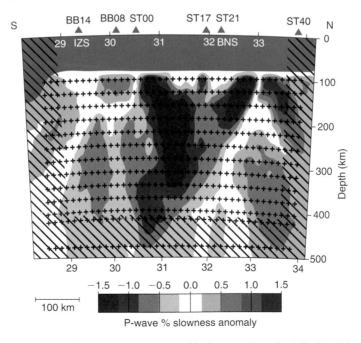

Plate 9.4 *(Bottom) Interpretive profile illustrating the possible downwelling of cold Indian lithosphere beneath central Tibet (redrawn from Tilmann et al., 2003, Science **300**, 1424–7, with permission from the AAAS). The image, which shows a moderate (2%) P-wave velocity contrast near 32°N with high (blue) velocities to the north and slow (red) velocities to the south, was obtained by a tomographic inversion of P-wave arrival times. IZS, Indus–Zangbo suture; BNS, Bangong–Nujiang suture.*

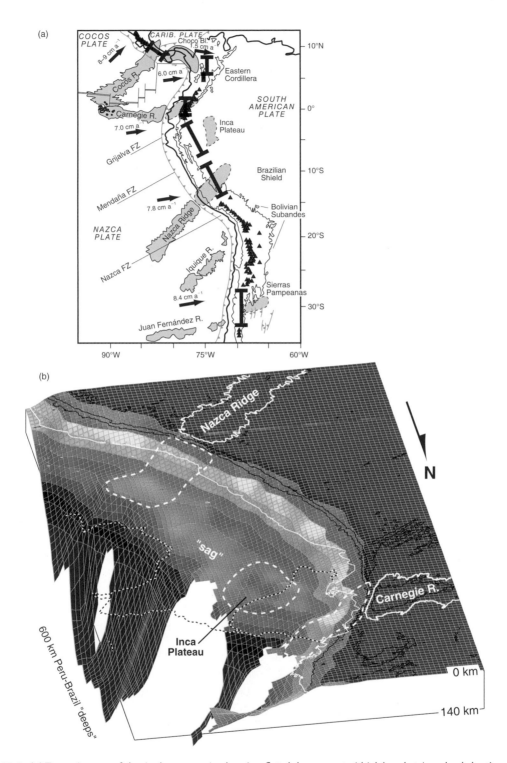

Plate 10.1 *(a) Tectonic map of the Andean margin showing flat slab segments (thick brackets) and subducting oceanic plateaux (gray). Red triangles, active volcanoes. Dashed lines indicate inferred position of subducted plateaux and ridges. Plate convergence vectors are based on the model of DeMets* et al. *(1990). (b) Three-dimensional view of the subducting Nazca plate surface determined from the gridding of earthquake hypocenters. View is to the south. Dashed lines show morphologic highs with intervening sag, corresponding to the estimated position of the subducted Nazca Ridge and Inca Plateau (images provided by M.-A. Gutscher and modified from Gutscher* et al., *2000, by permission of the American Geophysical Union. Copyright © 2000 American Geophysical Union).*

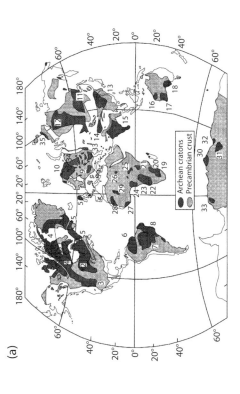

(a)

(b) 100–200 km depth

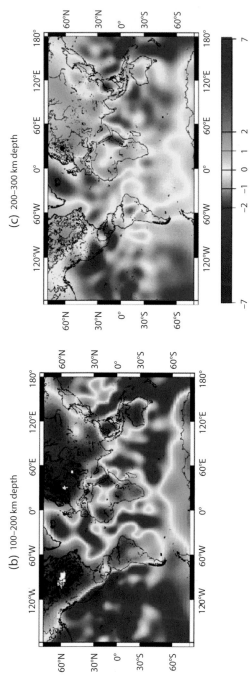

(c) 200–300 km depth

% V$_s$ Anomaly

−7 −2 −1 0 1 2 7

Plate 11.1 *Global maps showing (a) areas of Archean and Proterozoic crust (after Kusky & Polat, 1999, with permission from Elsevier) and seismic velocity anomalies at depths of (b) 100–200 km and (c) 200–300 km (image provided by S. King and modified from King, 2005, with permission from Elsevier). Cratons in (a) are as follows: 1, Slave; 2, Superior (including Abitibi); 3, Wyoming; 4, Kaminak (Hearne); 5, North Atlantic (Nain, Godthaab, Lewisian); 6, Guiana; 7, Central Brazil (Guapore); 8, Atlantic (São Francisco); 9, Ukranian; 10, Baltic (Kola); 11, Aldan; 12, Anabar; 13, Sino (North China and South China)-Korean; 14, Tarim; 15, Indian; 16, Pilbara; 17, Yilgarn; 18, Gawler; 19, Kaapvaal; 20, Tanzanian; 21, Zambian; 22, Angolan; 23, Kasai; 24, Gabon; 25, Kibalian; 26, Uweinat; 27, Liberian; 28, Maritanian; 29, Ouzzalian; 30, Napier; 31, Prince Charles Mountains; 32, Vestfold Hills; 33, Heimefront Ranges; 34, deeply buried Archean rocks of the East European Shield; 35, Tajmyr. S-wave velocity anomalies in (b) and (c) are from the tomographic model of Ritsema & van Heijst (2000).*

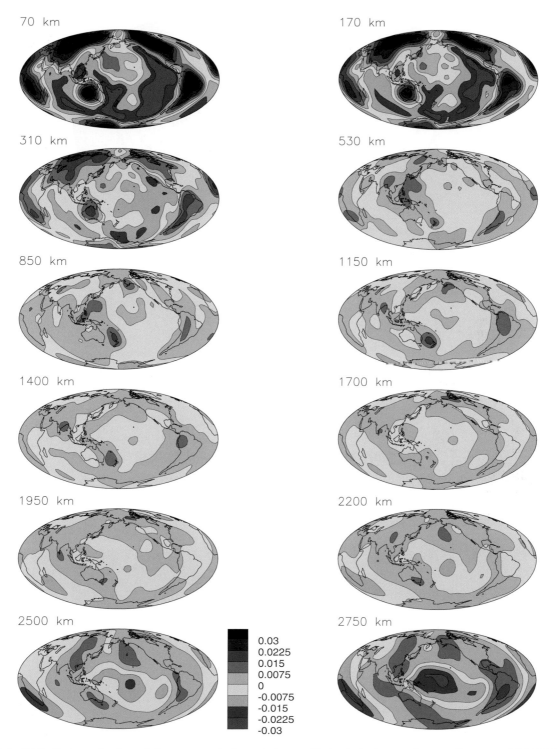

Plate 12.1 *Contoured values of shear wave velocity variations at 12 depths in the mantle according to model S16B30 of Masters et al. (1996). The variations are expressed as fractions of the global average model value at that depth (reproduced with permission from the Royal Society of London).*

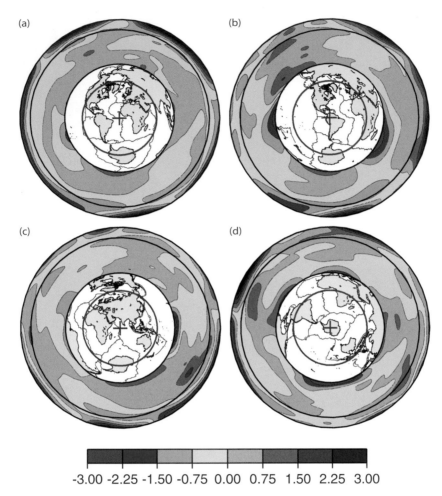

-3.00 -2.25 -1.50 -0.75 0.00 0.75 1.50 2.25 3.00

Plate 12.2 *Cross sections through the shear wave velocity model S16B30 of Masters* et al. *(1996). The variations are expressed as percentages of the global average model value at that depth. Within the mantle the 660 km discontinuity is indicated by a solid circle and within the core the intersection of the plane of the section with the Earth's surface is shown as a solid circle on an azimuthal equidistant projection of the globe (reproduced with permission from the Royal Society of London).*

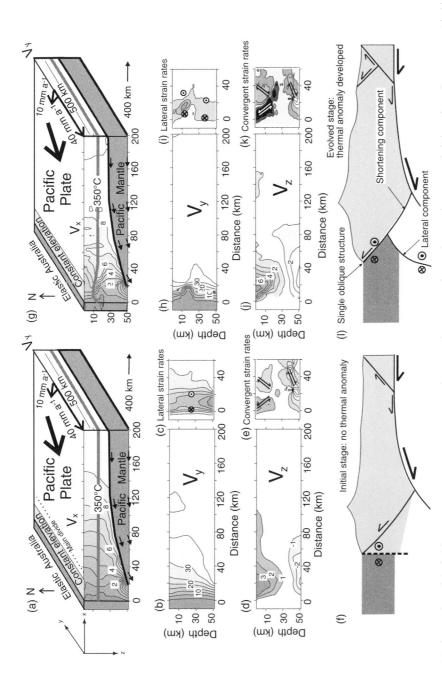

Figure 8.24 *Mechanical model of oblique compression between two plates involving a two-layered crust above mantle lithosphere (images provided by P. Koons and modified from Koons et al., 2003, with permission from the Geological Society of America). (a) Initial model where crust is dragged along its base at velocities of 40 mm a⁻¹ parallel to the y-axis (Vy) and 10 mm a⁻¹ parallel to the x-axis (Vx). Crustal rheology is horizontal and not yet perturbed by advection. (b) Vertical profile constructed parallel to x-axis showing the component of motion parallel to the y-axis (velocities in mm a⁻¹) and (c) plot of lateral strain rates. (d) Vertical profile parallel to the xz plane showing distribution of vertical motion and (e) plot of convergent strain rates. (f) In this model lateral and convergent components are accommodated on two separate structures. (g–k) Results with velocity profiles and strain rate analysis for a model that includes thermal weakening associated with exhumation and concentrated erosion. At this stage both lateral and convergent components of motion are accommodated along a single dipping structure in the upper crust (l) and may separate in the lower crust.*

eastward-dipping fault surface. In the lower crust, the two components separate, producing two zones of deformation (Fig. 8.24l). These results illustrate how an evolving thermal structure resulting from asymmetric erosion and exhumation stabilizes the lateral and convergent components of oblique collision along a single dipping fault. They also suggest that a partitioning of deformation onto separate strike-slip and dip-slip faults is favored where thermal weakening is absent.

8.7 MEASURING THE STRENGTH OF TRANSFORMS

Measures of the strength of continental transforms and large strike-slip faults provide a potentially useful means of testing models of continental rheology and evaluating the driving forces of continental deformation (Section 8.5.1). In many intraplate areas, the long-range (1000–5000 km) uniformity of stress orientations and their relative magnitudes inferred from measures of strain or displacement suggest that plate-driving forces provide the largest component of the total stress field (Zoback, 1992). Models of GPS-derived horizontal velocities in some regions, such as southern California, tend to support this view (McCaffrey, 2005). However, in other areas, such as the Basin and Range Province (Section 7.3), stresses caused by lateral variations in crustal buoyancy (Section 7.6.3) also appear to contribute significantly to the horizontal stress field (Sonder & Jones, 1999; Bennett et al., 2003).

There have been numerous attempts to evaluate the strength of the San Andreas Fault using various geologic and geophysical indicators (Zoback et al., 1987, Zoback, 2000). For some fault segments (Fig. 8.25), stress data suggest that the direction of maximum horizontal compression (σ_1, Section 2.10.1) lies at a high angle (β) to the fault zone. In central California these angles are as high as $\beta = 85°$. In southern California they are lower at $\beta = 68°$ (Townend & Zoback, 2004). These observations are problematic because classical theories of faulting (Section 2.10.2) cannot explain compression at high angles to a strike-slip fault with such a small component of convergence. Moreover, in the case

of the San Andreas Fault, a paradox exists in that heat flow observations (Lachenbruch & Sass, 1992) show no frictionally generated heat, so that the fault must slip in response to very low shear stresses.

One possible explanation of the high-angle stress directions in California is that the San Andreas is an extremely weak fault that locally reorients the regional stresses (Mount & Suppe, 1987; Zoback et al., 1987; Zoback, 2000). In this interpretation, shear stresses far from the fault are high and contained by the frictional strength of the crust, but shear stresses on planes parallel to the "weak" faults of the San Andreas system must be quite low. Consequently, the principal stresses become reoriented so as to minimize shear stresses on planes parallel to the San Andreas Fault. This requires a rotation such that the direction of maximum horizontal compressive stress (σ_1) becomes nearly orthogonal to the fault if the regional compression direction is at an angle in excess of 45° to the fault, which occurs at present. However, if this angle is less than 45°, the maximum horizontal compression is rotated into approximate parallelism with the fault. This latter type of rotation may have characterized the San Andreas Fault at some time in the past when relative plate motions were different than they are now.

This model of a weak continental strike-slip fault offers one explanation of conflicting geologic and geophysical data in California. However, alternative interpretations involving a strong or an intermediate-strength San Andreas Fault also have been proposed. These latter models are based on frictional theories of faulting, which suggest that σ_1 rotates to ~45° from the fault trace within a ~20–30-km-wide zone in the Big Bend region (Scholz, 2000). Scholz (2000) interpreted reports of high σ_1 angles in this area as representing local stresses related to folding instead of regional stresses. He also concluded that the presence and sense of the stress rotation fits predictions of a strong fault rather than a weak one. High fluid pressure (Section 8.6.3) is a possible mechanism for decreasing the strength of the fault and could explain some rotation of the stresses (Rice, 1992). Alternatively, the strength of the fault and the adjacent crust generally could be much lower than predicted by considerations of fault mechanics (Hardebeck & Michael, 2004).

These conflicting observations and interpretations concerning the strength of large strike-slip faults have yet to be resolved. In the case of the San Andreas Fault, part of the controversy may be related to different mechanical behaviors of the creeping versus locked

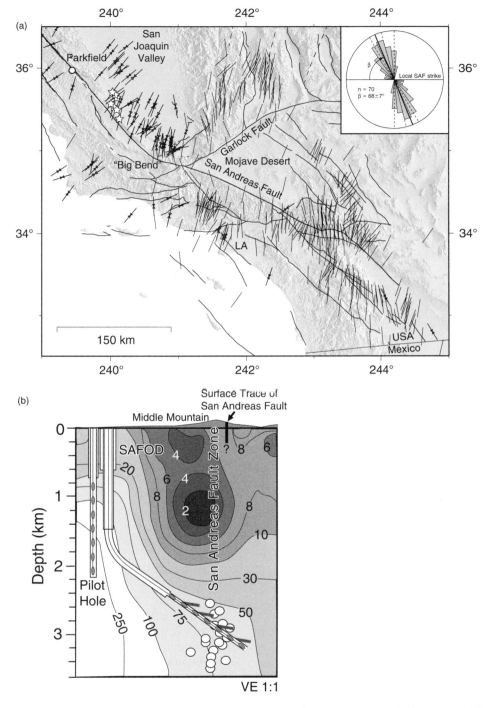

Figure 8.25 *(a) Maximum horizontal compression from southern California (image provided by J. Townend and M. Zoback and modified from Townend & Zoback, 2004, by permission of the American Geophysical Union. Copyright © 2004 American Geophysical Union). Stress determinations are as follows: inward-point arrows, borehole breakouts; stars, hydraulic fracturing experiments; plain straight lines, earthquake focal mechanisms. Inset summarizes the angle (β) between the maximum principal compressive stress and the local fault strike within 10 km of the San Andreas Fault (SAF). The angle of 68 ± 7° suggests a relatively low frictional strength for a 400-km-long fault segment. (b) Vertical profile showing location of SAFOD drill hole experiment near Parkfield California (after Hickman et al., 2004, with permission from the American Geophysical Union). Magnetotelluric resistivity readings are from Unsworth & Bedrosian (2004). White circles are earthquake hypocenters. Ovals in drill holes represent down-hole sensors. Contours show resistivity in ohm-meters.*

segments of the fault or to the different methods of inferring stresses. To resolve these problems independent measurements of principal stress orientations and magnitudes from within large, tectonically active faults are needed. The **S**an **A**ndreas **F**ault **O**bservatory at **D**epth (SAFOD) drilling program involves such measurements. This program involves drilling into the hypocentral zone of repeating M ≈ 2 earthquakes on a creeping segment of the San Andreas Fault near Parkfield, California (Fig. 8.1) at a depth of about 3 km. The goals include establishing an observatory in close prox-imity to these repeating earthquakes to obtain down-hole measurements of the physical and chemical conditions under which earthquakes occur and to exhume rock and fluid samples for laboratory analyses (Hickman *et al.*, 2004). Although there is still considerable uncertainty in the preliminary estimates of horizontal stress magnitudes, stress observations near the bottom of a 2.2-km-deep pilot hole (Fig. 8.25b) (Hickman & Zoback, 2004) and heat flow measurements (Williams *et al.*, 2004) suggest a locally weak San Andreas Fault in an otherwise strong crust.

9 | Subduction zones

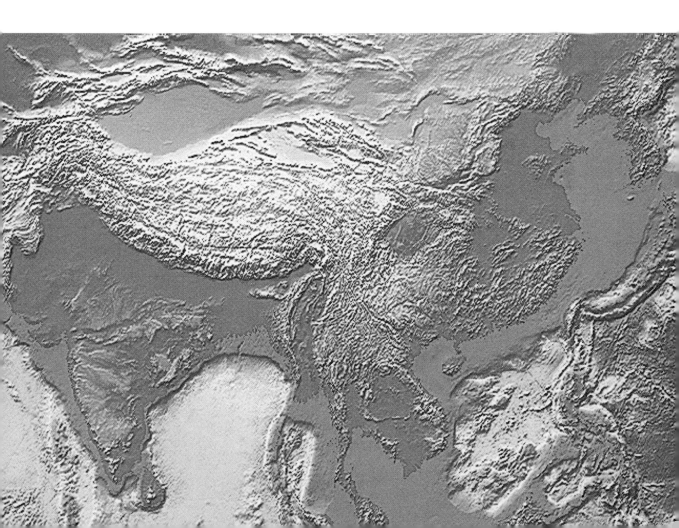

9.1 OCEAN TRENCHES

Oceanic trenches are the direct manifestation of under-thrusting oceanic lithosphere, and are developed on the oceanward side of both the island arcs and Andean-type orogens that form above subduction zones (Fig. 9.1). They represent the largest linear depressed features of the Earth's surface, and are remarkable for their depth and continuity. The Peru–Chile Trench is 4500 km long and reaches depths of 2–4 km below the surrounding ocean floor so that its base is 7–8 km below sea level. The trenches in the western Pacific are typically deeper than those of the eastern Pacific margin, the greatest trench depths, of 10–11 km, occurring in the Mariana and Tonga–Kermadec trenches. The main control on the maximum depth of a particular trench would appear to be the age of the oceanic lithosphere being sub-ducted, as this determines the depth to the oceanic crust entering the trench (Section 6.4). The striking contrast between trench depths in the east and west Pacific is largely explained therefore by the systematic difference in the age of the ocean floor in these areas (Plate 4.1 between pp. 244 and 245). Trenches are generally 50–100 km in width and in section form an asymmetric

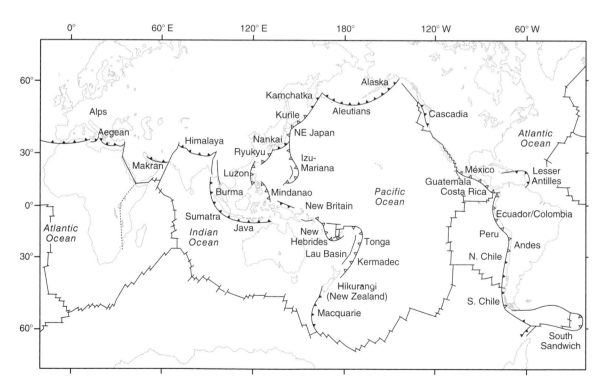

Figure 9.1 *The location of convergent plate margins (thin solid lines with barbs). Accretionary margins are indicated by solid barbs, and erosive margins by open barbs (Sections 9.6, 9.7). The thick solid lines are active spreading centers and include those in backarc basins (Section 9.10) (modified from Stern, 2002, and from Clift & Vanucchi, 2004, by permission of the American Geophysical Union. Copyright © 2002 and 2004 American Geophysical Union).*

V-shape with the steepest slope, of 8–20°, on the side opposite the underthrusting ocean floor. The sediment fill of trenches can vary greatly, from virtually nothing, as in the Tonga–Kermadec trench, to almost complete, as in the Lesser Antilles and Alaskan trenches because of the supply of sediment from adjacent continental areas. Trench depth is also reduced by the subduction of aseismic ridges (Section 10.2.2).

9.2 GENERAL MORPHOLOGY OF ISLAND ARC SYSTEMS

Island arc systems are formed when oceanic lithosphere is subducted beneath oceanic lithosphere. They are consequently typical of the margins of shrinking oceans such as the Pacific, where the majority of island arcs are located. They also occur in the western Atlantic, where the Lesser Antilles (Caribbean) and South Sandwich (Scotia) arcs are formed at the eastern margins of small oceanic plates isolated by transform faults against the general westward trend of movement.

All of the components of island arc systems are usually convex to the underthrusting ocean. This convexity may be a consequence of spherical geometry, as suggested by Frank (1968). If a flexible spherical shell, such as a table tennis ball, is indented an angle θ (Fig. 9.2), the indentation is a spherical surface with the same radius as the shell (R). The edge of the indentation is a circle whose radius r is given by $r = \frac{1}{2}R\theta$, where θ is in radians. If this theorem is applied to a plate on the Earth's surface, θ represents the angle of underthrusting of oceanic lithosphere, which averages about 45°. The radius of curvature of the trench and island arc on the Earth's surface is then about 2500 km. This value is in agreement with some, but not all, island arc systems. The general convexity of island arc systems is probably a consequence of spherical geometry, and deviations result from the oversimplification of this approach, in particular the fact that the conservation of surface area is not required by plate tectonics. Thus, for example, the angle of underthrusting at the Mariana arc is almost 90°, but it has one of the smallest radii of curvature (Uyeda & Kanamori, 1979).

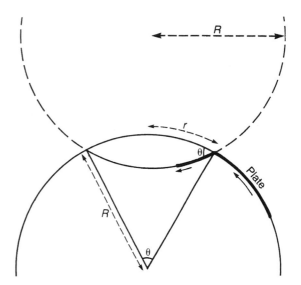

Figure 9.2 *Geometry of an indentation in a sphere of inextensible material (redrawn from Bott, 1982, by permission of Edward Arnold (Publishers Ltd).*

The generalized morphology of an island arc system is shown in Fig. 9.3, although not all components are present in every system. Proceeding from the oceanward side of the system, a flexural bulge about 500 m high occurs between 100 and 200 km from the trench. The forearc region comprises the trench itself, the accretionary prism, and the forearc basin. The accretionary prism is constructed of thrust slices of trench fill (flysch) sediments and possibly oceanic crust sediments that have been scraped off the downgoing slab by the leading edge of the overriding plate. The forearc basin is a region of tranquil, flat-bedded sedimentation between the accretionary prism and island arc. The island arc is made up of an outer sedimentary arc and an inner magmatic arc. The sedimentary arc comprises coralline and volcaniclastic sediments underlain by volcanic rocks older than those found in the magmatic arc. This volcanic substrate may represent the initial site of volcanism as the relatively cool oceanic plate began its descent. As the "cold" plate extended further into the asthenosphere the position of igneous activity moved backwards to its steady state location now represented by the magmatic arc. Processes contributing to the formation of the island arcs are discussed in Section 9.8 and 9.9. The island arc and remnant arc (backarc ridge), first recognized by Vening Meinesz (1951), enclose a

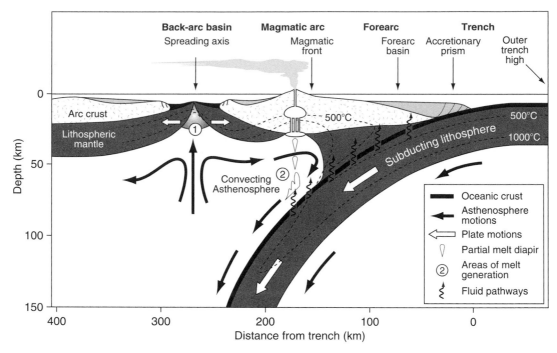

Figure 9.3 *Schematic section through an island arc system (modified from Stern, 2002, by permission of the American Geophysical Union. Copyright © 2002 American Geophysical Union).*

backarc basin (or marginal basin) behind the island arc. However not all backarc basins are formed by spreading above an active subduction zone, as indicated in Fig. 9.3 (Section 9.10).

9.3 GRAVITY ANOMALIES OF SUBDUCTION ZONES

Figure 9.4 shows a free air gravity anomaly profile across the Aleutian arc that is typical of most subduction zones. The flexural bulge of the downgoing lithosphere to seaward of the trench is marked by a positive gravity anomaly of about 500 g.u. (Talwani & Watts, 1974). The trench and accretionary prism are typified by a large negative anomaly of some 2000 g.u. amplitude which results from the displacement of crustal materials by sea water and low density sediments.

Conversely, the island arc is marked by a large positive anomaly. Isostatic anomalies over the trench and arc are large and exhibit the same polarity as the free air anomalies. These large anomalies result from the dynamic equilibrium imposed on the system by compression, so that the trench is forced down and the arc held up out of isostatic equilibrium by the forces driving the plates.

9.4 STRUCTURE OF SUBDUCTION ZONES FROM EARTHQUAKES

Subduction zones exhibit intense seismic activity. A large number of events occur on a plane that dips on average at an angle of about 45° away from the underthrusting oceanic plate (Fig. 9.5). The plane is known as a Benioff (or Benioff–Wadati) zone, after

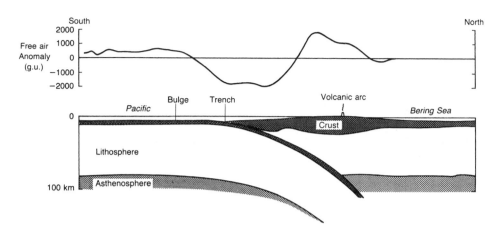

Figure 9.4 *Gravity anomalies of an oceanic subduction zone (after Grow, 1973, with permission from the Geological Society of America).*

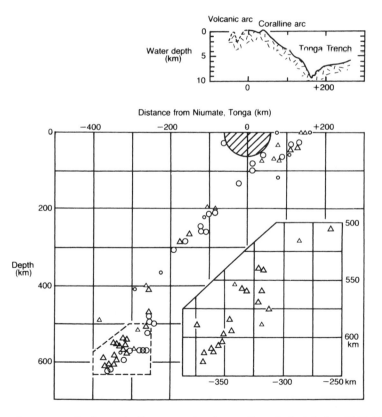

Figure 9.5 *Vertical section perpendicular to the Tonga arc showing earthquake foci during 1965. Circles, foci projected from within 0–150 km north of the section; triangles, from 0–150 km south. Exaggerated topography (13 : 1) above. Inset, enlargement of the region of deep earthquakes (redrawn from Isacks et al., 1969, with permission from the Geological Society of America).*

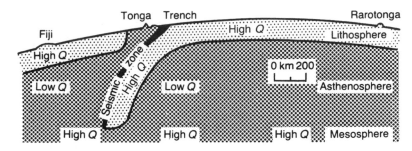

Figure 9.6 *Hypothetical section across the Tonga arc based on the attenuation of seismic waves (redrawn from Oliver & Isacks, 1967, by permission of the American Geophysical Union. Copyright © 1967 American Geophysical Union).*

its discoverer(s), and earthquakes on it extend from near the surface, beneath the forearc region, down to a maximum depth of about 670 km. Figure 9.5 shows a section through the Tonga–Kermadec island arc system with earthquake foci projected on to a vertical plane parallel to the direction of underthrusting. The foci can be seen to occur at progressively greater depths with increasing distance from the site of underthrusting at the Tonga Trench. Further information on the nature of the Benioff zone was obtained from a study of the body wave amplitudes from deep earthquakes (Fig. 9.6). Seismic arrivals at the volcanic islands of the arc, such as Tonga, were found to be of far greater amplitude than those recorded to the front or rear of the arc at stations such as Raratonga and Fiji. The differences in amplitude are usually described quantitatively in terms of the *Q*-factor, the inverse of the specific attenuation factor, and in general the higher the *Q*-factor the stronger the rock. High *Q* travel paths give rise to little attenuation, and vice versa. Seismic waves traveling up the length of the seismic zone appear to pass through a region of high *Q* (about 1000), while those traveling to lateral recorders pass through a more normal region of low *Q* (about 150). The Benioff zone thus appears to define the top of a high *Q* zone about 100 km thick. The Benioff zone had originally been interpreted as a large thrust fault between different crustal provinces. The seismic data allowed a new interpretation to be made in terms of a high *Q* belt of Pacific lithosphere underthrust into the mantle. This interpretation was refined by Barazangi & Isacks (1971), by the use of a local seismometer network in the region of the Tonga arc (Fig. 9.7). In addition to the previous results, a zone of very high attenuation

(extremely low *Q* of about 50) was defined in the uppermost mantle above the downgoing slab in a region about 300 km wide, stretching between the active island arc (Tonga) and backarc ridge (Lau Ridge). This implies that the mantle beneath the backarc basin (Lau basin) is much weaker than elsewhere or that the lithosphere is considerably thinner. The data have important ramifications for the origin of backarc basins and will be considered in more detail in Section 9.10.

Detailed investigations of the region above the subducting lithosphere have also been carried out using seismic tomography (Section 2.1.8). Plate 9.1 (between pp. 244 and 245) shows a section through the Tonga arc in which the subducting slab is clearly defined by a region of relatively high P-wave velocity. Above this there is a region of low velocities, beneath the Lau basin (see also Section 9.10), corresponding to the region of extremely low *Q* in Fig. 9.7. The lowest velocities occur beneath the Tonga arc volcanoes.

The earthquake activity associated with the downgoing slab occurs as a result of four distinct processes (Fig. 9.8). In region "a" earthquakes are generated in response to the bending of the lithosphere as it begins its descent. Bending, or downward flexure of the lithosphere, puts the upper surface of the plate into tension, and the normal faulting associated with this stress regime gives rise to the observed earthquakes, which occur to depths of up to 25 km (Christensen & Ruff, 1988).

Flexural bending of the lithosphere also gives rise to the topographic bulge present in the subducting plate on the oceanward side of the island arc. This regional rise of sea bed topography is located between 100–200 km from the trench axis and has an amplitude of

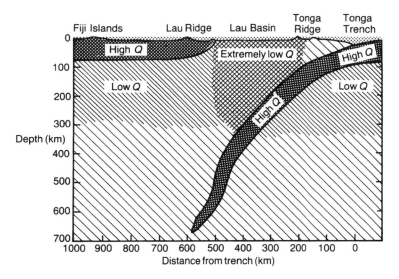

Figure 9.7 *Schematic section across the Tonga arc showing the zone of very high seismic attenuation beneath the Lau backarc basin (redrawn from Barazangi & Isacks, 1971, by permission of the American Geophysical Union. Copyright © 1971 American Geophysical Union).*

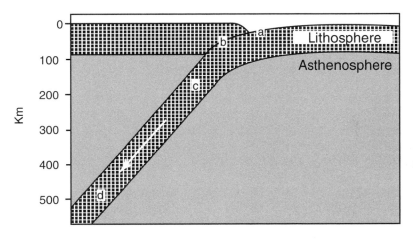

Figure 9.8 *Plate model of subduction zones; a, b, c, and d indicate regions of distinctive focal mechanisms.*

several hundred metres. Simple beam theory predicts that the presence of this bulge is a consequence of the downward deflection of the subducting plate (Fig. 9.9). However, closer investigation of lithospheric behavior in this environment indicates that the flexure is not completely elastic, and must involve considerable plastic (permanent) deformation (Fig. 9.10) (Turcotte *et al.*, 1978). Chapple & Forsyth (1979) deduced that the bending of a two layer elastic-perfectly plastic plate,

50 km thick, in which the upper 20 km are under tension and the lower 30 km under compression, fits most topographic profiles, and that the variations in these profiles are probably due to variations in the regional stress field.

Region "b" (Fig. 9.8) is characterized by earthquakes generated from thrust faulting along the contact between the overriding and underthrusting plates. Focal mechanism solutions for earthquakes associated with regions "a" and "b" of Fig. 9.8 are shown in Fig. 9.11,

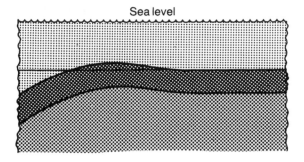

Figure 9.9 *Downbending of an elastic or elastic-perfectly plastic plate at a subduction zone (redrawn from Turcotte et al., 1978, with permission from Elsevier).*

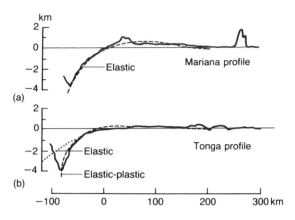

Figure 9.10 *Observed and theoretical profiles of lithosphere bending at a trench: (a) Mariana Trench, with an elastic lithosphere 29 km thick; (b) Tonga Trench, better modeled by an elastic-perfectly plastic plate 32 km thick (redrawn from Turcotte et al., 1978, with permission from Elsevier).*

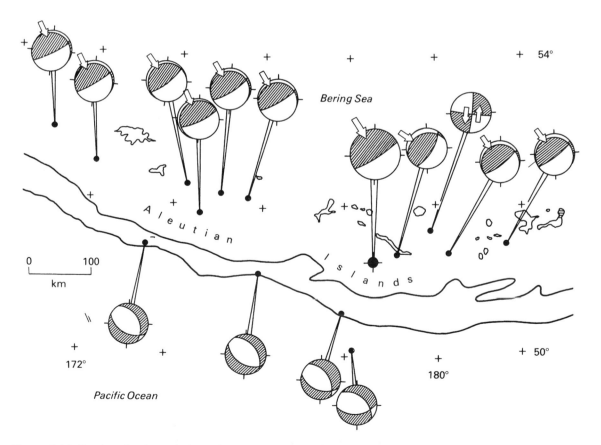

Figure 9.11 *Focal mechanism solutions of earthquakes in the Aleutian arc, compressional quadrant shaded (redrawn from Stauder, 1968, by permission of the American Geophysical Union. Copyright © 1968 American Geophysical Union).*

which represents the distribution of earthquake types around the Aleutian island arc (Stauder, 1968). The belt of earthquakes to the south of the islands is caused by normal faulting associated with the flexure of the top part of the Pacific Plate, which is underthrusting the Bering Sea in a northwesterly direction. The groups of earthquakes lying under or just to the south of the island chain are indicative of thrust faulting. The nodal planes dip steeply to the south and gently to the north. It is probable that the latter planes represent the fault planes, and that these earthquakes are generated by the relative movement between the Pacific and Bering Sea lithosphere. The single focal mechanism solution indicative of strike-slip movement is either on a sinistral strike-slip fault perpendicular to the island chain, as indicated on the diagram, or alternatively on a dextral strike-slip fault paralleling the island chain. In view of the oblique direction of underthrusting in this region, the latter interpretation is perhaps more likely to be correct (Section 5.3).

The earthquakes occurring in the Benioff zone in zone "c" (Fig. 9.8), at depths greater than the thickness of the lithosphere at the surface, are not generated by thrusting at the top of the descending plate, because the asthenosphere in contact with the plate is too weak to support the stresses necessary for extensive faulting. At these depths earthquakes occur as a result of the internal deformation of the relatively cold and hence strong descending slab of lithosphere. Hasegawa et al. (1978), making use of a local array of seismographs, identified two Benioff zones beneath the Japan arc that appear to merge down dip (Fig. 9.12). The arrival times of different seismic phases indicate that the upper of these zones corresponds to the crustal part of the descending slab, and the lower to the lithospheric mantle (Hasegawa et al., 1994).

Subsequently, double seismic zones, at depths between 70 and 200 km, have been documented in numerous well-studied subduction zones (Peacock, 2001), and it seems probable that they are a common feature of subduction zone seismicity. In some cases focal mechanism solutions for the upper zone earthquakes imply down-dip compression, and those for the lower zone earthquakes down-dip tension. This suggests that unbending of the downgoing plate may be important, the plate having suffered a certain amount of permanent, plastic deformation during its initial descent (Isacks & Barazangi, 1977). However the double seismic zones extend to depths well beyond the region of unbending of the downgoing plates. It is now thought that most of these earthquakes are triggered by metamorphic reactions involving dehydration; those in the upper zone associated with the formation of eclogite (Kirby et al., 1996), and those in the lower zone with the dehydration of serpentinite (Meade & Jeanloz, 1991). It is suggested that dehydration reactions generate high pore pressures along pre-existing fault planes in the subducting oceanic lithosphere, producing earthquakes

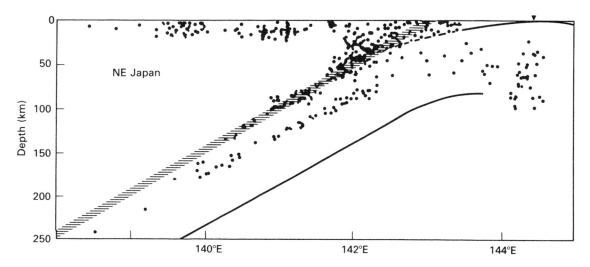

Figure 9.12 *Distribution of earthquakes beneath the northeastern Japan arc. Shaded line is probably the top of the descending lithosphere (redrawn from Hasegawa et al., 1978, with permission from Blackwell Publishing).*

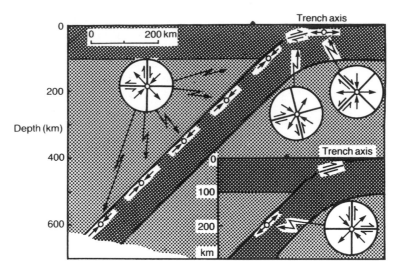

Figure 9.13 *Schematic focal mechanism solution distribution on a section perpendicular to an island arc. Inset shows alternative intermediate depth mechanism (redrawn from Isacks et al., 1969, with permission from the Geological Society of America).*

by brittle failure. The process is termed *dehydration embrittlement*.

Peacock (2001), using a detailed thermal model for the subduction zone beneath northeast Japan, has shown that the lower seismic zone (Fig. 9.12) migrates across the isotherms, from approximately 800 to 400°C, as the focal depths increase from 70 to 180 km. If these temperatures and implied pressures are plotted on a P–T diagram, the pressure/temperature values and negative slope are very analogous to those for the dehydration reaction serpentine to forsterite + enstatite + water. This strongly suggests that these earthquakes are the result of the dehydration of serpentinized mantle within the downgoing oceanic plate. This explanation assumes that the oceanic mantle is serpentinized to a depth of several tens of kilometers, whereas hydrothermal circulation and alteration at mid-ocean ridges is thought to be restricted to the crust. However, the normal faulting associated with the outer rise and bending of the oceanic lithosphere oceanward of the trench may well permit ingress of seawater and hydration of the lithosphere to depths of tens of kilometers (Peacock, 2001).

Below 300 km (zone "d" in Fig. 9.8) the earthquake mechanism is believed to be a result of the sudden phase change from olivine to spinel structure, producing *transformational* or *anticrack faulting*. This takes place

by rapid shearing of the crystal lattice along planes on which minute spinel crystals have grown (Green, 1994). At normal mantle temperatures this phase change occurs at a depth of approximately 400 km (Sections 2.8.5, 9.5). However, the anomalously low temperatures in the core of a downgoing slab enable olivine to exist metastably to greater depths, potentially until it reaches a temperature of about 700°C (Wiens et al., 1993). In old, rapidly subducting slabs this may, exceptionally, be at a depth of approximately 670 km, explaining the termination of subduction zone seismicity at this depth. It is also probable that a similar transformation from enstatite to ilmenite contributes to subduction zone seismicity in this depth range (Hogrefe et al., 1994). The phase changes that occur in the slab at a depth of approximately 700 km (Sections 2.8.5, 9.5) are thought to produce fine-grained materials that behave in a superplastic manner and thus cannot generate earthquakes (Ito & Sato, 1991).

The deep events of regions "c" and "d" (Fig. 9.8) are characterized by principal stress directions that are either parallel or orthogonal to the dip of the descending plate (Isacks et al., 1969) (Fig. 9.13). Consequently, the nodal planes determined by focal mechanism solutions do not correspond to the dip of the Benioff zone or a plane perpendicular to it. The principal stress directions show that the descending plate is thrown

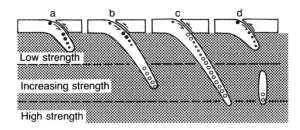

Figure 9.14 *A model of stress distributions in the descending lithosphere. Solid circles, extensional stress down dip; open circles, compressional stress down dip (redrawn from Isacks & Molnar, 1969, with permission from* Nature **223***, 1121–4. Copyright © 1969 Macmillan Publishers Ltd).*

into either down-dip compression or extension. Isacks & Molnar (1969) have suggested that the distribution of stress type in the seismic zone may result from the degree of resistance experienced by the plate during its descent, and Spence (1987) has described this resistance in terms of the net effect of ridge push and slab pull forces (Section 12.6). In Fig. 9.14a the plate is sinking through the asthenosphere because of its negative buoyancy and is thrown into down-dip tension as its descent is unimpeded. In Fig. 9.14b the bottom of the plate approaches the mesosphere, which resists descent and throws the leading tip into compression. As the plate sinks further (Fig. 9.14c), the mesosphere prevents further descent and supports the lower margin of the plate so that the majority of the seismic zone experiences compression. In Fig. 9.14d a section of the downgoing slab has decoupled so that the upper portion of the plate is thrown into tension and the lower portion into compression. A global summary of the stress directions determined from focal mechanism solutions (Isacks & Molnar, 1971) is shown in Fig. 9.15.

The stress distributions shown in Fig. 9.14b,d provide a possible explanation for the seismic gaps observed along the middle parts of the Benioff zone at certain trenches, such as the Peru–Chile Trench (Figs 9.15), where it is known that the slab is continuous (James & Snoke, 1990). A further type of seismic gap appears to be present in some island arcs at shallow depths. Figure 9.16 shows sections through the Benioff zone at the Aleutian–Alaska arc (Jacob *et al.*, 1977). There is a prominent gap in seismicity between the trench and a point about halfway towards the volcanic arc that becomes

progressively greater from west to east. The angle of underthrusting is very shallow in this region. The probable cause of this seismic gap and shallow underthrusting is the presence of copious quantities of terrigenous sediments within the trench that become increasingly abundant towards that section of the trench adjacent to Alaska. The unconsolidated nature of these sediments probably prevents any build-up of the strain energy necessary to initiate earthquakes, and their high positive buoyancy may force the subducting plate to descend at an anomalously shallow angle.

In reviewing the data for numerous subduction zones, Fukao *et al.* (2001) noted that subducted slabs are either deflected horizontally within or just beneath the transition zone, or penetrate the 660 km discontinuity and descend into the lower mantle (Plate 9.2 between pp. 244 and 245). Beneath Chile, the Aleutians, southern Kurile, and Izu-Bonin the slabs appear to flatten out within the transition zone, whereas beneath the Aegean, central Japan, Indonesia, and Central America they penetrate deep into the lower mantle. The slab beneath Tonga both flattens out within the transition zone and extends into the lower mantle (van der Hilst, 1995) (Plate 9.2e between pp. 244 and 245). There is no relationship between the age of a subducting slab and penetration into the lower mantle. Some researchers maintain that in places there is evidence for the slabs descending throughout the lower mantle to the core–mantle boundary (Section 12.8.2); others consider that there is little evidence for slab penetration beneath 1700 km depth (Kárason & van der Hilst, 2000). The possible implications of these tomographic results for convection in the mantle are considered in Section 12.9.

9.5 THERMAL STRUCTURE OF THE DOWNGOING SLAB

The strength and high negative buoyancy of subducting oceanic lithosphere and its capacity for sudden failure in the generation of earthquakes are consequences of its relatively low temperature with respect to normal mantle material at these depths. The subducting

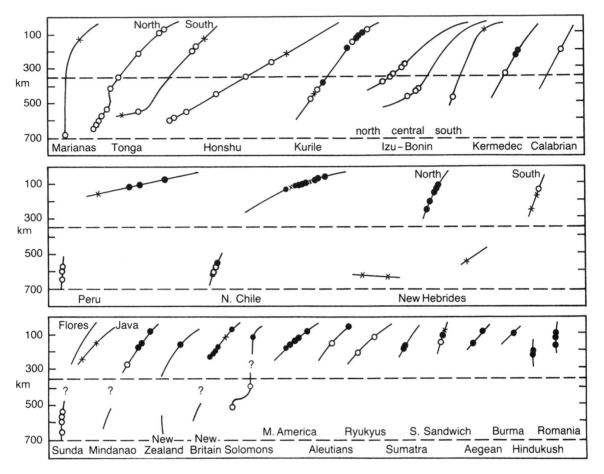

Figure 9.15 *Summary of the distribution of down dip stresses in Benioff zones. Open circles, events with compressional axis parallel to dip of zone; solid circles, events with tensional axis parallel to dip of zone; crosses, neither P- nor T-axis parallel to zone; solid lines, approximate form of seismic zone (redrawn from Isacks & Molnar, 1971, by permission of the American Geophysical Union. Copyright © 1971 American Geophysical Union).*

lithosphere can retain its separate thermal and mechanical identity to a considerable depth until sufficient heat has been transferred to it from the mantle to increase its temperature to that of its surroundings.

The variation of temperature within the sinking slab can be calculated from heat conduction equations provided that its thermal properties and boundary states are specified. The factors controlling the temperature distribution are:

1 the rate of subduction: the more rapid the descent the less time there is for absorption

of heat from the surrounding mantle by conduction;

2 the age and hence thickness of the descending slab: the thicker the slab the greater the time taken for it to equilibrate thermally with the surrounding asthenosphere;

3 frictional heating of the upper and lower surfaces of the slab as the descent of the slab is resisted by the asthenosphere;

4 the conduction of heat into the slab from the asthenosphere;

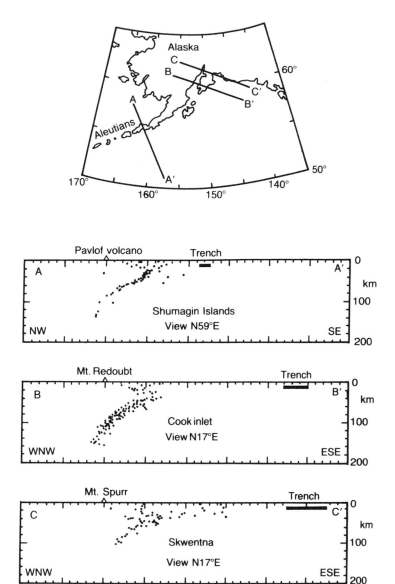

Figure 9.16 *Location map and cross-sections across the Aleutian arc showing earthquake foci (redrawn from Jacob et al., 1977, by permission of the American Geophysical Union. Copyright © 1977 American Geophysical Union).*

5 the adiabatic heating associated with compression of the slab as the pressure increases with depth;

6 the heat derived from radioactive decay of minerals in the oceanic lithosphere, likely to be small as oceanic plates are largely barren of radioactive minerals;

7 the latent heat associated with phase transitions of minerals to denser crystalline structures with depth: the principal phase changes experienced by the slab are the olivine–spinel transition at about 400 km depth which is exothermic, and the spinel–oxides transition at about 670 km, which is endothermic (Section 2.8.5).

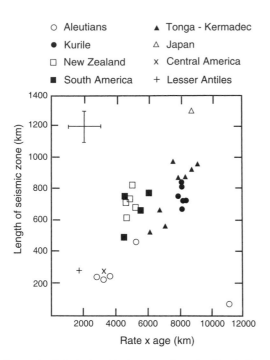

Figure 9.17 *Relationship between length of Benioff zone and the product of convergence rate and age. Approximate uncertainties given by error bars in upper left corner (redrawn from Molnar et al., 1979, with permission from Blackwell Publishing).*

Different solutions for the temperature distribution have been derived by various workers, depending on the assumptions made concerning the relative contributions of the above phenomena. Two models derived by Peacock & Wang (1999) and representing relatively cool and warm subducting lithosphere are shown in Plate 9.3 (between pp. 244 and 245). Although differing in detail, all such models indicate that the downgoing slab maintains its thermal identity to great depths and that, exceptionally, temperature contrasts up to 1000°C may exist between the core of the slab and normal mantle at a depth of 700 km.

As noted in Section 9.4, the length of the Benioff zone depends on the depth to which the subducting oceanic lithosphere maintains a relatively cold central core. Molnar *et al.* (1979) deduced that the downward deflection of isotherms, and hence the length of the seismic zone, is proportional to both the rate of subduction and the square of the thickness of the lithosphere. Lithosphere thickness is proportional to the square root of its age (Turcotte & Schubert, 2002) so that the length

of seismic zones should be proportional to the product of convergence rate and age. That this is generally so is illustrated by Fig. 9.17, and although there is considerable scatter the data appear to fit the relationship length (km) = rate (mm a^{-1}) $\times$ age (Ma)/10.

9.6 VARIATIONS IN SUBDUCTION ZONE CHARACTERISTICS

The age and convergence rate of the subducting oceanic lithosphere affect not only the thermal structure of the downgoing slab, and the length of the seismic zone, but a number of other characteristics of subduction zones. It can be seen from Fig. 9.15 that, although the dip of the Benioff zone is often approximately 45°, as typically illustrated, there is a great variation in dips, from 90° beneath the Marianas to 10° beneath Peru. It appears that the dip is largely determined by a combination of the negative buoyancy of the subducting slab, causing it to sink, and the forces exerted on it by flow in the asthenosphere, induced by the underthrusting lithosphere, which tend to uplift the slab. A higher rate of underthrusting produces a greater degree of uplift. Young oceanic lithosphere is relatively thin and hot; consequently it is more buoyant than older oceanic lithosphere. One would predict, therefore, that young subducting lithosphere, underthrusting at a high rate, will give rise to the shallowest dips, as in the case of Peru and Chile. It seems probable that the absolute motion of the overriding plate is also a contributing factor in determining the dip of the Benioff zone (Cross & Pilger, 1982).

Subduction zones with shallow dips have a stronger coupling with the overriding plate (Uyeda & Kanamori, 1979), giving rise to larger magnitude earthquakes in region "b" of Fig. 9.8. Shallow dips also restrict the flow of asthenosphere in the mantle wedge above the subduction zone, in extreme cases suppressing all supra subduction zone magmatism (Section 10.2.2), and in all cases giving rise to backarc compression rather than extension. Thus, Uyeda & Kanamori (1979) recognized two end-member types of subduction zone, which they referred to as Chilean and Mariana types (Fig. 9.18).

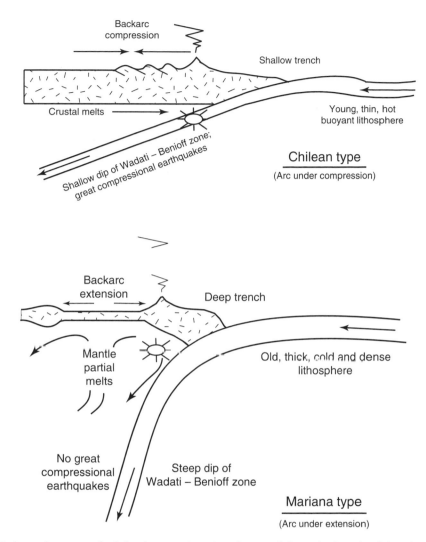

Figure 9.18 *End-member types of subduction zone based on the age of the underthrusting lithosphere and the absolute motion of the overriding plate (modified from Uyeda & Kanamori, 1979, and Stern, 2002, by permission of the American Geophysical Union. Copyright © 1979 and 2002 American Geophysical Union).*

Another first order variation in the nature of subduction zones is whether they are accretionary or erosive. Historically, oceanic trenches and magmatic arcs were considered to be the settings where material derived from the continental and oceanic crust is accreted to the margin of the overriding plate in the form of a wedge of sediments in the forearc region, and an edifice of igneous material in the magmatic arc. Increasingly however it has been realized that most of the oceanic crust and pelagic sediments is subducted into the mantle, and that, in approximately half of the convergent margins, some of the overriding plate is eroded and subducted. The process by which pelagic sediments on the downgoing plate are subducted is known as *sediment subduction* and the process whereby rock or sediment from the upper plate is subducted is termed *subduction erosion*. The latter may be derived from the base of the landward slope of the trench or from the underside of the upper plate. Moreover, the majority of the material accreted in the magmatic arc is thought to be derived from the mantle rather than subducted crust (Section 9.8). Thus, subduction zones have also been characterized as accretionary or erosive (Figs 9.1, 9.19). Examples of accretionary margins include the Nankai Trough and Barbados prisms (Section 9.7) (Saffer & Bekins, 2006); erosive prisms occur offshore of Costa Rica (Morris & Villinger, 2006) and Chile (Section 10.2.3).

On the basis of seismic reflection profiling data, it appears that the thickness of sediment on the oceanic plate entering a trench must exceed 400–1000 m for sediment to be scraped off and added to the accretionary prism. This implies that perhaps 80% of the pelagic sediments entering trenches is subducted, and that most of the sediment accreted in the forearc region is trench turbidites derived from continental material (von Huene & Scholl, 1991). The accretionary or nonaccretionary nature of a subduction zone will depend in part, therefore, on the supply of oceanic plate sediments and continentally derived clastic material to the trench. However, the causes of subduction erosion are very poorly understood (von Huene et al., 2004). Typically the thickness of trench sediments at accretionary margins exceeds 1 km (Saffer & Bekins, 2006). Other parameters that correlate with accreting margins are: orthogonal convergence rates of less than $76\,\mathrm{mm\,a^{-1}}$ and forearc bathymetric slopes of less than $3°$. In addition to the steeper slope of the forearc region at erosive margins, the forearc is characterized by subsidence, which reflects the thinning of the upper plate along its base. The amount of subsidence can be measured if drill cores are available from sedimentary sequences in this region. It is then possible to estimate the rate of erosion at the base of the forearc crust (Clift & Vannucchi, 2004).

9.7 ACCRETIONARY PRISMS

Where present, an accretionary prism forms on the inner wall of an ocean trench. The internal structure and construction of these features have been deduced from seismic reflection profiles and drilling at active subduction zones, and by the study of ancient subduction complexes now exposed on land.

Accretionary prisms develop where trench-fill turbidites (flysch), and some pelagic sediments, are scraped off the descending oceanic plate by the leading edge of the overriding plate, to which they become accreted. The Nankai Trough, located south of Japan (Fig. 9.20a), illustrates many of the structural, lithologic and hydrologic attributes of a large, active accretionary prism with a thick sedimentary section (Moore et al., 2001, 2005). Beneath the prism, the plate boundary is defined by a 20- to 30-m-thick, gently dipping fault or shear zone that separates a deformed sedimentary wedge above from a little-deformed section of subducted trench sediment, volcaniclastic rock, and basaltic crust below (Fig. 9.20b). This boundary, or *décollement*, develops in a weak sedimentary layer, typically a low permeability hemipelagic mud underlying stronger, more permeable trench turbidites. Above the décollement is a *fold and thrust belt* composed of listric thrust ramps that rise through the stratigraphic section forming imbricated arrays. These faults define wedge-shaped lenses that are internally folded and cleaved. At the base of the imbricate series, the décollement slopes downward toward the volcanic arc where it becomes progressively better developed. Away from the arc, it extends a short distance seaward of the *deformation front*, which is marked by the first small proto-thrusts and folds located inward from the trench. Farther seaward, the stratigraphic horizon that hosts the décollement is known as the incipient or *proto-décollement* zone where the incoming sedimentary section is only weakly deformed.

(a) Accretionary forearc

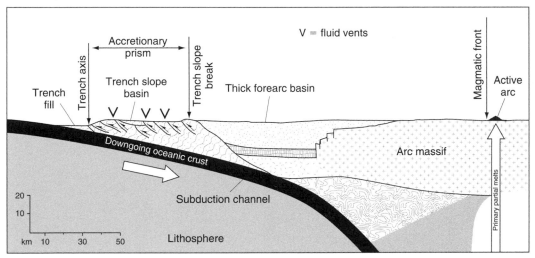

(b) Non-accretionary forearc

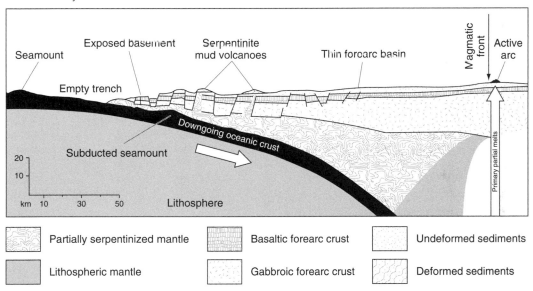

Figure 9.19 *Diagrams contrasting the characteristic features of (a) accretionary, and (b) nonaccretionary, convergent margins (redrawn from Stern, 2002, by permission of the American Geophysical Union. Copyright © 2002 American Geophysical Union). V, fluid vents.*

Seismic reflection data and the ages of deformed sediments suggest that the youngest faults in accretionary prisms occur at the deformation front and generally become older away from the trench (Moore *et al.*, 2001, 2005) (Fig. 9.20b). As shortening occurs, old thrust wedges gradually move upwards and are rotated toward the arc by the addition of new wedges to the toe of the prism. This process, called *frontal accretion*, causes older thrusts to become more steeply dipping with time and is responsible for the lateral growth of the prism. Lateral growth requires that the most intense deformation occurs at the oceanward base of the sedimentary pile,

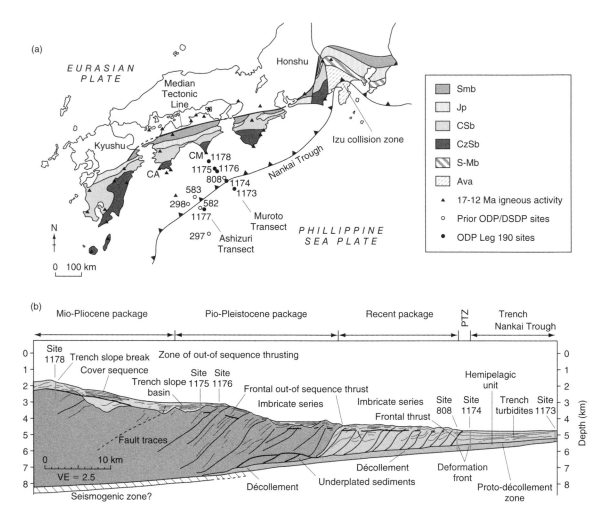

Figure 9.20 (a) Geologic map of the Nankai accretionary prism showing the Leg 190/196 Ocean Drilling Program (ODP, black circles) and previous ODP/Deep Sea Drilling Program (DSDP, open circles) drill sites (modified from Moore et al., 2001, by permission of the American Geophysical Union. Copyright © 2001 American Geophysical Union). CA, Cape Ashizuri; CM, Cape Muroto; Smb, Sanbagawa metamorphic belt; Jp, Jurassic accretionary prism; CSb, Cretaceous Shimanto belt; CzSb, Cenozoic Shimanto belt; S-Mb, Shimanto and Mineoka belts; Ava, Accreted volcanic arc; PTZ, protothrust zone. (b) Generalized cross-section of the Muroto transect showing main structural provinces (modified from Moore et al., 2005, by permission of the American Geophysical Union. Copyright © 2005 American Geophysical Union).

although some older thrusts may remain active during their rotation and some new thrusts may form and cut across older imbricate thrusts. These latter, cross cutting faults are termed *out-of-sequence thrusts* (Fig. 9.20b) because they do not conform to the common arcward progression of faulting. In addition to frontal accretion, some incoming material is carried downward past the deformation front where it is transferred, or *underplated*,

to the base of the prism by thrust faulting above the décollement. Unlike the off-scraped sediments at the toe, this underplated material may become deeply buried and undergo high pressure metamorphism (Section 9.9). Tectonic underplating, together with internal shortening, thickens the wedge and causes the slope of its upper surface to increase (Konstantinovskaia & Malavieille, 2005).

The top of an accretionary prism is defined by a relatively abrupt decrease in slope called the *trench slope break*. Between this break and the island arc, a forearc basin may develop, which is then filled with sediments derived from erosion of the volcanic arc and its substrate. This basin is a region of tranquil sedimentation where flat-lying units cover the oldest thrust slices in the wedge. Seaward of the forearc basin, on the trench slope, small pockets of sediment also accumulate on top of old thrust slices (Fig. 9.20b). The ages of these old slices, and their distance from the toe of the prism, provide a means of estimating lateral growth rates. For example, drilling at sites 1175 and 1176 in the Nankai prism has shown that trench slope sands unconformably overlie thrust slices that may be as young as 1 or 2 Ma (Moore *et al.*, 2001; Underwood *et al.*, 2003). Assuming steady state seaward growth, the distance of these thrust slices from the deformation front implies lateral growth rates as high as 40 km over the last 1 to 2 Myr. In comparison, the Middle America accretionary prism off the coast of Mexico has grown ~23 km in width over the past 10 Myr (Moore *et al.*, 1982) and the eastern Aleutian accretionary prism has grown 20 km in 3 Myr (von Huene *et al.*, 1998).

Erosion of the trench slope and other landward material commonly results in slump deposits and debris flows that can carry material as far as the trench, where it gets offscraped and recycled back into the wedge. At Site 1178 in the Nankai prism, the presence of thrust slices composed of Miocene turbidites indicates that the trench was accumulating large amounts of sediment derived from the erosion of rock exposed on Shikoku Island at that time (Moore *et al.*, 2005). Large (100- to 1000-m-long) blocks of slumped material, called *olistostromes*, remain semi-coherent during transport. This process provides much of the material that enables accretionary prisms to grow wider (Silver, 2000). Over time, erosion, deformation and sedimentary recycling result in a long-term circulation of material within the wedge (Platt, 1986). Offscraped material first moves down toward the base of the prism and then moves back toward the surface. This pattern results in a general increase in the metamorphic grade of rocks from the trench to the arc such that the oldest, high grade rocks are structurally highest and uplifted with respect to the younger deposits. The processes also may create a chaotic mixture of igneous, sedimentary and metamorphic rock types called a *mélange* (see also Section 10.6.1). Some of the oldest rock fragments in the mélange may record blueschist or eclogite facies metamorphism, indicating depths of burial of at least 30 km (Section 9.9).

The overall shape of accretionary prisms in profile approximates that of a tapered wedge, where the upper surface slopes in a direction opposite to that of the underlying décollement (Fig. 9.21a). Davis *et al.* (1983) and Dahlen (1990) showed that this tapered shape is required if the entire wedge moves together and the behavior of the system follows the Mohr–Coulomb fracture criterion (Section 2.10.2). The surface slope (α) is determined by the interplay between resistance to sliding on the décollement and the strength of the rock in the thrust wedge. Both of these latter two factors are strongly influenced by pore fluid pressure (λ), the dip of the basal décollement (β), and the weight of the overlying rock (Fig. 9.21b). Tectonic shortening and underplating thicken the wedge, thereby steepening the surface slope. If the surface slope becomes oversteepened, then various mechanical adjustments will occur until the slope decreases and a steady state is achieved. These adjustments may involve normal faulting and/or a lengthening of the décollement, and result from the same forces that drive the gravitational collapse of large topographic uplifts (Section 10.4.6). The mechanical behavior of the wedge also is especially sensitive to mass redistribution by surface erosion and deposition (Konstantinovskaia & Malavieille, 2005; Stolar *et al.*, 2006), which change topographic gradients and, at large scales, affect the thermal evolution of the crust (Section 8.6.3).

The results from drilling into active prisms have provided unequivocal evidence of the importance of fluid flow and changes in pore fluid pressure in accretionary prisms. Measurements of porosity, density, resistivity, and other physical characteristics suggest that accreted sediments descend so rapidly that they have no opportunity to dewater before burial (Silver, 2000; Saffer, 2003; Moore *et al.*, 2005). This process, and the low permeabilities that are typical of marine sediments, result in elevated pore pressures that reduce effective stress, lower the shear strength of rock (Section 2.10.2), and allow sliding on the décollement. Episodic fluid flow and the collapse of former flow paths also may allow the décollement to propagate laterally beneath the wedge (Ujiie *et al.*, 2003). These processes explain the generally small taper angles of most accretionary wedges, which can result only if the material within it is very weak and shear stresses on the décollement are very low (Davis *et al.*, 1983; Saffer & Bekins, 2002). High pore fluid pressure also explains

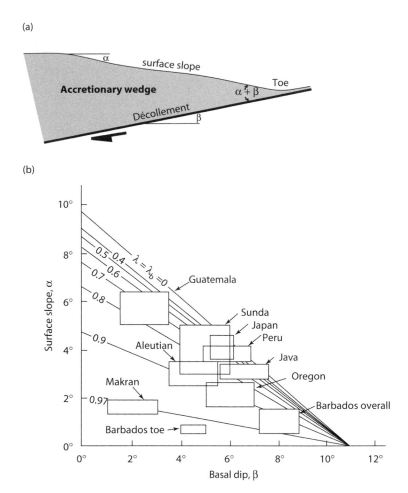

Figure 9.21 *(a) Schematic profile of a Coulomb wedge and (b) theoretical wedge tapers for various pore fluid pressure ratios (λ) for submarine accretionary prisms, assuming the pressure at the base is identical to that in the wedge (modified from Davis et al., 1993, by permission of the American Geophysical Union. Copyright © 1993 American Geophysical Union). Boxes in (b) indicate tapers of active wedges. Calculations involved a wedge sediment density of 2400 kg m⁻³.*

many other phenomena that are associated with prisms, including mud volcanoes and diapirs (Westbrook et al., 1984), and the development of unique chemical and biological environments at the leading edge of the prism (Schoonmaker, 1986; Ritger et al., 1987) (Fig. 9.19).

In addition to a mechanism by which pore fluid pressure increases by rapid burial, there also are competing mechanisms that decrease pore fluid pressure within a wedge. Fluids tends to flow along narrow, high permeability channels and exit to the décollement and the seafloor through vertical and lateral conduits (Silver,

2000; Morris & Villinger, 2006). Some of these conduits coincide with thrust faults overlying the décollement zone, whose high fracture permeability allows fluid to escape (Gulick et al., 2004; Tsuji et al., 2006). Fluid escape in this way implies that the décollement zone possesses a lower fluid pressure than its surroundings, a condition that is in apparent conflict with the evidence of high pore fluid pressures in this zone. However, the apparent conflict can be reconciled by models in which the fluid pressure in the décollement zone varies both spatially and temporally within the wedge. The nature of these variations, and their affect on the evolution of

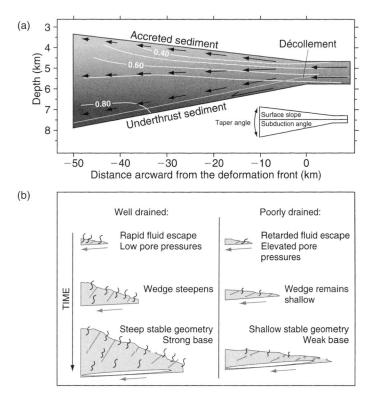

Figure 9.22 *(a) Schematic view of a numerical model of fluid flow within an accretionary prism and (b) cross-sections showing relationships between factors influencing accretionary wedge taper angle (modified from Saffer & Bekins, 2002, with permission from the Geological Society of America). Arrows in (a) represent approximate sediment velocities at the deformation front. Shading shows generalized porosity distribution, contours are modeled steady-state pore pressures (λ).*

the deforming wedge, are greatly influenced by factors such as the convergence rate and the stratigraphy, lithology, mineralogy, and hydrologic properties of the incoming sediments (Saffer, 2003).

The sensitivity of accretionary prisms to fluctuations in fluid flow and pore fluid pressure has been explored in detail using mechanical and numerical models. By combining a model of groundwater flow with critical taper theory (Fig. 9.22a), Saffer & Bekins (2002) concluded that low permeability, high pore pressure, and rapid convergence rates sustain poorly drained systems and result in shallow tapers, whereas high permeability, low pore pressure, and slow convergence result in well-drained systems and steep taper geometries (Fig. 9.22b). These authors also showed that the stratigraphic thickness and composition of the sediment that is incorporated into the wedge are among the most important factors governing pore fluid pressure in

wedges (Saffer & Bekins, 2006). Thick sedimentary sections give rise to large prisms that are able to sustain high pore fluid pressures and low stable taper angles (Fig. 9.23a). The results also suggest that prisms composed mostly of low permeability fine-grained sediment, such as northern Antilles (Barbados) and eastern Nankai (Ashizuri), will exhibit thin taper angles and those characterized by a high proportion of high permeability turbidites, such as Cascadia, Chile, and México, will have steep taper angles (Fig. 9.23b). This sensitivity to the physical properties of accreted and subducted sediment implies that any along-strike variation in sediment lithology or thickness strongly influences the geometry and mechanical behavior of accretionary prisms. Similarly, any variation in incoming sediment thickness or composition over time will force the accretionary complex to readjust until a new dynamic balance is reached.

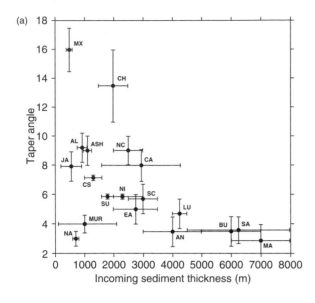

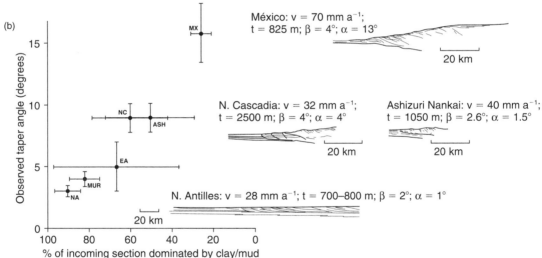

Figure 9.23 *Taper angles of active accretionary prisms plotted as a function of (a) thickness of incoming sediment and (b) lithology where the incoming sediment section has been sampled by drilling (modified from Saffer & Bekins, 2006, by permission of the American Geophysical Union. Copyright © 2006 American Geophysical Union). Horizontal error bars indicate uncertainty in lithology and vertical error bars indicate along-strike variations in taper angle. NA, northern Antilles; SA, southern Antilles; MUR, Nankai Muroto; AL, eastern Aleutians (160°W); EA, eastern Aleutians (148–150°W); CA, central Aleutians (172–176°W); NC, north Cascadia; SC, southern Cascadia; ASH, Nankai Ashizuri; MX, Mexico; JA, Java; CS, central Sumatra; SU, Sunda; CH, Chile; NI, Nicobar; AN, Andaman; LU, Luzon; BU, Burma; MA, Makran. Cross-sections in (b) are from Saffer & Bekins (2002). Plate convergence rate (v), incoming stratigraphic thickness (t), dip of the décollement (β), and surface slope (α) are indicated.*

9.8 VOLCANIC AND PLUTONIC ACTIVITY

Where subducting oceanic lithosphere reaches a depth of 65–130 km, volcanic and plutonic activity occurs, giving rise to an island arc or an Andean-type continental arc approximately 150–200 km from the trench axis (England *et al.*, 2004). The thickness of arc crust reflects both the age of the system and the type of crust on which the arc forms. Relatively young island arcs, such as the 3–4 Ma active part of the Mariana volcanic arc, may be underlain by a crust of 20 km thickness or less. Thin crust also generally occurs in settings where extension is dominant, such as in the Mariana arc system (Fig. 9.18b) (Kitada *et al.*, 2006). Mature island arcs, such as those in the Neogene Japanese arc system, generally show crustal thicknesses ranging from 30 km to 50 km because they have been constructed on older igneous and metamorphic rock (Taira, 2001). Continental arcs, including the Andes and the Cascades, are structurally the most complex of all arc systems because of the numerous structural and compositional heterogeneities that are intrinsic to continental lithosphere. In compressional continental settings (e.g. Figs 9.18a, Plate 10.1 (between pp. 244 and 245)), where substantial crustal thickening occurs, arc crust may reach thicknesses of 70–80 km (Section 10.2.4).

The types of volcanic rocks that occur in the supra-subduction zone environment generally form three volcanic series (Gill, 1981; Baker, 1982):

1 The low potassium *tholeiitic* series that is dominated by basaltic lavas associated with lesser volumes of iron-rich basaltic andesites and andesites.

2 The *calc-alkaline* series, dominated by andesites (Thorpe, 1982) that are moderately enriched in potassium, other incompatible elements, and the light rare earth elements. In continental arcs dacites and rhyolites are abundant, although they are subordinate to andesites.

3 The *alkaline* series that includes the subgroups of alkaline basalts and the rare, very high potassium-bearing (i.e. shoshonitic) lavas.

In general, the tholeiitic magma series is well represented above young subduction zones. These rocks have been interpreted as being derived by the fractional crystallization of olivine from a primary magma originating at relatively shallow mantle depths of 65–100 km. The calc-alkaline and alkaline series are encountered in more mature subduction zones, and appear to reflect magmas generated at depths greater than those that result in tholeiitic rocks. Calc-alkaline magmas, represented by andesite and basaltic andesite, are the most abundant of the volcanic series. Alkaline magmas exhibit the lowest abundance in island arcs and are more common in continental rifts and intraplate environments (Section 7.4.2).

Some island arcs exhibit spatial patterns in the distribution of the volcanic series. In the Japanese island arc system, for example, a compositional trend of tholeiite / calc-alkaline / alkaline volcanic rocks is apparent with increasing distance from the trench. This trend may reflect magmas derived from increasingly greater depths and / or differences in the degree of partial melting (Gill, 1981). A low degree of partial melting tends to concentrate alkalis and other incompatible elements into the small melt fraction (Winter, 2001), and could lead to an increase in alkalinity away from the trench due to a greater depth of melting or a decrease in the availability of water. However, there are many exceptions to this pattern in other arcs, indicating that differences in local conditions strongly influence magma compositions. The Izu-Bonin–Mariana arc system (Fig. 9.1), for example, shows compositional trends along the axis of the arc. From 35°N to 25°N latitude, volcanoes that form part of the Izu and Bonin arc segments are dominated by low and medium potassium rock suites (Fig. 9.24). The Mariana segment is dominated by medium potassium suites from 14°N to 23°N, and a shoshonitic province is found between the Mariana and Bonin segments (Stern *et al.*, 2003). This great spectrum of rock compositions reflects the diversity of processes involved in arc magmatism, including variations in the depth and degree of partial melting, magma mixing, fractionation, and assimilation (see also Section 7.4.2). In general, these observations indicate that the three volcanic series form a continuum of rock compositions and do not correspond to absolute magma types or source regions.

Mature arc systems, and especially continental arcs, typically include large, linear belts of plutonic rock called *batholiths*. These belts are so common in continental crust that they are widely used as indicators of ancient, now extinct convergent margins (Section 5.9). Occasionally the term *Cordilleran-type* batholith is used

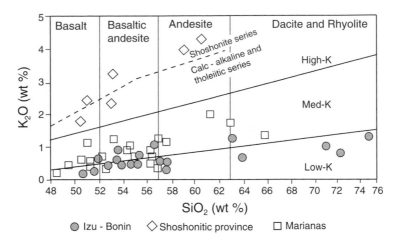

Figure 9.24 *Potassium-silica diagram for the mean composition of 62 volcanoes collected along the Izu–Bonin–Mariana arc system (modified from Stern et al., 2003, by permission of the American Geophysical Union. Copyright © 2003 American Geophysical Union).*

to describe large composite bodies of plutonic rock that were created above ocean–continent subduction zones. The majority of these batholiths are composed of hundreds to thousands of individual intrusions that range in composition from gabbro, tonalite and diorite to granodiorite and granite. Compositional similarities among many plutonic and some nearby volcanic rocks suggest that the former represent the crystallized residua of deep magma chambers that once fed shallow parts of the system. Their exposure in mature arcs results from prolonged periods of uplift and erosion.

One important, and highly controversial, area of research centers on the origin of the magmas supplying volcanic and plutonic complexes. Certainly the generation of the magmas must be linked in some way to the Benioff zone, as there is a very strong correlation between its depth and the systematic variation in volcanic rock composition and elemental abundances. Early models (e.g. Ringwood, 1975) suggested that the magmas were derived from melting of the top of the descending oceanic slab. However, this idea subsequently was rejected as a general model, in part because thermal models indicate that subducted lithosphere rarely becomes hot enough to melt (Peacock, 1991). In addition, petrologic and mineralogic evidence (Arculus & Curran, 1972) and helium isotope ratios (Hilton & Craig, 1989) indicate that the parental magmas originate by partial melting of asthenospheric mantle immediately overlying the descending plate. Karig & Kay

(1981), Davidson (1983), and Hilton & Craig (1989), among others, demonstrated that certain isotopic ratios require a large contribution from continent-derived sediments. Consequently, sediments from the trench must be carried down the subduction zone and incorporated into the asthenospheric melt (Plank & Langmuir, 1993). Most authors have concluded that the igneous crust of the subducting lithosphere contributes only very small amounts of melt, except, possibly, in special circumstances where young, hot lithosphere is subducted or warmed by mantle flow (Plate 9.3 between pp. 244 and 245). In this latter case, distinctive melt compositions such as *adakites* may be produced (Johnson & Plank, 1999; Yogodzinski, 2001; Kelemen *et al.*, 2003).

A major problem of arc magmatism is the source of the heat required for melting the asthenosphere above the descending slab. It was originally believed that this was derived solely by shear heating at the top of the slab. However, this is unlikely because the viscosity of the asthenosphere decreases with increasing temperature, and at the temperatures required for partial fusion the asthenosphere would have such a low viscosity that shear melting could not occur. Ringwood (1974, 1977) suggested that partial melting takes place at a relatively low temperature because of the high water vapor pressure resulting from the dehydration of various mineral phases in the downgoing slab. Indeed, the greater the amount of water present, the more the melting temperature of the mantle is reduced (Stolper & Newman, 1994). Thus, water acts as a primary agent that drives partial melting beneath arcs.

It is thought that as much as half of the water carried down into a subduction zone is released below the forearc region, partly into the crust, and also into the mantle producing serpentinite (Fig. 9.19) (Bostock *et al.*, 2002). Most of the water carried to great depths is sequestered in hydrous minerals in altered and metamorphosed crust, including serpentinite. With increasing pressure, hydrous basalt and gabbro are metamorphosed progressively to blueschist, then amphibolite, and then eclogite (Section 9.9). At each transformation water is released. Serpentinite is particularly effective in transporting water to great depth, but the extent to which the subducting lithosphere is serpentinized is unclear. Fast spread oceanic crust is thought to contain little or no serpentinite, but slow spread crust is known to contain some, perhaps as much as 10–20% (Carlson, 2001). However, as described above (Section 9.4), the lithospheric mantle in the downgoing slab may be hydrated to a depth of tens of kilometers as a result of the normal faulting associated with the bending of the plate as it approaches the subduction zone.

A generalized model of arc magmatism begins with the subduction of hydrated basalts beneath continental or oceanic lithosphere (Fig. 9.3). As the slab sinks through the mantle, heat is transferred to it from the surrounding asthenosphere and the basalt in the upper part of the slab begins to dehydrate through a series of metamorphic mineral reactions (Sections 9.4, 9.9). Sediments that have been subducted along with the basalt also dehydrate and may melt due to their low melting temperatures. The release of metamorphic fluids from the slab appears to be quite rapid, possibly occurring in as little as several tens of thousands of years (Turner & Hawkesworth, 1997). By contrast, the recycling of subducted sediment into the upper mantle may be slow (2–4 Ma). As heat is transferred to the slab, temperature gradients are established such that the asthenosphere in the vicinity of the slab becomes cooler and more viscous than surrounding areas, particularly near the upper part of the slab. This more viscous asthenosphere is then dragged down with the slab causing less viscous mantle to flow in behind it, as indicated in Fig. 9.3. It is the interaction of this downwelling mantle with aqueous fluids rising from the sinking slab that is thought to produce partial melting of the mantle. In addition, some melts may result from the upwelling of hot mantle material within the mantle wedge (Sisson & Bronto, 1998). If hot material rises quickly enough so that little heat is lost, the reduction in pressure may cause pressure release or decompression partial melting (see also Section 7.4.2).

A detailed study of the depth to the zone of seismicity and, hence, to the lithospheric slab directly beneath arc volcanoes has shown that, although these depths are consistent within a particular arc, they vary significantly from arc to arc within a range of 65–130 km (England *et al.*, 2004). Surprisingly these depths correlate not with the age or rate of underthrusting of the subducting lithosphere but inversely with the vertical rate of descent of the slab. England & Wilkins (2004) suggest that a high rate of descent increases the rate of flow in the mantle wedge and hence the rate at which hot mantle is drawn towards the corner of the wedge. This would produce higher temperatures, and hence melting, at a shallower depth than in the case of slow rates of descent.

Where sufficient partial melting occurs, probably $10 \pm 5\%$ (Pearce & Peate, 1995), the melt aggregates and begins to rise toward the base of the crust. As the magma moves into the crust it differentiates and may mix with either new, crust-derived melts or older melts, eventually forming the magmas that result in the calc-alkaline and alkaline series (Fig. 9.25). In the context of continental arcs, the generation of crust-derived melts appears to be common because the melting point of continental crust may be low enough to result in partial melting. Many of the Mesozoic Cordilleran-type batholiths of western North America (Tepper *et al.*, 1993), the Andes (Petford & Atherton, 1996), West Antarctica (Wareham *et al.*, 1997), and New Zealand (Tulloch & Kimbrough, 2003) contain chemically distinctive plutons that are thought to have originated from the partial melting of the lower continental crust. Tonalites, which are varieties of quartz diorite (see also Section 11.3.2), may be produced if the melting occurs at relatively high temperatures (~1100°C). Granodiorite may be produced if the melting occurs at cooler temperatures (~1000°C) and in the presence of sufficient quantities of water. Melts that move through a thick layer of continental crust may become enriched in incompatible elements before reaching the surface. These magmas also may lose some of their water content and begin to crystallize, with or without cooling. This latter process results in volcanic rocks that are characteristically fractionated, porphyritic, and wet. With time, the crust is thickened by overplating and underplating (Fig. 9.25).

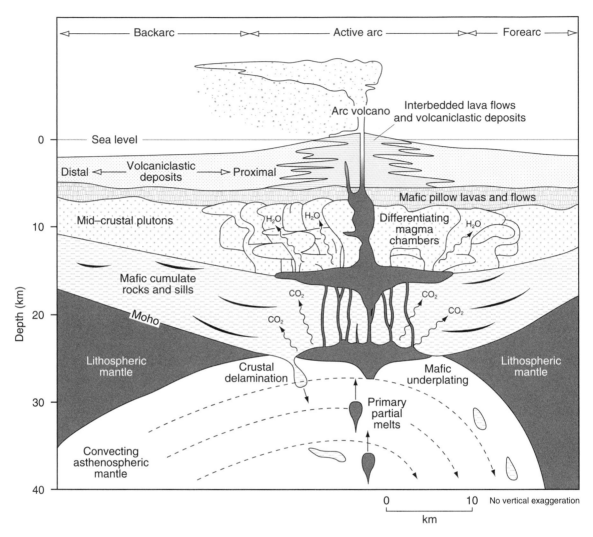

Figure 9.25 *Idealized section through an island arc illustrating the numerous processes involved in its construction. Similar processes may operate beneath Andean-type arcs (redrawn from Stern, 2002, by permission of the American Geophysical Union. Copyright © 2002 American Geophysical Union).*

Compression of the arc, such as that which occurs in the Chilean Andes (Fig. 9.18a), results in deformation that assists the thickening of arc crust (Section 10.2.4).

The mechanisms by which melts are transported through the mantle and crust are the source of a great deal of controversy. In general, transport processes operate on at least two different length scales (Petford *et al.*, 2000): the centimeter- to decimeter-scale segregation of melt near its source region and the kilometer-scale ascent of magma through the lithosphere to its final site of emplacement. Melt segregation from along grain boundaries probably involves porous flow mechanisms, assisted by ductile and brittle deformation. The ascent of melt appears to involve complex, nonvertical pathways from sources located at different depths. Schurr *et al.* (2003) identified regions of low Q (Section 9.4) from P-wave arrivals beneath the central Andes that reveal a variety of possible sources and ascent pathways for metamorphic fluids and partial melts (Plate 9.4(top) between pp. 244 and 245). A seismic reflection profile across the central Andes (Fig. 10.7) suggests the presence of fluids, including partial melt, at 20–30 km depth beneath the volcanic arc (ANCORP Working Group, 2003).

Measurements of disequilibria between short-lived uranium series isotopes in island arc lavas have suggested that melt ascent velocities from source to surface can be extremely rapid (10^3 m a^{-1}) (Turner et al., 2001). Such rates are much too fast for ascent to occur by grain-scale percolation mechanisms. Instead, melts probably separate into diapirs or form networks of low density conduits through which the flow occurs, either as dikes or as ductile shear zones. There is general agreement that deformation greatly enhances the rate of magma ascent. Laboratory experiments conducted by Hall & Kincaid (2001) suggest that buoyantly upwelling diapirs of melt combined with subduction-induced deformation in the mantle may create a type of channelized flow. Predicted transport times from source regions to the surface by channel flow range from tens of thousands to millions of years. It seems probable that a range of mechanisms is involved in the transport of magma from its various sources to the surface.

The emplacement of plutons and volcanic rock within or on top of the crust represents the final stage of magma transport. Most models of magma emplacement have emphasized various types of deformation, either in shear zones (Collins & Sawyer, 1996; Saint Blanquat et al., 1998; Brown & Solar, 1999; Marcotte et al., 2005) or in faults (e.g. Section 10.4.2), fractures and propagating dikes (Clemens & Mawer, 1992; Daczko et al., 2001). Some type of buoyant flow in diapirs also may apply in certain settings (e.g. Section 11.3.5). Various mechanisms for constructing plutons and batholiths are discussed by Crawford et al. (1999), Petford et al. (2000), Brown & McClelland (2000), Miller & Paterson (2001b), and Gerbi et al. (2002), among many others.

9.9 METAMORPHISM AT CONVERGENT MARGINS

As oceanic basalt is subducted at convergent margins, it undergoes a series of chemical reactions that both release water into the upper mantle wedge (Section 9.8) and increase the density of the subducting slab. These reactions involve specific metamorphic transformations that reflect the abnormally low geothermal gradients ($10°C$ km^{-1}) and the high pressures associated with the subduction zone environment (Section 9.5).

Prior to its subduction, oceanic basalt may exhibit low pressure (<0.6 GPa)/low temperature (<350°C) metamorphic mineral assemblages of the zeolite and prehnite-pumpellyite facies (Fig. 9.26). In some places greenschist facies minerals also may be present. In basalt, this latter facies typically includes chlorite, epidote and actinolite, which impart a greenish color to the rock (see also Section 11.3.2). This type of alteration of basalt results from the circulation of hot seawater in hydrothermal systems that develop near ocean ridges (Section 6.5).

As the altered basalt descends into a subduction zone, it passes through the pressure–temperature field of the blueschist facies (Fig. 9.26), which is characterized by the presence of the pressure-sensitive minerals glaucophane (a sodic blue amphibole) and jadeite (a pyroxene). A transitional zone, characterized by the presence of lawsonite, also may occur prior to the transformation to blueschist facies. Lawsonite is produced at temperatures below 400°C and at pressures of 0.3–0.6 GPa (Winter, 2001), conditions that are not yet high enough to produce glaucophane and jadeite. Lawsonite, along with glaucophane and other amphibole minerals, is an important host for water in subducting ocean crust.

One of the most important metamorphic reactions resulting in the dehydration and densification of subducting oceanic crust involves the transformation from the blueschist facies to the eclogite facies (Fig. 9.26). Eclogite is a dense, dry rock consisting mostly of garnet and omphacite (i.e. a variety of clinopyroxene rich in sodium and calcium). The exact depth at which eclogite facies reactions occur depends upon the pressures and temperatures in the subducting oceanic crust (Peacock, 2003). In relatively cool subduction zones, such as in northeast Japan (Plate 9.3a between pp. 244 and 245), the transformation may occur at depths of >100 km (Fig. 9.31). In relatively warm subduction zones, such as in southwest Japan (Plate 9.3b between pp. 244 and 245), the transformation may occur at depths as shallow as 50 km (Fig. 9.27). This transformation to eclogite enhances the negative buoyancy of the descending lithosphere and contributes to the slab-pull force acting on the subducting plate (Section 12.6).

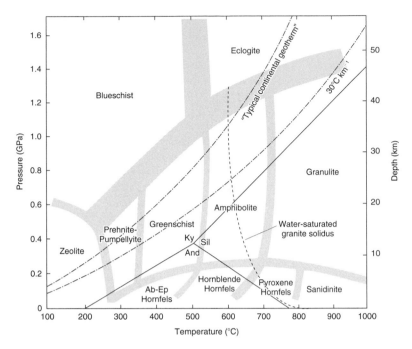

Figure 9.26 *Pressure–temperature diagram showing the approximate limits between the metamorphic facies (from Winter, John D.,* Introduction to Igneous and Metamorphic Petrology, *1st edition © 2001, p. 195. Reprinted by permission of Pearson Education Inc., Upper Saddle River, NJ). Example of an elevated (30°C km⁻¹) continental geotherm and stability ranges of three Al₂SiO₅ polymorphs commonly found in metamorphosed sedimentary rock (Ky, kyanite; And, andalusite; Sil, sillimanite) are shown for reference. Ab and Ep are albite and epidote, respectively.*

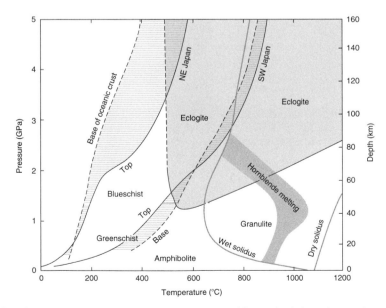

Figure 9.27 *Calculated pressure–temperature paths for the top and base of subducted oceanic crust beneath northeast and southwest Japan (after Peacock & Wang, 1999, with permission from* Science **286**, *937–9. Copyright by the AAAS, © 1999). Metamorphic facies and partial melting curves (dark gray lines) for basalt under wet and dry conditions are shown.*

Samples of blueschist and eclogite obtained from convergent margins (Section 9.7) provide important information on the physical and chemical conditions that occur within and above subducting lithosphere. Some of the first direct evidence of the conditions in the vicinity of the subduction zone décollement beneath a forearc has been provided by observations in the Mariana forearc. In this setting, large serpentine mud volcanoes up to 30 km in diameter and 2 km high occur in the forearc slope above an erosive margin (Fig. 9.19). In addition to serpentine, the volcanoes erupt slab-derived fluids and blueschist facies clasts that record the relatively cool temperatures of 150–250°C and pressures of 0.5–0.6 GPa (Maekawa et al., 1993). These determinations are consistent with thermal models of the slab–mantle interface where abnormally low geothermal gradients result from the rapid descent of cool oceanic lithosphere at trenches (Section 9.5) and from low to moderate levels of friction (Peacock, 1992). Samples of material obtained by drilling the Mariana mud volcanoes also provide evidence of the interactions among pore fluids, sediment and metamorphic rock that occur in an accretionary prism (Fryer et al., 1999). Similar material, known as *sedimentary serpentinite*, occurs in blueschist facies metamorphic belts preserved within continental crust. These belts commonly are interpreted to represent the suturing of ancient continental margins following the consumption of an intervening ocean (Sections 10.4.2, 11.4.3). Blueschist also is associated with ophiolitic suites (Ernst, 1973), lending support to the interpretation that some ophiolites formed in the forearc region of incipient subduction zones (Section 2.5).

In addition to the low temperature/high pressure type of metamorphism associated with subduction zones, some convergent margins also exhibit a type of regional metamorphism characterized by high temperatures (>500°C) and low to moderate pressures. This type of metamorphism commonly is associated with the high geothermal gradients that characterize magmatic arcs. Index minerals in metamorphosed sedimentary rocks, such as andalusite and sillimanite (Fig. 9.26), provide evidence of high temperatures in these regions. Temperature gradients of more than $25°C\,km^{-1}$ up to about $50°C\,km^{-1}$ result from the ascent of magmas generated where aqueous fluids from the subducted slab infiltrate the mantle wedge (Section 9.8). This type of metamorphism also is associated with the high differential stresses, deformation, and crustal thickening that accompany the formation of Andean-type

orogens (Section 10.2.5). Both associations affect large areas of the crust at convergent margins and thus reflect the large-scale thermal and tectonic disturbance associated with subduction and orogeny.

The most common groups of rocks associated with regional metamorphism belong to the greenschist, amphibolite and the granulite facies (Fig. 9.26). The transition from greenschist to amphibolite facies, like all metamorphic reactions, is dependent on the initial composition of the crust as well as the ambient pressure, temperature and fluid conditions. In metamorphosed basalt this transition may be marked by the change from actinolite to hornblende as amphibole is able to accept increasing amounts of aluminum and alkalis at high temperatures (>500°C) (Winter, 2001). At temperatures greater than 650°C amphibolite transforms into granulite. Granulites are highly diverse and may be of a low, medium or high pressure variety (Harley, 1989). In general, granulite facies rocks are characterized by the presence of anhydrous mineral assemblages such as orthopyroxene, clinopyroxene, and plagioclase.

If conditions at high temperatures are hydrous, then *migmatite* may form. Migmatite is a textural term that describes a mixed rock composed of both metamorphic and apparently igneous material. Proposed mechanisms for migmatite formation have included the partial melting of a rock, the injection of igneous (granitic or tonalitic) material into a rock, and the segregation of silicate material from a host during metamorphic rather than igneous activity. Migmatites are best developed in pelitic metasedimentary rocks, but also may occur in mafic rocks and granitoids. Brown et al. (1999) describe the structural and petrologic characteristics of migmatite derived from pelitic and basaltic rocks. Suda (2004) summarizes the formation and significance of migmatite in an intra-oceanic island arc setting. Klepeis et al. (2003) and Clarke et al. (2005) provide summaries of the tectonic setting and possible interpretations of a high-pressure (1.2–1.4 GPa) mafic granulite and associated migmatite belt located in Fiordland, New Zealand. These latter rocks represent the hot, lower crustal root of thick Cretaceous continental arc crust that has been exhumed during subsequent tectonic activity.

Attempts to place the evolution of the high pressure/low temperature and the high temperature/low pressure varieties of metamorphic rocks in the context of subduction zone processes are common in the scientific literature. One important early effort by Miyashiro (1961, 1972, 1973) led to the concept of

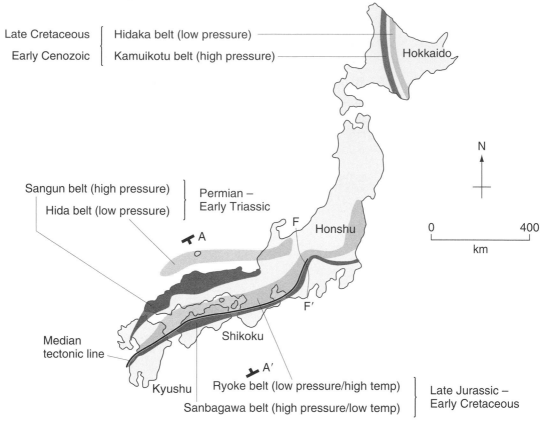

Figure 9.28 *Three paired metamorphic belts in Japan, F–F' is the Itoigawa–Shizuoka Line (from Miyashiro, 1972. Copyright 1972 by* American Journal of Science. *Reproduced with permission of* American Journal of Science *in the format Textbook via Copyright Clearance Center). Profile A–A' is shown in Fig. 9.29.*

paired metamorphic belts. On the Japanese islands of Hokkaido, Honshu, and Shikoku (Fig. 9.28), Miyashiro identified three pairs of metamorphic belts of different age that approximately parallel the trend of the modern Japanese subduction zone. Each of these belts consists of an outer zone of high pressure/low temperature blueschist and an inner belt of low pressure/high temperature rock. This spatial relationship and the similar age of each outer and inner belt led him to conclude that the belts formed together as a pair. After the introduction of plate tectonics, these paired belts were interpreted to be the result of underthrusting of oceanic crust beneath an island arc or continental crust (Uyeda & Miyashiro, 1974). The outer metamorphic belt was interpreted to develop near the trench due to the low geothermal gradient caused by subduction. The inner belt was interpreted to form in

the arc, some 100–250 km away, where geothermal gradients are high.

The application of the paired metamorphic belt model to Japan has allowed some investigators to infer the direction of subduction and plate motions at various times in the past. At present, Pacific lithosphere is subducted in a northwesterly direction beneath the Japan arc. The metamorphic polarity of the Sangun/Hida and Ryoke/Sanbagawa paired belts (Fig. 9.28) suggests that they were formed similarly, by underthrusting in a northwesterly direction. The Hidaka/Kamuikotu paired belt shows the opposite metamorphic polarity, and therefore may have formed during a different phase of plate movements when the direction of subduction was from the west of Japan. However, there are some discrepancies in this interpretation. For example, the Ryoke/Sanbagawa belts are much closer together than

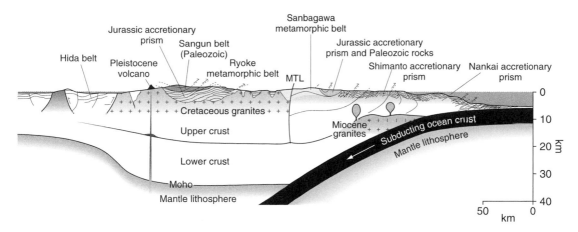

Figure 9.29 *Geologic cross-section of southwest Japan (modified from Taira, 2001,* Annual Review of Earth and Planetary Sciences *29, Copyright © 2001 Annual Reviews). Location of profile shown in Fig. 9.28. MTL, Median Tectonic Line.*

predicted by the model, and so it has been suggested that the boundary between them, called the Median Tectonic Line, experienced some 400 km of strike-slip movement (Section 5.3). This transcurrent movement has been confirmed by detailed mapping (Takagi, 1986) and indicates that strike-slip faulting was responsible for bringing the Sanbagawa and Ryoke belts into juxtaposition (Fig. 9.29).

Since the work of Miyashiro (1961, 1972, 1973), interpretations of paired metamorphic belts have been attempted in both island arc and Andean type settings around the Pacific margin (Fig. 9.30). The simplicity of these interpretations is appealing; however, in some examples, numerous inconsistencies exist. In the Atlantic region and in the Alps, many Phanerozoic metamorphic belts either lack one of the pairs or the contrast between them is unclear. These patterns, and the realization that many paired metamorphic belts did not form in their present positions, has led to skepticism about their overall significance. Brown (1998) summarized the evolution of thought that has led to a general demise of the concept of paired metamorphic belts in many convergent margins and orogens. One reason for this is that most metamorphic belts are no longer considered to be characterized by a single geothermal gradient, mainly because the rocks record an evolution across a range of geotherms through time. In addition, the recognition of suspect terranes and the importance of accretionary processes (Section 10.6) suggests that the tectonic units along these margins reflect a complex

array of processes that may or may not have accompanied subduction. Taira (2001) summarizes the importance of terrane collision for the evolution of Japan's metamorphic belts.

9.10 BACKARC BASINS

Backarc (or marginal) basins are relatively small basins of either oceanic or continental affinity that form behind the volcanic arc in the overriding plate of a subduction zone (Fig. 9.3). Oceanic varieties are most common in the western Pacific, but are also found in the Atlantic behind the Caribbean and Scotia arcs. In all of these settings, the basins reside on the inner, concave side of the island arc and many are bounded on the side opposite the arc by a backarc ridge (remnant arc). Most of these basins are associated with extensional tectonics and high heat flow, and the majority of oceanic varieties contain sea floor spreading centers where new oceanic crust is generated. In continental settings, extensional backarc basins have been described in the context of Andean-type convergent margins (Section 10.2). Some of the best preserved examples of this type formed along the western margin of South America during

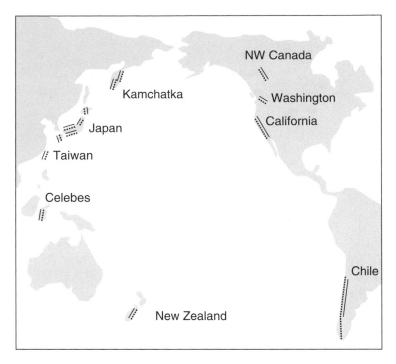

Figure 9.30 *Paired metamorphic belts in the circum-Pacific region. Dotted lines, high pressure belts; solid lines, low pressure belts (redrawn from Miyashiro, 1973, with permission from Elsevier)*

Mesozoic times (Section 10.2.1). An example of an active continental backarc basin is the extensional Taupo volcanic zone on the North Island of New Zealand (Stern, 1987; Audoine *et al.*, 2004).

Karig (1970) was one of the first to suggest that backarc basins form by the rifting of an existing island arc along its length, with the two halves corresponding to the volcanic and remnant arcs. This interpretation is based on observations in the Lau basin (Fig. 9.31), which lies west of the Tonga–Kermadec arc and is flanked on its western side by the Lau ridge. Karig (1970) concluded that extension was important during basin formation on the basis of the following observations: (i) the asymmetric cross-section of both the arc and ridge, which are mirrored across the center of the basin; (ii) the basin's topographic features, which are aligned parallel to both the arc and the ridge; (iii) the considerable sediment thickness present to the seaward side of the arc and landward side of the ridge and the absence of sediment within the basin; and (iv) the continuation of the arc–basin–ridge system to the south where it correlates with a zone of active backarc exten-

sion in the Taupo volcanic zone of New Zealand (Fig. 9.31). Further support for extension comes from the subsidence of remnant arcs as their dynamic support is removed after the development of backarc basins, earthquake focal mechanism solutions, and the segmented geometry of normal faults and spreading ridges, which also characterize continental rifts (Section 7.2) and mid-ocean ridges. The Woodlark Rift (Fig. 7.39b), which records the transition from rifting to sea floor spreading above a Neogene subduction zone (Section 7.8.2), illustrates this segmentation especially well.

In general, the composition of the crust in oceanic backarc basins is broadly similar to that of other ocean basins, although in some cases layer 1 is unusually thick. Net accretion rates are similar to those deduced for mid-ocean ridges, and range from approximately 160 mm a^{-1} in the northern Lau basin (Bevis *et al.*, 1995) to 70 mm a^{-1} in the East Scotia Sea (Thomas *et al.*, 2003) and 20–35 mm a^{-1} in the Mariana Trough (Martínez *et al.*, 2000). The crust in these settings commonly shows substantial thinning by normal faulting, although the total crustal thickness also depends upon the rate of

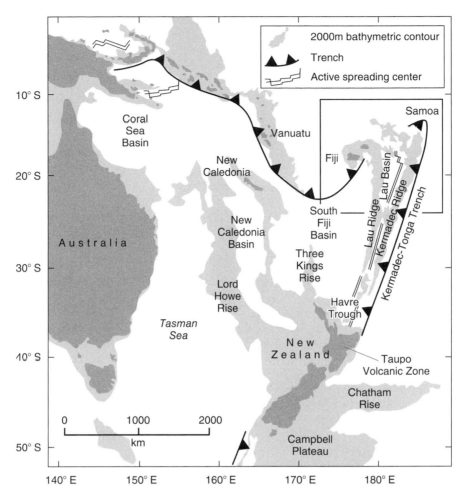

Figure 9.31 *Map showing the location of backarc basins in a part of the southwest Pacific, including the Lau basin, the South Fiji basin, the New Caledonian basin, and the Taupo volcanic zone (modified from Collins, 2002b, by permission of the American Geophysical Union. Copyright © 2002 American Geophysical Union). Box shows location of Fig. 9.32.*

magma addition and the age of the crust. In the Mariana Trough (Fig. 9.1), for example, crustal thicknesses range from 3.4 to 6.9 km, with the thinnest values corresponding to either slow spreading centers or magma-starved regions, and the thickest crust corresponding to regions of high magmatic activity (Kitada *et al.*, 2006).

These observations and the evidence for extension and sea floor spreading imply that backarc crust in oceanic settings is generated in a manner similar to that occurring at mid-ocean ridges (Section 6.10). However, there are many important differences in the processes that form basaltic crust in these two environments.

Structures corresponding to a mid-ocean ridge are not always present in backarc basins, and magnetic lineations may be poorly developed (Weissel, 1981). Those lineations that are present can be correlated with the magnetic polarity timescale (Section 4.1.6), although they tend to be shorter, of lower amplitude and less clearly defined than oceanic anomalies. In the southern Lau Basin, for example, individual magnetic anomaly lineations cannot be traced for more than 30 km along the strike of the basin (Fujiwara *et al.*, 2001). This short length probably reflects the small size of basement faults, which typically display segmented, en echelon patterns. Geochemical studies also indicate that backarc lavas commonly display greater compositional variations, including higher water contents, than mid-ocean ridge basalts (Taylor & Martínez, 2003). Many backarc lavas are chemically related to the lavas that form the adjacent island arc. These observations suggest that crustal accretion in backarc basins is strongly influenced by processes related to subduction (e.g. Kitada *et al.*, 2006).

Tomographic images of the mantle beneath active backarc spreading centers have confirmed the important linkages that exist between backarc crustal accretion and subduction. In one of the best-studied arc–backarc systems, Zhao *et al.* (1997) showed that very slow seismic velocities beneath the active Lau spreading center and moderately slow anomalies under the Tonga arc are separated at shallow (<100 km) depths in the mantle wedge but merge below 100 km to depths of 400 km (Plate 9.1 between pp. 244 and 245). The magnitude of the velocity anomalies is consistent with the presence of approximately 1% melt at depths of 30–90 km (Wiens & Smith, 2003). At greater depths the anomalies may result from the release of volatiles originating from the dehydration of hydrous minerals. These results indicate that backarc spreading is related to convective circulation in the mantle wedge and dehydration reactions in the subducting slab. They also suggest that backarc magma production is separated from the island arc source region within the depth range of primary magma production. By contrast, below 100 km, backarc magmas originate through mixing with components derived from slab dehydration and may help to explain some of the unique features in the petrology of backarc magmas relative to typical mid-ocean ridge basalts.

A wide variety of mechanisms has been postulated to explain the formation of backarc basins. One common view is that the extension and crustal accretion that characterize these environments occur in response to regional tensional stresses in the overriding lithosphere of the subduction zone (Packham & Falvey, 1971). These stresses may result from the trench suction force as the subducting slab steepens or "rolls-back" beneath the trench (Chase, 1978; Fein & Jurdy, 1986) (Section 12.6). Such a *roll-back* mechanism has been postulated to occur in subduction systems where the "absolute" direction of movement of the overriding plate is away from the trench (e.g. Figs. 10.9b,c, 10.37). Other sources of the tension could include convection in the upper mantle wedge induced by the descent of the

underthrusting slab (Hsui & Toksöz, 1981; Jurdy & Stefanick, 1983) or an increase in the angle of subduction with depth (Section 12.6). Although these and other mechanisms controlling the evolution of backarc basins are often debated, most authors agree that basin evolution is strongly influenced by the pattern of flow, partial melting and melt transport in the upper mantle wedge above a subduction zone. Geodynamic models increasingly have appealed to three-dimensional circulation patterns associated with trench migration and slab roll-back to explain the thermal evolution of the wedge and the production of melt within it (Kincaid & Griffiths, 2003; Wiens & Smith, 2003).

Martínez & Taylor (2002) developed a model of crustal accretion for the Lau basin that explains the mechanism of backarc magmatism and its relationship to magmatism in the Tonga arc. These authors observed that the various spreading centers in the basin (Fig. 9.32) display structural and compositional patterns that correlate with distance from the arc. As in most other intra-oceanic arc systems, the crust displays a general depletion of certain elements relative to mid-ocean ridge basalt that increases from the backarc toward the arc. In addition, the spreading center closest to the arc (the Valu Fa spreading ridge) displays a structure, depth and morphology indicating that it is characterized by an enhanced magma supply relative to other centers. Farther away from the arc, the East Lau and Central Lau spreading centers display diminished and normal magma supplies, respectively. Martínez & Taylor (2002) proposed that these variations result from the migration of magma source regions supplying the backarc spreading centers through the upper mantle wedge.

The model of Martínez & Taylor (2002) begins with the roll-back of the Pacific slab beneath the Tonga trench (large white arrow in Fig. 9.33a). This motion induces a compensating flow of mantle material beneath the Lau basin (small black arrows). As the mantle encounters water that is released from the subducting slab (Section 9.8), it partially melts, leaving a residual mantle depleted of a certain melt fraction. The stipple in Fig. 9.33a indicates the region of hydrated mantle. The region of partial melt is shown as the white background beneath the stippling. Flow induced by subduction drives the depleted layer toward the upper corner of the wedge where increased water concentrations from the slab promote additional melting. This region of enhanced melting (outlined area in Fig. 9.33a) supplies the Valu Fa spreading ridge close to the volcanic

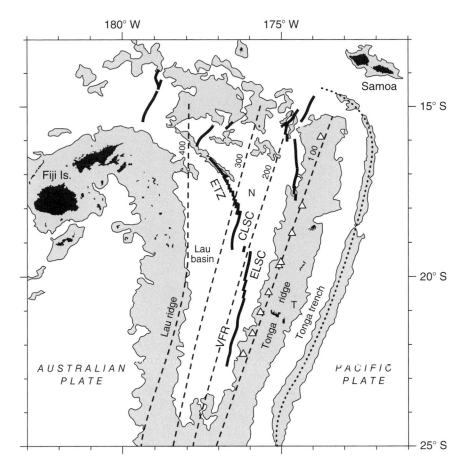

Figure 9.32 *Location map of the Lau basin showing the backarc spreading centers (heavy lines), trench axis (dotted line), arc volcanoes (white triangles), and contours of the subducted slab (dashed lines) labeled in km (after Martínez & Taylor, 2002, with permission from* Nature ***416**, 417–20. Copyright © 2002 Macmillan Publishers Ltd). N, Niuafo'ou plate; T, Tonga plate; VFR, Valu Fa ridge; ELSC, east Lau spreading center; CLSC, central Lau spreading center; ETZ, extensional transform zone.*

front, which receives melt that would otherwise supply the volcanic arc. The enhanced melting in this region also allows the depleted mantle (light gray in Fig. 9.33a) to remain weak enough to flow until it overturns and is carried back beneath the backarc basin as subduction proceeds. This return flow of depleted mantle results in diminished melt delivery to the East Lau spreading center farther from the volcanic arc because the melting regime is too far away to directly draw arc melts. Normal melting conditions occur at the Central Lau spreading ridge because this region overlies mantle that is farthest away from the volcanic front and from the effects of the slab. Consequently this latter spreading

center displays a crustal thickness, morphology, and geochemistry like those of a typical mid-ocean ridge.

Taylor & Martínez (2003) generalized this model to include other oceanic backarc basins, including the Mariana, Manus and East Scotia Sea basins. Variations in basalt geochemistry in these basins also can be explained by the migration of melt source regions in the mantle wedge and by differences in the extents and depths of partial melting. The geochemical data also suggest that the mantle source regions for the Lau and Manus backarc basins are hotter than those of the Mariana and Scotia due to faster rates of subduction. These increased rates appear to induce greater

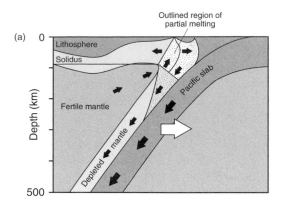

Figure 9.33 *Model for the formation of the Lau backarc basin (after Martínez & Taylor, 2002, with permission from Nature **416**, 417–20. Copyright © 2002 Macmillan Publishers Ltd). Large white arrow indicates the roll-back of the Tonga trench and Pacific slab. Large black arrows show the subduction component of the Pacific slab, and small black arrows show flow in the mantle wedge induced by the slab subduction and backarc spreading. Stippled gradient indicates region of hydrated mantle, with water concentration increasing eastward toward slab. Region of partial melting is indicated.*

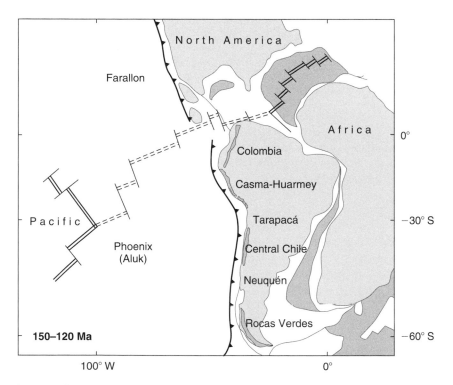

Figure 9.34 *Schematic plate reconstruction showing the location of early Mesozoic extensional backarc and marginal basins in South America (modified from Mpodozis & Allmendinger, 1993, with permission from the Geological Society of America). Marginal basins in Colombia and southern Chile (Rocas Verdes) are floored by oceanic crust. Casma-Huarmey in Peru and the central Chile basin are aborted marginal basins developed on thinned continental crust. Tarapacá and Neuquén are backarc basins with dominantly sedimentary infill.*

transport of heat within the mantle wedge (England & Wilkins, 2004). Thus, many of the fundamental differences between crustal accretion in backarc and mid-ocean ridge settings can be explained by the structure and dynamics of flow in the upper mantle wedge. However, it is important to realize that, in addition to processes related to subduction, it is probable that some backarc basins are influenced by the specific configurations of plate boundaries in their vicinity. An example of this may be the North Fiji basin where the backarc spreading direction is oriented at unusually low angles (10–30°) to the trend of the arc. Schellart *et al.* (2002) suggested that this unusual spreading direction could be related to an asymmetric opening of the basin and to collisional processes occurring along the plate boundary. The evolution of the Woodlark Rift (Section 7.8.2) also is strongly influenced by local boundary conditions, including rheological weaknesses in the lithosphere.

Variability in the structure and magmatic characteristics of backarc basins also is common in continental settings. Along the western margin of South America, for example, a series of extensional basins formed during a period of Mesozoic extension above a long-lived subduction zone (Fig. 9.34) (Section 10.2.1). In most of these basins, extension and backarc rifting occurred without the formation of a basaltic basin floor (Mpodozis & Allmendinger, 1993). Only in Colombia and southernmost Chile did extension proceed to a stage where complete rupture of continental crust (Section 7.5) occurred and oceanic-type spreading centers developed. A possible modern analogue of these oceanic basins is the Bransfield basin, located behind the South Shetland trench near the Antarctic Peninsula (Fig. 9.1). This latter basin is asymmetric and displays evidence for having opened by rift propagation through pre-existing arc crust beginning 4–5 million years ago (Barker *et al.*, 2003). Mora-Klepeis & McDowell (2004) discuss the geochemical signatures of rocks that record a similar transition from arc volcanism to rifting in the Baja California region of northwestern México.

Although common, not all continental backarcs are associated with extension or rifting. Many zones of ocean–continent convergence, including the modern Andean margin (Section 10.2), record shortening and orogenesis in the backarc environment. Regardless of the style of deformation they record, most continental backarcs are characterized by relatively thin, hot lithosphere (Hyndman *et al.*, 2005) (e.g. Fig. 10.7) whose properties greatly affect the mechanical evolution of the convergent margin (e.g. Sections 10.2.5, 10.4.6).

10 | Orogenic belts

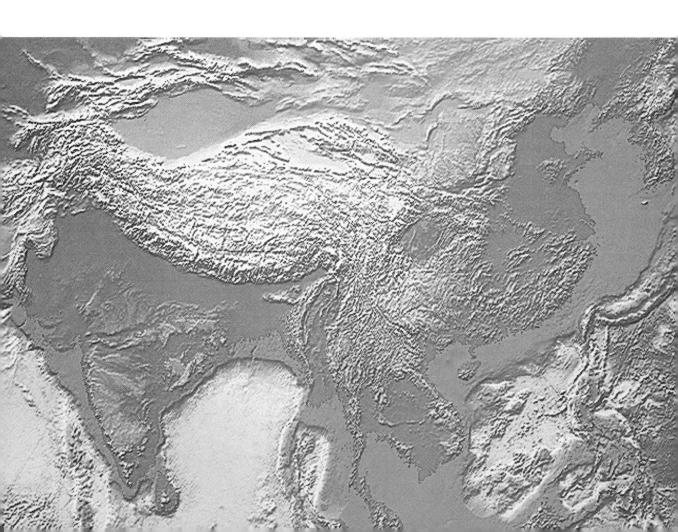

10.1 INTRODUCTION

Orogenic belts are long, commonly arcuate tracts of highly deformed rock that develop during the creation of mountain ranges on the continents. The process of building an orogen, or *orogenesis*, occurs at convergent plate margins and involves intra-plate shortening, crustal thickening, and topographic uplift. Ancient orogens, whose topography has been reduced or eliminated by erosion, mark the location of old, inactive plate margins and, thus, provide important information on past plate movements (e.g. Section 11.4.3).

The processes that control orogenesis vary considerably depending on the tectonic setting and the type of lithosphere involved in the deformation. *Noncollisional* or *Andean-type* orogens (Section 10.2) result from ocean–continent convergence where plate motions and other factors controlling subduction (Section 9.6) lead to compression within the overriding plate. *Collisional* orogens (Sections 10.4, 10.5) develop where a continent or island arc collides with a continental margin as a result of subduction. In these latter belts, the thickness and positive buoyancy of the colliding material inhibits its descent into the mantle and leads to compression and orogeny. The Himalayan–Tibetan belt and the European Alps represent orogens that form by continent–continent collision following the closure of a major ocean basin (i.e. *Himalaya-type*). Another variety where continental collision is highly oblique and did not involve ocean closure occurs in the Southern Alps of New Zealand (Sections 8.3.3, 8.6.3). Orogens that form by arc–continent collision include belts in Taiwan and the Timor–Banda arc region in the southwest Pacific.

Much of the internal variability displayed by both collisional and noncollisional orogens can be explained by differences in the strength and rheology of the continental lithosphere and by processes that influence these properties during orogenesis (Sections 10.2.5, 10.4.6). For example, where the continental lithosphere is relatively cool and strong, orogens tend to be comparatively narrow, ranging between 100 and 400 km wide. The Southern Alps of New Zealand (Fig. 8.2a) and the southern Andes near 40°S latitude (Fig. 10.1a) exhibit these characteristics. Conversely, where the continental lithosphere is relatively hot and weak, strain tends to delocalize and is distributed across zones that can be over a thousand kilometers wide. The central Andes near 20°S latitude (Fig. 10.1a) and the Himalayan–Tibetan orogen (Section 10.4) display these latter characteristics. Processes that change the strength and rheology of continental lithosphere during orogenesis commonly include magmatism, metamorphism, crustal melting, crustal thickening, sedimentation, and erosion.

The gradual accretion of continental fragments, island arcs, and oceanic material onto continental margins over millions of years is one of the primary mechanisms by which the continents have grown since Precambrian times (Sections 10.6, 11.4.2, 11.4.3). Most ancient and active orogens record many cycles of accretion and orogeny where distinctive assemblages of crustal material called *terranes* (Section 10.6.1) have collided and become attached to the continental margin. This process is augmented by other mechanisms of continental growth, including magma addition, sedimentation, and the creation and destruction of extensional basins (Section 10.6.3). Orogens that have grown significantly by these processes over long periods of time, often without ocean closure, generally have been termed *accretionary orogens*. Examples include the Paleozoic Altaids, which form much of northern China and Mongolia (Şengör & Natal'in, 1996); the western Cordillera of North America (Sections 10.6.2, 11.4.3); and the Lachlan Orogen of southeast Australia (Section 10.6.3). Pure accretionary orogens may lack evidence of a major continent–continent collision and consist of many small terranes and arc–continent collisions that have occurred along the margin of a long-lived ocean.

In this chapter, examples from South America, Asia, North America, Australia, and the southwest Pacific illustrate the diverse characteristics of orogens and the major mechanisms of orogenesis, including the evolution of compressional sedimentary basins.

10.2 OCEAN–CONTINENT CONVERGENCE

10.2.1 Introduction

One of the best-studied examples of an orogen that has formed by ocean–continent convergence lies in the

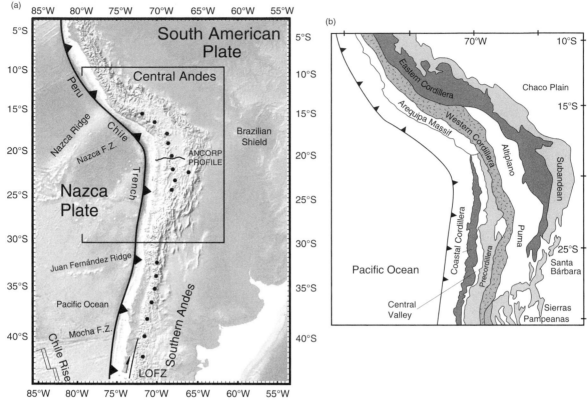

Figure 10.1 *(a) Shaded relief map of the central and southern Andes showing topographic features of the Nazca and South American plates. Map was constructed using the same topographic data and methods as in Fig. 7.1. Black dots are active volcanoes. LOFZ is the Liquiñe-Ofqui fault zone. Box shows location of Fig. 10.1b. ANCORP profile shown in Fig. 10.6. (b) Physiographic provinces of the central Andes (modified from Mpodozis et al., 2005, with permission from Elsevier).*

central Andes of Peru, Bolivia, northern Chile, and Argentina (Fig. 10.1). Here, the Andes exhibit the highest average elevations, the greatest width, the thickest crust, and the greatest amount of shortening in the orogen (Isacks, 1988; Allmendinger *et al.*, 1997; ANCORP Working Group, 2003). This central segment illustrates how many of the characteristic features of large orogens may form in the absence of continent–continent collision.

The Andean mountain chain, or *cordillera*, extends some 7500 km from Venezuela and Colombia in the north to Tierra del Fuego in the south. Along its length, the orogen displays a remarkable degree of diversity in structure, geologic history, and tectonic evolution. This diversity complicates determinations of the factors that control orogenesis within its different segments. Never-

theless, some common elements are evident that provide important boundary conditions on noncollisional orogenic processes. One of these constraints is that the active margin of South America was characterized by either a noncompressive or an extensional regime during the Late Jurassic and Early Cretaceous (Mpodozis & Ramos, 1989). At this time, most of the margin lay below sea level as a series of extensional backarc and marginal basins (Fig. 9.34) formed above a subduction zone (Dalziel, 1981; Mpodozis & Allmendinger, 1993; Mora *et al.*, 2006). This history shows that by itself subduction cannot account for the formation of Andean-type orogens. Rather, mountain building in this setting results only when ocean–continent convergence leads to compression in the overriding plate (Sections 9.6, 10.2.5).

In the Andes, compressional regimes have been established several times since the early Mesozoic, with the most recent phase beginning about 25–30 Ma (Allmendinger *et al.*, 1997). The beginning of this latest phase of compression has been interpreted to reflect two major processes: the trenchward acceleration of the South American plate (Pardo-Casas & Molnar, 1987; DeMets *et al.*, 1990; Somoza, 1998) and strong interplate coupling between the subducting oceanic lithosphere and the overriding continent (Jordan *et al.*, 1983; Gutscher *et al.*, 2000; Yáñez & Cembrano, 2004). One of the principal aims of tectonic studies in the Andes is to determine the origin of the highly variable response of the South American plate to this compression. This section provides a discussion of the first-order physical characteristics of the central and southern Andes that allow geoscientists to make inferences about the genesis of the mountain range.

10.2.2 Seismicity, plate motions, and subduction geometry

The general pattern of seismicity in the Andes is in accord with the eastward subduction of the Nazca plate beneath South America (Molnar & Chen, 1982). Geodetic data suggest that convergence velocities with respect to South America are 66–74 mm a^{-1} at the trench (Norabuena *et al.*, 1998; Angermann *et al.*, 1999; Sella *et al.*, 2002). These rates are slower than the 77–80 mm a^{-1} predicted by the NUVEL-1A model of plate motions (Section 5.8) and appear to reflect a deceleration from a peak of some 150 mm a^{-1} at 25 Ma (Pardo-Casas & Molnar, 1987; Somoza, 1998; Norabuena *et al.*, 1999). Currently, relative motion results in a variable component of trench-parallel displacement along the margin. In the central Andes, this component is minor and appears to be accommodated mostly within the subducted slab itself (Siame *et al.*, 2005). In the southern Andes, a moderate component of trench-parallel motion is accommodated by slip along major strike-slip faults (Cembrano *et al.*, 2000, 2002).

Focal mechanism solutions from shallow (≤70 km depth) earthquakes show that the South American plate is currently in compression (Fig. 10.2). Near the Peru–Chile Trench, some normal faulting occurs in response

to bending and other mechanical adjustments within the subducting oceanic lithosphere. Farther east thrust-type solutions are most abundant with some strike-slip motion (Gutscher *et al.*, 2000; Siame *et al.*, 2005). In general, the axes of maximum compressional stress are aligned with the plate motion vector, suggesting that plate boundary stresses are transmitted up to several hundred kilometers into the South American plate.

The distribution of earthquake hypocenters with depth indicates that the margin is divided into flat and steep subduction segments (Barazangi & Isacks, 1979; Jordan *et al.*, 1983). Beneath southern Peru and Bolivia, the Benioff zone dips about 30° (Fig. 10.3a,b). Beneath north-central Chile, it initially forms an angle of 30° to a depth of ~100 km and then dips at angles of 0–10° for several hundred kilometers (Fig. 10.3c). To allow subduction to take place at such different angles, either a lithospheric tear or a highly distorted down-going plate must accommodate the transitions between the flat and steep segments.

Above zones of flat subduction, shallow seismicity is more abundant and broadly distributed than over neighboring steep segments (Barazangi & Isacks, 1979; Jordan *et al.*, 1983). The seismic energy released in the upper plate above flat slabs is, on average, three to five times greater than in steep (>30°) segments between 250 and 800 km from the trench (Gutscher *et al.*, 2000). These differences suggest that flat slab segments are strongly coupled to the overlying continental plate (Section 9.6). The coupling appears to be controlled by the presence of a cool slab at shallow depths beneath the continental lithosphere, which strengthens the upper plate and enables it to transmit stresses over long distances.

In addition to influencing mechanical behavior, variations in the dip of the subducting plate affect patterns of volcanism. In the central Andes, where the slab dips steeply, Neogene volcanism is abundant (Plate 10.1a between pp. 244 and 245). By contrast, above the flat slab segments of north-central Peru and Chile (30°S latitude), significant Neogene volcanism is absent. These volcanic gaps and flat slab segments align with the location of partially subducted aseismic ridges. Gutscher *et al.* (2000) used relocated earthquake hypocenters (Engdahl *et al.*, 1998) below 70 km depth to generate a three-dimensional tomographic image of the subducted Nazca plate beneath the central and northern Andes (Plate 10.1b between pp. 244 and 245). The image shows two morphological highs that cor-

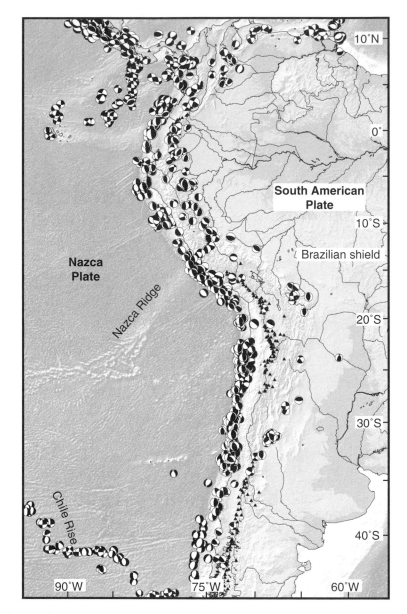

Figure 10.2 *Earthquake focal mechanism solutions in South America from the Harvard CMT catalogue for shallow (<70 km) earthquakes (1976–1999) (image provided by M.-A. Gutscher and modified from Gutscher et al., 2000, by permission of the American Geophysical Union. Copyright © 2000 American Geophysical Union). Shaded relief map is from data base of Smith & Sandwell (1997). Dark gray triangles are volcanoes.*

respond to the partially subducted Nazca Ridge and the fully subducted ridge beneath the Inca Plateau. A lithospheric tear may occur at the northwestern edge of the flat slab. These relationships support interpreta-

tions that the subduction of thick, buoyant oceanic lithosphere leads to flat subduction and can terminate magmatism by eliminating the asthenospheric wedge (Section 9.6).

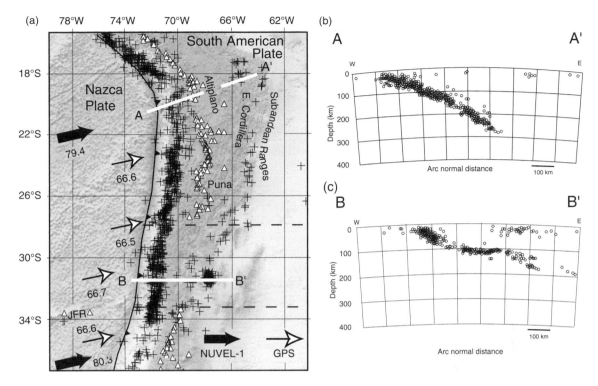

Figure 10.3 *(a) Shaded relief map of the central Andes showing the distribution of large to moderate earthquakes (crosses) and volcanoes (triangles) active since the Pliocene. Topographic databases are from Hastings & Dunbar (1998) and Smith & Sandwell (1997). Earthquakes are from the USGS National Earthquake Information Center, Preliminary Determination Epicenter catalogue (1973 to present) for shallow (≤70 km) events. Black and white arrows show relative convergence of the Nazca and South American plates from NUVEL-1 (DeMets et al., 1990, 1994) and continuous GPS observation (Kendrick et al., 1999; 2003) in mm a⁻¹. (b,c) Cross-sections showing depth distribution of relocated earthquakes (Engdahl et al., 1998) (images provided by L. Siame and modified from Siame et al., 2005, by permission of the American Geophysical Union. Copyright © 2005 American Geophysical Union). JFR, Juan Fernández Ridge.*

10.2.3 General geology of the central and southern Andes

The central Andes display two major mountain chains called the Western and Eastern cordilleras (Fig. 10.1b). Where the subducting Nazca plate dips steeply, south of latitude 15°S, the Western Cordillera contains the active volcanic arc. North of this latitude, where an active arc is absent, it is composed of Cenozoic extrusive rock. Paleozoic metasedimentary rock interfolded with Mesozoic-Cenozoic volcanic and sedimentary sequences comprise the Eastern Cordillera.

South of about latitude 15°S, the Western and Eastern cordilleras diverge around a large composite

plateau called the Altiplano-Puna (Fig. 10.1b). This orogenic plateau is 3.8–4.5 km high, 1800 km long, and 350–400 km wide (Isacks, 1988). Only the Tibetan Plateau (Section 10.4.2) is higher and wider. The Altiplano-Puna contains broad, internally drained areas of low relief and records little surface erosion. Its history of uplift began during the Miocene when plate convergence rates were at their peak (Allmendinger *et al.*, 1997). An initial stage of uplift coincided with a major ignimbrite flare-up and a period of intense crustal shortening that initially occurred in the Eastern and Western cordilleras (Allmendinger & Gubbels, 1996) and later migrated eastward into the sub-Andean zone and the Chaco foreland basin (Section 10.3.2). This shortening resulted in very thick, hot continental crust beneath the plateau (Section 10.2.4). Geodetic data indicate that shortening at the leading edge of the orogen now occurs at rates of 5–20 mm a^{-1} (Klotz *et al.*, 1999; Hindle *et al.*, 2002).

Between the Western Cordillera (volcanic arc) and the Peru–Chile Trench, elevations drop to depths of 7–8 km below mean sea level over a horizontal distance of only 60–75 km. This narrow forearc region suggests that part of the central Andean margin has been removed, either by strike-slip faulting or subduction erosion (von Huene & Scholl, 1991) (Section 9.6). The forearc includes two major belts of rock that are separated by a central valley filled with Cenozoic sediment. East of the valley, the Precordillera exposes Precambrian basement, Mesozoic sedimentary sequences, and Cenozoic intrusive and extrusive rock. The presence of this belt, which aligns with the Precambrian Arequipa Massif in southern Peru (Fig. 10.1b), indicates that the Andean orogen is founded on Precambrian continental crust. West of the central valley, the Coastal Cordillera is composed of early Mesozoic igneous rock that is a testament to the prolonged history of subduction along the margin. High-angle faults in the Coastal Cordillera, including the Atacama Fault System, record a long, complex history of normal, thrust, and strike-slip displacements (Cembrano *et al.*, 2005).

Near 20°S (Fig. 10.1a), where the orogen is >800 km wide, the backarc region records 300–350 km of Neogene shortening (Allmendinger *et al.*, 1997; McQuarrie, 2002). Most of this shortening occurs in the sub-Andean zone where combinations of thrust faults and folds deform sequences of Cenozoic, Mesozoic and Paleozoic rock in a *foreland fold and thrust belt* (see also Sections 9.7 and 10.3.4). East of the sub-Andean ranges, the 200-km-wide Chaco foreland basin is filled with at least 5 km of Neogene sediment on top of the Brazilian Shield. This basin provides an important record of Cenozoic uplift, erosion, and deposition in the central Andes (Section 10.3.2).

The Andean foreland records three different styles of tectonic shortening (Fig. 10.4): (i) thin-skinned fold and thrust belts that are detached within Paleozoic sedimentary sequences at depths of 7–10 km (Lamb *et al.*, 1997); (ii) thick-skinned fold and thrust belts with inferred detachments in Precambrian basement at 10–20 km depth; and (iii) foreland basement thrusts that appear to cut through the entire crust (Kley *et al.*, 1999). These different styles owe their origin partly to variations in the pre-Neogene lithospheric structure, temperature, and stratigraphy (Section 10.3.4). In addition, the deep-seated basement thrusts of the Pampeanas

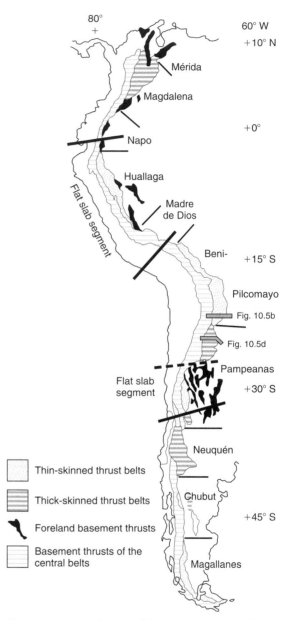

Figure 10.4 *Distribution of the segmented style of foreland deformation in the Andes (after Kley* et al., *1999, with permission from Elsevier). Flat slab segments are indicated.*

foreland (Fig. 10.5) correspond to a region of flat subduction, suggesting a possible causal relationship (Jordan *et al.*, 1983; Ramos *et al.*, 2002).

Alternations among the different styles of shortening along the strike of the orogen have produced a geologic segmentation of the Andean foreland. One of

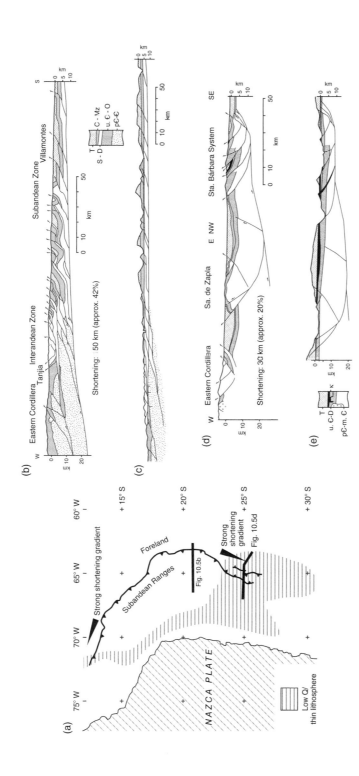

Figure 10.5 (a) Sketch map showing the transitions from thick to thin lithosphere under the Central Andes as determined from seismic wave attenuation data. Zone of thick lithosphere correlates with areas of strongest shortening gradient. Santa Bárbara system (SBS) is characterized by thick-skinned deformation. Thin-skinned thrust belt (b) and restored section (c) of the sub-Andean ranges, Bolivia using data from Baby et al. (1992), Dunn et al. (1995), and Kley (1996). Thick-skinned thrust belt (d) and restored section (e) of the Santa Bárbara system (all images modified from Kley et al., 1999, with permission from Elsevier). Location of profiles (b) and (d) also shown on Fig. 10.4.

the best-studied transitions occurs south of latitude 24°S. From north to south, a thin-skinned style of shortening in the sub-Andean belt (Fig. 10.5b,c) changes to a thick-skinned style of shortening in the Sierra de Santa Bárbara and northern Sierras Pampeanas (Fig. 10.5d,e). This change is accompanied by a decrease in the amount of shortening. A similar change in shortening magnitude occurs north of 14°S (Fig. 10.5a), implying that the present arcuate shape of the Central Andes either resulted from or has been accentuated by differences in the amount of shortening along the strike of the orogen (Isacks, 1988). The arcuate shape, or *orocline*, and the gradients in shortening also imply that the Central Andes have rotated about a vertical axis during the Neogene. GPS data, as well as paleomagnetic and geologic indicators, suggest that these rotations are counterclockwise in Peru and Bolivia north of the bend in the Central Andes, and clockwise to the south of the bend (Allmendinger et al., 2005).

Between 40° and 46°S latitude, the age of the subducting Nazca plate decreases from ~25 Ma at 38°S to essentially zero at 46°S, where the Chile Ridge is currently subducting (Herron et al., 1981; Cande & Leslie, 1986). Along this segment, convergence occurs at an angle of ~26° from the orthogonal to the trench (Jarrard, 1986). The oblique convergence has driven late Cenozoic deformation within a relatively narrow (300–400 km) orogen characterized by average elevations of <1 km (Montgomery et al., 2001). An active volcanic arc occurs north of the subducting ridge where the forearc is undergoing shortening. Inside the arc, dextral strike-slip faults of the 1000-km-long Liquiñe–Ofqui fault zone accommodate the trench-parallel component of relative plate motion (Cembrano et al., 2000, 2002) (Section 5.3). In the backarc region, shortening is relatively minor (<50 km) and controlled by the partial tectonic inversion (Section 10.3.3) of an extensional Mesozoic basin (Ramos, 1989; Kley et al., 1999). The Southern Andes, thus, is characterized by arc volcanism, relatively low relief, and deformation that is focused within a narrow transpressional (Section 8.2) orogen.

10.2.4 Deep structure of the central Andes

In 1996, geoscientists working on the **An**dean **C**ontinental **R**esearch **P**roject (ANCORP '96) completed a 400-km-long seismic reflection profile across the central Andes at 21°S latitude (Fig. 10.6). This profile, together with the results of geologic (Allmendinger et al., 1997; McQuarrie, 2002) and other geophysical studies (Patzwahl et al., 1999; Beck & Zandt, 2002), forms part of a >1000-km-long transect between the Pacific coast and the Brazilian craton (Fig. 10.7). Below the central Andean forearc, the seismic reflection profile shows east-dipping (~20°) packages of reflectors that mark the top of the subducting Nazca plate (ANCORP Working Group, 2003). Above and parallel to the slab are thick, highly reflective zones that indicate the presence of trapped fluids and sheared, hydrated mantle at the top of descending slab. Some diffuse seismicity in this region is probably related to dehydration embrittlement (Section 9.4). Sub-horizontal reflectors below the Coastal Cordillera may represent ancient intrusions that were emplaced during Mesozoic arc magmatism.

East of the forearc, converted (compressional-to-shear) teleseismic waves indicate that crustal thickness increases from about 35 km to some 70 km beneath the Western Cordillera and Altiplano (Yuan et al., 2000; Beck & Zandt, 2002). Crustal thickness also varies along the strike of the orogen, reaching a maximum of 75 km under the northern Altiplano and a minimum of 50 km under the Puna Plateau (Yuan et al., 2000, 2002). The lithosphere is 100–150 km thick below the Altiplano (Whitman et al., 1996) and several tens of kilometers thinner beneath the Puna. Lithospheric thinning beneath this latter segment explains the high elevation (~4 km) of the Puna above a relatively shallow Moho. The southward transition from thin-skinned to thick-skinned thrust faulting in this same region (Fig. 10.5) suggests that the removal of excess mantle lithosphere accommodates the westward underthrusting of the Brazilian Shield (McQuarrie et al., 2005).

Across the ANCORP '96 seismic reflection profile (Fig. 10.6), a distinct Moho is conspicuously absent. A broad transitional zone of weak reflectivity occurs at its expected depth. The cause of this diffuse character of the boundary appears to be related to active fluid-assisted processes, including the hydration of mantle rocks and the emplacement of magma under and into the lower crust. Most of the reflectivity across the profile is linked to petrologic processes involving the release, trapping, and/or consumption of fluids (ANCORP Working Group, 2003). These processes have produced a seismic reflection profile whose character contrasts with those collected across fossil mountain belts (Figs 10.33b, 10.34b, 11.15b,c) where seismic

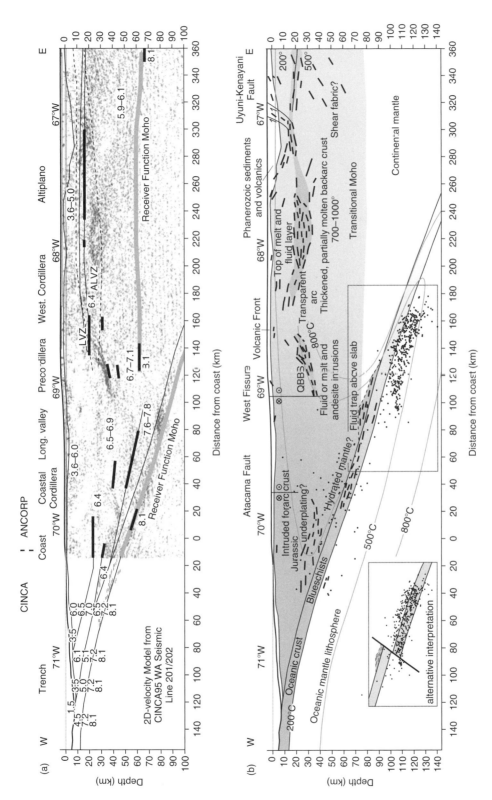

Figure 10.6 (a) Seismic results and (b) interpretation of depth-migrated ANCORP reflection data, including onshore wide-angle and receiver function results of Yuan et al. (2000) merged with offshore results of CINCA experiment (Patzwahl et al., 1999) (image provided by O. Oncken and modified from ANCORP Working Group, 2003, by permission of the American Geophysical Union. Copyright © 2003 American Geophysical Union). Location of section shown in Fig. 10.1a. Inset in (b) shows an alternative interpretation of the slab geometry and seismicity. ALVZ, Altiplano low velocity zone; QBBS, Quebrada Blanca Bright Spot.

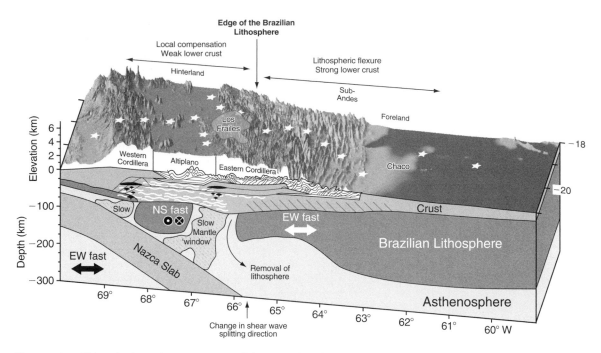

Figure 10.7 *Lithospheric-scale cross-section of the Central Andes at latitudes 18–20°S showing interpretations of the crust and mantle structure (image provided by N. McQuarrie and modified from McQuarrie et al., 2005, with permission from Elsevier). Fast upper mantle P-wave velocities (dark gray) and slow P-wave velocities (white and gray shades) are shown. White waves, crustal low-velocity zones. Cross-section shows basement thrust sheets beneath the Eastern Cordillera, cover rocks are white. White stars are data recording stations.*

reflection profiles can be interpreted only in terms of structure and lithologic contrasts.

Within the crust, seismic velocities indicate the presence of a 15- to 20-km-thick zone of low seismic wave speeds at depths of 14–30 km beneath the Western Cordillera and Altiplano-Puna (Yuan *et al.*, 2000, 2002). Average crustal V_p/V_s ratios of 1.77 beneath the plateau and peak values of 1.80–1.85 beneath the active volcanic arc suggest the presence of high crustal temperatures and widespread intra-crustal melting. Patches of high amplitude reflections and zones of diffuse reflectivity at some 20–30 km depth beneath the Precordillera and Altiplano suggest the presence of fluids rising through deep fault zones (ANCORP Working Group, 2003). One of the largest of these patches is the Quebrada Blanca Bright Spot (Fig. 10.6). Other low velocity zones and bright reflectors occur at mid-crustal depths beneath the Eastern Cordillera and backarc region. The presence of these features helps to explain differences in crustal thickness and in the altitude of the Altiplano and

Puna. Widespread fluid transport and partial melting of the middle and lower crust during Neogene shortening and plateau growth appears to have weakened the crust sufficiently to allow it to flow (Gerbault *et al.*, 2005). Similar features beneath the Tibetan Plateau (Section 10.4.5) suggest that orogenic plateaux in general involve very weak crust.

East of the Altiplano-Puna, receiver function determinations show a decrease in crustal thickness from 60 to 74 km beneath the Eastern Cordillera to about 30 km beneath the Chaco plain (Yuan *et al.*, 2000; Beck & Zandt, 2002). Mantle tomography images indicate that zones of low wave speed at 30 km depth extend through the lithosphere beneath Los Frailes ignimbrite field (Fig. 10.7), suggesting that the volcanism is rooted in the mantle and that the mantle lithosphere in this region has been altered or removed (Myers *et al.*, 1998). East of this low velocity zone, high seismic velocities in the shallow mantle, high Q (Section 9.4), and a change in the fast direction of shear-wave anisotropies suggest the presence of thick, strong, cold lithosphere of the Brazilian Shield (Polet *et al.*, 2000). Bouguer gravity anomalies show that lithospheric flexure (Sections 2.11.4, 10.3.2) supports part of the Eastern Cordillera and the sub-Andean zone (Watts *et al.*, 1995). Relationships between surface elevations and crustal thickness (Yuan *et al.*, 2002) indicate a lithospheric thickness of 130–150 km beneath the sub-Andean belt and much thicker mantle lithosphere farther east (Fig. 10.7).

10.2.5 Mechanisms of noncollisional orogenesis

Orogenesis at ocean–continent convergent margins initiates where two conditions are met (Dewey & Bird, 1970): (i) the upper continental plate is thrown into compression and (ii) the converging plates are sufficiently coupled to allow compressional stresses to be transmitted into the interior of the upper plate.

Studies of subduction zones in general suggest that the stress regime in the overriding plate is influenced by the rate and age of subducting oceanic lithosphere (Uyeda & Kanamori, 1979; Jarrard, 1986). High convergence rates and the underthrusting of young, thick, and/or buoyant lithosphere tend to induce compression, decrease slab dips, and enhance the transfer of compressional stresses (Section 10.2.2). However, although these factors may explain general differences between Chilean-type and Mariana-type subduction zones (Section 9.6), they do not explain the along-strike differences in the structure and evolution of the Andean orogen (Sections 10.2.3, 10.2.4). The Andean example shows that neither flat subduction nor the underthrusting of young and/or buoyant oceanic lithosphere control areas of maximum shortening and crustal thickening (Yáñez & Cembrano, 2004). From the Altiplano region northward and southward, there is a decrease in the total amount of crustal shortening and thickening with no direct correspondence to either the slab age (Jordan *et al.*, 1983; McQuarrie, 2002) or the convergence rate (Jordan *et al.*, 2001). These observations indicate that factors other than the geometry, rate, and age of subducting lithosphere control the response of the upper plate to compression. Among the most important of these other factors are: (i) the strength of inter-plate coupling at the trench and (ii) the internal structure and rheology of the continental plate.

1 *Interplate coupling at the trench.* Yáñez & Cembrano (2004) used a continuum mechanics approach to examine the effects of variable amounts of inter-plate coupling at the trench on upper plate deformation. These authors noted that patterns of seismicity in the Andes suggest that flat subduction controls some areas of strong inter-plate coupling (Section 10.2.2). However, the largest seismic energy release above flat segments occurs up to several hundred kilometers inland from the trench (Gutscher *et al.*, 2000). By contrast, seismicity at the Peru–Chile Trench is approximately equivalent in both flat and steep slab segments. This observation, and the lack of correlation between the amount of intra-plate shortening and the flat slab segments, suggests that the degree of inter-plate coupling at the trench may be equally or more important in controlling deformation of the upper plate.

To test this idea, Yáñez & Cembrano (2004) divided the South American plate into two tectonic domains that are characterized by different force balances: the forearc and the backarc-foreland. In the forearc, the age of the ocean crust and the convergent velocity control the strength of coupling across the ocean–continent interface. The strength of the coupling regulates the amount of deformation. The authors derived an empirical relationship between trench topography and the degree of coupling across the slipping interface using along-strike variations in the shape of the inner trench slope (Fig. 10.8a). This approach is based on the work of Wdowinski (1992) who suggested that after an equilibration period of 5–10 Ma, trench topography reflects the balance between the tectonic and buoyancy forces associated with subduction. Buoyancy forces associated with continental crust dominate the force balance if the strength of the plate interface is low, resulting in an upward movement of the forearc (Fig. 10.8b). Tectonic forces associated with the sinking of oceanic lithosphere dominate if the strength of the plate interface is high, causing downward movement of the forearc. By assuming the trench topography is in equilibrium with these forces, Yáñez &

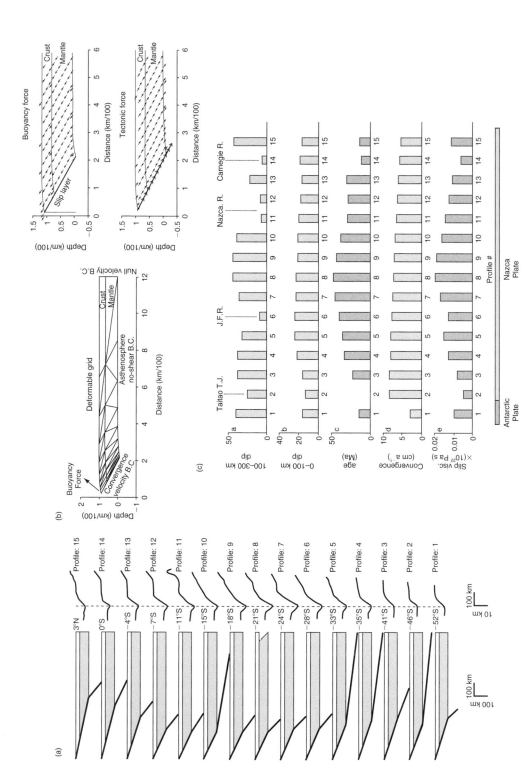

Figure 10.8 (a) Trench topography for 15 profiles of the Andes between 3°N and 56°S showing dip of the Benioff zone at 1:1 scale (left). Bathymetric profiles at right show a vertical exaggeration of 10. (b) Viscous model of trench topography showing mesh grid and boundary conditions (B.C.). Dynamic equilibrium of the trench is controlled by competing buoyancy (top-right) and tectonic (lower-right) forces. Velocity fields represented with arrows are for reference only. (c) Model results showing slab dip under asthenospheric wedge, near-trench slab dip angle, age of subducting slab at the trench, convergence velocity, slip layer viscosity for a layer of 10 km thickness (all images modified from Yáñez & Cembrano, 2004, by permission of the American Geophysical Union. Copyright © 2004 American Geophysical Union). T.J., triple junction; J.F.R., Juan Fernández Ridge.

Cembrano (2004) inferred the strength of the interface by finding the smallest displacement field for different ranges of slip parameters.

Figure 10.8c summarizes the results of the modeling. In the plots, the viscosity of the slip interface controls its strength. The slab dip, convergence rate, and the age of the subducting plate also are shown for comparison. The results indicate that the strongest inter-plate coupling occurs in the central Andes near latitude 21°S (Fig. 10.8a) where inner trench slopes are steepest and the age of subducted crust is oldest. Weak coupling occurs in the southern Andes south of 35°S where the age of ocean crust is significantly younger and trench slopes are gentle. For a constant convergent rate, the subduction of young oceanic crust and aseismic ridges results in weak coupling because the higher temperature of the oceanic lithosphere in these zones results in a thermal resetting of the ocean–continent interface.

In the backarc-foreland domain, deformation is controlled by the absolute velocity of the continental plate, its rheology, and the strength of inter-plate coupling at the trench (Yáñez & Cembrano, 2004). Strong coupling results in large amounts of compression in the backarc, which increases crustal shortening and thickening. Very weak coupling prevents backarc shortening. The rheology of the continental plate is governed by the strength of the mantle lithosphere and the temperature at the Moho. By varying the strength of coupling at the slip zone and incorporating a temperature- and strain rate-sensitive power-law rheology (Section 2.10.3), these authors reproduced several major features of the central and southern Andes. These include variations in the average topographic relief of the Andes, the observed shortening rate and crustal thickness in the Altiplano region, and block rotations (Section 10.2.3). The rotations are induced by differences in buoyancy forces caused by crustal thickness variations and in the strength of inter-plate coupling north and south of the Altiplano. Variations in the strength of inter-plate coupling also may explain differences in the degree of subduction erosion (Section 9.6) along the margin, although alternative models (e.g. von Huene et al., 2004) have been proposed.

In addition to the rate and age of subducting lithosphere, another factor that may control the strength of inter-plate coupling along the Peru–Chile Trench is the amount of surface erosion and deposition. Lamb & Davis (2003) postulated that the cold water current that flows along the coast of Chile and Peru inhibits water evaporation, resulting in little rainfall, small amounts of erosion, and minimal sediment transport into the trench. A dry, sediment-starved trench may result in a high degree of friction along the Nazca–South American plate interface, increasing shear stress, and leading to increased compression and uplift in the central Andes. By contrast, in the southern Andes where the flow of westerly winds, abundant rainfall, and the effects of glaciation result in high erosion rates, the Peru–Chile Trench is filled with sediment. The presence of large quantities of weak sediment in this region may reduce friction along the plate interface, effectively reducing the amount of shear stress and resulting in less topographic uplift and less intra-plate deformation.

2 *The structure and rheology of the continental plate.* Variations in the initial structure and rheology of the continental plate also can explain several first-order differences in the evolution of the central and southern Andes. Among these differences are the underthrusting of the Brazilian Shield beneath the Altiplano-Puna and major lithospheric thinning in the central Andes, and the absence of these features in the southern Andes.

Sobolev & Babeyko (2005) conducted a series of two-dimensional thermomechanical models (Fig. 10.9) that simulated deformation in the central and southern Andes using two different initial structures. The central Andes involve a thick felsic upper crust, a thin gabbroic lower crust, and a total thickness of 40–45 km. This configuration presumes that the crust already had been shortened prior to the start of deformation at 30–35 Ma

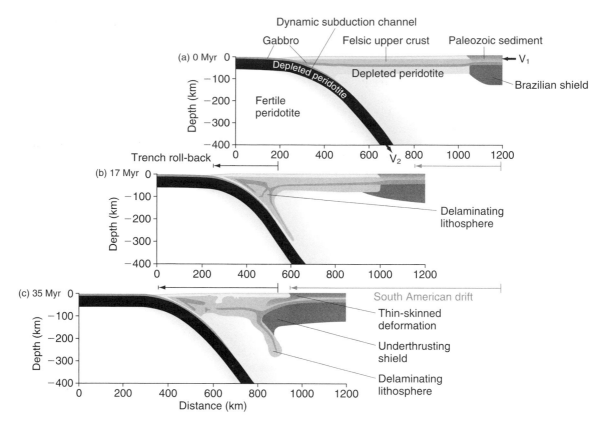

Figure 10.9 (a–c) Time snapshots showing the evolution of shortening for a mechanical model of the Central Andes (modified from Sobolev & Babeyko, 2005, with permission from the Geological Society of America).

(Allmendinger *et al.*, 1997; McQuarrie *et al.*, 2005). The southern Andes consist of an upper and lower crust of equal thickness and a total crustal thickness of 35–40 km. In all models, subduction initially occurs at a low-angle below a 100- to 130-km-thick continental lithosphere (Fig. 10.9a) and is free to move as subduction proceeds. The upper plate is pushed to the left (V_1), simulating the western drift of South American plate (Somoza, 1998). The slab is pulled from below at velocities (V_2) that conform to observations. A thin subduction channel simulates the plate interface where a frictional (brittle) rheology controls deformation at shallow depths and viscous flow occurs at deep levels. The depth of this change in rheology and the strength of the slip zone are regulated using a frictional coefficient.

The numerical experiment that best replicated the structure of the central Andes is shown in Fig. 10.9. In this model 58% of the westward drift of South America over a 35 Myr period is accommodated by roll-back (Section 9.10) of the Nazca plate at the trench, with the rest accommodated by intra-plate shortening (37%) and subduction erosion (5%). During shortening the crustal thickness doubles while the lower crust and mantle lithosphere become thinner by delamination (Fig. 10.9b). The delamination is driven by the transformation from gabbro to eclogite in the lower crust, which increases its density and allows it to peel off and sink into the mantle. Another possible mechanism for reducing lithospheric thickness is tectonic erosion driven by convective flow in the mantle (Babeyko *et al.*, 2002). These processes lead to an increase in temperature at the Moho, which weakens the crust and allows its lower part to flow. After 20–25 Myr in model time, tectonic shortening generates high topography between the magmatic arc and the Brazilian Shield (Fig. 10.10a). The large topographic gradients initiate flow in a weak middle and lower crust that evens out crustal thickness and the surface topography, forming a 4-km-high orogenic plateau after 30–35 Myr. The model also predicts mechanical failure of the wedge of Paleozoic sediments by thin-skinned thrust faulting in the foreland (Section 10.3.4) at 25 Myr model time, followed by underthrusting of the Brazilian Shield under the plateau (Fig. 10.9c). The failure of the foreland sediments marks a change in the mode of shortening from pure shear to simple shear deformation (e.g. Section 10.3.4, Fig. 10.12). Shortening reaches 300–350 km by 30–35 Myr, as indicated by the curve of filled circles in Fig. 10.10b.

These and other models allow the dominant process controlling tectonic shortening in the Andes to be the accelerating westward drift of the South American plate. However, to explain the major tectonic features of the central Andes high convergent velocities and strong inter-plate coupling must be accompanied by an initially thick, weak continental crust (Sobolev & Babeyko, 2005; McQuarrie et al., 2005). Geologic evidence

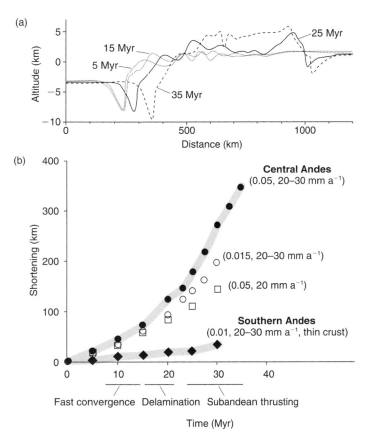

Figure 10.10 (a) Evolution of surface topography and (b) calculated shortening versus time for models using configuration shown in Fig. 10.9 (modified from Sobolev & Babeyko, 2005, with permission from the Geological Society of America). Note formation of high topography and then plateau during last 10 Ma in (a). Numbers near model results in (b) indicate subduction channel friction coefficient (first number) and western drift velocity of the South American plate (second group of numbers).

suggests that these conditions probably were only achieved in the central Andes, possibly as a consequence of high convergent rates, flat subduction, and/or the underthrusting of thick, buoyant oceanic crust. In addition, the mechanical failure of thick piles of sedimentary rock, continental underthrusting, and lithospheric thinning internally weaken the continental plate and influence its behavior as orogenesis proceeds.

10.3 COMPRESSIONAL SEDIMENTARY BASINS

10.3.1 Introduction

Sedimentary basins that either form or evolve in response to regional compression are common in orogenic belts. Among the most recurrent types are *foreland basins* (Section 10.3.2), which form as a direct result of the crustal thickening and topographic uplift that accompany orogenesis, and basins that initially form during a period of extension or transtension and later evolve during a period of subsequent compression. This latter process, called *basin inversion* (Section 10.3.3), also occurs in association with strike-slip faulting (Fig. 8.10) and is the mechanism by which old passive margin sequences deform during continental collision (Section 10.4.6). Any sedimentary basin in compression may develop a fold and thrust belt (Section 10.3.4) whose characteristics reflect the strength of the continental lithosphere and the effects of pre-existing stratigraphic and structural heterogeneities.

10.3.2 Foreland basins

In addition to topographic uplift, orogenesis commonly results in a region of subsidence called a *foreland basin* or *foredeep* (Dickinson, 1974). The *foreland* lies at the external edge of the orogen toward the undeformed continental interior (e.g. Fig. 10.7). If a volcanic arc is present, it coincides with the backarc region of the margin. Its counterpart, the *hinterland*, corresponds to

the internal zone of the orogen where the mountains are highest and rocks the most intensely deformed.

Foreland basins form where crustal thickening and topographic uplift create a mass of crust that is large enough to cause flexure (Section 2.11.4) of the continental craton. This flexure creates a depression that extends much farther into the surrounding craton than the margin of the thickened crust. It is bounded on one side by the advancing thrust front and on the other by a small flexural uplift called a *forebulge* (e.g. Fig. 10.18). The basin collects sedimentary material (molasse) that pours off the uplifting mountains as they experience erosion and as thrust sheets transport material onto the craton. Its stratigraphy provides an important record of the timing, paleogeography, and progressive evolution of orogenic events.

The shape of a foreland basin is controlled by the strength and rheology of the lithosphere. A low flexural rigidity, which characterizes young, hot and weak lithosphere, results in a narrow, deep basin. A high flexural rigidity, which characterizes old, cool and strong lithosphere, produces a wide basin with a better-developed forebulge (Flemings & Jordan, 1990; Jordan & Watts, 2005). Variations in the strength and temperature of the lithosphere can thus cause the character of the foreland basin to change along the strike of the orogen. Other factors such as inherited stratigraphic and structural inhomogeneities also influence basin geometry. In the Andes, an along-strike segmentation of the foreland partly coincides with variations in these properties and with the segmented geometry of the subducted Nazca plate (Section 10.2.3).

As a result of lithospheric flexure, the sediment thickness in a foreland basin decreases away from the mountain front to a feather edge on the forebulge (Flemings & Jordan, 1990; Gómez *et al.*, 2005). Close to the mountain range the sediments are coarse grained and deposited in a shallow water or continental environment; at the feather edge they are fine grained and often turbiditic. The sediments thus form a characteristic wedge-shaped sequence in profile whose stratigraphy reflects the subsidence history of the basin as it grows and migrates outwards during convergence. The stratigraphy is thus characterized by units that thin laterally, overstep older members, or may be truncated by erosion.

Belts of deformed sedimentary rock in which the layers are folded and duplicated by thrust faults are common in foreland basins. Like their counterparts in accretionary prisms (Section 9.7, Fig. 9.20) and in zones of transpression (Section 8.2, Fig. 8.8b), *foreland fold and thrust belts* form as the crust is shortened in a regime of compression (Fig. 10.5). During shortening, small sedimentary basins called *piggyback basins* may form on the top of moving thrust sheets.

10.3.3 Basin inversion

Many sedimentary basins record a reversal in the sense of motion on dip-slip faults at different stages in their evolution. This reversal is known as *inversion*. At present there is no universal definition of the process. However, the most common type refers to the compressional reactivation of pre-existing normal faults in sedimentary basins and passive margins that originally formed by extension or transtension (Turner & Williams, 2004). Fault reactivation changes the architecture of the basin and commonly results in the uplift of previously subsided areas and the exhumation of formerly buried rocks. Evidence for this type of inversion occurs at a wide range of scales in many different settings, including in collisional and noncollisional orogens and in regions of strike-slip faulting. At convergent margins the tectonic inversion of extensional backarc and intra-arc basins is an especially important process that accommodates crustal shortening, localizes contractional deformation, and results in an along-strike segmentation of the margin.

In many basins, a common criterion for recognizing fault-controlled inversion is the identification of the *null point* in vertical profiles or the *null line* in three dimensions. Figure 10.11 shows a cross-section illustrating the geometry of an inverted half graben in Indonesia (Turner & Williams, 2004). The profile shows a reactivated fault along which the net displacement changes from normal at its base to reverse near its top. The null point occurs where the net displacement along the fault is zero and divides the area displaying reverse displacement from that displaying normal displacement. As the magnitude of the inversion increases, the null point will migrate along the fault. The uplift and folding of synrift and postrift sediments also indicate that inversion has occurred by the compressional reactivation of a normal fault.

Basin inversion is caused by a variety of mechanisms. Continent–continent or arc–continent collision can result in compression, uplift, and fault reactivation. Changes in the rate and dip of subduction (Section 10.2.2, Fig. 9.18) also may cause basin inversion at ocean–continent convergent margins. In regions of strike-slip faulting, rapid reversals in the sense of motion on faults commonly occur between releasing bends and restraining bends (Section 8.2, Fig. 8.9). Isostatic, flexural, and thermal mechanisms also have been proposed to explain the uplift associated with basin inversion. However, many authors view these latter mechanisms as subordinate to external horizontal stresses that drive the compressional reactivation of faults.

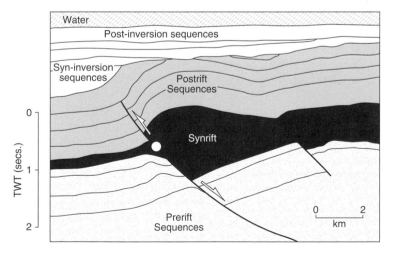

Figure 10.11 *Cross-section derived from a seismic reflection profile showing an inverted half graben from the East Java Sea Basin, Indonesia (redrawn from Turner & Williams, 2004, with permission from Elsevier). TWT is two-way-travel time of seismic reflections. White dot indicates null point.*

10.3.4 Modes of shortening in foreland fold-thrust belts

A common characteristic of fold and thrust belts is the presence of one or more *décollement* (or detachment) surfaces that underlie shortened sequences of sedimentary and volcanic rock (Section 9.7). The geometry of these surfaces tends to conform to the shape of the sedimentary and volcanic sections in which they form. In most foreland basins sedimentary sequences thin toward the foreland, resulting in décollements that dip toward the hinterland (Figs 10.5b, 10.7). In *thin-skinned* thrust belts (Section 10.2.4, Fig. 10.5b), the lowermost, or basal, décollement separates a laterally displaced sedimentary cover from an underlying basement that is still in its original position. In *thick-skinned* styles (Fig. 10.5d), the décollement surface cuts down through and involves the crystalline basement.

The development of thin- or thick-skinned styles of shortening commonly is controlled by the presence of inherited stratigraphic and structural heterogeneities in the crust. In the central Andean foreland, for example (Section 10.2.3), variations in the thickness and distribution of sedimentary sequences have been linked to different modes of Neogene shortening (Kley *et al.*, 1999). Thin-skinned styles preferentially occur in regions that have accumulated >3 km of sediment, where the low mechanical strength of the sequences localizes deformation above crystalline basement (Allmendinger & Gubbels, 1996). Thick-skinned styles tend to occur in regions where Mesozoic extensional basins have inverted (Sections 9.10, 10.3.3). As these latter basins experience the shift from extension to contraction, old normal faults involving basement rock reactivate (Turner & Williams, 2004; Saintot *et al.*, 2003; Mora *et al.*, 2006).

In many fold and thrust belts, and especially in thick-skinned varieties, shortening results in some faults that dip in a direction opposite to that of the basal décollement, creating a *doubly vergent thrust wedge* composed of forward-breaking and back-breaking thrusts. These doubly vergent wedges may occur at any scale, ranging from relatively small basement massifs (Fig. 10.5d) to the scale of an entire collisional orogen (Fig. 8.23b,d). Their bivergent geometry reflects a condition where the material on the upper part of an advancing thrust sheet or plate encounters resistance to continued forward motion (Erickson *et al.*, 2001; Ellis *et al.*, 2004). The resistance may originate from friction along the décol-

lement surface as the wedge thickens or as material moves over a thrust ramp (Section 9.7). It also may result where an advancing thrust sheet encounters a buttress made of a strong material, such as the volcanic arc in an accretionary prism (Section 9.7) or the boundary between a rigid plate and a weaker plate (Section 8.6.3). Buttresses also may result from a change in lithology across an old normal or thrust fault, from a thickening sequence of sedimentary rock, or from any other mechanism.

The lateral (across-strike) growth of thrust wedges (Section 9.7) and the involvement of deep levels in the deformation are controlled by the temperature and relative strengths of the shallow and deep crust. If the upper crust is strong and the deep crust relatively hot and weak, then shortening may localize into narrow zones and thick-skinned styles of deformation result (Ellis *et al.*, 2004; Babeyko & Sobolev, 2005). A weak middle and lower crust promotes ductile flow and inhibits the lateral growth of the thrust wedge. Deep crustal flow also tends to result in low critical tapers (Section 9.7) and a symmetric crustal structure that includes both forward- and back-breaking thrusts. During basin inversion, more normal faults tend to reactivate if the middle or lower crust is weak relative to the upper crust (Nemčok *et al.*, 2005; Panien *et al.*, 2005). By contrast, if the upper crust is weak and the deep crust is cool and strong, then shortening leads to a mechanical failure of upper crustal sequences and the orogen grows laterally by thin-skinned deformation. In scenarios involving a strong lower crust, thrust wedges tend to show high tapers, asymmetric styles (mostly forward-breaking thrusts), and rapid lateral growth.

A combination of these effects may explain why contractional deformation led to the rapid lateral growth of a foreland fold and thrust belt in the Central Andes and not in the Southern Andes (Section 10.2.3). Allmendinger & Gubbels (1996) recognized that deformation in these two regions involved two distinctive modes of shortening. In an older *pure shear* mode of shortening, deformation of the upper and lower crust occurred simultaneously in the same vertical column of rock. North of 23°S, this type of deformation was focused within the Altiplano. Later, during the Late Miocene the deformation migrated eastward, forming a thin-skinned foreland fold and thrust belt in the sub-Andean ranges while the middle and lower crust of the Altiplano continued to deform. This latter mode of shortening, where deformation in the upper crust and the deep crust are separated laterally, is known as *simple shear*.

The underthrusting of the Brazilian Shield beneath the Altiplano most likely drove the simple shear (Section 10.2.5). South of 23°S the pure shear mode of shortening lasted longer and was replaced by a thick-skinned thrust belt involving a mix of both pure and simple shear.

These differences in the style and mode of shortening along the strike of the Andes appear to be related to variations in the strength and temperature of the foreland lithosphere. Allmendinger & Gubbels (1996) postulated that the shallower basement and lack of a thick sedimentary cover in the Sierra Santa Bárbara ranges, south of latitude 23°S, precluded thin-skinned deformation and allowed deformation to remain in the thermally softened crust of the Puna for a long period of time. In addition, the mantle lithosphere beneath the Puna is significantly thinner than beneath the Altiplano, suggesting that the crust in the former is hotter and weaker.

To test this idea, Babeyko & Sobolev (2005) conducted a series of thermomechanical experiments where the cold, rigid lithosphere of the Brazilian Shield indented into the warm, soft lithosphere of the adjacent Altiplano-Puna. Figure 10.12a shows the model setup, which includes a thick plateau crust on the left and a three-layer crust on the right above mantle lithosphere. The three layer crust includes an 8-km-thick layer of Paleozoic sediments. The mechanical strength of this layer and the temperature of the foreland are the two main variables in the model. A Mohr–Coulomb elasto-plastic rheology simulates brittle deformation and a temperature- and strain rate-dependent viscoelastic rheology simulates ductile deformation. The whole system is driven by a constant shortening rate of $10\,mm\,a^{-1}$ applied to the right side of the model.

After 50 km of shortening, the models show distinctive modes of shortening. In the case where the Paleozoic sediments are strong (or absent) and lie on top of a cold strong Brazilian Shield (Fig. 10.12b), the crust and mantle deform together homogeneously in pure shear mode (Fig. 10.12d). No deformation occurs in the indenting foreland where cold temperatures inhibit lateral growth of a thrust wedge. In the case where the Paleozoic sediments are weak and the foreland cold and

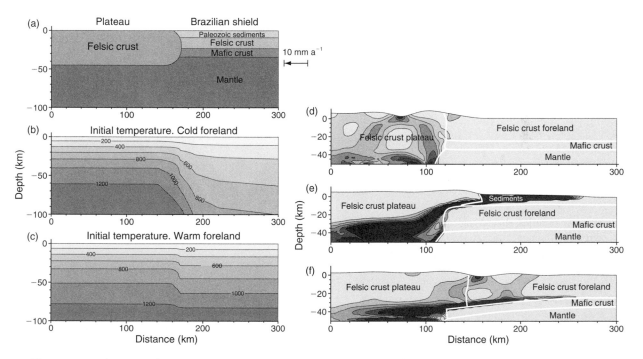

Figure 10.12 *(a–c) Initial setup and (d–f) results of numerical simulations of foreland deformation (after Babeyko & Sobolev, 2005, with permission from the Geological Society of America). (d–f) Accumulated finite strain after 50 km of shortening for three modes. White solid lines are boundaries of lithologic units. See text for explanation.*

strong (Fig. 10.12e), the foreland displays a simple shear thin-skinned mode of deformation. Underthrusting of the shield is accompanied by the eastward propagation of the thin-skinned thrust belt above a shallow décollement at 8–14 km depth and drives deformation in the lower crust beneath the plateau. This style conforms well to observations east of the Altiplano and north of 23°S. It also simulates the conditions of the Himalayan fold-thrust belt south of the Tibetan Plateau (Section 10.4.4). In the case where the Paleozoic sediments are weak and the foreland warm and weak (Fig. 10.12c), deformation in the foreland is thick-skinned with a deep décollement at ~25 km depth (Fig. 10.12f). This latter style conforms well to observations east of the Puna and south of 23°S and results because the foreland is weak enough to deform by buckling.

These observations and experiments illustrate that variations in lithospheric strength and rheology play an important role in controlling the tectonic evolution of compressional basins and fold-thrust belts. These effects are prominent at scales ranging from individual thrust sheets to the entire lithosphere.

10.4 CONTINENT– CONTINENT COLLISION

10.4.1 Introduction

Collisional mountain ranges form some of the most spectacular and dominant features on the surface of the Earth. Examples include the Himalayan–Tibetan orogen, the Appalachians, the Caledonides, the European Alps, the Urals (Section 11.5.5), the Southern Alps of New Zealand (Sections 8.3.3, 8.6.3), and many of the Proterozoic orogens (e.g. Section 11.4.3). The anatomy of these belts is highly diverse, in part due to differences in the size, shape, and mechanical strength of the colliding plates, and the effects of different precollisional tectonic histories. In addition, continental collision can range from being highly oblique, such as occurs on the South Island of New Zealand, to nearly orthogonal. These differences greatly influence the mechanisms of collisional orogenesis (Section 10.4.6).

The Himalayan–Tibetan orogen (Fig. 10.13) is one of the best places to study a large-scale continent–continent collision that followed the closure of a major ocean basin and formed an orogenic plateau. The active tectonics, diverse structure, and relatively well-known plate boundary history of this belt allow many tectonic relationships to be measured directly and provide important constraints on the driving mechanisms of deformation and the manner in which deformation is accommodated (Yin & Harrison, 2000). In addition, the immense size and high elevations of this orogen illustrate how mountain building and global climate are interrelated. These interactions form important elements of orogenesis in most, if not all, tectonic settings.

This section provides a discussion of four main aspects of the Himalayan–Tibetan orogen: (i) the relative motion of Indian and Eurasia and their tectonic history prior to collision; (ii) the nature of post-collisional convergent deformation as revealed by seismicity and geodetic data; (iii) the geologic history of the Himalaya and the Tibetan Plateau; and (iv) the deep structure of the orogen. Section 10.4.6 provides a discussion of the main factors controlling the mechanical evolution of the orogen.

10.4.2 Relative plate motions and collisional history

The Himalayan–Tibetan orogen was created mainly by the collision between India and Eurasia over the past 70–50 Myr (Yin & Harrison, 2000). The orogen is part of the greater Himalayan–Alpine system, which extends from the Mediterranean Sea in the west to the Sumatra arc of Indonesia in the east over a distance of >7000 km. This composite belt has evolved since the Paleozoic as the Tethyan oceans (e.g. Fig. 11.27) closed between two great converging landmasses: Laurasia in the north and Gondwana in the south (Şengör & Natal'in, 1996). Tethys may have been only a few hundred kilometers wide in the west but opened to the east to form an ocean that was at least several thousands of kilometers wide.

The India–Eurasia collision was brought about by the rifting of India from Africa and East Antarctica during the Mesozoic (Section 11.5.5) and by its migration northward as the intervening oceanic lithosphere was subducted beneath the Eurasian Plate. Magnetic

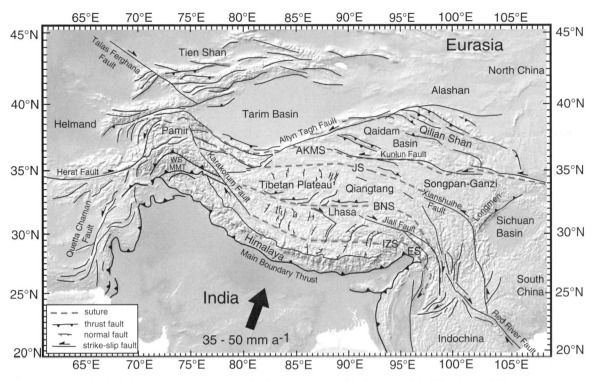

Figure 10.13 *Shaded relief map showing major faults and topographic features of the Himalayan–Tibetan orogen. Fault traces are from Hodges (2000), Yin & Harrison (2000), Tapponnier et al. (2001). WS, Western Himalayan Syntaxis; ES, Eastern Himalayan Syntaxis; MMT, Main Mantle Thrust; AKMS, Ayimaqin–Kunlun–Mutztagh suture; JS, Jinsha suture; BNS, Bangong–Nujiang suture; IZS, Indus–Zangbo suture. Map was constructed using the same topographic data and methods as in Fig. 7.1.*

anomalies in the Indian Ocean and paleomagnetic measurements from the Ninety-East Ridge and the Indian subcontinent record the northerly drift of the Indian plate and allow the reconstruction of its paleolatitude (Fig. 10.14). The data show a rapid decrease in the relative velocity between the Indian and Eurasian plates at 55–50 Ma. This time interval commonly is interpreted to indicate the beginning of the India–Eurasia collision. However, it is uncertain whether the decrease resulted from an increase in the resistance to continued motion of the India plate as it collided with Eurasia or if it simply reflects a sudden decrease in spreading rate along the mid-oceanic ridge south of India. This latter possibility allows the age of the initial contact between India and Eurasia to be older than 55–50 Ma.

Stratigraphic and sedimentological data provide additional information on the age and progressive evo-

lution of the India–Eurasia collision. Gaetani & Garzanti (1991) showed that marine sedimentation stopped and terrestrial deposition along the southern margin of Asia commenced at 55–50 Ma, which is in accord with interpretations of the age of the initial collision derived from magnetic anomalies. However, this observation in fact only constrains the youngest possible age of the onset of the collision because as much as 500–1000 km of the Indian passive continental margin has been underthrust beneath Asia, potentially eliminating the early record of the collision (Yin & Harrison, 2000). Beck *et al.* (1995) showed that trench and forearc material along the southern margin of the Eurasian plate near Pakistan was thrust onto the northern edge of India after 66 Ma and before 55 Ma. Willems *et al.* (1996) found changes in sedimentary facies and depositional patterns in south-central Tibet that suggest initial contact between some parts of India and Asia could

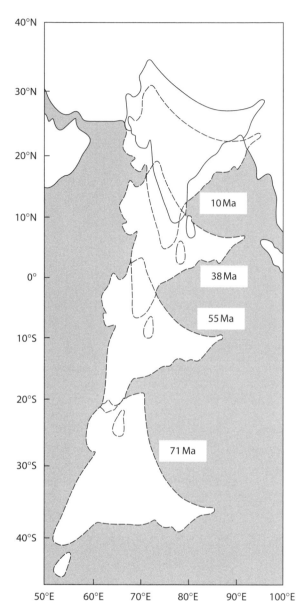

Figure 10.14 *Northward drift of India with respect to Asia from 71 Ma to the present, determined from magnetic lineations in the Indian and Atlantic oceans (redrawn from Molnar & Tapponnier, 1975,* Science **189***, 419–26, with permission from the AAAS).*

ity of India's northward drift from over 100 mm a^{-1} to about 50 mm a^{-1} or less. This latter time may mark the final stage of true continent–continent collision (Yin & Harrison, 2000).

Geologic observations in Tibet and China add important details to the sequence of events leading up to the India–Eurasia collision. The geology indicates that the main collision between India and Eurasia was preceded by the collision of several microcontinents, flysch complexes, and island arcs during Paleozoic and Mesozoic time. The collision and accretion of these terranes is marked by a series of suture zones (Fig. 10.13), some of which preserve ophiolites and blocks of high-pressure metamorphic rocks (Section 9.9). Some of these sutures expose relics of ultra-high-pressure (UHP) minerals such as coesite and microdiamond, commonly as inclusions in unreactive phases of zircon and garnet. The presence of these minerals, and the high pressures (2.5–4.0 GPa) under which they form, can reflect situations where a section of continental crust enters the subduction zone and descends to depths of 60–140 km before decoupling from the downgoing plate (Ernst, 2003; Harley, 2004). The mechanisms by which UHP and other high pressure metamorphic rocks are exhumed to the surface may involve contractional, extensional and/or strike-slip deformation accompanying the evolution of the plate boundary zone. Hacker *et al.* (2004) describe processes associated with the exhumation of UHP terranes in South China.

The Songpan–Ganzi terrane exposes thick Triassic flysch sequences that rest on top of Paleozoic marine sediments belonging to the passive margin of North China. These sequences were deposited, uplifted, and deformed during the Triassic collision between the North and South China blocks, forming the Ayimaqin–Kunlun–Mutztagh suture (Yin & Harrison, 2000). By the end of the Triassic (Fig. 10.15a), the Lhasa and Qiangtang terranes had rifted from Gondwana and began their journey toward Eurasia (Fig. 10.15b). The Qiangtang terrane collided with the Songpan–Ganzi by 140 Ma, forming the Jinsha suture. Continued convergence brought the Lhasa terrane into juxtaposition with Qiangtang and eventually welded the two fragments together, forming the Bangong–Nujiang suture. The formation of a new subduction zone beneath Lhasa (Fig. 10.15c) created an Andean-type orogen (Fig. 10.15d) and eventually resulted in the collision between India and Eurasia (Fig. 10.16e), forming the Indus–Zangbo suture. Continued convergence (Fig. 10.15f) resulted in intra-plate shortening and uplift, and is asso-

have occurred as early as 70 Ma. These relationships suggest that the initial collision may have begun as early as the Late Cretaceous. In general, most authors agree that all Tethyan oceanic lithosphere had disappeared by 45 Ma, and at ~36 Ma there was a decrease in the veloc-

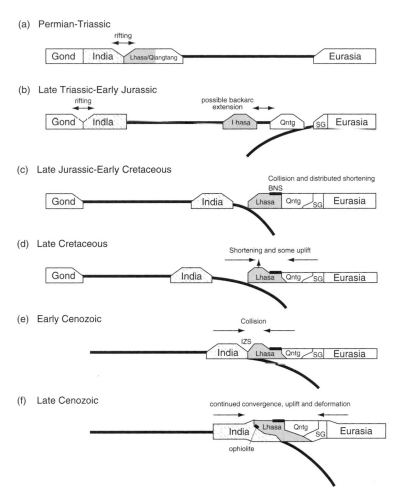

Figure 10.15 *Possible sequence of events in the evolution of the Himalayan–Tibetan orogen (modified from Haines et al., 2003, by permission of the American Geophysical Union. Copyright © 2003 American Geophysical Union). Interpretation incorporates relationships developed by Allègre et al. (1984) and Yin & Harrison (2000). BNS, Bangong–Nujiang suture; SG, Songpan–Ganzi terrane; Gond, Gondwana; Qntg, Qiantang terrane; IZS, Indus–Zangbo suture.*

ciated with a new plate boundary that is beginning to form in the Indian Ocean (Van Orman *et al.*, 1995). The Indus–Zangbo suture now forms the southern boundary of the Tibetan Plateau (Fig. 10.13), which lies more than 5000 m above mean sea level and covers an area of more than a million square kilometers.

This history shows that the Himalayan–Tibetan orogen is built upon a collage of exotic material that became welded to the Eurasian Plate before the main India–Eurasia collision (Şengör & Natal'in, 1996; Yin & Harrison, 2000). This type of sequential amalgamation of microcontinents and other material during prolonged subduction is characteristic of accretionary orogens (Section 10.6.2) and represents one of the most efficient mechanisms of forming supercontinents (Section 11.5). This history also resulted in a hot, weak Eurasia continental plate prior to its collision with India.

10.4.3 Surface velocity fields and seismicity

Since about 50 Ma, continued convergence between India and Eurasia at a slowed rate has caused India to penetrate some 2000 km into Asia (Dewey *et al.*, 1989;

Johnson, 2002). This motion created a zone of active deformation that stretches ~3000 km north of the Himalayan mountain chain (Fig. 10.13). Global Positioning System (GPS) measurements show that India is moving to the northeast at a rate of some 35–38 mm a^{-1} relative to Siberia (Larson *et al.*, 1999; Chen *et al.*, 2000; Shen *et al.*, 2000; Wang *et al.*, 2001). This rate is considerably slower than the long-term rates of 45–50 mm a^{-1} estimated from global plate motion models (DeMets *et al.*, 1994), which is typical of the short-term interseismic strain rates measured using geodetic data (e.g. Section 8.5).

The geodetic data suggest that deformation within the Tibetan Plateau and its margins absorbs more than 90% of the relative motion between the India and Eurasia plates, with most centered on a 50-km-wide region of southern Tibet (Wang *et al.*, 2001). Internal shortening of the plateau accounts for more than one-third of the total convergence. An additional component of shortening is accommodated north of the Tibetan Plateau in Pamir, Tien Shan, Qilian Shan, and elsewhere, although the rates are not well known in these areas.

South of the Kunlun Fault (Fig. 10.13), the surface velocity field shows that the Tibetan Plateau is extruding eastward relative to both India and Asia (Fig. 10.16). This motion, where slices of crust move laterally out of the way of colliding plates by slip on strike-slip faults, is termed *lateral escape*. The movement also involves the rotation of material around a curved belt in Myanmar called the eastern Himalayan syntaxis. The term *syntaxis* refers to the abrupt changes in trend that occur on either side of the Himalaya in Myanmar and Pakistan where mountain ranges strike at nearly right angles to the trend of the Himalaya. East of the plateau, North China and South China are moving to the east-southeast at rates of 2–8 mm a^{-1} and 6–11 mm a^{-1} relative to stable Eurasia, respectively.

A GPS velocity profile across the Tibetan Plateau (Fig. 10.16a) is mostly linear parallel to the predicted direction of the India–Eurasia collision (N21°E), except for a high gradient across the Himalaya at the southern end of the plateau (Wang *et al.*, 2001). This mostly linear trend suggests that the shortening across the plateau is broadly distributed; otherwise significant deviations across individual fault zones would be expected. However, this generally continuous style of deformation appears to be restricted mostly to the plateau itself. Rigid block-like motion appears to characterize regions to the north and northeast of the plateau, including the

Tarim Basin and the North and South China blocks. These observations, and geologic data, suggest that the northward growth of the orogen was not a smooth, continuous process, but occurred in an irregular series of steps. In a direction orthogonal (N111°E) to the convergence direction, horizontal motion increases steadily northward from the Himalaya across the Tibetan Plateau (Fig. 10.16b), reflecting the eastward motion of the latter with respect to both India and Eurasia. At its northern margin velocities decrease rapidly as a result of left-lateral strike-slip motion on the Kunlun and other faults (Wang *et al.*, 2001). The Longmen Shan (Fig. 10.13) moves eastward with the South China block (Burchfiel, 2004).

Earthquake focal mechanism solutions, compiled for the period 1976–2000 by Liu & Yang (2003), reveal the style of active faulting in the Himalayan–Tibetan orogen (Fig. 10.17). Zones of concentrated thrust faulting occur along both the northern, southern, and eastern margins of the Tibetan Plateau. Within the Himalaya, thrust faulting is prevalent. South of the Himalaya (Fig. 10.18), intra-plate earthquakes and other geophysical evidence indicate that the Indian plate flexes and slides beneath the Himalaya, where it lurches northward during large earthquakes (Bilham *et al.*, 2001). The overall pattern of the deformation is similar to that which occurs at ocean–continent convergence zones where an oceanic plate flexes downward into a subduction zone. North of the Himalaya, normal faulting and east–west extension dominate southern and central Tibet. Strike-slip faulting dominates a region some 1500 km wide north of the Himalaya and extending eastward into Indo-China. Farthest from the mountain chain is a region of crustal extension and normal faulting extending from the Baikal Rift of Siberia to the northern China Sea. Active strike-slip faulting also occurs in the western Himalayan syntaxis and eastern Himalayan syntaxis in Pakistan and in Myanmar, respectively. South of the syntaxis in Pakistan, movement along north-striking faults is dominantly sinistral; south of the one in Myanmar it is mostly dextral. These opposite senses of motion on either side of India are compatible with the northward penetration of India into southern Asia.

These observations indicate that convergence between India and Eurasia is accommodated by combinations of shortening, east–west extension, strike-slip faulting, lateral escape, and clockwise rotations. In addition, uplift of the high elevations of the Tibetan Plateau by Miocene time (Blisniuk *et al.*, 2001; Kirby *et al.*, 2002)

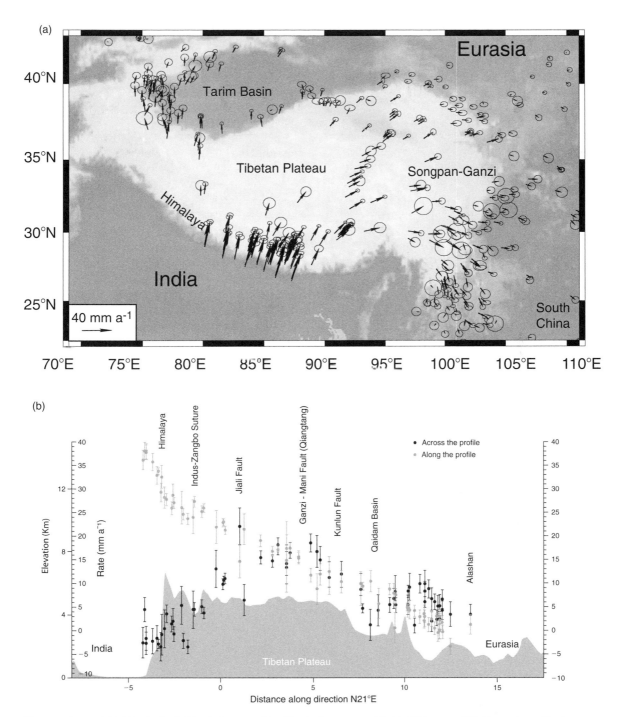

Figure 10.16 *(a) GPS velocity field relative to stable Siberia (image provided by Y. Yang and M. Liu and modified from Liu & Yang, 2003, by permission of the American Geophysical Union. Copyright © 2003 American Geophysical Union), combining data from Chen et al. (2000), Larson et al. (1999), and Wang et al. (2001). Error ellipses are 95% confidence. Velocity scale is shown in lower left corner. (b) GPS velocity profile across the Tibetan Plateau in the direction N21°E (after Wang et al., 2001, Science **294**, 574–7, with permission from the AAAS). Black diamonds represent the component of velocity perpendicular to the profile, light gray diamonds represent the component of the velocity parallel to the profile.*

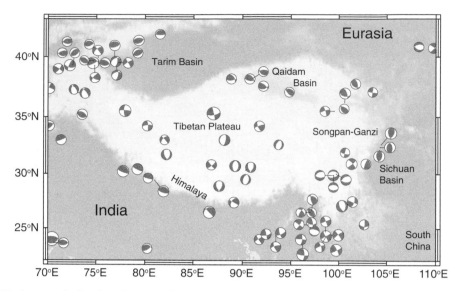

Figure 10.17 *Earthquake focal mechanism solutions showing the predominant east–west crustal extension in the Tibetan Plateau (image provided by Y. Yang and M. Liu and modified from Liu & Yang, 2003, by permission of the American Geophysical Union. Copyright © 2003 American Geophysical Union). Data are events with magnitude >5.5 and depth <33 km from the Harvard catalogue (1976–2000).*

indicates that significant vertical uplift occurred after India collided with Asia. Currently, the Himalaya are uplifting rapidly at rates between 0.5 and 4 mm a⁻¹ and experience very high rates of erosion along their southern flank (Hodges *et al.*, 2001).

10.4.4 General geology of the Himalaya and Tibetan Plateau

The Himalaya are composed of three large, imbricated thrust slices and related folds separated by four major fault systems (Figs 10.19, 10.20). These imbricated thrusts, which occupy a section about 250–350 km wide, appear to accommodate approximately one-third to one-half of the ~2000 or more kilometers of post-collisional shortening between India and Eurasia (Besse & Courtillot, 1988; DeCelles *et al.*, 1998). At the base of the stack the mostly buried Main Frontal Thrust lies along the topographic front of the mountain range (Wesnousky *et al.* 1999). This fault is the youngest and most active fault in the mountain range and carries rock of the Himalaya southward into a flexural foredeep (Section 10.3.2) called the Ganga foreland basin. The

Ganga basin contains over 5 km of Miocene–Pliocene terrigenous sedimentary sequences overlain by late Pleistocene alluvium (DeCelles *et al.*, 2001). The northern part of this basin, which forms the Himalayan foothills, defines a 10- to 25-km-wide physiographic province commonly referred to as the Sub-Himalaya.

Above and to the north of the Main Frontal Thrust is the Main Boundary Thrust (Fig. 10.19). This latter fault system dips gently to the north and appears to have been active mostly during the Pleistocene, although slip on it may have initiated during the Late Miocene–Pliocene (Hodges, 2000). The fault carries Precambrian–Mesozoic low-grade schist and unmetamorphosed sedimentary rock of the Lesser (or Lower) Himalaya southward over the Sub-Himalaya. The Lesser Himalaya form a zone at elevations between about 1500 and 3000 m. Above the Lesser Himalaya, high-grade gneisses and granitic rocks of the Greater (or Higher) Himalaya are carried southward along the Main Central Thrust (DeCelles *et al.*, 2001). This latter thrust accommodated significant shortening during the Early Miocene and Pliocene, and appears inactive in most places today (Hodges, 2000).

The Greater Himalaya, which reach altitudes of over 8000 m, consist of Precambrian gneiss overlain by

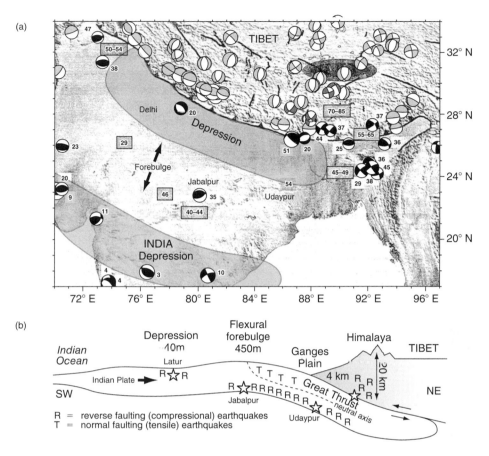

Figure 10.18 (a) Earthquake focal mechanism solutions in India and southern Tibet (modified from Jackson et al., 2004, with permission from the Geological Society of America). Numbers represent depths. Black solutions are from events that occur within the Indian craton, light gray solutions are at depths of 10–15 km. Depths highlighted by a box are Moho depths from receiver function studies. Ellipse in Tibet is the high velocity anomaly imaged by Tilmann et al. (2003) and shown in Plate 9.4 (bottom) (between pages 244 and 245). (b) Schematic cross-section showing flexure of Indian lithosphere as it is underthrust to the north beneath Tibet (modified from Bilham, 2004). At the crest of the flexural bulge the surface of the Indian plate is in tension (T) and its base is in compression (R).

Paleozoic and Mesozoic sedimentary rock of Tethyan origin. These rocks have been thrust southward for a distance exceeding 100 km. The unit includes migmatite and amphibolite grade metamorphic rocks intruded by light-colored granitic bodies of Miocene age called leucogranite (Hodges et al., 1996; Searle et al., 1999). The migmatite and leucogranite have originated by the partial melting of the lower crust beneath Tibet (Le Fort et al., 1987) and are absent north of the Greater Himalaya.

The progressive decrease in the age of thrusting from north to south within the Himalaya defines a fore-land-propagating fold-thrust system. At depth, each of three main thrusts of the system merges downward into a common décollement called the Main Himalayan Thrust (Fig. 10.20). Seismic reflection and velocity profiles (Zho et al., 1991; Nelson et al., 1996) show that the décollement continues beneath the Greater Himalaya where it disappears beneath southern Tibet amid a zone of weak reflectivity thought to represent a zone of partially molten rock (Section 10.4.5).

Bounding the top of the thrust stack at the surface is a system of normal faults that form the South Tibetan Detachment System (Burchfiel et al., 1992). The basal

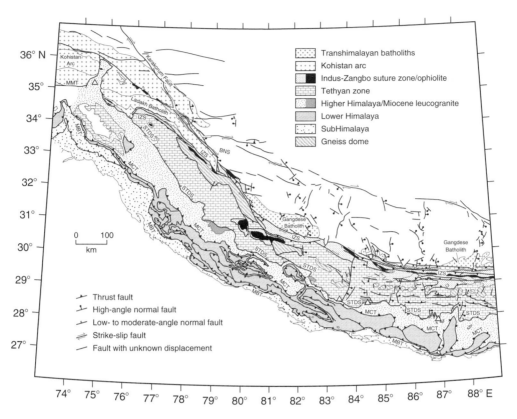

Figure 10.19 *Geologic map of the Himalaya (modified from Hodges, 2000, with permission from the Geological Society of America). BNS, Bangong–Nujiang suture; IZS, Indus–Zangbo suture; MBT, Main Boundary Thrust; MCT, Main Central Thrust; MMT, Main Mantle Thrust; STDS, South Tibetan Detachment System.*

detachment dips gently-moderately to the north and separates the high-grade gneisses of the Greater Himalaya from low-grade Cambrian–Eocene rocks of the Tethyan zone (Fig. 10.19). These latter rocks were deposited on the passive margin of northern India prior to its collision with Eurasia. The basal detachment records Miocene and, possibly, Pliocene north-directed normal displacements of at least 35–40 km that occurred contemporaneously with south-directed motion on the Main Central Thrust (Hodges, 2000). In its hanging wall, Tethyan rocks are dissected by complex arrays of splay faults (Fig. 10.19) whose cumulative displacement probably approaches that of the basal detachment (Searle, 1999).

Throughout the Tethyan zone are a discontinuous series of metamorphic culminations called gneiss domes. The most extensively studied of these is the Kangmar gneiss dome, which forms part of an antiform

cored by Precambrian metamorphic basement surrounded by a mantle of less metamorphosed Carboniferous–Triassic rocks (Burg *et al.*, 1984). A few of the largest gneiss domes preserve eclogite-facies metamorphic assemblages that are overprinted by amphibolite-facies assemblages (Guillot *et al.*, 1997). The domes are dissected by normal faults and bear some resemblance to the extensional metamorphic core complexes in the western USA and elsewhere (Section 7.3). However, their origin is not well understood and several different mechanisms have been proposed to explain them, including thrust faulting and folding in addition to normal faulting and lower crustal flow.

At the northern end of the Tethyan Zone the Indus–Zangbo suture separates rocks that once formed part of the Indian Plate from Paleozoic–Mesozoic rocks of the Lhasa terrane (Section 10.4.2). The suture is defined by a deformed mixture of components derived from

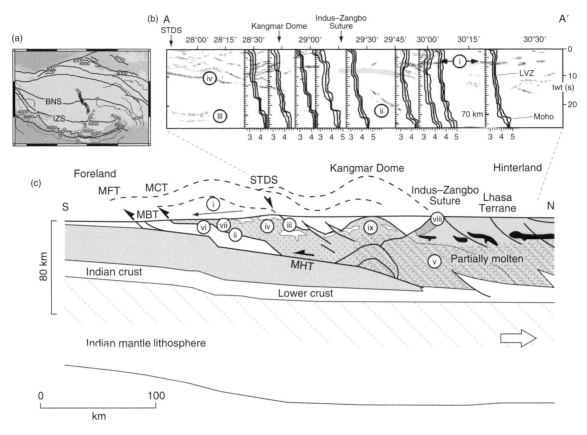

Figure 10.20 *(a) Map showing location of INDEPTH profiles (modified from Xie et al., 2001, by permission of the American Geophysical Union. Copyright © 2001 American Geophysical Union). Right-way-up triangles, INDEPTH I, II; upside-down triangles, INDEPTH III. (b) Composite of INDEPTH seismic information, including S-wave velocity models derived from waveform modeling of broad-band earthquake data (error bars shown around central profiles) and wide-angle reflection data beneath and north of the Indus–Zangbo suture (modified from Nelson et al., 1996, Science **274**, 1684–8, with permission from the AAAS). LVZ, midcrustal low velocity zone; STDS, South Tibetan Detachment; i, reflections interpreted to represent fluids at 15–20 km depth; ii, steep reflection in the lower crust interpreted to represent thrusting; iii, Moho at 75 km depth; iv, fault that accommodates underthrusting of India beneath Tibet. (c) Interpretive cross-section of the central Himalaya and southern Tibet (section provided by C. Beaumont and modified from the compilation of Beaumont et al., 2004, by permission of the American Geophysical Union. Copyright © 2004 American Geophysical Union). Section incorporates observations from Nelson et al. (1996), Hauck et al. (1998), and DeCelles et al. (2002). Roman numerals are explained in Section 10.4.6. MFT, Main Frontal Thrust; MHT, Main Himalayan Thrust. Other abbreviations as in Fig. 10.19.*

both the Indian and Eurasia plates, as well as Tethyan ophiolites and blueschist (Section 9.9). The ophiolites are not continuous, and in places are replaced by sediment deposited in a forearc environment. South-dipping thrusts and strike-slip faults deform these rock units. North of the suture, the Paleozoic–Mesozoic sedimentary rocks that form most of southern Tibet are intruded

by the Cretaceous–Eocene Gangdese batholith of the Transhimalayan zone (Fig. 10.19). This batholith formed along an ocean–continent convergent plate margin in response to northwards underthrusting of Tethyan oceanic lithosphere prior to the India–Eurasia collision (Fig. 10.15c,d). In the western Himalaya the equivalent unit is an island arc that formed within the Tethys

Ocean in mid-Cretaceous times. The Bangong–Nujiang suture separates this unit from the Karakorum granite batholith on its northern side (Fig. 10.19).

North of the Indus–Zangbo suture, active normal faulting and east–west extension are dominant. This style of deformation has formed a series of young rift basins that trend approximately north–south. At most these basins record extension of a few tens of kilometers. Most are filled with Pliocene and younger conglomerates and appear to have formed since the Miocene. Some are associated with major strike-slip faults, such as the Jiali Fault, and may represent pull-apart basins (Section 8.2). These observations and geochronologic data suggest that the east–west extension is either younger than or outlasted the north–south extension recorded by the South Tibetan Detachment System (Harrison et al., 1995). Late Cenozoic intrusive and extrusive activity also occurs in southern and central Tibet (Chung et al., 2005).

Between the Bangong–Nujiang suture and the Qaidam Basin (Fig. 10.13) are three major mid–late Cenozoic fold-thrust belts. All three are associated with the development of a foreland basin (Yin & Harrison, 2000). The cumulative amount of shortening accommodated by these belts is poorly constrained but may reach several hundreds of kilometers. At the northern margin of Tibet deformation is partitioned between active folding and thrusting and several major active strike-slip faults, including the Altyn Tagh and Kunlun faults. Along the former fault, left-lateral strike-slip motion is transferred to as much as 270 km of northeast–southwest shortening in the Qilian Shan. Farther east and southeast of the Qilian Shan, the shortening direction turns east–west where motion on east-striking strike-slip faults is transferred onto active north-striking thrust faults in the Longmen Shan (Burchfiel, 2004). This latter mountain range also records Mesozoic shortening and rises more than 6 km above the rigid, virtually undeformed Sichuan Basin, forming one of the steepest fronts along the Tibetan Plateau (Clark & Royden, 2000). To the south of the basin, many of the major strike-slip faults, including the Jiali and Xianshuihe faults, are curved. These faults rotate clockwise around the eastern Himalayan syntaxis relative to South China (Wang et al., 1998).

North of the plateau active shortening also occurs in the Tien Shan and the Altai ranges of northern China and Mongolia. The deformation in these regions appears to be controlled mostly by pre-existing strength heterogeneities in the Eurasian lithosphere.

10.4.5 Deep structure

Velocity models and tomographic images derived from studies of Rayleigh surface waves (Section 2.1.3) show that the crust and uppermost mantle of the Indian Shield are characterized by high seismic velocities (Mitra et al., 2006). This characteristic suggests that the subcontinent is composed of relatively cool, strong lithosphere. At the northern edge of the Shield, comparatively low velocities occur beneath the Gangetic plains as a result of the molasse sediments and alluvial cover in the Ganga foredeep. Low velocities also characterize the thick crust beneath the Himalaya and Tibetan Plateau. South of the Himalaya broad-band teleseismic data indicate that crustal thickness ranges from 35 to 44 km (Mitra et al., 2005). This variability partially reflects the flexure of the Indian plate (Fig. 10.18), as it is underthrust to the north beneath Eurasia (Section 10.4.3).

Below the Himalaya, seismic reflection and shear wave velocity profiles (Fig. 10.20b) show a well-defined Moho at 45 km depth that descends as a single smooth surface to depths of 70–80 km beneath southern Tibet (Nelson et al., 1996; Schulte-Pelkum et al., 2005). A crustal décollement surface above the Moho dips northward from 8 km below the Sub-Himalaya to a mid-crustal depth of 20 km beneath the Greater Himalaya. Above the décollement a strongly (20%) anisotropic layer characterized by fast seismic velocities has formed in response to localized shearing. Slightly north of the Greater Himalaya the lower crust of the Indian Shield shows a high velocity region that may contain eclogite. These observations suggest that the upper and middle parts of the Indian crust detach along the base of the shear zone and are incorporated into the Himalaya while the lower crust continues its descent under southern Tibet (Fig. 10.21). This conclusion is consistent with gravity measurements that predict an increase in Moho depth beneath the Greater Himalaya (Cattin et al., 2001).

The deep structure of the Tibetan Plateau has been studied using passive and active source seismic surveys, magnetotelluric measurements, and surface geologic studies as part of an interdisciplinary project called INDEPTH (InterNational DEep Profiling of Tibet and the Himalaya). The geophysical data indicate that the reflection Moho beneath Tibet is rather diffuse (Fig. 10.20b), similar to that observed on seismic reflection profiles across the plateaux of the Central

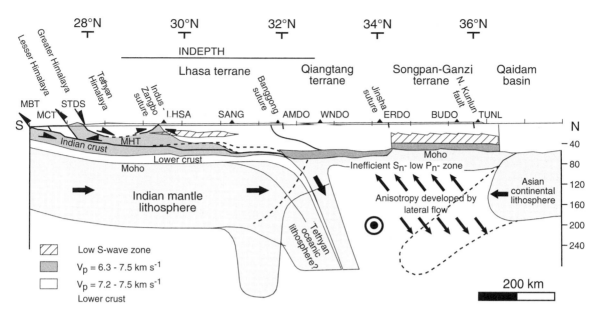

Figure 10.21 *Representative north–south cross-section of the Himalayan–Tibetan orogen (image provided by C. Beaumont and modified from compilation of Beaumont* et al., *2004, by permission of the American Geophysical Union. Copyright © 2004 American Geophysical Union). Section incorporates observations from Owens & Zandt (1997), DeCelles* et al. *(2002), Johnson (2002), Tilmann* et al. *(2003), Haines* et al. *(2003). Triangles denote seven seismic recording stations. (LHASA, SANG, AMDO, WNDO, ERDO, BUDO, TUNL). Fault abbreviations as in Figs 10.19 and 10.20. Bulls-eye symbol denotes lateral movement of material out of the plane of the page.*

Andes (Fig. 10.6). Like the Central Andes, the southern part of Tibet is characterized by low velocity zones in the crust and bands of bright intra-crustal reflections at 15–20 km depth that result from either a concentration of aqueous fluids or the presence of partial melt (Nelson *et al.*, 1996; Makovsky & Klemperer, 1999). Low values of Q (~90) in this region are consistent with abnormally high temperatures as well as partially molten crust (Xie *et al.*, 2004). Magnetotelluric data, which measure subsurface electrical resistivity (see also Section 8.6.3), are particular sensitive to the presence of interconnected fluids in a rock matrix. Unsworth *et al.* (2005) found low resistivity along at least 1000 km of the southern margin of the Tibetan Plateau, suggesting that it is characterized by weak, low viscosity crust. This weak zone is confined on its southern side by the faulted Indian crust of the Greater Himalaya and is underlain by stiff Indian mantle (Rapine *et al.*, 2003).

In central Tibet, teleseismic data and receiver functions provide information on the crustal structure and mechanisms of deformation below the Bangong–Nujiang suture (Ozacar & Zandt, 2004). Strong (>10%)

seismic anisotropy in the upper crust shows a fabric that trends WNW–ESE parallel to both the suture and younger strike-slip faults. Seismic anisotropy (18%) also occurs at 24–32 km depth in the middle crust. This latter zone shows a near horizontal and gently dipping fabric that suggests mid-crustal flow in a north–south direction. The seismic properties of the lower crust and upper mantle also change across the suture (McNamara *et al.*, 1997; Huang W. *et al.*, 2000), although the suture itself has little geophysical expression (Haines *et al.*, 2003). In the lower crust, some north-dipping reflections may represent ductile thrust slices or wedges (Fig. 10.20c) and active-source seismic data show that the Moho shallows by up to 5 km on the northern side of the boundary (Haines *et al.*, 2003). These observations, and the relatively small amount of shortening recorded in the upper crust of Tibet, suggest that the upper crust is mechanically decoupled from the underlying layers across a weak ductilely flowing middle crust.

Surface wave studies (Curtis & Woodhouse, 1997) and P_n and S_n wave observations (McNamara *et al.*,

1997; Zho et al., 2001) of the upper mantle indicate that fast mantle velocities occur beneath southern Tibet and slow mantle velocities occur north of the Bangong–Nujiang suture (Fig. 10.21). These differences suggest the presence of cold, strong mantle beneath southern Tibet and anomalously warm, weak mantle beneath central and northern Tibet. The pattern may indicate that Indian lithosphere has been underthrust to at least a point beneath the center of the Tibetan Plateau. However, this interpretation is in conflict with estimates of the total amount of convergence and shortening of the lithosphere since the collision began. Estimates of the total convergence (~2000 km) derived from magnetic anomalies, paleomagnetic studies, and estimates of the minimum amount of post-collisional shortening (Johnson, 2002) suggest that cold Indian lithosphere also may occur beneath northern Tibet.

High resolution tomographic images of the upper mantle may help to resolve this discrepancy. Tilmann et al. (2003) interpreted the presence of a subvertical, high velocity zone located south of the Bangong–Nujiang suture between 100 km and 400 km depth (Plate 9.4(bottom) between pp. 244 and 245). This subvertical zone may represent downwelling Indian mantle lithosphere. The additional Indian lithosphere helps account for the total amount of shortening in the Himalayas and Tibet. The downwelling also may explain the presence of warm mantle beneath northern and central Tibet, which would flow upwards to counterbalance a deficit in asthenosphere caused by the downwelling. The occurrence of calc-alkaline-type volcanic rocks in southern and central Tibet may support this interpretation by requiring a portion of continental crust to have been underthrust into the mantle beneath Tibet from the north and south (Yin & Harrison, 2000). Nevertheless, the mechanisms by which Indian lithosphere shortens and is underthrust beneath Tibet remain controversial.

At the northern and northwestern margin of Tibet, the Moho abruptly shallows to depths of 50–60 km across the Altyn Tagh Fault and beneath the Tarim Basin (Wittlinger et al., 2004). The Moho also appears to shallow across the Jinsha suture beneath the Songpan–Ganzi terrane (Fig. 10.21). From the receiver functions it is impossible to distinguish whether the Moho is part of Indian or Eurasian lithosphere. Relatively thick (60 km) crust occurs beneath the Tien Shan and gradually thins to the north to an average of 42 km beneath the Shield of Eurasia (Bump & Sheehan, 1998). The thick crust beneath the Tien Shan is consistent with evidence of crustal shortening in this region (Section 10.4.3).

10.4.6 Mechanisms of continental collision

Like all other major zones of continental deformation (e.g. Sections 7.6, 8.6, 10.2.5), the evolution of collisional orogens is governed by the balance among regional and local forces, the strength and rheology of the continental lithosphere, and by processes that change these parameters over time. To determine how interactions among these factors control the development of the Himalayan–Tibetan orogen, geoscientists have developed physical and analogue models of continental collision. This section provides a discussion of the main results and different approaches used in this field of study.

1 *Precollisional history.* The strength and rheology of the continental lithosphere at the start of continental collision is governed by the pre-collisional history of the two colliding plates. In the case of the Himalayan–Tibetan orogen, millions of years of subduction, arc magmatism, terrane accretion, and crustal thickening along the southern margin of Eurasia (Section 10.4.2) weakened the lithosphere. During the India–Eurasia collision, the many suture zones, thick flysch sequences, and other weak zones that characterized Eurasia allowed deformation to extend deep into the interior of the continent (Yin & Harrison, 2000; Tapponnier et al., 2001).

Unlike Eurasia, the relatively cool and deeply rooted Precambrian shield of India resulted in a relatively strong plate that resisted shortening during collision. The generally high mechanical strength and high elastic thickness of the Indian lithosphere led to its underthrusting beneath southern Tibet (Section 10.4.3). An exception to its generally high strength is the sediment that was deposited on the passive continental margin of northern India from the Early Proterozoic to Paleocene. During collision, these weak sequences failed and were scraped off the downgoing plate, forming the Himalayan fold and thrust belt.

2 *Continental underthrusting.* The underthrusting, or subduction, of continental lithosphere beneath another continental plate is one of the most important mechanisms that accommodates convergence in zones of continental collision. The rheology of the two plates and the degree of mechanical coupling between them control shortening and the evolution of stresses within the overriding plate. In the Himalayan–Tibetan orogen, the underthrusting of Indian continental lithosphere drives intra-plate shortening at the leading edge of the Indian plate and in Tibet, and, possibly, also farther north into Asia. The resultant shortening has generated crust that is up to 70–80 km thick (Section 10.4.5) and has contributed to the uplift and growth of the Tibetan Plateau. Like its counterpart in the central Andes (Section 10.2.4), the plateau is associated with high crustal temperatures and widespread intra-crustal melting that have weakened the crust sufficiently to allow it to flow. This process has decoupled the Tibetan crust from the underlying convergent motions and has altered the dynamics of the orogen. Although geophysical observations show that Indian lithosphere is underthrust to at least a point beneath central Tibet, interpretations differ on how this process is accommodated (Dewey *et al.*, 1989; Yin & Harrison, 2000; Johnson, 2002). The main problem is that the underthrusting requires the removal or displacement of Asian lithosphere from under Tibet (Section 10.4.5). Several mechanisms may alleviate this problem, including the downturning of Indian mantle lithosphere beneath the Bangong–Nujiang suture (Figs 10.21, Plate 9.4(bottom) (between pp. 244 and 245), the convective removal or delamination of the lithospheric mantle beneath Tibet (England & Houseman, 1988; Molnar *et al.*, 1993), the southward subduction of Asian mantle (Willett & Beaumont, 1994), and the removal of Asian mantle by strike-slip faulting during the lateral escape of Tibet (Section 10.4.3).

Although the role of these various processes remains uncertain, it seems likely that a combination of mechanisms accommodates shortening beneath Tibet.

3 *Indentation, lateral escape, and gravitational collapse.* A comparison between the total amount of convergence between India and Eurasia since they collided and estimates of the total amount of shortening accommodated by fold-thrust belts in the orogen has yielded a shortening deficit ranging anywhere from 500 km to over 1200 km (Dewey *et al.*, 1989; Johnson, 2002). This deficit has led to numerous attempts to explain how the convergence not accounted for by folding and thrusting has been accommodated. A leading hypothesis involves the indentation of India into Asia and the lateral escape of eastern Tibet (Section 10.4.3).

Indentation is the process by which a rigid block presses into and deforms a softer block during convergence. The theory of indentation originally was developed by mechanical engineers to predict the configuration of lines of maximum shear stress, or slip lines, in deforming plastic materials. In geologic applications, the slip lines correspond to dextral and sinistral strike-slip faults whose pattern is controlled by the shape of the indenter and by lateral constraints placed on the plastic medium (Tapponnier & Molnar, 1976; Tapponnier *et al.*, 1982).

In one pioneering application, Tapponnier *et al.* (1982) explored the effects of indentation as a rigid 50-mm-wide block (India) penetrates into a softer block (Asia) made of laminated plasticine. Figure 10.22 shows two evolutionary sequences where the plasticine is either bilaterally confined at the two edges parallel to the motion of the indenter (Fig. 10.22a–c) or unilaterally confined at only one of these edges (Fig. 10.22d–f). The bilaterally confined case produces a symmetric pattern of slip lines ahead of a "dead triangle" that rapidly welds to the indenter. The penetration proceeds by the creation of numerous, short-lived, dextral and sinistral faults near the triangle's apex. The unilateral

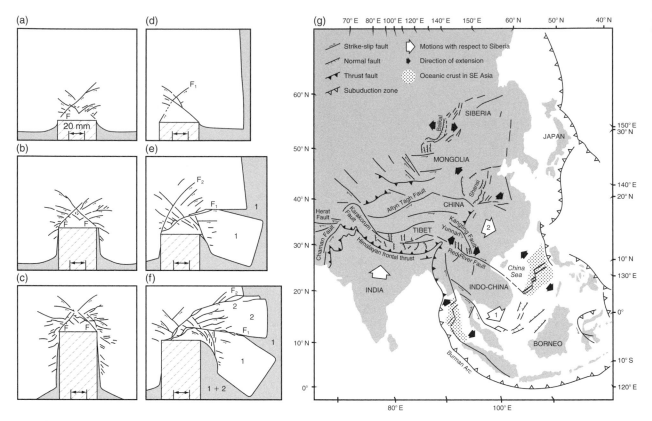

Figure 10.22 *(a–f) Indentation experiments on Plasticine and (g) schematic map illustrating extrusion tectonics in eastern Asia (redrawn from Tapponnier et al., 1982, with permission from the Geological Society of America). F, major fault. Numbers associated with arrows in (g) are extrusion phases: 1 ~ 50–20 Ma; 2 ~ 20–0 Ma.*

case generates an asymmetric pattern where faults that allow displacement towards the free edge predominate, such as F_1. The block translated sideways rotates about 25° clockwise and is followed by the extrusion of a second block along another sinistral fault, F_2, which allows a continued rotation of the first block by up to about 40°. Pull-apart basins (Section 8.2) develop along the sinistral faults because of their irregular geometry. As these movements progress, a gap grows between the indenter and extruded plasticine. Tapponnier *et al.* (1982) suggested that these results explain the dominance of sinistral offsets in China (Fig. 10.22g). The pull-apart structures may be analogous to the extensional regimes in Shansi, Mongolia, and Baikal. The Altyn Tagh Fault may correlate with the major dislocation F_2, and the Red River Fault with F_1. The comparison also suggests that indentation causes the curvature of fault systems located east of Tibet. Finally, lateral extrusion between and to the southeast of the Altyn Tagh and Red River faults results in extension that resembles patterns observed in the South China Sea and the Gulf of Thailand (Fig. 10.22g).

Since its development in the late 1970s and early 1980s, the indentation model of continental collision has evolved considerably. Although the model of Tapponnier *et al.*

(1982) explains the general pattern and distribution of strike-slip faulting in eastern Tibet and southeast Asia, it has been less successful at explaining other aspects of the deformation. One problem is that it predicts lateral displacements of hundreds to a thousand kilometers on the large strike-slip faults within Tibet. However, estimates of the magnitude of slip on major strike-slip faults, so far, have failed to confirm the extremely large magnitudes of displacement. The Altyn Tagh Fault, for example, may only have 200 km of left-lateral slip and the Xianshuihe Fault about 50 km (Yin & Harrison, 2000). These observations suggest that while lateral escape is occurring, it may occur on a smaller scale than originally predicted.

Another problem with the application shown in Fig. 10.22 is that it does not predict, or take into account, the effects of variations in crustal thickness during deformation. In addition, the region of east–west extension and normal faulting in Tibet has no analogue in the model. One possible explanation for the extension is that it results from gravitational buoyancy forces associated with the great thickness and high elevations of the plateau. In this view, an excess in gravitational potential energy enhanced by the presence of a buoyant crustal root and the possible removal of mantle lithosphere by convective erosion or delamination (see also Section 10.2.5) drives the *gravitational collapse* of the overthickened crust (Dewey, 1988; England & Houseman, 1989). Lateral gradients in gravitational potential energy may help the plateau spread out and move laterally toward the eastern lowlands where it interacts with other lithospheric elements. Whereas other origins for this extension also have been proposed, quantitative considerations of these forces suggest that the evolution of the plateau depends as much upon buoyancy forces and local boundary conditions as it does on indentation or stresses arising at the edges of the Indian and Eurasian plates (Royden, 1996; Liu & Yang, 2003). The gravitational collapse of over-thickened continental crust also explains the evolution

of orogens after convergence stops where, in many areas, it has been linked to the formation of extensional metamorphic core complexes (Section 7.3).

To account for these effects, investigators have simulated the deformation of Asia using a viscous sheet that deforms in response to both the edge forces arising from continental collision and the internal forces generated by differences in crustal thickness (England & Houseman, 1989; Robl & Stüwe, 2005a). Rather than modeling displacements on individual faults, these continuum models simulate deformation as a zone of distributed flow between two colliding plates. Most predict that a zone of shortening and thickening crust grows in front of and, with the appropriate boundary conditions, to the side of an advancing indenter. The results suggest that the zone of deformation directly related to the India–Eurasia collision is much smaller than that predicted by the plasticine models and that other regions of deformation in Southeast Asia are unrelated to the local and tectonic forces arising from the collision. Instead, deformation in these latter regions may result from regional tectonic stress fields related to the plate boundaries located south and east of Asia.

Continuum models of indentation, in general, have been successful at explaining the asymmetry of deformation in Asia, including the lateral escape of eastern Tibet. They also are well suited for examining the effects of variations in lithospheric strength and rheology on the style of deformation observed in India and Asia. Robl & Stüwe (2005a, 2005b), for example, explored the effects of variations in the shape, convergence angle, and rheology of a continental indenter on both lateral and vertical strain patterns in Asia during lateral escape. These authors investigated the sensitivity of a deforming viscous sheet to indentation involving combinations of these parameters. An especially interesting aspect of their application is the investigation of how buoyancy forces arising from crustal thickening are balanced by edge forces from indentation.

In the experiments of Robl & Stüwe (2005a) Asia is modeled as a viscous sheet consisting of a regular square mesh with 3200 triangles (Fig. 10.23a). The eastern, western, and northern boundaries of the mesh are rigid and cannot move. These constraints simulate the effects of the Tarim Basin to the north and Pamir to the west. At the southern boundary an indenter of width ($D/2$) and length (ω) moves northward into the mesh with a velocity scaled to be 50 mm a^{-1}. This motion results in deformation and thickening that is distributed between the indenter and the foreland to the north (Fig. 10.23b). Two important variables include the viscosity contrast (η) between the indenter and the foreland and the angle (α) between the indenter front and the direction of indentation. All materials are described using a power law rheology with exponent (n), which describes how strain rates are related to stress (Section 2.10.3).

The effect of indenter shape on the distribution of deformation is best illustrated in simulations where the indenter is strong. Viscosity contrasts of $\eta = 1000$ and $\eta = 100$ simulate this condition. The results show that for an indentation angle of $\alpha = 45°$ and a strong, viscous indenter, deformation localizes along the interface between the colliding blocks and the horizontal velocity field is highly asymmetric. Crustal thickening is at a maximum north of the western tip and slightly less in front of the northeastern edge (Fig. 10.24a,b). A band of eastward-moving material develops on the northeast side of the indenter (Fig. 10.25a). These asymmetries contrast with the symmetric patterns that surround rectangular indenters with high viscosities (Figs 10.23b, 10.25b). For a low viscosity indenter ($\eta = 2$ or 3), the indenter angle plays only a minor role. In these latter cases, the indenter accommodates most of the shortening and thickening, with the pattern becoming progressively more symmetric and delocalized through time (Fig. 10.24c,d). Figure 10.25c and d show that the horizontal velocity field for a rectangular

indenter is similar to that with an indenter angle of $\alpha = 45°$. These results indicate that, for the given set of boundary

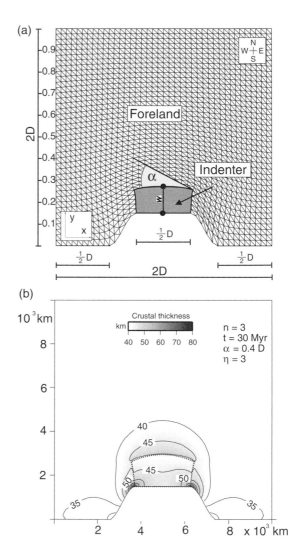

Figure 10.23 *Geometry and boundary conditions of a finite element model of indention (image provided by J. Robl and modified from Robl & Stüwe, 2005a, by permission of the American Geophysical Union. Copyright © 2005 American Geophysical Union). (a) Regular mesh consisting of 3200 triangles. Shaded region is the indenter and light region is the foreland. (b) Typical model result scaled for a length scale of D = 5000 km and an indentation velocity of 50 mm a^{-1}. Gray scale indicates crustal thickening distributed between indenter and foreland.*

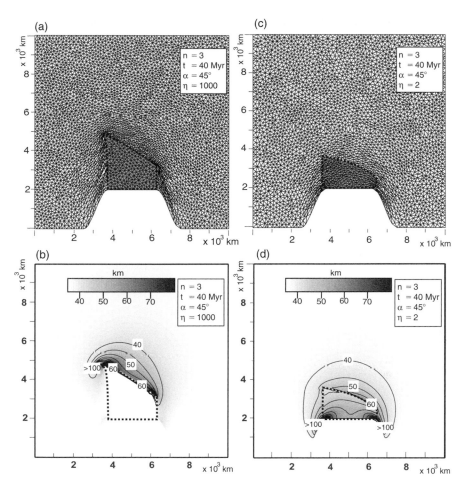

Figure 10.24 *Finite element model showing the influence of viscosity contrast on the evolution of oblique indenters (image provided by J. Robl and modified from Robl & Stüwe, 2005a, by permission of the American Geophysical Union. Copyright © 2005 American Geophysical Union). In both model runs, the geometry was identical, and n = 3 and α = 45°. Bold dotted line is the outline of the indenter. (a,b) Viscosity contrast η = 1000. (c), (d) Viscosity contrast η = 2. (a) and (c) show finite element mesh after 40 Myr, (b) and (d) show corresponding diagrams contoured for crustal thickness.*

conditions, lateral escape of the crust increases with indenter angle for relatively strong indenter rheologies and simulates the patterns of displacement observed in eastern Tibet.

In situations where Asian lithosphere is especially viscous and strong, lateral escape results mainly from horizontal compression as blocks move out of the way of the rigid indenter. In these cases, buoyancy forces arising from crustal thickening contribute little to the horizontal velocity field because

the thickening tends to be either highly localized or inhibited by the high strength of the material. As the strength of Asia decreases, the magnitude and distribution of crustal thickening increase and gravitational buoyancy forces become increasingly important. The numerical simulations of Robl & Stüwe (2005a) and others (Liu & Yang, 2003) suggest that buoyancy forces developing in weak thick crust such as that in Tibet enhance the rate of lateral escape.

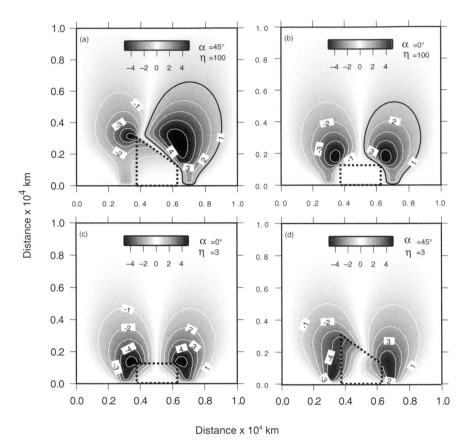

Figure 10.25 *Results from finite element modeling showing the instantaneous lateral displacement field during indentation for two different indentor rheologies and two indentor angles (α) (image provided by J. Robl and modified from Robl & Stüwe, 2005a, by permission of the American Geophysical Union. Copyright © 2005 American Geophysical Union). (a) and (b) show a viscosity contrast between indentor and foreland of η = 100; (c) and (d) show a contrast of η = 3. Indentor angles of α = 0° and α = 45° are represented. Grayscale bar shows horizontal velocity. Contour interval is 1 mm a⁻¹. Eastward moving area is largest for the oblique indentor shown in (a).*

A three-dimensional viscoelastic model developed by Liu & Yang (2003) illustrates how various possible driving forces and a rheological structure involving both vertical and lateral variations influence deformation patterns in Tibet and its surrounding regions. This model, like most others, involves a rigid Indian plate that collides with a deformable Eurasian continent at a constant velocity relative to Eurasia. The two plates are coupled across a fault zone that simulates the Main Boundary Thrust (Fig. 10.26a). On the eastern and southeastern sides of the model,

boundary conditions are assigned to simulate the lateral escape of the crust. On the west, the effects of a spring or roller simulate the lateral resisting force of a rigid block in Pamir. On the northern side of the model boundary conditions approximate the resistance to motion by the rigid Tarim Basin. The top surface approximates the real topography and the bottom lies at 70 km depth. A vertical topographic load is included by calculating the weight of rock columns in each surface grid of the finite element model. An isostatic restoring force is applied

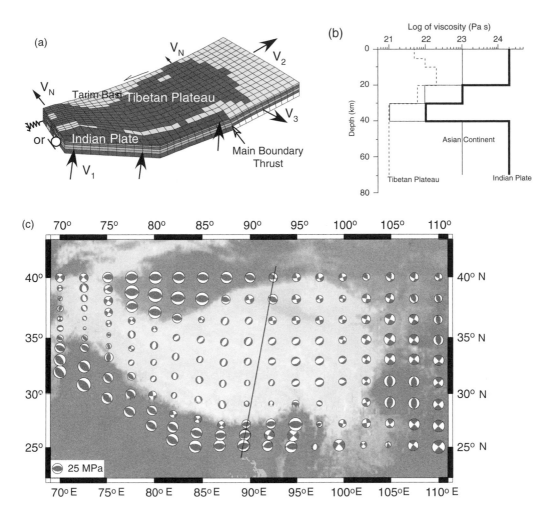

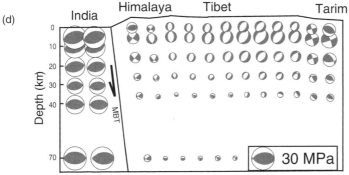

Figure 10.26 *(a,b) Initial conditions and (c,d) results of a finite element model of the Himalayan–Tibetan orogen (images provided by Y. Yang and M. Liu and modified from Liu & Yang, 2004, by permission of the American Geophysical Union. Copyright © 2004 American Geophysical Union). The Main Boundary Thrust is simulated by a weak zone that is adjusted to reflect the degree of mechanical coupling between the Indian plate and the Eurasian continent. Effective viscosity profiles in (b) correspond to the Tibetan Plateau (dashed line), the Indian plate (thick line), and the rest of the Eurasian continent (thin line). (c) Shows the predicted stresses at 10 km depth. Each symbol is a lower hemisphere stereonet projection of the three-dimensional stress state, similar to that used for earthquake focal mechanisms. (d) Shows variation in predicted stresses with depth. The depth variation is related to the rheological model used in (b). Scale shown in lower right.*

to the bottom of the model. Unlike many other models, this experiment also incorporates lateral variations in rheology using three crustal blocks, including a stiff Indian plate, a weak Tibetan Plateau, and an intermediate-strength Asian continent north of the plateau. In addition, six layers of material with different effective viscosities represent vertical variations in the rheology of these blocks (Fig. 10.26b).

Within this framework, Liu & Yang (2003) considered that combinations of the following forces contribute to the present state of stresses in the Himalaya and Tibetan Plateau: (i) a horizontal compressive force resulting from the collision of India with Asia; (ii) buoyancy forces resulting from isostatically compensated topography; (iii) basal shear on the Eurasia plate as India slides beneath Tibet; and (iv) horizontal forces originating from the pull of subduction zones located south and east of Asia. The stress field is constrained by GPS data, earthquake focal mechanisms, topography, and other observations. Figure 10.26c and d show the predicted stresses in the upper crust (at 10 km depth) using the velocity boundary conditions based on geodetic data: a uniform convergence rate of 44 mm a^{-1} toward N20°E at the Himalayan front (V$_1$), 7 mm a^{-1} to the east on the east side (V$_2$), and 10 mm a^{-1} to the southeast on the southeast side of the model (V$_3$). A velocity of 20 mm a^{-1} to the north (V$_N$) occurs at the western side of the model and decreases to zero on the eastern side. Higher convergent rates lead to enhanced mechanical coupling between the Eurasian and Indian plates, although this effect can be offset by a Main Boundary Thrust Zone that is mechanically weak.

The model results suggest that the surface velocity field and regime of deformation in the orogen (Section 10.4.3) reflect a mechanical balance between gravitational buoyancy, the indenting Indian plate, and the specific geometry and the boundary conditions of the plateau. Crustal thickening and topographic uplift are enhanced by the presence of the Tarim Basin, which acts as a rigid back-stop at the northern end of the model. To obtain the observed east–west extension and the high elevations of Tibet, the Tibetan crust must be very weak. The model suggests that the force balance evolves through time as the crust deforms and thickens. When the plateau is 50% lower than its present elevation of nearly 5 km, strike-slip and reverse faulting dominate the plateau region. Significant crustal extension occurs when the plateau reaches 75% of its present elevation. The model also suggests that although far field extensional forces may enhance the collapse of the plateau they are not required. Basal shear also enhances the extensional regime in the Himalaya and southern Tibet while increasing shortening in northern Tibet. This latter effect results because basal shear relieves the compressive (indentation) stresses that balance the buoyancy forces driving extension at the southern edge of Tibet. This leads to a decrease in compressive stress in the upper crust, which enhances extension. North of the Indus–Zangbo suture, the basal shear adds to the horizontal compression, resulting in increased shortening.

4 *Lower crustal flow and ductile extrusion.* The simple numerical and analogue experiments of indentation described above illustrate the sensitivity of deformation in collisional belts to local boundary conditions and variations in lithospheric rheology. A particularly interesting group of numerical experiments has explored the effects of weak, flowing middle and lower crust on the dynamics of continental collision. This condition of weak crust is in good agreement with geologic and geophysical observations indicating that the middle crust beneath Tibet is hot, fluid-rich, and/or partially molten (Section 10.4.5).

Royden (1996) and Ellis *et al.* (1998) showed that a vertical stratification of the lithosphere into strong and weak layers influences the degree of strain localization during convergence. Where the lower crust is relatively strong and resists flow, the crust tends to couple to the underlying mantle during shortening and results in a relatively narrow zone of localized strain at the surface. This effect may explain the relatively narrow width, triangular shape, and lack of a high orogenic plateau in the Eastern Alps and the Southern Alps of New Zealand. By contrast, where the lower crust is relatively weak and flows easily, the crust decouples from the mantle and results in diffuse strain. This latter effect may apply to Tibet and the central Andes where low viscosity zones have developed in the deep crust during crustal thickening and wide, steep-sided plateaux have formed above the weak zones (Sections 10.2.4, 10.4.5).

A vertical decoupling of the lithosphere as a result of ductile flow in a weak lower crust is well illustrated along the northern and eastern margins of Tibet. In these regions balanced cross-sections show that thrust faults sole out into décollement surfaces in the middle crust (Yin & Harrison, 2000). A comparison of geodetic data (Fig. 10.16a) and geologic observations has indicated that the lateral motion of the crust in the Longmen Shan region of eastern Tibet is mostly accommodated by lower crustal flow with little faulting occurring at the surface (Burchfiel, 2004). Northwest of the Sichuan Basin topography is anomalously high compared to the rest of Tibet. Clark & Royden (2000) and Clark et al. (2005) explained these relationships as a result of dynamic pressure resulting from the lateral flow of a partially molten lower crust as it encounters the strong crust and upper mantle of the Sichuan Basin. At the western margin of the basin, the flowing lower crust diverts northeastward along a rheologically weak crustal corridor that coincides with the Paleozoic–Mesozoic Qinling suture. The response of the upper crust to this flow may include dynamic uplift and strike-slip faulting, resulting in the anomalously high topography of eastern Tibet compared to its central and southern sectors.

The strain-softening effects of continental underthrusting coupled with enhanced surface erosion also can result in strain localization that alters the dynamics of orogenesis. An excellent example of this process occurs in the Southern Alps of New Zealand (Section 8.6.3). Strain-softening feedbacks also have contributed significantly to the tectonic evolution of the Himalayan fold and thrust belt and southern Tibet where Indian lithosphere is underthrust to the north beneath Eurasia. Hodges (2000) summarized nine principal geologic and tectonic features of this relatively narrow zone that require explanation in any quantitative model of the orogen. These features include (Fig. 10.20c): (i) rapid erosion of the southern flank of the Himalaya; (ii) shortening on the Main Central Thrust (MCT) system and thrust faults to the south; (iii) extension on the South Tibetan Detachment (STD) system; (iv) high-grade metamorphism and crustal melting in the Greater Himalaya; (v) crustal melting in the middle crust beneath Tibet; (vi) juxtaposition of contrasting lithologies across the MCT; (vii) an inverted metamorphic sequence where high-grade rocks are thrust over the Lesser Himalaya along the MCT, (viii) the position of the Indus–Zangbo suture; and (ix) normal faults accommodating north–south extension in the southern Tibetan Plateau.

To determine how enhanced erosion coupled with continental underthrusting may explain these principle features, Beaumont et al. (2001, 2004) constructed thermomechanical models involving combinations of two related processes. The first process is a channel flow of ductile middle to lower crust. Channel flow involves the lateral movement of partially molten crust in a narrow zone bounded above and below by shear zones. These authors used this type of flow to explain the progressive growth of the Tibetan Plateau. The second process is the ductile extrusion of high-grade metamorphic rocks between coeval normal-sense and thrust-sense shear zones. This latter process is used to explain the exhumation of the Greater Himalaya rocks along the southern flank of the mountain range. In the models these two processes are linked through the effects of surface denudation (i.e. the removal of surface material) that is focused along the southern edge of a plateau and the presence of low viscosity, partially molten crust beneath Tibet. Variations in crustal thickness between the high plateau and the Ganga foreland, the rate of denudation and upper crustal strength also affect the style of the deformation. The models are relatively insensitive to channel heterogeneities and to variations in the behavior of the mantle lithosphere beneath the modeled plateau.

The thermomechanical models of Beaumont *et al.* (2004) consist of a vertical plane divided into crust and mantle layers (Fig. 10.27a). A passive marker grid and numbered vertical markers track the progressive deformation of the model during convergence. The suture (S) marks the position where Indian lithosphere is subducted beneath Eurasia and descends into the mantle at a constant velocity (V_p) and a constant dip angle (θ). This point is allowed to migrate during convergence. The basal velocity condition drives flow in the upper plate. The crust consists of upper and middle quartzo-felspathic layers overlying a dry granulitic lower crust that are modeled using a viscous-plastic power-law rheology. The initial thermal structure (Fig. 10.27b) shows two radioactive layers (A_1, A_2) that provide internal heat to the crust. The lithosphere-asthenosphere boundary is defined to be at the 1350°C isotherm. Given a basal heat flux of $q_m = 20\,\mathrm{mW\,m^{-2}}$, a surface heat flux of $q_s = 71.25\,\mathrm{mW\,m^{-2}}$, and a surface temperature (T_s) of 0°C with no heat flux through the sides of the model, the Moho temperature is 704°C. Other important model properties include an extra increment of viscous weakening in the crust, which simulates the presence of a small amount of partial melt, and surface denudation scaled to 1.0–20 mm a^{-1}.

Figure 10.27c–e show the results of a model that provides an internally consistent explanation of the large-scale geometry and tectonic features of the Himalaya and southern Tibet. This model incorporates a convergence rate of 50 mm a^{-1} and advancing subduction, which mimics the manner in which precollisional suture zones wrap around the rigid India indenter as it penetrates into Eurasia. Surface denudation also requires the suture (S) to advance, which is modeled at a rate of $V_s = 25\,\mathrm{mm\,a^{-1}}$. Although S moves during the model, the results in Fig. 10.27c–e are shown with a fixed point "S" to keep the size of the diagrams manageable. Advancing subduction requires the removal of Eurasian lithosphere, which also is modeled by subduction. Indian lower crust is subducted along with its underlying mantle lithosphere. No displacements occur out of the plane of the model.

As Indian lithosphere is underthrust beneath southern Tibet, channel flow initiates by the development of partially molten material in the mid-lower crust beneath the plateau (Fig. 10.27c,d). Coeval thrust- and normal-sense shear zones develop across the lower and upper parts of the channel, respectively. These shear zones are interpreted to correspond to the Main Central Thrust and the South Tibetan Detachment Fault. The channel propagates through the converging crust. Efficient erosion at the southern edge of the plateau leads to a coupling between the channel flow and surface denudation. Denudation causes the surface position of the suture (S) to migrate toward India relative to the mantle (Fig. 10.27c–e) because it creates an imbalance in the flux of crustal material through the model. The final position of the suture after 51–54 Ma mimics the position of the Indus–Zangbo suture within the Tibetan Plateau.

The coupling between channel flow and surface denudation eventually leads to the ductile extrusion and exhumation of hot material in the channel between the coeval thrust and normal faults (Fig. 10.27d,e). The exhumation exposes the high-grade metamorphic rocks and *migmatite* (i.e. a mixed rock consisting of both metamorphic and igneous components) of the Greater Himalaya. The provenance of the channel material is derived from two sources. Initially, melt weakening in the middle crust occurs just to the south of point "S" (Fig. 10.27c). Later, as the suture is advected southward, material is derived from

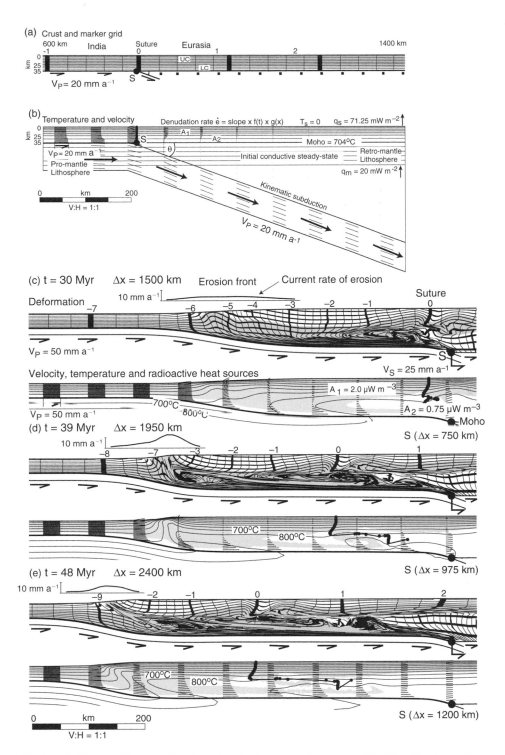

Figure 10.27 (a,b) Initial conditions and (c–e) results of a thermomechanical model of the Himalayan–Tibetan orogen (images provided by C. Beaumont and modified from Beaumont et al., 2004, by permission of the American Geophysical Union. Copyright © 2004 American Geophysical Union). Passive marker grid and mechanical layers are shown in (a). Initial thermal structure, radioactive layers, isotherms, and instantaneous velocity vectors are shown in (b). Top panels in (c–e) show deformed marker grid, bottom panels show evolved thermal structure. Heavy line with dots represents position of model suture (vertical marker 0), S is the mantle suture, whose position is tracked by a horizontal distance (Δx). Plots above profiles show the distribution and rate of slope-dependent erosion across model surface. The amount of convergence, which progresses from 1500 km to 2400 km, also is marked by horizontal distance Δx.

the Eurasian side of the suture. This process predicts that channel crust south of the Indus–Zangbo suture will show Indian crustal affinities, whereas channel material north of the suture will have Eurasian crustal affinities in a manner consistent with geologic observations (Section 10.4.4). Other similar models predict the formation of gneiss domes similar to those observed in the Greater Himalaya.

10.5 ARC–CONTINENT COLLISION

Orogenic belts that result from the collision between an island arc and a continent typically are smaller than those that form by continent–continent collision (Dewey & Bird, 1970). Arc–continent collision also tends to be relatively short-lived because it usually represents an intermediate step during the closure of a contracting ocean basin. Active examples of this type of orogen occur in Taiwan (Huang C.-Y. et al., 2000, 2006), Papua New Guinea (Wallace et al., 2004), and the Timor–Banda arc region north of Australia (Audley-Charles, 2004). These belts provide important information on the mechanisms by which continents grow, including by the accretion of terranes (Section 10.6.3).

The sequence of events that occurs during arc–continent collision begins as the island arc approaches a continent by the consumption of an intervening ocean. Collision begins when the continental margin is driven below the inner wall of the trench. At this point the positive buoyancy of continental lithosphere slows the rate of underthrusting and may lock the trench. If the continental margin is irregular or lies at an angle with respect to the island arc, the timing of arc–continent collision may vary along the strike of the orogen. Once collision begins, the forearc region and accretionary wedge are uplifted and deformed as thrust faults carry slices of flysch and oceanic crust onto the continental plate. If the two plates continue to converge, a new trench may develop on the oceanward (or backarc) side of the island arc.

The Timor–Banda arc region provides an example of an arc–continent collision in its early stages of development. Prior to 3 Ma, oceanic lithosphere of the Indo-

Australian plate subducted northward beneath the Eurasian plate at the Java Trench (Fig. 10.28a). This subduction created the Banda volcanic arc and a north dipping Benioff zone that extends to depths of at least 700 km. Between 3 Ma and 2 Ma, subduction brought Australian continental lithosphere in contact with the Banda forearc, part of which was thrust southward over the colliding Australian continental margin and is now well exposed on Timor (Harris et al., 2000; Hall, 2002). The downgoing Australian continental slope choked the subduction zone and created a fold and thrust belt (Fig. 10.28b) that has deformed both the forearc sequences and the structurally lower unsubducted cover sequences of the Australian continental margin. The Australian sequences include pre-rift Late Jurassic to Permian sedimentary rocks of a Gondwana cratonic basin, and younger post-rift Late Jurassic to Pliocene continental margin deposits that accumulated on the rifted continental slope and shelf (Audley-Charles, 2004). Within the adjacent volcanic arc north of Timor, volcanism has stopped on the islands of Alor, Wetar, and Romang. West of the tectonic collision zone volcanism is still occurring on the islands of Flores, Sumbawa and Lombok, north of the triangular Savu-Wetar forearc basin (Fig. 10.28a).

In eastern Indonesia, east of the Australian–Timor collision zone, seismicity patterns provide evidence of the past northward subduction of Indian oceanic lithosphere beneath the Banda Sea (Milsom, 2001). Figure 10.28a shows the inferred position of the former Banda trench, which represents the eastward continuation of the Java trench before it was obliterated by its collision with Australian continental lithosphere. The distribution of earthquake hypocenters beneath the Wetar Strait and Banda arc marks the location of the descending continental lithosphere to below depths of 300 km (Engdahl et al., 1998). Earthquake records suggest that the upper and lower plates of the subduction zone in the Timor region are now locked (McCaffrey, 1996; Kreemer et al., 2000). North of the Banda arc, Silver et al. (1983) discovered two north-directed thrust faults (the Wetar and Flores thrusts) that appear to represent the precursors of a new subduction zone that is forming in response to the collision (Fig. 10.28a,b).

An example of an oblique arc–continent collision occurs in Taiwan and its offshore regions. This belt is especially interesting because an oblique angle of convergence between the Luzon arc and the Eurasia continental margin has resulted in a progressive younging of the collision zone from north to south (Fig.

(a)

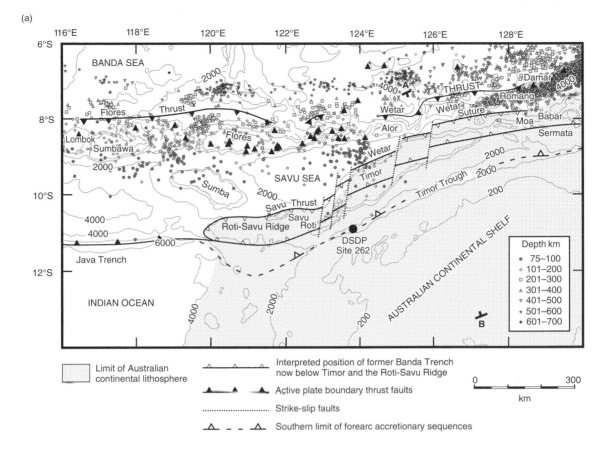

Limit of Australian continental lithosphere

△——△——△ Interpreted position of former Banda Trench now below Timor and the Roti-Savu Ridge

▲▲▲▲ Active plate boundary thrust faults

.... Strike-slip faults

△ — — △ Southern limit of forearc accretionary sequences

0   300
km

(b)

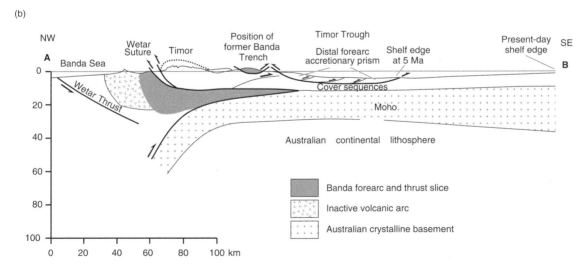

Figure 10.28 *(a) Tectonic map and (b) interpretive cross-section of the Australia–Banda arc collision zone (images provided by M. Audley-Charles and R. Hall and modified from Audley-Charles, 2004, with permission from Elsevier). Triangles are active volcanoes. Hypocenters for events below 75 km depth show the north-dipping Australian lithospheric slab and are based on the data set of Engdahl et al. (1998). Geologic section incorporates data from Hughes et al. (1996), Richardson & Blundell (1996), Harris et al. (2000), Hall & Wilson (2000), and Hall (2002). Australian continental shelf edge is marked by the 200 m contour. Dotted pattern in (a) illustrates that Australian continental lithosphere underlies the islands of Savu, Roti, Timor, Moa, Sermata and Babar. Shelf edge at 5 Ma is estimated from Deep Sea Drilling Program drilling site 262 (black circle).*

10.29). This geometry has allowed geoscientists to use spatial variations in the patterns of deformation, uplift, and sedimentation to piece together the progressive evolution of an oblique collision. C.-Y. Huang *et al.* (2000, 2006) used this approach to propose four stages of arc–continent collision beginning with intra-oceanic subduction and evolving through initial and advanced stages before the arc and forearc collapse and subside.

Off southern Taiwan, near latitude 21°N (Fig. 10.29), subduction of South China Sea oceanic lithosphere beneath the Philippine Sea plate results in volcanism and has formed an accretionary prism and forearc basin (Figs 10.29, 10.30a). The Hengchun Ridge/Kaoping slope and North Luzon Trough represent these two tectonic elements, respectively. Farther north, near 22°N, the North Luzon Trough narrows at the expense of an expanding accretionary prism (Fig. 10.29). In this latter region arc–continent collision began about 5 Ma and resulted in the formation of a suture between the arc and prism. The suture records both convergent and sinistral strike-slip motion (Malavieille *et al.*, 2002) and separates two zones of contrasting structural vergence. To the east, forearc sequences have been thrust eastward toward the arc, forming the Huatung Ridge (Fig. 10.29). To the west, forearc material on the Asian continental slope and South China Sea basin is carried westward within a growing accretionary prism. On the Hengchun Peninsula, Miocene slates and turbidites of the prism have been uplifted and exposed. These and other observations suggest that the initial stage of oblique arc–continent collision involves the following processes (Fig. 10.30b):

1 uplift and erosion of the accretionary prism and the continued deposition of forearc basin sequences;

2 waning arc volcanism and the build-up of fringing reefs on inactive volcanic islands;

3 arc subsidence, strike-slip faulting, and the development of intraarc pull-apart basins;

4 suturing, clockwise rotation, and shortening of forearc sequences to form a syn-collisional fold and thrust belt.

North of the Huatung Ridge, near 23°N, arc–continent collision has reached an advanced stage (Huang *et al.*, 2006). Here, collision since the Plio-Pleistocene has resulted in the west-directed thrusting and accretion of the Luzon arc and forearc sequences onto the accretionary wedge and Asian continent (Fig. 10.30c). These events have led to the uplift and exhumation of the underthrust Eurasian continental crust in the Coastal Range of eastern Taiwan. The last stage in the collision/accretion process is recorded north of about 24°N where the collapse and subsidence of the accreted arc and forearc has occurred over the last one or two million years (Fig. 10.30d), possibly as a result of the northward subduction of the northernmost Coastal Range at the Ryukyu Trench (Fig. 10.29). C.-Y. Huang *et al.* (2000) postulated that the Longitudinal Valley–Chingshui faults mark the collapsed trace of the arc where it approaches the subduction zone. This sequence of events suggests that orogens formed by arc–continent collision can progress rapidly through the initial stage of collision to an advanced stage and even collapse of the arc and forearc in only a few million years.

10.6 TERRANE ACCRETION AND CONTINENTAL GROWTH

10.6.1 Terrane analysis

Many orogens are composed of a collage of fault-bounded blocks that preserve geologic histories unrelated to those of adjacent blocks. These units are known as *terranes* and may range in size from a few hundreds to thousands of square kilometers. Terranes usually are classified into groups according to whether they are native or exotic to their adjacent continental cratons (e.g. Section 11.5.5). Exotic (or allochthonous) terranes are those that have moved relative to adjacent bodies and, in some cases, have traveled very great distances. For example, paleomagnetic investigations have demonstrated that some terranes have a north–south component of motion of several thousand kilometers (Beck, 1980; Ward *et al.*, 1997) and have undergone rotations of up to 60° (Cox, 1980; Butler *et al.*, 1989). The boundaries of terranes may be normal, reverse, or strike-slip

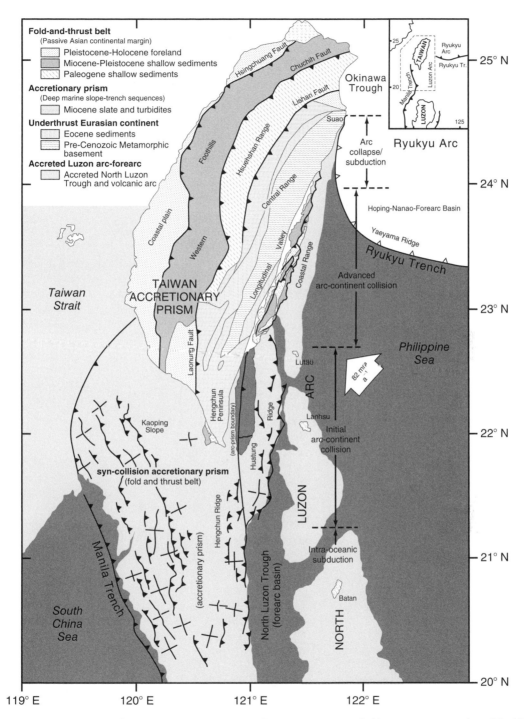

Figure 10.29 *Tectonic map of the Taiwan arc–continent collision (images provided by C.-Y. Huang and modified from Huang C.-Y. et al., 2000, with permission from Elsevier).*

(a) **Intra-oceanic subduction** (15–11 Ma to Recent)

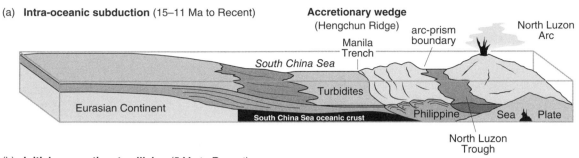

(b) **Initial arc-continent collision** (5 Ma to Recent)

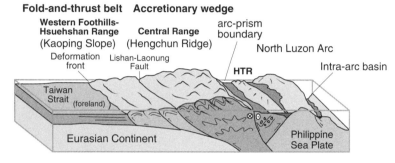

(c) **Advanced arc-continent collision** (2 Ma to Recent)

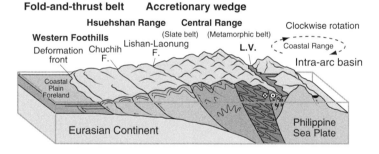

(d) **Arc collapse/subduction** (1–2 Ma to Recent)

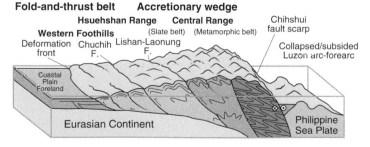

Figure 10.30 *Tectonic evolution of the Taiwan arc–continent collision (images provided by C.-Y. Huang and modified from Huang C.-Y. et al., 2000, with permission from Elsevier). (a) Intra-oceanic subduction. (b) Initial arc–continent collision. (c) Advanced arc–continent collision. (d) Arc collapse, subsidence, and subduction. LV, Longitudinal Valley, HTR, Huatung Ridge.*

faults; occasionally they may preserve thin ophiolites, blueschist, or highly deformed flysch. Terranes are *'suspect'* if there is doubt about their paleogeography with respect to adjacent terranes or to the continental margin (Coney *et al.*, 1980; Howell, 1989).

The identification and analysis of terranes is one of the most useful approaches to determining the long-term evolution of orogens, the mechanisms of continental growth, and the origin of the constituent components of continental lithosphere. Terrane recognition is based on contrasts in detailed stratigraphic and structural histories, although in many cases these have been destroyed or modified by younger events. Similarly the original nature of the bounding faults of many terranes may be obscured by metamorphism, igneous activity, or deformation. Consequently, in order to determine whether the geologic histories of adjacent terranes are compatible with their present spatial relationships, very detailed and comprehensive structural, geochemical and isotopic investigations are necessary (e.g. Keppie & Dostal, 2001; Vaughan *et al.*, 2005). In practice several criteria are used to distinguish the identity of terranes, including contrasts in the following:

1 the provenance, stratigraphy, and sedimentary history;

2 petrogenetic affinity and the history of magmatism and metamorphism;

3 the nature, history, and style of deformation;

4 paleontology and paleoenvironments;

5 paleopole position and paleodeclination.

The rock associations that make up terranes tend to be similar among orogens. Consequently, investigators have grouped them into several general types (Jones *et al.*, 1983; Vaughan *et al.*, 2005):

1 Turbidite terranes characterized by thick piles of land-derived sediment that are transported offshore by density currents and deposited in a deep marine environment. The sequences commonly are siliciclastic and may also be calcareous. Most of these terranes have been metamorphosed and imbricated by thrust faulting during or after accretion; some may preserve a crystalline basement. Three main varieties occur:

(a) turbidites forming part of an accretionary prism in a forearc setting (Section 9.7)

(b) turbidites forming part of an accretionary prism in a forearc setting with a minor proportion of basaltic rock;

(c) turbidites that escaped being incorporated into an accretionary prism.

2 Tectonic and sedimentary *mélange* terranes consisting of a heterogeneous assembly of altered basalt and serpentinite, chert, limestone, graywacke, shale, and metamorphic rock fragments (including blueschist) in a fine-grained, highly deformed, and cleaved mudstone matrix. These terranes commonly are associated with flysch, turbidite terranes, and collision-subduction zone assemblages (Section 9.7), and may occur along the boundaries between other terranes.

3 Magmatic terranes, which may be predominantly mafic or felsic according to the environment in which they form. Mafic varieties commonly include ophiolites, pillow basalts associated with pelagic and volcanogenic sediment, subaerial flood basalts, sheeted dikes, and plutonic complexes. This category may represent rock generated by seafloor spreading, LIP formation (Section 7.4.1), arc volcanism, ocean islands, and fragments of basement derived from backarc and forearc basins. In some cases oceanic fragments are associated with overlying sedimentary sequences charting travel from deep sea to continental margin environments. Felsic varieties commonly include calc-alkaline plutonic rock and dispersed fragments of old continental crust.

4 Nonturbiditic clastic, carbonate, or evaporite sedimentary terranes, which fall into two categories:

(a) well-bedded, shallow marine fluvial or terrestrial sequences; such as those deposited on continental margins and shallow basins;

(b) massive limestones, such as those scraped off the tops of seamounts as they become incorporated into accretionary prisms.

5 Composite terranes, which consist of a collage of two or more terranes of any variety that amalgamated prior to accretion onto a continent. Examples of this type of

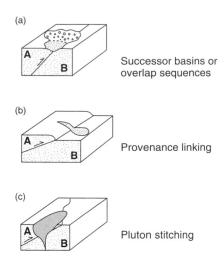

(a) Successor basins or overlap sequences

(b) Provenance linking

(c) Pluton stitching

Figure 10.31 *Geologic relationships that help establish the timing of terrane amalgamation and accretion (redrawn from Jones et al., 1983).*

terrane include the Intermontane and Insular Superterranes of the Canadian Cordillera (Fig. 10.33a) and Avalonia (Figs 10.34; 11.24b).

The chronological sequence of terrane accretion onto a continent can be determined from geologic events that postdate accretion and link adjacent terranes (Fig. 10.31). These include the deposition of sediments across terranes boundaries (Fig. 10.31a), the appearance of sediments derived from an adjacent terrane (Fig. 10.31b), and the "stitching" together of terranes by plutonic activity (Fig. 10.31c).

Following the identification of the terranes that comprise an orogen, a variety of analytical tools may be applied to determine whether they are exotic or native to the adjacent cratons. In addition to paleomagnetic, structural, and paleontologic studies; these include the application of isotope geochemistry and geochronology to determine the thermal evolution, provenance history, and crustal sources of the terranes. The most commonly used provenance techniques include the dating and geochemical characterization of zircon using U, Pb, and Hf isotope compositions (e.g. Gehrels, 2002; Hervé *et al.*, 2003; Griffin *et al.*, 2004). Zircon is a highly refractory mineral that commonly occurs in granitoids and sedimentary rock and may

preserve isotopic evidence of multiple phases of igneous and metamorphic growth. Comparisons of the age spectra from detrital zircon populations collected from sedimentary and metasedimentary rocks are especially useful for determining provenance history (e.g. Sections 3.3, 11.1). Analyses of the composition and petrologic evolution of xenoliths carried to the surface from great depths provide another important means of probing the deep roots of terranes to determine their age, sources, and tectonic evolution (e.g. Section 11.3.3).

10.6.2 Structure of accretionary orogens

One of the most fully investigated belts of accreted terranes is the Cordillera of western North America (Fig. 10.32). The distribution of terranes in this region forms a zone some 500 km wide that makes up about 30% of the continent (Coney *et al.*, 1980). Most of the terranes in the Cordillera accreted onto the margin of ancestral North America during Mesozoic times (Coney, 1989). Some also experienced lateral translations along strike-slip faults. This latter process of *dispersal*, where accreted terranes become detached and are redistributed along the margin, is still occurring today as active strike-slip faults dismember and transport terranes within Canada, the USA, and México.

Two composite cross-sections across the Canadian Cordillera (Fig. 10.33a) illustrate the large-scale tectonic structure of a major accretionary orogen. The sections were constructed by combining deep seismic reflection and refraction data from the Lithoprobe **S**lave-**North**ern **C**ordillera **L**ithospheric **E**volution (SNORCLE) and Southern Cordillera transects, with geologic information and the results of other geophysical surveys (Clowes *et al.*, 1995, 2005; Hammer & Clowes, 2007). Figure 10.33b shows a part of the Northern Cordillera, where subduction has ceased and the western side of the orogen is marked by a zone of active strike-slip faulting. Figure 10.33c shows a part of the Southern Cordillera, where subduction is still occurring. These transects elucidate the youngest part of a four billion year history of subduction, arc–continent collision, and terrane accretion along the western margin of North America (Clowes *et al.*, 2005) (see also Section 11.4.3).

Following the amalgamation of the Canadian Shield during the Proterozoic (Section 11.4.3), a number of

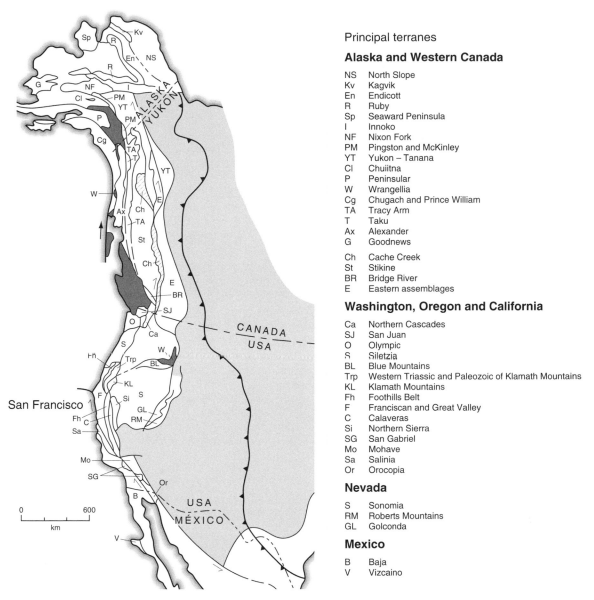

Principal terranes

Alaska and Western Canada

NS North Slope
Kv Kagvik
En Endicott
R Ruby
Sp Seaward Peninsula
I Innoko
NF Nixon Fork
PM Pingston and McKinley
YT Yukon – Tanana
Cl Chuiitna
P Peninsular
W Wrangellia
Cg Chugach and Prince William
TA Tracy Arm
T Taku
Ax Alexander
G Goodnews

Ch Cache Creek
St Stikine
BR Bridge River
E Eastern assemblages

Washington, Oregon and California

Ca Northern Cascades
SJ San Juan
O Olympic
S Siletzia
BL Blue Mountains
Trp Western Triassic and Paleozoic of Klamath Mountains
KL Klamath Mountains
Fh Foothills Belt
F Franciscan and Great Valley
C Calaveras
Si Northern Sierra
SG San Gabriel
Mo Mohave
Sa Salinia
Or Orocopia

Nevada

S Sonomia
RM Roberts Mountains
GL Golconda

Mexico

B Baja
V Vizcaino

Figure 10.32 *Generalized map of suspect terranes in western North America. Stippled ornament, North American cratonic basement; barbed line, eastern limit of Cordilleran Mesozoic–Cenozoic deformation; solid ornament, Wrangellia; diagonal ornament, Cache Creek terrane (redrawn from Coney et al., 1980, with permission from* Nature ***288**, 329–33. Copyright © 1980 Macmillan Publishers Ltd).*

rifting events between 1.74 Ga and the Middle Devonian created thick passive margin sequences that were deposited on top of Proterozoic crust of the North American craton (Thorkelson *et al.*, 2001). These sequences occupy the central part of SNORCLE Line 2B/21 (Fig. 10.33a,b). During the Middle Jurassic, a composite

terrane, called the Intermontane Superterrane, began to accrete onto the continental margin. The collision shortened the passive margin sequences and translated them eastward, resulting in a major foreland fold and thrust belt that now forms most of the eastern Cordillera. West of the foreland, the Omineca belt consists of

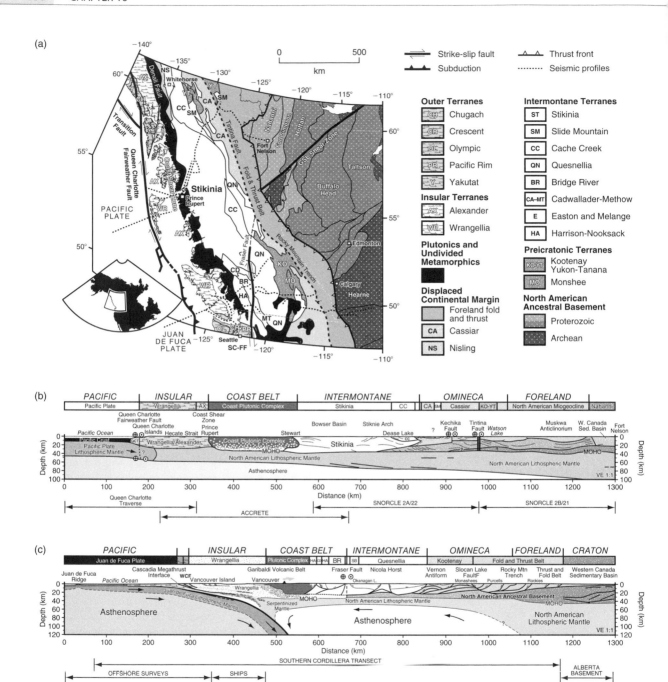

Figure 10.33 *(a) Tectonic terrane map and interpreted cross-sections of the (b) northern and (c) southern Canadian Cordillera based on lithospheric velocities, seismic reflection data (solid black lines), and geologic information (images provided by R. Clowes and P. Hammer and modified from Hammer and Clowes, 2007, with permission from the Geological Society of America). Data sources: Queen Charlotte Traverse – Spence & Asudeh (1993), Dehler & Clowes (1988); Accrete – Morozov et al. (1998, 2001), Hammer et al. (2000); SNORCLE – Hammer & Clowes (2004), Cook et al. (2004), Welford et al. (2001); Offshore surveys – Rohr et al. (1988), Hasselgren & Clowes (1995), Drew & Clowes (1990); SHIPS – Ramachandran et al. (2006); Southern Cordillera Transect – Clowes et al. (1987), Hyndman et al. (1990), Clowes et al. (1995), Varsek et al. (1993), Cook et al. (1992); Alberta Basement – Chandra & Cumming (1972), Lemieux et al. (2000). QCT, Queen Charlotte Transform; WCF, West Coast Fault.*

highly deformed and metamorphosed rocks of Middle Jurassic age that represent the suture zone created by the Intermontane–North American collision. A key result of the Northern Cordillera transect (Fig. 10.33b) is that most of the accreted terranes are relatively thin flakes of crust with a vertical extent of less than 10 km. Crustal thickness is unusually low and almost uniform across the entire Cordillera, ranging between 33 and 36 km (Clowes et al., 2005). Lithospheric thickness also is unusually thin and gradually thickens to the east beneath the Precambrian shield. These observations demonstrate that many accreted terranes lack the thick mantle roots that characterize most continental cratons (Section 11.3.1) (Plate 11.1 between pp. 244 and 245).

During the Late Cretaceous, the North American Cordillera again grew westward as another composite terrane, called the Insular Superterrane, accreted to the margin. This composite assemblage was exotic to North America and consisted mostly of two island arc terranes: Alexander and Wrangellia. The latter terrane, which is named after the Wrangell Mountains in Alaska (Jones et al., 1977), is particularly well studied and comprises upper Paleozoic island arc rocks overlain by thick, subaerial lava flows, and capped by a Triassic carbonate sequence. This distinctive geology has allowed investigators to identify several fragments of the terrane that are now scattered along some 2500 km of the Cordillera, occupying a latitudinal spread of almost 24° (Fig. 10.32). However, some paleomagnetic data suggest that the original spread was 4°, implying that a large amount of post-accretion fragmentation and dispersal has occurred. The paleolatitude of the fragments is centered on 10° N or S (the hemisphere is unknown due to uncertainties in the polarity of the Earth's magnetic field during the Triassic) and is in accord with a tropical environment suggested by their geology. It thus appears that Wrangellia may have originated in the western Pacific in Triassic times near the present position of New Guinea. Following its formation, it appears to have traversed the Pacific as a complete entity and accreted to North America where it was subsequently fragmented and translated to its present locations by strike-slip faulting.

The arrival of the Insular Superterrane deformed the interior of the North American continent and formed a major part of the Coast belt. Prior to and during the amalgamation, subduction beneath the margin formed the Coast Plutonic Complex (Hutchison, 1982; Crawford et al., 1999; Andronicos et al., 2003) during a major period of crustal growth by magma addition. The ACCRETE marine seismic transect across the Coast belt (Hollister & Andronicos, 1997) (Fig. 10.33b) shows the presence of layered, high velocity intrusions and a thinner than average (32 km) continental crust (Morozov et al., 1998; Hammer et al., 2000). At its western margin, the lithospheric-scale Coast Shear Zone (Klepeis et al., 1998) separates rocks of the Coast Plutonic Complex from those of the Alexander terrane, which lies on continental crust only 25 km thick. The Queen Charlotte Fault, which forms the boundary between the Pacific and North American plates, coincides with the western boundary of the orogen. This transform, and the Denali Fault in the western Yukon and southeast Alaska, represent the only major active strike-slip faults in the transect.

Strike-slip displacements also accommodated some relative motion between the accreted terranes and North America in the Canadian Cordillera. One of the most prominent of these zones is the Tintina Fault, which now forms the boundary between the accreted terranes to the west and ancestral North America to the east. This fault is a major lithospheric-scale structure that may record several hundred kilometers of dextral displacement since the Paleocene (Clowes et al., 2005). Other major strike-slip displacements are more speculative. For example, some paleomagnetic data suggest that, from 90 Ma to 50 Ma, many of the terranes in southeast Alaska and British Columbia were displaced several thousand kilometers northward parallel to the margin from a latitude near present day Baja, California (Umhoefer, 1987; Irving et al., 1996). This interpretation is known as the Baja–British Columbia or *Baja–BC hypothesis*. However, the great magnitude of the postulated displacements has been disputed mainly because the faults along which the terranes may have moved great distances have not been found (e.g. Cowan et al., 1997). Many correlations of stratigraphy and structures across faults suggest that the displacements are much less than those indicated by some paleomagnetic data. Numerous attempts to resolve these conflicting observations have been proposed, including: (i) tests of the paleomagnetic data using other types of data (Housen & Beck, 1999; Keppie & Dostal, 2001); (ii) determining reasons why strike-slip faults that record large translations are unlikely to be preserved (Umhoefer, 2000); and (iii) evaluating alternative correlations of units across terranes (Johnston, 2001). It seems probable that only through an interdisciplinary approach that combines paleomagnetic data, plate motions, paleontologic data, and geologic evidence will the history of large-magnitude terrane translation in western North America be resolved.

In general, these observations from the Canadian Cordillera and elsewhere suggest that ancient accretionary orogens are characterized by the following (Clowes *et al.*, 2005):

1 An extremely heterogeneous seismic velocity structure in the crust, produced by both thin-skinned and thick-skinned (Section 10.3.4) deformation, with the majority of terranes consisting of thin (<10 km thick) crustal flakes and lacking the thick mantle roots that characterize most continental cratons. There are exceptions to this 'thin flake' pattern, such as the Stikinia terrane (Fig. 10.33b), which exhibits a full crustal extent. Thick-skinned belts commonly display crustal-scale tectonic wedges characterized by a complex pattern of indentations and interfingering.

2 Observed crustal thicknesses are unusually low (33–36 km) compared to global averages (Section 2.4.3) except for averages in zones of continental extension.

3 The Moho remains mostly flat regardless of the age of crustal accretion or the age at which the last major tectonic deformation occurred. Lateral changes in crustal thickness tend to be gradual, with abrupt variations occurring at major terrane boundaries.

4 The dispersal of terranes by strike-slip faulting is an important process that occurs in most orogens. Subtle variations in seismic velocity and/or crustal thickness typically occur across these faults.

The structure of the Southern Cordillera (Fig. 10.33c), where subduction is occurring, provides some additional information on the mechanisms that result in many of the characteristics of the northern transect at the lithospheric scale. This southern part of the margin shows shortening and crustal thickening in the forearc region and an active volcanic arc within the Coast belt. The mantle lithosphere shows evidence of hydrothermal alteration (serpentinization) in the upper mantle wedge beneath the arc and substantial thinning for several hundred kilometers toward the interior of the continent. This thinning of the lithosphere in the backarc region is similar to that observed in other ocean–continent convergent margins (e.g. Fig. 10.8) and appears to reflect processes closely associated with subduction. These processes could include thinning by delamination or tectonic erosion driven by convective flow in the mantle (Section 10.2.5).

At scales smaller than that of the transects shown in Fig. 10.33b and c, the structure of ancient accretionary orogens provides a record of the processes involved in terrane accretion, including subduction and the formation of crustal-scale wedges. For example, seismic reflection data collected across the Appalachian orogen in Newfoundland provide an image (Fig. 10.34) of an Ordovician-Devonian collisional zone that resulted when several exotic terranes accreted onto the margin of Laurentia (Hall *et al.*, 1998; van der Velden *et al.*, 2004). Prior to the collision, thick sequences of sedimentary rock were deposited on a passive continental margin located outboard of the craton. These sequences record the stretching, thinning and eventual rupture (Sections 7.2, 7.7) of Proterozoic continental lithosphere as the Iapetus Ocean opened during the Late Proterozoic and Early Cambrian. This rifting event was followed by a series of terrane collisions and accretionary cycles that formed the Paleozoic orogenies of the Appalachian Mountains (Section 11.5.4). Many of the accreted terranes, such as Meguma and Avalonia, were microcontinents and composite terranes rifted from northwestern Gondwana during the Early Ordovician (Section 11.5.5, Fig. 11.24a).

The seismic reflection data (Fig. 10.34b) show prominent reflectivity at deep levels of the Appalachian crust that taper westward and merge with a well-defined Moho (van der Velden *et al.*, 2004). The shape and character of these reflections suggest that they mark the location of an old Ordovician–Devonian subduction zone. A similar feature occurs beneath the Canadian Shield (Fig. 11.15b), suggesting that the preservation of ancient subduction channels may be relatively common. Above and to the east of the paleosubduction zone are a series of dipping thrust faults and tectonic wedges composed of interlayered slices of the amalgamated terranes. Some reflections are truncated by a near vertical strike-slip fault that cuts through the entire crust.

These and other relationships observed in the Appalachians and the northern Canadian Cordillera show that ancient accretionary orogens tend to preserve the large-scale tectonics structures and lithologic contrasts associated with terrane accretion and dispersal. By contrast, active orogens such as the Andes, the Himalayan–Tibetan orogen, and the southern Canadian Cordillera produce seismic reflection profiles whose

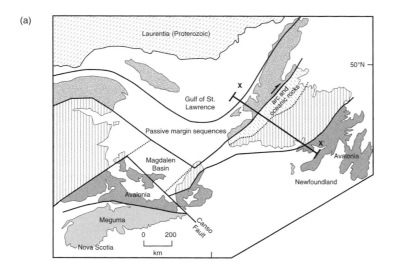

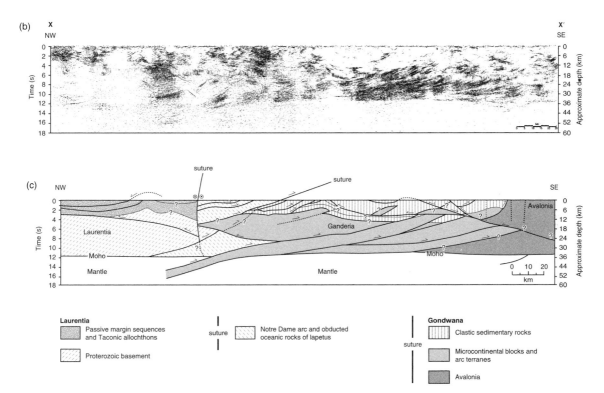

Figure 10.34 (a) Tectonic provinces of Newfoundland in eastern Canada and plot (b), and interpretation (c), of seismic reflection data (after van der Velden et al., 2004, with permission from the Geological Society of America). Migrated data are plotted with no vertical exaggeration and an assumed average velocity of 6 km s⁻¹.

deep structure exhibits the effects of processes related to the release, trapping, and consumption of fluids above a subduction zone (Figs 10.6a, 10.20b).

10.6.3 Mechanisms of terrane accretion

Observations from the North American Cordillera, the Appalachians, and many other ancient orogens suggest that the accretion and dispersal of terranes involves processes that are similar to those that occur in modern orogens. The regimes of active arc–continent collision in the southwest Pacific (Section 10.5), for example, offer excellent analogues for how a variety of tectonic and sedimentary terranes originate and are emplaced onto continental margins. In general, as the subduction of oceanic lithosphere brings thick sequences of continental, oceanic, and island arc material into contact with the trench, their positive buoyancy chokes the subduction zone. Once the collision begins, the forearc and accretionary wedge are uplifted and are carried, or obducted, onto the continental margin by thrust faults. As subduction slows or stops, a new trench may develop on the oceanward side of the old one (Section 10.5) and the process of accretion may begin again.

Many exotic terranes appear to originate during rifting events associated with the formation and break-up of the supercontinents (e.g. Fig. 11.24). Others may owe their origin to the abundant oceanic ridges, rises, and plateaux that make up about 10% of the area of the present ocean basins (Ben-Avraham et al., 1981). Most of these topographic highs represent extinct island arcs, submerged microcontinents, and LIPs (Section 7.4.1). As these features are brought to a trench, their positive buoyancy also may inhibit their subduction and allow them to be accreted as exotic terranes.

In addition to the processes described elsewhere in this chapter, two additional mechanisms of terrane accretion and continental growth deserve further mention: the obduction of ophiolites (Section 2.5); and continental growth by magmatism, sedimentation, and the formation and destruction of backarc, intraarc, and forearc basins.

The presence of ophiolitic assemblages in orogens provides an important marker of accretionary tectonic processes (Sections 2.5, 10.4.3, 11.4.3). As indicated in Section 2.5, models of ophiolite obduction tend to be quite variable, in part due to the diversity of the envi-

ronments in which these assemblages can form and the way in which they are uplifted and emplaced in the upper crust. In one postulated model, Wakabayashi & Dilek (2000) described how ophiolitic material in a backarc environment might become entrapped in a forearc setting prior to its obduction. This model is interesting because it explains how changes in the location or polarity of subduction can result in the capture of material that originally formed in an environment different than the one in which it is emplaced. This mechanism also may occur at larger scales, where it can result in the formation of a marginal sea by the entrapment of oceanic crust. Several of the present marginal seas for which there is no convincing evidence for backarc spreading (Section 9.10), such as the eastern Caribbean and Bering Sea, may have formed in this way (Ben-Avraham et al., 1981; Cooper et al., 1992).

In the Wakabayashi & Dilek (2000) model, the Coast Range ophiolite of western North America forms behind a Mesozoic island arc located offshore and above a subduction zone that dips to the west (Fig. 10.35a). Later, the island arc collides with the continent and a new east-dipping subduction zone initiates, capturing the ophiolite in the developing forearc (Fig. 10.35b,c). Ophiolite obduction subsequently occurs in a forearc setting when layers of the crust become detached and uplifted as a result of compression (Fig. 10.35d). The compression may result from any number of mechanisms, including the arrival of buoyant material at the trench.

In addition to the collision and accretion of exotic terranes, significant continental growth may occur by magma addition and sedimentation. An example of an accretionary orogen that has grown by more than 700 km mainly by these latter mechanisms is the Middle Paleozoic Lachlan orogen of southeastern Australia (Foster & Gray, 2000; Collins, 2002a; Glen, 2005). This orogen lacks many of the features that characterize major collisional orogens, such as exotic terranes, the development of high topography, deep-seated thrust faults, and exposures of high pressure metamorphic rocks. Instead, it is dominated by a huge volume of granitoid rock (Fig. 10.36), volcanic sequences, and extensive low-grade quartz-rich turbidites that overlie thinned continental crust and mafic lower crust of oceanic affinity (Fergusson & Coney, 1992). Like the Mesozoic-Cenozoic Andes, it records a history of ocean–continent convergence that lasted some 200 million years and involved many cycles of extension and contraction (Foster et al., 1999). Large (up to 1000 km-

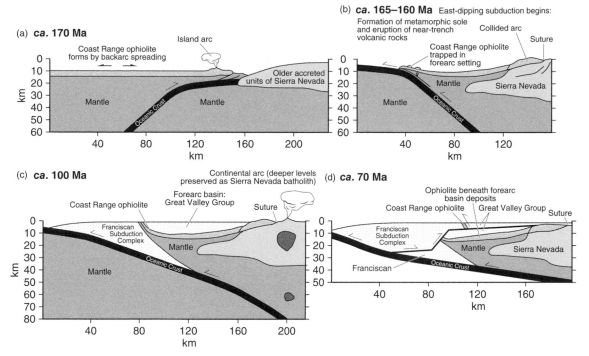

Figure 10.35 *Possible evolution of the Coast Range ophiolite in a backarc setting offshore of California and its subsequent emplacement in a forearc setting (after Wakabayashi & Dilek, 2000, with permission from the Geological Society of America). (a) Coast Range ophiolite forms behind a Mesozoic island arc. (b,c) Island arc collides with the continent and a new east-dipping subduction zone initiates, capturing the ophiolite in forearc. (d) Ophiolite obduction occurs in a forearc setting.*

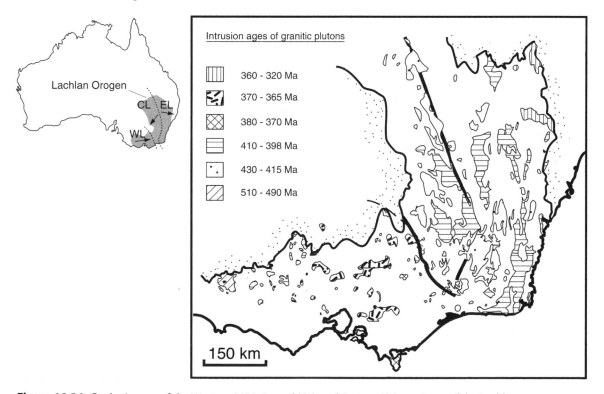

Figure 10.36 *Geologic map of the Western (WL), Central (CL), and Eastern (EL) provinces of the Lachlan orogen showing distribution of granitoids and their intrusion ages (redrawn with permission from Foster & Gray, 2000, Annual Review of Earth and Planetary Sciences **28**. Copyright © 2000 Annual Reviews).*

wide) extensional basins floored by basalt and gabbro were created behind one or more island arcs that eventually accreted onto the continental margin (Glen, 2005). Between the volcanic rocks are the accreted parts of a huge submarine sediment dispersal system that developed along the Gondwana margin during the early Paleozoic. Diachronous pulses of contractional and strike-slip deformation followed each extensional cycle, generating upright folds and overprinting cleavages in a series of thrust wedges in the upper 15 km of the crust. This style of shortening did not lead to the development of a well-defined foreland basin nor a foreland fold and thrust belt of the type seen in the central Andes (Fig. 10.5) and the Himalaya (Figs 10.19, 10.20). Instead, it was controlled by the thick (10 km) succession of turbidites and locally high geothermal gradients. These rela-

tionships suggest that orogenesis and crustal growth in the Lachlan orogen were dominated by magmatism and the recycling of continental detritus during cycles of extension and contraction that lasted from Late Ordovician through early Carboniferous times.

Cycles of backarc and intra-arc extension, such as those that occurred in the Lachlan orogen, generate thin, hot lithosphere that may localize deformation during subsequent phases of contraction, collision, and orogeny (Hyndman *et al.*, 2005). Collins (2002b) illustrated this process in a model of orogenesis involving the formation and closure of autochthonous backarc basins (Section 9.10) above a long-lived subduction zone. The model begins with a zone of intra-arc extension that evolves in response to the roll back (Section 9.10) of a subducting slab (Fig. 10.37a). This setting

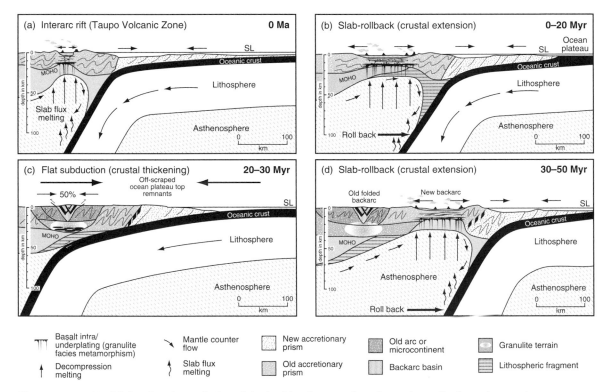

Figure 10.37 *Model showing the evolution of the Lachlan Orogen of southeast Australia through accretionary tectonics involving the creation and destruction of backarc basins above an ocean–continent subduction zone (after Collins, 2002a, with permission from the Geological Society of America). (a) Intra-arc extension due to the roll-back of a subducting slab. (b) Backarc basin and remnant arc form. (c) Subduction zone flattens and the upper plate of the orogen is thrown into compression. Contraction and crustal thickening are focused in the thermally softened backarc. (d) Extension is re-established and a new arc–backarc system forms.*

is analogous to the present day Taupo volcanic zone of the North Island, New Zealand. As the arc splits apart and migrates away from the trench, a backarc basin and remnant arc form (Fig. 10.37b), leading to subsidence and crustal thinning. Decompression melting (Section 9.8) in the upper mantle wedge beneath the backarc region generates basaltic crust as mafic magma underplates and intrudes the thinned crust. Next, the subduction zone flattens and the upper plate of the orogen is thrown into compression, possibly as a result of the arrival of an oceanic plateau or island arc at the subduction zone (Fig. 10.37c). This stage also may be analogous to the regime of flat slab subduction and contraction that characterizes part of the Andes (Section 10.2.2). Contractional deformation and crustal thickening are focused in the thermally softened backarc region. The contraction closes the

backarc basins and may lead to the accretion of the arc and forearc onto the continental margin. If a thick sequence of sediment has infilled the basin, a hot short-lived (~10 Ma) narrow orogen forms. Once the oceanic plateau has subducted, extension is re-established and a new arc–backarc system forms along the margin (Fig. 10.37d).

Models of accretionary orogens such as this, while speculative, illustrate how some continental margins may grow in the absence of major collisional events. Another example of a margin that appears to have grown by accretionary mechanisms is preserved in the Mesozoic history of Baja, California (Busby, 2004). Here, as in the Lachlan orogen, extension above a subduction zone created buoyant arc, forearc, and ophiolite terranes that accreted onto the upper plate during convergence, resulting in significant continental growth.

11 Precambrian tectonics and the supercontinent cycle

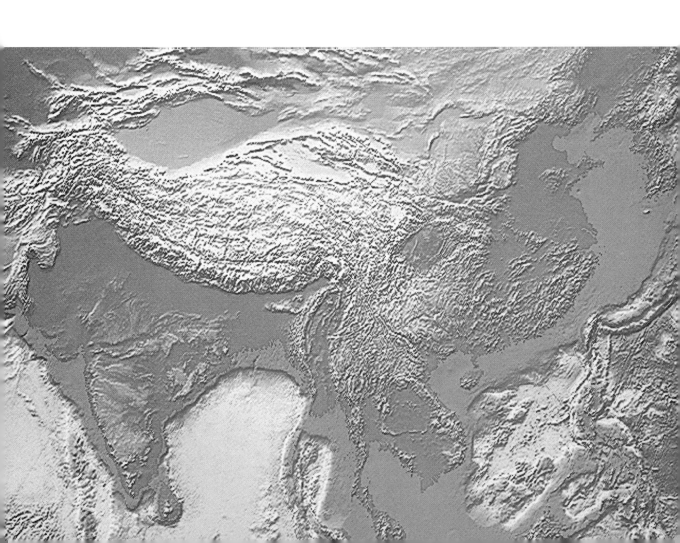

11.1 INTRODUCTION

The relatively flat, stable regions of the continents contain remnants of Archean crust that formed some 4.4 to 2.5 billion years ago (Plate 11.1a between pp. 244 and 245). The formation of these cratonic nucleii marks the transition from an early Earth that was so hot and energetic that no remnants of crust were preserved, to a state where crustal preservation became possible. Most of the cratons are attached to a high velocity mantle root that extends to depths of at least 200 km (King, 2005) (Plate 11.1b,c between pp. 244 and 245). These cratonic roots are composed of stiff and chemically buoyant mantle material (Section 11.3.1) whose resistant qualities have contributed to the long-term survival of the Archean continental lithosphere (Carlson *et al.*, 2005).

The beginning of the Archean Eon approximately coincides with the age of the oldest continental crust. A conventional view places this age at approximately 4.0 Ga, which coincides with the age of the oldest rocks found so far on Earth: the Acasta gneisses of the Slave craton in northwestern Canada (Bowring & Williams, 1999). However, >4.4 Ga detrital zircon minerals found in the Yilgarn craton of Western Australia (Wilde *et al.*, 2001) suggest that some continental crust may have formed as early as 4.4–4.5 Ma, although this interpretation is controversial (Harrison *et al.*, 2005, 2006; Valley *et al.*, 2006). Because evidence for continental crust and the ages of the oldest known rocks and minerals continually are being pushed back in time, the Archean has no defined lower boundary (Gradstein *et al.*, 2004). The end of the Archean, marking the beginning of the Proterozoic Eon, approximately coincides with inferred changes in the tectonic style and the petrologic characteristics of Precambrian rocks. It is these inferences that are central to a debate over the nature of tectonic activity in Precambrian times. Among the most important issues are whether some form of plate tectonics was operating in the early Earth and, if so, when it began. Current evidence (Sections 11.3.3, 11.4.3) suggests that plate tectonic mechanisms, including subduction, were occurring at least by 2.8–2.6 Ga and possibly much earlier (van der Velden *et al.*, 2006; Cawood *et al.*, 2006).

In considering the nature of Precambrian tectonic processes, three approaches have been adopted (Kröner, 1981; Cawood *et al.*, 2006). First, a strictly uniformitar-ian approach is taken in which the same mechanisms of plate tectonics that characterize Phanerozoic times are applied to the Precambrian cratons. This approach is common in the interpretation of Proterozoic belts, although it also has been applied to parts of the Archean cratons. Second, a modified uniformitarian approach can be postulated in which plate tectonic processes in the Precambrian were somewhat different from present because the physical conditions affecting the crust and mantle have changed throughout geologic time. This approach has been used in studies of both Archean and Early Proterozoic geology. Third, alternatives to plate tectonic mechanisms can be invoked for Precambrian times. This latter, nonuniformitarian approach most often is applied to the Early and Middle Archean. Each of these three approaches has yielded informative results.

11.2 PRECAMBRIAN HEAT FLOW

One of the most important physical parameters to have varied throughout geologic time is heat flow. The majority of terrestrial heat production comes from the decay of radioactive isotopes dispersed throughout the core, mantle, and continental crust (Section 2.13). Heat flow in the past must have been considerably greater than at present due to the exponential decay rates of these isotopes (Fig. 11.1). For an Earth model with a K/U ratio derived from measurements of crustal rocks, the heat flow in the crust at 4.0 Ga would have been three times greater than at the present day and at 2.5 Ga about two times the present value (Mareschal & Jaupart, 2006). For K/U ratios similar to those in chondritic meteorites, which are higher than those in crustal rocks, the magnitude of the decrease would have been greater.

The increased heat flow in Archean times implies that the mantle was hotter in the younger Earth than it is today. However, how much hotter and whether a hotter mantle caused young continental lithosphere to be much warmer than at present is uncertain. This uncertainty arises because there is no direct way to determine the ratio of heat loss to heat produc-

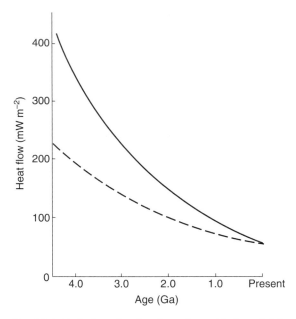

Fig. 11.1 *Variation of surface heat flow with time. Solid line, based on a chondritic model; dashed line, based on a K/U ratio derived from crustal rocks (redrawn from McKenzie & Weiss, 1975, with permission from Blackwell Publishing).*

tion in the early Earth. If the heat loss mostly occurred by the relatively inefficient mechanism of conduction then the lithosphere would have been warmer. However, if the main mechanism of heat loss was convection beneath oceanic lithosphere, which is very effective at dissipating heat, then the continental lithosphere need not have been much hotter (Lenardic, 1998). Clarifying these aspects of the Archean thermal regime is essential in order to reconstruct tectonic processes in the ancient Earth and to assess whether they were different than they are today.

Another part of the challenge of determining the Precambrian thermal regime is to resolve an apparent inconsistency that comes from observations in the crust and mantle parts of Archean lithosphere. Geologic evidence from many of the cratons, including an abundance of high temperature/low pressure metamorphic mineral assemblages and the intrusion of large volumes of granitoids (Section 11.3.2), suggest relatively high (500–700 or 800°C) temperatures in the crust during Archean times, roughly similar to

those which occur presently in regions of elevated geotherms. By contrast, geophysical surveys and isotopic studies of mantle nodules suggest that the cratonic mantle is strong and cool and that the geotherm has been relatively low since the Archean (Section 11.3.1). Some of the most compelling evidence of cool mantle lithosphere comes from thermobarometric studies of silicate inclusions in Archean diamonds, which suggest that temperatures at depths of 150–200 km during the Late Archean were similar to the present-day temperatures at those depths (Boyd et al., 1985; Richardson et al., 2001). Although geoscientists have not yet reconciled this apparent inconsistency, the relationship provides important boundary conditions for thermal models of Archean and Proterozoic tectonic processes.

In addition to allowing estimates of ancient mantle geotherms, the evidence from mantle xenoliths indicates that the cool mantle roots beneath the cratons quickly reached their current thickness of ≥200 km during Archean times (Pearson et al., 2002; Carlson et al., 2005). This thickness is greater than that of old oceanic lithosphere but much thinner than it would be if the lithosphere simply had cooled from above by conduction since the Archean (Sleep, 2005). Progressive thickening by conductive cooling also can be ruled out because the mantle roots do not display an age progression with depth (James & Fouch, 2002; Pearson et al., 2002). Instead, the relatively small thickness and long-term preservation of the cratonic roots indicate that they must have been kept thin by convective heat transfer from the underlying mantle (Sleep, 2003). Once the cratonic roots stabilized, the heat supplied to the base of the lithosphere from the rest of the mantle must have been balanced by the heat that flows upward to the surface. In this model, a chemically buoyant layer of lithosphere forms a highly resistant lid above the convecting mantle, allowing it to maintain nearly constant thickness over time. These considerations illustrate how the formation and long-term survival of the cool mantle roots beneath the cratons has helped geoscientists constrain the mechanisms of heat transfer during Precambrian times.

Differences in the inferred mechanism of heat loss from the Earth's interior have resulted in contrasting views about the style of tectonics that may have operated during Precambrian times (e.g. Hargraves, 1986; Lowman et al., 2001; van Thienen et al., 2005). A conventional view suggests that an increased heat supply in the Archean mantle could be dissipated by increasing the length of ocean ridge systems or by increasing the rate of plate production with respect to the present (Bickle, 1978). Hargraves (1986) concluded that heat loss through the oceanic lithosphere is proportional to the cube root of the total length of the mid-ocean ridge. Assuming a nonexpanding Earth (Section 12.3), the increased rate of plate production implies a similar increase in plate subduction rate. These computations suggest that some form of plate tectonics was taking place during the Precambrian at a much greater rate than today. The fast rates suggest an image of the solid surface of the early Earth where the lithosphere was broken up into many small plates that contrasts with the relatively few large plates that exist presently. This interpretation is consistent with the results of numerical models of mantle convection, which show that small plates are capable of releasing more heat from the Earth's interior than large plates (Lowman *et al.*, 2001).

More recent calculations have disputed this conventional view, at least for the Late Archean. Van Thienen *et al.* (2005) suggested that the increased heat flux from the Archean mantle could have been dissipated by thinning the lithosphere and thereby increasing the heat flow through the lithosphere. These authors concluded that for a steadily (exponentially) cooling Earth, plate tectonics is capable of removing all the required heat at a plate tectonic rate comparable to or lower than the current rate of operation. This result is contrary to the notion that faster spreading would be required in a hotter Earth to be able to remove the extra heat (e.g. Bickle, 1978). It also suggests that reduced slab pull and ridge push forces in a hotter mantle would result in a slower rate of plate tectonics compared to the modern Earth. Korenaga (2006) showed that a more sluggish style of plate tectonics during Archean times satisfies all the geochemical constraints on the abundance of heat-producing elements in the crust and mantle and the evidence for a gradual cooling of the mantle with time in the framework of whole mantle convection. This result removes the thermal necessity of having extensive ocean ridges and/or rapid spreading and subduction. It must be appreciated, however, that thermal conditions during Archean times are quite conjectural, so that these and other alternative interpretations remain speculative.

11.3 ARCHEAN TECTONICS

11.3.1 *General characteristics of cratonic mantle lithosphere*

A defining characteristic of the cratonic mantle lithosphere is a seismic velocity that is faster than normal subcontinental mantle to depths of at least 200 km and locally to depths of 250–300 km (Plate 11.1b,c between pp. 244 and 245). Many Proterozoic belts lack these fast velocity anomalies at similar depths. In addition, Archean cratons are characterized by the lowest surface heat flow of any region on Earth, with a heat flux that is lower than adjacent Proterozoic and Phanerozoic crust by some $20\,mW\,m^{-2}$ (Jaupart & Mareschal, 1999; Artemieva & Mooney, 2002). Isotopic age determinations and Re-Os studies of mantle nodules (Pearson *et al.*, 2002; Carlson *et al.*, 2005) confirm that the mantle roots are Archean in age and indicate that most have remained thermally and mechanically stable over the past 2–3 Ga. These observations indicate that the roots of the Archean cratons are cool, strong, and compositionally distinct from the surrounding mantle.

In the ocean basins, the base of the oceanic lithosphere is marked by a strong decrease in the velocity of shear waves at depths generally less than 100 km beneath the crust (Sections 2.8.2, 2.12). Similar low velocity zones occur under tectonically active continental regions, such as the Basin and Range Province, but beneath the stable cratons low velocity zones are either extremely weak or entirely absent (Carlson *et al.*, 2005). Consequently the base of the continental lithosphere is not well defined by seismological data. With increasing depth the high seismic velocities of the stable lithosphere gradually approach those of the convecting mantle across a broad, ill-defined transition zone below 200 km. Thermal modeling and geochemical data from mantle xenoliths have helped to define the location of the lithosphere–asthenosphere boundary. The results suggest that the base of the continental lithosphere is deepest (~250 km) in undisturbed cratonic areas and shallowest (~180 km) beneath Phanerozoic rifts and orogens (O'Reilly *et al.*, 2001). This determination is in general agreement with seismic observations.

In addition to being cool and strong, studies of mantle xenoliths indicate that the Archean mantle roots are chemically buoyant and highly depleted in incompatible elements (O'Reilly *et al.*, 2001; Pearson *et al.*, 2002). When mantle melting occurs elements such as calcium, aluminum and certain radiogenic elements are concentrated into and extracted by the melt whereas other elements, particularly magnesium, selectively remain behind in the solid residue. Those elements that concentrate into the melt are known as *incompatible* (Section 2.4.1). Both buoyancy and chemical depletion are achieved simultaneously by partial melting and melt extraction, which, in the case of the mantle lithosphere, has left behind a residue composed of Mg-rich harzburgites, lherzolites, and peridotite (O'Reilly *et al.*, 2001). Eclogite also appears to be present in the cratonic lithosphere. However, high velocity bodies consistent with large, dense masses of eclogite have not been observed in the continental mantle (James *et al.*, 2001; Gao *et al.*, 2002). An inventory of mantle xenoliths from the Kaapvaal craton suggests that eclogite reaches abundances of only 1% by volume in the continental mantle (Schulze, 1989). These characteristics have resulted in the mechanical and thermal stability of the cratons for up to three billion years (Section 11.4.2).

11.3.2 General geology of Archean cratons

Archean cratons expose two broad groups of rocks that are distinguished on the basis of their metamorphic grade: *greenstone belts* and *high grade gneiss terrains* (Windley, 1981). Both groups are intruded by large volumes of granitoids. Together these rocks form the Archean *granite-greenstone belts*. The structure and composition of these belts provide information on the origin of Archean crust and the evolution of the early Earth.

The greenstones consist of metavolcanic and metasedimentary rocks that exhibit a low pressure (200–500 MPa), low temperature (350–500°C) regional metamorphism of the greenschist facies. Their dark green color comes from the presence of minerals that typically occur in altered mafic (i.e. Mg- and Fe-rich) igneous rock, including chlorite, actinolite, and epidote. Three main stratigraphic groups are recognized within greenstone belts (Windley, 1981). The lower group is composed of tholeiitic and komatiitic lavas. *Komatiites*,

named after the Komati Formation in the Barberton Greenstone belt of the Kaapvaal craton, South Africa (Viljoen & Viljoen, 1969), are varieties of Mg-rich basalt and ultramafic lava that occur almost exclusively in Archean crust. The high Mg content (>18 wt% MgO) of these rocks (Nisbet *et al.*, 1993; Arndt *et al.*, 1997) commonly is used to infer melting temperatures that are higher than those of modern basaltic magmas (Section 11.3.3). The central group contains intermediate and felsic volcanic rocks whose trace and rare earth elements are similar to those found in some island arc rocks. The upper group is composed of clastic sediments, such as graywackes, sandstones, conglomerates, and banded iron formations (BIFs). These latter rocks are chemical-sedimentary units consisting of iron oxide layers that alternate with chert, limestone, and silica-rich layers (see also Section 13.2.2).

High-grade gneiss terrains typically exhibit a low pressure, high temperature (>500°C) regional metamorphism of the amphibolite or granulite facies (Section 9.9). These belts form the majority of the area of Archean cratons. A variety of types commonly are displayed, including quartzofeldspathic gneiss of mostly granodiorite and tonalite composition, layered peridotite-gabbro-anorthosite or leucogabbro-anorthosite complexes, and metavolcanic amphibolites and metasediment (Windley, 1981). Peridotite (Sections 2.4.7, 2.5) is an ultramafic rock rich in olivine and pyroxene minerals. Leucogabbro refers to the unusually light color of the gabbroic rock due to the presence of plagioclase. *Anorthosites* are plutonic rocks consisting of >90% plagioclase and have no known volcanic equivalents. These latter rocks occur exclusively in Archean and Proterozoic crust. Most authors view Archean anorthosites as having differentiated from a primitive magma, such as a basalt rich in Fe, Al and Ca elements or, possibly, a komatiite (Winter, 2001). High-grade gneiss terrains are highly deformed and may form either contemporaneously with, structurally below, or adjacent to the low-grade greenstone belts (Percival *et al.*, 1997).

The granitoids that intrude the greenstones and high-grade gneisses form a compositionally distinctive group known as *tonalite-trondhjemite-granodiorite*, or *TTG*, suites (Barker & Arth, 1976). Tonalites (Section 9.8) and trondhjemites are varieties of quartz diorite that typically are deficient in potassium feldspar. These igneous suites form the most voluminous rock associations in Archean crust and represent an important step in the formation of felsic continental crust from the primordial mantle (Section 11.3.3).

11.3.3 The formation of Archean lithosphere

The distinctive composition and physical properties of the stiff, buoyant mantle roots beneath the cratons (Section 11.3.1) result from the chemical depletion and extraction of melts from the primitive mantle. These two processes lowered the density and increased the viscosity of the residue left over from partial melting and resulted in a keel that consists mostly of high-Mg olivine and high-Mg orthopyroxene (O'Reilly et al., 2001; Arndt et al., 2002). Both of these components are absent in fertile (undepleted) mantle peridotite and are rare in the residues of normal mantle melting, such as that which produces modern oceanic crust and oceanic islands. Consequently, most workers have concluded that the distinctive composition is related to unusually high degrees (30–40%) of mantle melting over a range (4–10 GPa) of mantle pressures (Pearson et al., 2002; Arndt et al., 2002). High-degree partial melting of mantle peridotite produces magma of komatiitic (Section 11.3.2) composition and a solid residue that is very similar to the composition of Archean lithospheric mantle (Herzberg, 1999; Arndt et al., 2002).

One radiometric system that has been of considerable use in determining when melt extraction and Archean root formation occurred involves the decay of ^{187}Re to ^{188}Os (Walker et al., 1989; Carlson et al., 2005). The key feature of this isotopic system is that Os is compatible during mantle melting and Re is moderately incompatible. Consequently, any residue left behind after melt extraction will have a lower Re and a higher Os concentration than in either the mantle melts or the fertile mantle. This characteristic allows Re-Os isotope analyses of mantle xenoliths to yield information on the age of melt extraction. The data from mantle xenoliths show that the oldest melting events are Early–Middle Archean in age. Significant amounts of lithospheric mantle also formed in Late Archean times and are associated with voluminous mafic magmatism (Pearson et al., 2002).

Although high-degree partial melting undoubtedly occurred, this process alone cannot explain the origin of the Archean mantle lithosphere. The main reason for this conclusion is that the abundance of komatiite found in the Archean crustal record is much too low to balance the amount of highly depleted peridotite found in the cratonic mantle (Carlson et al., 2005). This imbalance suggests that either a large proportion of komati-itic magma never reached the surface or other processes must have contributed to root formation. One of these processes is an efficient sorting mechanism that concentrated the unusual components of the Archean mantle lithosphere at the expense of all others (Arndt et al., 2002). The most likely driving force of the sorting is the buoyancy and high viscosity of high-Mg olivine and orthopyroxene, although exactly how this happened is uncertain. The density and viscosity of these minerals depends upon their Mg–Fe ratios and water content, respectively; which are lower in Archean mantle lithosphere compared to normal asthenospheric mantle. Arndt et al. (2002) considered three processes that could have resulted in the mechanical segregation and accumulation of a layer of buoyant, viscous mantle near the Earth's surface during Archean times. First, upwelling buoyant residue in the core of a mantle plume could have separated from the cooler, denser exterior and accumulated during ascent (Fig. 11.2a). In this model, melting begins at high pressure (~200 km depth) and continues to shallow depths, by which point melt volumes are high and the dense residue of early, low-degree melting is swept away by mantle flow. Second, buoyant residue could have segregated slowly as material was transported down subduction zones and recycled through the mantle in convection cells (Fig. 11.2b). Third, some subcontinental lithosphere could be the remnants of an initial crust that crystallized in an Archean magma ocean that formed during the final stages of Earth accretion (Fig. 11.2c). In all three cases, buoyant, viscous material rises and is separated from higher density residue during mantle flow. Whether some combination of these or other processes helped to form the cratonic keels is highly speculative. Nevertheless, they illustrate how several different mechanisms could have concentrated part of the residue of mantle melting into a near-surface layer.

In addition to high-degree partial melting and efficient sorting, most authors also have concluded that the formation and evolution of mantle lithosphere involved a multi-stage history involving many tectonic and magmatic events (James & Fouch, 2002). However, opinions are divided over whether root construction preferentially involved the underthrusting and stacking of subducted oceanic slabs (Carlson et al., 2005), the accretion and thickening of arc material (Lee, 2006), or the extraction of melt from hot mantle plumes (Wyman & Kerrich, 2002). By applying a range of criteria some geologic studies have made compelling cases that ancient mantle plumes played a key role in the

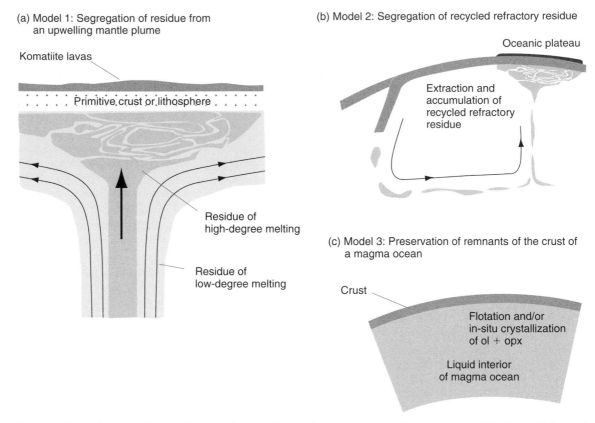

(a) Model 1: Segregation of residue from an upwelling mantle plume

Komatiite lavas

Primitive crust or lithosphere

Residue of high-degree melting

Residue of low-degree melting

(b) Model 2: Segregation of recycled refractory residue

Oceanic plateau

Extraction and accumulation of recycled refractory residue

(c) Model 3: Preservation of remnants of the crust of a magma ocean

Crust

Flotation and/or in-situ crystallization of ol + opx

Liquid interior of magma ocean

Fig. 11.2 *(a–c) Three possible mechanisms that could allow the segregation and accumulation of high-Mg olivine and orthopyroxene near the surface of the Earth (after Arndt et al., 2002, with permission from the Geological Society of London).*

evolution of Archean lithosphere (Tomlinson & Condie, 2001; Ernst *et al.*, 2005). Data from seismic profiles, geochronologic studies, and isotopic analyses indicate that many roots were affected by large pulses of mafic magmatism during the Late Archean (Wyman & Kerrich, 2002; James & Fouch, 2002). Other studies, however, have emphasized a subduction zone setting to explain the evolution of Archean mantle lithosphere. Most of the cratons display evidence for the significant modification of cratonic roots by terrane collisions and thickening during at least some stage in their history (James & Fouch, 2002; Schmitz *et al.*, 2004). In support of a subduction zone mechanism, a Late Archean (2.8–2.6 Ga) fossil subduction zone (Fig. 11.3) has been found within the Abitibi craton in northern Canada using seismic data (Calvert & Ludden, 1999; van der Velden *et al.*, 2006). Nevertheless, it is important to recognize

that Archean mantle roots probably resulted from more than one tectonic environment and that no single setting or event is applicable to all cases.

The distinctive rock associations that comprise granite-greenstone belts (Section 11.3.2) provide another important means of evaluating the mechanisms that contributed to the formation and evolution of Archean lithosphere. One of the key questions to answer is whether the komatiitic and tholeiitic lavas that form the majority of the greenstones formed in environments that were broadly similar to modern tectonic environments. For example, if these lavas loosely represent the Archean equivalent of modern mid-ocean ridge basalts, as is commonly believed, then they might be used to infer that much of the volcanism in Archean times involved the creation and destruction of ocean crust (Arndt *et al.*, 1997). However, one of the problems

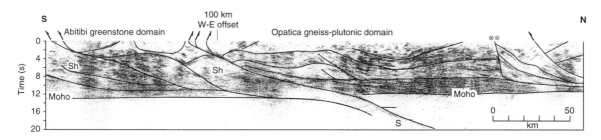

Fig. 11.3 *Seismic reflection profile of the Opatica–Abitibi belt in the Superior Province of northern Canada (modified from van der Velden et al., 2006, by permission of the American Geophysical Union. Copyright © 2006 American Geophysical Union). Interpretation is modified from Calvert et al. (1995), Lacroix & Sawyer (1995), and Calvert & Ludden (1999). S, fossil subduction zone; Sh, shingle reflections suggesting imbricated material in the middle crust.*

with these comparisons is that no chemically unaltered, complete example of Archean ocean crust is preserved. In addition, the Archean mantle was hotter by some amount than the modern mantle (Section 11.2), which undoubtedly influenced the compositions, source depths, and patterns of the volcanism (Nisbet *et al.*, 1993). These problems have complicated interpretations of the processes that produced and recycled Archean crust and how they may differ from those in modern environments.

Most authors have concluded that the high magnesium contents and high degrees of melting associated with the formation of komatiites reflect melting temperatures (1400–1600°C) that are higher than those of modern basaltic magmas (Nisbet *et al.*, 1993). Exactly how much hotter, however, is problematic. Parman *et al.* (2004) proposed a subduction-related origin for these rocks similar to that which produced boninites in the Izu-Bonin-Mariana island arc (Fig. 11.4). *Boninites* are high-Mg andesites that are thought to result from the melting of hydrous mantle in anomalously hot forearc regions above young subduction zones (Crawford *et al.*, 1989; Falloon & Danyushevsky, 2000). If the komatiites were produced by the melting of hydrous mantle, then the depth of melting could have been relatively shallow, as in subduction zones, and the Archean mantle need only be slightly hotter (~100°C) than at present (Grove & Parman, 2004). In this interpretation, shallow melting and subduction result in the formation and thickening of highly depleted mantle lithosphere that some time later is incorporated into the cratonic mantle below a continent.

Alternatively, if the source rocks of komatiites were dry then high ambient temperatures in the Archean

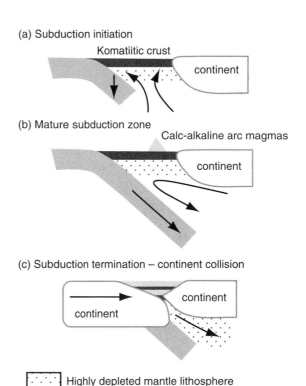

Fig. 11.4 *Conceptual model for the generation of komatiites and cratonic mantle by partial melting in a subduction zone (after Parman et al., 2004. Copyright © 2004 Geological Society of South Africa). (a) Partial melting produces komatiitic magma in a forearc setting. (b) Mature subduction cools and hydrates residual mantle. (c) Obduction of komatiitic crust occurs during collision.*

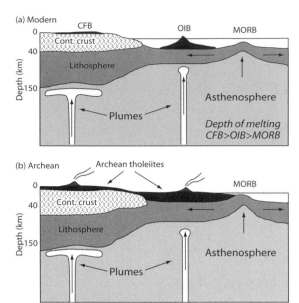

Fig. 11.5 *Model of komatiitic and tholeiitic basalt formation involving mantle plumes (after Arndt et al., 1997, by permission of Oxford University Press). Model shows the influence of lithospheric thickness on depth of melting where CFB is continental flood basalt, OIB oceanic island basalt, and MORB mid-ocean ridge basalt.*

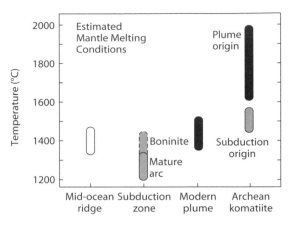

Fig. 11.6 *The range of mantle melt generation temperatures estimated for various modern tectonic settings compared to temperatures inferred for komatiite melt generation by a plume model (black filled oval) and a subduction model (gray filled oval) (after Grove & Parman, 2004, with permission from Elsevier).*

mantle would have caused melting to begin at depths that were much greater than occurs in subduction zones, possibly in upwelling mantle plumes or at unusually deep levels (~200 km) beneath mid-ocean ridges. The greater depths of melting would produce large volumes of basalt and oceanic crust that was much thicker (20–40 km) than it is today (Bickle *et al.*, 1994). Evidence of large volumes of mafic magma and high eruption rates have suggested that oceanic plateaux and continental flood basalts are the best modern analogues for such thick mafic crust and invites comparisons with Phanerozoic LIPs (Section 7.4.1) (Arndt *et al.*, 1997, 2001). In this latter context, the differences between the modern and ancient rocks are explained by variations in the depth of melting and in the effects of a thick overriding lithosphere (Fig. 11.5). These and a variety of other models (Fig. 11.6) illustrate how information on the depth and source of the melting that produced komatiites has important consequences for both the tectonic setting and the thermal evolution of the early Earth.

A variety of tectonic models also have been postulated for the origin of Archean continental crust. Windley (1981) noted the geologic and geochemical similarities between Archean tonalite-trondhjemite-granodiorite (TTG) suites and exhumed granitoids associated with Andean-type subduction zones (Section 9.8). He considered this to be an environment in which voluminous quantities of tonalite can be produced, and concluded that this represents a reasonable analogue for the formation of these rocks in Archean times. Subsequent work has led to a general consensus that these subduction models are applicable to the Late Archean. However, their applicability to Early and Middle Archean times when thick oceanic crust may have inhibited subduction is more controversial. As an alternative to subduction, Zegers & van Keken (2001) postulated that TTG suites formed by the removal and sinking of the dense, lower part of thick oceanic plateaux. The peeling away, or *delamination*, of a dense eclogite root results in uplift, extension, and partial melting to produce TTG suite magmas. This process could have returned some oceanic material into the mantle and may have accompanied collisions among oceanic terranes in Early–Middle Archean times. However, the possible absence of subduction creates a problem in that, assuming a nonexpanding Earth (Section 12.3), a high rate of formation of oceanic lithosphere during these times must have been accompanied by a mechanism by which oceanic lithosphere also was destroyed at high rates.

An important aspect of the origin of TTG suites is the type of source material that melted to produce the magma. Early petrologic studies suggested that these magmas could result from the partial melting of subducted oceanic crust in the presence of water (Martin, 1986). However, more recent work has emphasized other sources, including the lower crust of arcs and the base of thick oceanic plateaux (Smithies, 2000; Condie, 2005a). The importance of the source material is illustrated by a two-stage model proposed by Foley *et al.* (2003). This model envisages that during the Early Archean, oceanic crust was too thick to be subducted as a unit, and so its lowermost parts delaminated and melted (Fig. 11.7a). These lower roots are inferred to have been pyroxenites that were produced by the metamorphism of ultramafic cumulate layers. The melting of the pyroxenite did not favor the generation of TTG

melts but produced basaltic melts instead. As the oceanic crust cooled and became thinner, a point was reached in the Late Archean when the entire crust could subduct (Fig. 11.7b). At this time hydrothermally altered crust, such as amphibolite, was introduced into subduction zones and led to the widespread formation of TTG suites. This model supports the view that the formation of the earliest continental crust requires subduction and the melting of a hydrous mafic source.

11.3.4 Crustal structure

Granite-greenstone belts display a variety of structural styles and outcrop patterns, many of which also occur in Phanerozoic orogens (Kusky & Vearncombe, 1997). Those common to both Archean and Phanerozoic belts include large tracts of metamorphosed igneous and

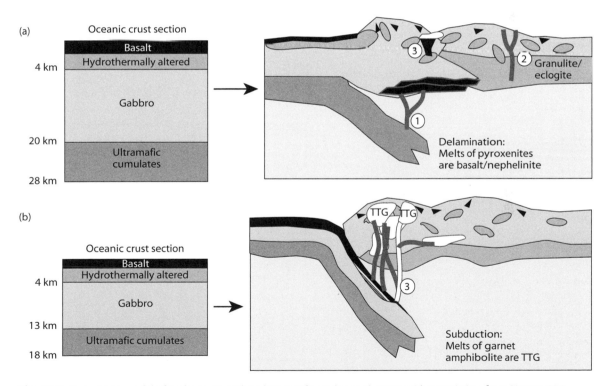

Fig. 11.7 *Two-stage model of Archean crustal evolution (after Foley et al., 2003, with permission from Nature* **421**, *249–52. Copyright © 2003 Macmillan Publishers Ltd). (a) Early Archean delamination of thick oceanic crust and the melting of pyroxenites (1). Local melting of lower crust (2) and garnet amphibolite (3) may also occur to produce small volumes of felsic magma. (b) Late Archean whole-crust subduction and the large-scale melting of garnet amphibolite to produce TTG suites (3).*

sedimentary rock that are deformed by thrust faults and strike-slip shear zones (e.g. Figs 10.13, 10.19). Another pattern, commonly referred to as a dome-and-keel architecture, occurs exclusively in Archean crust. This latter structural style forms the focus of the discussion in this section.

Dome-and-keel provinces consist of trough-shaped or synclinal keels composed of greenstone that surround ellipsoidal and ovoid-shaped domes composed of gneiss, granitoid, and migmatite (Section 9.8). The contacts between domes and keels commonly are high-grade ductile shear zones. Marshak *et al.* (1997) distinguished between two types of these provinces. One type has keels composed of greenstones and their associated metasedimentary strata (Section 11.3.2) and domes composed of granitoid rock that is similar in age or slightly younger than the greenstones. The other type has domes of mostly gneissic and migmatitic basement rock and keels composed of greenstones that are younger than the dome rocks.

The Eastern Pilbara craton of northwestern Australia (Fig. 11.8) is one of the oldest and best preserved examples of a granite-greenstone belt and dome-and-keel province with a history spanning 3.72–2.83 Ga (Collins *et al.*, 1998; Van Kranendonk *et al.*, 2002, 2007). The craton exposes nine granitoid domes with diameters ranging from 35 km to 120 km. Studies of seismic refraction data and gravity and magnetic anomalies (Wellman, 2000) indicate that the margins of the domes are generally steep and extend to mid-crustal depths of ~14 km. Despite their simple outlines, the internal structure of the domes is complex. Each contains remnants of 3.50–3.43 Ga TTG suite granitoids (Section 11.3.2) that are intruded by younger (3.33–2.83 Ga) more potassic igneous suites (Fig. 11.9a). The domes display compositional zonations and variable degrees of deformation. In many cases, the youngest bodies are located in the cores of the domes with older, more deformed granitoids at the margins. This internal structure indicates that each dome was constructed through the emplacement of many intrusions over hundreds of millions of years and that deformation accompanied the magmatism.

Between the granitoid domes are synclinal tracts of greenstone composed of dipping volcanic and sedimentary sequences up to 23 km thick (Van Kranendonk

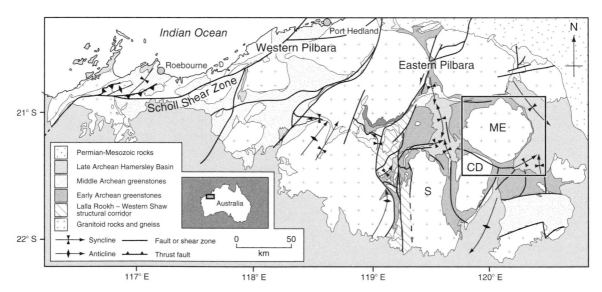

Fig. 11.8 *Geologic map of the Pilbara granite-greenstone belt (modified from Zegers & van Keken, 2001, with permission from the Geological Society of America, with additional structural information from Van Kranendonk et al., 2007) showing the typical ovoid pattern of TTG suite granitoids surrounded by greenstones belts. ME, Mt. Edgar Dome; CD, Corunna Downs Dome; S, Shaw Dome. Black box shows location of Fig. 11.9.*

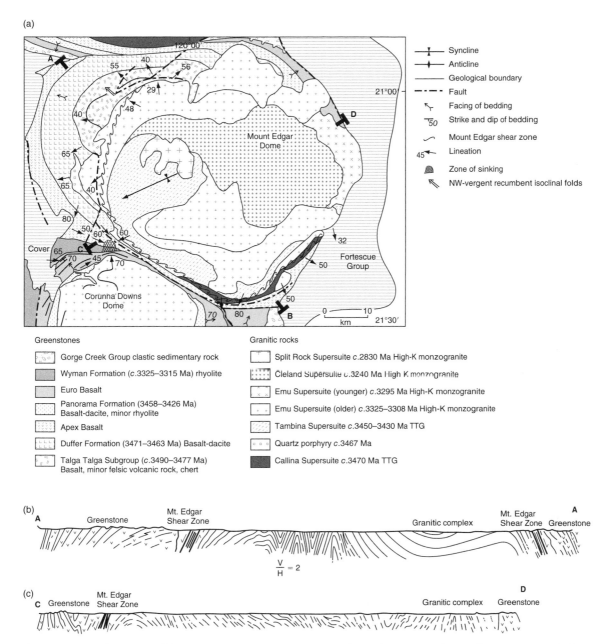

Fig. 11.9 *(a) Geologic map of the Mt. Edgar Dome and granitic complex in the Eastern Pilbara (after Van Kranendonk et al., 2007, with permission from Blackwell Publishing). Map shows the internal structure of the domes and the radial distribution of 3.32–3.30 Ga metamorphic mineral elongation lineations (arrows) that converge in a vertical zone of sinking between the Mt. Edgar, Carunna Downs and Shaw (not shown) domes. These features are contemporaneous with the arcuate shear zones that formed along granite-greenstone contacts (Fig. 11.8). (b,c) Cross-sections showing the trends of foliation surfaces and shear zones within the Mt. Edgar Dome (after Collins, 1989, with permission from Elsevier).*

et al., 2002). Successive groups of these strata were deposited in autochthonous basins that developed on synclines of older greenstones. Episodes of felsic volcanism in these belts accompanied emplacement of the granitoid domes. The degree of metamorphism and the age of the strata gradually decrease away from the deformed margins of the domes and toward the cores of synclines where the greenstones are only weakly deformed. These weakly deformed areas preserve the delicate Archean *stromatolites* and other evidence of early life (Buick, 2001). The geometry of the synclines between the domes creates a high amplitude (~15 km), long wavelength (120 km) dome-and-keel structure that developed throughout the entire history of the Eastern Pilbara.

The contacts between the granitoid domes and the greenstones in the Eastern Pilbara vary from being intrusive to unconformities, ring faults and high grade shear zones. The shear zones and ring faults are concentric about the domes and generally display steep to subvertical orientations (Figs 11.8, 11.9b,c). Many of these shear zones, including the Mt. Edgar Shear Zone, formed during the period 3.32–3.30 Ga (Van Kranendonk *et al.*, 2007). The central part of the craton contains a 5- to 15-km-wide zone of ductile deformation called the Lalla Rookh–Western Shaw structural corridor (Fig. 11.8). This zone formed during a period (~2.94 Ga) of contraction and is characterized by multiple generations of folds and ductile rock fabrics (Van Kranendonk & Collins, 1998).

11.3.5 Horizontal and vertical tectonics

The origin of the unique dome-and-keel architecture of the Archean cratons (Section 11.3.4) is important for understanding the nature of Archean tectonics. In general, interpretations can be divided into contrasting views about the relative roles of vertical and horizontal displacements in producing this pattern. The Eastern Pilbara craton in western Australia illustrates how vertical and horizontal tectonic models have been applied to explain the dome-and-keel structural style. During this discussion, it is important to keep in mind that the crustal structure, as illustrated by the Pilbara example, is the product of multiple episodes of deformation, metamorphism, and pluton emplacement rather than a single tectonic episode.

Vertical tectonic models describe the diapiric rise of hot granitoid domes as the result of a partial convective overturning of the middle and upper crust. Collins *et al.* (1998) and Van Kranendonk *et al.* (2004) used strain patterns, a dome-side-up/greenstone-side-down sense of displacement in shear zones, and other features to link the formation of dome-and-keel structures to a sinking of the greenstones. The process begins with the emplacement of hot TTG suite (Section 11.3.2) granitoids into an older greenstone succession (Fig. 11.10a). Domes are initiated at felsic volcanic centers due to a laterally uneven emplacement of TTG magma. After a hiatus of several tens of millions of years, the emplacement of thick piles of basalt on top of less dense granitoids creates an inverted crustal density profile (Fig. 11.10b). The magmatism also buries the granitoids to mid-crustal depths where they partially melt due to the build up of radiogenic heat and, possibly, the advection of heat from mantle plume activity. Thermal softening and a reduction in mid-crustal viscosity facilitates the sinking of the greenstones, which then squeezes out the underlying partial melts into rising, high-amplitude granitoid domes (Fig. 11.10c). The convective overturning depresses geotherms in the greenstone tracts, resulting in local cooling and the preservation of kyanite-bearing metamorphic rocks, which equilibrate at moderate-low pressures (~600 MPa) and temperatures (500°C). This model explains the formation of the dome-and-keel structure without rigid plates or plate boundary forces and is similar to the sinking or *sagduction* models proposed for the formation of dome-and-keel structures in the Dharwar craton of India (Chardon *et al.*, 1996).

Horizontal tectonic models for the Eastern Pilbara propose that the greenstones were affected by one or more periods of horizontal contraction and extension (Blewitt, 2002). In these interpretations, the contraction results from episodes of Early Archean collision (Sections 10.4, 10.5) and terrane accretion (Section 10.6). Periods of horizontal extension result in the formation of crustal detachments and the emplacement of the granitoid domes. Kloppenburg *et al.* (2001) used observations of multiple cross cutting fabrics and unidirectional patterns of stretching lineations to suggest that the Mt. Edgar Dome initially formed as an extensional metamorphic core complex (Sections 7.3, 7.6.3, 7.6.6). An initial period of terrane collision and thrusting prior to 3.32 Ga thickens the Early Archean Warrawoona Greenstone Belt and buries granitoid basement to mid-crustal levels where it partially melts. Partial melting

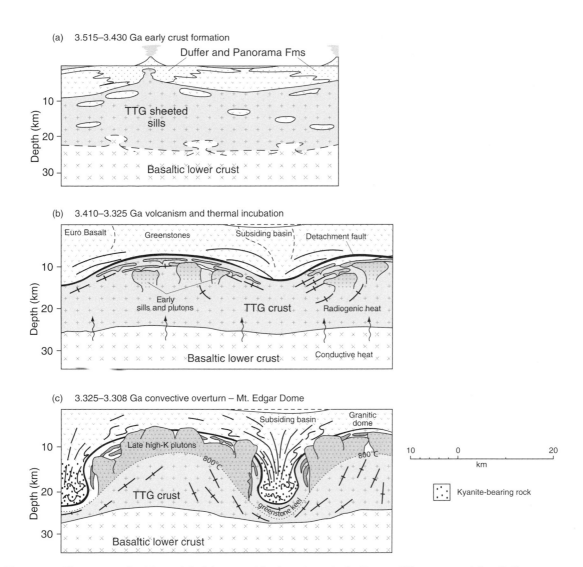

Fig. 11.10 *Three-stage diapiric model of dome-and-keel provinces in the Eastern Pilbara craton (after Collins et al., 1998, with permission from Pergamon Press, Copyright Elsevier 1998; the age-spans of the stages are from Van Kranendonk et al., 2007).*

facilitates the extensional collapse of the thickened crust, forming detachment faults (i.e. Mt Edgar shear Zone, Fig. 11.11a) similar to those found in Phanerozoic core complexes (e.g. Figs. 7.14b, 7.39c). The density inversion created by dense greenstones overlying buoyant, partially molten basement triggers the rise of granitoid domes at 3.31 Ga by solid state flow during extension (Fig. 11.11b). This extension is accommodated by displacement on the Mt. Edgar shear zone and by lateral strike-slip motion in a transfer zone within

the gneissic basement. Normal-sense displacements drop greenstones down between the rising domes. Emplacement of the domes steepens the detachments and changes the geometry of the system so that its structure no longer resembles that of Phanerozoic core complexes (Fig. 11.11c). Steepening during periods of shortening provides an alternative explanation of the near vertical sides of the granitoid domes.

The application of both vertical and horizontal models to Archean cratons involves numerous uncer-

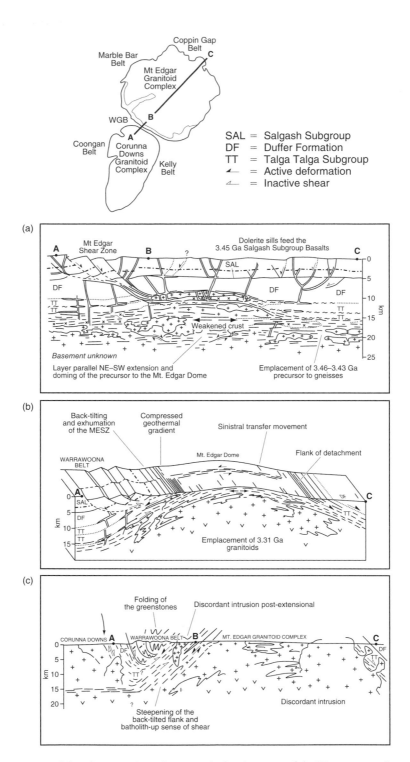

Fig. 11.11 *Cartoon summarizing the tectonic and magmatic development of the Warrawoona Greenstone Belt (WGB) and Mt. Edgar Dome by horizontal extension (after Kloppenburg et al., 2001, with permission from Elsevier). (a) Pre 3.33 Ga gabbro/diorite and dolerite intrusions, NE–SW extension, and doming of the Mt. Edgar Granitoid Complex. (b) Differential extension on the Mt. Edgar Shear Zone (MESZ) and lateral motion along a transfer zone within the granitoid complex at 3.31 Ga. (c) Final localized normal displacements and steepening of the MESZ followed by discordant intrusions of post-extensional plutons.*

tainties. Problems with diapiric models commonly include uncertainties surrounding the timing of convective overturn and how an inverted density profile could be maintained or periodically established over a 750 million year history without thrust faulting (Van Kranendonk *et al.*, 2004). How the stiff rheology of granitoids allows diapirism also is unclear. Problems with horizontal tectonic models may include a lack of evidence of large-scale tectonic duplication of the greenstones by thrusts in some areas and uncertainties surrounding how the formation of metamorphic core complexes could produce the distinctive ovoid patterns of the granitoids. Horizontal tectonic models also commonly encounter difficulty explaining the kinematics and horseshoe-shaped geometry of shear zones that border many granitoid domes (Marshak, 1999).

A comparison of the evolution of various Archean cratons has suggested that aspects of both horizontal and vertical tectonic processes occurred in different places and at different times. Hickman (2004) highlighted numerous tectonic and metamorphic differences between the Eastern and Western parts of the Pilbara craton prior to ~2.95 Ga. He showed that, unlike the more or less autochthonous dome-and-keel structure of the Eastern Pilbara, the Western Pilbara preserves a series of amalgamated terranes (Section 10.6.1) that are separated by a series of thrusts and strike-slip shear zones (Fig. 11.8) and involved episodes of horizontal compression that resemble a Phanerozoic style of plate tectonics. These differences suggest that both vertical and horizontal tectonics played an important role during the formation of the Pilbara craton.

11.4 PROTEROZOIC TECTONICS

11.4.1 General geology of Proterozoic crust

Proterozoic belts display two groups of rocks that are distinguished on the basis of their metamorphic grade and deformation history. The first group consists of thick sequences of weakly deformed, unmetamorphosed sedimentary and volcanic rocks that were deposited in large basins on top of Archean cratons. The second group is composed of highly deformed,

high-grade metamorphic rocks that define large orogenic belts. Both these groups contain distinctive suites of igneous rocks.

The most common lithologic assemblage in the weakly deformed parts of Early–Middle Proterozoic crust are quartzite-carbonate-shale sequences that reach thicknesses of some 10 km (Condie, 1982b). Quartz-pebble conglomerates and massive, cross-bedded sandstones also are common. Many of these sequences are intercalated with banded iron formations, cherts, and volcanic rocks. Other rock types that are either rare or absent in Archean belts appeared at this time, including extensive evaporites, phosphorous-rich sedimentary sequences, and red bed deposits (Section 3.4). These latter rocks generally are interpreted to have accumulated in stable, shallow water environments after 2.0 Ga. The appearance and the preservation of such thick sequences of sedimentary rock has been interpreted to reflect the stabilization of Precambrian continental crust during Proterozoic times (Eriksson *et al.*, 2001, 2005) (Section 11.4.2). In the Pilbara region of northwest Australia (Fig. 11.8) the deposition of 2.78–2.45 Ga coarse clastic sedimentary rocks and volcanic sequences in a shallow platform environment in the Hamersley Basin (Trendall *et al.*, 1991) reflects this stabilization. By 1.8 Ga, the existence of large, stable landmasses and free oxygen in the Earth's atmosphere allowed all of the well-known sedimentary environments that characterize Phanerozoic times to develop (Eriksson *et al.*, 2005).

The highly deformed regions of Proterozoic crust are divisible into two types (Kusky & Vearncombe, 1997). The first type consists of thick sedimentary sequences that were deformed into linear fold-and-thrust belts similar to those in Phanerozoic orogens (Figs 10.5, 10.19). The second type consists of high-grade gneisses of the granulite and upper amphibolite facies. Some of the largest and best known of these latter belts form the ~1.0 Ga Grenville provinces of North America, South America, Africa, Antarctica, India, and Australia (Fig. 11.19). Other belts (Fig. 11.12) evolved during the period 2.1–1.8 Ga (Zhao *et al.*, 2002). These orogens contain large ductile thrust zones that separate distinctive terranes. Some contain ophiolites (Section 2.5) that resemble Phanerozoic examples except for the lack of highly deformed mantle-derived rocks at their bases in ophiolites older than ~1 Ga (Moores, 2002). The presence of these features reflects the importance of subduction, collision, and terrane accretion along Proterozoic continental margins (Carr *et al.*, 2000; Karlstrom *et al.*, 2002).

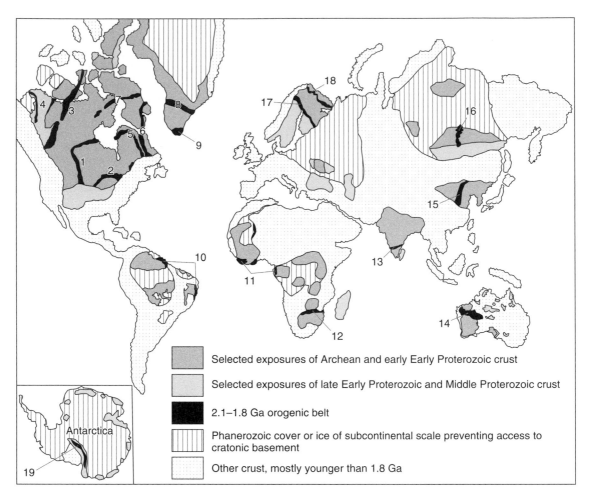

Fig. 11.12 *Global distribution of 2.1–1.8 Ga orogenic belts showing selected areas of Archean and Early Proterozoic basement (after Zhao et al., 2002, with permission from Elsevier). Orogens labeled as follows: 1, Trans-Hudson; 2, Penokean; 3, Taltson-Thelon; 4, Wopmay; 5, Cape Smith–New Quebec; 6, Torngat; 7, Foxe; 8, Nagssugtoqidian; 9, Makkovikian–Ketilidian; 10, Transamazonian; 11, Eburnian; 12, Limpopo; 13, Moyar; 14, Capricorn; 15, Trans-North China; 16, Central Aldan; 17, Svecofennian; 18, Kola-Karelian; 19, Transantarctic.*

A comparison of igneous rocks in Archean and Proterozoic belts indicates a progressive change in the bulk composition of the crust through time (Condie, 2005b). During the Early Archean, basaltic rocks were most abundant (Section 11.3.2). Later, the partial melting of these rocks in subduction zones or at the base of oceanic plateaux produced large volumes of TTG suite granitoids (Sections 11.3.2, 11.3.3). By 3.2 Ga granites first appeared in the geologic record and were produced in large quantities after 2.6 Ga.

This compositional trend from basalt to tonalite to granite generally is attributed to an increase in the importance of subduction and crustal recycling during the transition from Late Archean to Early Proterozoic times.

Large swarms of mafic dikes were emplaced into Archean cratons and their cover rocks during the Late Archean–Early Proterozoic and onwards. One of the best exposed examples of these is the 1.27 Ga MacKenzie dike swarm of the Canadian Shield, which exhibits dikes that fan out over a 100° arc and extend for more than 2300 km (Ernst *et al.*, 2001). Some of these shield regions also contain huge sills and layered intrusions of mafic and ultramafic rock that occupy hundreds to thousands of square kilometers. These intrusions provide information on the deep plumbing systems of Precambrian magma chambers and on crust–mantle interactions. Three of the best known examples are the ~1.27 Ga Muskox intrusion in northern Canada (Le Cheminant & Heaman, 1989; Stewart & DePaolo, 1996), the ~2.0 Ga Bushveld complex in South Africa (Hall, 1932; Eales & Cawthorn, 1996), and the ~2.7 Ga Stillwater complex in Montana, USA (Raedeke & McCallum, 1984; McCallum, 1996). Unlike the layered igneous suites of the Archean high-grade gneiss terrains (Section 11.3.2), these intrusions are virtually undeformed.

Anorthosite massifs (Section 11.3.2) emplaced during Proterozoic times also differ from the Archean examples. Proterozoic anorthosites are associated with granites and contain less plagioclase than the Archean anorthosites (Wiebe, 1992). These rocks form part of an association known as anorthosite-mangerite-charnockite-granite (AMCG) suites. *Charnockites* are high temperature, nearly anhydrous rocks that can be of either igneous or high-grade metamorphic origin (Winter, 2001). The source of magma and the setting of the anorthosites are controversial. Most studies interpret them as having crystallized either from mantle-derived melts that were contaminated by continental crust (Musacchio & Mooney, 2002) or as primary melts derived from the lower continental crust (Schiellerup *et al.*, 2000). Current evidence favors the former model. Some authors also have suggested that these rocks were emplaced in rifts or backarc environments following periods of orogenesis, others have argued that they are closely related to the orogenic process (Rivers, 1997). Their emplacement represents an important mechanism of Proterozoic continental growth and crustal recycling.

11.4.2 Continental growth and craton stabilization

Many of the geologic features that comprise Protero-zoic belts (Section 11.4.1) indicate that the continental lithosphere achieved widespread tectonic stability during this Eon. Tectonic stability refers to the general resistance of the cratons to large-scale lithospheric recy-cling processes. The results of seismic and petrologic studies (Sections 11.3.1, 11.3.3) and numerical modeling (Lenardic *et al.*, 2000; King, 2005) all suggest that com-positional buoyancy and a highly viscous cratonic mantle explain why the cratons have been preserved for billions of years. These properties, and isolation from the deeper convecting mantle, have allowed the mantle lithosphere to maintain its mechanical integrity and to resist large-scale subduction, delamination and/or erosion from below. Phanerozoic tectonic processes have resulted in some recycling of continental litho-sphere (e.g. Sections 10.2.4, 10.4.5, 10.6.2), however the scale of this process relative to the size of the cratons is small.

The cores of the first continents appear to have reached a sufficient size and thickness to resist being returned back into the mantle by subduction or delam-ination some 3 billion years ago. Collerson & Kamber (1999) used measurements of Nb/Th and Nb/U ratios to infer the net production rate of continental crust since 3.8 Ga. This method exploits differences in the behavior of these elements during the partial melting and chemical depletion of the mantle. The different ratios potentially provide information on the extent of the chemical depletion and the amount of continental crust that was present on Earth at different times. This work and the results of isotopic age determinations (Fig. 11.13) suggest that crust production was episodic with rapid net growth at 2.7, 1.9, and 1.2 Ga and slower growth afterward (Condie, 2000; Rino *et al.*, 2004). Each of these pulses may have been short, lasting ≤100 Ma

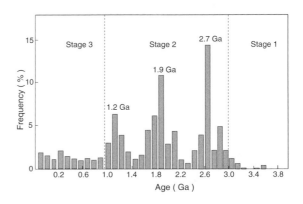

Fig. 11.13 *Plot showing the distribution of U-Pb zircon ages in continental crust (after Condie, 1998, with permission from Elsevier).*

(Condie, 2000). On the basis of available data, Condie (2005b) concluded that 39% of the continental crust formed in the Archean, 31% in the Early Proterozoic, 12% in the Middle–Late Proterozoic, and 18% in the Phanerozoic.

Two of the most important mechanisms of Late Archean and Early Proterozoic continental growth and cratonic root evolution were magma addition (Section 9.8) and terrane accretion (Section 10.6). Several authors (e.g. Condie, 1998; Wyman & Kerrich, 2002) have suggested that the ascent of buoyant mafic material in mantle plumes may have initiated crust formation and may have either initiated or modified root formation during periods of rapid net growth (Section 11.3.3). Schmitz *et al.* (2004) linked the formation and stabilization of the Archean Kaapvaal craton in South Africa to subduction, arc magmatism, and terrane accretion at 2.9 Ga. In this and most of the other cratons, isotopic ages from mantle xenoliths and various crustal assemblages indicate that chemical depletion in the mantle lithosphere was coupled to accretionary processes in the overlying crust (Pearson *et al.*, 2002). This broad correspondence is strong evidence that the crust and the underlying lithospheric mantle formed more or less contemporaneously and have remained mechanically coupled since at least the Late Archean. A progressive decrease in the degree of depletion in the lithospheric mantle since the Archean (Fig. 11.14) indicates that the Archean–Proterozoic boundary represents a major shift in the nature of lithosphere-forming processes, with more gradual changes occurring during the Phanerozoic (O'Reilly *et al.*, 2001). The most obvious driving mechanism of this change is the secular cooling of the Earth (Section 11.2). In addition, processes related to subduction, collision, terrane accretion, and magma addition also helped to form and stabilize the continental lithosphere.

Whereas these and many other investigations have identified some of the processes that contributed to the formation and stabilization of Archean cratons, numerous questions still remain. Reconciling the composition of craton roots determined from petrologic studies with the results of seismic velocity studies is problematic (King, 2005). There are many uncertainties about how stability can be achieved for billions of years without suffering mechanical erosion and recycling in the presence of subduction and mantle convection. Another problem is that the strength of mantle materials required to stabilize craton roots in numerical experiments exceeds the strength estimates of these materials

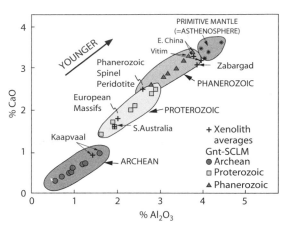

Fig. 11.14 *CaO-Al₂O₃ plot showing the range of subcontinental lithospheric mantle (SCLM) compositions for selected cratons that have been matched with ages of the youngest tectonothermal events in the overlying crust (after O'Reilly* et al.*, 2001, with permission from the Geological Society of America). Compositions have been calculated from garnet xenocrysts (Gnt). Xenolith averages shown for comparison. Plot shows that newly formed subcontinental lithospheric mantle has become progressively less depleted in Al and Ca contents from Archean through Proterozoic and Phanerozoic time. Garnet peridotite xenoliths from young extensional areas (e.g. eastern China, Vitim in the Baikal region of Russia, and Zabargad Island in the Red Sea) are geochemically similar to primitive mantle, indicating very low degrees of melt depletion.*

derived from laboratory measurements (Lenardic *et al.*, 2003). These issues, and the extent to which the cratonic mantle interacts with and influences the pattern of mantle convection, presently are unsolved. Improved resolution of the structure, age and geochemical evolution of the continental crust and lithospheric mantle promise to help geoscientists resolve these problems in the future.

11.4.3 Proterozoic plate tectonics

Early tectonic models of Proterozoic lithosphere envisaged that the Archean cratons were subdivided by mobile belts in which deformation was wholly ensialic, with no rock associations that could be equated with ancient ocean basins. These interpretations where Proterozoic orogenies occurred far from continental margins have since fallen out of favor. Most studies now indicate that Proterozoic orogens (Fig. 11.12) evolved along the margins of lithospheric plates by processes that were similar to those of modern plate tectonics.

One of the best-studied examples of an Early Proterozoic orogen that formed by plate tectonic processes lies between the Slave craton and the Phanerozoic Canadian Cordillera in northwestern Canada. This region provides a record of nearly 4 billion years of lithospheric development (Clowes *et al.*, 2005). Deep seismic reflection data collected as part of the Lithoprobe SNORCLE (**S**lave-**Nor**thern **C**ordillera **L**ithospheric **E**volution) transect of the Canadian Shield (see also Section 10.6.2) provide evidence of a modern, plate tectonic-style of arc–continent collision, terrane accretion, and subduction along the margin of the Archean Slave craton between 2.1 and 1.84 Ga (Cook *et al.*, 1999). These processes formed the Early Proterozoic Wopmay Orogen (Fig. 11.15a) and resulted in continental growth through the addition of a series of magmatic arcs, including the Hottah and Fort Simpson terranes and the Great Bear magmatic arc.

Final assembly of the Slave craton occurred by ~2.5 Ma. Cook *et al.* (1999) suggested that low-angle seismic reflections beneath the Yellowknife Basin (Fig. 11.15b) represent surfaces that accommodated shortening during this assembly. Some of these reflections project into the upper mantle and represent the remnants of an east-dipping Late Archean subduction zone. Following amalgamation of the craton, the Hottah terrane formed as a magmatic arc some distance outboard of the ancient continental margin between 1.92 and 1.90 Ga. During the Calderian phase (1.90–1.88 Ga) of the Wopmay Orogen this arc terrane collided with the Slave craton, causing compression, shortening, and the eastward translation of exotic material (Fig. 11.15c). In the seismic profile (Fig. 11.15b), the accreted Proterozoic crust displays gently folded upper crustal layers overlying reflectors that appear to be thrust slices above detachment faults that flatten downward into the Moho. Remnants of the old, east-dipping subduction zone associated with the collision are still visible today as reflections that project to 200 km or more beneath the Slave craton.

Once accretion of the Hottah terrane terminated, the subduction of oceanic lithosphere to the east beneath the continental margin created the 1.88–1.84 Ga Great Bear Magmatic arc and eventually led to the collision of the Fort Simpson terrane some time before 1.71 Ga (Fig. 11.15c,d). Mantle reflections that record subduction and shortening during this arc–continent collision dip eastward beneath the Great Bear magmatic arc from the lower crust to depths of 100 km (Fig. 11.15b,c,d). Where the mantle reflections flatten into the lower crust, they merge with west-dipping crustal reflections, producing a lithospheric-scale accretionary wedge that displays imbricated thrust slices. This faulted material and the underthrust lower crust represent part of a Early Proterozoic subduction zone that bears a remarkable resemblance to structures that record subduction and accretion within the Canadian Cordillera (Fig. 10.33) and along the Paleozoic margin of Laurentia (Fig. 10.34). Seismic refraction and wide-angle reflection data (Fernández-Viejo & Clowes, 2003) indicate the presence of unusually high velocity (7.1 km s^{-1}) lower crust and unusually low velocity (7.5 km s^{-1}) upper mantle in this zone (Fig. 11.16c) compared to other parts of this section (Fig. 11.16a,b,d). This observation indicates that the effects of collision, subduction, and the accompanying physical changes in rocks of the mantle wedge remain identifiable 1.84 billion years after they formed.

In western Australia, distinctive patterns of magnetic anomalies provide direct evidence for the collision and suturing of the Archean Yilgarn and Pilbara cratons beginning by ~2.2 Ga (Cawood & Tyler, 2004). The Capricorn Orogen (Fig. 11.17a,b) is composed of Early Proterozoic plutonic suites, medium- to high-grade metamorphic rocks, a series of volcano-sedimentary basins, and the deformed margins of the two Archean cratons. Late Archean rifting and the deposition of passive margin sequences at the southern margin of the Pilbara craton is recorded by the basal sequences of the Hamersley Basin. Following rifting between the cratons, several major pulses of contractional deformation and metamorphism took place during the intervals 2.00–1.96 Ga, 1.83–1.78 Ga, and 1.67–1.62 Ga. These events resulted in basin deformation and the juxtaposition of cratons of different age and structural trends (Fig. 11.17b,c). The episodic history of rifting followed by multiple episodes of contraction and collision corresponds to at least one and probably two Wilson cycles (Section 7.9) involving the opening and closing of Late Archean–Early Proterozoic ocean basins (Cawood & Tyler, 2004). The presence of similar collisional orogens in Laurentia, Baltica, Siberia, China, and India suggests that the early to mid-Early Proterozoic marks a period

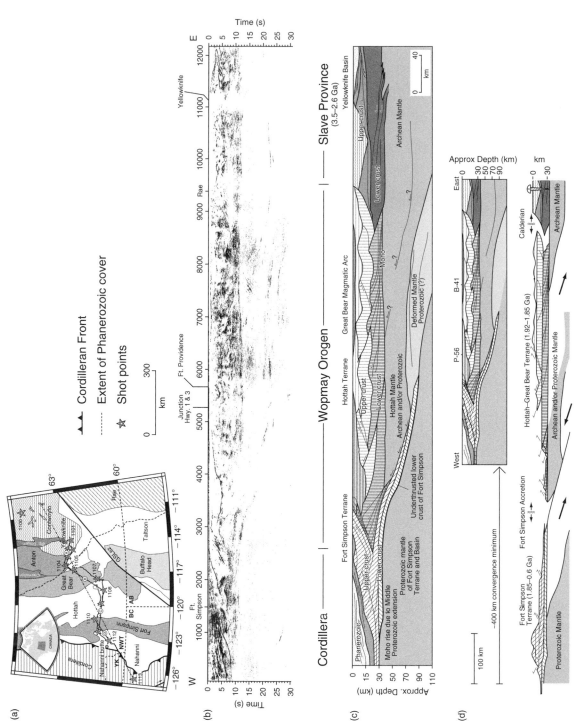

Fig. 11.15 (a) Map showing tectonic elements of the Wopmay Orogen and location of the SNORCLE transect (after Fernández-Viejo & Clowes, 2003, with permission from Blackwell Publishing). Straight black lines (circled letters) show division used for interpretation of crustal structure. Orogen includes Coronation, Great Bear, Hottah, Fort Simpson and Nahanni domains. BC, British Colombia; AB, Alberta; YK, Yukon; NWT, Northwest Territories; GSLsz, Great Slave Lake shear zone. (b) Seismic profile, (c) interpretation and (d) reconstruction of the Wopmay Orogen and Slave Province (modified from Cook et al., 1999, by permission of the American Geophysical Union. Copyright © 1999 American Geophysical Union). Reconstruction is made along major faults (bold black lines) and shows a minimum estimate.

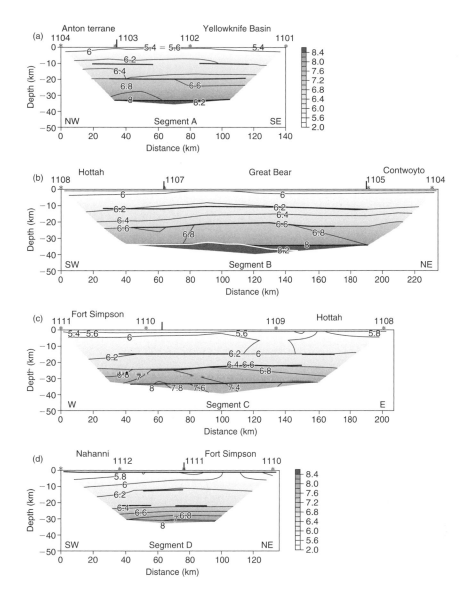

Fig. 11.16 *P-wave velocity model for (a) Slave province, (b) eastern part of the Wopmay Orogen and transition to Slave Province, (c) Hottah and Fort Simpson terranes, and (d) Fort Simpson terrane and Nahanni domain (after Fernández-Viejo & Clowes, 2003, with permission from Blackwell Publishing). Line segments shown in Fig. 11.15a. Contours are drawn at 0.2 km s^{-1} increments. Heavy black lines are locations from which wide-angle reflections were observed.*

of supercontinent assembly by plate tectonic processes (Fig. 11.12) (Section 11.5.4).

The appearance of the metamorphic products of subduction and continental collision during the Early Proterozoic, including eclogites and other >1 GPa high-pressure metamorphic assemblages (e.g. Sections 9.9,

10.4.2), represents an important marker of the onset of tectonic processes similar to those seen in the Phanerozoic Earth (O'Brien & Rötzler, 2003; Collins *et al.*, 2004; Brown, 2006). Phanerozoic eclogites commonly preserve evidence of having been partially subducted to depths greater than ~50 km and then rapidly exhumed

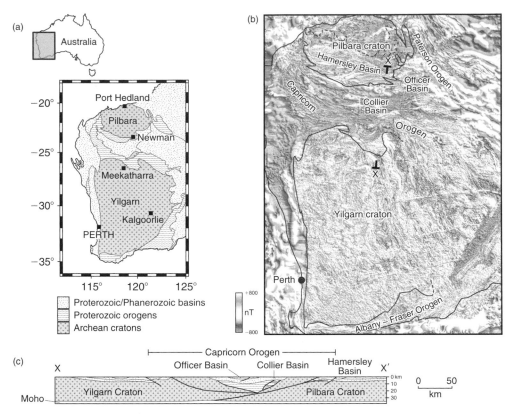

Fig. 11.17 *(a) Tectonic map (after Hackney, 2004, with permission from Elsevier) and (b) magnetic anomaly image emphasizing gradients in total magnetic intensity (after Kilgour & Hatch, 2002, with permission from Geoscience Australia, image provided by M. Van Kranendonk, Geological Survey of Western Australia). Magnetic anomaly image shows total magnetic intensity measured in nanoteslas (nT) compiled from airborne, marine and land-based geophysical surveys. The distinctive magnetic anomaly patterns from the Pilbara and Yilgarn cratons reflect different structural trends that resulted from Precambrian plate tectonic processes. (c) Interpretive cross-section of the Capricorn Orogen showing the sutured cratons (after Cawood & Tyler, 2004, with permission from Elsevier).*

(Fig. 11.18) (Collins *et al.*, 2004). In the modern Earth, these unusual conditions are met in subduction–accretion complexes and at the sites of continental collision where relatively cold crust is buried to subcrustal depths. For these subducted rocks to return to the Earth's surface with preserved eclogite facies mineral assemblages, they must be exhumed rapidly before tectonically depressed isotherms can re-equilibrate and overprint the assemblages with higher temperature granulite facies minerals.

The oldest examples of in-situ eclogites (i.e. rocks other than xenoliths) include 1.80 Ga and 2.00 Ga varieties (<1.8 GPa, 750°C) from the North China craton and the Proterozoic orogens surrounding the Tanzanian craton (Zhao *et al.*, 2001; Möller *et al.*, 1995). Eclogites

recording conditions of ~1.2 GPa and 650–700°C at 1.9–1.88 Ga also occur in the Lapland Granulite Belt of Finland (Tuisku & Huhma, 1998). The Aldan Shield in Siberia (Smelov & Beryozkin, 1993) and the Snowbird tectonic zone between the Rae and Hearne cratons of Canada (Baldwin *et al.*, 2003) preserve retrogressed eclogites of 1.90 Ga. The absence of Archean eclogite facies rocks suggests that before Early Proterozoic times either the conditions to produce such rocks did not exist, the processes to exhume them at a sufficient rate to preserve eclogite-facies mineral assemblages did not exist, or all pre-existing examples have been obliterated by subsequent tectonic events.

The presence of ophiolitic assemblages in Precambrian orogens provides another possible marker of tec-

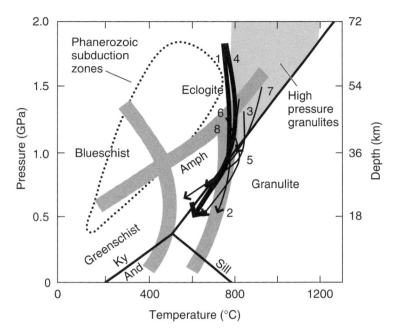

Fig. 11.18 *Pressure–temperature plot showing the published paths of the moderate temperature eclogite and high-pressure granulite facies metamorphic rocks older than 1.5 Ga (after Collins et al., 2004, with permission from Elsevier). 1, Usagaran Belt, Tanzania; 2, Hengshan Belt, China; 3, Sanggan Belt, China; 4, Ubendian Belt, Tanzania; 5, Jianping Belt, China; 6, Sare Sang, Badakhshan Block, Afghanistan; 7, Snowbird tectonic zone, Canada; 8, Lapland Granulite Belt. Thick arrows refer to Usagaran/Ubende eclogites. Field for Phanerozoic subduction zone metamorphism and metamorphic facies indicated by thick gray curves. Aluminosilicate polymorph fields plotted for reference. And, andalusite; Ky, kyanite; Sill, sillimanite; Amph, amphibolite.*

tonic processes similar to those operating during the Phanerozoic (de Wit, 2004; Parman *et al.*, 2004). Some Archean greenstone belts have been interpreted as ophiolites, although these interpretations typically are controversial. The oldest unequivocal examples are Early Proterozoic in age and support interpretations that seafloor spreading and associated ocean crust formation was an established mechanism of plate tectonics by this time. One of the best preserved and least equivocal of the Early Proterozoic examples is the Purtuniq Complex (Scott *et al.*, 1992; Stern *et al.*, 1995) of the Trans-Hudson Orogen between the Hearne and Superior cratons of northern Canada (Plate 11.1a between pp. 244 and 245). Other examples occur in the Arabian-Nubian Shield (Kröner, 1985) and the Yavapai-Mazatzal orogens of the southwestern USA (Condie, 1986). The presence of these features suggests that complete Wilson cycles (Section 7.9) of ocean opening and closure and ophiolite obduction were occurring by at least 2.0 Ga.

Together these observations strongly suggest that plate tectonic mechanisms became increasingly important after the Late Archean. Most authors link this development to the increased stability of continental crust during this Eon (Section 11.4.2). Nevertheless, it is important to realize that many uncertainties about Proterozoic tectonics remain. The age of formation of the rock units and the timing of regional metamorphism, deformation, and cooling are poorly known in many regions. The sources of magmatism and the amounts and mechanisms of crustal recycling also commonly are unclear. In addition, there continues to be a need for high resolution paleomagnetic and geochronologic data to enable accurate reconstructions of the continents and oceans during Proterozoic as well as Archean times (Section 11.5.3). These data are crucial for determining the tectonic evolution of Proterozoic lithosphere and for detailed comparisons between Proterozoic and Phanerozoic orogens.

11.5 THE SUPERCONTINENT CYCLE

....

11.5.1 Introduction

Geologic evidence for the repeated occurrence of continental collision and rifting since the Archean has led to the hypothesis that the continents periodically coalesced into large landmasses called supercontinents. The best known of the supercontinents include Gondwana (Fig. 3.4) and Pangea (Fig. 11.27), which formed in the latest Proterozoic and late Paleozoic times, respectively. Other supercontinents, such as Rodinia and Laurussia, also have been proposed for Late Proterozoic and late Paleozoic times, respectively. Processes in the mantle that may have led to their assembly and dispersal are discussed in Section 12.11.

11.5.2 Pre-Mesozoic reconstructions

Paleogeographic maps for the Mesozoic and Cenozoic can be computed by the fitting together of continental margins or oceanic lineations of the same age on either side of an ocean ridge (Chapters 3, 4). The location of the paleopoles can be determined from paleomagnetic measurements (Section 3.6) and so the only unknown in these reconstructions is the zero meridian of longitude. These combined techniques cannot be used for reconstructions prior to the Mesozoic because *in situ* oceanic crust is lacking.

Methods of quantifying plate motions in pre-Mesozoic times involve the use of paleomagnetic data coupled with high-precision geochronology. Ancient plate edges, although somewhat distorted, are marked by orogenic belts and ophiolite assemblages (Sections 2.5, 11.4.3), which indicate ancient sutures between welded continents and accreted terranes. Evidence provided from the past distributions of flora and fauna and indicators of paleoclimate also aid these plate reconstructions (Sections 3.4, 3.5). For a particular time the paleomagnetic pole for each ancient plate is rotated to an arbitrary single magnetic pole and the continents on the plate are re-projected using the same Eulerian rotation. The continents are then successively shifted along fixed latitudes, that is, rotated about the magnetic pole, until the overlap of continental margins is minimized. Although the paleomagnetic data do not provide a unique sequence of reconstructions, they clearly indicate the gross trends of plate movements during ancient times. More detailed inferences on the evolution of particular regions are then made from their geology viewed in terms of plate tectonic mechanisms.

The application of paleomagnetic methods for the Precambrian is less straightforward than for Phanerozoic times for three main reasons (Dunlop, 1981). First, the error limits of isotopic ages typically are larger. Second, isotopic and magnetic records may be partially reset during metamorphism to different degrees, and the distinction between pre- and post-orogenic isotopic and magnetic overprints can be difficult. Third, overprints occur during post-orogenic cooling and uplift, and the temperatures at which isotopic systems close and magnetizations stabilize are different, so that the dates may be younger or older than the magnetizations by intervals of tens of millions of years. However, even given these uncertainties and the gaps in the paleomagnetic record arising from the lack of suitable samples of certain ages, the data allow investigators to test the validity of paleogeographic reconstructions for pre-Mesozoic times based on the geologic record on the continents.

11.5.3 A Late Proterozoic supercontinent

Similarities between the Late Proterozoic geologic record in western Canada and eastern Australia (Bell & Jefferson, 1987; Young, 1992) and between the southwestern USA and East Antarctica suggest that these areas were juxtaposed during Late Proterozoic times (Dalziel, 1991, 1995; Moores, 1991; Hoffman, 1991) (Fig. 11.19a). This seemingly radical suggestion was referred to as the SWEAT (**S**outh **W**est US and **E**ast An**T**arctica) hypothesis. The widespread Grenville orogenic belts, that immediately pre-date the Late Proterozoic, suggest that many other continental fragments can be added to this reconstruction to form a Late Proterozoic supercontinent called *Rodinia* (Fig. 11.19a). Laurentia (North America and Greenland) forms the core of the supercontinent and is flanked to the north by East Antarctica. The reconstruction shows that the North

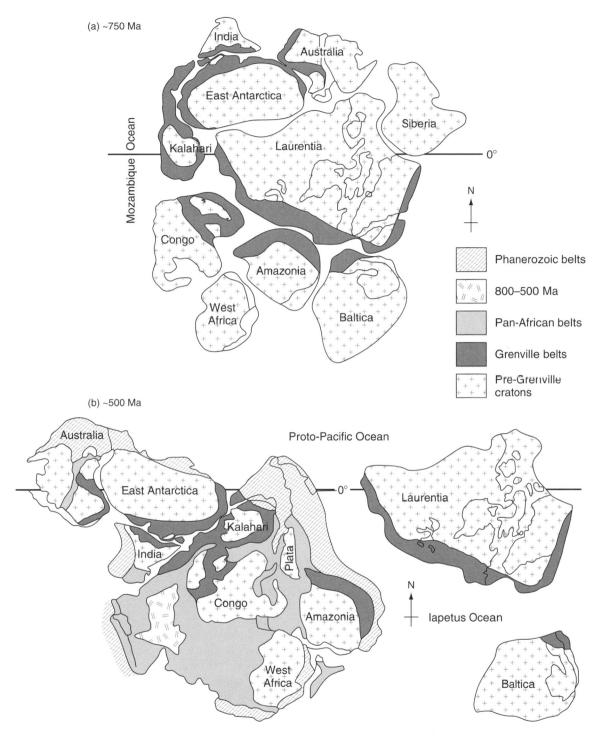

Fig. 11.19 *(a) Reconstruction of the Late Proterozoic supercontinent Rodinia. (b) Late Cambrian paleogeography after the break-up of Rodinia and the formation of Gondwana (after Hoffman, 1991, with permission from* Science ***252**, 1409–12 with permission from the AAAS).*

American Grenville Province continues directly into East Antarctica, and similar belts of this age can be traced over most of the Gondwana fragments. The age of the oldest sedimentary rock associated with break-up, and the provinciality of certain animal groups across the split, suggest that the supercontinent fragmented at about 750 Ma (Storey, 1993). During fragmentation the blocks now making up East Gondwana (East Antarctica, Australia, and India) moved anticlockwise, opening the proto-Pacific Ocean (Panthalassa), and collided with the blocks of West Gondwana (Congo, West Africa, and Amazonia). The intervening Mozambique Ocean closed by the pincer-like movements of these blocks and Gondwana was created when they collided to form the Mozambique belt of East Africa and Madagascar. Gondwana then rotated clockwise away from Laurentia about 200 Ma later. Southern Africa was located at the pivot of these movements and Baltica moved independently away from Laurentia, opening the Iapetus Ocean, which subsequently closed during the assembly of Pangea (Section 11.5.5). Figure 11.19b shows a postulated configuration at 500 Ma.

The first paleomagnetic test of the SWEAT hypothesis was carried out by Powell et al. (1993) who showed that paleomagnetic poles at 1055 Ma and at 725 Ma for Laurentia and East Gondwana are in agreement when repositioned according to the Rodinia reconstruction, thereby lending support to the hypothesis. Between 725 Ma and the Cambrian the APWPs diverge, suggesting that East Gondwana broke away from Laurentia after 725 Ma. The only fragment of Rodinia for which a detailed Apparent Polar Wanderer (APW) path can be defined for the period 1100–725 Ma is Laurentia (McElhinny & McFadden, 2000). This, therefore, has been used as a reference path against which repositioned paleomagnetic poles from other Rodinian fragments can be compared. However, many of the tests were hindered by a lack of high quality geochronology. As new data were collected, the existence of a Late Proterozoic supercontinent gained acceptance, although numerous modifications have been proposed (Dalziel et al., 2000b; Karlstrom et al., 2001; Meert & Torsvik, 2003). There is now considerable geologic and paleomagnetic evidence that, except for Amazonia, the cratons of South America and Africa were never part of Rodinia, although they probably were close to it (Kröner & Cordani, 2003). Newer models also indicate the piecemeal assembly of Rodinia beginning with Grenville-age collisions in eastern Canada and Australia at 1.3–1.2 Ga, followed by an Amazonia–Laurentia collision at 1.2 Ga

(Tohver et al., 2002), the majority of assembly between 1.1 and 1.0 Ga, and minor collisions between 1.0 and 0.9 Ga (Li et al., 2008). Most current models of Rodinia also show a fit between the cratons at 750 Ma that differs substantially from the older hypotheses (Wingate et al., 2002). Torsvik (2003) published a model (Fig. 11.20) that summarizes some of these changes. The position of the continents suggests that the break-up of Rodinia had begun by 850 or 800 Ma with the opening of the proto-Pacific ocean between western Laurentia and Australia-East Antarctica. The emplacement of mafic dike swarms in western Laurentia at 780 Ma may reflect this fragmentation (Harlan et al., 2003). The position of Australia-East Antarctica also suggests that India was not connected to East Antarctica until after ~550 Ma. This model emphasizes that the internal geometry of Rodinia probably changed repeatedly during the few hundred million years it existed.

The differences among the new and old models of Rodinia illustrate the controversial and fluid nature of Precambrian reconstructions. Numerous uncertainties in the relative positions of the continents exist, with the paleolatitudes of only a few cratons being known for any given time. It also must be remembered that paleomagnetic methods give no control on paleolongitude (Section 3.6), so that linear intercratonic regions whose strike is directed towards the Eulerian pole used to bring the cratons into juxtaposition are not constrained to have had any particular width. For these reasons, most reconstructions rely on combinations of many different data sets, including geological correlations based on orogenic histories, sedimentary provenance, the ages of rifting and continental margin formation, and the record of mantle plume events (Li et al., 2008).

Another controversial aspect of the Rodinia supercontinent concerns the effect of its dispersal on past climates. Some studies suggest that as Rodinia fragmented the planet entered an icehouse or *snowball Earth* state in which it was intermittently completely covered by ice (Evans, 2000; Hoffman & Schrag, 2002). The geologic evidence for this intermittent but widespread glaciation includes glacial deposits of Late Proterozoic age that either contain carbonate debris or are directly overlain by carbonate rocks indicative of warm waters. In addition, paleomagnetic data suggest that during at least two Late Proterozoic glacial episodes ice sheets reached the equator. One possible explanation of these observations is that periods of global glaciation during the Late Proterozoic were controlled by anomalously low atmospheric CO_2 concentrations (Hyde et al., 2000;

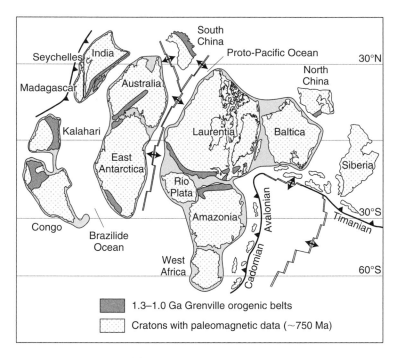

Fig. 11.20 *Reconstruction of Rodinia at ~750 Ma (after Torsvik, 2003, with permission from* Science *300, 1379–81, with permission from the AAAS).*

Donnadieu *et al.*, 2004). During break-up, the changing paleogeography of the continents may have led to an increase in runoff, and hence consumption of CO_2, through continental weathering that decreased atmospheric CO_2 concentrations (Section 13.1.3). The extreme glacial conditions may have ended when volcanic outgassing of CO_2 produced a sufficiently large greenhouse effect to melt the ice. The resulting "hothouse" would have enhanced precipitation and weathering, giving rise to the deposition of carbonates on top of the glacial deposits during sea-level (Hoffman *et al.*, 1998). Alternatively, these transitions may have resulted mainly from the changing configuration of continental fragments and its effect on oceanic circulation (Sections 13.1.2, 13.1.3). Whichever view is correct, these interpretations suggest that the break-up of Rodinia triggered large changes in global climate. However, the origin, extent, and termination of the Late Proterozoic glaciations, and their possible relationship to the supercontinental break-up, remains an unresolved and highly contentious issue (Kennedy *et al.*, 2001; Poulsen *et al.*, 2001).

11.5.4 Earlier supercontinents

The origin of the first supercontinent and when it may have formed are highly speculative. Bleeker (2003) observed that there are about 35 Archean cratons today (Plate 11.1a between pp. 244 and 245) and that most display rifted margins, indicating that they fragmented from larger landmasses. Several possible scenarios have been envisioned for the global distribution of the cratons during the transition from Late Archean to Early Proterozoic times (Fig. 11.21). These possibilities include a single supercontinent, called *Kenorland* by Williams *et al.* (1991) after an orogenic event in the Canadian Superior Province, and the presence of either a few or many independent aggregations called *supercratons*. Bleeker (2003) concluded that the degree of geologic similarity among the exposed cratons favors the presence of several transient, more or less independent supercratons rather than a single supercontinent or many small dispersed landmasses. He defined a

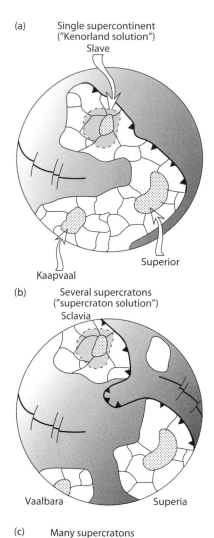

(a) Single supercontinent
("Kenorland solution")
Slave

Kaapvaal

Superior

(b) Several supercratons
("supercraton solution")
Sclavia

Vaalbara

Superia

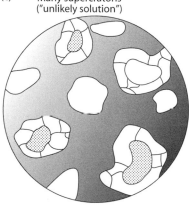

(c) Many supercratons
("unlikely solution")

Fig. 11.21 *Cartoons representing possible craton configurations during Late Archean–Early Proterozoic times. Three well-known cratons (Slave, Superior and Kaapvaal) are shown shaded in (a). These cratons may have been spawned by the larger supercratons shown in (b) (after Bleeker, 2003, with permission from Elsevier).*

minimum of three supercratons, *Sclavia*, *Superia*, and *Vaalbara*, that display distinct amalgamation and break-up histories (Fig. 11.21b). The *Sclavia* supercraton appears to have stabilized by 2.6 Ga. Confirmation of these tentative groupings awaits the collection of detailed chronostratigraphic profiles for each of the 35 Archean cratons.

Diachronous break-up of the supercratons defined by Bleeker (2003) occurred during the period 2.5–2.0 Ga, spawning the 35 or more independently drifting cratons. Paleomagnetic evidence supports the conclusion that significant differences in the paleolatitudes existed between at least several of these fragments during the Early Proterozoic (Cawood *et al.*, 2006). Following the break-up the cratons then appear to have amalgamated into various supercontinents. Hoffman (1997) postulated a Middle Proterozoic supercontinent called *Nuna*, which Bleeker (2003) considered to represent the first true supercontinent. Zhao *et al.* (2002) also recognized that most continents contain evidence for 2.1–1.8 Ga orogenic events (Section 11.4.3) (Fig. 11.12). They postulated that these orogens record the collisional assembly of an Early–Middle Proterozoic supercontinent called *Columbia* (Fig. 11.22). These studies, while still speculative, suggest that at least one supercontinent formed prior to the final assembly of Rodinia and after the Archean cratons began to stabilize during the Late Archean.

11.5.5 Gondwana–Pangea assembly and dispersal

The assembly of Gondwana began immediately following the break-up of Rodinia in Late Proterozoic times. According to the SWEAT hypothesis (Section 11.5.3) West Gondwana formed when many small ocean basins that surrounded the African and South American cratons closed during the opening of the proto-Pacific Ocean, creating the Pan-African orogens (Fig. 11.19b). Subsequent closure of the Mozambique Ocean resulted in the collision and amalgamation of West Gondwana with the blocks of East Gondwana. This amalgamation may have created a short-lived Early Cambrian supercontinent called *Pannotia*. The existence of this supercontinent is dependent on the time of rifting between Laurentia and Gondwana (Cawood *et al.*, 2001). Models of Pannotia (Fig. 11.23a) are based mostly on geologic evidence that Laurentia and Gondwana were attached

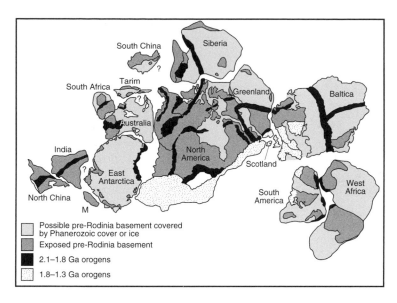

Fig. 11.22 *Reconstruction of the postulated Early–Middle Proterozoic supercontinent Columbia (after Zhao et al., 2002, with permission from Elsevier). M, Madagascar. Early Proterozoic orogens (2.1–1.8 Ga) are identified in Fig. 11.12.*

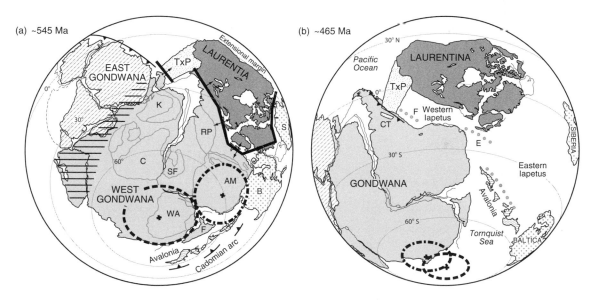

Fig. 11.23 *Postulated reconstructions of (a) Pannotia at ~545 Ma and (b) the rifting of Laurentia and Gondwana at ~465 Ma emphasizing the paleogeography of Laurentia relative to Gondwana (after Dalziel, 1997, with permission from the Geological Society of America). Crosses with 95% confidence circles shown in (a) with a dashed line and a dashed-dotted line indicate paleomagnetic poles for Laurentia and Gondwana, respectively. Horizontal lines in (a) denote limit of the Mozambique orogenic belt; thick solid black line marks the location of a Laurentia–Gondwana rift. In (b) paleomagnetic poles are for Laurentia + Baltica + Siberia + Avalonia (cross with dashed confidence circle) and Gondwana (cross with dashed-dotted confidence circle). Cuyania (CT) has accreted onto the Gondwana margin. Abbreviations: C, Congo; K, Kalahari; WA, West Africa; AM, Amazonia; RP, Río de la Plata; SF, São Francisco; S, Siberia; B, Baltica; TxP, the hypothetical Texas Plateau; F, Famatina arc; E, Exploits arc.*

or in close proximity at the end of the Late Proterozoic (Dalziel, 1997). However, the paleomagnetic poles for these two landmasses do not overlap, suggesting that an alternative configuration where Laurentia is separated from Gondwana at this time also is possible (Meert & Torsvik, 2003).

Most models suggest that the break-up of Pannotia began with the latest Proterozoic or Early Cambrian opening of the Iapetus Ocean as Laurentia rifted away from South America and Baltica (Figs 11.19b, 11.23b). Subduction zones subsequently formed along the Gondwana and Laurentia margins of Iapetus, creating a series of volcanic arcs, extensional backarc basins, and rifted continental fragments. As the ocean closed this complex assemblage of terranes accreted onto the margins of both Laurentia and Gondwana. The provenance of these terranes provides a degree of control on the relative longitudes and paleogeography of these two continents prior to the Permo-Carboniferous assembly of Pangea (Dalziel, 1997).

The Early Paleozoic accreted terranes of Laurentia and Gondwana are classified into groups according to whether they are native or exotic to their adjacent cratons (Keppie & Ramos, 1999; Cawood, 2005). Those native to Laurentia include the Notre Dame–Shelburne Falls (Taconic) and Lough–Nafooey volcanic arcs (Figs 10.34, 11.24a), which formed near and accreted onto Laurentia during Early–Middle Ordovician times. These collisions were part of the Taconic Orogeny in the Appalachians (Karabinos et al., 1998), the Grampian Orogeny in the British Isles, and the Finnmarkian Orogeny in Scandinavia. During the same period, the Famatina arc terrane (Fig. 11.23b), of Gondwana affinity, formed near and accreted onto the western margin of South America (Conti et al., 1996).

Terranes exotic to Laurentia include Avalonia, Meguma, Carolina, and Cadomia (Fig. 11.24a). These

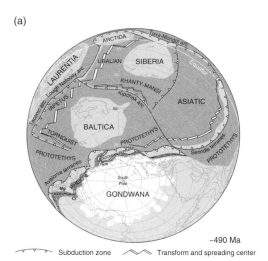

(a)

~490 Ma

⌄⌄⌄ Subduction zone   ⋀⋁⋀ Transform and spreading center

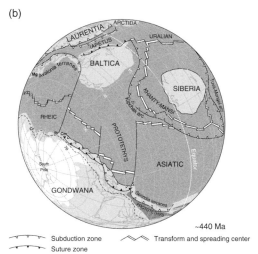

(b)

~440 Ma

⌄⌄⌄ Subduction zone   ⋀⋁⋀ Transform and spreading center
⌄⌄⌄ Suture zone

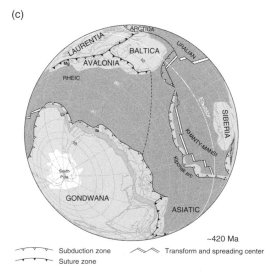

(c)

~420 Ma

⌄⌄⌄ Subduction zone   ⋀⋁⋀ Transform and spreading center
⌄⌄⌄ Suture zone

Fig. 11.24 *Postulated Paleozoic plate reconstructions for (a) 490 Ma, (b) 440 Ma, and (c) 420 Ma emphasizing the paleogeography of terranes derived from northern Gondwana and the opening of the Rheic Ocean (images provided by G.M. Stampfli and modified from von Raumer et al., 2003, and Stampfli & Borel, 2002, with permission from Elsevier). Interpretations incorporate the dynamics of hypothesized convergent, divergent and transform plate boundaries. Labeled terranes in (a) are: Mg, Meguma; Cm, Cadomia; Ib, Iberia; Cr, Carolina.*

and other terranes rifted from northwestern Gondwana in the Early Ordovician and later accreted onto the Laurentian margin, forming part of the Silurian-Devonian Acadian and Salinic orogens in the northern Appalachians and the Caledonides of Baltica and Greenland (Figs 11.24c, 11.25). Cuyania, an exotic terrane located in present day Argentina (Fig. 11.23b), rifted from southern Laurentia during Early Cambrian times and later accreted onto the Gondwana margin (Dalziel, 1997). These tectonic exchanges suggest that at least two different plate regimes existed in eastern and western Iapetus during the Paleozoic with subduction zones forming along parts of both Gondwana and Laurentia (Fig. 11.24a). Although the geometry of the plate boundaries is highly speculative, the interpretation of distinctive plate regimes explains the piecemeal growth of both continents by terrane accretion prior to the assembly of Pangea.

The rifting of the Avalonia terranes from Gondwana in the Late Cambrian and Early Ordovician led to the opening of the Rheic Ocean between the Gondwana mainland and the offshore crustal fragments (Fig. 11.24a,b). After the closure of Iapetus and the accretion of Avalonia, the Rheic Ocean continued to exist between Laurentia and Gondwana, although its width is uncertain (Fig. 11.24c). During these times a new series of arc terranes rifted from the Gondwana margin, resulting in the opening of the Paleotethys Ocean (Fig. 11.26a). The opening of Paleotethys and the closure of the Rheic Ocean eventually resulted in the accretion of these Gond-

wana-derived terranes onto Laurentia followed by a continent–continent collision between Laurussia and Gondwana (Fig. 11.26b). This latter collision produced the Permo-Carboniferous Alleghenian and Variscan orog-

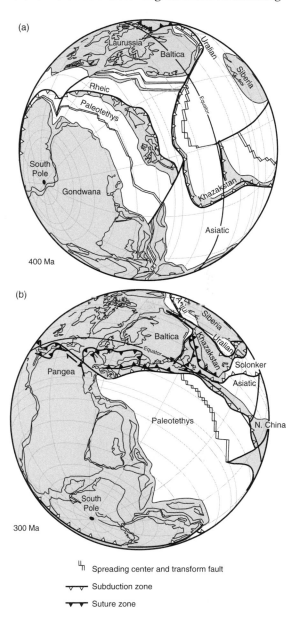

Fig. 11.26 *Postulated Paleozoic plate reconstructions for (a) 400 Ma and (b) 300 Ma (images provided by G.M. Stampfli and modified from Stampfli & Borel, 2002, with permission from Elsevier). In (a) The Rheic Ocean closes as Paleotethys opens. In (b) Gondwana has collided with Laurussia creating the European Variscides and Alleghenian Orogen.*

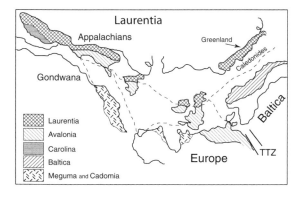

Fig. 11.25 *Late Paleozoic reconstruction showing the Silurian–Devonian Appalachian (Acadian and Salinic)–Caledonian orogens (after Keller & Hatcher, 1999, with permission from Elsevier). TTZ is the Teisseyre-Tornquist zone, representing a major crustal boundary between Baltica and southern Europe.*

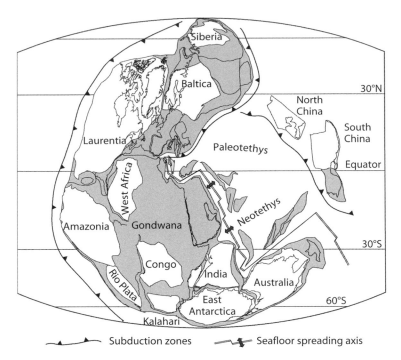

Fig. 11.27 *Reconstruction of Pangea at 250 Ma (after Torsvik, 2003, with permission from Science* **300**, *1379–81, with permission from the AAAS). Major cratons are shown.*

enies in North America, Africa, and southwest Europe. Collisions in Asia, including the suturing of Baltica and Siberia to form the Ural Orogen at ~280 Ma, resulted in the final assembly of Pangea. The supercontinent at the height of its extent at ~250 Ma is shown in Fig. 11.27.

Like its assembly, the fragmentation of Pangea was heterogeneous. Break-up began in the mid-Jurassic with the rifting of Lhasa and West Burma from Gondwana and the opening of the central Atlantic shortly after 180 Ma (Lawver *et al.*, 2003). Magnetic anomalies indicate that by 135 Ma the southern Atlantic had started to open. Rifting between North America and Europe began during the interval 140–120 Ma. Africa and Ant-

arctica began to separate by 150 Ma. Australia also began to rift from Antarctic by 95 Ma with India separating from Antarctica at about the same time. These data indicate that the majority of Pangea break-up occurred during the interval 150–95 Ma. Small fragments of continental crust such as Baja California and Arabia continue to be rifted from the continental remnants of Pangea. As with the older supercontinents, the break-up of Pangea was accompanied by the closure of oceans, such as Paleotethys and Neotethys (Fig. 11.27), and by collisions, including those that occur presently in southern Asia (Fig. 10.13), southern Europe, and Indonesia (Fig. 10.28).

12

The mechanism of plate tectonics

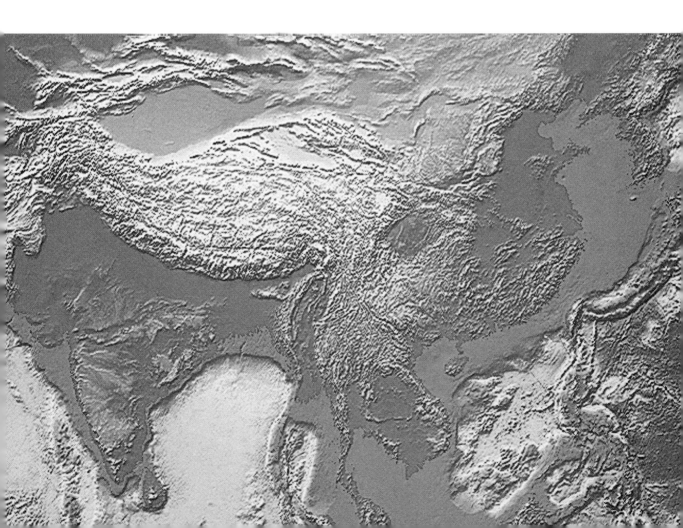

12.1 INTRODUCTION

The mechanism behind the motion of plates is still controversial. Older theories for the origin of the major structural features of the Earth's surface that relied on the supposed contraction or expansion of the Earth have now been discounted. The most likely mechanism of heat transfer from depth appears to be convection. The form of this convection and the manner in which the thermal energy is utilized in driving the plates are discussed in this chapter.

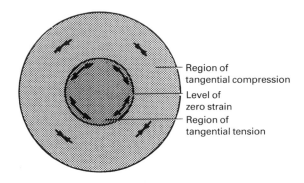

Figure 12.1 *Contracting Earth model.*

12.2 CONTRACTING EARTH HYPOTHESIS

In the 19th century it was believed that, since its formation, the Earth had been cooling due to heat loss by thermal conduction. Computations by Lord Kelvin on the rate of cooling of an initially molten Earth provided the first, erroneous, estimate of the age of the Earth of 100 Ma. As a corollary, it was suggested that the accompanying contraction of the Earth on cooling might provide a mechanism for mountain building. It was estimated that the circumference of the Earth had decreased by 200–600 km since the Earth's accretion. The discovery of radioactivity at the end of the 19th century negated much of the early work as it provided a precise method of dating rocks and also demonstrated that the Earth possesses its own internal sources of heat.

The contraction hypothesis envisaged that the central region of the Earth underwent more rapid cooling and contraction than the outer part and was placed in a state of tangential tension. Above a horizon of no strain, the outer shell of the Earth was then subjected to tangential compression as it collapsed inwards upon the shrinking center (Fig. 12.1). The lithosphere is too thick to respond to this compression by buckling, but would yield by thrust faulting, producing mountain belts by the stacking of thrust slices.

A contracting Earth is no longer recognized as a possible mechanism for tectonic activity for two convincing reasons:

1 The Earth is not cooling sufficiently rapidly to be consistent with contraction, and modern

evaluations of cooling rates imply a total contraction of only a few tens of kilometers. Consequently, the contraction hypothesis cannot account for the many thousands of kilometers of crustal shortening which must have occurred in mountain belts throughout geologic time.

2 The hypothesis implies that the lithosphere is everywhere in compression, and cannot provide an explanation for phenomena that must have originated in tensional regimes, such as normal faults, ocean ridges, and rift valleys.

12.3 EXPANDING EARTH HYPOTHESIS

The expanding Earth hypothesis was first proposed in the 1920s and was subsequently adopted by several geologists as the mechanism behind the break-up of continents, the formation of continental rifts, and the presence of extensional features such as normal faults (Carey, 1976, 1988). Their proposal was that the continental lithosphere was originally continuous over the surface of an Earth of smaller radius and that, as the Earth expanded and its surface area increased, the continental lithosphere fragmented and dispersed, while mantle material welled up into the consequent gaps to form the oceans. Independent evidence for the expanding Earth hypothesis appeared to be provided by certain theoretical physicists, who suggested that the universal gravitational constant was decreasing with time as the universe expanded and its constituent matter became more widely dispersed. Gravitational forces are responsible for binding the Earth into a spherical form, and since the gravitational constant directly controls the magnitude of the force of attraction between masses, its decrease would imply a progressive relaxation of the binding forces and an increase in the Earth's radius.

The most recent versions of the expanding Earth hypothesis correlate the period of rapid expansion with the break-up and fragmentation of Pangea in the past 200 Ma. These argue that continental reconstructions can be arranged more accurately on a globe of smaller radius, and propose that during this period the surface area of the Earth increased by a factor of 2.5 implying an increase in the radius from 63% of its present value and a mean radial expansion rate of about $12\,mm\,a^{-1}$.

There are two methods available that can be used to test the expanding Earth hypothesis directly.

12.3.1 Calculation of the ancient moment of inertia of the Earth

The moment of inertia of a rigid body about a given axis is defined as Σmr^2, where m is the mass of each small element of the body and r the distance of the element from the axis of rotation. The moment of inertia of a uniform sphere is given by $2MR^2/5$, where M is the mass of the sphere and R its radius. Newton's laws of mechanics for linear motion state that the momentum (mass × velocity) of a system is conserved unless an external force acts upon it. These laws apply equally to angular (rotational) motion, in that angular momentum (moment of inertia × angular velocity) is conserved unless the system is acted upon by an external torque.

Since the Earth's mass remains constant, any determination of the ancient moment of inertia of the Earth would allow a calculation of its ancient radius and so demonstrate if any expansion had occurred.

The theory behind any such determination is complicated by the fact that momentum is conserved within a system comprising the Earth and the Moon. Throughout geologic time the angular momentum inherited from the fragments that accreted to form the Earth and Moon has been progressively partitioned between the two bodies, by a mechanism known as tidal interaction, in such a way as to reduce the rotational energy of the system. At present the stage has been reached in which the Moon spins very slowly and must consequently lie at a greater distance from the Earth than in the past, so that momentum is conserved in its orbital motion. The tidal interaction of the Moon on the Earth is similarly causing the latter's angular rotation to decelerate. Since the number of rotations in a complete orbit of the Sun determines the number of days in a year, the year in the past would have consisted of more days than at present. This also implies that the length of the day has progressively increased.

Transfer of angular momentum from Earth to Moon thus causes an increase in the length of the day. A further contributor to this phenomenon would be an increase in the Earth's moment of inertia, which would allow angular momentum to be conserved by a slower rate of rotation. Knowledge of the length of the lunar month would allow an estimate to be made of the lunar contribution to the Earth's rotational deceleration and allow any change in its moment of inertia to be isolated.

Information on the rotational history of the Earth–Moon system has been provided from a detailed examination of fossil organisms whose patterns of growth are strongly affected by diurnal effects. In particular, certain rugose corals of Middle Devonian age (390 Ma) have been shown to exhibit epithecal banding which can be attributed to daily, monthly, and yearly cycles of growth (Scrutton, 1967). Such studies have indicated that the Middle Devonian year comprised 400 ± 7 days, and was divided into 13 lunar months of 30.5 days. The average increase in the length of day up to present times is $20\,s\,Ma^{-1}$.

The length of the lunar month in Devonian times allows an estimate to be made of the Moon's angular momentum at that time, and hence the deceleration of the Earth's rotation resulting from tidal friction. The deceleration not accounted for in this way can be used to provide an estimate of the Devonian moment of inertia of the Earth, which is found to be 99.4–99.9% of its present value. Given the uncertainties in the calculation, the moment of inertia does not seem to have altered significantly. The expansion of the Earth required to cause continental drift implies that the Devonian moment of inertia would have to have been only 94% of its present value. Consequently, such rapid expansion can be ruled out.

12.3.2 Calculation of the ancient radius of the Earth

A rather less involved method of testing the expanding Earth hypothesis entails determining the paleoradius of the Earth using paleomagnetic techniques (Egyed, 1960).

The method involves selecting sampling sites of the same age, on the same paleomeridian and differing as much as possible in paleolatitude. They must also be on a landmass that has been stable since the time the sites acquired their primary remanent magnetizations (Fig. 12.2). Determining the paleolatitudes (ϕ_1, ϕ_2) of the sites then provides the angle originally subtended at the center of the Earth ($\phi_1 + \phi_2$). The known separation of the sites (d) can then be used to calculate the paleoradius of the Earth (R_a) according to the relationship $R_a = d/(\phi_1 + \phi_2)$, where angles are expressed in radians. However, it is rare to find two paleomagnetic sampling sites on the same paleomeridian so, in practice, this method is of limited applicability. Ward (1963) devised a more general *minimum dispersion method* that facilitates an analysis of arbitrarily distributed sampling sites. The dispersion of paleomagnetic poles from sites of the same age and known relative paleogeographic position is calculated, using the Fisher (1953) method for disper-

sion on a sphere, for different values of the Earth's radius. The radius for which the dispersion of the poles is a minimum is taken to be the best estimate of the paleoradius. McElhinny *et al.* (1978) analyzed the paleomagnetic data available at that time using this method. They found that for the past 400 Ma the average paleoradius has been 102 ± 2.8% of the present radius. A small contraction or very slight expansion of the Earth could be tolerated by this analysis, but the very large increase in radius required by the expanding Earth hypothesis can be ruled out. Additional analyses by McElhinny & McFadden (2000) produced very similar results.

The expanding Earth hypothesis clearly does not stand up to direct testing. Also, indirectly the hypothesis cannot account for presently observable phenomena. If continental drift results from this mechanism there would be no necessity for subduction zones for the consumption of oceanic lithosphere, and no explanation is provided for extensive zones affected by collisional tectonics. The majority of plates are presently spreading in an east–west sense. If such a pattern results from an expanding Earth it would imply a progressive increase in the size of the equatorial bulge, which is not occurring. An expansion of the Earth would imply the existence of extensive zones subjected to membrane stresses as plates attempt to adjust to the increasing radius of curvature of the Earth, and these do not exist. Finally, the theory does not provide a mechanism for the continental drift that is known to have occurred in pre-Mesozoic times (Section 11.5).

12.4 IMPLICATIONS OF HEAT FLOW

The average vertical thermal gradient at the Earth's surface is about 25°C km^{-1}. If this gradient remained constant with depth, the temperature at a depth of 100 km would be 2500°C. This temperature is in excess of the melting temperature of mantle rocks at this depth, and so a fluid layer is implied. Such a molten layer does not exist because S waves are known to propagate through this region (Section 2.1.3). Two possibilities exist in explanation of this phenomenon: first, that heat sources are concentrated above a depth of

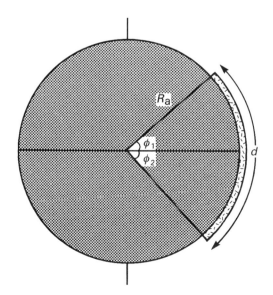

Figure 12.2 *Parameters used in estimating the paleoradius of the Earth from paleomagnetic data.*

100 km; and second, that a more efficient mechanism than conduction operates below this depth whereby heat is transferred at a much lower thermal gradient. These processes can be distinguished by considering the variation in heat flow over the Earth's surface in conjunction with the variation in content of radioactive minerals of different crustal types.

Heat flow generally decreases with the age of the crust (Sclater *et al.*, 1980). Within the oceans heat flow decreases from the ocean ridges to the flanking basins and it has been shown (Section 6.4) that this cooling correlates with a progressive thickening of the oceanic lithosphere and an increase in water depth. Similarly, the heat flow of backarc basins (Section 9.10) decreases with age, with the presently active basins exhibiting the greatest heat flow. Within continental regions the heat flow generally decreases with increasing time since the last tectonic event. Consequently, Precambrian shields are characterized by the lowest heat flow and young mountain belts by the highest.

The representation of the global pattern of heat flow is difficult because the density of the observations is highly variable so that the location of contours can be greatly biased by only a small number of measurements. Chapman & Pollack (1975) overcame the problem of limited observations in some areas by predicting the heat flow in those areas on the basis of the correlation of heat flow with the age of the oceanic lithosphere and the age of the last tectonic event to affect continental crust. In Fig. 12.3 their results are presented by a spherical harmonic analysis of the heat flow measured or predicted in $5° \times 5°$ grid areas of the globe. This procedure imparts a certain smoothing of the true pattern, so that variations with wavelengths of less than about 3300 km are not represented. Figure 12.3 illustrates the high heat flow associated with the ocean ridge system and the youngest marginal basins of the western Pacific. Low heat flow values are associated with old oceanic crust and with Precambrian shields.

Histograms of heat flow measurements from oceans and continents are presented in Fig. 12.4. The greater dispersion of the oceanic values reflects variability arising from localized extreme values at the crests of ocean ridges. By contrast, there are fewer extreme high or low values present in the continental values. The mean of oceanic heat flow measurements is $67 \, mW \, m^{-2}$. However, this only represents the heat loss by conduction, and ignores the heat reaching the surface by the discharge of hot fluids such as water and lava. It is now recognized that the hydrothermal contribution accounts for about a quarter of the global heat loss, and that the average oceanic heat flow is $101 \, mW \, m^{-2}$. The mean continental heat flow is $65 \, mW \, m^{-2}$, including the small contribution from lavas. The global average heat flow is $87 \, mW \, m^{-2}$ (Pollack *et al.*, 1993).

The majority of the heat escaping at the Earth's surface originates from the decay of long-lived radioactive isotopes of uranium, thorium and potassium (Section 2.13) which have half-lives of the same order as the age of the Earth. These isotopes are relatively

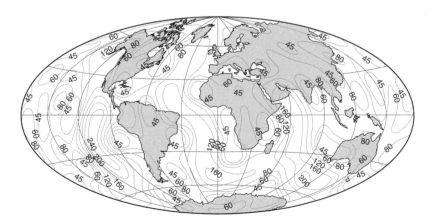

Figure 12.3 *Pattern of global heat flow represented by spherical harmonic analysis. Contour interval 40 mW m^{-2} (after Chapman & Pollack, 1975).*

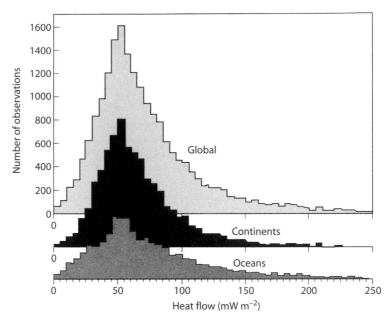

Figure 12.4 *Comparison of the heat flow from continents and oceans (redrawn from Pollack et al., 1993, by permission of the American Geophysical Union. Copyright © 1993 American Geophysical Union).*

enriched in the upper continental crust, and it has been estimated that their decay contributes 18–38 mW m^{-2} to the observed heat flow (Pollack & Chapman, 1977). Consequently up to about 60% of the heat flow in continental regions may be generated within the upper 10–20 km of the crust. The oceanic crust, however, is virtually barren of radioactive isotopes, and only about 4 mW m^{-2} can be attributed to this source. Over 96% of the oceanic heat flow must originate from beneath the crust, and so different processes of heat supply must act beneath continents and oceans (Sclater & Francheteau, 1970).

Thus, a large proportion of the continental heat flow is from sources concentrated at a shallow depth, and only a small sub-crustal component is required. Conversely, the majority of oceanic heat flow must originate at sub-crustal levels. Because of the melting problems discussed above, this heat must be transported under the influence of a low thermal gradient. The mechanism of heat transfer by convection is the only feasible process conforming to these constraints. Therefore, although heat transfer by conduction takes place within the rigid lithosphere, heat transfer by convection must predominate in the sublithospheric

mantle. Indeed, conduction cannot occur to any great depth as the rate of heat transfer by this mechanism is much slower than required. The feasibility and form of such convection is discussed in the following sections.

12.5 CONVECTION IN THE MANTLE

12.5.1 The convection process

The nature of convective flow in the mantle is problematic. Analytical solution is difficult because of the complex rheological structure, including the presence of a transition zone (Section 2.8.5), the presence of heat sources within the convecting layer as well as beneath it, the influence of an overlying rigid lithosphere on the pattern of convection, and the fact that the convecting layer has the form of a spherical shell.

However, as a result of advances in numerical simulations and analogue modeling, and constraints on the pattern of convection supplied by seismic tomography and past and present plate motions, it is now possible to derive considerable information on the convective process.

Convection in a fluid involves heat transport by motion of the fluid caused by positive or negative buoyancy of some of the fluid, that is, horizontal density contrasts or gradients within it. The latter are typically produced by more dense downwellings from a cold boundary layer or less dense upwellings from a hot boundary layer, but they may also be of compositional origin. Indeed one tends to think of a convecting fluid layer as being heated from below and

cooled from above, in which case there is a hot thermal boundary layer at its base and a cold thermal boundary layer at the top (Fig. 12.5a). However, it is possible that one of these boundary layers may be weak or absent. In addition, the fluid layer may be heated from within (Fig. 12.5b,c). In Fig. 12.5b the lower boundary layer is missing and the fluid is heated internally. The cold dense fluid sinking from the top boundary layer drives convection and the upwelling is passive rather than buoyant; fluid has to move upwards to create space for the sinking cold fluid. The mantle is probably more like Fig. 12.5c in that it is heated from below, by heat flowing from the core, and from within by radioactivity. In Fig. 12.5 if the temperature of the lower boundary is fixed in each case then the

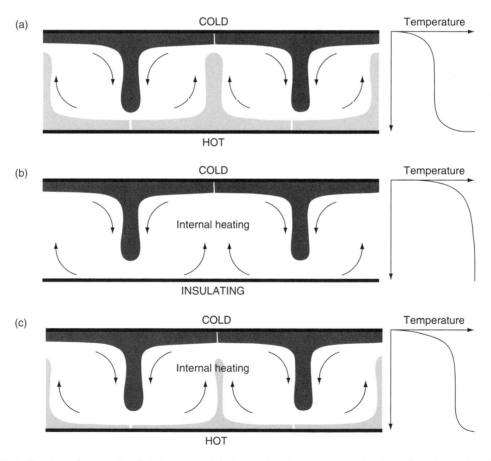

Figure 12.5 *Sketches of convecting fluid layers, and their associated temperature–depth profiles, illustrating the varying nature of the lower thermal boundary layer depending on the way in which the fluid layer is heated (from Davies, 1999. Copyright © Cambridge University Press, reproduced with permission).*

temperature profiles will be as shown to the right of the figure. If there is no heating from below the temperature in the interior of the fluid will be the same as that at the base of the fluid layer (Fig. 12.5b). If there is some heating from below, in addition to internal heating (Fig. 12.5c), then the interior of the convecting fluid will have an intermediate temperature between cases (a) and (b). This results in a greater drop in temperature across the upper boundary layer, and a lower drop in temperature across the lower boundary layer, compared to case (a). This effect of internal heating, whereby the top boundary layer is strengthened and the bottom boundary layer weakened, may therefore be applicable to the mantle.

The effect of internal heating and the lack of a lower thermal boundary layer is illustrated in Fig. 12.6. This shows the results of two numerical models with parameters appropriate to the mantle. The three frames on the left relate to a model with heating from below and no internal heating and those on the right to a model with internal heating and no lower boundary layer. In the first case one can clearly see cold sinking columns and hot rising columns analogous to Fig. 12.5a. In the right hand case only downwellings are apparent and the upwellings are passive and widely distributed (*cf.* Fig. 12.5b).

Although instructive, these models probably do not accurately simulate convection in the mantle as they assume uniform viscosity throughout the convecting layer, a parallel-sided rather than a spherical layer, thermal convection alone and no phase changes within the fluid. In the Earth's mantle it is now thought that the viscosity increases with depth and that buoyancy is in part created by compositional variations. As discussed in Section 12.9 these two factors appear to stabilize the convective pattern for hundreds of millions of years, whereas the convective patterns developed in the models of Fig. 12.6 are clearly unstable over this period of time.

12.5.2 Feasibility of mantle convection

In order to gain insight into the feasibility and nature of convection within a spherical, rotating Earth, it is convenient to assume that the mantle approximates a Newtonian fluid. Although this assumption may be erroneous, it does allow simple calculations to be made on the convective process.

The condition for the commencement of thermal convection is controlled by the magnitude of the dimensionless Rayleigh number (R_a), which is defined as the ratio of the driving buoyancy forces to the resisting effects of the viscous forces and thermal diffusion.

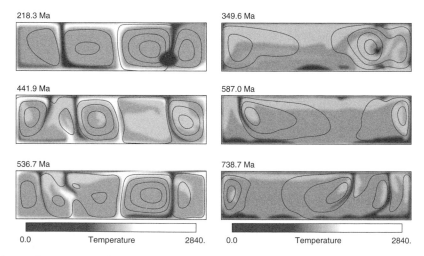

Figure 12.6 *Frames from numerical models illustrating (a) convection in a layer heated from below and (b) convection in a layer heated internally and with no heat from below (from Davies, 1999. Copyright © Cambridge University Press, reproduced with permission).*

$$R_a = \alpha\beta\rho g d^4 / k\eta$$

where α is the coefficient of thermal expansion, β the superadiabatic temperature gradient (the gradient in excess of that expected to be associated with the increasing pressure), ρ the density of the fluid, g the acceleration due to gravity, d the thickness of the convecting fluid, k the thermal diffusivity (the ratio of the thermal conductivity to the product of density and specific heat at constant volume), and η the dynamic viscosity (Section 2.10.3). For convection in the mantle, the Rayleigh number corresponding to the onset of convection is approximately 10^3. This corresponds to the minimum temperature gradient required for convection to occur. For the actual temperature gradient the Rayleigh number is of the order of 10^6 or greater. This implies very favorable conditions for convection and, as a consequence, thin boundary layers compared to the total layer thickness.

The nature of the flow in a convecting fluid can be judged by the magnitude of the Reynolds number (Re), which allows discrimination between laminar and turbulent flow. Re is defined:

$$Re = vd / v$$

where v is the velocity of flow and v is the kinematic viscosity (the ratio of the dynamic viscosity, η, to density). Taking $v = 200\,mm\,a^{-1} = 6 \times 10^{-9}\,ms^{-1}$, $d = 3000\,km = 3 \times 10^6\,m$ and $v = 2 \times 10^{17}\,m^2\,s^{-1}$, $Re = 9 \times 10^{-20}$. This very low value indicates that viscous forces dominate and hence the flow is laminar. The effect of the Earth's rotation on convection can be judged by the magnitude of the Taylor number (T), which is defined:

$$T = (2wd^2 / v)^2$$

where w is the angular velocity of rotation. Putting $w = 7.27 \times 10^{-5}\,rad\,s^{-1}$ and other values as above, $T \approx 4 \times 10^{-17}$. A value of T less than unity implies no significant effect of rotation on convection and so the Earth's rotation should have no effect on the pattern of mantle convection.

The efficiency of convection is measured by the Nusselt number (Nu), which is the ratio of the total heat transferred to that transferred by thermal conduction alone. Elder (1965) computed experimentally the relationship between Nu and R_a. He found that at values of R_a appropriate to marginal convection Nu is unity and very little heat is transferred by convection. At R_a values 10^6 or greater, appropriate to the mantle, Nu is about 100, indicating the predominance of heat transfer by convection.

12.5.3 The vertical extent of convection

The mantle transition zone (Section 2.8.5) may well influence the nature or even the vertical extent of convection in the mantle. If this zone represents a change in chemical composition, then it implies that convection currents do not cross it. In this case separate layers of convective circulation would occur above and below the transition zone, with heat transported by conduction across a thermal boundary layer within all or part of the transition zone.

The nature of the mantle transition zone is equivocal, but the majority view appears to be that it represents a region in which solid state phase changes take place, whereby the mineralogy of mantle material changes to higher pressure forms with depth, rather than representing a change in chemical composition (Section 2.8.5). For example, Watt & Shankland (1975) have shown, from an inversion of velocity–density data, that the mean atomic weight of the mantle shows no change across the transition zone. If this is the case, convection currents could cross the transition zone, as long as the phase changes take place very rapidly, and convection cells would then be mantle wide. The phase changes would have two important effects on convection, as they are temperature and pressure dependent and involve latent heat. In the case of olivine to spinel the change from low pressure to high pressure forms takes place at shallower than average depths in the cold descending currents and at greater than average depths in the hot ascending currents. Consequently, low-density minerals are created deeper on ascent and denser, high-pressure forms at shallower depths on descent. Their positive and negative buoyancies respectively then help to drive the convection cells. The phase change is also associated with a release or absorption of latent heat, the high- to low-pressure reaction being exothermic and the low to high-pressure reaction being endothermic. This causes steepening of the thermal gradient across the transition zone, so that the temperature in the lower mantle is 100–150°C higher than if the zone did not exist.

Tackley *et al.* (1993) have numerically modeled mantle convection in three dimensions with an endo-thermic phase change at the base of the transition zone. They suggest that cold downwelling material accumu-lates above 660 km and then periodically flushes into the lower mantle. This fits well with the results of seismic tomographic imaging of subduction zones, which sug-gests that some slabs flatten out within the transition zone and others penetrate the base of the zone and descend into the lower mantle (Section 9.4; (Plate 9.2 between pp. 244 and 245).

Thus, the transition zone may not be a barrier to mantle-wide convection, and a number of workers have presented evidence in accord with this premise. Kanasewich (1976) noted an organized distribution of plates, in which the Pacific and African plates are approximately circular with the smaller plates having an approximately elliptical form and arranged sys-tematically between these two large plates. Kanasewich attributed this organization to convection that is mantle-wide. Davies (1977) conducted model experi-ments and concluded that only extreme viscosity contrasts would restrict convection to the upper mantle, and maintained that such contrasts do not exist. Elsasser *et al.* (1979) employed a scaling analy-sis in which the depth of convection is derived as a function of known parameters, and concluded that this depth is consistent with convection throughout the entire mantle. The topography on the base of the mantle transition zone has an amplitude of about 30 km (Shearer & Masters, 1992), which is an order of magnitude lower than predicted for a chemical, rather than a phase, change at this depth. Morgan & Shearer (1993) derived the buoyancy distribution in the mantle from seismic tomographic maps and concluded that there must be significant flow between the lower and upper mantle. However, other work, summarized by van Keken *et al.* (2002), suggests that the geochemical and isotopic pattern of trace ele-ments found in oceanic volcanic rocks supports a model in which portions of the mantle have been chemically isolated for much of Earth history. This would suggest that the mixing implied by whole mantle convection has not occurred, and that layered convection is more likely. However, in the light of the geophysical evidence for mantle-wide convection many geochemists have derived models in which dis-tinct chemical reservoirs can be preserved within this context (e.g. Tackley, 2000; Davies *et al.*, 2002). It would seem, therefore, that convective circulation is most likely to be mantle-wide and not constrained by the transition zone.

12.6 THE FORCES ACTING ON PLATES

In order to understand the structural styles and tectonic development of plate margins and interiors, it is necessary to consider the nature and magnitude of all the forces that act on plates. Forsyth & Uyeda (1975) solved the inverse problem of determining the relative magnitude of plate forces from the observed motions and geometries of plates. Since the present velocities of plates appear to be constant, each plate must be in dynamic equilibrium, with the driving forces being balanced by inhibiting forces. Forsyth & Uyeda (1975) used the corollary of this, that the sum of the torques on each plate must be zero, to deter-mine the relative size of the forces on the 12 plates which they assumed make up the Earth's surface. The asthenosphere's role in this scenario was considered to be essentially passive. A similar set of computations based on a similar method, and providing similar results, was made by Chapple & Tullis (1977). The following description of forces is based on the exten-sions of the work of Forsyth & Uyeda (1975) made by Bott (1982).

At ocean ridges the ridge push force F_{RP} (Fig. 12.7) acts on the edges of the separating plates. This derives from the buoyancy of the hot inflowing material causing the elevation of the ridge and hence an additional hydrostatic head at shallow depths which acts on the thinner lithosphere at the ridge crest. It may also arise from the cooling and thickening of the oceanic litho-sphere away from the ridge (Section 6.4), which exerts a pull on the ridge region. Hence, it is basically a grav-itational force. The ridge-push force may be two or three times greater if a mantle plume (Section 5.5) underlies the ridge (Bott, 1993), because of the increased pressure in the asthenosphere at the ridge crest. The separation of plates at ocean ridges is opposed by a minor ridge resistance R_R that originates in the brittle upper crust and whose existence is demonstrated by earthquake activity at ridge crests. The resisting forces are small so that the net effect is the presence of a driving force.

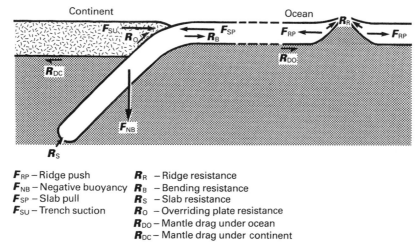

Figure 12.7 *Some of the forces acting on plates (developed from Forsyth & Uyeda, 1975, by Bott, 1982, reproduced by permission of Edward Arnold (Publishers) Ltd).*

Beneath plate interiors a mantle drag force acts on the base of both the oceanic and continental lithospheres if the velocity of the underlying asthenosphere differs from that of the plate. If the asthenosphere velocity exceeds that of the plate, mantle drag enhances the plate motion (F_{DO}, F_{DC}), but if the asthenosphere velocity is lower, as shown in Fig. 12.7, the mantle drag tends to resist plate movement (R_{DO}, R_{DC}). Mantle drag beneath continents is about eight times the drag beneath oceans; this may be due to the increased thickness of the subcontinental lithosphere beneath cratonic areas (Sleep, 2003).

At subduction zones the major force acting on plates results from the negative buoyancy (F_{NB}) of the cold, dense slab of descending lithosphere. Part of this vertical force is transmitted to the plate as the slab-pull force F_{SP}. The density contrast, and hence F_{NB}, is greatly enhanced at depths of 300–400 km where the olivine–spinel transition occurs in the slab. F_{SP} is opposed by a slab resistance (R_S), which mainly acts on the leading edge of the descending plate where it is five to eight times greater than the viscous drag on its upper and lower surfaces. Underthrusting involves a downward flexure of the lithosphere in response to F_{NB}, and since it behaves in an elastic manner in the top few tens of kilometers flexure is opposed by a bending resistance (R_B). A further resistance to motion at subduction zones is the friction between the two plates. This overriding

plate resistance (R_O) is expressed in the intense earthquake and tectonic activity observed at shallow depths at destructive plate margins. The downgoing slab achieves a terminal velocity when F_{SP} is nearly balanced by $R_S + R_O$. If F_{SP} exceeds $R_S + R_O$, the slab descends at greater than the terminal velocity and throws the slab into tension at shallow depths. If F_{SP} is less than $R_S + R_O$ the slab is thrown into compression. The balance between driving and resistive forces may thus control the distribution of stress types, as revealed by earthquake focal mechanism solutions, within downgoing slabs (Section 9.4).

In the region on the landward side of subduction zones the overriding lithosphere is thrown into tension by the trench suction force (F_{SU}). There are several possible causes of this force (Fig. 12.8):

1 It may arise because the angle of subduction becomes progressively greater with depth (Fig. 12.8a). Tension would then arise as the overriding plate collapses toward the trench.

2 The tension could result from the "roll-back" of the underthrusting plate (Fig. 12.8b). That is, the downgoing slab retreats from the overriding plate.

3 Tension could be generated by secondary convective flow in the region overlying the downgoing slab (Fig. 12.8c). This would require

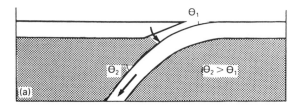

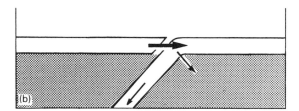

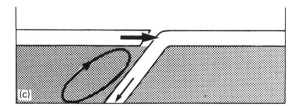

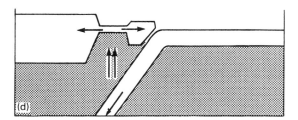

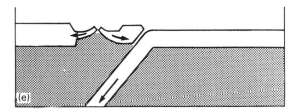

Figure 12.8 *Possible sources of the trench suction force (after Forsyth & Uyeda, 1975, with permission from Blackwell Publishing).*

4   Tension may arise from any of several mechanisms proposed for the formation of backarc basins on the landward side of subduction zones (Fig. 12.8d) as described in Section 9.10. However, once backarc spreading commences the landward plate becomes decoupled from the trench system (Fig. 12.8e).

When two plates of continental lithosphere are brought into contact after the complete consumption of an intervening ocean at a subduction zone, the resistance to any further motion is known as collision resistance. The mechanism of this resistance is complex because it takes place both at the suture between the plates and within the overriding plate (Sections 10.4.3, 10.4.6). Finally, transform fault resistance affects conservative plate margins in both continental and oceanic areas. The resistance acts parallel to the faults and gives rise to earthquakes with a strike-slip mechanism (Section 2.1.5) confined to a shallow depth. More complex resistance is encountered where the fault trend is sinuous so that motion is not purely strike-slip (Section 8.2).

The relative magnitude of the forces acting on plates and their relevance to the driving mechanism of plate tectonics will be discussed in Section 12.7.

12.7 DRIVING MECHANISM OF PLATE TECTONICS

The energy available to drive plate motions is the heat generated in the core and mantle that is brought to the surface by convection in the mantle. It now remains to consider the manner in which this thermal energy is employed in driving the lithospheric plates. The proposal by Morgan (1971, 1972b) that plates are driven by the horizontal flow of material brought to the base of the lithosphere by hotspots was discounted initially (Chapple & Tullis, 1977), as the lateral flow would probably be equal in all horizontal directions and thus would not apply a directional force to the plates. Two models have been proposed. The classical, or mantle drag, model considers that the upper, cool, boundary layer of the convecting system is represented by the upper part of the asthenosphere, and that plates are driven by the viscous drag of the asthenosphere on their bases. By

a relatively high geothermal gradient giving rise to a relatively low viscosity in the asthenosphere (Section 12.5.2).

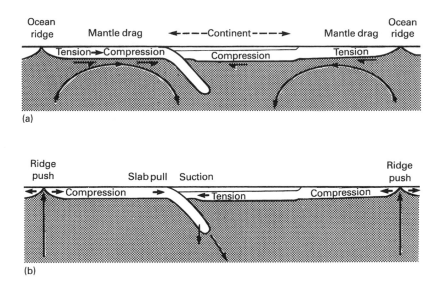

Figure 12.9 *Two concepts of plate driving mechanism: (a) cellular convection, with the cells exerting a mantle drag on the lithosphere; (b) Orowan–Elsasser-type convection, with plates driven by edge forces (redrawn from Bott, 1982, by permission of Edward Arnold (Publishers) Ltd).*

contrast, the edge-force model recognizes the lithosphere itself as the upper, cool boundary layer of the convection cells and proposes that the plates are driven by forces applied to their margins. The two models thus differ in the importance placed on the various forces acting on plates (Section 12.6) described by Forsyth & Uyeda (1975).

12.7.1 Mantle drag mechanism

Mantle drag was the first driving mechanism to be proposed, and envisages plate motion in response to the viscous drag exerted on the base of the lithosphere by the lateral motion of the top of mantle convection cells in the asthenosphere (Fig. 12.9a). The convection cells would consequently rise beneath oceanic ridges and descend beneath trenches, being largely absent beneath continental regions. This mechanism predicts that the oceanic lithosphere would be in a state of tension at the ocean ridges and compression at the trenches.

Because of their relationship to accretive and destructive plate margins, the horizontal dimensions of the convection cells powering mantle drag would be expected to be about half the width of an ocean, that is, 3000 km. This great lateral extent implies that the cells should have a relatively simple form. It is consequently difficult to explain how cells of simple geometry could drive plates with irregularly shaped margins, such as the Mid-Atlantic Ridge at equatorial latitudes where it is offset along a suite of transform faults. Also, the constant geometry of the convection cells cannot explain the relative movements between plate margins, such as is happening between the Mid-Atlantic and Carlsberg ridges. The large horizontal dimensions of the cells cannot account for the movements of small plates, such as the Caribbean and Philippine plates, which can hardly be powered by their own individual convective systems.

It would therefore seem that the classical mantle drag mechanism is not the main process causing the mobility of plates. It is possible, however, that our views on mantle drag are biased by the fact that the present continents are dispersed. Ziegler (1993) argues that mantle drag may have been a significant mechanism during supercontinent break-up and, indeed, Phanerozoic plate motions appear to require this mechanism (Section 12.11).

12.7.2 Edge-force mechanism

In this mechanism the oceanic lithosphere represents the top of the convection system, and the plates move in response to forces applied to their edges (Fig. 12.9b). The mechanism was first proposed by Orowan (1965) and Elsasser (1969, 1971) and is sometimes referred to as Orowan–Elsasser-type convection (Davies & Richards, 1992).

Only a small percentage of the energy supplied from the mantle is available to drive the plates, but this fraction is adequate to power the present plate motions (Bott, 1982). The energy is utilized by the lithosphere to drive the plates in several ways. The ridge-push force (Section 12.6) originates from the uplift of the ridge crest caused by the anomalously hot asthenosphere beneath it. This provides a lateral push to the rear of accreting oceanic lithosphere. The slab-pull force (Section 12.6) arises from the negative buoyancy of the downgoing slab at trenches, and is assisted by phase changes to denser forms that affect minerals in the slab at increased pressure. The slab-pull force is potentially some four times larger than the ridge-push force, although in practice much of this force is probably utilized in overcoming slab resistance (Chapple & Tullis, 1977). The trench suction force (Section 12.6) originates from the geometry of the downgoing slab and also provides a significant driving force.

The edge-force mechanism can account for many phenomena more satisfactorily than the mantle drag mechanism, in particular:

1 It is more acceptable thermodynamically and is much more effective in transporting heat from the mantle.

2 It is consistent with the observed pattern of intraplate stress. As discussed in Section 12.7.1, the mantle drag mechanism implies tension at ocean ridges and compression at trenches. The edge-force mechanism would give rise to the opposite stress configuration, and this is in accord with the stress regime indicated by focal mechanism solutions of intraplate earthquakes.

3 It is reconcilable with the present plate motions, in particular with the observations of Forsyth & Uyeda (1975) that:

 (a) plate velocity is independent of plate area (Fig. 12.10a). If mantle drag were operative it would be expected that the greatest velocities would be experienced by plates with the greatest area over which the mantle drag would act;

 (b) plates attached to downgoing slabs move more rapidly than other plates (Fig.

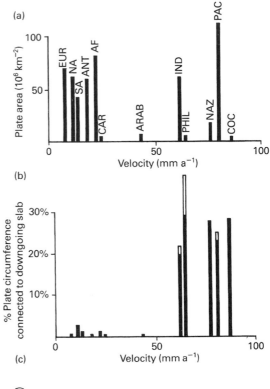

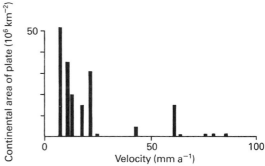

Figure 12.10 *Correlations of plate parameters with plate velocity: (a) plate area; (b) plate circumference connected to downgoing slab (open bar, total length; filled bar, effective length); (c) continental area of plate (redrawn from Forsyth & Uyeda, 1975, with permission from Blackwell Publishing).*

12.10b). This is in accord with the slab-pull force being greater than other forces affecting the plates;

 (c) plates with a large area of continental crust move more slowly (Fig. 12.10c). This implies that mantle drag inhibits the motion of such plates rather than driving them.

The mechanism also provides a reasonable explanation of the motions of small plates.

Consequently, the edge-force mechanism of plate movement appears to be much more successful in explaining all observed phenomena, and has been adopted by most workers, certainly for present-day plate motions.

12.8 EVIDENCE FOR CONVECTION IN THE MANTLE

12.8.1 Introduction

A fundamental axiom of plate tectonics is that oceanic lithosphere is formed from mantle material at mid-ocean ridge crests and returned to the mantle in subduction zones. Thus, plate creation, movement and destruction provide evidence for convection in the mantle. There must be downwellings in the mantle associated with subduction zones and upwelling beneath mid-ocean ridge crests. However, beyond this, plate tectonics provides no evidence for the location of the return flow in the mantle, or the depth extent of convection other than the seismicity associated with subducting slabs (Section 9.4). One must turn therefore to other lines of evidence for information on the pattern of convection in the deep mantle.

12.8.2 Seismic tomography

Much important information on the three-dimensional structure of the mantle has been supplied by seismic tomography (Section 2.1.8). Convection is driven by lateral differences in temperature and density. These variables affect seismic velocity, which typically decreases with decreasing density and increasing temperature (Dziewonski & Anderson, 1984). By mapping velocities in the mantle it is possible to infer the differences in temperature and density that are a consequence of convection. Also, by mapping seismic anisotropy both vertically and laterally it is possible to obtain estimates of the direction of mantle flow.

The first three-dimensional seismic velocity models for the mantle derived by the tomographic technique were published in the early 1980s (Woodhouse & Dziewonski, 1984). Since then there have been great improvements in data quality, geographic coverage, and processing techniques, and the resolution of subsequent models has greatly improved. However, many of the essential features of the velocity variations were apparent in the earliest models. Plate 12.1 (between pp. 244 and 245) shows the variations in the shear wave velocity at 12 depths in the mantle according to model S16B30 of Masters et al. (1996). It is immediately apparent that the greatest variations occur near the top and bottom of the mantle, presumably within or in the vicinity of the thermal boundary layers. Within the top 200 km the perturbations are very closely related to surface tectonic features. Ocean ridges, the rifts of northeast Africa, and the active backarc basins of the western Pacific are all underlain by anomalously low velocity mantle. Continental areas in general, and shield areas in particular, are underlain by the highest velocities, and older oceanic crust by relatively high velocities. These variations essentially reflect the different thermal gradients and hence the thickness of the lithosphere in these areas (Section 11.3.1). Between 200 and 400 km most of these generalizations still apply but the velocity contrasts are lower. A notable exception is the mantle beneath the backarc basins where the slow velocities at shallower depths have been replaced by near zero anomalies. In the transition zone (e.g. 530 km) the variations are in general quite small and the correlation with surface features has largely broken down. Again an exception is the mantle beneath the backarc basins of the western Pacific, which, at this depth, is characterized by high velocities presumably associated with cold subducted lithosphere. In the lower mantle (depths greater than 660 km) the variations in shear wave velocity are generally quite small (less than ±1.5%), but a persistent feature is a ring of higher than average velocities beneath the rim of the Pacific. This becomes particularly marked in the lowest 400 km of the mantle (e.g. depths of 2500 and 2750 km). At depths greater than 2000 km, large regions of anomalously low velocity occur beneath the central Pacific, and beneath southern Africa and part of the South Atlantic.

Plate 12.2 (between pp. 244 and 245) shows four cross-sections through the shear wave velocity model of Masters *et al.* (1996), each on a plane passing through the center of the Earth. Three of these sections are longitudinal sections, that is, the planes also pass through the north and south poles; the fourth is an equatorial section. Note that in each cross-section the great circle showing the intersection of the plane of the section with the Earth's surface is the smallest circle on the diagram. Plate 12.2a (between pp. 244 and 245) clearly illustrates the way in which the high velocities associated with continental areas, such as North America and Eurasia, and the low velocities associated with mid-ocean ridges, such as the East Pacific Rise and the mid-Indian Ocean ridge, only extend to depths of 200–400 km within the upper mantle. The section in Plate 12.2b (between pp. 244 and 245) passes through the central Pacific and southern Africa and reveals the low velocity regions in the lowermost mantle beneath these areas, and the way in which they have their greatest extent at the core–mantle boundary. In this section it is also noteworthy that beneath Alaska higher than average velocities extend from the surface to the core, and that beneath parts of the Pacific and to the south of South Africa low velocities extend from the surface to the core–mantle boundary. Plate 12.2c (between pp. 244 and 245) shows that low velocities also exist from the surface to the core beneath the region of the Azores and the Canary Islands in the North Atlantic. As the sections shown in Plate 12.2a–c (between pp. 244 and 245) all pass through both poles they all have low velocity regions in the upper mantle associated with the Arctic ridge and high velocities in the upper mantle beneath the continent of Antarctica. The equatorial section of is particularly instructive and revealing in that it not only passes through the low velocity regions in the lowermost mantle beneath southern Africa and the central Pacific but also shows that the high velocity regions associated with subduction beneath South America and the Indonesian region extend continuously through the transition zone and the lower mantle to the core–mantle boundary. Moreover it illustrates that these two pairs of features, which may represent hot upwellings and cold downwellings respectively, are approximately diametrically opposite to each other.

As discussed in Section 2.10.6, measurements of seismic anisotropy in the mantle can yield information on the pattern of flow. Depending on the deformation mechanism and the minerals involved crystal lattices can be preferentially aligned causing seismic waves to propagate with different velocities in different directions. The preferential alignment of olivine by flow in the upper mantle for example gives rise to the highest seismic velocities in the flow direction (Karato & Wu, 1993). Studies of seismic anisotropy in the upper mantle reveal flow directions that are in general parallel to plate motions with indications of vertical flow beneath mid-ocean ridges and in the vicinity of subduction zones (Park & Levin, 2002).

Most of the lower mantle is isotropic. This is probably because under the temperature, pressure and deformation mechanism pertaining in the lower mantle the minerals present, such as perovskite and magnetowüstite, are effectively isotopic (Karato *et al.*, 1995). In the lowermost mantle, the D'' layer, seismic anisotropy has been observed (Section 2.10.6). It is thought to reflect deformation due to horizontal flow in general, but at the base of the low velocity regions beneath the central Pacific and southern Africa there are indications of vertical flow suggesting the onset of upwelling (Panning & Romanowicz, 2004).

12.8.3 Superswells

The most pronounced features in the lower mantle revealed by seismic tomography are two extensive regions of low velocity beneath the south Atlantic and southern Africa, and the central and southwest Pacific (Plates 12.1, 12.2 between pp. 244 and 245). These correlate with anomalously high elevation of the Earth's surface in these areas. Indeed the width of the topographic swell in each case, several thousand kilometers, is so large that they have been termed *superswells* (McNutt & Judge, 1990; Nyblade & Robinson, 1994). This is in contrast to the topographic swells associated with hot spots that are typically less than 1000 kilometers across. However the elevated topography and bathymetry of superswells cannot be explained by anomalously high temperatures and/or low density rock types in the lithosphere and asthenosphere beneath these regions (Ritsema & van Heijst, 2000). The only plausible explanation is that they are dynamically supported by major upwellings of hot material in the lower mantle (Hager *et al.*, 1985; Lithgow-Bertelloni & Silver, 1998). These hot, low velocity regions, defined by seismic tomography, appear to rise from the thermal boundary layer at the core–mantle boundary (Plate 12.2 between pp. 244 and 245) (Section 12.8.4).

Just as upwellings in the mantle produce regional uplift of the Earth's surface, downwellings produce regional subsidence (Gurnis, 2001). The most notable example of depressed crust at the present day is the Indonesian region. This is situated above anomalously high seismic velocities in the transition zone and upper part of the lower mantle (Plate 12.2b between pp. 244 and 245) that probably reflect a confluence of downgoing lithospheric slabs. Seismic tomography can only map regions of low and high velocity, and hence possible upwellings and downwellings in the mantle, at the present day. However, evidence from the geologic record for regional scale elevation and subsidence of the Earth's crust may indicate that a particular area has been underlain by a major mantle upwelling or deep subducting slabs in the past. Originally it was assumed that changes in sea level, causing major marine transgressions and regressions on continental crust, were synchronous worldwide, away from areas of active tectonism. However, as more data accumulated it became clear that this was not so, although an obvious explanation was lacking. It is now apparent that elevation and subsidence of the lithosphere associated with convection in the mantle, could provide an explanation for what were previously some very enigmatic observations.

Denver, Colorado in the central USA has an elevation of 1.6 km but is underlain by Cretaceous sediments typical of shallow water deposition. At that time the Farallon plate, the eastern flank of the East Pacific Rise in the northeast Pacific, was being subducted beneath western and central North America and is thought to have caused depression of the crust above it. With the progressive elimination of the East Pacific Rise in the northeast Pacific throughout the late Cenozoic, the Farallon plate has become detached and continues to sink eastwards, allowing the buoyancy of the crust of the western and central USA to reassert itself, thereby causing the uplift of the Colorado region. Van der Hilst *et al.* (1997), using seismic tomography, imaged the sinking Farallon plate 1600 km beneath the eastern USA. Similar anomalous vertical movements of parts of Australia since the early Cretaceous are thought to be due to the influence of downwellings created by subduction zones, initially to the east of Australia, and more recently to the north (Gurnis *et al.*, 1998).

12.8.4 The D″ layer

It has long been recognized that the greatest contrasts in physical properties and chemical composition within the Earth occur at the core–mantle boundary and that this is almost certainly the location of a thermo-chemical boundary layer (Section 2.8.6). Initially, seismologists were unable to detect any layering in the lower mantle and referred to it as Layer D (Bullen, 1949). Subsequently it was realized that a layer at the base of the mantle, perhaps 2–300 km thick, has distinctive, if variable, characteristics; typically lower seismic velocities or a lower velocity gradient than in the lower mantle above. Hence the lower mantle is now divided into two seismologic layers D′ and D″. With further refinements in seismologic techniques, studies of seismic waves reflected, refracted and diffracted at the core–mantle boundary have revealed remarkable details of the complexity and lateral variability of layer D″. The geographic distribution of earthquakes and seismologic observatories is such that not all parts of the layer can be studied in the same degree of detail. Clearly for such a remote layer, that is now thought to have vertical and horizontal variability analogous to that of the lithosphere, this poses quite a challenge for future seismologic studies.

Figure 12.11 illustrates the picture that is emerging of the nature of layer D″ for three very different regions for which detailed studies have been possible: beneath central America, Hawaii, and southern Africa. The upper boundary of the layer is characterized by a velocity discontinuity. Below this there may be an increase or decrease in the seismic velocities, particularly the shear wave velocity, or a decrease in the velocity gradient with depth. A velocity increase is most marked beneath regions where there are subducting slabs such as Central America (Fig. 12.11a). In a 5- to 50-km-thick layer immediately above the core–mantle boundary there is often a zone of ultra-low seismic velocities, with decreases in the shear wave velocity of 10–50%. This implies partial melting with more than 15% melt (Thybo *et al.*, 2003). These ultra-low velocity zones (ULVZ) are most extensively developed beneath major hotspots such as Hawaii (Fig. 12.11b) and beneath the superswells, and inferred upwellings, of the central Pacific and southern Africa (Fig. 12.11c). Unlike the variations in seismic velocity in the main part of the lower mantle, that are thought to be largely due to temperature differences, the marked

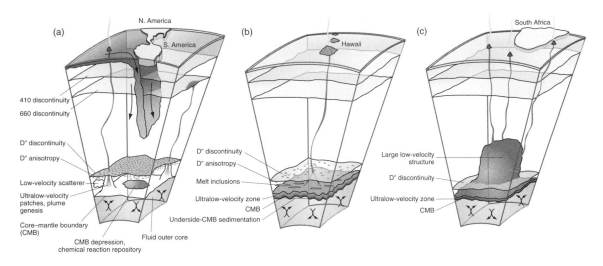

Figure 12.11 *Sections through the Earth's interior beneath regions centered on (a) central America, (b) Hawaii, and (c) South Africa, illustrating variations in the nature of the D″ layer (reproduced from Garnero, 2004,* Science ***304****, 834–6, with permission from the AAAS).*

lateral and vertical variations within layer D″ may be caused by variations in chemical composition, mineralogic phase changes and/or varying degrees of partial melting, in addition to temperature differences. Compositional variations may be due to the mixing of molten iron from the core with the perovskite of the mantle to form new high-pressure minerals (Section 2.8.6). It is thought that this is most likely to occur in the ULVZs where it is facilitated by higher temperatures, partial melting, and low viscosity. The result would be a chemically distinct, high-density layer but with a low viscosity. A phase change in perovskite to a denser and strongly anisotropic form is an interesting possibility as some parts of the D″ layer exhibit a marked anisotropy. It is thought that this anisotropy may be induced by subducted slabs beneath downwellings and by shear flow beneath upwellings.

It seems likely that the slabs of subducted lithosphere that sink into the lower mantle affect the nature of the D″ layer beneath them, most notably its temperature. This in turn will modulate the flow of heat from the core which will influence convection in the core and the nature of the Earth's magnetic field, and determine where flow may occur within and above layer D″.

12.9 THE NATURE OF CONVECTION IN THE MANTLE

The evidence for convective flow in the mantle, from seismic tomography and studies of the regional elevation and subsidence of the Earth's surface, strongly suggests that there are two main driving forces for this convection. The negative buoyancy of cold subducting lithosphere would appear to determine the main sites of downwelling, and the positive buoyancy of hot, low viscosity material originating in the lowermost, D″, layer of the mantle determines the upwellings. These two complementary modes of convection in the mantle have been termed the plate and plume modes, respectively (Davies, 1999). Both have their origins in thermal boundary layers: the plate mode in the lithosphere immediately beneath the Earth's surface, and the plume mode in the D″ layer of the mantle, immediately above the core–mantle boundary. As Davies (1993) has aptly put it, the plate mode is crucial in cooling the mantle, by the creation of oceanic lithosphere, and the plume mode releases heat from the core. The heat released by the plate mode is thought to be much greater than that released from the core as the mantle is heated internally by radioactivity. One might expect therefore that the plate mode is dominant. These two very different modes of convection need not necessarily be strongly coupled. However it is noteworthy that the two major upwellings at the present day, beneath southern Africa and the south central Pacific, are at the centers of the expanding ring of subduction zones around what was Gondwana and the contracting ring of subduction zones around the Pacific respectively, and hence distant from the cooling effect of the subducting slabs that appear to extend to the core–mantle boundary ((Plate 12.2 (between pp. 244 and 245), Fig. 12.12). It is also striking that these two active upwellings do not correspond directly to mid-ocean ridges. This is consistent with the interpretation of the upwelling beneath ridges

being entirely passive. Meguin & Romanowicz (2000) and Montelli *et al.* (2004b) note that there is evidence in their mantle tomographic models for lateral flow in the upper mantle from the African upwelling to the Atlantic and Indian Ocean ridges, and from the Pacific upwelling to the East Pacific Rise. If so this would complete the elusive route of the return flow from subduction zones to mid-ocean ridges, or at least provide one such route.

The scale, or wavelength, of this gross pattern of convection in the mantle is greater than that predicted by analogue experiments and early numerical models assuming a Rayleigh number greater than 10^6. It transpires that this is because these models assumed uniform viscosity throughout the convecting layer. In the Earth's mantle the viscosity varies with both temperature and pressure. For the relevant temperature gradient in the mantle the effect of increasing pressure with depth almost certainly means that the viscosity of the lower mantle is significantly greater than that of the upper mantle. Bunge *et al.* (1997) investigated three-

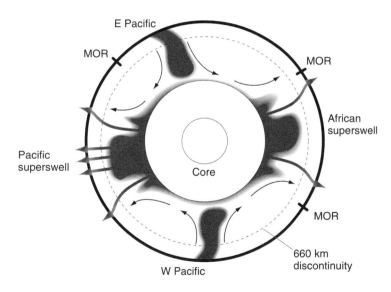

Figure 12.12 *Cartoon showing an approximately equatorial section through the Earth and illustrating the possible relationship of subduction zones, superswells, plumes, and mid-ocean ridges (MOR) to the gross pattern of circulation in the mantle. Note that deep-seated or primary plumes, such as Afar, Reunion, Tristan, Hawaii, Easter, and Louisville, are peripheral to the superswells, and that secondary plumes are common above the Pacific superswell. The mid-ocean ridges are a passive response to the plate separation and not systematically related to the main convective pattern.*

dimensional spherical convection models of the mantle in which the viscosity of the lower mantle was 30 times that of the upper mantle. They found that not only was the wavelength of the resulting convection greater but that long linear downwellings formed from the upper boundary layer; both effects making the pattern of convection very comparable to that deduced for the mantle. The convective pattern also had greater temporal stability.

Researchers also have investigated the effect on mantle convection of the endothermic phase change at a depth of 660 km, the base of the transition zone. For plausible physical characteristics of this phase change the results suggest that it might inhibit but not prevent the passage of upwellings and downwellings through it. This is consistent with the results of seismic tomography that indicate that the transition zone has some effect but that it is not sufficient to impede whole mantle convection (Montelli *et al.*, 2004b).

The chemical heterogeneity of layer D″ (Section 12.8.4) means that it acts as a thermochemical, rather than a thermal boundary layer. Indeed where it is hottest it is essentially a thermal boundary layer over a chemical boundary layer, the ultra-low velocity zone (ULVZ). Upwellings of the low viscosity, low density thermal boundary layer at these points entrain the low viscosity but higher density chemical boundary layer to a height of 50–100 km depending on the strength of the upwelling (Fig. 12.13). Analogue experiments (Davaille, 1999) indicate that the nature of the upwelling depends on the ratio of the stabilizing chemical density anomaly to the destabilizing thermal density anomaly. If this is greater than 1, a plume-like upwelling forms; if it is approximately 0.5, thermals (broad upwellings or domes) are produced. In either case the entrainment of the dense chemical boundary layer is thought to stabilize the location of the plume or thermal upwelling (Jellinek & Manga, 2004). However, as a result of the greater stability ratio, plumes will tend to be very long-lived.

If this general picture of convection in the mantle is correct the roles of subduction zones and a chemical boundary layer at the base of the mantle are crucial in determining the pattern and nature of the convection. Indeed it could be argued that the location of subduction zones is most fundamental in that they not only determine downwellings occur but also where the boundary layer at the core–mantle boundary is hottest, and hence where upwellings occur. However, subduction zones are transient features within the context of

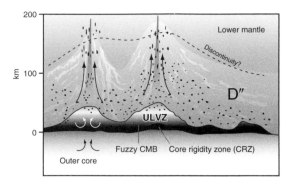

Figure 12.13 *Cartoon of the D″ layer where it is hotter than its average temperature. These regions include an ultra-low-velocity zone (ULVZ), thought to be characterized by partial melt and chemical heterogeneity, chemical and melt scatterers throughout and, possibly, the points of origin of plumes (redrawn, with permission, from Garnero, 2000.* Annual Review of Earth and Planetary Sciences, **28**. *Copyright © 2000, Annual Reviews).*

geologic time. Within the supercontinent cycle (Section 11.5) there are times when subduction zones are initiated, as a result of continental break-up, and terminated by continent–continent collision. Such events could initiate changes in the gross pattern of convection in the mantle and even change the distribution of mass within the Earth causing a change in the location of the rotational axis, that is, the axis about which the moment of inertia is a maximum. This would be particularly true if the initial development of subduction zones includes a build-up of subducted material in the transition zone that ultimately avalanches down into the lower mantle. Such True Polar Wander (Section 5.6) is thought to have occurred between 130 and 50 Ma ago (Besse & Courtillot, 2002), a time period bracketed by the break-up of Pangea, with the initiation of subduction zones, and the collision of India with Eurasia and a major change in the rate of subduction in this zone. The change in direction of the Hawaiian–Emperor seamount chain, and the change in the relative motion between the Pacific and Indo-Atlantic hotspot reference frames (Section 5.5) also occurs at the time of the Indian collision, 40–50 Ma ago. Thus, these too might reflect the consequent changes in the thermal regime and pattern of convection in the mantle, and, hence, the relative positions of the two major convection cells within the African and Pacific hemispheres.

12.10 PLUMES

Certain volcanic hotspots at the Earth's surface appear to be essentially fixed with respect to the Earth's deep interior and to provide an absolute reference frame for plate motions for the past 40 Ma (Section 5.5). The fixed nature of hotspots such as Hawaii, first suggested by Wilson (1963), led Morgan (1971) to propose that they are located over plumes of mantle material upwelling from the lower mantle or even the core–mantle boundary. The plume hypothesis has been, and continues to be, controversial because it has proven difficult to provide unequivocal evidence of such plumes (Foulger & Natland, 2003). There is now, however, a growing body of evidence, from both modeling and observational data, that some hotspots at the Earth's surface may be fed by narrow plumes of high temperature, low viscosity material rising from essentially fixed points on the core–mantle boundary. There is also theoretical and empirical evidence that the source material for other hotspots may be derived from much shallower depths, within the mantle transition zone or the uppermost part of the lower mantle, or even from immediately beneath the lithosphere; the latter being a passive response to various forms of lithospheric break-up (Anderson, 2000). The suggestion that there are three types of hotspot, in terms of their depth of origin, has been deduced by Courtillot *et al.* (2003), mostly by a consideration of the roles of the three potential thermal boundary layers in the mantle. However, there is substantial support for this from the results of seismic tomography (Montelli *et al.*, 2004a, 2004b). Hotspots that are underlain by low seismic velocities in the upper mantle only appear to be limited in number. Examples in Fig. 5.7 are Bowie, Cobb, Galapagos, East Australia and, surprisingly perhaps, Iceland, although it is underlain by a very large upper mantle anomaly (Montelli *et al.*, 2004b). Iceland is anomalous also in terms of geochemical indicators of deep mantle origin, notably the ratios of $^3He/^4He$ and $^{186}Os/^{187}Os$ in the lavas (Foulger & Pearson, 2001; Brandon, 2002). Montelli *et al.* (2004b) describe the tomographic anomaly beneath Yellowstone as being virtually nonexistent.

Laboratory experiments indicate that the peaks that develop on the ULVZ in the D″ layer, where it is sufficiently dense, form where ridges between embayments in the surface of the ULVZ meet at an elevated point or arête (Jellinek & Manga, 2004). As a result the upwelling of the thermal boundary layer that produces these peaks is focused into a narrow, cylindrical conduit. The temperature difference between this plume and the surrounding mantle is probably 200–300°C, implying more than two orders of magnitude reduction in the viscosity across the boundary layer between them. If partial melt is present in the upwelling thermal boundary layer this would also lower the viscosity. Partial melt entrained from the ULVZ may be required to explain the osmium isotopic ratios in certain hotspot lavas that are thought to indicate that the source of the osmium is the outer core (Brandon *et al.*, 1998).

Numerical and analogue models of these hot, low viscosity plumes, originating in the deep mantle, suggest that plume shape and mobility are controlled by the magnitude of the viscosity contrast with the surrounding mantle (Kellogg & King, 1997; Lowman *et al.*, 2004; Lin & van Keken, 2006). As the contrast increases, the plume conduit becomes narrower and its head becomes broad and mushroom-shaped as hot material is able to move upward more efficiently (Fig. 12.14). This model, with a mushroom-shaped plume head and a long, thin tail extending to the depth of origin, has achieved widespread application. Nevertheless, numerical models also predict a great variety of plume shapes and sizes in cases where density contrasts due to chemical variations in the lowermost mantle are incorporated into models of plume formation (Section 12.9) (Farnetani & Samuel, 2005; Lin & van Keken, 2006).

In general, the model of a narrow, mushroom-shaped plume fits well with the initial expression of some hotspots in terms of continental flood basalts or oceanic plateaux, reflecting the arrival of the plume head beneath a thinned lithosphere, and the subsequent trace of the hotspot, in the form of a volcanic ridge or line of volcanoes, produced by the tail. Courtillot *et al.* (2003) suggest that these hotspots be termed *primary* hotspots (Section 5.5). They also suggested that the lifespan of primary hotspots might be approximately 140 Ma. Those initiated within the past 100 Ma, such as Afar and Reunion, are still active; those that are 100 to 140 Ma old, that is Louisville and Tristan, might be failing, and those that formed more than 140 Ma ago, such as Karoo and Siberia, have no active trace. Theoretical arguments predict that such large plume heads and long-lived tails must originate in a thermal boundary layer at great depth, presumably layer D″ at the

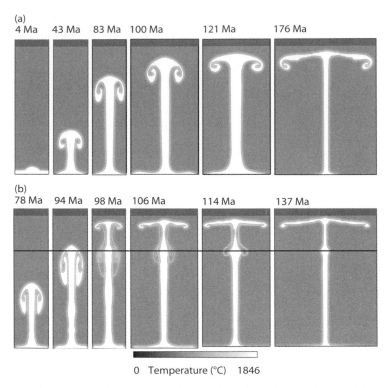

(a)
4 Ma 43 Ma 83 Ma 100 Ma 121 Ma 176 Ma

(b)
78 Ma 94 Ma 98 Ma 106 Ma 114 Ma 137 Ma

0 Temperature (°C) 1846

Figure 12.14 *Sequences from numerical models, scaled approximately to the mantle, in which a plume grows from a thermal boundary layer. In (a) the viscosity is a function of temperature, and in (b) the viscosity also increases by a factor of 20 at 700 km depth. In (b) the plume slows and thickens through the 700 km discontinuity but then narrows and speeds up in the low viscosity upper layer (from Davies, 1999. Copyright © Cambridge University Press, reproduced with permission).*

core–mantle boundary. It has been estimated that a major plume may be fed for 100 Ma from a volume of layer D″ only tens of kilometers thick and 500–1000 km in diameter.

If major upwellings, such as those beneath southern Africa and the south Pacific, reach the base of the transition zone they may well form a thermal boundary layer at this depth from which *secondary* plumes may originate (Brunet & Yuen, 2000; Courtillot *et al.*, 2003). These would be relatively short-lived and without initial flood basalts but may well account for the hotspots on the south Pacific superswell such as the Society and Cook-Austral islands, Samoa, Pitcairn, and Caroline (Fig. 5.7) (Adam & Bonneville, 2005). By contrast, there are no plumes within the southern

African superswell although, as in the Pacific, there are several potential deep mantle, or *primary*, plumes around it (Figs 5.7, 12.11c). This contrast is also reflected in the marked difference in the seismic velocity anomalies in the upper mantle beneath the two areas (Plate 12.2b,d between pp. 244 and 245). The differing characteristics of the African and Pacific superswells may arise from the fact that the south Pacific upwelling is the remnant of the Cretaceous superplume in this area (Section 5.7). The uplift of southern Africa was also initiated in the mid-Cretaceous, suggesting that major mantle upwellings or superplumes may have a life cycle analogous to, and perhaps related to, the life cycle of the assembly and break-up of supercontinents.

12.11 THE MECHANISM OF THE SUPERCONTINENT CYCLE

The assembly and dispersal of the supercontinents reflect interactions between continental lithosphere and processes operating in the mantle. The first type of interaction involves the broad upwellings and down-wellings that define mantle convection cells (Section 12.9). The second is related to the possible impingement of deep mantle plumes (Section 12.10) on the base of continental lithosphere.

Numerical simulations have provided an important means of investigating the possible relationships between mantle convection patterns and plate motions. Gurnis (1988) suggested that, during periods of dispersal, the continents tend to aggregate over cold down-wellings in the mantle, where they act as an insulating blanket. The mantle consequently heats up, altering the convection pattern, and the supercontinent rifts apart in response to the resulting tension. The continental fragments then move toward the new cold down-wellings resulting from the changed convective regime. Gurnis emphasized the fact that the continents, except Africa, are currently moving to cold regions of the mantle, which are characterized by few hotspots and high seismic velocities. It appears that about 200 Ma ago, Pangea was positioned over what is today the upwelling beneath southern Africa. Since Africa has moved only slowly with respect to the hotspot reference frame, it seems that Pangea may have been situated over this upwelling prior to break-up, in accord with the model. It would thus appear that a positive feedback exists between patterns of mantle convection and the formation of the supercontinents.

The results of experiments also suggest that several mechanisms produce convection patterns that promote the growth and dispersal of supercontinents. The insulating properties of large masses of continental lithosphere create mantle upwelling beneath their interiors (Gurnis, 1988; Zhong & Gurnis, 1993; Guillou & Jaupart, 1995). Large plates also prevent the mantle beneath them from being cooled by subduction, which

further promotes upwelling (Lowman & Jarvis, 1999). Sites of downwelling may be controlled by the intrinsic buoyancy of continental lithosphere, which tends to concentrate subduction zones along continental margins. This effect was illustrated by Lowman & Jarvis (1996, 1999) who showed that the collision of two continents at a site of downwelling can trigger a reorganization of the convection pattern, leading to downwelling at the margins and upwelling beneath their interiors (Fig. 12.15). These authors also showed that slab-pull and trench suction forces (Section 12.6) probably were as important as mantle upwelling in the break-up of the supercontinents.

Another important process that affects the relationship between patterns of mantle convection and plate motions is internal heating (Section 12.5.1). Lowman et al. (2001, 2003) showed that, in internally heated models, plate motion is characterized by episodic reversals in direction as mantle circulation patterns change from clockwise to counterclockwise and vice versa. These reversals are caused by the trapping and build-up of heat and buoyancy forces in the interior of convection cells, which destabilizes the convection pattern. The results of modeling suggest that the downwelling of cold material at one edge of a plate can entrain hot material that is trapped below the plate and drag it into the lower mantle. The hot, buoyant material then begins to ascend as the drag of the cold downwelling wanes. The ascent of hot material pushes the plate laterally and induces new cold downwelling on the other side of the plate, beginning a new cycle of upwelling and plate motion in the opposite direction. This type of feedback relationship between plate motion and internally heated mantle convection may explain why some plates suddenly change direction on timescales of some 300 Myr.

Many geologic investigations (e.g. Hill, 1991; Storey, 1995; Dalziel et al., 2000a) have demonstrated time–space relationships among LIPs, hotspots, and supercontinental fragmentation. Nevertheless, the role of hot spots or upwelling deep mantle plumes during continental break-up is uncertain. Thermal buoyancy forces due to mantle upwellings and tractions at the base of the lithosphere caused by convecting asthenosphere may contribute to a horizontal deviatoric tension that is sufficient to break continental lithosphere (Section 7.5). Lowman & Jarvis (1999) showed that tensile stresses in the interior of supercontinents depend on the size of the plate, the Rayleigh number of mantle

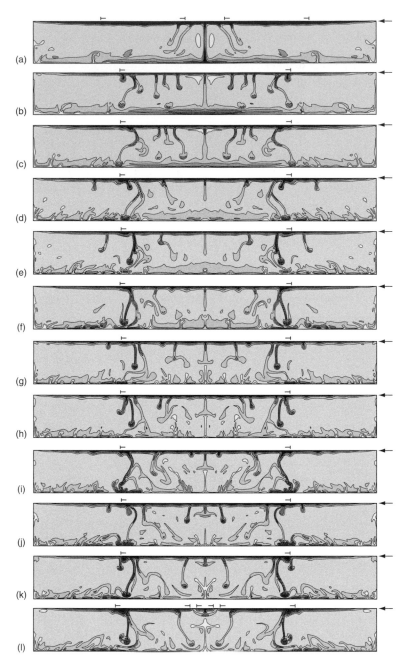

Figure 12.15 *Temperature fields in a numerical model of whole mantle convection that incorporates two continents 5800 km wide (modified from Lowman & Jarvis, 1999, by permission of the American Geophysical Union. Copyright © 1999 American Geophysical Union). Dark and light shading represent cool and warm temperatures, respectively; markings at the top indicate the locations of the continental margins. The continents collide at the model symmetry plane between panels (a) and (b), forming a supercontinent of width 11,600 km. As subduction (dark downwellings) between the two continents ceases, new subduction zones form along the continent margins. Eventually a central upwelling of warm material forms beneath the supercontinent. The supercontinent rifts between panels (k) and (l) some 600 Myr after its formation.*

convection (Section 12.5.2), the viscosity profile of the mantle, and the amount of radioactive heat present. In addition, their models suggest that, given an internally heated mantle, stresses generated at subduction zones also may be sufficiently large to cause rifting in a stationary supercontinent.

Some geologic data suggest that plume-related magmatism coincided with the assembly, rather than the break-up, of the supercontinents. Hanson *et al.* (2004) showed that large-scale magmatic events occurred within continental interiors during the Proterozoic assembly of Rodinia (Section 11.5.3). These authors also concluded that the impingement of mantle upwellings on the base of continental lithosphere prob-ably occurred independently of the supercontinent cycle. Isley & Abbott (2002) used a series of plume proxies, including massive dike swarms, high-Mg extrusive rocks (e.g. Section 11.3.2), flood basalts, and layered intrusions, to identify mantle plume events through time. At least two global scale events coincided with continental assembly in Late Archean and Proterozoic times. From these relationships, it seems that there may be two types of mantle plume events, those associated with supercontinental break-up and those associated with their formation (Condie, 2000). These studies highlight the intriguing but uncertain relationships between mantle plumes and the supercontinent cycle.

13 Implications of plate tectonics

13.1 ENVIRONMENTAL CHANGE

13.1.1 Changes in sea level and seawater chemistry

The sedimentary record in continental areas is characterized by marine transgressions and regressions due to changes in sea level throughout geologic time. One of the highest sea level stands occurred in late Cretaceous time when, for example, the very pure marine limestone, Chalk, was deposited throughout much of northwest Europe.

Major changes in sea level, of 100 m or more, are difficult to explain, except during ice ages, when large volumes of fresh water are locked up in land-bound ice sheets. However, for much of geologic time, there were no major glaciations, and yet there were major changes in sea level. The concepts of sea floor spreading, hot spots, and plumes provide plausible mechanisms to resolve this problem. The water depth above oceanic crust formed solely by sea floor spreading is related to the age of the crust (Section 6.4), younger crust occurring at shallower depths. Such crust has an essentially uniform thickness of 6–7 km (Section 2.4.4). However, if this crust is thickened, as a result of enhanced igneous activity above a hot spot or plume, the water depth will be shallower than that predicted by the age/depth relationship. Exceptionally, as in the case of Iceland and the Azores, the volcanic edifice rises above sea level. Thus, enhanced rates of sea floor spreading, hot spot or plume activity can produce elevated ocean floor that will displace the water upwards and cause a rise in sea level. During the Cretaceous period, for example, the high sea level stand might well be due to exceptionally high rates of sea floor spreading and plume activity, as discussed in Section 5.7.

Changes in the net rate of formation of oceanic crust, as a result of changes in spreading rates and/or the total length of actively spreading ridges, are a very effective way of changing the proportion of young, elevated ocean floor, and hence producing, in the long term, changes in sea level. Variations in net accretion rate also imply changes in the amount of igneous and hydrothermal activity at spreading centers that will have implications for the chemistry of seawater. Interactions

between the circulating seawater and the hot basaltic rock at ridge crests are thought to remove magnesium and sodium from the seawater and to release calcium ions from the rock. It is also possible that the sulfate ion is removed from the water when it encounters the oxic conditions at or near the sea floor. These changes would predict that the Mg/Ca, SO_4/Cl, and Na/K ratios in seawater decrease during periods of high rates of formation of oceanic crust and hydrothermal activity.

Stanley & Hardie (1999) suggest that such changes in seawater chemistry are reflected in the mineralogy of marine evaporites and carbonate sediments throughout the Phanerozoic. They assume that a first order sea level curve may be used as a proxy for the rate of production of oceanic crust, and hence the variation in hydrothermal brine flux, throughout the past 550 Ma (Fig. 13.1). From this the temporal variation in the

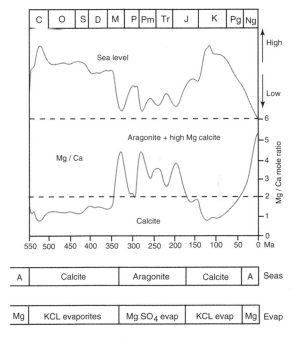

Fig. 13.1 *Variation in the Mg/Ca ratio in seawater, calculated by Hardie, 1996, from an assumed curve of long term changes in sea level, and (below) summaries of the mineralogy of nonskeletal carbonates, and marine evaporites, illustrating the correlation with the predicted changes in the Mg/Ca ratio in seawater during the past 550 Ma (based on figure 2 in Stanley & Hardie, 1999).*

Mg/Ca ratio for seawater is calculated (Hardie, 1996). During the resulting periods of low Mg/Ca ratio, associated with high sea level stands, nonskeletal carbonates are composed of (low magnesium) calcite, and marine evaporites are characterized by late forming KCl (sylvite), and an absence of Mg salts. By contrast, the periods of high Mg/Ca ratio are characterized by nonskeletal carbonate deposits composed of high magnesium calcite and aragonite (a polymorph of calcite), and marine evaporites in which $MgSO_4$ formed during the final stages of evaporation. The former periods have been termed periods of "calcite seas," and are thought to be associated with high pCO_2 and high surface temperatures; i.e. a "Greenhouse Earth" such as that which probably characterized the Cretaceous. The periods of high Mg/Ca ratio have been designated as periods of "aragonite seas." These appear to correlate with times of low pCO_2, and low surface temperatures, and include ice ages, i.e. an "Icehouse Earth".

Variations in pCO_2 in the atmosphere in the geologic past are thought to have been largely due to the outgassing of CO_2 from volcanic activity. Thus eustatic changes in sea level, changes in seawater chemistry, and variations in the concentration of CO_2 in the Earth's atmosphere in the past might all be related to variations in the rates of sea floor spreading and plume activity.

13.1.2 Changes in oceanic circulation and the Earth's climate

Two of the most significant influences on the Earth's climate are the concentration of greenhouse gases in the atmosphere (Sections 5.7, 13.1.1), and the extent, distribution, and bottom topography of the oceans. The configuration of the ocean basins affects the transport of heat in the oceans, by surface currents and deep-water circulation, thereby affecting the temperature and moisture content of the atmosphere over oceanic areas. Surface currents are essentially wind driven, and, therefore, largely determined by the circulation of the atmosphere. The rotation of the Earth, and the concentration of incoming solar radiation within the tropics, produces surface easterly (trade) winds at low latitudes, westerlies at intermediate latitudes, and easterlies at high latitudes (greater than $60°$).

If the Earth's surface was entirely covered by an ocean, the resulting westerly directed, equatorial ocean current, and the intermediate latitude easterly directed, circumpolar currents, would bracket irregular "gyres," circulating clockwise in the northern hemisphere, and anticlockwise in the southern hemisphere. In this situation the world-encircling equatorial and circumpolar currents would tend to inhibit the transfer of heat by surface currents from low to high latitudes, and the temperature gradient between the equator and the poles would be accentuated. As a consequence, sea ice might form in the polar oceans. However, land masses with north–south trending shorelines in low and intermediate latitudes, will deflect the equatorial and circumpolar currents, to the right in the northern hemisphere, and to the left in the southern hemisphere, thereby intensifying the gyres, and transferring heat from the tropics to higher latitudes by means of western boundary currents. In this scenario the temperature gradient is reduced. A classic example at the present day is the Gulf Stream of the western North Atlantic, which warms the air above the ocean in the extreme North Atlantic, thereby ameliorating the climate of Iceland and northwest Europe. The opening or closing of gateways for the equatorial or circumpolar currents, as a result of continental drift, can, therefore, have pronounced effects on the Earth's climate (Smith & Pickering, 2003).

During the past 200 Ma the supercontinent of Pangea has progressively rifted apart. The resulting fragments have drifted across the face of the globe, such that a continuous tropical seaway, the neo-Tethys, was formed, and subsequently closed, and a southern ocean gradually opened up around Antarctica (Figs 13.2–13.7). By the mid-Cenozoic, a complete southern circumpolar current came into existence, which isolated and insulated Antarctica, and was probably instrumental in triggering the first major build-up of the Antarctic ice cap (Kennett, 1977).

At the beginning of the Mesozoic Era, 250 Ma ago, the supercontinent of Pangea extended from pole to pole (Fig. 13.2), without extensive polar landmasses in either hemisphere. Strong western boundary currents off the eastern shores of Pangea would have transported warm water to high latitudes, preventing the formation of ice-sheets and warming the east-facing coasts relative to the west. The interior of the supercontinent would have had strong seasonal extremes. By 160 Ma (Fig. 13.3) a low latitude east–west seaway had started to open up, between what is now North America

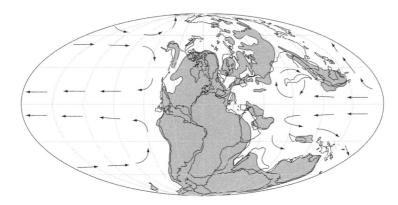

Fig. 13.2 *Possible circulation pattern of surface ocean currents during the early Triassic (245 Ma). Figs 13.2–13.7. Continental reconstructions, and present and paleo-shorelines, are from Smith et al., 1994 (Copyright © Cambridge University Press, reproduced with permission). Land areas shaded. Any indication of a surface equatorial counter current has been omitted, because of the uncertainties surrounding its existence and location in the geologic past. The currents shown in Figs 13.3, 13.5, 13.6, and 13.7 are based in part on Haq, 1989.*

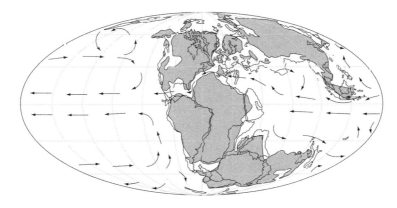

Fig. 13.3 *As for Fig. 13.2, for the Late Jurassic (160 Ma).*

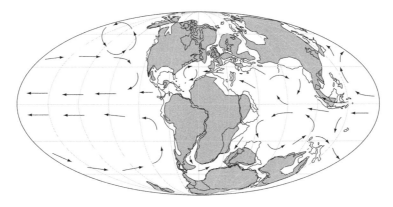

Fig. 13.4 *As for Fig. 13.2, for the Early Cretaceous (130 Ma).*

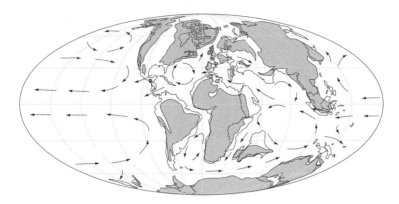

Fig. 13.5 As for Fig.13.2, for the mid-Cretaceous (95 Ma).

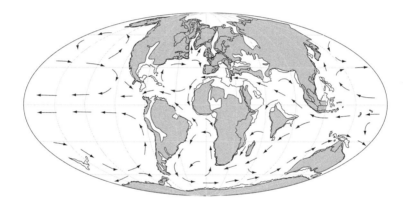

Fig. 13.6 As for Fig. 13.2, for the mid-Paleocene (60 Ma).

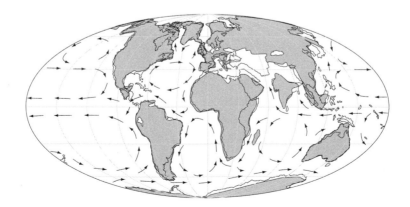

Fig. 13.7 As for Fig. 13.2, for the mid-Oligocene (30 Ma).

and northwest Africa, as a consequence of the first phase of rifting of the supercontinent. This was initiated about 180 Ma ago. Thus the "Tethyan embayment" in Pangea (Fig. 13.2) was extended to the west to facilitate a circum-global equatorial current. This meant that some tropical waters were heated to a higher temperature before turning northwards and southwards to warm higher latitudes. In this way the whole Earth became warmer and the temperature gradient from the equator to the poles was further reduced.

The separation of Antarctica from Africa, which started about 165 Ma ago, was the first stage in the break-up of Gondwana (Fig. 13.4). This was followed at about 125 Ma by the rifting apart of South America and Africa, which started in the south and propagated northwards. This, coupled with the complex fracture zone pattern in the equatorial Atlantic region, due to transform faulting, meant that the gateway between the North and South Atlantic did not open up until about 95 Ma (Fig. 13.5) (Poulsen et al., 2001). The initial changes in the deep-water circulation, resulting from the opening of this gateway, may explain the "anoxic event" that produced the widespread black shales in adjacent areas at that time (Poulsen et al., 2001). By 95 Ma India had separated from Antarctica and a major Southern Ocean was opening up south of Africa and India. However, in the late Cretaceous, and even in the early Cenozoic (Fig. 13.6), the circum-equatorial current still existed, and the surface water in the high latitude oceans was still very much warmer than it is today.

Throughout the Cenozoic, Africa, India, and Australia continued to drift northwards, away from Antarctica, thereby enlarging the southern and Indian Oceans, and ultimately forming the Alps and the Himalayas as a result of the collision of Africa and India with Eurasia (Section 10.4.1). By 30 Ma (Fig. 13.7), the Tethyan seaway was effectively closed, and the Southern Ocean completely encircled Antarctica, as a result of the opening of gateways south of Tasmania, and in the Drake Passage, south of South America. These results of continental drift, gave rise to major changes in the near surface oceanic circulation. There was no longer a complete circum-equatorial current, and a pronounced circum-polar current was established in the Southern Ocean. Thus the equatorial water became less warm, and Antarctica was insulated from the warmer water circulating in the major southern hemisphere gyres of the Pacific, Atlantic, and Indian Oceans. A change in oxygen isotope ratios in the tests of planktonic and

benthic microfossils (Shackleton & Kennett, 1975), and the first major build-up of ice on Antarctica, coincided with these developments, and appear to mark a transition from a Greenhouse to an Icehouse Earth. The change in oxygen isotope values is particularly pronounced and well documented, and is essentially coincident with the Eocene–Oligocene boundary (Fig. 13.8). This is also the time of the opening of the gateway south of Tasmania (Exon et al., 2002). The full opening of the Drake Passage is less well constrained, but was probably shortly after this (Livermore et al., 2004). Oxygen isotope ratios and a drop in sea level of 40 m suggest that during the early Oligocene the volume of ice in Antarctica built up to perhaps as much as one-half of its present volume. This and subsequent increases in ice volume, and changes in sea level, gave rise to an emergence of land areas, and a major reduction in the area of shallow seas on continental crust (cf. Figs 13.6, 13.7).

Following a period of warming and deglaciation in the late Oligocene (Fig. 13.8), additional major increases in the volume of ice on Antarctica, and associated drops in sea level, are thought to have occurred in the mid Miocene and at the end of the Miocene. The drop in sea level associated with the increase in ice volume at 6 Ma may explain the isolation, and subsequent desiccation, of the Mediterranean Sea, as a result of the exposure of the sill at the Strait of Gibraltar (Van Couvering et al., 1976), and would have restricted the flow of water through the ocean gateway between North and South America. However, additional tectonic movements were required before a complete land bridge formed, about 3 Ma ago; as determined from the interchange of mammals between North and South America (Marshall, 1988). The gradual formation of the Isthmus of Panama would have led to the intensification of the Gulf Stream, and ultimately, perhaps, to the formation of the northern hemisphere ice-sheets (Haug & Tiederman, 1998).

The warm waters of the Gulf Stream would have given rise to more warm and moist air, and hence more precipitation, in relatively high latitudes in the North Atlantic area. The geographic distribution of ice sheets is determined not only by cold ambient temperatures, but also by the availability of precipitation. The Plio-Pleistocene ice-sheets of the northern hemisphere were restricted to Greenland, northern North America, and northwest Europe for this reason. Similarly, the occurrence of tropical rain forests is determined not only by high temperatures, but also by the delivery of

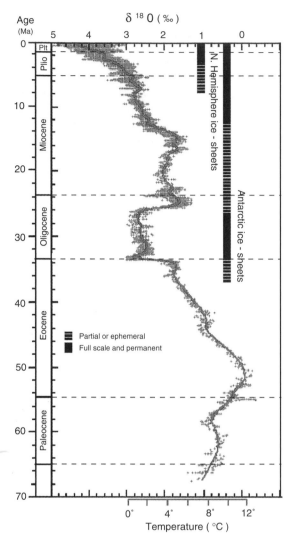

Fig. 13.8 *Global deep sea oxygen isotope record compiled from measurements on benthic fauna from numerous Deep Sea Drilling Project and Ocean Drilling Project cores. Fitted curve is a smoothed five point running mean. The temperature scale relates to an ice-free ocean and only applies therefore to the time prior to the onset of large scale glaciation in Antarctica (approx. 35 Ma). Much of the subsequent variability in the δ¹⁸O records reflects changes in Antarctic and Northern Hemisphere ice volume. When seawater evaporates, molecules containing the lighter isotope ¹⁶O evaporate more readily. Thus when atmospheric water vapor precipitates as snow in polar regions, ¹⁸O depleted water becomes sequestered in the polar ice caps and the proportion of ¹⁸O in seawater increases (part of figure 2 in Zachos et al., 2001, reproduced from Science **292**, 686–93, with permission from the AAAS).*

precipitation by warm equatorial currents. The tropical coal forests of the Carboniferous, for example, formed at the western end of the equatorial Tethyan embayment in the embryonic supercontinent of Pangea (Fig. 3.9). At the present day the most extensive areas of tropical rain forest are in the Amazon basin and the archipelago of southeast Asia, areas warmed by the main westward directed equatorial currents of today's oceans. The progressive cooling of the Earth's climate during the last 50 Ma, particularly in higher latitudes, led to a general reduction in the amount of precipitation, and an increase in aridity. Thus formerly forested areas in high latitudes were turned to tundra, and in temperate latitudes to grassland. As a consequence of the major cooling about 6 Ma, even some low latitude, tropical forests were converted to savannah. This is thought to have had a profound effect on mammalian, and, ultimately, human evolution.

During "Greenhouse Earth" conditions the oceans are warm throughout, with very little deep-water circulation. As a consequence the bottom waters become deoxygenated and there is the potential for the preservation of organic material and hence the formation of black shale deposits (Sections 3.4, 5.7). In as much as there is vertical mixing, it is probably triggered by regional changes in the salinity, and hence the density, of seawater in the tropics. Weak circulation, and the preservation of organic matter, meant that there was much less upwelling of nutrients compared to the present oceans. Thus, the overall fertility of the Cretaceous oceans was low, but the potential for the ultimate formation of oil, from Cretaceous marine source rocks, was high. In "Icehouse Earth" conditions cold, dense water forms in polar regions, sinks, and flows towards the equator, thereby creating a relatively vigorous, and certainly very significant, deep water circulation. The cooling that marked the transition from a Greenhouse to an Icehouse Earth, at about the Eocene–Oligocene boundary, probably enabled sea ice to form around the margin of Antarctica for the first time. During the formation of the sea ice much of the salt content of the seawater is expelled, increasing the density of the seawater beneath the ice. This cold, dense water would then sink to the ocean floor, and flow northwards, as it does at the present day.

One of the enigmas of the late Cenozoic cooling of the Earth is the relatively sudden build-up of ice in Antarctica in the Mid-Miocene (Fig. 13.8). One interesting and remarkable possibility is that it was caused by a change in the topography of the sea floor in the extreme North Atlantic, as a consequence of tectonic processes

(Schnitker, 1980). It is likely that the Greenland–Iceland–Faroes Ridge had subsided sufficiently at this time for cold water from the Arctic to spill over this sill and sink towards the ocean floor. Although cold and saline, it is not as dense as the Antarctic bottom water moving northwards. As a consequence, the Arctic water travels south at an intermediate depth, and is ultimately deflected towards the surface off Antarctica. Here it is "warm," relative to the surrounding seawater, and creates more moisture laden air, and hence enhanced precipitation over Antarctica. This model again emphasizes the importance of precipitation, in addition to sub-zero temperatures, in facilitating the build-up of an ice sheet.

13.1.3 *Land areas and climate*

The extent, distribution, and topography of land areas also affect the Earth's climate. Land heats up and cools down more rapidly than the sea. The daily cycle of sea and land breezes in coastal areas is a well-known consequence of this. A similar phenomenon on a longer, seasonal timescale, and affecting a larger geographic area, is the monsoonal climate of India and the Arabian Sea. In the northern summer the large landmass of southern Asia heats up, and the air rising above this creates a low pressure area and draws in moisture laden air from the northwest Indian Ocean – the southwest monsoon. In the winter, cold, dense air over the cold land area creates high pressure, and gives rise to the dry, northeast monsoon, which blows from land to sea. Seasonal heating and cooling of the air over the Sahara produces a similar, but smaller scale, effect over central Africa and the equatorial Atlantic in the Gulf of Guinea, where a similar monsoon regime pertains. These two monsoonal areas account for the tropical rain forests of central Africa, and Burma, Sri Lanka, and parts of India.

The albedo of land areas is variable depending on the type, or lack, of vegetative cover, but it is typically higher than that of sea areas, which have a low albedo. The distribution of land and sea, and its affect on the Earth's albedo in the past, might be expected to have produced an appreciable effect on climate, but as yet this is poorly understood. Ice or snow covered land or sea has a high albedo, and clearly is significant, not least in that it provides a positive feedback mechanism: the greater the extent of the ice and/or snow, the greater the degree of cooling. Mountains, even in low latitudes, can be covered with permanent or seasonal snow, thereby increasing the Earth's albedo. However, the formation of mountain belts may affect the climate in a more substantial way, by changing the rate of weathering at the Earth's surface, which in turn affects the amount of carbon dioxide in the atmosphere.

The weathering of carbonates exposed on land, by a weak carbonic acid solution, formed by the dissociation of carbon dioxide from the atmosphere, or soil, in rainwater, produces calcium and bicarbonate ions that are then transported to the ocean by rivers. In the oceans the weathering reaction is reversed: calcium carbonate is secreted by organisms, to produce their tests, which, if preserved after the death of the organism, form carbonates on the sea floor.

$$CaCO_3 + CO_2 + H_2O \leftrightarrow Ca^{2+} + 2HCO_3^- \quad \text{(eq. 1)}$$

The carbon dioxide so released ultimately returns to the atmosphere. Thus the carbon fixed in the carbonates on land is redeposited on the sea floor, with no net change in the CO_2 content of the atmosphere. The weathering of silicate rocks by carbonic acid, however, has important differences. A simplified weathering reaction may be expressed as:

$$\text{Silicate mineral} + 2CO_2 + \text{water} \rightarrow$$
$$2HCO_3^- + \text{clay mineral} + \text{cation(s)} \quad \text{(eq. 2)}$$

In the ocean the HCO_3^- ions combine with Ca^{2+}, as in the reverse of equation 1, to form calcium carbonate. In this case, two molecules of CO_2 are removed from the atmosphere, for every one molecule returned to the atmosphere when $CaCO_3$ is formed in the ocean. Increased weathering of silicate rocks could, therefore, draw down the CO_2 content of the atmosphere, and be a possible cause of global cooling (Raymo & Ruddiman, 1992).

As a result of the most recent phase of continental drift, the Cenozoic was characterized by a major episode of mountain building, notably throughout the Alpine–Himalayan belt, and culminating in the uplift of the Tibetan Plateau in the Late Cenozoic. The elevation of mountains would have greatly increased physical and chemical weathering processes, particularly as they concentrate rainfall on their windward flanks. The elevation of the Tibetan Plateau, for example, is thought to have greatly intensified the southwest monsoon, bringing much heavier rainfall, and causing much more intense weathering, on the southern slopes of the Himalaya. The elevation of Tibet and surrounding areas is particularly important because, although this represents just 5% of the Earth's land area, 23% of the global flux of dissolved material in rivers is derived from rivers with a source in the Tibetan/Himalayan region. There is some indication from the fauna and flora, and the sedimentary record of northern India, that there was a major intensification of the southwest monsoon about 8 Ma ago (An *et al.*, 2001). The question remains, however, whether this correlates with the uplift of the Tibetan Plateau (Section 10.4.3). Current models for this uplift, which may involve the convective removal of thickened lithosphere beneath Tibet (Section 10.4.6), imply that the final uplift phase may have been relatively sudden, in geologic terms. Attempts to obtain an independent estimate of the timing of this uplift, using paleobotanical evidence or the dating of fault systems, have proved to be inconclusive, with some results confirming the 8 Ma date, but others indicating a date of 14–15 Ma for the final uplift of Tibet (Spicer *et al.*, 2003). Spicer *et al.* suggest that a possible explanation for this is that the uplift occurred progressively from south to north over a period of 6–7 million years.

Greatly enhanced weathering of silicate rocks in the late Miocene, would have removed CO_2 from the Earth's atmosphere and might well account for the pronounced global cooling revealed by oxygen isotope studies at or near the Miocene–Pliocene boundary, i.e. at about 6 Ma ago (Fig. 13.8). As indicated above this may have produced effects that led, ultimately, to the initiation of the Ice Ages approximately 3 Ma ago.

Thus plate tectonic processes influence all the major factors that are currently thought to determine the Earth's long-term changes in climate. The concentration of CO_2 in the atmosphere, at any particular point in time, is thought to be determined largely by the amount of volcanism at that time. Thus the exceptionally high levels of CO_2 associated with the "Greenhouse Earth" of the Cretaceous period are related to superplume activity, and high rates of sea floor spreading and subduction, all three giving rise to enhanced volcanic activity. Conversely, systematic decreases in plume activity, and plate accretion and destruction, would cause global cooling. However, the periods of pronounced global cooling during the past 50 Ma are not associated with decreases in volcanism (Fig. 5.13). It seems probable therefore that one needs to invoke the other potential impacts of plate tectonic processes on the Earth's climate, notably changes in oceanic circulation and the consequences of mountain building, and enhanced weathering, to explain the mid-Cenozoic transition to an "Icehouse Earth," and eventually the triggering of the Ice Ages of the past 3 Ma.

13.2 ECONOMIC GEOLOGY

13.2.1 Introduction

The application of plate tectonic theory to the exploration of economically viable mineral and hydrocarbon deposits is a common approach in the field of economic geology. Plate tectonics has provided exploration geologists with a framework to which they can relate the specific environments and spatial relationships of economic deposits (Rona, 1977; Bierlein *et al.*, 2002; Richards, 2003). Studies of this kind have increased as the search for small and covered deposits becomes progressively more important. This approach has led to a classification of economic deposits according to plate tectonic processes. Many of the observations that support this classification (listed below) are discussed by Mitchell & Garson (1976, 1981), Rona (1977), Tarling (1981), Hutchinson (1983), Sawkins (1984), and Evans (1987).

1 autochthonous deposits directly related to magmatism at plate margins and interiors;

2 allochthonous deposits related to plate margin magmatism;

3 deposits related to sedimentary basins formed by plate motions;

4 deposits related to climate and to changes in paleolatitude resulting from plate motions.

Whereas plate tectonic theory has been useful for understanding the origin and evolution of economically viable deposits, alternative approaches also have been employed, especially with regard to mineral deposits. One area of current research involves investigating the potential links among the formation of ore deposits, the evolution of large igneous provinces (LIPs, Section 7.4.1), and the effects of deep mantle plumes (Ernst *et al.*, 2005). The formation of LIPs and the rise of deep mantle plumes may involve tectonic and magmatic activities that operate independently of plate motions. In addition, the understanding of mineral deposits in Archean cratons is complicated by the possibility of unique geologic or tectonic process operating in the early Earth (Section 11.1). These differences require models that incorporate the unique aspects of Archean geology and tectonics (Herrington *et al.*, 1997).

13.2.2 Autochthonous and allochthonous mineral deposits

The various plate tectonic environments in which many metalliferous deposits are found are shown in Fig. 13.9. The initial rifting of a continent includes the emplacement of alkaline and peralkaline igneous rocks and the establishment of high geothermal gradients (Sections 7.4.2, 7.2, respectively). Ore minerals are generated from this magmatism and from the large-scale circulation of hydrothermal fluids that are energized by it. One group of igneous rocks frequently associated with extensive mineralization includes *carbonatites*. These are unusual rocks composed of more than 50% carbonate minerals that form ring complexes within alkaline rocks. The important elements found in this environment are phosphorus (as apatite), niobium (pyrochlore), rare earths (monazite, bastnaesite), copper, uranium, thorium, and zircon. Also found are magnetite, fluorite, barite, strontianite, and vermiculite. Carbonatites may also contribute to the sodium carbonate, chloride, and fluoride found in vast quantities in the lakes of the East African Rift system, although it is possible that these derive from weathering of the alkaline rocks. Directly related to the magmatism are porphyry and vein-type molybdenum deposits associated with subalkaline granites, copper-nickel deposits associated with mafic intrusions, and hydrothermal copper deposits. Within the sediments related to rifting, stratiform copper deposits of great volume are associated with specific shale or sandstone horizons. These disseminated ores are believed to form during the first marine transgression into the continental interior and overlie red-bed horizons, and are probably derived from the copper-rich rift basalts in response to the elevated heat flow of the rift. Carbonate hosted lead-zinc-barite ores are also found in the intracratonic rifts and rifted continental margins (Section 7.7), typified by the deposits of the Upper Mississippi Valley of North America. Attempts also have been made to link a wide range of mineral deposits that are related to magmatism in continental rift environ-

ments to the effects of rising mantle plumes (Pirajno, 2004).

With the development of a narrow ocean basin between the rifted continental fragments, new mineral deposits are created at the mid-ocean ridge. The present day example of this environment is the Red Sea. Here 13 pools of hot brines (Fig. 13.10) have been located along the central ridge where it is intersected by transform faults. These contain zinc-copper-lead sediments of possible economic value, for example the Atlantis II Deep, which contains sulfide layers with zinc contents of up to 20% which are 20 m thick and cover an area of over 50 km². It is generally agreed that the metals are of volcanic origin and have been concentrated into brines by the thermally induced circulation of seawater through the volcanic rocks and thick evaporite sequences found in this region (Cowan & Cann, 1988). As the ocean basin evolves, these deposits may become buried by sediment and reappear in collisional orogens where the tectonism obscures their original setting. Also associated with this advanced phase of rifting are sediment-hosted massive sulfide deposits which occur in thick continentally derived clastic sediments on passive continental margins. They comprise single or multiple lenses of pyrite, galena, and sphalerite ores with minor silver and copper. These are not common and probably reflect the influence of metal-rich-formation waters powered by long-lived geothermal systems.

As an ocean basin continues to grow, contemporaneous mineralization takes place at the mid-ocean ridge, and has been observed at certain locations along the Pacific (Corliss *et al.*, 1979), Atlantic (Scott *et al.*, 1973), and Indian Ocean ridges. The mineralization is of hydrothermal origin and its location depends upon the availability of oceanic crust of high permeability overlying the magma chamber which allows fluids to percolate with relative ease. Hydrothermal processes of low intensity lead to the formation of ferromanganese nodules, and encrustations of iron and manganese on pillow basalts at the layer 1–layer 2 interface. Higher intensity hydrothermal activity has been observed at some locations, such as on the East Pacific Rise where discharge is of two types. *Black smokers* are vents where pyrrhotite particles are discharged, producing ores which may be zinc- or iron-rich and containing lesser amounts of cobalt, lead, silver, and cadmium. At *white smokers* little sulfide material is discharged, and the main precipitate is barite.

In open ocean conditions ferromanganese nodules and encrustations form on top of basalt or sediments

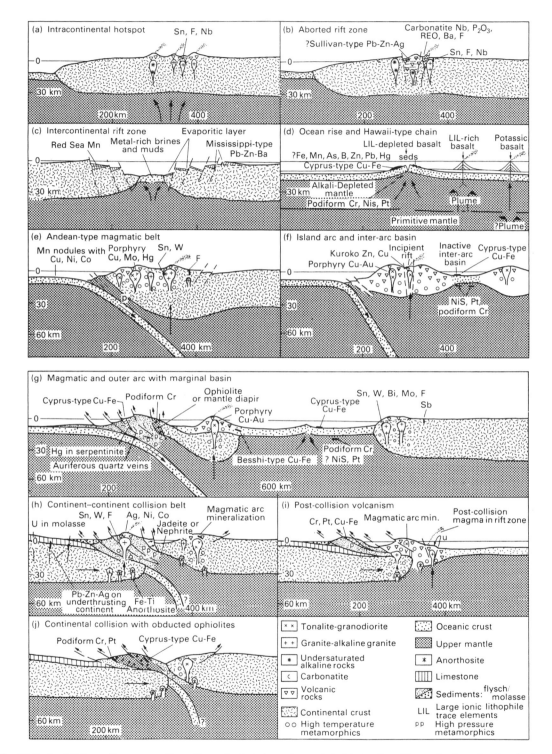

Fig. 13.9 *(a–j) Schematic cross-sections through plate boundary-related tectonic settings of mineralization (redrawn from Mitchell & Garson, 1976).*

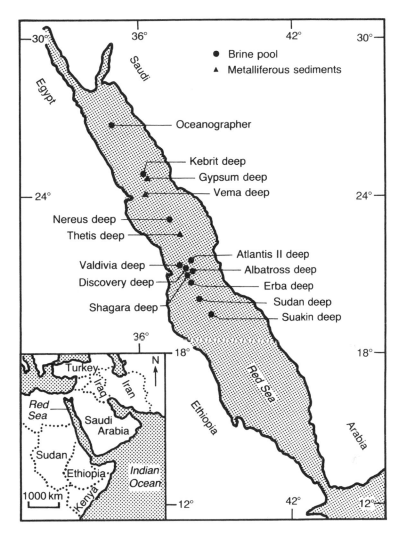

Fig. 13.10 *Locations of hot brine pools and metalliferous sediments in the Red Sea (redrawn from Bignell* et al., *1976, with permission from the Geological Association of Canada).*

where strong ocean currents prevent the accumulation of clastic sediments. These deposits are of hydrous origin and accumulate slowly, sometimes forming extensive pavements. As well as iron and manganese, they also contain smaller amounts of copper, nickel, and cobalt.

In addition to the exhalative processes of mineralization described above, ore bodies may form within the oceanic lithosphere as it is created. Much of our knowledge of these deposits is derived from studies of ophiolites (Section 2.5), which are interpreted as allochthonous slices of oceanic or backarc basin lithosphere tectonically emplaced within continental crust during collisional orogenesis. One of the most intensively studied bodies of this type is the Troodos complex of Cyprus (Fig. 13.11), and other examples include ophiolites of northwestern Newfoundland, the Semail Ophiolite of Oman, and Ergani Maden in Turkey. At high levels in the lithosphere, massive sulfide deposits (marcasite, chalcopyrite and sphalerite) occur on top of or within the pillow lavas of layer 2. It has been suggested that these sulfides formed either in a manner similar to the brines of the Red Sea or by precipitation from hydro-thermal solutions which became enriched in metals by circulating within the volcanic rocks. The deeper, plutonic portions of ophiolites contain economic deposits of chromite, which occur as podiform bodies and tabular masses within the harzburgites and dunites of the upper lithospheric mantle. These deposits may have formed by partial melting of primitive mantle material or by crystal fractionation within the magma chamber underlying ocean ridges (Section 6.10). Similarly associated with the magma chamber are nickel and platinum sulfides. The model for mineralization of the oceanic lithosphere derived from ophiolite studies is shown in Fig. 13.12 (Rona, 1984).

The formation of ore deposits composed of nickel, copper and platinum group elements also can be linked to the mafic and ultramafic magmatism that results in the formation of LIPs (Section 7.4.1). Examples of these types of deposit include the 250 Ma Noril'sk deposit in Siberia, which currently produces 70% of the world's palladium, and the Proterozoic Bushveld intrusion (Section 11.4.1) of South Africa, which produces large quantities of platinum and chrome (Naldrett, 1999).

Exploration models that integrate the characteristics of these deposits with the formation of LIPs rely on detailed information about the sources of the magma and the deep plumbing systems that transport it through the crust (Pirajno, 2004).

Several forms of mineralization are present in subduction zone environments, their types depending upon whether the overriding lithosphere is continental or oceanic. Hedenquist & Lowenstern (1994) have reviewed the role of magmas in the formation of hydrothermal ores in such environments. The most important mineralizations are the porphyry coppers. These are relatively rare, low grade deposits that are gold-rich and molybdenum-poor when associated with island arcs and gold-poor and molybdenum-rich in Andean-type mountain ranges. They are found, for example, in the Andes themselves, as well as in the Philippines, Taiwan, Puerto Rico, and the Ryuku and Burman arcs. The broad uniformity of these deposits worldwide suggests that the controls on their location are related to the primary subduction-related magmatism and do not require the presence of any specific crustal or magma type. Magma may be emplaced and porphyry coppers

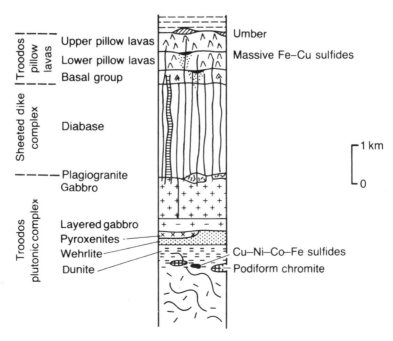

Fig. 13.11 *Mineralization in the Troodos Ophiolite (redrawn from Searle & Panayiotou, 1980, with permission from the Ministry of Agriculture and Natural Resources, Cyprus).*

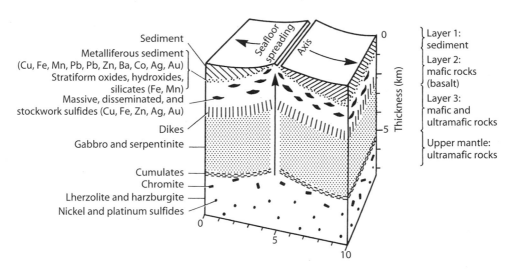

Fig. 13.12 *Schematic block diagram showing the potential distribution of mineral deposits in the oceanic lithosphere (redrawn from Rona, 1984, with permission from Elsevier).*

may form anywhere along the volcanic arc, but large deposits will most likely be formed where magma ascent is concentrated over a prolonged period of time. Richards (2003) reviews many of the large-scale magmatic and tectonic processes leading to the formation of porphyry deposits at convergent margins.

Another important class of deposits found associated with oceanic subduction zones (Fig. 13.13) is stratiform massive sulfides of zinc, lead and copper known, after their type area of occurrence in Japan, as Kuroko-type ores. These ores also are known as volcanic-hosted or volcanogenic massive sulfide (VMS) deposits. They reflect deposition in a shallow marine environment and occur interbedded with pyroclastics and silicic calc-alkaline lavas. Many appear to occur during a late stage of volcanic arc evolution. Halbach *et al.* (1989) suggest that they formed in a backarc basin (Section 9.10), and cite the Okinawa Trough as a modern analogue. They may have been deposited by saline submarine hot springs arising from the separation of aqueous ore fluids during the final stages of magmatic fractionation or from the leaching of older volcanic rocks. Kuroko-type ores may be incorporated into continents during continent–island arc collisions, such as at Río Tinto in Spain, Umm Samiuki in Egypt, and the Buchans mine in Newfoundland.

There also exist other forms of stratiform massive sulfides that differ in their depositional environment from Cyprus- or Kuroko-type. They are associated with intermediate to basic volcanic rocks with carbonaceous mudstones, clastic limestones, or quartzites, all of which suggest a deep water environment unlike ocean ridges, ocean basins, or island arcs. They have been termed Besshi-type deposits. They may have formed in a trench or a tensional environment, but their origin remains, as yet, obscure.

There are several types of deposit that are specific to Andean-type subduction. These include stratabound copper sulfide deposits, such as are found in Chile, which are closely related to episodic calcalkaline volcanism and occur within porphyritic andesite lavas. The principal minerals are chalcosite, bornite, and chalcopyrite, and they contain significant amounts of silver. The intercalation of these deposits with shallow marine and terrestrial deposits suggests their formation in small lagoons. Tin and tungsten mineralization occurs in the eastern Andes of Peru and Bolivia on the landward side of the porphyry copper belt. It appears to be derived from the same Benioff zone region as the magmas, and may owe its existence to the anomalously shallow dip of the subduction zone in this region (Section 10.2.2).

In the backarc environment of Andean-type subduction zones in the Pacific there are granite belts that contain deposits of tin and tungsten with lesser molybdenum, bismuth, and fluorite. The origin of the tin, in particular, is controversial. Tin is present in only minute

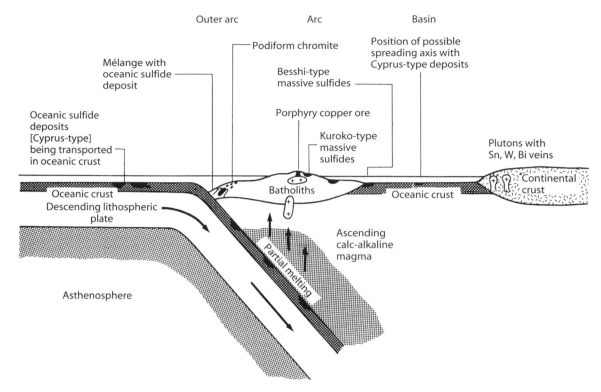

Fig. 13.13 *Development and emplacement of mineral deposits in a subduction-related setting (redrawn from Evans, 1987, using data from Sillitoe, 1972a, 1972b, with permission from the Economic Geology Publishing Co. and the Institute of Mining and Metallurgy).*

quantities in the oceanic crust, and is similarly absent in island arcs. An oceanic origin of the tin appears unlikely. One hypothesis is that tin is derived from deep in a Benioff zone which is migrating away from a continent during backarc spreading. The fluorine originating at these levels would extract tin from deep levels of still hot granite plutons and deposit it at the surface in their vicinity. Another hypothesis is that the generation of tin requires the presence of thick continental crust, such as is present in the tin belts of the Andes, Alaska and upper Myanmar, and a shallow dipping Benioff zone, which acts as a source of heat and volatiles. In this case the tin may originate from pre-existing concentrations in the lower continental crust.

Within ensialic backarc basins vein-type gold and silver deposits are common, such as are found in the Great Basin of Nevada. These are associated with andesites, dacites, and rhyolites, and pre-date the major episode of faulting. They probably originate in magmatically associated brines. In backarc basins that form above oceanic subduction zones the crust is similar to oceanic, although generated in a different fashion, and so mineral deposits would not be expected to differ greatly from those in oceanic crust. The mineralization in these settings may be similar to that formed during the early development of a spreading ridge, and thus may be related to magmatic and exhalative volcanic processes.

Zones of continental collision and terrane accretion (Sections 10.4, 10.6) also host a wide range of metalliferous deposits. These belts may display allochthonous terranes containing mineral associations that formed during the early stages of crustal accretion, such as ophiolites, ferromanganese nodules, subduction-related deposits, and mineralization related to the early stages of rifting. Deposits that originate during the continental collision also are present. Solomon (1990) noted that, in the southwestern Pacific rim, porphyry copper-gold deposits mostly form after a reversal of arc polarity following a collisional event. Sometimes associated with porphyry coppers are mercury deposits (as cinnabar or quicksilver), which may have originated in a similar manner. Granite bodies commonly are emplaced during and after a collisional event. Associated with these granites are tin-tungsten deposits of cassiterite and wolframite and, in some cases, vein-type deposits of uranium. This mineralization, like the granites, may be derived from the partial melting of the lower continental crust.

The Paleozoic Lachlan Orogen of southeast Australia illustrates the types of base and precious metals that form and are preserved in long-lived accretionary orogens (Section 10.6.3). Bierlein et al. (2002) describe orogenic gold deposits that evolved within developing accretionary wedges while major porphyry copper-gold deposits formed in an oceanic island arc located offshore of the Pacific margin of Gondwana. As deformed oceanic sequences, volcanic arcs, and microcontinents accreted onto the Australian margin, sediment-hosted copper-gold and lead-zinc deposits formed in short-lived intra-arc basins, while volcanogenic massive sulfide deposits were produced in forearc regions. Compression leading to the inversion (Section 10.3.3) of these basins also triggered pulses of orogenic gold mineralization. This and other studies (Groves et al., 2003) illustrate that gold-rich deposits can form during any stage of orogenic evolution.

Oceanic transform faults are favorable environments for mineralization because they may be associated with high heat flow and provide highly fractured and permeable conduits for both the downward percolation of seawater and the upward migration of mineralizing fluids. Iron sulfide concretions have been reported from the Romanche Fracture Zone of the equatorial Atlantic which may have originated by this mechanism. The brine pools of the Red Sea appear to be located where transform faults intersect the central ridge, and it is possible that the metals ascend along these faults. Indeed, base metal deposits are found along the continental continuation of the faults. A similar mechanism has been postulated for the brines of the Salton Sea, California. It is probable that the ultramafic intrusions occurring in fracture zones (Section 6.12) contain high proportions of nickel, cobalt, and copper.

For mineralization in the Archean cratons, analogies with the plate tectonic settings of some Phanerozoic deposits are possible. For example, many Archean greenstone belts host volcanogenic massive sulfides (Kuroko-type), copper-zinc-lead sulfides, and gold deposits that also occur throughout the Phanerozoic record. However, many aspects of Archean metallogenesis require further investigation. Porphyry coppers, which typically have a clear association with subduction zone environments, are extremely rare in the Archean, except for a few controversial examples (Herrington et al., 1997). In addition, nickel-sulfide deposits hosted by komatiites in Archean greenstone belts (Section 11.3.2) have no modern analogues. Some studies (de Ronde et al., 1997) have suggested that fluid circulation in the Archean occurred at a larger scale than during other times in Earth's history, which would have influenced the formation of hydrothermal ore deposits. These features may reflect fundamentally different tectonic and/or crustal processes operating during the Archean compared to Phanerozoic times (Section 11.3).

Banded iron formations (BIFs) are common in Archean cratons (Section 11.3.2), although they also occur in rocks as young as Devonian. These rocks contain magnetite, hematite, pyrite, siderite, and other iron-rich silicates. Two main types have been identified (Pirajno, 2004). An Algoma type is associated with volcanic sequences in backarc environments. A Superior type is associated with sedimentary sequences deposited on the continental shelves of rifted continental margins. The development of BIFs on a global scale during Late Archean and Early Proterozoic times also may reflect a period of enhanced mantle plume activity.

Proterozoic mineral deposits are widely interpreted as forming in plate tectonic environments, particularly those related to divergent plate margins and subduction zones (Gaál & Schulz, 1992). Possible exceptions to this approach may include massif-type anorthosite complexes, which are associated with iron-titanium deposits of magnetite and ilmenite. These magma-hosted ore deposits may have originated during episodes of lower crustal melting (Section 11.4.1). Some studies have related such magmatism to the break-up of supercontinents, to zones of continental rifting, and to mantle plumes (Pirajno, 2004).

Another type of magma-hosted deposit includes diamonds that occur in kimberlite pipes. Kimberlites consist of small potassic, ultramafic intrusions that originate from the mantle. These intrusions occur in virtually every Archean craton as well as throughout the Phanerozoic. In some areas, such as in parts of North and South America, there is good evidence that the majority of kimberlites were generated during times of enhanced hot spot or mantle plume activity (Sections 5.5, 5.7), possibly associated with the break-up of the supercontinent Gondwana. The relationships among kimberlite magmatism, supercontinent assembly and dispersal, the relative stability of the crust and mantle, and diamond productivity are discussed further by Heaman *et al.* (2003).

13.2.3 Deposits of sedimentary basins

The majority of fossil fuels are found within sedimentary basins whose formation can be related directly or indirectly to plate motions. In addition to the sedimentary environment, quite stringent conditions are necessary for the development and preservation of these resources.

There are four principal criteria which must be met for the development of petroleum and gas, hereinafter referred to as hydrocarbons: layers rich in organic matter within the sedimentary succession; a source of heat applied for a time sufficient for the maturation of organic materials into hydrocarbons; permeable pathways which allow movement of the hydrocarbons; and a porous reservoir whose top is sealed by impermeable capping beds.

The main source of the disseminated organic matter, or kerogens, in sediments is plankton. The abundance of plankton is controlled by climate, the quantity of nutrients available, and water body geometry. The first two factors are latitude dependent, and the majority of oil basins originate at low latitudes. The latitude is obviously affected by the north–south component of plate motion, while the plate configuration at any given time determines water body geometry. Organic material is especially abundant along continental margins where there is major river runoff into large deltas.

The preservation of kerogens requires conditions which are not oxidizing. These are achieved along continental slopes where the production of organic matter exceeds the availability of free oxygen to convert them to carbon dioxide, and in closed anoxic basins. It follows that the shales and mudstones produced in such environments are the most common source rocks as they

have the ability to absorb kerogens and remove them from the effects of free oxygen.

The temperature experienced by the kerogens after burial is critical, and depends on the local geothermal gradient. Temperatures of 70–85°C are required to develop liquids and 150–175°C for dry gas. It is also important that a critical exposure time to these temperatures is exceeded, so the basin must be free from tectonism and uplift during this period.

After formation, the hydrocarbons undergo primary migration from the fine-grained source rocks and secondary migration as they concentrate and accumulate in a reservoir of high porosity. Migration occurs because of the buoyancy of the hydrocarbons, and it follows that all hydrocarbon accumulations are allochthonous. There are several types of oil trap, including anticline, fault, stratigraphic, unconformity, and lithological traps, which have the effect of providing a capping to the reservoir with an impermeable cover which prevents further upward movement.

Plate tectonics controls the locations of reservoirs in that it is responsible for the formation and preservation of the sedimentary basins in which hydrocarbons are generated and trapped. These include:

1 intracratonic basins formed by hotspot activity, Paris and Michigan basins;

2 basins associated with continental rifting, e.g. the Gulf of Aden, Red Sea;

3 aulacogens (Section 7.1), e.g. the North Sea;

4 passive continental margin basins, e.g. the Gabon Basin;

5 ensialic backarc basins, e.g. the Oriente Basin of Ecuador and Peru;

6 marginal seas, e.g. the Andaman Sea;

7 accretionary prisms, e.g. the coastal oil-fields of Ecuador and Peru;

8 forearc basins, e.g. the Cook Inlet of southern Alaska;

9 pull-apart basins associated with strike-slip faults (Section 8.2), e.g. the Los Angeles Basin, western USA (Moody, 1973);

10 foreland basins (Section 10.3.2) of orogens, e.g. the Aquitaine Basin, southwest France,

11 tensional basins associated with indentation tectonics (Section 10.4.6), e.g. southern Asia and Tibet.

Not only can plate tectonics create the habitat of hydrocarbon deposits, it can also explain why certain regions are particularly rich in these deposits. A large proportion of the Earth's hydrocarbon reserves are located in the Middle East, and the evolution and preservation of these deposits has been a consequence of a specific pattern of plate interactions (Irving *et al.*, 1974).

During Mesozoic and early Cenozoic times two large embayments existed on the continental shelf of the Afro-Arabian continent on the southern side of the Tethys Ocean (Figs 13.5, 13.6). Such embayments around the Tethys, which also included the Gulf of México and the Persian Gulf, may have been connected via the proto-Mediterranean Sea, or Tethys-Atlantic seaway, which was situated at low latitudes. At about 100 Ma the rate of spreading of the seaway increased, maximizing the development of hydrocarbon source rocks because of the formation of extensive, warm, shallow seas to which were supplied large quantities of nutrients from the spreading center. When the seaway subsequently began to close following the development of a subduction zone at its north margin, the geometry of the plate movements was such as to protect the Persian Gulf from major tectonism. This arose because the rapid northerly motion of the Indian Plate absorbed most of the energy associated with the collision with the Eurasian Plate. The Gulf of México was similarly protected by northeastward motion of the Greater Antilles.

Coal is a combustible sedimentary rock containing in excess of 50% by weight of carbonaceous material. It is formed by the decomposition, compaction, and diagenesis of an accumulation of terrestrial and freshwater plant debris. Coals thus appear in the geological record from Devonian times when the first plants appeared.

In order to prevent the total destruction of the vegetable matter by biochemical decomposition, very wet conditions are required to stop the decay by the accumulation of toxic waste products. The conditions under which this process occurs are controlled by climate and topography. Normally a warm, wet climate is required to promote luxuriant growth, and this should be under the condition of constant standing water. Although, in regions of high rainfall, peat forms in upland areas, it is rarely preserved due to the erosion experienced in this environment. The prime conditions for coal formation are those of flat, low-lying ground invaded by swamps with stagnant water. The slow

sinking of these regions preserves the organic layers by progressive burial.

The process of coalification refers to the physical and chemical changes experienced by the organic matter after burial in response to rising temperature and pressure. On compression, water and volatiles are expelled and the deposit becomes enriched in carbon. The degree of coalification is reflected in coal of different ranks, varying from the low rank lignite to high rank anthracite.

Plate tectonics affects coal formation in that it controls the latitude of a region (Section 3.4) and creates the environments necessary for the preservation of organic matter, of which the most important are passive continental margins (Section 7.7). Deltas formed on such margins produce the most favorable conditions for coal formation, and swamps can develop on a regional scale. Present day examples include the Niger, Amazon, and Mississippi deltas, and ancient examples the Carboniferous coals of North America and northwest Europe. Intracratonic deltas, such as the Rhine, are similarly productive and are likely to be preserved due to their stable surroundings. Coal deposits are also found in aulacogens and ensialic backarc basins. The tectonism associated with collisional orogens provides an environment whereby coals increase in grade by high-pressure metamorphism.

13.2.4 Deposits related to climate

It has already been stated that the formation of hydrocarbon and coal deposits is dependent upon both climate and special conditions of sedimentation. There are certain deposits, however, which appear to be related solely to climate. Since climate is largely dependent upon latitude, north–south plate motion can be considered as controlling the formation of such deposits. They include laterites and evaporites.

The most important lateritic deposit is nickel laterite, which results from the extreme weathering of the ultramafic parts of ophiolite bodies under tropical conditions. The original nickel content of fresh peridotite becomes enriched by a factor of about seven under the influence of such weathering by percolating ground water. These deposits are becoming increasingly important sources of nickel, and are exploited in the southwestern Pacific and the northern Caribbean.

A similar deposit is bauxite, a residual deposit enriched in aluminum hydroxide, which provides the vast majority of the world's aluminum. This forms by the *in situ* weathering of aluminosilicate minerals on stable peneplaned topography in a wet tropical climate by the intense leaching of alkalis and silica. Bauxite only forms within 30° of the equator and requires high rainfall and high ambient temperatures. It is mined in Jamaica, northern Australia, and China.

Evaporites form in an arid climate by the evaporation of seawater in semi-isolated basins which receive periodic marine influxes. They cannot develop by the evaporation of a single isolated body of water, as this could not explain the vast observed thicknesses of evaporite deposits. The sequence of minerals precipitated is calcium carbonate and sulfate, sodium chloride, and finally magnesium or potassium minerals. Evaporites are important commercially in the chemical industry, particularly for the potash salts. They are also important in the generation of hydrocarbon traps because, being of low density, they mobilize after burial and rise through the sedimentary layers. Such halokinesis provides fault traps along the sides of rising salt masses and anticlinal traps in the layers above the masses which are folded during the ascent. This is an important process in the North Sea and Gulf of México, for example, which are underlain by salt deposits of Permian age.

13.2.5 Geothermal power

Geothermal energy can be effectively utilized for power generation when the vertical thermal gradient is several times its mean value of about $25°C\,km^{-1}$, producing near-surface temperatures above $180°C$. This condition is achieved at constructive and destructive plate margins, as exemplified by the geothermal power plants in Iceland and the North Island of New Zealand, respectively. Anomalously high geothermal gradients are also present in intra-plate areas where they are frequently associated with granitic plutons. The normal geothermal gradient can be utilized for lower energy power generation, such as for space heating, wherever a thick pile of permeable sediments allows the circulation of fluids to depths of several kilometers. An example of this type is the Paris Basin, where space heating for over 20,000 dwellings is provided by deep fluid circulation.

13.3 NATURAL HAZARDS

The most obvious natural hazards resulting from tectonic activity are earthquakes and volcanic eruptions. However both may cause tidal waves either directly, or indirectly, by triggering major slides or slumps from steep slopes at the continental shelf edge or on volcanic islands. The largest tidal waves, or tsunamis, are caused by earthquakes on faults that displace the ocean floor, typically in the vicinity of ocean trenches and associated with subduction of ocean floor (Fig. 9.5). A recent example was the South Asia tsunami arising from the Indonesian earthquake of December 26, 2004. Volcanic eruptions on volcanic islands, particularly if they are explosive, as in the case of island arc volcanoes, can also produce very damaging tidal waves. The eruption of Santorini in the Aegean, in the 17th century BC, that severely damaged that part of the Minoan civilization on Crete, and the eruption of Krakatoa in Indonesia in 1883, are classic examples. Detailed mapping of the ocean floor has revealed major debris slides off continental slopes, such as the Storega slide off Norway, and around volcanic islands such as those of the Hawaiian chain. As most of these slides are pre-historic one cannot be sure that they were all triggered by earthquakes or igneous activity but it seems very likely that they were.

Several hundred thousand earthquakes are recorded by the Global Seismograph Network every year. Most of them are so minor that they go undetected by people and less than 1%, approximately 1000, cause damage. Earthquakes are most common on or in the vicinity of plate boundaries and in the other zones of deformation shown in Fig. 5.10. Fortunately damaging, large magnitude earthquakes are relatively rare, on average one or two a year, and they are confined to these major earthquake zones. In any one geographic area therefore the interval between large, damaging earthquakes may be several hundred years. The frequency of occurrence is inversely proportional to the magnitude, thus small magnitude earthquakes are very common particularly in the main earthquake zones. Earthquakes also occur in plate interiors, away from the main earthquake areas, as a result of the stresses and strains set up within plates by plate driving forces. Such earthquakes often occur on pre-existing faults that have an appropriate orientation with respect to the present day stress field. The recurrence interval between earthquakes in a particular intra-plate region is typically very long – hundreds or thousands of years. The long intervals between earthquakes in a within-plate region and between very large magnitude earthquakes in the main earthquake regions means that populations are often surprised and ill-prepared when an earthquake does occur.

Major volcanic eruptions are even less frequent but unlike the situation for earthquakes the techniques for providing useful predictions of volcanic eruptions are showing greater promise and should increasingly save lives. The nature of volcanic eruptions depends on the chemistry of the magma involved. The relatively silica- and volatile-rich magmas of island arc volcanoes give rise to explosive eruptions as a result of their relatively low temperature, high viscosity, and high gas content. Such volcanoes give rise to large volumes of ash, slow moving, viscous lavas, and at worst turbulent clouds of superheated gases and incandescent ash known as *nuees*

ardentes. Silica- and volatile-poor tholeiitic magmas, such as those of Iceland and Hawaii, give rise to quiet extrusion of low viscosity lavas that flow readily at their source but slow as they cool and fill topographic depressions. Affected populations can usually avoid advancing lava flows but not clouds of incandescent ash. The pumice and ash that build up on the slopes of explosive volcanoes can also be remobilized by torrential rains that are triggered by the cloud of dust particles that accompanies the eruption. The resulting mudflows then sweep down the flanks with great speed, gathering more material and ultimately engulfing towns and villages in their path.

Devastating as the natural hazards associated with earthquakes and volcanoes are they are an inevitable consequence of the dynamic state of the Earth's interior. Without such processes the Earth would not be such a distinctive planet, not only in terms of its surface features and the concentration of energy and mineral resources at and near its surface, but also, in all probability, in terms of the origin and evolution of life.

Review questions

1. a. Summarize the subdivisions of the Earth's silicate mantle and metallic core that have been deduced from seismologic studies. Indicate briefly the type of seismologic study or observation that provides the evidence for each layer that you mention.
   b. Suggest a plausible temperature distribution within the Earth that is consistent with the physical state and rheological properties of these different layers.

2. a. Describe the way in which the seismic "first motion" at a number of seismologic observatories across the world can be used to obtain a focal mechanism solution for an earthquake of sufficiently large magnitude.
   b. Explain how, for near-surface earthquakes, such solutions can be related to a simple theory of faulting and hence to the principal stress directions in the vicinity of the earthquake and the sense of motion on the fracture.

3. Review critically the various lines of geophysical evidence suggesting that there is a layer of low strength (the asthenosphere) in the upper mantle. How, if at all, does the depth of this layer vary beneath different geological provinces?

4. In the late 1960s four aspects of seismology provided the basis for the formulation of the concept of plate tectonics:
   a. more precise epicentral and focal depth determinations;
   b. the study of surface waves and free oscillations;
   c. attenuation studies; and
   d. the determination of focal mechanism solutions.

   Explain briefly the nature and significance of each of these contributions. Show how a nearly complete statement of plate tectonics can be derived from seismology alone.

5. a. Explain the underlying assumptions of the paleomagnetic method for determining the paleolatitude and orientation of a continental area. Outline the various stages involved in carrying out such a paleomagnetic study.

   b. Describe two ways in which studies based on the remanent magnetization of rocks have contributed to the verification of continental drift and the formulation of plate tectonics.

6. a. Define the term transform fault and illustrate the six theoretical possibilities for active, right-lateral transform faults which do not terminate in triple junctions. Indicate the way in which the geometry of each type evolves with time and, where possible, cite a recent or present day example.
   b. Explain the significance of transform faults when deducing the relative motion between adjacent lithospheric plates.

7. Explain the principles upon which the direction and rate of relative motion between lithospheric plates are deduced for the different types of plate boundary developed at the Earth's surface. Review briefly the types of argument used to convert these "relative" motions into "absolute" plate motions with respect to the Earth's deep interior. To what extent do the resulting models for absolute plate motions differ?

8. a. Briefly outline the basic assumptions involved in making a paleomagnetic study.
   b. Explain qualitatively how the global paleomagnetic dataset for rocks formed during the past 200 Ma may be analysed to test:
      i. the geocentric dipole model for the Earth's magnetic field;
      ii. the hotspot model for absolute plate motions; and
      iii. the suggestion that much of the Western Cordillera of North America consists of "displaced" or "suspect" terranes.

   In each case state any assumptions and limitations of the method.

9. Explain the following plate tectonic concepts, illustrating your answer with specific examples:
   a. failed arms or rifts;
   b. triple junctions;

c. backarc spreading; and

d. Benioff plane reversal.

10 Explain the main structural differences between volcanic and nonvolcanic rifted continental margins. Discuss how these two types of margin form in terms of continental break-up processes.

11 In what ways do the main principles of plate tectonics fail to explain how continental crust deforms? What types of data are used to determine how large zones of continental deformation behave?

12 The evolution of continental rifts is thought to be controlled by a number of competing mechanisms, including lithospheric stretching, strain-induced weakening, and magma intrusion. Explain how these mechanisms influence rift evolution.

13 Explain how earthquake focal mechanism solutions can give an indication of the stress field in the vicinity of the earthquake focus. Cite examples of focal mechanism solutions and inferred maximum and minimum principal compressive stress directions for earthquakes occurring:

a. at plate boundaries; and

b. within plates.

14 What are the controls on the geometry of foreland basins and modes of shortening in foreland fold and thrust belts?

15 Describe the generalized morphology of a subduction zone, explaining all the features described in terms of a subduction model.

16 Formulate a possible "life cycle" for an ocean basin within the framework of continental drift, sea floor spreading, and plate tectonics. Give present day examples to illustrate various stages in your proposed cycle.

17 Describe the physical and chemical properties of Archean cratonic mantle lithosphere and explain how these characteristics have contributed to its long term survival. Discuss at least two models that explain how the Archean lithosphere may have formed.

18 Provide a detailed description of the igneous rock suites that are expected in different tectonic environments. How are the Archean suites similar to and different from those that characterize Phanerozoic times?

19 How and why might the Early Archean tectonic regime have differed from that of the present?

20 Some geologists and geophysicists claim that studies of the structural setting, internal structure, and composition of ophiolites are highly relevant to a better understanding of the processes occurring in subduction zones and at ocean ridge crests. Describe the evidence derived from ophiolites and provide a critical review of these claims.

21 Compare and contrast the properties and mechanical behavior of continental versus oceanic lithosphere.

22 Explain why studies of isostasy are important in considering the Earth's dynamic behavior.

23 Explain the processes that may be involved in the formation of orogenic plateaus at convergent margins.

24 Summarize the observations that geoscientists have used to determine when plate tectonics first began on planet Earth.

25 Describe the data and techniques used in making continental reconstructions.

26 In what ways have studies of terrestrial heat flow contributed to plate tectonic theory?

27 Explain the strengths and limitations of paleoclimatological techniques for determining past plate motions and geographies.

28 Describe the evidence for plate tectonic processes being operative in Proterozoic times.

29 Explain the types of metamorphism that are expected within the different provinces of convergent margins.

30 Discuss the role of supercontinents within the basic framework of plate tectonics.

31 Summarize the main characteristics of Large Igneous Provinces (LIPs) and the principal observations that constrain their possible origins. Explain why these features are important in the tectonic evolution of the planet.

32 Describe the physical changes to a convergent margin that occur as a result of arc–continent collision

33 Evaluate the pros and cons of applying the basic tenets of plate tectonics to explain the origin and evolution of economically viable mineral deposits.

34 Describe the principle of terrane analysis as applied to the continents and explain what it has revealed about the mechanisms by which the continents have grown since Precambrian times.

35 Describe the principal factors that control deformation within the upper plate of convergent margins. Which combination of factors explains the differences in the width of the Andean orogen in the Central and Southern Andes?

36 Explain the mechanisms that allow continental transforms to accomplish the very large displacements that are observed on them.

37 Explain the evidence for and against indentation tectonics as applied to southeast Asia.

38 Discuss the evidence for convection in the mantle.

39 Outline and discuss the data which have led to current theories of the causes of plate motions.

40 Explain the various mechanisms of ductile creep in the mantle.

41 Explain how fluid flow and variations in pore fluid pressure influence the structure and mechanical evolution of accretionary prisms?

42 Describe the principal characteristics of accretionary orogens and how they form?

43 Discuss dehydration processes in subduction zones and how they influence the tectonic evolution of convergent margins.

44 Explain how concentrated surface erosion and the exhumation of deep crustal rocks influence the mechanical evolution of orogens.

45 a. Explain how pull-apart basins and localized regions of uplift form along continental strike-slip faults.
   b. Explain the various ways in which deformation may be accommodated in zones of transtension and transpression. Provide simple sketches to illustrate your answer.
   c. Compare and contrast the deep structure of the San Andreas and the Alpine fault systems. How do models of these continental transforms differ?

46 Describe at least two unsolved problems in our understanding of the processes that accommodate continental collision and discuss ways in which these problems may be resolved in the future.

47 The following table lists data obtained from two tectonic provinces in North America. Calculate the heat flow at each site and, using graphs or calculations as appropriate, discuss the significance of the results:

Site	Heat production of surface rocks (10^{-6} W m^{-3})	Thermal conductivity (W m^{-1} °K^{-1})	Temperature gradient (°K km^{-1})
Cenozoic province (last thermal event at 20 Ma)			
A1	4.28	6.6	15.0
A2	5.50	4.9	22.5
A3	5.90	3.8	30.0
A4	3.78	3.9	23.9
Caledonian province (last thermal event at 400 Ma)			
B1	2.14	2.0	25.0
B2	2.90	3.0	18.5
B3	3.57	3.0	20.2
B4	4.83	4.2	16.0

48 Assuming that Airy's mechanism for isostatic compensation is perfectly obeyed, show that the thickness of continental crust at any point is a linear function of the topographic height at that point.

The Bouguer gravity anomaly map for a mountainous area of Utopia was corrected for isostatic compensation using an Airy model. In general the anomalies on the resulting isostatic anomaly map were negligible, with the exception of one area where a broad band of negative anomalies persisted. This was interpreted as arising from compensation for topography which has since been removed by erosion. If this residual isostatic anomaly had an amplitude of -1000 g.u. and the area is broad, estimate the excess thickness of the root. To what thickness of eroded crustal material does this correspond?

($G = 6.67 \times 10^{-11}$ m^3 kg^{-1} s^{-2}; 1 g.u. $= 10^{-6}$ m s^{-2};
Mean density of crust $= 2.85$ Mg m^{-3};
Density of upper mantle $= 3.35$ Mg m^{-3}).

49 The thermoremanent magnetization of the Palisade Sill, which overlooks the Hudson River and the city of New York, is directed towards the north and has an inclination of $25.6°$ downwards. Calculate the implied paleomagnetic pole position (in present day coordinates), and the geographic latitude of the sill at the time of its intrusion, approximately 190 Ma.

Assuming that the difference between the present and the paleolatitude of the sill is due to the drift of North America with respect to the geographic pole, calculate the implied average rate of drift during the past 190 Ma. Is this an absolute, maximum, or minimum average rate? Explain. (Present latitude and longitude of sill = 40.5°N, 75°W; 1 degree of latitude = 111 km.)

50 In the plate tectonic situation illustrated below three plates (A, B, and C) are separated by a symmetrically spreading ridge at R and a subduction zone at T. If the rate of spreading at R is x mm a^{-1} per ridge flank, and the rate of subduction at T is y mm a^{-1}, describe the way in which the situation evolves with time for each of the following five cases:

a. y less than x;
b. y equal to x;
c. y between x and $2x$;
d. y equal to $2x$; and
e. y greater than $2x$.

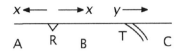

Hint: Consider plate C to be fixed and assign velocities to A, R, and B, relative to C, in terms of x and y.

References

Abers, G.A. & Roecker, S.W. (1991) Deep structure of an arc–continent collision: earthquake relocation and inversion for upper mantle P and S wave velocities beneath Papua New Guinea. *J. geophys. Res.* **96**, 6379–401.

Abers, G.A., Mutter, C.Z. & Fang, J. (1997) Earthquakes and normal faults in the Woodlark–D'Entrecasteaux rift system, Papua New Guinea. *J. geophys. Res.* **102**, 15301–17.

Abers, G.A. *et al.* (2002) Mantle compensation of active metamorphic core complexes at Woodlark rift in Papua New Guinea. *Nature* **418**, 862–5.

Achauer, U. & Masson, F. (2002) Seismic tomography of continental rifts revisited: from relative to absolute heterogeneities. *Tectonophysics* **358**, 17–37.

Adam, C. & Bonneville, A. (2005) Extent of the south Pacific superswell. *J. geophys. Res.* **110**, B09408, doi:10.1029/2004JB003465.

Afonso, J.C. & Ranalli, G. (2004) Crustal and mantle strengths in continental lithosphere: is the jelly sandwich model obsolete? *Tectonophysics* **394**, 221–32.

Aki, K., Christofferson, A. & Husebye, E. (1977) Determination of three-dimensional seismic structures of the lithosphere. *J. geophys. Res.* **82**, 277–96.

Allègre, C.J. *et al.* (1984) Structure and evolution of the Himalaya–Tibet orogenic belt. *Nature* **307**, 17–22.

Allen, C.R. (1981) The modern San Andreas Fault. *In* Ernst, G. (ed.) *The Geotectonic Development of California*, pp. 511–34. Prentice Hall, Englewood Cliffs, NJ.

Allmendinger, R.W. (1992) Fold and thrust tectonics of the western United States exclusive of the accreted terranes. *In* Burchfiel, B.C., Lipman, P.W. & Zoback, M.L. (eds) *The Cordilleran Orogen: conterminous US. Geology of North America*, **G3**, pp. 583–607. Geological Society of America, Boulder, CO.

Allmendinger, R.W. & Gubbels, T. (1996) Pure and simple shear plateau uplift, Altiplano–Puna, Argentina and Bolivia. *Tectonophysics* **259**, 1–14.

Allmendinger, R.W. *et al.* (1983) Cenozoic and Mesozoic structure of the eastern Basin and Range Province, Utah, from COCORP seismic-reflection data. *Geology* **11**, 532–6.

Allmendinger, R.W. *et al.* (1986) Phanerozoic tectonics of the Basin and Range–Colorado Plateau transition from COCORP data and geological data; a review. *In* Baraxangi, M. & Brown, L.D. (eds) *Reflection Seismology; the continental crust*, pp 257–67. Geodynamics Series. American Geophysical Union, Washington, DC.

Allmendinger, R.W. *et al.* (1997) The evolution of the Altiplano–Puna plateau of the central Andes. *Annu. Rev. Earth Planet. Sci.* **25**, 139–74.

Allmendinger, R.W. *et al.* (2005) Bending the Bolivian orocline in real time. *Geology* **33**, 905–8.

Al-Zoubi, A. & ten Brink, U. (2002) Lower crustal flow and the role of shear in basin subsidence: an example from the Dead Sea basin. *Earth planet. Sci. Lett.* **199**, 67–79.

An, Z. *et al.* (2001) Evolution of Asian monsoons and uplift of the Himalaya–Tibetan plateau since Late Miocene times. *Nature* **411**, 62–6.

ANCORP Working Group (2003) Seismic imaging of a convergent continental margin and plateau in the central Andes. *J. geophys. Res.* **108**, 23–28.

Anderson, D.L. (1982) Hot spots, polar wander, Mesozoic convection and the geoid. *Nature* **297**, 391–3.

Anderson, D.L. (1994) Superplumes or supercontinents? *Geology* **22**, 39–42.

Anderson, D.L. (2000) The thermal state of the upper mantle: no role for mantle plumes. *Geophys. Res. Lett.* **27**, 3623–6.

Anderson, D.L. (2007) *New Theory of the Earth*, 2nd edn. Cambridge University Press, Cambridge, UK.

Anderson, D.L. & Natland, J.H. (2005) A brief history of the plume hypothesis and its competitors: concept and controversy. *In* Foulger, G.R. *et al.* (eds) *Plates, Plumes, & Paradigms. Geol. Soc. Amer. Sp. Paper* **388**, 119–45.

Anderson, D.L., Tanimoto, T. & Zhang, Y. (1992) Plate tectonics and hotspots: the third dimension. *Science* **256**, 1645–51.

Anderson, E.M. (1951) *The Dynamics of Faulting*, 2nd edn. Oliver and Boyd, Edinburgh.

Andrews, J.A. (1985) True polar wander: an analysis of Cenozoic and Mesozoic palaeomagnetic poles. *J. geophys. Res.* **90**, 7737–50.

Andronicos, C.L. *et al.* (2003) Strain partitioning in an obliquely convergent orogen, plutonism, and synorogenic collapse: Coast Mountains Batholith, British Columbia, Canada. *Tectonics* **22**, 1012, doi:10.1029/2001TC001312.

Angelier, J. (1994) Fault slip analysis and palaeostress reconstruction. *In* Hancock, P.L. (ed.) *Continental Deformation*, pp. 53–100. Pergamon Press, Tarrytown, NY.

Angermann, D., Klotz, J. & Reigber, C. (1999) Space–geodetic estimation of the Nazca–South America angular velocity. *Earth planet. Sci. Lett.* **171**, 329–34.

Arculus, R.J. & Curran E.B. (1972) The genesis of the calc-alkaline rock suite. *Earth planet. Sci. Lett.* **15**, 255–62.

Argus, D.F. &. Gordon, R.G. (2001) Present tectonic motion across the Coast Ranges and San Andreas fault system in California. *Bull. geol. Soc. Am.* **113**, 1580–92.

Arndt, N.T., Albaréde, F. & Nisbet, E.G. (1997) Mafic and ultramafic volcanism. *In* De Wit, M.J. & Ashwal, L.D. (eds) *Greenstone Belts*, pp. 398–420. Clarendon Press, Oxford.

Arndt, N.T., Bruzak, G. & Reischmann, T. (2001) The oldest continental and oceanic plateaus: geochemistry of basalts and komatiites of the Pilbara Craton, Australia. *In* Ernst, R.E. &

Buchan, K.L. (eds) *Mantle Plumes: their identification through time. Geol. Soc. of Amer. Sp. Paper* **352**, 359–87.

Arndt, N.T., Lewin, E. & Albarède, F. (2002) Strange partners: formation and survival of continental crust and lithospheric mantle. *In* Fowler, C.M.R., Ebinger, C.J. & Hawkesworth, C.J. (eds) *The Early Earth: physical, chemical and biological development. Spec. Pub. geol. Soc. Lond.* **199**, 91–103.

Artemieva, I.M. & Mooney, W.D. (2002) On the relations between cratonic lithosphere thickness, plate motions, and basal drag. *Tectonophysics* **358**, 211–31.

Ashby, M.F. & Verrall, R.A. (1977) Micromechanisms of flow and fracture, and their relations to the rheology of the upper mantle. *Phil. Trans. Roy. Soc. Lond. A* **288**, 59–95.

Attoh, K., Brown, L., Guo, J. & Heanlein, J. (2004) Seismic stratigraphic record of transpression and uplift on the Romanche transform margin, offshore Ghana. *Tectonophysics* **378**, 1–16.

Atwater, T. (1970) Implications of plate tectonics for the Cenozoic tectonic evolution of Western North America. *Bull. geol. Soc. Am.* **81**, 3513–36.

Atwater, T. (1989) Plate tectonic history of the northeast Pacific and western North America. *In* Winterer, E.L., Hussong, D.M. & Decker, R.W. (eds) *The Eastern Pacific and Hawaii. The Geology of North America* **N**, pp. 21–72. Geological Society of America, Boulder, CO.

Audley-Charles, M.G. (2004) Ocean trench blocked and obliterated by Banda forearc collision with Australian proximal continental slope. *Tectonophysics* **389**, 65–79.

Audoine, E., Savage, M.K. & Gledhill, K. (2004) Anisotropic structure under a back arc spreading region, the Taupo Volcanic Zone, New Zealand. *J. geophys. Res.* **109**, B11305, doi:10.1029/2003JB002932.

Auzende, J.-M. *et al.* (1989) Direct observation of a section through slow-spreading oceanic crust. *Nature* **337**, 726–9.

Axen, G.J. (2004) Mechanics of low-angle normal faults. *In* Karner, G.D. *et al.* (eds) *Rheology and Deformation of the Lithosphere at Continental Margins*, pp. 46–91. Columbia University Press, New York.

Axen, G.J. & Bartley, J.M. (1997) Field test of rolling hinges: existence, mechanical types, and implications for extensional tectonics. *J. geophys. Res.* **102**, 20515–37.

Aydin, A. & Page, B.M. (1984) Diverse Pliocene–Quaternary tectonics in a transform environment, San Francisco Bay region, California. *Bull. geol. Soc. Am.* **95**, 1303–17.

Ayele, A., Stuart, G.W. & Kendall, J.-M. (2004) Insights into rifting from shear-wave splitting and receiver functions: an example from Ethiopia. *Geophys. J. Int.* **157**, 354–62.

Babeyko, A.Y. & Sobolev, S.V. (2005) Quantifying different modes of the late Cenozoic shortening in the central Andes. *Geology* **33**, 621–4.

Babeyko, A.Y. *et al.* (2002) Numerical models of the crustal scale convection and partial melting beneath the Altiplano–Puna plateau. *Earth planet Sci. Lett.* **199**, 373–88.

Baby, P. *et al.* (1992) Geometry and kinematic evolution of passive roof duplexes deduced from cross section balancing: example from the foreland thrust system of the southern Bolivian Subandean Zone. *Tectonics* **11**, 523–36.

Baines, A.G. *et al.* (2003) Mechanism for generating the anomalous uplift of oceanic core complexes: Atlantis Bank, southwest Indian Ridge. *Geology* **31**, 1105–8.

Baker, P.E. (1982) Evolution and classification of orogenic volcanic rocks. *In* Thorpe, R.S. (ed.) *Andesites*, pp. 11–23. Wiley, Chichester.

Baldock, G. & Stern, T. (2005) Width of mantle deformation across a continental transform: evidence from upper mantle (Pn) seismic anisotropy measurements. *Geology* **33**, 741–4.

Baldwin, J.A., Bowring, S.A. & Williams, M.L. (2003) Petrological and geochronological constraints on high pressure, high temperature metamorphism in the Snowbird tectonic zone, Canada. *J. metam. geol.* **21**, 81–98.

Baldwin, S.L. *et al.* (2004) Pliocene eclogite exhumation at plate tectonic rates in eastern Papua New Guinea. *Nature* **431**, 263–7.

Ballard, R.D. & van Andel, T.H. (1977) Morphology and tectonics of the inner rift valley at lat 36°50′N on the mid-Atlantic ridge. *Bull. geol. Soc. Am.* **88**, 507–30.

Banks, R.J., Parker, R.L. & Huestis, S.P. (1977) Isostatic compensation on a continental scale: local versus regional mechanisms. *Geophys. J. Roy. astr. Soc.* **51**, 431–52.

Barazangi, M. & Dorman, J. (1969) World seismicity maps compiled from ESSA, Coast and Geodetic Survey epicenter data, 1961–1967. *Bull. seism. Soc. Am.* **59**, 369–80.

Barazangi, M. & Isacks, B. (1971) Lateral variations of seismic wave attenuation in the upper mantle above the inclined earthquake zone of the Tonga island arc: deep anomaly in the upper mantle. *J. geophys. Res.* **76**, 8493–516.

Barazangi, M. & Isacks, B. (1979) Subduction of the Nazca plate beneath Peru: evidence from spatial distribution of earthquakes. *Geophys. J. Roy. astr. Soc.* **57**, 537–55.

Barker, D.H.N. *et al.* (2003) Backarc basin evolution and cordilleran orogenesis: insights from new ocean-bottom seismograph refraction profiling in Bransfield Strait, Antarctica. *Geology* **31**, 107–10.

Barker, F. & Arth, J.G. (1976) Generation of trondhjemitic-tonalitic liquids and Archaean bimodal trondhjemite-basalt suites. *Geology* **4**, 596–600.

Barker, P.F. (1979) The history of ridge crest offset at the Falklands–Agulhas Fracture Zone from a small circle geophysical profile. *Geophys. J.* **59**, 131–45.

Barnes, P.M. *et al.* (2001) Rapid creation and destruction of sedimentary basins on mature strike-slip faults: an example from the offshore Alpine fault, New Zealand. *J. struct. Geol.* **23**, 1727–39.

Barnes, P.M., Sutherland, R. & Delteil, J. (2005) Strike-slip structure and sedimentary basins of the southern Alpine Fault, Fiordland, New Zealand. *Bull. geol. Soc. Am.* **117**, 411–35, doi:10.1130/B25458.1.

Basile C. & Allemand, P. (2002) Erosion and flexural uplift along transform faults. *Geophys. J. Int.* **151**, 646–53.

Bastow, I.D., Stuart, G.W., Kendall, J.–M. & Ebinger, C.J. (2005) Upper-mantle seismic structure in a region of incipient continental breakup: northern Ethiopian rift. *Geophys. J. Int.* **162**, 479–93.

Båth, M. (1979) *Introduction to Seismology*. Birkhäuser Verlag, Basel.

Batiza, R., Melson, W.G. & O'Hearn, T. (1988) Simple magma supply geometry inferred beneath a segment of the Mid-Atlantic Ridge. *Nature* **335**, 428–31.

Batt, G.E. & Braun, J. (1999) The tectonic evolution of the Southern Alps, New Zealand: insights from fully thermally coupled dynamical modelling. *Geophys. J. Int.* **136**, 403–20.

Batt, G.E. *et al.* (2004) Cenozoic plate boundary evolution in the South Island of New Zealand: new thermochronological constraints. *Tectonics* **23**, TC4001, doi:10.1029/2003TC001527.

Beanland, S. & Clark, M.M. (1994) The Owens Valley fault zone, eastern California, and surface faulting associated with the 1872 earthquake. *US Geol. Surv. Bull.* **1982**, 1–29.

Beaumont, C. *et al.* (1996) The continental collision zone, South Island, New Zealand: comparison of geodynamical models and observations. *J. geophys. Res.* **101**, 3333–59.

Beaumont, C. *et al.* (2001) Himalayan tectonics explained by extrusion of a low-viscosity crustal channel coupled to focused surface denudation. *Nature* **414**, 738–42.

Beaumont, C. *et al.* (2004) Crustal channel flows: 1. Numerical models with applications to the tectonics of the Himalayan–Tibetan orogen. *J. geophys. Res.* **109**, B06406, doi:10.1029/2003JB002809.

Beavan, J. *et al.* (1999) Crustal deformation during 1994–98 due to oblique continental collision in the central Southern Alps, New Zealand, and implications for seismic potential of the Alpine fault. *J. geophys. Res.* **104**, 25233–55.

Bechtel, T.D. *et al.* (1990) Variations in effective elastic thickness of the North American lithosphere. *Nature* **343**, 636–8.

Beck, M.E. Jr (1980) The palaeomagnetic record of plate margin tectonic processes along the western edge of North America. *J. geophys. Res.* **85**, 7115–31.

Beck, R.A. *et al.* (1995) Stratigraphic evidence for an early collision between northwest India and Asia. *Nature* **373**, 55–8.

Beck, S.L. & Zandt, G. (2002) The nature of orogenic crust in the central Andes. *J. geophys. Res.* **107**, 2230, doi:10.1029/2000JB000124.

Becker, T.W., Hardebeck, J.L. & Anderson, G. (2005) Constraints on fault slip rates of the southern California plate boundary from GPS velocity and stress inversions. *Geophys. J. Int.* **160**, 634–50.

Behn, M.D., Lin, J. & Zuber, M.T. (2002) A continuum mechanics model for normal faulting using a strain–rate softening rheology: implications for thermal and rheological controls on continental and oceanic rifting. *Earth planet. Sci. Lett.* **202**, 725–40.

Bell, R. & Jefferson, C.W. (1987) An hypothesis for an Australian–Canadian connection in the Late Proterozoic and the birth of the Pacific Ocean. *In Proceedings of Pacific Rim Congress, 1987*, pp. 39–50. Australian Institute of Mining and Metallurgy, Parkville, Victoria.

Ben-Avraham, Z. *et al.* (1981) Continental accretion and orogeny. From oceanic plateaus to allochthonous terranes. *Science* **213**, 47–54.

Ben-Avraham, Z., Nur, A. & Jones, D. (1982) The emplacement of ophiolites by collision. *J. geophys. Res.* **87**, 3861–7.

Bennett, R.A., Davis, J.L. & Wernicke, B.P. (1999) Present-day pattern of Cordilleran deformation in the western United States. *Geology* **27**, 371–4.

Bennett, R.A. *et al.* (2003) Contemporary strain rates in the northern Basin and Range province from GPS data. *Tectonics* **22**, 1008, doi:10.1029/2001TC1355.

Benoit, M.H., Nyblade, A.A. & Pasyanos, M.E. (2006) Crustal thinning between the Ethiopian and East African plateaus from modeling Rayleigh wave dispersion. *Geophys. Res. Lett.* **33**, L13301, doi:10.1029/2006GL025687.

Besse, J. & Courtillot, V. (1988) Paleogeographic maps of the continents bordering the Indian Ocean since the Early Jurassic. *J. geophys. Res.* **93**, 11791–808.

Besse, J. & Courtillot, V. (1991) Revised and synthetic polar wander paths of the African, Eurasian, North American and Indian plates and true polar wander since 200 Ma. *J. geophys. Res.* **96**, 4029–50.

Besse, J. & Courtillot, V. (2002) Apparent and true polar wander and the geometry of the geomagnetic field over the past 200 Ma. *J. geophys. Res.* **107**, B11, doi:10.1029/2000JB000050.

Bevis, M. *et al.* (1995) Geodetic observations of very rapid convergence and back arc extension in the Tonga arc. *Nature* **374**, 249–51.

Bickle, M.J. (1978) Heat loss from the Earth: a constraint on Archaean tectonics from the relation between geothermal gradients and the rate of plate production. *Earth planet. Sci. Lett.* **40**, 301–15.

Bickle, M.J., Nisbet, E.G. & Martin, A. (1994) Archean greenstone belts are not oceanic crust. *J. Geol.* **102**, 121–38.

Bicknell, J.D. *et al.* (1988) Tectonics of a fast spreading center – a deep-tow and seabeam survey at EPR 19°30′S. *Marine geophys. Res.* **9**, 25–45.

Bierlein, F.P., Gray, D.R. & Foster, D.A. (2002) Metallogenic relationships to tectonic evolution – the Lachlan Orogen, Australia. *Earth planet. Sci. Lett.* **202**, 1–13.

Bignell, R.D., Cronan, D.S. & Tooms, J.S. (1976) Red Sea metalliferous brine precipitates. *In* Strong, D.F. (ed.) *Metallogeny and Plate Tectonics. Geol. Assoc. Can. Spec. Paper* **14**, pp. 147–79. St. John's, NL.

Bilham, R. (2004) Earthquakes in India and the Himalaya; tectonics, geodesy and history. *Ann. Geophys.* **47**, 839–58.

Bilham, R. *et al.* (1999) Secular and tidal strain across the Main Ethiopian rift. *Geophys. Res. Lett.* **26**, 2789–92.

Bilham, R., Gaur, V.K. & Molnar, P. (2001) Himalayan seismic hazard. *Science* **293**, 1442–4.

Bird, R.T. *et al.* (1998) Plate tectonic reconstructions of the Juan Fernandez microplate: transformation from internal shear to rigid rotation. *J. geophys. Res.* **103**, 7049–67.

Birt, C. *et al.* (1997) A combined interpretation of the KRISP '94 seismic and gravity data: evidence for a mantle plume beneath the East African plateau. *Tectonophysics* **278**, 211–42.

Blackman, D.K. *et al.* (1998) Origin of extensional core complexes: evidence from the Mid-Atlantic Ridge at Atlantis Fracture Zone. *J. geophys. Res.* **103**, 21315–33.

Bleeker, W. (2003) The late Archean record: a puzzle in ca. 35 pieces. *Lithos* **71**, 99–134.

Blewitt, R.S. (2002) Archaean tectonic processes: a case for horizontal shortening in the North Pilbara Granite–Greenstone Terrane, Western Australia. *Precambrian Res.* 113, 87–120.

Blisniuk, P.M. *et al.* (2001) Normal faulting in central Tibet since at least 13.5 Myr ago. *Nature* 412, 628–32.

Boillot, G. & Froitzheum, N. (2001) Non-volcanic rifted margins, continental break-up and the onset of seafloor spreading: some outstanding questions. *In* Wilson, R.C.L. *et al.* (eds) *Non-volcanic Rifting of Continental Margins: a comparison of evidence from land and sea. Spec. Pub. geol. Soc. Lond.* 187, 9–30.

Bokelmann, G.H.R. & Beroza, G.C. (2000) Depth-dependent earthquake focal mechanism orientation: evidence for a weak zone in the lower crust. *J. geophys. Res.* 105, 21683–95.

Bonatti, E. (1978) Vertical tectonism in oceanic fracture zones. *Earth planet. Sci. Lett.* 37, 369–79.

Bonatti, E. & Crane, K. (1984) Oceanic fracture zones. *Sci. Am.* 250, 36–47.

Bonatti, E. & Honnorez, J. (1976) Sections of the Earth's crust in the equatorial Atlantic. *J. geophys. Res.* 81, 410–6.

Bonatti, E. *et al.* (1977) Easter volcanic chain (southeast Pacific): a mantle hot line. *J. geophys. Res.* 82, 2457–78.

Bos, A.G. & Spakman, W. (2005) Kinematics of the southwestern US deformation zone inferred from GPS motion data. *J. geophys. Res.* 110, B08405, doi:10.1029/2003JB002742.

Bos, B. & Spiers, W. (2002) Frictional-viscous flow of phyllosilicate-bearing fault rock; microphysical model and implications for crustal strength profiles. *J. geophys. Res.* 107, B2, 2028, doi:10.1029/2001JB000301.

Bostock, M.G., Hyndman, R.D., Rondenay, S. & Peacock, S.M. (2002) An inverted continental Moho and serpentinization of the forearc mantle. *Nature*, 417, 536–8.

Bott, M.H.P. (1967) Solution of the linear inverse problem in magnetic interpretation with application to oceanic magnetic anomalies. *Geophys. J. Roy astr. Soc.* 13, 313–23.

Bott, M.H.P. (1982) *The Interior of the Earth, its structure, constitution and evolution*, 2nd edn. Edward Arnold, London.

Bott, M.H.P. (1993) Modelling the plate driving mechanism. *J. geol. Soc. Lond.* 150, 941–51.

Bowring, S.A. & Williams, I.S. (1999) Priscoan (4.00–4.03 Ga) orthogneisses from northwestern Canada. *Contrib. Mineral. Petrol.* 134, 3–16.

Boyd, F.R., Gurney, J.J. & Richardson, S.H. (1985) Evidence for a 150–200 km thick Archaean lithosphere from diamond inclusion thermobarometry. *Nature* 315, 387–9.

Brace, W.F. & Kohlstedt, D.L. (1980) Limits on lithospheric stress imposed by laboratory experiments. *J. geophys. Res.* 85, 6248–52.

Brandon, A.D. (2002) ^{186}Os–^{187}Os systematics of Gorgona komatiites and Iceland picrites (abstract). *Geochim. cosmochim. Acta* 66, 100.

Brandon, A.D. *et al.* (1998) Coupled ^{186}Os and ^{187}Os evidence for core–mantle interaction. *Science* 280, 1570–3.

Brett, R. (1976) The current status of speculations on the composition of the core of the Earth. *Rev. Geophys. Space Phys.* 14, 375–83.

Briggs, J.C. (1987) *Biogeography and Plate Tectonics, developments in palaeontology and stratigraphy 10*. Elsevier, Amsterdam.

Brown, E.H. & McClelland, W.C. (2000) Pluton emplacement by sheeting and vertical ballooning in part of the southeast Coast plutonic complex, British Columbia. *Bull. geol. Soc. Am.* 112, 708–19.

Brown, M. (1998) Unpairing of metamorphic belts: *P–T* paths and a tectonic model for the Ryoke Belt, southwest Japan. *J. metam. geol.* 16, 3–22.

Brown, M. (2006) Duality of thermal regimes is the distinctive characteristics of plate tectonics since the Neoarchean. *Geology* 34, 961–4.

Brown, M. & Solar, G.S. (1999) The mechanism of ascent and emplacement of granite magma during transpression; a syntectonic granite paradigm. *Tectonophysics* 312, 1–33.

Brozena, J.M. & White, R.S. (1990) Ridge jumps and propagations in the South Atlantic Ocean. *Nature* 348, 149–52.

Brunet, D. & Yuen, D.A. (2000) Mantle plumes pinched in the transition zone. *Earth planet. Sci. Lett.* 178, 13–27.

Buck, W.R. (1988) Flexural rotation of normal faults. *Tectonics* 7, 959–73.

Buck, W.R. (1991) Modes of continental lithospheric extension. *J. geophys. Res.* 96, 20161–78.

Buck, W.R. (1993) Effect of lithospheric thickness on the formation of high- and low-angle normal faults. *Geology* 21, 933–36.

Buck, W.R. (2004) Consequences of asthenospheric variability on continental rifting. *In* Karner, G.D. *et al.* (eds) *Rheology and Deformation of the Lithosphere at Continental Margins*, pp. 1–30. Columbia University Press, New York.

Buck, W.R. *et al.* (1988) Thermal consequences of lithospheric extension: pure and simple. *Tectonics* 7, 213–34.

Buck, W.R., Lavier, L.L. & Poliakov, A.N.B. (1999) How to make a rift wide. *Phil. Trans. Roy. Soc. Lond. A* 357, 671–93.

Buick, R. (2001) Life in the Archaean. *In* Briggs, D.E.G. & Crowther, P.R. (eds) *Palaeogeology II*, pp. 13–21. Blackwell, Oxford.

Bull, W.B., & Cooper, A.F. (1986) Uplifted marine terraces along the Alpine Fault, New Zealand. *Science* 240, 804–5.

Bullard, E.C., Everett, J.E. & Smith, A.G. (1965) The fit of the continents around the Atlantic. *Phil. Trans. Roy. Soc. Lond. A* 258, 41–51.

Bullen, K.E. (1949) Compressibility–pressure hypothesis and the Earth's interior. *Geophys. J. R. Astron. Soc.* 5, 355–68.

Bump, H.A. & Sheehan, A.F. (1998) Crustal thickness variations across the northern Tien Shan from teleseismic receiver functions. *Geophys. Res. Lett.* 25, 1055–8.

Bunge, H.-P., Richards, M.A. & Baumgardner, J.R. (1997) A sensitivity study of three-dimensional spherical mantle convection at 10^8 Rayleigh number: effects of depth-dependent viscosity, heating mode, and an endothermic phase change. *J. geophys. Res.* 102, 11991–12007.

Burchfiel, B.C. (2004) New technology; new geological challenges. *GSA Today* 14, 4–9.

Burchfiel, B.C. *et al.* (1992) The south Tibetan detachment system, Himalayan orogen: extension contemporaneous with and parallel to shortening in a collisional mountain belt. *Geol. Soc. Am. Sp. Paper* 269, 1–41.

Burg, J.P. *et al.* (1984) Himalayan metamorphism and deformations in the North Himalayan belt, southern Tibet, China. *Earth planet Sci. Lett.* **69**, 391–400.

Bürgmann, R. *et al.* (2006) Resolving vertical tectonics in the San Francisco Bay Area from permanent scatterer InSAR and GPS analysis. *Geology* **34**, 221–4.

Busby, C. (2004) Continental growth at convergent margins facing large ocean basins: a case study from Mesozoic convergent-margin basins of Baja California, Mexico. *Tectonophysics* **392**, 241–77.

Butler, R. *et al.* (1989) Discordant paleomagnetic poles from the Canadian Coast Plutonic Complex: regional tilt rather than large-scale displacement? *Geology* **17**, 691–94.

Butler, R. *et al.* (2004) The Global Seismograph Network surpasses its design goal. *EOS Trans. Amer. Geophys. Un.* **85**, 225–9.

Byerlee, J.D. (1978) Friction of rocks. *Pure Appl. Geophys.* **116**, 615–26.

Bystricky, M. (2003) Mantle flow revisited. *Science* **301**, 1190–1.

Calvert, A.J. (1995) Seismic evidence for a magma chamber beneath the slow-spreading Mid-Atlantic Ridge. *Nature* **377**, 410–14.

Calvert, A.J. & Ludden, J.N. (1999) Archean continental assembly in the southeastern Superior Province of Canada. *Tectonics* **18**, 412–29.

Calvert, A.J. *et al.* (1995) Archaean subduction inferred from seismic images of a mantle suture in the Superior Province. *Nature* **375**, 670–4.

Campbell, A.C. *et al.* (1988) Chemistry of hot springs on the Mid-Atlantic Ridge. *Nature* **335**, 514–9.

Cande, S.C. & Kent, D.V. (1992) A new geomagnetic polarity time scale for the late Cretaceous and Cenozoic. *J. geophys. Res.* **97**, 13917–51.

Cande, S.C. & Kent, D.V. (1995) Revised calibration of the geo-magnetic polarity timescale for the Late Cretaceous and Cenozoic. *J. geophys. Res.* **100**, 6093–5.

Cande, S.C. & Leslie, R.B. (1986) Late Cenozoic tectonic of the Southern Chile Trench. *J geophys. Res.* **91**, 471–96.

Cande, S.C. & Stock, J.M. (2004) Pacific–Antarctic–Australia motion and the formation of the Macquarie Plate. *Geophys. J. Int.* **157**, 399–414.

Cande, S.C. *et al.* (1989) *Magnetic Lineations of the World's Ocean Basins*, pp. 13, 1 sheet. American Association of Petroleum Geologists, Tulsa, Oklahoma.

Cande, S.C. *et al.* (1999) Cenozoic motion between East and West Antarctica. *Nature* **404**, 145–50.

Cann, J.R. (1970) New model for the structure of the ocean crust. *Nature* **226**, 928–30.

Cann, J.R. (1974) A model for oceanic crustal structure developed. *Geophys. J. Roy. astr. Soc.* **39**, 169–87.

Cann, J.R. *et al.* (1997) Corrugated slip surfaces formed at ridge–transform intersections on the Mid-Atlantic Ridge. *Nature* **385**, 329–32.

Cannat, M. *et al.* (1995) Thin crust, ultramafic exposures, and rugged faulting patterns at the Mid-Atlantic Ridge (22°–24°N). *Geology* **23**, 49–52.

Carbotte, S.M. & MacDonald, K.C. (1994) The axial topographic high at intermediate and fast spreading ridges. *Earth planet. Sci. Lett.* **128**, 85–97.

Caress, D.W., Menard, H.W. & Hey, R.N. (1988) Eocene reorganization of the Pacific–Farallon spreading center north of the Mendocino fracture zone. *J. geophys. Res.* **93**, 2813–38.

Caress, D.W., Burnett, M.S. & Orcutt, J.A. (1992) Tomographic image of the axial low velocity zone at 12°50′N on the East Pacific Rise. *J. geophys. Res.* **97**, 9243–64.

Carey, S.W. (1958) A tectonic approach to continental drift. *In* Carey, S.W. (ed.) *Continental drift: a symposium,* pp. 177–355. Univ. of Tasmania, Hobart.

Carey, S.W. (1976) *The Expanding Earth.* Elsevier, Amsterdam.

Carey, S.W. (1988) *Theories of the Earth and Universe.* Stanford University Press, Stanford, CA.

Carlson, R.L. (2001) The abundance of ultramafic rocks in the Atlantic Ocean crust. *Geophys. J. Int.* **144**, 37–48.

Carlson, R.W., Pearson, D.G. &. James, D.E. (2005) Physical, chemical, and chronological characteristics of continental mantle. *Rev. Geophys.* **43**, RG1001, doi:10.1029/2004RG000156.

Carr, S.D. *et al.* (2000) Geologic transect across the Grenville orogen of Ontario and New York. *Can. J. Earth Sci.* **37**, 193–216.

Carter, W.E. & Robertson, D.S. (1986) Studying the earth by very-long-baseline interferometry. *Sci. Am.* **255**(5), 44–52.

Catchings, R.D. & Mooney, W.D. (1991) Basin and Range and upper mantle structure northwest to central Nevada. *J. geophys. Res.* **96**, 6247–67.

Cattin, R. *et al.* (2001) Gravity anomalies, crustal structure and thermo-mechanical support of the Himalaya of central Nepal. *Geophys. J. Int.* **147**, 381–92.

Cawood, P.A. (2005) Terra Australis Orogen: Rodinia breakup and development of the Pacific and Iapetus margins of Gondwana during the Neoproterozoic and Paleozoic. *Earth Sci. Rev.* **69**, 249–79.

Cawood, P.A. & Suhr, G. (1992) Generation and obduction of ophiolites: constraints from the Bay of Islands complex, western Newfoundland. *Tectonics* **11**, 884–97.

Cawood, P.A. & Tyler, I.M. (2004) Assembling and reactivating the Proterozoic Capricorn Orogen: lithotectonic elements, orogenies, and significance. *Precambrian Res.* **128**, 201–18.

Cawood, P.A., McCausland, P.J.A. & Dunning, G.R. (2001) Opening Iapetus: constraints from the Laurentian margin in Newfoundland. *Bull. geol. Soc. Am.* **113**, 443–53.

Cawood, P.A. *et al.* (2003) Source of the Dalradian Supergroup constrained by U-Pb dating of detrital zircon and implications for the East Laurentian margin. *J. geol. Soc. Lond.* **160**, 231–46.

Cawood, P.A., Kröner, A. & Pisarevsky, S. (2006) Precambrian plate tectonics: criteria and evidence. *GSA Today* **16**, 4–11.

Cembrano, J. *et al.* (2000) Contrasting nature of deformation along an intra-arc shear zone, the Liquiñe–Ofqui fault zone, southern Chilean Andes. *Tectonophysics* **319**, 129–49.

Cembrano, J. *et al.* (2002) Late Cenozoic transpressional ductile deformation north of the Nazca–South America–Antarctica triple junction. *Tectonophysics* **354**, 289–314.

Cembrano, J. *et al.* (2005) Fault zone development and strain partitioning in an extensional strike-slip duplex: a case study from the Mesozoic Atacama fault system, Northern Chile. *Tectonophysics* **400**, 105–25.

Chandra, N.N. & Cumming. G.L. (1972) Seismic refraction studies in Western Canada. *Can. J. Earth Sci.* **9**, 1099–109.

Chapman, D.S. & Pollack, H.N. (1975) Global heat flow: a new look. *Earth planet. Sci. Lett.* **28**, 23–32.

Chapple, W.M. & Forsyth, D.W. (1979) Earthquakes and bending of plates at trenches. *J. geophys. Res.* **84**, 6729–49.

Chapple, W.M. & Tullis, T.E. (1977) Evaluation of the forces that drive plates. *J. geophys. Res.* **82**, 1967–84.

Chardon, D., Choukroune, P. & Jayananda, M. (1996) Strain patterns, décollement and incipient sagducted greenstones terrains in the Archaean Dharwar craton (south India). *J. struct. Geol.* **18**, 991–1004.

Chase, C.G. (1978) Plate kinematics: the Americas, east Africa, and the rest of the world. *Earth planet. Sci. Lett.* **37**, 355–68.

Chen, Y. & Morgan, W.J. (1990) Rift valley/no rift valley transition at mid-ocean ridges. *J. geophys. Res.* **95**, 17571–81.

Chen, Z. *et al.* (2000) Global positioning system measurements from eastern Tibet and their implications for India/Eurasia intercontinental deformation. *J. geophys. Res.* **105**, 16215–27.

Chesley, J.T., Rudnick, R.L. & Lee, C.-T. (1999) Re–Os systematics of mantle xenoliths from the East African Rift: age, structure, and history of the Tanzanian craton. *Geochim. cosmochim. Acta* **63**, 1203–17.

Choukroune, P., Francheteau, J. & Le Pichon, X. (1978) *In situ* structural observations along Transform Fault A in the FAMOUS area, Mid-Atlantic Ridge. *Bull. geol. Soc. Am.* **89**, 1013–29.

Christensen, D.H. & Ruff, L.J. (1998) Seismic coupling and outer rise earthquakes. *J. geophys. Res.* **93**, 13421–44.

Christensen, N.I. & Fountain, D.M. (1975) Constitution of the lower continental crust based on experimental studies of seismic velocities in granulite. *Bull. geol. Soc. Am.* **86**, 227–36.

Christensen, N.I. & Mooney, W.D. (1995) Seismic velocity structure and the composition of the continental crust: a global view. *J. geophys. Res.* **100**, 9761–88.

Christensen, N.I. & Salisbury, M.H. (1972) Seafloor spreading, progressive alteration of layer 2 basalts, and associated changes in seismic velocities. *Earth planet. Sci. Lett.* **15**, 367–75.

Christie-Blick, N. & Biddle, K.T. (1985) Deformation and basin formation along strike-slip faults. In Biddle, K.T. & Christie-Blick, N. (eds) *Strike-slip Deformation, Basin Formation, and Sedimentation. Soc. Econ. Pal. Mineral. Spec. Pub.* **37**, 1–35.

Christodoulidis, D.C. *et al.* (1985) Observing tectonic plate motions and deformations from satellite laser ranging. *J. geophys. Res.* **90**, 9249–63.

Chulick, G.S. & Mooney, W.D. (2002) Seismic structure of the crust and uppermost mantle of North America and adjacent oceanic basins: a synthesis. *Bull. seis. Soc. Am.* **92**, 2478–92.

Chung, S.-L. *et al.* (2005) Tibetan tectonic evolution inferred from spatial and temporal variations in post-collisional magmatism. *Earth Sci. Rev.* **68**, 173–96.

Clague, D.A. & Dalrymple, G.B. (1989) Tectonics, geochronology and origin of the Hawaiian–Emperor chain. In Winterer, E.L., Hussong, D.M. & Decker, R.W. (eds) *The Eastern Pacific and Hawaii. The Geology of North America* N, pp. 188–217. Geological Society of America, Boulder, CO.

Clark, M.K. & Royden, L.H. (2000) Topographic ooze: building the eastern margin of Tibet by lower crustal flow. *Geology* **28**, 703–6.

Clark, M.K., Bush, J.W.M. & Royden, L.H. (2005) Dynamic topography produced by lower crustal flow against rheological strength heterogeneities bordering the Tibetan Plateau. *Geophys. J. Int.* **162**, 575–90.

Clark, T.A. *et al.* (1987) Determination of relative site motions in the western United States using Mark III Very Long Baseline Interferometry. *J. geophys. Res.* **92**, 12741–50.

Clarke, G.L. *et al.* (2005) Roles for fluid and/or melt advection in forming high-P mafic migmatites, Fiordland, New Zealand. *J. metam. geol.*, **23**, 557–67.

Clarke, P.J. *et al.* (1998) Crustal strain in central Greece from repeated GPS measurements in the interval 1989–1997. *Geophys. J. Int.* **135**, 195–214.

Clemens, J. D. & Mawer, C.K. (1992) Granitic magma transport by fracture propagation. *Tectonophysics* **204**, 339–60.

Clift, P. & Vannucchi, P. (2004) Controls on tectonic accretion versus erosion in subduction zones: implications for the origin and recycling of continental crust. *Rev. Geophys.* **42**, RG2001, doi:10.1029/2003RG000127.

Clouard, V. & Bonneville, A. (2001) How many Pacific hotspots are fed by deep mantle plumes? *Geology* **29**, 695–8.

Clowes, R. M. *et al.* (1987) Lithoprobe–Southern Vancouver Island: Cenozoic subduction complex imaged by deep seismic reflection. *Can. J. Earth Sci.* **24**, 31–51.

Clowes, R.M. *et al.* (1995) Lithospheric structure in the southern Canadian Cordillera from a network of seismic refraction lines. *Can. J. Earth Sci* **32**, 1485–513.

Clowes, R.M. *et al.* (2005) Lithospheric structure in northwestern Canada from Lithoprobe seismic refraction and related studies: a synthesis. *Can. J. Earth Sci.* **42**, 1277–93.

Coe, R.S., Hongre, L. & Glatzmaier, G.A. (2000) An examination of simulated geomagnetic reversals from a paleomagnetic perspective. *Phil. Trans. Roy. Soc. Lond. A* **358**, 1141–70.

Coffin, M.F. & Eldholm, O. (1994) Large igneous provinces: crustal structure, dimensions, and external consequences. *Rev. Geophys.* **32**, 1–36.

Cohen, S.C. & Smith, P.E. (1985) LAGEOS scientific results: introduction. *J. geophys. Res.* **90**, 9217–20.

Collerson, K.D. & Kamber, B.S. (1999) Evolution of the continents and the atmosphere inferred from Th–U–Nb systematics of the depleted mantle. *Science* **282**, 1519–22.

Collette, B.J. (1979) Thermal contraction joints in spreading seafloor as origin of fracture zones. *Nature* **251**, 299–300.

Collier, J.S. & Singh. S.C. (1997) Detailed structure of the top of the melt body beneath the East Pacific Rise at 9°40′N from waveform inversion of seismic reflection data. *J. geophys. Res.* **102**, 20287–304.

Collins, A.S. *et al.* (2004) Temporal constraints on Palaeoproterozoic eclogite formation and exhumation (Usagaran Orogen, Tanzania). *Earth planet. Sci. Lett.* **224**, 175–92.

Collins, W.J. (1989) Polydiapirism of the Archean Mount Edgar Batholith, Pilbara Block, Western Australia. *Precambrian Res.* **43**, 41–62.

Collins, W.J. (2002a) Hot orogens, tectonic switching, and creation of continental crust. *Geology* **30**, 535–8.

Collins, W.J. (2002b) Nature of extensional accretionary orogens. *Tectonics* **21**, doi:1029/2000TC001272.

Collins, W.J. & Sawyer, E.W. (1996) Pervasive granitoid magma transfer through the lower–middle crust during non-coaxial compressional deformation. *J. metam. geol.* **14**, 565–79.

Collins, W.J., Van Kranendonk, M.J. & Teyssier, C. (1998) Partial convective overturn of Archaean crust in the east Pilbara Craton, Western Australia: driving mechanisms and tectonic implications. *J. struct. Geol.* **20**, 1405–24.

Condie, K.C. (1982a) *Plate Tectonics and Crustal Evolution.* Pergamon Press, Oxford.

Condie, K.C. (1982b) Early and middle Proterozoic supracrustal successions and their tectonic settings. *Amer. J. Sci.* **282**, 341–57.

Condie, K.C. (1986) Geochemistry and tectonic setting of early Proterozoic supracrustal rocks in the southwestern United States. *J. Geol.* **94**, 845–64.

Condie, K.C. (1998) Episodic continental growth and supercontinents: a mantle avalanche connection? *Earth planet. Sci. Lett.* **163**, 97–108.

Condie, K.C. (2000) Episodic continental growth models: afterthoughts and extensions. *Tectonophysics* **322**, 153–62.

Condie, K.C. (2005a) TTGs and adakites: are they both slab melts? *Lithos* **80**, 33–44.

Condie, K.C. (2005b) *Earth as an Evolving Planetary System.* Elsevier, Amsterdam.

Coney, P.J. (1989) Structural aspects of suspect terranes and accretionary tectonics in western North America. *J. struct. Geol.* **11**, 107–25.

Coney, P.J. & Harms, T.A. (1984) Cordilleran metamorphic core complexes: Cenozoic extensional relics of Mesozoic compression. *Geology* **12**, 550–4.

Coney, P.J., Jones, P.L. & Monger, J.W.H. (1980) Cordilleran suspect terranes. *Nature* **288**, 329–33.

Conti, C.M. *et al.* (1996) Paleomagnetic evidence of an early Paleozoic rotated terrane in Northwest Argentina; a clue for Gondwana–Laurentia interaction? *Geology* **24**, 953–6.

Cook, F.A. (2002) Fine structure of the continental reflection Moho. *Bull. geol. Soc. Am.,* **114**, 64–79.

Cook, F.A. *et al.* (1992) Lithoprobe crustal reflection cross-section of the southern Canadian Cordillera I: foreland thrust and fold belt to Fraser River fault. *Tectonics* **11**, 12–35.

Cook, F.A. *et al.* (1999) Frozen subduction in Canada's Northwest Territories: Lithoprobe deep lithospheric reflection profiling of the western Canadian Shield. *Tectonics* **18**, 1–24.

Cook, F.A. *et al.* (2004) Precambrian crust beneath the Mesozoic northern Canadian Cordillera discovered by Lithoprobe seismic reflection profiling. *Tectonics* **23**, doi:10.1029/2002TC001412.

Cooper, A.K. *et al.* (1992) Evidence for Cenozoic crustal extension in the Bering Sea region. *Tectonics* **11**, 719–31.

Corliss, J.B. *et al.* (1979) Submarine thermal springs on the Galapagos Rift. *Science* **203**, 1073–83.

Corti, G. *et al.* (2005) Active strike-slip faulting in El Salvador, Central America. *Geology* **33**, 989–92.

Courtillot, V. & Besse, J. (1987) Magnetic field reversals, polar wander and core–mantle coupling. *Science* **237**, 1140–7.

Courtillot, V. *et al.* (2003) Three distinct types of hotspots in the Earth's mantle. *Earth planet. Sci. Lett.* **205**, 295–308.

Cowan, D., Brandon, M. & Garver, J. (1997) Geologic tests of hypotheses for large coastwise displacements – a critique illustrated by the Baja British Columbia controversy. *Amer. J. Sci.* **297**, 117–173.

Cowan, J. & Cann, J. (1988) Supercritical two-phase separation of hydrothermal fluids in the Troodos ophiolite. *Nature* **333**, 259–61.

Cox, A. (1980) Rotation of microplates in western North America. *In* Strangway, D.W. (ed.) *The Continental Crust and its Mineral Deposits.* Geol. Assoc. Canada. Spec. Paper **20**, pp. 305–21.

Cox, A. & Hart, R.B. (1986) *Plate Tectonics. How it works.* Blackwell Scientific Publications, Oxford.

Cox, A., Doell, R.R. & Dalrymple, G.B. (1964) Geomagnetic polarity epochs. *Science* **143**, 351–2.

Cox, A.V., Dalrymple, G.B. & Doell, R.R. (1967) Reversals of the Earth's magnetic field. *Sci. Am.* **216**, 44–54.

Crawford, A.J., Falloon, T.J. & Green, D.G. (1989) Classification, petrogenesis, and tectonic setting of boninites. *In* Crawford, A.J. (ed.) *Boninites and Related Rocks*, pp. 1–49. Unwin-Hyman, London.

Crawford, M.L. *et al.* (1999) Batholith emplacement at mid-crustal levels and its exhumation within an obliquely convergent margin. *Tectonophysics* **312**, 57–78.

Creer, K.M. (1965) Palaeomagnetic data from the Gondwanic continents. *In* Blackett, P.M.S., Bullard, E. & Runcorn, S.K. (eds) *A Symposium on Continental Drift. Phil. Trans. Roy. Soc. Lond. A* **258**, 27–90.

Crittenden, M.D., Jr, Coney, P.J. & Davis, G.H. (1980) Cordilleran metamorphic core complexes. *Geol. Soc. Am. Mem.* **153**, 490pp.

Cross, T.A. & Pilger, R.H. (1982) Controls of subduction geometry, location of magmatic arcs, and tectonics of arc and back-arc regions. *Bull. geol. Soc. Am.* **93**, 545–62.

Crough, S.T. (1979) Hotspot epeirogeny. *Tectonophysics* **61**, 321–33.

Curtis, A. & Woodhouse, J.H. (1997) Crust and upper mantle shear velocity structure beneath the Tibetan plateau and surrounding regions from interevent surface wave phase velocity inversion. *J. geophys. Res.* **102**, 11789–813.

d'Acremont, E. et al. (2005) Structure and evolution of the eastern Gulf of Aden conjugate margins from seismic reflection data. *Geophys. J. Int.* **160**, 869–90.

Daczko, N.R., Clarke, G.L. & Klepeis, K.A. (2001) Transformation of two-pyroxene hornblende granulite to garnet granulite involving simultaneous melting and fracturing of the lower crust, Fiordland, New Zealand. *J. metam. geol.* **19**, 549–62.

Dahlen, F.A. (1990) Critical taper model of fold-and-thrust belts and accretionary wedges. *Annu. Rev. Earth planet. Sci.* **18**, 55–99.

Dahlen, F.A. & Barr, T.D. (1989) Brittle frictional mountain building: 1. Deformation and mechanical energy budget. *J. geophys. Res.* **94**, 3906–22.

Dalziel, I.W.D. (1981) Back-arc extension in the Southern Andes; a review and critical reappraisal. *Phil. Trans. Roy. Soc. Lond. A* **300**, 319–35.

Dalziel, I.W.D. (1991) Pacific margins of Laurentia and East Antarctica–Australia as a conjugate rift pair: evidence and implications for an Eocambrian supercontinent. *Geology* **19**, 598–601.

Dalziel, I.W.D. (1995) Earth before Pangea. *Sci. Am.* **272**, 38–43.

Dalziel, I.W.D. (1997) Neoproterozoic–Paleozoic geography and tectonics: review, hypothesis, environmental speculation. *Bull. geol. Soc. Am.* **109**, 16–42.

Dalziel, I.W.D., Lawver, L.A. & Murphy, J.B. (2000a) Plumes, orogenesis, and supercontinental fragmentation. *Earth planet. Sci. Lett.* **178**, 1–11.

Dalziel, I.W.D., Mosher, S. & Gahagan, L.M. (2000b) Laurentia–Kalahari collision and the assembly of Rodinia. *J. Geol.* **108**, 499–513.

Davaille, A. (1999) Simultaneous generation of hotspots and superswells by convection in a homogeneous planetary mantle. *Nature* **402**, 756–60.

Davey, F.J. *et al.* (1995) Crustal reflections from the Alpine Fault zone, South Island, New Zealand. *NZ J. Geol. Geophys.* **38**, 601–4.

Davidson, J.P. (1983) Lesser Antilles isotopic evidence of the role of subducted sediment in island arc magma genesis. *Nature* **306**, 253–5.

Davies, D. (1968) A comprehensive test ban. *Sci. J. Lond.* Nov. 1968, 78–84.

Davies, G.F. (1977) Whole mantle convection and plate tectonics. *Geophys. J. Roy. astr. Soc.* **49**, 459–86.

Davies, G.F. (1993) Cooling the core and mantle by plume and plate flows. *Geophys. J. Int.* **115**, 132–46.

Davies, G.F. (1999) *Dynamic Earth: plates, plumes and mantle convection.* Cambridge University Press, Cambridge, 458pp.

Davies, G.F. & Richards, M.A. (1992) Mantle convection. *J. Geol.* **100**, 151–206.

Davies, J.H., Brodholt, J.P. & Wood. B.J. (eds) (2002) Chemical reservoirs and convection in the Earth's mantle. *Phil. Trans. Roy. Soc. Lond. A* **360**, 2361–648.

Davis, D., Suppe, J. & Dahlen, F.A. (1983) Mechanics of fold-and-thrust belts and accretionary wedges. *J. geophys. Res.* **88**, 1153–72.

Davis, M. & Kusznir, N. (2002) Are buoyancy forces important during the formation of rifted margins? *Geophys. J. Int.* **149**, 524–33.

Davis, M. & Kusznir, N. (2004) Depth-dependent lithospheric stretching at rifted continental margins. *In* Karner, G.D. *et al.* (eds) *Rheology and Deformation of the Lithosphere at Continental Margins*, pp. 31–45. Columbia University Press, New York.

Davis, P.M. & Slack, P.D. (2002) The uppermost mantle beneath the Kenya dome and relation to melting, rifting and uplift in East Africa. *Geophys. Res. Lett.* **29**, doi:10.1029/2001GL013676.

Dean, S.M. *et al.* (2000) Deep structure of the ocean–continent transition in the southern Iberia Abyssal Plain from seismic refraction profiles. II. The IAM–9 transect at 40°20′N. *J. geophys. Res.* **105**, 5859–86.

DeCelles, P.G. *et al.* (1998) Eocene–early Miocene foreland basin development and the history of Himalayan thrusting, western and central Nepal. *Tectonics* **17**, 741–65.

DeCelles, P.G., Robinson, D.M., & Zandt, G. (2002) Implications of shortening in the Himalayan fold-thrust belt for uplift of the Tibetan Plateau. *Tectonics* **21**, 1062, doi:10.1029/2001TC001322.

DeCelles, P.G. *et al.* (2001) Stratigraphy, structure, and tectonic evolution of the Himalayan fold-thrust belt in western Nepal. *Tectonics* **20**, 487–509.

Dehler, S.A. & Clowes, R.M. (1988) The Queen Charlotte Islands refraction project. Part I. The Queen Charlotte Fault Zone. *Can. J. Earth Sci.* **25**, 1857–70.

DeLong, S.E., Dewey, J.F. & Fox, P.J. (1977) Displacement history of oceanic fracture zones. *Geology* **5**, 199–201.

DeMets, C. (2001) A new estimate for present-day Cocos–Caribbean plate motion: implications for slip along the Central American volcanic arc. *Geophys. Res. Lett.* **28**, 4043–6.

DeMets, C. & Dixon, T.H. (1999) New kinematic models for Pacific–North America motion from 3 Ma to present: 1. Evidence for steady motion and biases in the NUVEL–1A model. *Geophys. Res. Lett.* **26**, 1921–4.

DeMets, C. et al. (1990) Current plate motions. *Geophys. J. Int.* **101**, 425–78.

DeMets. C. et al. (1994) Effect of recent revisions to the geomagnetic time scale on estimates of current plate motions. *Geophys. Res. Lett.* **21**, 2191–94.

de Ronde, C.E.J., Channer, D.M. DeR. & Spooner, E.T.C. (1997) Archaean Fluids. *In* De Wit, M.J. & Ashwal, L.D. (eds) *Greenstone Belts*, pp. 309–35. Clarendon Press, Oxford.

DESERT Group (2004) The crustal structure of the Dead Sea Transform. *Geophys. J. Int.* **156**, 655–81.

Detrick, R.S. *et al.* (1987) Multi-channel seismic imaging of a crustal magma chamber along the East Pacific Rise. *Nature* **326**, 35–41.

Detrick, R.S. *et al.* (1990) No evidence from multi-channel reflection data for a crustal magma chamber in the MARK area on the Mid-Atlantic Ridge. *Nature* **347**, 61–4.

Detrick, R.S., White R.S. & Purdy, G.M. (1993a) Crustal structure of North Atlantic fracture zones. *Rev. Geophys.* **31**, 439–57.

Detrick, R.S. *et al.* (1993b) Seismic structure of the southern East Pacific Rise. *Science* **259**, 499–503.

Detrick, R. *et al.* (1994) *In situ* evidence for the nature of the seismic layer 2/3 boundary in oceanic crust. *Nature* **370**, 288–90.

Dewey, J.F. (1969) Evolution of the Appalachian/Caledonian orogen. *Nature* **222**, 124–9.

Dewey, J.F. (1976) Ophiolite obduction. *Tectonophysics* **31**, 93–120.

Dewey, J.F. (1988) Extensional collapse of orogens. *Tectonics* **7**, 1123–39.

Dewey, J.F. & Bird, J.M. (1970) Mountain belts and the new global tectonics. *J. geophys. Res.* **75**, 2625–47.

Dewey, J.F. & Burke, K.C.A. (1974) Hot spots and continental break-up: implications for collisional orogeny. *Geology* **2**, 57–60.

Dewey, J.F., Cande, S. & Pitman, W.C. (1989) Tectonic evolution of the India–Eurasia collision zone. *Ecol. Geol. Helv.* **82**, 717–34.

de Wit, M.J. (2004) Archean Greenstone Belts do contain fragments of ophiolites. *In* Kusky, T.M. (ed.) *Precambrian Ophiolites and Related Rocks. Developments in Precambrian Geology*, **13**, pp. 599–613. Elsevier, Amsterdam.

Dick, H.B.J., Lin, J. & Schouten, H. (2003) An ultraslow-spreading class of oceanic ridge. *Nature* **426**, 405–12.

Dickinson, W.R. (1971) Plate tectonic models of geosynclines. *Earth planet. Sci. Lett.* **10**, 165–74.

Dickinson, W.R. (1974) Plate tectonics and sedimentation. *In* Dickinson, W.R. (ed.) *Tectonics and Sedimentation. Soc. econ. Pal. Mineral. Spec. Pub.* **22**, 1–27.

Dickinson, W.R. (2002) The Basin and Range Province as a composite extensional domain. *Int. geol. Rev.* **44**, 1–38.

Dietz, R.S. & Holden, C. (1970) The breakup of Pangaea. *Sci. Am.* **223**, 30–41.

Dietz, R.S. (1961) Continental and ocean basin evolution by spreading of the sea floor. *Nature* **190**, 854–7.

Dillon, J.T. & Ehlig, P.L. (1993) Displacement on the southern San Andreas fault. *In* Powell, R.E., Weldon, R.J.I. & Matti, J.C. (eds) *The San Andreas Fault System: displacement, palinspastic reconstruction, and geologic evolution. Geol. Soc. Am. Mem.* **178**, 199–216.

Dixon, T.H. (1991) An introduction to the Global Positioning System and some geological applications. *Rev. Geophys.* **29**, 249–76.

Dixon, T.H., Norabuena, E. & Hotaling, L. (2003) Paleoseismology and Global Positioning System; earthquake-cycle effects and geodetic versus geologic fault slip rates in the Eastern California shear zone. *Geology* **31**, 55–8.

Donnadieu, Y. *et al.* (2004) A "snowball Earth" climate triggered by continental break-up through changes in runoff. *Nature* **428**, 303–6.

Dooley, T. & McClay, K.R. (1997) Analog modelling of strike-slip pull-apart basins. *Bull. Am. Assoc. Petroleum Geols.* **81**, 804–26.

Drew, J.J. & Clowes, R.M. (1990) A re-interpretation of the seismic structure across the active subduction zone of western Canada – CCSS Workshop Topic I, onshore–offshore data set. *In* Green, A.G. (ed.) *Studies of Laterally Heterogeneous Structures Using Seismic Refraction And Reflection Data. Proceedings of the 1987 Commission on Controlled Source Seismology Workshop, Geol. Surv. Canada*, Paper 89–13, pp. 115–32.

Driscoll, N.W. & Karner, G.D. (1998) Lower crustal extension across the Northern Carnarvon basin, Australia: evidence for an eastward dipping detachment. *J. geophys. Res.* **103**, 4975–91.

du Toit, A.L. (1937) *Our Wandering Continents*. Oliver & Boyd, Edinburgh.

Duclos, M. *et al.* (2005) Mantle tectonics beneath New Zealand inferred from SKS splitting and petrophysics. *Geophys. J. Int.* **163**, 760–4.

Dugda, M.T. *et al.* (2004) Crustal structure in Ethiopia and Kenya from receiver function analysis: implications for rift development in eastern Africa. *J. geophys. Res.* **110**, doi:10.1029/2004JB003065.

Duncan, R.A. & Richards, M.A. (1991) Hotspots, mantle plumes, flood basalts, and true polar wander. *Rev. Geophys.* **29**, 31–50.

Dunlop, D.J. (1981) Palaeomagnetic evidence for Proterozoic continental development. *Geophys. J. Roy. astr. Soc.* **301A**, 265–77.

Dunn, J.F., Hartshorn, K.G. & Hartshorn, P.W. (1995) Structural styles and hydrocarbon potential of the Sub-Andean Thrust Belt of Southern Bolivia. *In* Tankard, A.J., Suárez–Soruco, R. & Welsink, H.J. (eds) *Petroleum Basins of South America. Amer. Assoc. Pet. geol. Mem.* **62**, pp. 523–43. Tulsa, Oklahoma.

Dziak, R.P. & Fox, C.G. (1999) The January 1998 earthquake swarm at Axial Volcano, Juan de Fuca Ridge: evidence for submarine volcanic activity. *Geophys. Res. Lett.* **26**, 3429–32.

Dziewonski, A.M. (1984) Mapping the lower mantle: determination of lateral heterogeneity up to degree and order 6. *J. geophys. Res.* **89**, 5929–52.

Dziewonski, A.M. & Anderson, D.L. (1984) Seismic tomography of the Earth's interior. *Am. Sci.* **72**, 483–94.

Eales, H.V. & Cawthorn, R.G. (1996) The Bushveld Complex. *In* Cawthorn R.G. (ed.) *Layered Intrusions*, pp. 181–229. Elsevier, Amsterdam.

Ebinger, C.J. (2005) Continental break-up: the East African perspective. *Astro. Geophys.* **46**, 216–21.

Ebinger C.J. & Casey, M. (2001) Continental breakup in magmatic provinces: an Ethiopian example. *Geology* **29**, 527–30.

Ebinger, C.J. & Ibrahim, A. (1994) Multiple episodes of rifting in Central and East Africa: a reevaluation of gravity data. *Geol. Rundsch.* **83**, 689–702.

Ebinger, C.J. & Sleep, N.H. (1998) Cenozoic magmatism throughout east Africa resulting from impact of a single plume. *Nature* **395**, 788–91.

Ebinger, C.J. *et al.* (1999) Extensional basin geometry and the elastic lithosphere. *Phil. Trans. Roy. Soc. Lond. A* **357**, 741–65.

Eckstein, Y. & Simmons, G. (1978) Measurements and interpretation of terrestrial heat flow in Israel, *Geothermics* **6**, 117–42.

Edgar, N.T. (1974) Acoustic stratigraphy in the deep oceans. *In* Bur, C.A. & Drake, C.L. (eds) *The Geology of Continental Margins*, pp. 243–6. Springer-Verlag, Berlin.

Edwards, R.A., Whitmarsh, R.B. & Scrutton, R.A. (1997) The crustal structure across the transform continental margin off Ghana, eastern equatorial Atlantic. *J. Geophys. Res.* **102**, 747–72.

Egyed, L. (1960) Some remarks on continental drift. *Geofis. Pura Appl.* **45**, 115–16.

Elder, J.W. (1965) Physical processes in geothermal areas. *In* Lee, W.M.K. (ed.) *Terrestrial Heat Flow. Geophys. Monogr. Ser.* **8**, pp. 211–39. American Geophysical Union, Washington, DC.

El-Isa, Z. *et al.* (1987a) A crustal structure study of Jordan derived from seismic refraction data. *Tectonophysics* **138**, 235–53.

El-Isa, Z., Mechie, J. & Prodehl, C. (1987b) Shear velocity structure of Jordan from explosion seismic data. *Geophys. J. Roy. astr. Soc.* **90**, 265–81.

Ellis, S. *et al.* (1998) Continental collision including a weak zone: the vise model and its application to the Newfoundland Appalachians. *Can. J. Earth Sci.* **35**, 1323–46.

Ellis, S., Schreurs, G. & Panien, M. (2004) Comparisons between analogue and numerical models of the thrust wedge development. *J. struct. Geol.* **26**, 1659–75.

Elsasser, W.M. (1969) Convection and stress propagation in the upper mantle. *In* Runcorn, S.K. (ed.) *The Application of Modern Physics to the Earth and Planetary Interiors*, pp. 223–46. Wiley-Interscience, London.

Elsasser, W.M. (1971) Sea-floor spreading as thermal convection. *J. geophys. Res.* **76**, 1101–12.

Elsasser, W.M., Olsen, P. & Marsh, B.D. (1979) The depth of mantle convection. *J. geophys. Res.* **84**, 147–55.

Elthon, D. (1981) Metamorphism in oceanic spreading centres. *In* Emiliani, C. (ed.) *The Oceanic Lithosphere. The Sea 7*, pp. 285–303. Wiley, New York.

Elthon, D. (1991) Geochemical evidence for formation of the Bay of Islands ophiolite above a subduction zone. *Nature* **354**, 140–3.

Engdahl, E.R., van der Hilst, R. & Buland, R. (1998) Global teleseismic earthquake relocation with improved travel times and procedures for depth determination. *Bull. seis. Soc. Am.* **88**, 722–43.

Engebretson, D.C., Cox, A. & Gordon, K.G. (1985) Relative motion between oceanic and continental plates in the Pacific Basin. *Geol. Soc. Am. Sp. Paper* **206**.

Engeln, J.F. *et al.* (1988) Microplate and shear zone models for oceanic spreading center reorganisations. *J. geophys. Res.* **93**, 2839–56.

England, P. (1983) Constraints on extension in the continental lithosphere. *J. geophys. Res.* **88**, 1145–52.

England, P. & Molnar, P. (1997) Active deformation of Asia: from kinematics to dynamics. *Science* **278**, 647–50.

England, P. & Wilkins, C. (2004) A simple analytical approximation to the temperature structure in subduction zones. *Geophys. J. Int.* **159**, 1138–54.

England, P., Engdahl, R. & Thatcher, W. (2004) Systematic variation in the depth of slabs beneath arc volcanoes. *Geophys. J. Int.* **156**, 377–408.

England, P.C. & Houseman, G. (1988) The mechanics of the Tibetan plateau. *Phil. Trans. Roy. Soc. Lond. A* **326**, 301–19.

England, P.C. & Houseman, G. (1989) Extension during continental convergence, with application to the Tibetan Plateau. *J. geophys. Res.* **94**, 17561–79.

Erickson, S.G., Strayer, L.M. & Suppe, J. (2001) Initiation and reactivation of faults during movements over a thrust-fault ramp: numerical mechanical models. *J. struct. Geol.* **23**, 11–23.

Eriksson, P.G. *et al.* (2001) An introduction to Precambrian basins: their characteristics and genesis. *Sed. Geol.* **141–142**, 1–35.

Eriksson, P.G. *et al.* (2005) Patterns of sedimentation in the Precambrian. *Sed. Geol.* **176**, 17–42.

Ernst, R.E., Grosfils, E.B. & Mege, D. (2001) Giant Dike Swarms: Earth, Venus and Mars. *Ann. Rev. Earth planet. Sci.* **29**, 489–534.

Ernst, R.E., Buchan, K.L. & Campbell, I.H. (2005) Frontiers in Large Igneous Province research. *Lithos* **79**, 271–97.

Ernst, W.G. (1973) Blueschist metamorphism and P–T regimes in active subduction zones. *Tectonophysics* **17**, 255–72.

Ernst, W.G. (2003) High-pressure and ultrahigh-pressure metamorphic belts – subduction, recrystallization, exhumation, and significance for ophiolite study. *In* Dilek, Y. & Newcomb, S. (eds) *Ophiolite Concept and the Evolution of Geological Thought. Geol. Soc. Am. Sp. Paper* **373**, 365–84.

Evans, A.M. (1987) *An Introduction to Ore Geology*, 2nd edn. Blackwell Scientific Publications, Oxford.

Evans, D. (2000) Stratigraphic, geochronological, and paleomagnetic constraints upon the Neoproterozoic climatic paradox. *Amer. J. Sci.* **300**, 347–433.

Evans, R.L. *et al.* (1999) Asymmetric electrical structure in the mantle beneath the East Pacific Rise at 17°S. *Science* **286**, 752–6.

Exon, N. *et al.* (2002) Drilling reveals climatic consequences of Tasmanian Gateway opening. *EOS Trans. Amer. Geophys. Un.* **83**, pp. 253, 258–9.

Eyles, N. (1993) Earth's glacial record and its tectonic setting. *Earth Sci. Rev.* **35**, 1–248.

Falloon, T.J. & Danyushevsky, L.V. (2000) Melting of refractory mantle at 1.5, 2 and 2.5 GPa under anhydrous and H_2O undersaturated conditions: implications for the petrogenesis of high-Ca boninites and the influence of subduction components on mantle melting. *J. Petrol.* **41**, 257–83.

Farnetani, C.G. & Samuel, H. (2005) Beyond the thermal plume paradigm. *Geophys. Res. Lett.* **32**, L07311, doi:10.1029/2005GL022360.

Faul, U.H., Fitzgerald, J.D. & Jackson, I. (2004) Shear wave attenuation and dispersion in melt-bearing olivine polycrystals: 2. Microstructural interpretation and seismological implications. *J. geophys. Res.* **109**, BO6202, doi:10.1029/2003JB002407.

Fein, J.B. & Jurdy, D.M. (1986) Plate motion controls on back-arc spreading. *Geophys. Res. Lett.* **12**, 545–8.

Fergusson, C.L. & Coney, P.J. (1992) Implications of a Bengal Fan-type deposit in the Paleozoic Lachlan fold belt of southeastern Australia. *Tectonophysics* **214**, 417–39.

Fernandes, R.M.S. et al. (2004) Angular velocities of Nubia and Somalia from continuous GPS data: implications on present-day relative kinematics. *Earth planet. Sci. Lett.* **222**, 197–208.

Fernández-Viejo G.F. & Clowes, R.M. (2003) Lithospheric structure beneath the Archaean Slave Province and Proterozoic Wopmay orogen, northwestern Canada, from a LITHOPROBE refraction/wide-angle reflection survey. *Geophys. J. Int.* **153**, 1–19.

Ferris, A. *et al.* (2006) Crustal structure across the transition from rifting to spreading: the Woodlark rift system of Papua New Guinea. *Geophys. J. Int.* **166**, 622–34.

Fisher, R.A. (1953) Dispersion on a sphere. *Proc. Roy. Soc. London* **A217**, 295–305.

Fleitout, L. & Froidevaux, C. (1982) Tectonics and topography for a lithosphere containing density heterogeneities. *Tectonics* **1**, 21–56.

Flemings, P.B. & Jordan, T.E. (1990) Stratigraphic modeling of foreland basins; interpreting thrust deformation and lithosphere rheology. *Geology* **18**, 430–4.

Flesch, L.M. *et al.* (2000) Dynamics of the Pacific–North American plate boundary in the western United States. *Science* **287**, 834–6.

Flower, M.F.J. (1981) Thermal and kinematic control on ocean ridge magma fractionation: contrasts between Atlantic and Pacific spreading axes. *J. geol. Soc. Lond.* **138**, 695–712.

Flower, M.F.J. & Dilek, Y. (2003) Arc–trench rollback and forearc accretion: 1. A collision–induced mantle flow model for Tethyan ophiolites. *In* Dilek, Y. & Robinson. P.T. (eds) *Ophiolites in Earth History. Spec. Pub. geol. Soc. Lond.* **218**, 21–41.

Foley, S.F., Buhre, S. & Jacob, D.E. (2003) Evolution of the Archaean crust by delamination and shallow subduction. *Nature* **421**, 249–52.

Force, E.R. (1984) A relation among geomagnetic reversals, sea floor spreading rate, paleoclimate, and black shales. *EOS Trans. Amer. Geophys. Un.* **65**, 18–19.

Forsyth, D.W. (1975) The early structural evolution and anisotropy of the oceanic upper mantle. *Geophys. J. Roy. astr. Soc.* **43**, 103–62.

Forsyth, D.W. (1977) The evolution of the upper mantle beneath mid-ocean ridges. *Tectonophysics* **38**, 89–118.

Forsyth, D.W. (1992) Finite extension and low-angle normal faulting. *Geology* **20**, 27–30.

Forsyth, D.W. & Uyeda, S. (1975) On the relative importance of the driving forces of plate motion. *Geophys. J. Roy. astr. Soc.* **43**, 163–200.

Foster, A.N. & Jackson, J.A. (1998) Source parameters of large African earthquakes: implications for crustal rheology and regional kinematics. *Geophys. J. Int.* **134**, 422–48.

Foster, A.N. *et al.* (1997) Tectonic development of the northern Tanzanian sector of the East African Rift system. *J. geol. Soc. Lond.* **154**, 689–700.

Foster, D.A. & Gray, D.R. (2000) Evolution and structure of the Lachlan fold belt (orogen) of Eastern Australia. *Ann. Rev. Earth planet. Sci.* **28**, 47–80.

Foster, D.A., Gray, D.R. & Bucher, M. (1999) Chronology of deformation within the turbidite–dominated Lachlan orogen: implications for the tectonic evolution of eastern Australia and Gondwana. *Tectonics* **18**, 452–85.

Foulger, G.R. & Natland, J.H. (2003) Is "hotspot" volcanism a consequence of plate tectonics? *Science* **300**, 921–2.

Foulger, G.R. & Pearson, D.G. (2001) Is Iceland underlain by a plume in the lower mantle? Seismology and helium isotopes. *Geophys. J. Int.* **145**, F1–F5.

Fowler, C.M.R. (1976) Crustal structure of the Mid-Atlantic ridge crest at 37°N. *Geophys. J. Roy. astr. Soc.* **47**, 459–91.

Fowler, C.M.R. (2005) *The Solid Earth: an introduction to global geophysics*, 2nd edn. Cambridge University Press, Cambridge.

Fox, P.J. & Stroup, J.B. (1981) The plutonic foundation of the oceanic crust. *In* Emliani, C. (ed.) *The Oceanic Lithosphere, The Sea*, 7, pp. 119–218. Wiley, New York.

Fox, P.J. *et al.* (1976) The geology of the Oceanographer Fracture Zone: a model for fracture zones. *J. geophys. Res.* **81**, 4117–28.

Frakes, L.A. (1979) *Climates throughout Geologic Time.* Elsevier, Amsterdam.

Francheteau, J. (1983) The oceanic crust. *Sci. Am.* **249**, 68–84.

Frank, F.C. (1968) Curvature of island arcs. *Nature* **220**, 363.

Frankel, H. (1988) From continental drift to plate tectonics. *Nature* **335**, 127–30.

Fryer, P., Wheat, C.G. & Mottl, M.J. (1999) Mariana blueschist mud volcanism: implications for conditions within the subduction zone. *Geology* **27**, 103–6.

Fuis, G.S. *et al.* (2001) Crustal structure and tectonics from the Los Angeles basin to the Mojave Desert, southern California. *Geology* **29**, 15–8.

Fuis, G.S. *et al.* (2003) Fault systems of the 1971 San Fernando and 1994 Northridge earthquakes, southern California: relocated aftershocks and seismic images from LARSE II. *Geology* **31**, 171–74.

Fujiwara, T., Yamazaki, T. & Joshima, M. (2001) Bathymetry and magnetic anomalies in the Havre Trough and southern Lau Basin: from rifting to spreading in back-arc basins. *Earth Planet. Sci. Lett.* **185**, 253–64.

Fukao, Y., Widiyantoro, S. & Obayashi, M. (2001) Stagnant slabs in the upper and lower mantle transition region. *Rev. Geophys.* **39**, 291–323.

Furman, T. *et al.* (2004) East Africa Rift System (EARS) plume structure: insights from Quaternary mafic lavas of Turkana, Kenya. *J. Petrol.* **45**, 1069–88.

Fyfe, W.S. & Londsdale, P. (1981) Ocean floor hydrothermal activity. *In* Emiliani, C. (ed.) *The Oceanic Lithosphere, The Sea* 7, pp. 589–638. Wiley, New York.

Gaál, G. & Schulz, K.J. (eds) (1992) Precambrian metallogeny related to plate tectonics. *Precambrian Res.* **58**, 1–446.

Gaetani, M. & Garzanti, E. (1991) Multicyclic history of the northern India continental margin (northwestern Himalaya). *Bull. Am. Assoc. Petroleum Geols.* **75**, 1247–46.

Gaina, C., Müller, R.D. & Cande, S.C. (2000) Absolute plate motion, mantle flow and volcanics at the boundary between the Pacific and Indian Ocean mantle domains since 90 Ma. *In* Richard, M., Gordon, R.G. & van der Hilst, R.O. (eds) *The History and Dynamics of Global Plate Motions. Geophys. Monogr. Ser.* **121**, pp. 189–210. American Geophysical Union, Washington, DC.

Galanis Jr, S.P. *et al.* (1986) Heat flow at Zerqa Ma'in and Zara and a geothermal reconnaissance of Jordan. *USGS Open-File Report* **86–631**, 110pp.

Gans, P.B. (1987) An open system, two-layer crustal stretching model for the Eastern Great Basin. *Tectonics* **6**, 1–12.

Gao, S. *et al.* (1998) Chemical composition of the continental crust as revealed by studies in east China. *Geochim. cosmochim. Acta* **62**, 1959–75.

Gao, S.S., Silver, P.G. & Liu, K.H. (2002) Mantle discontinuities beneath southern Africa. *Geophys. Res. Lett.* **29**, 1491.

Gao, W. *et al.* (2004) Upper mantle convection beneath the central Rio Grande rift imaged by P and S wave tomography. *J. geophys. Res.* **109**, B03305, doi:10.1029/2003JB002743.

Garfunkel, Z., Zak, I. & Freund, R. (1981) Active faulting in the Dead Sea Rift. *Tectonophysics*, **80**, 1–26.

Garland, G.D. (1979) *Introduction to Geophysics*, 2nd edn. W.B. Saunders, Philadelphia, PA.

Garnero, E.J. (2000) Heterogeneity of the lowermost mantle. *Ann. Rev. Earth planet. Sci.* **28**, 509–37.

Garnero, E.J. (2004) A new paradigm for the Earth's core–mantle boundary. *Science* **304**, 834–6.

Garnero, E.J. *et al.* (1998) Ultralow velocity zone at the core–mantle boundary. *Geodyn. Ser.* **28**, 319–34.

Gass, I.G. (1980) The Troodos massif; its role in the unravelling of the ophiolite problem and its significance in the understanding of constructive plate margin processes. *In* Panayistou, A. (ed.) *Ophiolites*, pp. 23–35. Geol. Surv. Cyprus.

Gehrels, G. (2002) Detrital zircon geochronology of the Taku terrane, southeast Alaska. *Can. J. Earth Sci.* **39**, 921–31.

Gente, P. *et al.* (1995) Characteristics and evolution of the segmentation of the Mid-Atlantic Ridge between 20°N and 24°N during the last 10 million years. *Earth planet. Sci. Lett.* **129**, 55–71.

Gerbault, M., Davey, F. & Henrys, S. (2002) Three-dimensional lateral crustal thickening in continental oblique collision: an example from the Southern Alps, New Zealand. *Geophys. J. Int.* **150**, 770–9.

Gerbault M., Martinod, J. & Hérail, G. (2005) Possible orogeny-parallel lower crustal flow and thickening in the Central Andes. *Tectonophysics* **399**, 59–72.

Gerbi, C., Johnson, S.E. & Paterson, S.R. (2002) Implications of rapid, dike-fed pluton growth for host-rock strain rates and emplacement mechanisms. *J. struct. Geol.* **26**, 583–94.

Gilbert, H.J. & Sheehan, A.F. (2004) Images of crustal variations in the intermountain west. *J. Geophys. Res.* **109**, B03306, doi:10:1029/2003JB002730.

Gill, J.B. (1981) *Orogenic Andesites and Plate Tectonics.* Springer-Verlag, Berlin.

Ginzburg, A. *et al.* (1979a) A seismic study of the crust and upper mantle of the Jordan–Dead Sea Rift and their transition toward the Mediterranean Sea. *J. geophys. Res.* **84**, 1569–82.

Ginzburg, A. *et al.* (1979b) Detailed structure of the crust and upper mantle along the Jordan–Dead Sea Rift. *J. geophys. Res.* **84**, 5605–12.

Glatzmaier, G.A. & Roberts, P.H. (1995) A three-dimensional self-consistent computer simulation of a geomagnetic field reversal. *Nature* **377**, 203–9.

Glatzmaier, G.A. *et al.* (1999) The role of the Earth's mantle in controlling the frequency of geomagnetic reversals. *Nature* **401**, 885–90.

Glen, R.A. (2005) The Tasmanides of eastern Australia. *In* Vaughan, A.P.M., Leat, P.T. & Pankhurst, R.J. (eds) *Terrane Processes at the Margins of Gondwana. Spec. Pub. geol. Soc. Lond.* **246**, 23–96.

Godfrey, N.J. *et al.* (2002) Lower crustal deformation beneath the central Transverse Ranges, southern California. *J. geophys. Res.* **107**, doi:10.1029/2001JB000354.

Goes, S. & van der Lee, S. (2002) Thermal structure of the North American uppermost mantle inferred from seismic tomography. *J. geophys. Res.* **107**, 2050, doi:10.1029/2000JB000049.

Goldsworthy, M., Jackson, J. & Haines, J. (2002) The continuity of active fault systems in Greece. *Geophys. J. Int.* **148**, 596–618.

Gómez, E. *et al.* (2005) Development of the Colombian foreland-basin system as a consequence of diachronous exhumation of the Northern Andes. *Bull. geol. Soc. Am.* **117**, 1272–92.

Gordon, R.G. (1995) Present plate motions and plate boundaries. *In Global Earth Physics: A Handbook of Physical Constants. AGU Reference Shelf* 1, pp. 66–87. American Geophysical Union, Washington, DC.

Gordon, R.G. (1998) The plate tectonic approximation: plate non-rigidity, diffuse plate boundaries, and global plate reconstructions. *Ann. Rev. Earth planet. Sci.* **26**, 615–42.

Gordon, R.G. (2000) Diffuse oceanic plate boundaries: strain rates, vertically averaged rheology, and comparisons with narrow plate boundaries and stable plate interiors. *In* Richards, M.A., Gordon, R.G. & van der Hilst, R.D. (eds) *The History and Dynamics of Plate Motions. Geophys. Monogr. Ser.* **121**, pp. 143–59. American Geophysical Union, Washington, DC.

Gordon, R.G. & Stein, S. (1992) Global tectonics and space geodesy. *Science* **256**, 333–42.

Gradstein, F.M., Ogg, J.G. & Smith, A.G. (eds) (2004) *A Geologic Time Scale 2004.* Cambridge University Press, Cambridge, 610pp.

Green, H.W. (1994) Solving the paradox of deep earthquakes. *Sci. Am.* **271**, 50–7.

Green, W.V., Achauer, U. & Meyer, R.P. (1991) A three dimensional seismic image of the crust and upper mantle beneath the Kenya rift. *Nature* **354**, 199–203.

Griffin, W.L. *et al.* (2004) Archean crustal evolution in the northern Yilgarn Craton: U-Pb and Hf-isotope evidence from detrital zircons. *Precambrian Res.* **131**, 231–82.

Gripp, A.E. & Gordon, R.G. (2002) Young tracks of hotspots and current plate velocities. *Geophys. J. Int.* **150**, 321–61.

Grove, T.L. & Parman, S.W. (2004) Thermal evolution of the Earth as recorded by komatiites. *Earth planet. Sci. Lett.* **219**, 173–87.

Groves, D.I. *et al.* (2003) Gold deposits in metamorphic belts: overview of current understanding, outstanding problems, future research, and exploration significance. *Econ. Geol.* **98**, 1–29.

Grow, J.A. (1973) Crustal and upper mantle structure of the central Aleutian arc. *Bull. geol. Soc. Am.* **84**, 2169–92.

Guillot, S. *et al.* (1997) Eclogitic metasediments from the Tso Morari area (Ladakh, Himalaya): evidence for continental subduction during India–Asia convergence. *Contrib. Mineral. Petrol.* **128**, 197–212

Guillou, L. & Jaupart, C. (1995) On the effect of continents on mantle convection. *J. geophys. Res.* **100**, 24217–38.

Gulick, S.P.S. *et al.* (2004) Three-dimensional architecture of the Nankai accretionary prism's imbricate thrust zone off Cape Muroto, Japan: prism reconstruction via en echelon thrust propagation. *J. Geophys. Res.* **109**, B02105, doi:10.1029/2003JB002654.

Gurnis, M. (1988) Large-scale mantle convection and the aggregation and dispersal of supercontinents. *Nature* **332**, 695–9.

Gurnis, M. (2001) Sculpting the Earth from inside out. *Sci. Am.* **284**, 40–47.

Gurnis, M., Müller, R.D. & Moresi, L. (1998) Cretaceous vertical motion of Australia and the Australian–Antarctic discordance. *Science* **279**, 1499–1504.

Gutscher, M.-A. *et al.* (2000) Geodynamics of flat subduction: seismicity and tomographic constraints from the Andean margin. *Tectonics* **19**, 814–33.

Hacker, B.R., Ratschbacher, L. & Liou, J.G. (2004) Subduction, collision and exhumation in the ultrahigh-pressure Qinling–Dabie Orogen. *In* Malpas, J. *et al.* (eds) *Aspects of the Tectonic Evolution of China. Spec. Pub. geol. Soc. Lond.* **226**, 157–75.

Hackney, R. (2004) Gravity anomalies, crustal structure and isostasy associated with the Proterozoic Capricorn Orogen, Western Australia. *Precambrian Res.* **128**, 219–36.

Hager, B.H. *et al.* (1985) Lower mantle heterogeneity, dynamic topography and the geoid. *Nature* **313**, 541–5.

Haines, S.S. *et al.* (2003) INDEPTH III seismic data: from surface observations to deep crustal processes in Tibet. *Tectonics* **22**, 1001, doi:10.1029/2001TC001305.

Halbach, P. *et al.* (1989) Probable modern analogue of Kuroko-type massive sulphide deposits in the Okinawa Trough back-arc basin. *Nature* **338**, 496–9.

Hall, A.L. (1932) The Bushveld igneous complex of the Central Transvaal. *Mem. Geol. Surv. S. Afr.* **28**, 544pp.

Hall, J., Marillier, F. & Dehler, S. (1998) Geophysical studies of the structure of the Appalachian orogen in the Atlantic borderlands of Canada. *Can. J. Earth Sci.* **35**, 1205–221.

Hall, J.K. (1993) The GSI digital terrain model (DTM) project completed. *Curr. Res. Geol. Surv. Isr.* **8**, 47–50.

Hall, P.S. & Kincaid, C. (2001) Diapiric flow at subduction zones: a recipe for rapid transport. *Science* **292**, 2472–5.

Hall, R. (2002) Cenozoic geological and plate tectonic evolution of SE Asia and the SW Pacific: computer-based reconstructions, model and animations. *J. Asian Earth Sci.* **20**, 353–431.

Hall, R. & Wilson, M.E.J. (2000) Neogene sutures in eastern Indonesia. *J. Asian Earth Sci.* **18**, 781–808.

Hallam, A. (1972) Continental drift and the fossil record. *Sci. Am.* **227**, 56–66.

Hallam, A. (1973a) *A Revolution in the Earth Sciences.* Clarendon Press, Oxford.

Hallam, A. (1973b) Provinciality, diversity and extinction of Mesozoic marine invertebrates in relation to plate movements. *In* Tarling, D.H. & Runcorn, S.C. (eds) *Implications of Continental Drift to the Earth Sciences*, **1**, pp. 287–94. Academic Press, London.

Hallam, A. (1975) Alfred Wegener and the hypothesis of continental drift. *Sci. Am.* **232**, 88–97.

Hallam, A. (1981) Relative importance of plate movements, eustasy, and climate in controlling major biogeographical changes since the early Mesozoic. *In* Nelson, G. & Rosen, D.E. (eds) *Vicariance Biogeography, a critique*, pp. 303–40. Columbia University Press, New York.

Hammer, P.T.C. & Clowes, R.M. (2004) Accreted terranes of northwestern British Columbia, Canada: lithospheric velocity structure and tectonics. *J. geophys. Res.* **109**, B06305, doi:10.1029/2003JB002749.

Hammer, P.T.C. & Clowes, R.M. (2007) Lithospheric-scale structures across the Alaskan and Canadian Cordillera: comparisons and tectonic implications. *In* Sears, J., Harms, T. & Evenchick, C. (eds) *Whence the Mountains? Inquiries into the Evolution of Orogenic Systems: A volume in honor of Raymond A. Price. Geol. Soc. Am. Sp. Paper* **433**, 99–116.

Hammer, P.T.C., Clowes, R.M. & Ellis, R.M. (2000) Crustal structure of NW British Columbia and SE Alaska from seismic wide-angle studies: Coast Plutonic Complex to Stikinia. *J. geophys. Res.* **105**, 7961–81.

Hammond, W.C. & Thatcher, W. (2004) Contemporary tectonic deformation of the Basin and Range province, western United States: 10 years of observation with the Global Positioning System. *J.geophys. Res.* **109**, B08403, doi:10.1029/2003JB002746.

Handy, M.R. & Brun, J.-P. (2004) Seismicity, structure and strength of the continental lithosphere. *Earth planet. Sci. Lett.* **223**, 427–41.

Hanson, R.E. *et al.* (2004) Coeval large-scale magmatism in the Kalahari and Laurentian cratons during Rodinia assembly. *Nature* **304**, 1126–9.

Haq, B.U. (1989) Paleoceanography: a synoptic overview of 200 million years of ocean history. *In* Haq, B.U. & Millman, J.D. (eds) *MarineGgeology and Oceanography of Arabian Sea and Coastal Pakistan*, pp. 201–31. Van Nostrand Reinhold, New York.

Hardebeck, J.L. & Michael, A.J. (2004) Stress orientations at intermediate angles to the San Andreas Fault, California. *J. geophys. Res.* **109**, B11303, doi:10.1029/2004JB003239.

Hardie, L.A. (1996) Secular variation in sea water chemistry: an explanation for the coupled secular variation in the mineralogy of marine limestones and potash evaporites over the past 600 Ma. *Geology* **24**, 279–83.

Harding, T.P. (1974) Petroleum traps associated with wrench faults. *Bull. Am. Assoc. Petroleum Geols.* **58**, 1290–304.

Harding, T.P. (1985) Seismic characteristics and identification of negative flower structures, positive flower structures and positive structural inversion. *Bull. Am. Assoc. Petroleum Geols.* **69**, 582–600.

Hargraves, R.B. (1986) Faster spreading or greater ridge length in the Archean? *Geology* **14**, 750–2.

Harlan, S.S. *et al.* (2003) Gunbarrel mafic magmatic event: a key 780-Ma time marker for Rodinia plate reconstructions. *Geology* **31**, 1053–6.

Harley, S.L. (1989) The origin of granulites: a metamorphic perspective. *Geol. Mag.* **126**, 215–47.

Harley, S.L. (2004) Extending our understanding of ultrahigh temperature crustal metamorphism. *J. Mineral. Petrol. Sci.* **99**, 140–58.

Harper, J.F. (1978) Asthenosphere flow and plate motions. *Geophys. J. Roy. astr. Soc.* **55**, 87–110.

Harris, R.A. *et al.* (2000) Thermal history of Australian passive margin cover sequences accreted to Timor during Late Neogene arc–continent collision, Indonesia. *J. Asian Earth Sci.* **18**, 47–69.

Harrison, C.G.A. & Bonatti, E. (1981) The oceanic lithosphere. *In* Emiliani. C. (ed.) *The Oceanic Lithosphere. The Sea* **7**, pp. 21–48. Wiley, New York.

Harrison, C.G.A. & Sclater, J.G. (1972) Origin of the disturbed magnetic zone between the Murray and Molokai fracture zones. *Earth planet. Sci. Lett.* **14**, 419–27.

Harrison, T.M., McKeegan, K.D. & Le Fort, P. (1995) Detection of inherited monazite in the Manaslu leucogranite by $^{208}Pb/^{232}Th$ ion microprobe dating: crystallization age and tectonic significance. *Earth planet. Sci. Lett.* **133**, 271–82.

Harrison, T.M. *et al.* (2005) Heterogeneous Hadean Hafnium: evidence of continental crust at 4.4 to 4.5 Ga. *Science* **310**, 1947–50.

Harrison, T.M. *et al.* (2006) Response to comment on "Heterogeneous Hadean Hafnium: evidence of continental crust at 4.4 to 4.5 Ga". *Science* **312**, 1139b.

Hartog, R. & Schwartz, S.Y. (2001) Depth-dependent mantle anisotropy below the San Andreas fault system: apparent splitting parameters and waveforms. *J. geophys. Res.* **106**, 4155–67.

Hasegawa, A., Umino, N. & Takagi, A. (1978) Doubleplaned seismic zone and upper-mantle structure in the northeastern Japan arc. *Geophys. J. Roy. astr. Soc.* **54**, 281–96.

Hasegawa, A., Horiuchi, S. & Umino, N. (1994) Seismic structure of the northeastern Japan convergent margin: a synthesis. *J. geophys. Res.* **99**, 22295–311.

Hastings, D.A. & Dunbar, P.K. (1998) Development and assessment of the Global Land One-km base digital elevation model (GLOBE), *ISPRS Arch.* **32**, pp. 218–21. *Int. Soc. Photogramm. Remote Sens.* (ISPRS), Stuttgart, Germany.

Hauck, M.L. *et al.* (1998) Crustal structure of the Himalayan orogen at ~90° east longitude from Project INDEPTH deep reflection profiles. *Tectonics* **17**, 481–500.

Haug, G.H. & Tiederman, R. (1998) Effect of the formation of the Isthmus of Panama on Atlantic Ocean thermohaline circulation. *Nature* **393**, 673–6.

Hauksson, E. (1987) Seismotectonics of the Newport–Inglewood fault zone in the Los Angeles basin, southern California. *Bull. seis. Soc. Am.* **77**, 539–61.

Hauksson, E. (1994) The 1991 Sierra Madre earthquake sequence in Southern California: seismological and tectonic analysis. *Bull. seis. Soc. Am.* **84**, 1058–74.

Hauksson, E. *et al.* (1988) The 1987 Whittier Narrows earthquake in the Los Angeles metropolitan area, California. *Science* **239**, 1409–12.

Hauksson, E., Jones, L.M. & Hutton, K. (1995) The 1994 Northridge earthquake sequence in California: seismological and tectonic aspects. *J. geophys. Res.* **100**, 12335–55.

Hauser, E. *et al.* (1987) Crustal structure of eastern Nevada from COCORP deep seismic reflection data. *Bull. geol. Soc. Am.* **99**, 833–44.

Heaman, L.M., Kjarsgaard, B.A. & Creaser, R.A. (2003) The timing of kimberlite magmatism in North America: implications for global kimberlite genesis and diamond exploration. *Lithos* **71**, 153–84.

Heaton, T.H. (1982) The 1971 San Fernando earthquake: a double event? *Bull. seis. Soc. Am.* **72**, 2037–62.

Hedenquist, J.W. & Lowenstern, J.B. (1994) The role of magmas in the formation of hydrothermal ore deposits. *Nature* **370**, 519–27.

Heirtzler, J.R. *et al.* (1968) Marine magnetic anomalies, geomagnetic field reversals and motions of the ocean floor and continents. *J. geophys. Res.* **73**, 2119–36.

Heirtzler, J.R., Le Pichon, X. & Baron, J.G. (1966) Magnetic anomalies over the Reykjanes Ridge. *Deep Sea Res.* **13**, 427–43.

Helffrich, G.R. & Wood, B.J. (2001) The Earth's mantle. *Nature* **412**, 501–7.

Hendrie, D. *et al.* (1994) A quantitative model of rift basin development in the northern Kenya Rift: evidence for the Turkana region as an "accommodation zone" during the Palaeogene. *Tectonophysics* **236**, 409–38.

Henrys, S.A. *et al.* (2004) Mapping the Moho beneath the Southern Alps continent–continent collision, New Zealand, using wide-angle reflections. *Geophys. Res. Lett.* **31**, L17602, doi:10.1029/2004GL020561.

Henstock, T.J. & Levander, A. (2000) Lithospheric evolution in the wake of the Mendocino triple junction: structure of the San Andreas Fault system at 2 Ma. *Geophys. J. Int.* **140**, 233–47.

Herring, T.A. *et al.* (1986) Geodesy by radio interferometry: evidence for contemporary plate motion. *J. geophys. Res.* **91**, 8341–7.

Herrington, R.J., Evans, D.M. & Buchanan, D.L. (1997) Metallogenic aspects. *In* De Wit, M.J. & Ashwal, L.D. (eds) *Greenstone Belts*, pp. 177–219. Clarendon Press, Oxford.

Herron, E.M. (1972) Sea-floor spreading and the Cenozoic history of the East–Central Pacific *Bull. geol. Soc. Am.* **83**, 1671–92.

Herron, E.M., Cande, S.C. & Hall, B.R. (1981) An active spreading center collides with a subduction zone, a geophysical survey of the Chile margin triple junction. *Geol. Soc. Am. Mem.* **154**, 683–701.

Herron, T.J., Stoffa, P.L. & Buhl, P. (1980) Magma chamber and mantle reflections – East Pacific Rise. *Geophys. Res. Lett.* **7**, 989–92.

Hervé, F., Fanning, C.M. & Pankhurst, R.J. (2003) Detrital zircon age patterns and provenance of the metamorphic complexes of southern Chile. *J. S. Am. Earth Sci.* **16**, 107–23.

Herzberg, C. (1999) Formation of cratonic mantle as plume residues and cumulates. *In* Fei, Y., Bertka, C. & Mysen, B.O. (eds) *Mantle Petrology: field observations and high pressure experimentation*, pp. 241–57, Geochemical Society, Houston, TX.

Hess, H.H. (1962) History of ocean basins. *In Petrologic Studies: a volume in honor of A.F. Buddington*, pp. 599–620. Geological Society of America, New York.

Hetland, E.A. & Hager, B.H. (2004) Relationship of geodetic velocities to velocities in the mantle. *Geophys. Res. Lett.* **31**, L17604, doi:10.1029/2004GL020691.

Hey, R.N. (1977) A new class of pseudofaults and their bearing on plate tectonics: a propagating rift model. *Earth planet. Sci. Lett.* **37**, 321–5.

Hey, R.N., Dunnebier, F.K., & Morgan, W.J. (1980) Propagating rifts on mid-ocean ridges. *J. geophys. Res.* **85**, 3647–58.

Hey, R.N. *et al.* (1986) Sea Beam/Deep-Tow investigation of an active propagating rift system, Galapagos 95.5°W. *J. geophys. Res.* **91**, 3369–94.

Hey, R.N. *et al.* (1988) Changes in direction of seafloor spreading revisited. *J. geophys. Res.* **93**, 2803–11.

Hickman, A.H. (2004) Two contrasting granite-greenstone terranes in the Pilbara craton, Australia: evidence for vertical and horizontal tectonic regimes prior to 2900 Ma. *Precambrian Res.* **131**, 153–72.

Hickman, S. & Zoback, M. (2004) Stress orientations and magnitudes in the SAFOD pilot hole. *J. geophys. Res.* **31**, L15S12, doi:10.1029/2004GL020043.

Hickman, S., Zoback, M. & Ellsworth, W. (2004) Introduction to special section: preparing for the San Andreas Fault Observatory at Depth. *Geophys. Res. Lett.* **31**, L12S01, doi:10.1029/2004GL020688.

Hill, R.I. (1991) Starting plumes and continental breakup. *Earth planet Sci. Lett.* **104**, 398–416.

Hilton, D.R. & Craig, H. (1989) A helium isotope transect along the Indonesian archipelago. *Nature* **342**, 906–8.

Hindle, D. *et al.* (2002) Consistency of geologic and geodetic displacements during Andean orogenesis. *Geophys. Res. Lett.* **29**, doi:10.1029/2001GL013757.

Hirth, G. & Kohlstedt, D. (2003) Rheology of the uper mantle and the mantle wedge: a view from experimentalists. *In* Eiler, J. (ed.) *Inside the Subduction Factory. Geophys. Monogr. Ser.* **138**, pp. 83–106. American Geophysical Union, Washington, DC.

Hodges, K.V. (2000) Tectonics of the Himalaya and southern Tibet from two perspectives. *Bull. geol. Soc. Am.* **112**, 324–50.

Hodges, K.V., Parrish, R.R. & Searle, M.P. (1996) Tectonic evolution of the central Annapurna Range, Nepalese Himalayas. *Tectonics* **15**, 1264–91.

Hodges, K.V., Hurtado, J.M. & Whipple, K.X. (2001) Southward extrusion of Tibetan crust and its effect on Himalayan tectonics. *Tectonics* **20**, 799–809.

Hoffman, P.F. (1991) Did the breakout of Laurentia turn Gondwanaland inside out? *Science* **252**, 1409–12.

Hoffman, P.F. (1997) Tectonic genealogy of North America. *In* Van der Pluijm, B.A. & Marshak, S. (eds) *Earth Structure and Introduction to Structural Geology and Tectonics*, pp. 459–64. McGraw Hill, New York.

Hoffman, P.F. & Schrag, D.P. (2002) The snowball Earth hypothesis: testing the limits of global change. *Terra Nova* **14**, 129–55.

Hoffman, P.F. et al. (1998) A Neoproterozoic snowball Earth. *Science* **281**, 1342–6.

Hofmann, A.W. (1997) Mantle geochemistry: the message from oceanic volcanism. *Nature* **385**, 219–29.

Hofmann, C. *et al.* (1997) Timing of the Ethiopian flood basalt event and implications for plume birth and global change. *Nature* **389**, 838–41.

Hofmann, C., Feraud, G. & Courtillot, V. (2000) ^{40}Ar/^{39}Ar dating of mineral separates and whole rocks from the Western Gnats lava pile: further constraints on the duration and age of the Deccan Traps. *Earth planet. Sci. Lett.* **180**, 13–28.

Hofstetter, R. & Beth, M. (2003) The Afar Depression: interpretation of the 1960–2000 earthquakes. *Geophys. J. Int.* **155**, 715–32.

Hogrefe, A. et al. (1994) Metastability of estatite in deep subducting lithosphere. *Nature* **372**, 351–3.

Holbrook, W.S. et al. (1996) Crustal structure of a transform boundary: San Francisco Bay and the central California continental margin. *J. geophys. Res.* **101**, 22311–34.

Hole, J.A. *et al.* (2000) Three-dimensional seismic velocity structure of the San Francisco Bay area. *J. geophys. Res.* **105**, 13859–74.

Hollister, L.S. & Andronicos, C.L. (1997) A candidate for the Baja British Columbia fault system in the Coast Plutonic Complex. *GSA Today* **7**, 1–7.

Holmes, A. (1928) Radioactivity and Earth movements. *Trans. Geol. Soc. Glasgow* **18**, 559–606.

Hopper, J.R. & Buck, W.R. (1996) The effect of lower crustal flow on continental extension and passive margin formation. *J. geophys. Res.* **101**, 20175–94.

Hopper, J.R. *et al.* (2003) Structure of the SE Greenland margin from seismic reflection and refraction data: implications for nascent spreading center subsidence and asymmetric crustal accretion during North Atlantic opening. *J. geophys. Res.* **108**, 2269, doi:10.1029/2002JB001996.

Hopper, J.R. *et al.* (2004) Continental breakup and the onset of ultraslow seafloor spreading off Flemish Cap on the Newfoundland rifted margin. *Geology* **32**, 93–6.

Housen, B.A. & Beck, M.E. (1999) Testing terrane transport: an inclusive approach to the Baja B.C. controversy. *Geology* **27**, 1143–46.

Howell, D.G. (1989) *Tectonics of Suspect Terranes: mountain building and continental growth.* Chapman & Hall, London.

Hsui, A.T. & Toksöz, M.N. (1981) Back-arc spreading: trench migration, continental pull or induced convection? *Tectonophysics* **74**, 89–98.

Huang, C.-Y. *et al.* (2000) Geodynamic processes of Taiwan arc–continent collision and comparison with analogs in Timor, Papua New Guinea, Urals and Corsica. *Tectonophysics* **325**, 1–21.

Huang, C.-Y., Yuan, P.B. & Tsao, S.-J. (2006) Temporal and spatial records of active arc–continent collision in Taiwan: a synthesis. *Bull. geol. Soc. Am.* **118**, 274–88.

Huang, W. *et al.* (2000) Seismic polarization anisotropy beneath the central Tibetan Plateau. *J. geophys. Res.* **105**, 27979–89.

Hughes, B.D. *et al.* (1996) Detailed processing of seismic reflection data from the frontal part of the Timor trough accretionary wedge, eastern Indonesia. *In* Hall, R., & Blundell, D. (eds) *Tectonic Evolution of Southeast Asia. Spec. Pub. geol. Soc. Lond.* **106**, 75–83.

Huismans, R.S. & Beaumont, C. (2003) Symmetric and asymmetric lithospheric extension: relative effects of frictional-plastic and viscous strain softening. *J. geophys. Res.* **108**, 2496, doi:10.1029/2002JB002026.

Huismans, R.S. & Beaumont, C. (2007) Roles of lithospheric strain softening and heterogeneity in determining the geometry of rifts and continental margins. *In* Karner, G.D., Manatschal, G. & Pinhiero, L.M. (eds) *Imaging, Mapping and Modelling Continental Lithosphere Extension and Breakup. Spec. Pub. geol. Soc. Lond.* **282**, 107–34.

Huismans, R.S., Podladchikov, Y.Y. & Cloetingh, S. (2001) Transition from active to passive rifting: relative importance of asthenospheric doming and passive extension of the lithosphere. *J. geophys. Res.* **106**, 11271–91.

Hurley, P.M. (1968) The confirmation of continental drift. *Sci. Am.* **218**(4), 52–64.

Hutchinson, C.S. (1983) *Economic Deposits and their Tectonic Setting.* MacMillan Press, London.

Hutchison, W.W. (1982) Geology of the Prince Rupert–Skeena map area, British Columbia. *Mem. Geol. Surv. Can.* **394**, 1–116.

Hyde, W.T. et al. (2000) Neoproterozoic "snowball Earth" simulations with a coupled climate/ice-sheet model. *Nature* **405**, 425–9.

Hyndman, R.D., Currie, C.A. & Mazzotti, S.P. (2005) Subduction zone backarcs, mobile belts, and orogenic heat. *GSA Today* **15**, doi:10.1130/1052-5173(2005)15<4:SZBMBA>2.0CO;2.

Hyndman, R.D. et al. (1990) The northern Cascadia subduction zone at Vancouver Island: seismic structure and tectonic history. *Can. J. Earth Sci* **27**, 313–29.

Ibs-von Seht, M. et al. (2001) Seismicity, seismotectonics and crustal structure of the southern Kenya Rift – new data from the Lake Magadi area. *Geophys. J. Int.* **146**, 439–53.

Irving, E., Emslie, R.F. & Ueno, H. (1974) Upper Proterozoic paleomagnetic poles from Laurentia and the history of the Grenville structural province. *J. Geophys. Res.* **79**, 5491–502.

Irving, E., North, F. & Couillard, R. (1974) Oil, climate and tectonics. *Can. J. Earth Sci.* **11**, 1–17.

Irving, E. et al. (1996) Large (1000–4000 km) northward movements of tectonic domains in the northern Cordillera, 83 to 45 Ma. *J. geophys. Res.* **101**, 17901–16.

Isacks, B. & Molnar, P. (1969) Mantle earthquake mechanisms and the sinking of the lithosphere. *Nature* **223**, 1121–4.

Isacks, B. & Molnar, P. (1971) Distribution of stresses in the descending lithosphere from a global survey of focal mechanism solutions of mantle earthquakes. *Rev. Geophys. Space Phys.* **9**, 103–74.

Isacks, B., Oliver, J. & Sykes, L.R. (1968) Seismology and the new global tectonics. *J. geophys. Res.* **73**, 5855–99.

Isacks, B., Sykes, L.R. & Oliver, J. (1969) Focal mechanisms of deep and shallow earthquakes in the Tonga–Kermadec region and tectonics of island arcs. *Bull. geol. Soc. Am.* **80**, 1443–69.

Isacks, B.L. (1988) Uplift of the central Andean plateau and bending of the Bolivian orocline. *J. geophys. Res.* **93**, 3211–31.

Isacks, B.L. & Barazangi, M. (1977) Geometry of Benioff zones: lateral segmentation and downward bending of the subducted lithosphere. *In* Talwani, M. & Pitman, W.C. III (eds) *Island Arcs, Deep Sea Trenches and Backarc Basins*, pp. 99–114. American Geophysical Union, Washington, DC.

Isley, A.E. & Abbott, D.H. (2002) Implications of the temporal distribution of high-Mg magmas for mantle plume volcanism through time, *J. Geol.* **110**, 141–58.

Ito, E. & Sato, H. (1991) Aseismicity in the lower mantle by superplasticity of the descending slab. *Nature* **351**, 140–1.

Jackson, H.R. (2002) Seismic refraction profiles in the Gulf of Saint Lawrence and implications for extent of continuous Grenville lower crust. *Can. J. Earth Sci.* **39**, 1–17.

Jackson, J. (2002) Strength of continental lithosphere: time to abandon the jelly sandwich? *GSA Today* **12**, 4–10.

Jackson, J. (2004) Velocity fields, faulting, and strength on the continents. *In* Karner, G.D. et al. (eds) *Rheology and Deformation of the Lithosphere at Continental Margins*, pp. 31–45. Columbia University Press, New York.

Jackson, J.A., Haines, A. & Holt, W. (1992) The horizontal velocity field in the deforming Aegean Sea region determined from the moment tensors of earthquakes. *J. geophys. Res.* **97**, 17657–84.

Jackson, J.A. et al. (2004) Metastability, mechanical strength and the support of mountain belts. *Geology*, **32**, 625–8.

Jacob, K.H., Nakamura, K. & Davies, J.N. (1977) Trench–volcano gap along the Alaska–Aleutian arc: facts and speculations on the role of terrigenous sediments for subduction. *In* Talwani, M. & Pitman, W.C. III (eds) *Island Arcs, Deep Sea Trenches and Back-arc Basins. Maurice Ewing Series* I, pp. 243–58. American Geophysical Union, Washington, DC.

Jacobs, J.A. (1991) *The Deep Interior of the Earth.* Chapman & Hall, London.

Jacobs, J.A. (1994) *Reversals of the Earth's Magnetic Field.* Cambridge University Press, Cambridge.

James, D.E. & Fouch, M.J. (2002) Formation and evolution of Archaean cratons: insights from southern Africa. *In* Fowler, C.M.R., Ebinger, C.J. & Hawkesworth, C.J. (eds) *The Early Earth: physical, chemical and biological development. Spec. Pub. geol. Soc. Lond.* **199**, 1–26.

James, D.E. & Snoke, J.A. (1990) Seismic evidence for continuity of the deep slab beneath central and eastern Peru. *J. geophys. Res.* **95**, 4789–5001.

James, D.E. et al. (2001) Tectospheric structure beneath southern Africa. *Geophys. Res. Lett.* **28**, 2485–8.

James, E.W., Kimbrough, D.L. & Mattinson, J.M. (1993) Evaluation of displacements of pre-Tertiary rocks on the northern San Andreas fault using U-Pb zircon dating, initial SR, and common Pb isotopic ratios. *In* Powell, R.E., Weldon, R.J.I. & Matti, J.C. (eds) *The San Andreas Fault System: displacement, palinspastic reconstruction, and geologic evolution. Geol. Soc. Am. Mem.* **178**, 257–71.

Jarrard, R.D. (1986) Relations among subduction parameters. *Rev. Geophys.* **24**, 217–84.

Jaupart, C. & Mareschal, J.C. (1999) The thermal structure and thickness of continental roots. *Lithos* **48**, 93–114.

Jellinek, A.M. & Manga, M. (2004) Links between long-lived hot spots, mantle plumes, D″, and plate tectonics. *Rev. Geophys.* **42**, 3002.

Jennings, C.W. (1994) Fault activity map of California and adjacent areas. *Geol. Data Map* **6**, Calif. Dep. of Conserv. Div. Mines and Geol., Sacramento, CA.

Johnson, M.C. & Plank, T. (1999) Dehydration and melting experiments constrain the fate of subducted sediments. *Geochem. Geophys. Geosyst.* **1**, doi:10.1029/1999GC000014.

Johnson, M.R.W. (2002) Shortening budgets and the role of continental subduction during the India–Asia collision. *Earth Sci. Rev.* **59**, 101–23.

Johnston, S.T. (2001) The great Alaskan terrane wreck: reconciliation of paleomagnetic and geologic data in the northern Cordillera. *Earth planet. Sci. Lett.* **193**, 259–72.

Jones, C.H. & Phinney, R.A. (1998) Seismic structure of the lithosphere from teleseismic converted arrivals observed at small arrays in the southern Sierra Nevada and vicinity, California. *J. geophys. Res.* **103**, 10065–90.

Jones, C.H. *et al.* (1992) Variations across and along a major continental rift: an interdisciplinary study of the Basin and Range Province, Western USA. *Tectonophysics* **213**, 57–96.

Jones, D.G., Silberling, N.J. & Hillhouse, J. (1977) Wrangellia – a displaced terrane in northwestern North America. *Can. J. Earth Sci.* **14**, 2565–77.

Jones, D.L. *et al.* (1983) Recognition, character and analysis of tectonostratigraphic terranes in western North America. *In* Hashimoto, M. & Uyeda, S. (eds) *Accretion Tectonics in the circum-Pacific Regions*, pp. 21–35. Terra Scientific, Tokyo.

Jones, E.J.W. (1999) *Marine Geophysic.* Wiley, Chichester, England.

Jordan, T.A. & Watts, A.B. (2005) Gravity anomalies, flexure and the elastic thickness structure of the India–Eurasia collisional system. *Earth planet. Sci. Lett.* **236**, 732–50.

Jordan, T.E. *et al.* (1983) Andean tectonics related to geometry of subducted Nazca plate. *Bull. geol. Soc. Am.* **94**, 341–61.

Jordan, T.E. *et al.* (2001) Extension and basin formation in the southern Andes caused by increased convergence rate: a mid-Cenozoic trigger for the Andes. *Tectonics* **20**, 308–24.

Jurdy, D.M. & Stefanick, M. (1983) Flow models for back-arc spreading. *Tectonophysics* **99**, 191–206.

Kamp, P.J.J., Green, P.F. & Tippett, J.M. (1992) Tectonic architecture of the mountain front–foreland basin transition, South Island, New Zealand, assessed by fission track analysis. *Tectonics* **11**, 98–113.

Kanasewich, E.R. (1976) Plate tectonics and planetary convection. *Can. J. Earth Sci.* **13**, 331–40.

Karabinos, P. *et al.* (1998) Taconic orogeny in the New England Appalachians: collision between Laurentia and the Shelburne Falls arc. *Geology* **26**, 215–18.

Kárason, H. & van der Hilst, R.D. (2000) Constraints on mantle convection from seismic tomography. *In* Richards, M.A., Gordon, R. & van der Hilst, R.D. (eds) *The History and Dynamics of Global Plate Motion*. Geophys. Monogr. Ser. **121**, pp. 277–88. American Geophysical Union, Washington, DC.

Karato, S.-I. (1998) Seismic anisotropy in the deep mantle, boundary layers and the geometry of mantle convection. *Pure Appl. Geophys.* **151**, 565–87.

Karato, S.-I. & Wu, P. (1993) Rheology of the upper mantle: a synthesis. *Science* **260**, 771–8.

Karato, S.-I., Zhang, S. & Wenk, H.-R. (1995) Superplasticity in the Earth's lower mantle: evidence from seismic anisotropy and rock physics. *Science* **270**, 458–61.

Karig, D.E. (1970) Ridges and basins of the Tonga–Kermadec island arc system. *J. geophys. Res.* **75**, 239–54.

Karig, D.E. & Kay, R.W. (1981) Fate of sediments on the descending plate at convergent margins. *Phil. Trans. Roy. Soc. Lond. A* **301**, 233–51.

Karl, D.M. *et al.* (1988) Loihi Seamount, Hawaii: a mid-plate volcano with a distinctive hydrothermal system. *Nature* **335**, 532–5.

Karlstrom, K.E. & Williams, M.L. (1998) Heterogeneity of the middle crust: implications for strength of continental lithosphere. *Geology* **26**, 815–18.

Karlstrom, K.E. *et al.* (2001) Long-lived (1.8–1.0 Ga) convergent orogen in southern Laurentia, its extensions to Australia and Baltica, and implications for refining Rodinia. *Precambrian Res.* **111**, 5–30.

Karlstrom, K.E. *et al.* (2002) Structure and evolution of the lithosphere beneath the Rocky Mountains: initial results from the CD-ROM Experiment. *GSA Today* **12**, 4–10.

Karson, J.A. (2002) Geologic structure of the uppermost oceanic crust created at fast- to intermediate-rate spreading centers. *Annu. Rev. Earth planet. Sci.* **30**, 347–84.

Karson, J.A. *et al.* (1987) Along-axis variations in seafloor spreading in the MARK area. *Nature* **328**, 681–5.

Karson, J.A. *et al.* (2002) Structure of Uppermost fast-spread oceanic crust exposed at the Hess Deep Rift: implications for subaxial processes at the East Pacific Rise. *Geochem. Geophys. Geosyst.* **3**, doi: 10.1029/2001GC000155.

Karson, J.A. *et al.* (2006) Detachment shear zone of the Atlantis Massif core complex, Mid-Atlantic Ridge, 30°N. *Geochem. Geophys. Geosyst.* **7**, Q06016, doi:10.1029/2005GC001109.

Kaula, W.M. (1975) Absolute plate motions by boundary velocity minimizations. *J. geophys. Res.* **80**, 244–8.

Kay, R, Hubbard, N.J. & Gast, P.W. (1970) Chemical characteristics and origin of oceanic ridge volcanic rocks. *J. geophys. Res.* **75**, 1585–613.

Kearey, P., Brooks, M. & Hill, I. (2002) *Introduction to Geophysical Exploration*, 3rd edn. Blackwell Publications, Oxford.

Keen, C. & Tramontini, C. (1970) A seismic refraction survey on the Mid-Atlantic ridge. *Geophys. J. Roy. astr. Soc.* **20**, 473–91.

Keir, D. *et al.* (2006) Strain accommodation by magmatism and faulting as rifting proceeds to breakup: seismicity of the northern Ethiopian rift. *J. geophys. Res.* **111**, B05314, doi:10.1029/2005JB003748.

Kelemen, P.B., Yogodzinski, G.M. & Scholl, D.W. (2003) Along-strike variation in lavas of the Aleutian arc: genesis of high Mg andesite and implications for continental crust. *In* Eiler, J. (ed.) *Inside the Subduction Factory. Geophys. Monogr. Ser.* **138**, pp. 223–76. American Geophysical Union, Washington, DC.

Keller, R.G. & Hatcher, Jr, R.D. (1999) Some comparisons of the structure and evolution of the southern Appalachian–Ouachita orogen and portions of the Trans-European Suture Zone region. *Tectonophysics* **314**, 43–68.

Kellogg, L.H. & King, S.D. (1997) The effect of temperature dependent viscosity on the structure of new plumes in the mantle: results of a finite element model in a spherical axisymmetrical shell. *Earth planet. Sci. Lett.* **148**, 13–26.

Kelly, D.S. *et al.* (2001) An off-axis hydrothermal vent field near the Mid-Atlantic Ridge at 30°N. *Nature* **412**, 145–9.

Kelly, D.S., Baross, J.A. & Delaney, J.R. (2002) Volcanoes, fluids and life at mid-ocean ridge spreading centers. *Annu. Rev. Earth planet. Sci.* **30**, 385–491.

Kendall, J.-M. *et al.* (2005) Magma-assisted rifting in Ethiopia. *Nature* **433**, 146–8.

Kennedy, M.J., Christie-Blick, N. & Prave, A.R. (2001) Carbon isotopic composition of Neoproterozoic glacial carbonates as a test of paleoceanographic models for snowball Earth phenomena. *Geology* **29**, 1135–8.

Kennett, B.L.N. (1977) Towards a more detailed seismic picture of the oceanic crust and mantle. *Marine geophys. Res.* **3**, 7–42.

Kennett, B.L.N., & Orcutt, J.A. (1976) A comparison of travel time inversions for marine refraction profiles. *J. geophys. Res.* **81**, 4061–70.

Kennett, B.L.N., Engdahl, E.R. & Buland, R. (1995) Constraints on seismic velocities in the Earth from traveltimes. *Geophys. J. Int.* **122**, 108–24.

Kennett, J.P. (1977) Cenozoic evolution of Antarctic glaciation, the circum-Antarctic current, and their impact on global paleoceanography. *J. geophys. Res.* **82**, 3843–60.

Kent, D.V. & Gradstein, F.M. (1985) A Cretaceous and Jurassic geochronology. *Bull. geol. Soc. Am.* **96**, 1419–27.

Kent, D.V. & Gradstein, F.M. (1986) A Jurassic to recent chronology. *In* Vogt, P.R. & Tucholke, B.E. (eds) *The Western North Atlantic region. The Geology of North America* 1, pp. 45–50. Geological Society of America, Boulder, CO.

Kent, G.M. *et al.* (1994) Uniform accretion of oceanic crust south of the Garrett transform at 14°15′S on the East Pacific Rise. *J. geophys. Res.* **99**, 9097–116.

Kent, G.M., Harding, A.J. & Orcutt, J.A. (1990) Evidence for a smaller magma chamber beneath the East Pacific Rise at 9°30′N. *Nature* **344**, 650–3.

Keppie, J.D. & Dostal, J. (2001) Evaluation of the Baja controversy using paleomagnetic and faunal data, plume magmatism, and piercing points. *Tectonophysics* **339**, 427–42.

Keppie, J.D. & Ramos, V.A. (1999) Odyssey of terranes in the Iapetus and Rheic oceans during the Paleozoic. *In* Ramos, V.A. & Keppie, J.D. (eds) *Laurentia–Gondwana Connections before Pangea. Geol. Soc. Am. Sp. Paper* **336**, 267–76.

Kidd, R.G.W. (1977) A model for the process of formation of the upper oceanic crust. *Geophys. J. Roy. astr. Soc.* **50**, 149–83.

Kieffer, B. *et al.* (2004) Flood and shield basalt from Ethiopia: magmas from the African superswell. *J. Petrol.* **45**, 793–834.

Kilgour, B. & Hatch, L. (compilers) (2002) *Magnetic Anomaly Images of the Australian Region* [Digital Dataset]. Geoscience Australia, Canberra.

Kincaid C. & Griffiths R.W. (2003) Laboratory models of the thermal evolution of the mantle during rollback subduction. *Nature*, **425**, 58–62.

King, S.D. (2005) Archean cratons and mantle dynamics. *Earth planet. Sci. Lett.* **234**, 1–14.

Kirby, S.H., Engdahl, E.R. & Denlinger, R. (1996) Intermediate depth intraslab earthquakes and arc volcanism as physical expressions of crustal and upper mantle metamorphism in subducting slabs. *In* Bebout, G.E. *et al.* (eds) *Subduction: top to bottom. Geophys. Monogr. Ser.* **96**, pp. 195–214. American Geophysical Union, Washington, DC.

Kirby, E. *et al.* (2002) Late Cenozoic evolution of the eastern margin of the Tibetan Plateau: inferences from ^{40}Ar/^{39}Ar and (U-Th)/He thermochronology. *Tectonics* **21**, 1001, doi:10.1029/2000TC001246.

Kitada, K. *et al.* (2006) Distinct regional differences in crustal thickness along the axis of the Mariana Trough, inferred from gravity anomalies. *Geochem. Geophys. Geosyst.* **7**, Q04011, doi:10.1029/2005GC001119.

Klein, E.M. *et al.* (1988) Isotope evidence of a mantle convection boundary at the Australian–Antarctic Discordance. *Nature* **333**, 623–9.

Kleinrock, M.C. & Hey, R.N. (1989) Detailed tectonics near the tip of the Galapagos 95.5°W propagator: how the lithosphere tears and a spreading axis develops. *J. geophys. Res.* **94**, 13801–38.

Klepeis, K.A., Crawford, M.L. & Gehrels, G. (1998) Structural history of the crustal-scale Coast shear zone near Portland Canal, Coast Mountains orogen, southeast Alaska and British Columbia. *J. struct. Geol.* **20**, 883–904.

Klepeis, K.A, Clarke, G.L. & Rushmer, T. (2003) Magma transport and coupling between deformation and magmatism in the continental lithosphere. *GSA Today* **13**, 4–11.

Klepeis, K.A. *et al.* (2004) Processes controlling vertical coupling and decoupling between the upper and lower crust of orogens: results from Fiordland, New Zealand. *J. struct. Geol.* **26**, 765–91.

Klepeis, K.A., King, D., De Paoli, M., Clarke, G.L. & Gehrels, G. (2007) Interaction of strong lower and weak middle crust during lithospheric extension in western New Zealand. *Tectonics* **26**, TC4017.

Kley, J. (1996) Transition from basement-involved to thin skinned thrusting in the Cordillera Oriental of southern Bolivia. *Tectonics* **15**, 763–75.

Kley, J., Monaldi, C.R. & Salfity, J.A. (1999) Along-strike segmentation of the Andean foreland: causes and consequences. *Tectonophysics* **301**, 75–94.

Kloppenburg, A., White, S.H. & Zegers, T.E. (2001) Structural evolution of the Warrawoona Greenstone Belt and adjoining granitoid complexes, Pilbara craton, Australia: implications for Archaean tectonic processes. *Precambrian Res.* **112**, 107–47.

Klosko E.R. *et al.* (1999). Upper mantle anisotropy in the New Zealand region. *Geophys. Res. Lett.* **26**, 1497–500.

Klotz, J. *et al.* (1999) GPS-derived deformation of the central Andes including the 1995 Antofagasta Mw = 8.0 earthquake. *Pure Appl. Geophys.* **154**, 709–30.

Knittle, E. & Jeanloz, R (1991) Earth's core–mantle boundary: results of experiments at high pressures and temperatures. *Science* **251**, 1438–43.

Kohler, M.D. (1999) Lithospheric deformation beneath the San Gabriel Mountains in the Southern California Transverse Ranges. *J. geophys. Res.* **104**, 15025–41.

Kohler, M.D. & Eberhart-Phillips, D. (2003) Intermediate-depth earthquakes in a region of continental convergence: South Island, New Zealand. *Bull. seis. Soc. Am.* **93**, 85–93.

Kono, M. & Roberts, P.H. (2002) Recent geodynamo simulations and observations of the geomagnetic field. *Rev. geophys.* **40**(4), doi:101029/2000RG000102, 1013.

Konstantinovskaia, E. & Malavieille, J. (2005) Erosion and exhumation in accretionary orogens: experimental and geological approaches. *Geochem. Geophys. Geosyst.* **6**, Q02006, doi:10.1029/2004GC000794.

Koons, P.O. (1987) Some thermal and mechanical consequences of rapid uplift; an example from the Southern Alps, New Zealand. *Earth planet. Sci. Lett.* **86**, 307–19.

Koons, P. *et al.* (2003) Influence of exhumation on the structural evolution of transpressional plate boundaries: an example from the Southern Alps, New Zealand. *Geology* **31**, 3–6.

Koons, P.O. *et al.* (1998) Fluid flow during active oblique convergence: a Southern Alps model from mechanical and geochemical observations. *Geology* **26**, 159–62.

Korenaga, J. (2006) Archean geodynamics and the thermal evolution of the Earth. *In* Benn, K., Mareschal, J.C., & Condie, K.C. (eds) *Archean Geodynamics and Environments. Geophys. Monogr. Ser.* **164**, pp. 7–32. American Geophysical Union, Washington, DC.

Kreemer, C. *et al.* (2000) Active deformation in eastern Indonesia and the Philippines from GPS and seismicity data. *J. geophys. Res.* **105**, 663–80.

Kröner, A. (1981) Precambrian plate tectonics. *In* Kröner, A. (ed.) *Precambrian Plate Tectonics*, pp. 57–90. Elsevier, Amsterdam.

Kröner, A. (1985) Ophiolites and the evolution of tectonic boundaries in the late Proterozoic Arabian–Nubian Shield of northeast Africa and Arabia. *Precambrian Res.* **27**, 277–300.

Kröner, A. & Cordani, U. (2003) African, southern Indian and South American cratons were not part of the Rodinia supercontinent: evidence from field relationships and geochronology. *Tectonophysics* **375**, 325–52.

Kurtén, B. (1969) Continental drift and evolution. *Sci. Am.* **220**, 54–65.

Kusky, T.M. & Polat, A. (1999) Growth of granite–greenstone terranes at convergent margins, and stabilization of Archean cratons. *Tectonophysics* **305**, 43–73.

Kusky, T.M. & Vearncombe, J.R. (1997) Structural Aspects. *In* de Wit, M.J. & Ashwal, L.D. (eds) *Greenstone Belts*, pp. 91–124. Clarendon Press, Oxford, UK.

Kusznir, N.J. & Bott, M.H.P. (1976) A thermal study of the formation of oceanic crust. *Geophys. J. Roy. astr. Soc.* **47**, 83–95.

Kusznir, N.J. & Park, R.G. (1987) The extensional strength of the continental lithosphere: its dependence on geothermal gradient, and composition and thickness. *In* Coward, M.P., Dewey, J.F. & Hancock, P.L. (eds) *Continental Extensional Tectonics. Spec. Pub. geol. Soc. Lond.* **28**, 35–52.

Kusznir, N.J., Hunsdale, R. & Roberts, A.M. (2004) Timing of depth-dependent lithosphere stretching on the S. Lofoten rifted margin offshore mid-Norway: pre-breakup or post-breakup? *Basin Research* **16**, 279–96.

Lachenbruch, A.H. & Sass, J.H. (1992) Heat flow from Cajon Pass, fault strength, and tectonic implications. *J. geophys. Res.* **97**, 4995–5015.

Lacroix, S. & Sawyer, E.W. (1995) An Archean fold-thrust belt in the northwestern Abitibi greenstone belt: structural and seismic evidence. *Can. J. Earth Sci.* **32**, 97–112.

Lamb, A. & Davis, P. (2003) Cenozoic climate change as a possible cause for the rise of the Andes. *Nature* **425**, 792–7.

Lamb, S. *et al.* (1997) Cenozoic evolution of the central Andes in Bolivia and northern Chile. *In* Burg, J.-P. & Ford, M. (eds) *Orogeny through Time. Spec. Pub.geol. Soc. Lond.* **121**, 237–64.

Langmuir, C.H., Bender, J.B. & Batiza, R. (1986) Petrological and tectonic segmentation of the East Pacific Rise, 5°30′–14°30′N. *Nature* **322**, 422–9.

Larson, K.M. *et al.* (1999) Kinematics of the India–Eurasia collision zone from GPS measurements. *J. geophys. Res.* **104**, 1077–93.

Larson, R.L. (1991a) Latest pulse of Earth: evidence for a mid-Cretaceous superplume. *Geology* **19**, 547–50.

Larson, R.L. (1991b) Geological consequences of superplumes. *Geology* **19**, 963–6.

Larson, R.L. (1995) The mid-Cretaceous superplume episode. *Sci. Am.* **272**, 66–70.

Larson, R.L. & Pitman, W.C. III (1972) World-wide correlation of Mesozoic magnetic anomalies, and its implications. *Bull. geol. Soc. Am.* **83**, 3645–61.

Larson, R.L. *et al.* (1992) Roller-bearing tectonic evolution of the Juan Fernandez microplate. *Nature* **356**, 571–6.

Latin, D., Norry, M.J. & Tarzey, R.J.E. (1993) Magmatism in the Gregory Rift, East Africa: evidence for melt generation by a plume. *J. Petrol.* **34**, 1007–27.

Lavier, L.L. & Buck, W.R. (2002) Half graben versus large-offset low-angle normal fault: importance of keeping cool during normal faulting. *J. geophys. Res.* **107**, 2122, doi:10.1029/2001JB000513.

Lavier, L.L. & Manatschal, G. (2006) A mechanism to thin the continental lithosphere at magma-poor margins. *Nature* **440**, 324–8.

Lavier, L.L., Buck, W.R. & Poliakov, A.N.B. (1999) Self-consistent rolling hinge model for the evolution of large-offset low-angle normal faults. *Geology* **27**, 1127–30.

Lavier, L.L., Buck, W.R. & Poliakov, A.N.B. (2000) Factors controlling normal fault offset in an ideal brittle layer. *J. geophys. Res.* **105**, 23 431–42.

Lawver, L.A. & Müller, R.D. (1994) Iceland hotspot track. *Geology* **22**, 311–4.

Lawver, L.A. *et al.* (2003) The PLATES 2003 Atlas of Plate Reconstruction (750 Ma to Present Day). PLATES Progress Report No. 280–0703. *University of Texas Institute for Geophysics Technical Report* No. 190. University of Texas Press, Houston, Texas.

Lazar, M., Ben-Avraham, Z. & Schattner, U. (2006) Formation of sequential basins along a strike-slip fault – geophysical observations from the Dead Sea basin. *Tectonophysics* **421**, 53–69.

Le Cheminant, A.N. & Heaman, L.M. (1989) Mackenzie igneous event, Canada; Middle Proterozoic hotspot magmatism associated with ocean opening. *Earth planet. Sci. Lett.* **96**, 38–48.

Lee, C.-T.A. (2006) Geochemical/petrologic constraints on the origin of cratonic mantle. *In* Benn, K., Mareschal, J.C., & Condie, K.C. (eds) *Archean Geodynamics and Evironments. Geophys. Monogr. Ser.* **164**, pp. 89–114. American Geophysical Union, Washington, DC.

Le Fort, P. *et al.* (1987) Crustal generation of Himalayan leucogranites. *Tectonophysics* **134**, 39–57.

LeGrand, H.E. (1988) *Drifting Continents and Shifting Theories.* Cambridge University Press, Cambridge, UK.

Leitner, B. *et al.* (2001) A focused look at the Alpine fault, New Zealand: seismicity, focal mechanisms, and stress observations, *J. geophys. Res.* **106**, 2193–220.

Lemieux, S., Ross, G.M. & Cook, F.A. (2000) Crustal geometry and tectonic evolution of the Archean crystalline basement beneath the southern Alberta Plains, from new seismic reflection and potential field studies. *Can. J. Earth Sci.* **37**, 1473–91.

Lenardic, A. (1998) On the partitioning of mantle heat loss below oceans and continents over time and its relationship to the Archaean paradox. *Geophys. J. Int.* **134**, 706–20.

Lenardic, A., Moresi, L.N. & Muehlhaus, H. (2003) Longevity and stability of cratonic lithosphere; insights from numerical simulations of coupled mantle convection and continental tectonics. *J. geophys. Res.* **108**, 2303, doi:10.1029/2002JB001859.

Lenardic, A. *et al.* (2000) What the mantle sees; the effect of continents on mantle heat flow. *In* Richards, M.A., Gordon, R.G. & van der Hilst, R.D. (eds) *The History and Dynamics of Global Plate Motions. Geophys. Monogr. Ser.* **121**, pp. 95–112. American Geophysical Union, Washington, DC.

Le Pichon, X. (1968) Sea-floor spreading and continental drift. *J. geophys. Res.* **73**, 3661–97.

Le Pichon, X., Francheteau, J. & Bonnin, J. (1973) *Plate Tectonics.* Elsevier, Amsterdam.

le Roex, A.P., Späth, A. & Zartman, R.E. (2001) Lithospheric thickness beneath the southern Kenya Rift: implications from basalt geochemistry. *Contrib. Mineral. Petrol.* **142**, 89–106.

Li, L. *et al.* (2004) Stress measurements of deforming olivine at high pressure. *Phys. Earth planet. Int.* **143–144**, 357–67.

Li, Z.X. *et al.* (2008) Assembly, configuration, and break-up history of Rodinia: a synthesis. *Precambrian Res.* **160**, 179–210.

Lin, J *et al.* (1990) Evidence from gravity data for focused magmatic accretion along the Mid-Atlantic Ridge. *Nature* **344**, 627–32.

Lin, S.-C. & van Keken, P.E. (2006) Dynamics of thermochemical plumes: 1. Plume formation and entrainment of a dense layer. *Geochem. Geophys. Geosyst.* **7**, Q02006, doi:10.1029/2005GC001071.

Lister, C.R.B. (1980) Heat flow and hydrothermal circulation. *Ann. Rev. Earth planet. Sci.* **8**, 95–117.

Lister, G.S., Etheridge, M.A. & Simons, P.A. (1986) Detachment faulting and the evolution of passive continental margins. *Geology* **14**, 246–50.

Lithgow-Bertelloni, C. & Silver, P.G.(1998) Dynamic topography, plate driving forces and the Africal superswell. *Nature* **395**, 269–72.

Little, T.A., Holcombe, R.J. & Ilg, B.R. (2002) Kinematics of oblique collision and ramping inferred from microstructures and strain in middle crustal rocks, central Southern Alps, New Zealand. *J. struct. Geol.* **24**, 219–39.

Liu, M. & Yang, Y. (2003) Extensional collapse of the Tibetan Plateau: results of three-dimensional finite element modelling. *J. geophys. Res.* **108**, 2361, doi:10.1029/2002JB002248.

Livermore, R., Vine, F.J. & Smith, A.G. (1984) Plate motions and the geomagnetic field – II. Jurassic to Tertiary. *Geophys. J. Roy. astr. Soc.* **79**, 939–61.

Livermore, R. *et al.* (2004) Shackleton fracture zone: no barrier to early circumpolar ocean circulation. *Geology* **32**, 797–800.

Lottes, A.L. & Rowley, D.B. (1990) Early and Late Permian reconstructions of Pangaea. *In* McKerrow, W.S. & Scotese, C.R. (eds) *Paleozoic Paleogeography and Biogeography. Geol. Soc. Lond. Mem.* **12**, 383–95.

Louden, K.E. & Chian, D. (1999) The deep structure of nonvolcanic rifted continental margins. *Phil. Trans. Roy. Soc. Lond. A* **357**, 767–805.

Louie, J.N. *et al.* (2004) The northern Walker Lane refraction experiment: Pn arrivals and the northern Sierra Nevada root. *Tectonophysics* **388**, 253–69.

Lowman, J.P. & Jarvis, G.T. (1996) Continental collisions in wide aspect ratio and high Rayleigh number two-dimensional mantle convection models. *J. geophys. Res.* **101**, 25 485–97.

Lowman, J.P. & Jarvis, G.T. (1999) Effects of mantle heat source distribution on supercontinent stability. *J. geophys. Res.* **104**, 12,733–46.

Lowman, J.P., King, S.D. & Gable, C.W. (2001) The influence of tectonic plates on mantle convection patterns, temperature and heat flow. *Geophys. J. Int.* **146**, 619–36.

Lowman, J.P., King, S.D. & Gable, C.W. (2003) The role of the heating mode of the mantle in intermittent reorganization of the plate velocity field. *Geophys. J. Int.* **152**, 455–67.

Lowman, J.P., King, S.D. & Gable, C.W. (2004) Steady plumes in viscously stratified, vigorously convecting, three-dimensional numerical mantle convection models with mobile plates. *Geochem. Geophys. Geosyst.* **5**, Q01L01, doi:10.1029/2003GC000583.

Lowrie, A., Smoot, C. & Bartiza, R. (1986) Are oceanic fracture zones locked and strong or weak? New evidence for volcanic activity and weakness. *Geology* **14**, 242–5.

Lysack, S. (1992) Heat flow variations in continental rifts. *Tectonophysics* **208**, 309–23.

Lyzenga, G.A. *et al.* (1986) Tectonic motions in California inferred from very long baseline interferometry observations, 1980–4. *J. geophys. Res.* **91**, 9473–87.

Macdonald, K.C. & Fox. P.J. (1983) Overlapping spreading centres: new accretion geometry on the East Pacific Rise. *Nature* **302**, 55–8.

Macdonald, K.C. (1982) Mid-ocean ridges: fine scale tectonic, volcanic and hydrothermal processes within a plate boundary zone. *Ann. Rev. Earth planet. Sci.* **10**, 155–90.

Macdonald, K.C. *et al.* (1988) A new view of the mid-ocean ridge from the behaviour of ridge-axis discontinuities. *Nature* **335**, 217–25.

Macdonald, K.C., Sheirer, D.S. & Carbotte, S. (1991) Mid-ocean ridges: discontinuities, segments and giant cracks. *Science* **253**, 986–94.

Macdonald, R. *et al.* (2001) Plume–lithosphere interactions in the generation of the basalts of the Kenya Rift, East Africa. *J. Petrol.* **42**, 877–900.

Mackenzie, G.D., Thybo, H. & Maguire, P.K.H. (2005) Crustal velocity structure across the Main Ethiopian Rift: results from two-dimensional wide-angle seismic modelling. *Geophys. J. Int.* **162**, 994–1006.

Mackwell, S.J., Zimmerman, M.E. & Kohlstedt, D.L. (1998) High-temperature deformation of dry diabase with application to tectonics on Venus. *J. geophys. Res.* **103**, 975–84.

MacLeod, C.J. *et al.* (2002) Direct evidence for oceanic detachment faulting at the Mid-Atlantic Ridge, 15°45′N. *Geology* **30**, 879–82.

Maekawa, H. *et al.* (1993) Blueschist metamorphism in an active subduction zone. *Nature* **364**, 520–3.

Mahmoud, S. *et al.* (2005) GPS evidence for northward motion of the Sinai Block: implications for E. Mediterranean tectonics. *Earth planet. Sci. Lett.* **238**, 217–24.

Makovsky, Y. & Klemperer, S.L. (1999) Measuring the seismic properties of Tibetan bright spots: evidence for free aqueous fluids in the Tibetan middle crust. *J. geophys. Res.* **104**, 10,795–825.

Makris, J. *et al.* (1983) Seismic refraction profiles between Cyprus and Israel and their interpretation. *Geophys. J. R. astr. Soc.* **75**, 575–91.

Malavieille, J. *et al.* (2002) Arc–continent collision in Taiwan: new marine observations and tectonic evolution. *Geol. Soc. Am. Sp. Paper* **358**, 187–211.

Marcotte, S.B. *et al.* (2005) Intra-arc transpression in the lower crust and its relationship to magmatism in a Mesozoic magmatic arc. *Tectonophysics* **407**, 135–63.

Mareschal, J.-C. & Jaupart, C. (2006) Archean thermal regime and stabilization of the cratons. *In* Benn, K., Mareschal, J.C., & Condie, K.C. (eds) *Archean Geodynamics and Evironments. Geophys. Monogr. Ser.* **164**, pp. 61–73. American Geophysical Union, Washington, DC.

Marshak, S. (1999) Deformation style way back when: thoughts on the contrasts between Archean/Paleoproterozoic and contemporary orogens. *J. struct.Geol.* **21**, 1175–82.

Marshak, S. *et al.* (1997) Dome-and-keel provinces formed during Paleoproterozoic orogenic collapse-core complexes, diapirs, or neither?: examples from the Quadrilátero Ferrífero and the Penokean orogen. *Geology* **25**, 415–18.

Marshall, L.G. (1988) Land mammals and the great American interchange. *Amer. Sci.* **76**, 380–8.

Martin, H. (1981) The late Palaeozoic Gondwana glaciation. *Geol. Rund.* **70**, 480–98.

Martin, H. (1986) Effects of steeper Archean geothermal gradient on geochemistry and subduction-zone magmas. *Geology* **14**, 753–6.

Martínez, F. & Taylor, B. (2002) Mantle wedge control on back-arc crustal accretion. *Nature* **416**, 417–20.

Martínez, F., Fryer, P. & Becker, N. (2000) Geophysical characteristics of the Mariana Trough, 11°50′N–13°40′N. *J. Geophys. Res.* **105**, 16,591–607.

Martínez, F., Goodliffe, A.M. & Taylor, B. (2001) Metamorphic core complex formation by density inversion and lower-crust extrusion. *Nature* **411**, 930–4.

Martínez-Díaz, J.J. *et al.* (2004) Triggering of destructive earthquakes in El Salvador. *Geology* **32**, 65–8.

Marty B., Pik, R. & Gezahegan, Y. (1996) Helium isotopic variations in Ethiopian plume lavas: nature of magmatic sources and limit on lower mantle contribution. *Earth planet. Sci. Lett.* **144**, 223–37.

Marvin, U.B. (1973) *Continental Drift: the evolution of a concept.* Smithsonian Institution, Washington, DC.

Mascle, J. & Blarez, E. (1987) Evidence for transform margin evolution from the Ivory Coast–Ghana continental margin. *Nature* **326**, 378–81.

Mason, R. (1985) Ophiolites. *Geology Today* **1**, 136–40.

Mason, R.G. & Raff, A.D. (1961) Magnetic survey off the west coast of North America, 32°N latitude to 42°N latitude. *Bull. geol. Soc. Am.* **72**, 1259–66.

Massonnet, D. & Feigl, K. (1998) Radar interferometry and its application to changes in the Earth's surface. *Rev. Geophys.* **36**, 441–500.

Masters, G. *et al.* (1996) A shear velocity model of the mantle. *Phil. Trans. Roy Soc. London A* **354**, 1385–411.

Maxwell, A.E. *et al.* (1970) Deep sea drilling in the South Atlantic. *Science* **168**, 1047–59.

McCaffrey, R. (1996) Slip partitioning at convergent plate boundaries of SE Asia. *In* Hall, R. & Blundell, D.J. (eds) *Tectonic Evolution of Southeast Asia. Spec. Pub. geol. Soc. Lond.* **106**, 3–18.

McCaffrey, R. (2005) Block kinematics of the Pacific–North America plate boundary in the southwestern United States from inversion of GPS, seismological, and geologic data. *J. geophys. Res.* **110**, B07401, doi:10.1029/2004JB003307.

McCallum, I.S. (1996) The Stillwater Complex. *In* Cawthorn, R. G. (ed.) *Layered Intrusions*, pp. 441–83. Elsevier, Amsterdam.

McClay, K. & Bonora, M. (2001) Analog models of restraining stepovers in strike-slip fault systems. *Bull. Am. Assoc. Petroleum Geols.* **85**, 233–60.

McElhinny, M.W. & McFadden, P.L. (2000) *Paleomagnetism: continents and oceans.* Academic Press, San Diego.

McElhinny, M.W., Taylor, S.R. & Stevenson, D.J. (1978) Limits to the expansion of Earth, Moon, Mars and Mercury and to changes in the gravitational constant. *Nature* **271**, 316–21.

McKenzie, D.P. (2003) Estimating T_e in the presence of internal loads. *J. geophys. Res.* **108**, doi:10.1029/2002JB001766.

McKenzie, D.P. & Morgan, W.J. (1969) Evolution of triple junctions. *Nature* **224**, 125–33.

McKenzie, D.P. & Parker, R.L. (1967) The North Pacific: an example of tectonics on a sphere. *Nature* **216**, 1276–80.

McKenzie, D.P. & Sclater, J.G. (1971) The evolution of the Indian Ocean since the late Cretaceous. *Geophys. J. Roy. astr. Soc.* **24**, 437–528.

McKenzie, D.P. & Weiss, N. (1975) Speculations on the thermal and tectonic history of the earth. *Geophys. J. Roy. astr. Soc.* **42**, 131–74.

McLennan, S.M. & Taylor, S.R. (1996) Heat flow and the chemical composition of continental crust. *J. Geol.* **104**, 369–77.

McNamara, D.E. *et al.* (1997) Upper mantle velocity structure beneath the Tibetan Plateau from Pn travel time tomography. *J. geophys. Res.* **102**, 493–505.

McNutt, M.K. & Judge, A.V. (1990) The superswell and mantle dynamics beneath the south Pacific. *Science* **248**, 969–75.

McQuarrie, N. (2002) The kinematic history of the central Andean fold-thrust belt, Bolivia: implications for building a high plateau. *Bull. geol. Soc. Am.* **114**, 950–63.

McQuarrie N. *et al.* (2005) Lithospheric evolution of the Andean fold-thrust belt, Bolivia, and the origin of the central Andean plateau. *Tectonophysics* **399**, 15–37.

Meade, B.J. & Hager, B.H. (2005) Block models of crustal motion in southern California constrained by GPS measurements. *J. geophys. Res.* **110**, B03403, doi:10.1029/2004JB003209.

Meade, C. & Jeanloz, R. (1991) Deep focus earthquakes and recycling of water into the Earth's mantle. *Science* **252**, 68–72.

Mechie, J. *et al.* (1997) A model for the structure, composition and evolution of the Kenya Rift. *Tectonophysics* **278**, 95–119.

Mechie, J. *et al.* (2005) Crustal shear velocity structure across the Dead Sea Transform from 2-D modeling of DESERT explosion seismic data. *Geophys. J. Int.* **160**, 910–24.

Meert, J.G. & Torsvik, T.H. (2003) The making and unmaking of a supercontinent: Rodinia revisited. *Tectonophysics* **375**, 261–88.

Meguin, C. & Romanowicz, B. (2000) The three-dimensional shear velocity structure of the mantle from the inversion of body, surface and higher waveforms. *Geophys. J. Int.* **143**, 709–28.

MELT seismic team (1998) Imaging the deep seismic structure beneath a mid-ocean ridge: the MELT experiment. *Science* **280**, 1215–8.

Menard, H.W. (1964) *Marine Geology of the Pacific.* McGraw-Hill, New York.

Menard, H.W. & Atwater, T. (1968) Changes in direction of sea floor spreading. *Nature* **219**, 463–7.

Menard, H.W. & Atwater, T. (1969) Origin of fracture zone topography. *Nature* **222**, 1037–40.

Menzies, M.A. *et al.* (2002) Characteristics of volcanic rifted margins. *In* Menzies, M.A. *et al.* (eds) *Volcanic Rifted Margins.* Geol. Soc. Am. Sp. Paper **362**, 1–14.

Merrill, R.T., McElhinny, M.W. & McFadden, P.L. (1996) *The Magnetic Field of the Earth: paleomagnetism, the core and the deep mantle.* Academic Press, San Diego.

Michael, P.J. *et al.* (2003) Magmatic and amagmatic sea floor spreading at the ultraslow-spreading Gakkel Ridge, Arctic Basin. *Nature* **423**, 956–61.

Miller, E.L. *et al.* (1999) Rapid Miocene slip on the Snake Range–Deep Creek Range fault system, east-central Nevada. *Bull. geol. Soc. Am.* **111**, 886–905.

Miller, R.B. & Paterson, S.R. (2001a) Influence of lithological heterogeneity, mechanical anisotropy, and magmatism on the rheology of an arc, North Cascades, Washington. *Tectonophysics* **342**, 351–70.

Miller, R.B. & Paterson, S.R. (2001b) Construction of mid-crustal sheeted plutons; examples from the North Cascades, Washington. *Bull. Geol. Soc. Am.* **113**, 1423–42.

Milsom, J. (2001) Subduction in eastern Indonesia: how many slabs? *Tectonophysics* **338**, 167–78.

Minshull, T.A. (2002) The break-up of continents and the formation of new ocean basins. *Phil. Trans. R. Soc. Lond. A* **360**, 2839–52.

Minster, J.B. & Jordan, T.H. (1978) Present-day plate motions. *J. geophys. Res.* **83**, 5331–54.

Mitchell, A.H.G. & Garson, M.S. (1976) Mineralization at plate boundaries. *Minerals Sci. Engineering* **8**, 129–69.

Mitchell, A.H.G. & Garson, M.S. (1981) *Mineral Deposits and Global Tectonic Settings.* Academic Press, London.

Mitchell, A.H.G. & Reading, H.G. (1969) Continental margins, geosynclines, and ocean floor spreading. *J. Geol.* **77**, 629–46.

Mitchell, A.H.G. & Reading, H.G. (1986) Sedimentation and tectonics. *In* Reading, H.G. (ed.) *Sedimentary Environments and Facies*, pp. 471–519. Blackwell Scientific Publications, Oxford.

Mitra, S. *et al.* (2005) Crustal structure and earthquake focal depths beneath northeastern India and southern Tibet. *Geophys. J. Int.* **160**, 227–48.

Mitra, S. *et al.* (2006) Variation of Rayleigh wave group velocity dispersion and seismic heterogeneity of the Indian crust and uppermost mantle. *Geophys. J. Int.* **164**, 88–98.

Miyashiro, A. (1961) Evolution of metamorphic belts. *J. Petrol.* **2**, 277–311.

Miyashiro, A. (1972) Metamorphism and related magmatism in plate tectonics. *Am. J. Sci.* **272**, 629–56.

Miyashiro, A. (1973) Paired and unpaired metamorphic belts. *Tectonophysics* **17**, 241–54.

Mohr, P.A. & Zanettin, B. (1988) The Ethiopian flood basalt province. *In* MacDougall, J.D. (ed.) *Continental Flood Basalts*, pp. 63–110. Kluwer Academic Publishers, Dordrecht, Netherlands.

Mohriak, W.U. & Rosendahl, B.R. (2003) Transform zones in the South Atlantic rifted continental margins. *In* Storti, F., Holdsworth, R.E. & Salvini, F. (eds) *Intraplate Strike-slip Deformation Belts.* Spec. Pub. geol. Soc. Lond. **210**, 211–28.

Möller, A. *et al.* (1995) Evidence for a 2 Ga subduction zone: eclogites in the Usagarian belt of Tanzania, *Geology* **23**, 1067–70.

Molnar, P. & Chen, W.-P. (1982) Seismicity and mountain building. *In* Hsü, K.G. (ed.) *Mountain Building Processes*, pp. 41–57. Academic Press, London.

Molnar, P. & Tapponnier, P. (1975) Cenozoic tectonics of Asia: effects of a continental collision. *Science* **189**, 419–26.

Molnar, P., Freedman, D. & Shih, J.S.F. (1979) Lengths of intermediate and deep seismic zones and temperatures in downgoing slabs of lithosphere. *Geophys. J. Roy. astr. Soc.* **56**, 41–54.

Molnar, P., England, P. & Martinod, J. (1993), Mantle dynamics, uplift of the Tibetan Plateau, and the Indian monsoon. *Rev. Geophys.* **31**, 357–96.

Montelli, R. *et al.* (2004a) Finite-frequency tomography reveals a variety of plumes in the mantle. *Science* **303**, 338–43.

Montelli, R. *et al.* (2004b) Global P and PP traveltime tomography: rays versus waves. *Geophys. J. Int.* **158**, 637–54.

Montési, L.G.J. & Zuber, M.T. (2002) A unified description of localization for application to large-scale tectonics. *J. geophys. Res.* **107**, 2045, doi:10.1029/2001JB000465.

Montési, L.G.J. & Zuber, M.T. (2003) Spacing of faults at the scale of the lithosphere and localization instability: 1. Theory. *J. geophys. Res.* **108**, 2110, doi:10.1029/2002JB001923.

Montgomery, D.R., Balco, G. & Willett, S.D. (2001) Climate, tectonics, and the morphology of the Andes. *Geology*, **29**, 579–82.

Moody, J.D. (1973) Petroleum exploration aspects of wrench–fault tectonics. *Am. Assoc. Pet. Geol.* **57**, 449–76.

Mooney, W.D., Laske, G. & Masters, T.G. (1998) CRUST 5.1: a global crustal model at $5 \times 5°$. *J. geophys. Res.* **103**, 727–47.

Moore, G.F. *et al.* (2001) New insights into deformation and fluid flow processes in the Nankai Trough accretionary prism: results of Ocean Drilling Program Leg 190. *Geochem. Geophys. Geosyst.* **2**, doi:10.1029/2001GC000166.

Moore, G.F. *et al.* (2005) Legs 190 and 196 synthesis: deformation and fluid flow processes in the Nankai Trough accretionary prism. *In* Mikada, H. *et al.* (eds) *Proceedings of the Ocean Drilling Program. Scientific Results.* **190/196**, pp. 1–26. College Station, Texas.

Moore, J.C. *et al.* (1982) Geology and tectonic evolution of a juvenile accretionary terrane along a truncated convergent margin: synthesis of results from Leg 66 of the Deep Sea Drilling Project, southern Mexico. *Bull. geol. Soc. Am.* **93**, 847–61.

Moores, E.M. (1982) Origin and emplacement of ophiolites. *Rev. Geophys. Space Phys.* **20**, 737–60.

Moores, E.M. (1991) South west US–East Antarctic (SWEAT) connection: a hypothesis. *Geology* **19**, 325–8.

Moores, E.M. (2002) Pre–1 Ga (pre-Rodinian) ophiolites: their tectonic and environmental implications. *Bull. geol. Soc. Am.* **114**, 80–95.

Moores, E.M. & Vine, F.J. (1971) The Troodos Massif, Cyprus, and other ophiolites as oceanic crust; evaluation and implications. *Phil. Trans. Roy. Soc. Lond. A* **268**, 443–66.

Mora, A. *et al.* (2006) Cenozoic contractional reactivation of Mesozoic extensional structures in the Eastern Cordillera of Colombia. *Tectonics* **25**, 2010, doi:10.1029/2005TC001854.

Mora-Klepeis, G. & McDowell, F.W. (2004) Late Miocene calc-alkalic volcanism in northwestern Mexico: an expression of rift or subduction-related magmatism? *JS Amer. Earth Sci.* **17**, 297–310.

Morgan, J.P. & Chen, Y.J. (1993) Dependence of ridge-axis morphology on magma supply and spreading rate. *Nature* **364**, 706–8.

Morgan, J.P. & Shearer, P.M. (1993) Seismic constraints on mantle flow and topography of the 660-km discontinuity: evidence for whole-mantle convection. *Nature* **365**, 506–11.

Morgan, W.J. (1968) Rises, trenches, great faults and crustal blocks. *J. geophys. Res.* **73**, 1959–82.

Morgan, W.J. (1971) Convection plumes in the lower mantle. *Nature* **230**, 42–3.

Morgan, W.J. (1972a) Plate motions and deep mantle convection. *Geol. Soc. Am. Mem.* **132**, 7–22.

Morgan, W.J. (1972b) Deep mantle convection plumes and plate motions. *Bull. Am. Assoc. Petroleum Geols.* **56**, 203–13.

Morgan, W.J. (1981) Hot spot tracks and the opening of the Atlantic and Indian Oceans. In Emiliani, C. (ed.) *The Oceanic Lithosphere. The sea* 7, pp. 443–87. Wiley, New York.

Morgan, W.J. (1983) Hotspot tracks and the early rifting of the Atlantic. *Tectonophysics* **94**, 123–39.

Morozov, I.B. *et al.* (1998) Wide-angle seismic imaging across accreted terranes, southeastern Alaska and western British Columbia. *Tectonophysics* **299**, 281–96.

Morozov, I.B. *et al.* (2001) Generation of new continental crust and terrane accretion in Southeastern Alaska and Western British Columbia: constraints from P- and S-wave wide-angle seismic data (ACCRETE). *Tectonophysics* **341**, 49–67.

Morris, J.D. & Villinger, H.W. (2006) Leg 205 synthesis: subduction fluxes and fluid flow across the Costa Rica convergent margin. In Morris, J.D., Villinger, H.W. & Klaus, A. (eds) *Proceedings of the Ocean Drilling Program. Scientific Results*, **205**, pp. 1–54. College Station, TX.

Mount, V.S. & Suppe, J. (1987) State of stress near the San Andreas Fault: implications for wrench tectonics. *Geology* **15**, 1143–6.

Mpodozis, C. & Allmendinger, R.W. (1993) Extensional tectonics, Cretaceous Andes, northern Chile (27°S). *Bull. geol. Soc. Am.* **105**, 1462–77.

Mpodozis, C. & Ramos, V. (1989) The Andes of Chile and Argentina. In Ericksen, G.E., Cañas Pinochet, M.T. & Reinemund, J.A. (eds) *Geology of the Andes and its Relation to Hydrocarbon and Mineral Resources. Circum–Pacific Council for Energy and Mineral Resources Earth Science Series* **11**, pp. 59–90. Houston, TX.

Mpodozis, C. *et al.* (2005) Late Mesozoic to Paleogene stratigraphy of the Salar de Atacama Basin, Antofagasta, Northern Chile: implications for the tectonic evolution of the Central Andes. *Tectonophysics* **399**, 125–54.

Müller, B. *et al.* (1997) Short-scale variation of tectonic regimes in the western European stress province north of the Alps and Pyrenees. *Tectonophysics* **275**, 199–219.

Müller, R.D. *et al.* (1997) Digital isochrons of the world's ocean floor. *J. geophys. Res.* **102**, 3211–14.

Müller, R.D., Royer, J.Y. & Lawver, L.A. (1993) Revised plate motions relative to the hotspots from combined Atlantic and Indian Ocean hotspot tracks. *Geology* **21**, 275–8.

Murakami, M. *et al.* (2004) Post-perovskite phase transition in $MgSiO_3$. *Science* **304**, 855–8.

Musacchio, G. & Mooney, W.D. (2002) Seismic evidence for a mantle source for mid-Proterozoic anorthosites and implications for model soft crustal growth. In Fowler, C.M.R., Ebinger, C.J. & Hawkesworth, C.J. (eds) *The Early Earth: physical, chemical and biological development. Spec. Pub. geol. Soc. Lond.* **199**, 125–34.

Mutter, J., Talwani, M. & Stoffa, P. (1982) Origin of seaward-dipping reflectors in ocean crust off the Norwegian margin by subaerial seafloor spreading. *Geology* **10**, 353–7.

Mutter, J.C. *et al.* (1988) Magma distribution across ridge axis discontinuities on the East Pacific Rise from multichannel seismic images. *Nature* **336**, 156–8.

Myers, S. *et al.* (1998) Lithospheric scale structure across the Bolivian Andes from tomographic images of velocity and attenuation for P and S waves. *J. geophys. Res.* **103**, 21,233–52.

Nafe, J.E. & Drake, C.L. (1963) Physical properties of marine sediments. In Hill, M.N. (ed.) *The Earth Beneath the Sea. The sea* 3, pp. 794–815. Interscience Publishers, New York.

Nagel, T.J. & Buck, W.R. (2004) Symmetric alternative to asymmetric rifting models. *Geology* **32**, 937–40.

Naldrett, A.J. (1999) World-class Ni-Cu-PGE deposits: key factors in their genesis. *Miner. Depos.* **34**, 227–40.

Naylor, M. *et al.* (2005) A discrete element model for orogenesis and accretionary wedge growth. *J. geophys. Res.* **110**, B12403, doi:10.1029/2003JB002940.

Neev, D., & Hall, J.K. (1979) Geophysical investigations in the Dead Sea. *Sedim. Geol.* **23**, 209–38.

Nelson, K.D. *et al.* (1996) Partially molten middle crust beneath southern Tibet: synthesis of project INDEPTH results. *Science* **274**, 1684–88.

Nemčok, M., Schamel, S. & Gayer, R. (2005) Thrustbelts, Structural Architecture, Thermal Regimes and Petroleum Systems. Cambridge University Press, New York.

Newman, R. & White, N. (1997) Rheology of the continental lithosphere inferred from sedimentary basins. *Nature* **385**, 621–4.

Nicolas, A. (1989) *Structure of Ophiolites and Dynamics of Oceanic Lithosphere.* Kluwer Academic Publishers, Dordrecht.

Nicolas, A., Boudier, F. & Ildefonse, B. (1994) Evidence from the Oman ophiolite for active mantle upwelling beneath a fast-spreading ridge. *Nature* **370**, 51–3.

Niell, A.E. *et al.* (1979) Comparison of a radiointerferometric differential baseline measurement with conventional geodesy. *Tectonophysics* **52**, 49–58.

Nielsen T.K. & Hopper, J.R. (2002) Formation of volcanic rifted margins: are temperature anomalies required? *Geophys. Res. Lett.* **29**, 2022, doi:10.1029/2002GL015681.

Nielsen T.K. & Hopper, J.R. (2004) From rift to drift: mantle melting during continental breakup. *Geochem. Geophys. Geosyst.* **5**, Q07003, doi:10.1029/2003GC000662.

Niemi, N.A. *et al.* (2004) BARGEN continuous GPS data across the eastern Basin and Range province, and implications for fault system dynamics. *Geophys. J. Int.* **159**, 842–62.

Nisbet, E.G. & Fowler, C.M.R. (1978) The Mid-Atlantic Ridge at 37 and 45°N: some geophysical and petrological constraints. *Geophys. J. Roy. astr. Soc.* **54**, 631–60.

Nisbet, E.G. *et al.* (1993) Constraining the potential temperature of the Archean mantle: a review of the evidence from komatiites. *Lithos* **30**, 291–307.

Norabuena, E.O. *et al.* (1998) Space geodetic observations of Nazca–South America convergence across the central Andes. *Science* **279**, 358–62.

Norabuena, E.O. *et al.* (1999) Decelerating Nazca–South America and Nazca Pacific plate motions. *Geophys. Res. Lett.* **26**, 3405–8.

Norris, R.J. & Cooper, A.F. (2001) Late Quaternary slip rates and slip partitioning on the Alpine Fault, New Zealand. *J. struct. Geol.* **23**, 507–20.

Norris, R.J., Koons, P.O. & Cooper, A.F. (1990) The obliquely-convergent plate boundary in the South Island of New Zealand: implications for ancient collisional zones. *J. struct. Geol.* **12**, 715–25.

Norton, I.O. (1995) Plate motion in the North Pacific: the 43 Ma nonevent. *Tectonics* **14**, 1080–94.

Norton, I.O. & Sclater, J.G. (1979) A model for the evolution of the Indian Ocean and the break up of Gondwanaland. *J. geophys. Res.* **84**, 6803–30.

Nunns, A.G. (1983) Plate tectonic evolution of the Greenland–Scotland Ridge and surrounding areas. *In* Bott, M.H.P. *et al.* (eds) *Structure and Development of the Greenland–Scotland Ridge. NATO Conference Series IV*, **8**, pp. 11–30. Plenum Press, London.

Nyblade, A.A. & Robinson, S.W. (1994) The African superswell. *Geophys. Res. Lett.* **21**, 765–68.

O'Brien, P.J. & Rötzler, J. (2003) High pressure granulites: formation, recovery of peak conditions and implications for tectonics. *J. metam. geol.* **21**, 3–20.

O'Connell, R.J. & Budiansky, B. (1977) Viscoelastic properties of fluid-saturated cracked solids. *J. geophys. Res.* **82**, 5719–35.

O'Reilly, S.Y. *et al.* (2001) Are lithospheres forever? Tracking changes in subcontinental lithospheric mantle through time. *GSA Today* **11**, 4–10.

Okino, K. *et al.* (2004) Development of oceanic detachment and asymmetric spreading at the Australian–Antarctic Discordance. *Geochem. Geophys. Geosyst.* **5**, Q12012, doi:10.1029/2004GC000793.

Oldow, J.S. (2003) Active transtensional boundary zone between the western Great Basin and Sierra Nevada block, western US Cordillera. *Geology* **31**, 1033–6.

Oliver, J. (1982) Tracing surface features to great depths: a powerful means for exploring the deep crust. *Tectonophysics* **81**, 257–72.

Oliver, J. & Isacks, B. (1967) Deep earthquake zones, anomalous structures in the upper mantle, and the lithosphere. *J. geophys. Res.* **72**, 4259–75.

Opdyke, N.D. & Channel, J.E.T. (1996) *Magnetic Stratigraphy*. Academic Press, San Diego.

Opdyke, N.D., Burckle, L.H. & Todd, A. (1974) The extension of the magnetic time scale in sediments of the central Pacific Ocean. *Earth planet. Sci. Lett.* **22**, 300–6.

Opdyke, N.D. *et al.* (1966) Palaeomagnetic study of Antarctic deep sea cores. *Science* **154**, 349–57.

Oreskes, N. (1999) *The Rejection of Continental Drift: theory and method in American earth science*. Oxford University Press, New York.

Oreskes, N. (2001) (ed.) *Plate Tectonics: an insider's history of the modern theory of the Earth*. Westview press, Boulder, CO.

Orowan, E. (1965) Convection in a non-Newtonian mantle, continental drift, and mountain building. *Phil. Trans. Roy. Soc. Lond. A* **258**, 284–313.

Owens, T.J. & Zandt, G. (1997) Implications of crustal property variations for models of Tibetan plateau evolution. *Nature* **387**, 37–43.

Ozacar, A.A. & Zandt, G. (2004) Crustal seismic anisotropy in central Tibet: implications for deformational style and flow in the crust. *Geophys. Res. Lett.* **31**, L23601, doi:10.1029/2004GL021096.

Özalaybey, S. & Savage, M.K. (1995) Shear-wave splitting beneath western United States in relation to plate tectonics. *J. geophys. Res.* **100**, 18 135–49.

Packham, G.H. & Falvey, D.A. (1971) An hypothesis for the formation of marginal seas in the Western Pacific. *Tectonophysics* **11**, 79–110.

Pakiser, L.C. (1963) Structure of the crust and upper mantle in the western United States. *J. geophys. Res.* **68**, 5747–56.

Pancha, A., Anderson, J.G. & Kreemer, C. (2006) Comparison of seismic and geodetic scalar moment rates across the Basin and Range Province. *Bull. seis. Soc. Am.* **96**, 11–32.

Panien, M., Schreurs, G. & Pfiffner, A. (2005) Sandbox experiments on basin inversion: testing the influence of basin orientation and basin fill. *J. struct. Geol.* **27**, 433–45.

Panning, M. & Romanowicz, B. (2004) Inferences on flow at the base of Earth's mantle based on seismic anisotropy. *Science* **303**, 351–3.

Pardo-Casas, F. & Molnar, P. (1987) Relative motion of the Nazca (Farallon) and South American plates since Late Cretaceous time. *Tectonics* **6**, 233–48.

Park, J. & Levin, V. (2002) Seismic anisotropy: tracing plate dynamics in the mantle. *Science* **296**, 485–9.

Park, R.G. (1983) *Foundations of Structural Geology*. Blackie, London & Glasgow.

Park, Y. & Nyblade, A.A. (2006) P-wave tomography reveals a westward dipping low velocity zone beneath the Kenya Rift. *Geophys. Res. Lett.* **33**, L07311, doi:10.1029/2005GL025605.

Park, S.K. & Wernicke, B. (2003) Electrical conductivity images of Quaternary faults and Tertiary detachments in the California Basin and Range. *Tectonics* **22**, 1030, doi:10.1029/2001TC001324.

Parman, S.W., Grove, T.L. & Dann, J.C. (2004) A subduction origin for komatiites and cratonic lithospheric mantle. *S. Afr. J. Geol.* **107**, 107–18.

Parsons, B. & McKenzie, D.P. (1978) Mantle convection and the thermal structure of the plates. *J. geophys. Res.* **83**, 4485–96.

Parsons, B. & Sclater, J.G. (1977) An analysis of the variation of ocean floor bathymetry and heat flow with age. *J. geophys. Res.* **82**, 803–27.

Parsons, T. *et al.* (1996) Crustal structure of the Colorado Plateau, Arizona: application of new long-offset seismic data analysis techniques. *J. geophys. Res.* **101**, 11,173–94.

Patzwahl, R. *et al.* (1999) Two-dimensional velocity models of the Nazca plate subduction zone between 19.5°S and 25°S from wide angle seismic measurements during the CINCA95 project. *J. geophys. Res.* **104**, 7293–317.

Peacock, S.M. (1991) Numerical simulation of subduction zone pressure–temperature–time paths: constraints on fluid production and arc magmatism. *Philos. Trans. R. Soc. London Ser. A* **335**, 341–53.

Peacock, S.M. (1992) Blueschist-facies metamorphism, shear heating, and P–T–t paths in subduction shear zones. *J. geophys. Res.* **97**, 17 693–707.

Peacock, S.M. (2001) Are the lower planes of double seismic zones caused by serpentine dehydration in subducting oceanic mantle? *Geology* **29**, 299–302.

Peacock, S.M. (2003) Thermal structure and metamorphic evolution of subducting slabs. *In* Eiler, J. (ed.) *Inside the Subduction Factory.* Geophys. Monogr. Ser. **138**, pp. 7–22. American Geophysical Union, Washington, DC.

Peacock, S.M. & Wang, K. (1999) Seismic consequences of warm versus cool subduction zone metamorphism: examples from northeast and southwest Japan. *Science* **286**, 937–9.

Pearce, J.A. (1980) Geochemical evidence for the genesis and eruptive setting of lavas from Tethyan ophiolites. *In* Panayiotou, A. (ed.) *Ophiolites,* pp. 261–72. Geol. Surv., Cyprus.

Pearce, J.A. & Peate, D.W. (1995) Tectonic implications of the composition of volcanic arc magmas. *Annu. Rev. Earth Planet. Sci.* **23**, 251–85.

Pearson, D.G. *et al.* (2002) The development of lithosperic keels beneath the earliest continents: time constraints using PGE and Re–Os isotope systematics. *In* Fowler, C.M.R., Ebinger, C.J. & Hawkesworth, C.J. (eds) *The Early Earth: physical, chemical and biological development.* Spec. Pub. geol. Soc. Lond. **199**, 65–90.

Peltier, W.R. & Andrews, J.T. (1976) Glacial–isostatic adjustment – I. The forward problem. *Geophys. J. Roy. astr. Soc.* **46**, 605–46.

Percival, J.A. *et al.* (1997) Tectonic evolution of associated greenstone belts and high-grade terrains. *In* de Wit, M.J. & Ashwal, L.D. (eds) *Greenstone Belts,* pp. 398–420. Clarendon Press, Oxford.

Perez-Gussinge, M. & Watts, A.B. (2005) The long-term strength of Europe and its implications for plate forming processes. *Nature* **436**, 381–4.

Petford, N. & Atherton, M.P. (1996) Na-rich partial melts from newly underplated basaltic crust: the Cordillera Blanca batholith, Peru. *J. Petrol.* **37**, 1491–521.

Petford, N. *et al.* (2000) Granite magma formation, transport and emplacement in the Earth's crust. *Nature* **408**, 669–73.

Petit, C. & Ebinger, C. (2000) Flexure and mechanical behavior of cratonic lithosphere: gravity models of the East African and Baikal rifts. *J. geophys. Res.* **105**, 19,151–62.

Petronotis, K.E. & Gordon, R.G. (1999) A Maastrichtian paleomagnetic pole for the Pacific plate from a skewness analysis of marine magnetic anomaly 32. *Geophys. J. Int.* **139**, 227–47.

Pickup, S.L.B. *et al.* (1996) Insight into the nature of the ocean–continent transition off West Iberia from a deep multichannel seismic reflection profile. *Geology* **24**, 1079–82.

Pilger, R.H. Jr (1982) The origin of hotspot traces: evidence from eastern Australia. *J. geophys. Res.* **87**, 1825–34.

Piper, J.D.A. (1987) *Palaeomagnetism and the Continental Crust.* Open University Press, Milton Keynes, UK.

Pirajno, F. (2004) Hotspots and mantle plumes: global intraplate tectonics, magmatism and ore deposits. *Mineral. Petrol.* **82**, 183–216.

Pitman, W.C. III & Heirtzler, J.R. (1966) Magnetic anomalies over the Pacific–Antarctic ridge. *Science* **154**, 1164–71.

Pitman, W.C. III & Talwani, M. (1972) Sea-floor spreading in the North Atlantic. *Bull. geol. Soc. Am.* **83**, 619–46.

Plank, T. & Langmuir, C.H. (1993) Tracing trace elements from sediment input to volcanic output at subduction zones. *Nature* **362**, 739–43.

Planke, S. *et al.* (2000) Seismic volcanostratigraphy of large-volume basaltic extrusive complexes on rifted margins. *J. geophys. Res.* **105**, 19 335–51.

Platt, J.P. (1986) Dynamics of orogenic wedges and the uplift of high-pressure metamorphic rocks. *Bull. geol. Soc. Am.* **97**, 1037–53.

Plumstead, E.P. (1973) The enigmatic Glossopteris flora and uniformitarianism. *In* Tarling, D.H. & Runcorn, S.K. (eds) *Implications of Continental Drift to the Earth Sciences,* I, pp. 413–24. Academic Press, London.

Polet, J. *et al.* (2000) Shear wave anisotropy beneath the Andes from the BANJO, SEDA, and PISCO experiments. *J. geophys. Res.* **105**, 6287–304.

Pollack, H.N. & Chapman, D.S. (1977) The flow of heat from the Earth's interior. *Sci. Am.* **237**, 60–76.

Pollack, H.N., Hunter, S.J. & Johnson, J.R. (1993) Heat flow from the Earth's interior: analysis of the global data set. *Rev. Geophys.* **31**, 267–80.

Poulsen, C.J. *et al.* (2001) Response of the mid-Cretaceous global ocean circulation to tectonic and CO_2 forcings. *Paleoceanography,* **16**, 576–92.

Powell, C. McA. *et al.* (1993) Paleomagnetic constraints on timing of the Neoproterozoic breakup of Rodinia and the Cambrian formation of Gondwana. *Geology* **21**, 889–92.

Prawirodirdjo, L. & Bock, Y. (2004) Instantaneous global plate motion model from 12 years of continuous GPS observations. *J. geophys. Res.* **109**, B8, B08405, doi:10.1029/2003JB002944.

Prescott, W.H. *et al.* (2001) Deformation across the Pacific–North America plate boundary near San Francisco, California. *J. geophys. Res.* **106**, B4, 6673–82.

Prevot, M. *et al.* (2000) Evidence for a 20° tilting of the Earth's rotation axis 110 million years ago. *Earth planet. Sci. Lett.* **179**, 517–28.

Pritchard, M.E. & Simons, M. (2004) Surveying volcanic arcs with satellite radar interferometry: the Central Andes, Kamchatka, and beyond. *GSA Today* **14**, 4–11.

Purdy, G.M. (1987) New observations of the shallow seismic structure of young oceanic crust. *J. geophys. Res.* **92**, 9351–62.

Purdy, G.M. & Detrick, R.S. (1986) The crustal structure of the Mid-Atlantic Ridge at 23°N from seismic reflection studies. *J. geophys. Res.* **91**, 3739–62.

Raedeke, L.D. & McCallum, I.S. (1984) Investigations of the Stillwater Complex: Part II. Petrology and petrogenesis of the Ultramafic series. *J. Petrol.* **25**, 395–420.

Raff, A.D. & Mason, R.G. (1961) Magnetic survey off the west coast of North America, 40°N latitude to 52°N latitude. *Bull. geol. Soc. Am.* **72**, 1267–70.

Rainbird, R.H., Hamilton, M.A. & Young, G.M. (2001) Detrital zircon geochronology and provenance of the Torridonian, NW Scotland. *J. geol. Soc. Lond.* **158**, 15–27.

Ramachandran, K., Hyndman, R.D. & Brocher, T.M. (2006) Regional P wave velocity structure of the Northern Cascadia Subduction Zone. *J. geophys. Res.* **111**, B12301, doi:10.1029/2005JB004108.

Ramos, V.A. (1989) Foothills structure in northern Magallanes Basin, Argentina. *Amer. Assoc. Pet. geol.* **73**, 887–903.

Ramos, V.A., Cristallini, E.O. & Pérez, D.J. (2002) The Pampean flat-slab of the Central Andes. *J. S Am. Earth Sci.* **15**, 59–78.

Ranalli, G. (1995) *Rheology of the Earth*, 2nd edn Chapman & Hall, London.

Ranalli, G. (2001) Mantle rheology: radial and lateral viscosity variations inferred from microphysical creep laws. *J. Geodyn.* **32**, 425–44.

Ranalli, G. & Murphy, D.C. (1987) Rheological stratification of the lithosphere. *Tectonophysics* **132**, 281–96.

Ranero, C.R. & Reston, T.J. (1999) Detachment faulting at ocean core complexes. *Geology* **27**, 983–6.

Rapine, R. *et al.* (2003) Crustal structure of northern and southern Tibet from surface wave dispersion analysis. *J. geophys. Res.* **108**, doi:10.1029/2001JB000445.

Raymo, M.E. & Ruddiman, W.F. (1992) Tectonic forcing of Late Cenozoic climate. *Nature* **359**, 117–22.

Reston, T.J., Krawczyk, C.M. & Klaeschen, D. (1996) The S reflector west of Galicia (Spain): evidence from prestack depth migration for detachment faulting during continental breakup. *J. geophys. Res.* **101**, 8075–92.

Rice, J.R. (1992) Fault stress states, pore pressure distributions, and the weakness of the San Andreas fault. *In* Evans, B. & Wong, T.F. (eds) *Fault Mechanics and Transport Properties of Rocks*, pp. 475–504. Academic Press, New York.

Richards, J.P. (2003) Tectono-Magmatic Precursors for Porphyry Cu-(Mo-Au) Deposit Formation. *Econ. Geol.* **98**, 1515–33.

Richardson, A.N. & Blundell, D.J. (1996) Continental collision in the Banda arc. *In* Hall, R. & Blundell, D.J. (eds) *Tectonic Evolution in Southeast Asia. Spec. Pub. geol. Soc. Lond.* **106**, 47–60.

Richardson, S.H. *et al.* (2001) Archean subduction recorded by Re-Os isotopes in eclogitic sulfide inclusions in Kimberley diamonds. *Earth planet. Sci. Lett* **191**, 257–66.

Ringwood, A.E. (1974) The petrological evolution of island arc systems. *J. geol. Soc. Lond.* **130**, 183–204.

Ringwood, A.E. (1975) *Composition and Petrology of the Earth's Mantle.* McGraw-Hill, New York.

Ringwood, A.E. (1977) Petrogenesis in island arc systems. *In* Talwani, M. & Pitman, W.C. III (eds) *Island Arcs, Deep Sea Trenches and Back-arc Basins. Maurice Ewing Series* **I**, pp. 311–24. American Geophysical Union, Washington, DC.

Rino, S. *et al.* (2004) Major episodic increase of continental crustal growth determined from zircon ages of river sands; implications for mantle overturns in the Early Precambrian. *Phys. Earth planet. Interiors* **146**, 369–94.

Ritger, S., Canson, B. & Snegge, E. (1987) Methane-derived authigenic carbonates formed by subduction-induced pore-water expulsion along the Oregon/Washington margin. *Bull. geol. Soc. Am.* **98**, 147–56.

Ritsema, J. & van Heijst, H.J. (2000) New seismic model of the upper mantle beneath Africa. *Geology* **28**, 63–6.

Ritsema, J., van Heijst, H.J. & Woodhouse, J.H. (1999) Complex shear wave velocity structure imaged beneath Africa and Iceland. *Science* **286**, 1925–8.

Ritsema, J. *et al.* (1998) Upper mantle seismic velocity structure beneath Tanzania: implications for the stability of cratonic roots. *J. geophys. Res.* **103**, 21 201–14.

Rivers, T. (1997) Lithotectonic elements of the Grenville Province: review and tectonic implications. *Precambrian Res.* **86**, 117–54.

Roberts, A., Lundin, E.R. & Kusznir, N.J. (1997) Subsidence of the Vøring Basin and the influence of the Atlantic continental margin. *J. Geol. Soc. Lond.* **154**, 551–7.

Robl, J. & Stüwe, K. (2005a) Continental collision with finite indenter strength: 1. Concept and model formulation. *Tectonics* **24**, TC4005, doi:10.1029/2004TC0011727.

Robl, J. & Stüwe, K. (2005b) Continental collision with finite indenter strength: 2. European eastern Alps. *Tectonics* **24**, TC4014, doi:10.1029/2004TC001741.

Rogers, A.M. *et al.* (1991) The seismicity of Nevada and some adjacent parts of the Great Basin. *In* Slemmons, D.B. *et al.* (eds) *Neotectonics of North America. Decade Map*, **1**, 153–84. Geological Society of America, Boulder, CO.

Romanowicz, B. (2003) Global mantle tomography: progress status in the past 10 years. *Ann. Rev. Earth Planet. Sci.* **31**, 303–28.

Romm, J. (1994) A new forerunner for continental drift. *Nature* **367**, 407–8.

Rona, P.A. (1977) Plate tectonics, energy and mineral resources: basic research leading to payoff. *EOS Trans. Amer. Geophys. Un.* **58**, 629–39.

Rona, P.A. (1984) Hydrothermal mineralization at seafloor spreading centres. *Earth Sci. Rep.* **20**, 1–104.

Rona, P.A. & Richardson, E.S. (1978) Early Cenozoic global plate reorganization. *Earth planet. Sci. Lett.* **40**, 1–11.

Royden, L. (1996) Coupling and decoupling of the crust and mantle in convergent orogens: implications for strain partitioning in the crust. *J. geophys. Res.* **101**, 17 679–705.

Royden, L. & Keen, C.E. (1980) Rifting process and thermal evolution of the continental margin of eastern Canada determined from subsidence curves. *Earth planet. Sci. Lett.* **51**, 343–61.

Rudnick, R.L. (1992) Xenoliths – samples of the lower continental crust. In Fountain, D.M., Arculus, R. & Kay, R.W. (eds) *Continental Lower Crust*, pp. 269–316. Elsevier, Amsterdam.

Rudnick, R.L. & Fountain, D.M. (1995) Nature and composition of the continental crust: a lower crustal perspective. *Rev. Geophys.* **33**, 267–309.

Rudnick, R.L. & Gao, S. (2003) The Composition of the Continental Crust. In Rudnick, R.L. (ed.) *The Crust*, pp. 1–64. Holland, H.D. & Turekian, H.K. (eds) *Treatise on Geochemistry*, **3**. Elsevier–Pergamon, Oxford.

Rupke, N.A. (1970) Continental drift before 1900. *Nature* **227**, 349–50.

Ruppel, C. (1995) Extensional processes in continental lithosphere. *J. geophys. Res.* **100**, 24 187–215.

Rusby, R.I. & Searle, R.C. (1995) A history of the Easter Island microplate, 5.25 Ma to present. *J. geophys. Res.* **100**, 12 617–40.

Sabadini, R. & Yuen, D.A. (1989) Mantle stratification and long-term polar wander. *Nature* **339**, 373–5.

Sabadini, R., Yuen, D.A. & Boschi, E. (1982) Polar wandering and the forced responses of a rotating, multi-layered, viscoelastic planet. *J. geophys. Res.* **81**, 2885–903.

Saffer, D.M. (2003) Pore pressure development and progressive dewatering in underthrust sediments at the Costa Rican subduction margin: comparison with northern Barbados and Nankai. *J. geophys. Res.* **108** (B5), 2261, doi:10.1029/2002JB001787.

Saffer, D.M. & Bekins, B.A. (2002) Hydrologic controls on the mechanics and morphology of accretionary wedges and thrust belts. *Geology* **30**, 271–4.

Saffer, D.M. & Bekins, B.A. (2006) An evaluation of factors influencing pore pressure in accretionary complexes: implications for taper angle and wedge mechanics. *J. geophys. Res.* **111**, B04101, doi:10.1029/2005JB003990.

Saint Blanquat, M. et al. (1998) Transpressional kinematics and magmatic arcs In Holdsworth, R.E, Strachan, R.A. & Dewey, J.F. (eds) *Continental Transpressional and Transtensional Tectonics. Spec. Pub. geol. Soc. Lond.* **135**, 327–40.

Saintot, A. et al. (2003) Structures associated with inversion of the Donbas fold belt (Ukraine and Russia). *Tectonophysics* **373**, 181–207.

Salisbury, M.H. & Christensen, N.I. (1978) The seismic velocity structure of a traverse through the Bay of Islands ophiolite complex, Newfoundland, an exposure of oceanic crust and upper mantle. *J. geophys. Res.* **83**, 805–17.

Sass, J.H. et al. (1994) Thermal regime of the southern Basin and Range Province: 1. Heat flow data from Arizona and the Mojave Desert of California and Nevada. *J. geophys. Res.* **99**, 22,093–119.

Savage, J.C. (2000) Viscoelastic-coupling model for the earthquake cycle driven from below. *J. geophys. Res.* **105**, B11, 25,525 32.

Savage, J.C. & Burford, R.O. (1973) Geodetic determination of relative plate motion in central California. *J. geophys. Res.* **78**, 832–45.

Savage, J.C. et al. (2004a) Strain accumulation across the Coast Ranges at the latitude of San Francisco, 1994–2000. *J. geophys. Res.* **109**, B03413, doi:10.1029/2003JB002612.

Savage, J.C., Svarc, J.L & Prescott, W.H. (2004b) Interseismic strain and rotation rates in the northeast Mojave domain, eastern California. *J. geophys. Res.* **109**, B02406, doi:10.1029/2003JB002705.

Savage, M.K. (1999) Seismic anisotropy and mantle deformation: what have we learned from shear wave splitting? *Rev. Geophys.* **37**, 65–106.

Savage, M.K. & Sheehan, A.F. (2000) Seismic anisotropy and mantle flow from the Great Basin to the Great Plains, western United States. *J. geophys. Res.* **105**, 13,715–34.

Sawkins, F.J. (1984) *Metal Deposits in Relation to Plate Tectonics.* Springer-Verlag, Berlin.

Sawyer, T.L. (1999) Assessment of contractional deformation rates of the Mt. Diablo fold and thrust belt, eastern San Francisco Bay Region, Northern California. *USGS National Earthquake Hazards Reduction Program.* Final Technical Report No. 98-HQ-GR-1006, pp. 1–53. US Geol. Survey, Reston, Virginia.

Sayers, J. et al. (2001) Nature of the continent–ocean transition on the non-volcanic rifted margin of the central Great Australian Bight. In Wilson, R.C.L., et al. (eds) *Non-volcanic Rifting of Continental Margins: a comparison of evidence from land and sea. Spec. Pub. geol. Soc. Lond.* **187**, 51–76.

Schellart, W.P., Lister, G.S. & Jessell, M.W. (2002) Analogue modeling of arc and backarc deformation in the New Hebrides arc and North Fiji Basin. *Geology* **30**, 311–14.

Scherwath, M. et al. (2002) Pn anisotropy and distributed upper mantle deformation associated with a continental transform. *Geophys. Res. Lett.* **29**, doi:10.1029/2001GLO14179.

Scherwath, M. et al. (2003) Lithospheric structure across oblique continent collision in New Zealand from wide-angle P wave modeling. *J. geophys. Res.* **108**, 2566, doi:10.1029/2002JB002286.

Scherwath, M. et al. (2006) Three-dimensional lithospheric deformation and gravity anomalies associated with oblique continental collision in South Island, New Zealand. *Geophys. J. Int.* **167**, 906–16.

Schiellerup, H. et al. (2000) Re-Os isotopic evidence for a lower crustal origin of massif-type anorthosites. *Nature* **405**, 781–4.

Schilling, J.-G., Anderson, R.N. & Vogt. P. (1976) Rare earth, Fe and Ti variations along the Galapagos spreading centre and their relationship to the Galapagos mantle plume. *Nature* **261**, 108–13.

Schmitz, M.D. et al. (2004) Subduction and terrane collision stabilize the western Kaapvaal craton tectosphere 2.9 billion years ago. *Earth planet. Sci. Lett.* **222**, 363–76.

Schnitker, D. (1980) North Atlantic oceanography as a possible cause of Antarctic glaciation and eutrophication. *Nature* **284**, 615–16.

Scholz, C.H. (1998) Earthquakes and friction laws. *Nature* **391**, 37–42.

Scholz, C.H. (2000) Evidence for a strong San Andreas fault. *Geology* **28**, 163–6.

Schoonmaker, J. (1986) Clay mineralogy and diagenesis of sediments from deformation zones in the Barbados accretionary prism. *In* Moore, J.C. (ed.) *Synthesis of Structural Fabrics in Deep Sea Drilling Project Cores from Forearcs. Geol. Soc. Am. Mem.* **166**, 105–16.

Schroeder, T. & John, B.E. (2004) Strain localization on an oceanic detachment fault system, Atlantis Massif, 30°N, Mid-Atlantic Ridge. *Geochem. Geophys. Geosyst.* **5**, Q11007, doi:10.1029/2004GC000728.

Schubert, G., Turcotte, D.L. & Olsen P. (2001) *Mantle Convection in the Earth and Planets*, pp. 940. Cambridge University Press, Cambridge.

Schulte-Pelkum, V. *et al.* (2005) Imaging the Indian subcontinent beneath the Himalaya. *Nature.* **435**, doi:10.1038/nature03678.

Schulze, D.J. (1989) Constraints on the abundance of eclogite in the upper mantle. *J. geophys. Res* **94**, 4205–12.

Schurr, B. *et al.* (2003) Complex patterns of fluid and melt transport in the central Andean subduction zone revealed by attenuation tomography. *Earth planet. Sci. Lett.* **215**, 105–19.

Sclater, J.G. & Francheteau, J. (1970) The implications of terrestrial heat flow observations on current tectonic and geochemical models of the crust and upper mantle of the Earth. *Geophys. J. Roy. astr. Soc.* **20**, 509–42.

Sclater, J.G., Jaupart, C. & Galson, D. (1980) The heat flow through oceanic and continental crust and the heat loss of the Earth. *Rep. Geophys. Space Phys.* **18**, 269–311.

Scotese, C.R., Gahagan, L.M. & Larson, R.L. (1988) Plate tectonic reconstructions of the Cretaceous and Cenozoic ocean basins. *Tectonophysics* **155**, 27–48.

Scott, D.J., Helmstaedt, H. & Bickle, M.J. (1992) Purtuniq ophiolite, Cape Smith belt, northern Quebec, Canada: a reconstructed section of early Proterozoic oceanic crust. *Geology* **20**, 173–6.

Scott, M.R. *et al.* (1973) Hydrothermal manganese in the median valley of the Mid-Atlantic Ridge. *EOS Trans. Amer. Geophys. Un.* **54**, 244.

Scrutton, C.T. (1967) Absolute time data from palaeontology. *In* Runcorn, S.K. *et al.* (eds) *International Dictionary of Geophysics I*, p. 1. Pergamon Press, Oxford.

Scrutton, R.A. (1979) On sheared passive continental margins. *Tectonophysics* **59**, 293–305.

Searle, D.L. & Panayiotou, A. (1980) Structural implications in the evolution of the Troodos massif, Cyprus. *In* Panayiotou, A. (ed.) *Ophiolites*, pp. 50–60. Geol. Surv. Cyprus.

Searle, M.P. (1999) Extensional and compressional faults in the Everest–Lhotse massif, Khumbu Himalaya, Nepal. *J. geol. Soc. Lond.* **156**, 227–40.

Searle, M.P. *et al.* (1999) Age of crustal melting, emplacement and exhumation history of the Shivling leucogranite, Garhwal Himalaya. *Geol. Mag.* **136**, 513–25.

Searle, R.C. (1983) Multiple, closely spaced transform faults in fast-slipping fracture zones. *Geology* **11**, 607–10.

Searle, R.C. *et al.* (1989) Comprehensive sonar imagery of the Easter microplate. *Nature* **341**, 701–5.

Sella, G.F., Dixon, T.H. & Mao, A. (2002) REVEL: a model for Recent plate velocities from space geodesy. *J. geophys. Res.* **107**, 2081, doi:10.1029/2000JB000033.

Sempéré, J.-C., Purdy, G.M. & Schouten, H. (1990) Segmentation of the Mid-Atlantic Ridge between 24°N and 30°40'N. *Nature* **344**, 427–9.

Şengör A.M.C. & Natal'in B.A. (1996) Turkic-type orogeny and its role in the making of the continental crust. *Ann. Rev. Earth planet. Sci.* **24**, 263–337.

Shackleton, N.J. & Kennett, J.P. (1975) Paleotemperature history of the Cenozoic and the initiation of Antarctic glaciation: oxygen and carbon isotope analyses in DSDP Sites 277, 270 and 281. *In* Kennett, J.P. & Houtz, R.E. (eds) *Initial Reports of the Deep Sea Drilling Project* **29**, pp. 743–56. US Government Printing Office, Washington, DC.

Shackleton, R.M., Dewey, J.F. & Windley, B.F. (1988) *Tectonic Evolution of the Himalayas and Tibet.* Royal Society, London.

Shamir, G., Eyal, Y. & Bruner I. (2005) Localized versus distributed shear in transform plate boundary zones: the case of the Dead Sea Transform in the Jericho Valley. *Geochem. Geophys. Geosyst.* **6**, Q05004, doi:10.1029/2004GC000751.

Shearer, P.-M. & Masters, G.M. (1992) Global mapping of topography on the 660-km discontinuity. *Nature* **355**, 791–6.

Shen, Z.-K. *et al.* (2000) Contemporary crustal deformation in east Asia constrained by Global Positioning System measurements. *J. geophys. Res.* **105**, 5721–34.

Shen-Tu, B., Holt, W.E. & Haines, A.J. (1998) Contemporary kinematics of the Western United States determined from earthquake moment tensors, very long baseline interferometry, and GPS observations. *J. geophys. Res.* **103**, 18,087–117.

Siame, L.L. *et al.* (2005) Deformation partitioning in flat subduction setting: case of the Andean foreland of western Argentina (28°S–33°S). *Tectonics* **24**, TC5003, doi:10.1029/2005TC001787.

Sibson, R.H. (1990) Conditions for fault-valve behaviour. *In* Knipe, R.J. & Rutter, E.H. (eds) *Deformation Mechanisms, Rheology and Tectonics. Spec. Pub. geol. Soc. Lond.* **54**, 15–28.

Sillitoe, R.H. (1972a) Formation of certain massive sulphur deposits at sites of sea-floor spreading. *Trans. Inst. Min. Metall.* **81**, B141–8.

Sillitoe, R.H. (1972b) A plate tectonic model for the origin of porphyry copper deposits. *Econ. Geol.* **67**, 184–97.

Silver, E.A. (2000) Leg 170: synthesis of fluid–structural relationships of the Pacific margin of Costa Rica. *In* Silver, E.A., Kimura, G. & Shipley, T.H. (eds) *Proceedings of the Ocean Drilling Program, Scientific Results*, **170**, pp. 1–11. College Station, TX.

Silver, E.A. *et al.* (1983) Back-arc thrusting in the eastern Sunda arc, Indonesia; a consequence of arc–continental collision. *J. geophys. Res.* **88**, 7429–48.

Singh, S.C. *et al.* (1998) Melt to mush variations in crustal magma properties along the ridge crest at the southern East Pacific Rise. *Nature* **394**, 874–8.

Sinha, M.C. *et al.* (1998) Magmatic processes at slow spreading ridges: implications of the RAMESSES experiment at 57°45′N on the Mid-Atlantic ridge. *Geophys. J. Int.* **135**, 731–45.

Sinton, J.M. & Detrick, R.S. (1992) Mid-ocean ridge magma chambers. *J. geophys. Res.* **97**, 197–216.

Sisson, T.W. & Bronto, S. (1998) Evidence for pressure-release melting beneath magmatic arcs from basalt at Galunggung, Indonesia. *Nature* **391**, 833–86.

Sleep, N.H. (1975) Formation of oceanic crust: some thermal constraints. *J. geophys. Res.* **80**, 4037–42.

Sleep, N.H. (2003) Survival of Archean cratonal lithosphere. *J. geophys. Res.* **108**, 2302, doi:10.1029/2001JB000169.

Sleep, N.H. (2005) Evolution of continental lithosphere. *Annu. Rev. Earth planet. Sci.* **33**, 369–93.

Smelov, A.P. & Beryozkin, V.I. (1993) Retrograded eclogites in the Olekma granite–greenstone region, Aldan Shield. *Precambrian Res.* **62**, 419–30.

Smith, A.G. (1999) Gondwana: its shape, size and position from Cambrian to Triassic times. *J. African Earth Sci.* **28**, 71–97.

Smith, A.G. & Hallam, A. (1970) The fit of the southern continents. *Nature* **225**, 139–44.

Smith, A.G. & Pickering, K.T. (2003) Oceanic gateways as a critical factor to initiate icehouse Earth. *J. Geol. Soc. London* **160**, 337–40.

Smith, A.G., Smith, D.G. & Funnell, B.M. (1994) *Atlas of Mesozoic and Cenozoic coastlines.* Cambridge University Press, Cambridge.

Smith, D.E. *et al.* (1985) Global plate motion results from satellite laser ranging (abstract). *EOS Trans. Amer. Geophys. Un.* **66**, 848.

Smith, D.K. & Cann, J.R. (1993) Building the crust at the Mid-Atlantic Ridge. *Nature* **365**, 707–15.

Smith, M. & Mosley, P. (1993) Crustal heterogeneity and basement influence on the development of the Kenya rift, East Africa. *Tectonics* **12**, 591–606.

Smith, W.H.F. & Sandwell, D.T. (1997) Global seafloor topography from satellite altimetry and ship depth soundings. *Science* **277**, 1957–62.

Smithies, R.H. (2000) The Archean tonalite–tondhjemite–granodiorite (TTG) series is not an analogue of Cenozoic adakite. *Earth planet. Sci. Lett.* **182**, 115–25.

Smithson, S.B. & Brown, S.K. (1977) A model for the lower continental crust. *Earth planet. Sci. Lett.* **35**, 134–44.

Snider, A. (1858) *La Création et ses Mystères Dévoilés.* Frank and Dentu, Paris.

Snow, J.K. & Wernicke, B. (2000) Cenozoic tectonism in the central Basin and Range: magnitude, rate, and distribution of upper crustal strain. *Amer. J. Sci.* **300**, 659–719.

Sobolev, S.V. & Babeyko, A.Y. (2005) What drives orogeny in the Andes? *Geology* **33**, 671–20.

Sobolev, S.V. *et al.* (2005) Thermo-mechanical model of the Dead Sea transform. *Earth planet. Sci. Lett.* **238**, 78–95.

Solomon, M. (1990) Subduction, arc reversal, and the origin of porphyry copper-gold deposits in island arcs. *Geology* **18**, 630–3.

Solomon, S.C., Sleep, N.H. & Richardson, R.M. (1975) On the forces driving plate tectonics: inferences from absolute plate velocities and intraplate stress. *Geophys. J. Roy. astr. Soc.* **42**, 769–801.

Somoza, R. (1998) Updated Nazca (Farallon)–South America relative motions during the last 40 My: implications for mountain building in the central Andean region. *JS Am. Earth Sci.* **11**, 211–15.

Sonder, L.J. & Jones, C.H. (1999) Western United States extension: how the West was widened. *Annu. Rev. Earth planet. Sci.* **27**, 417–62.

Spada, G., Ricard, Y. & Sabadini, R. (1992) Excitation of true polar wander by subduction. *Nature* **360**, 452–4.

Späth A., le Roex, A.P. & Opiyo-Akech, N. (2001) Plume–lithosphere interaction and the origin of continental rift-related alkaline volcanism – the Chyulu Hills Volcanic Province, southern Kenya. *J. Petrol.* **42**, 765–87.

Spence, G.D. & Asudeh, I. (1993) Seismic velocity structure of the Queen Charlotte Basin beneath Hecate Strait. *Can. J. Earth Sci.* **30**, 787–805.

Spence, W. (1987) Slab pull and the seismotectonics of subducting lithosphere. *Rev. Geophys.* **25**, 55–69.

Spicer, R.A. *et al.* (2003) Constant elevation of southern Tibet over the past 15 million years. *Nature* **421**, 622–4.

Spitzak, S. & DeMets, C. (1996) Constraints on present day plate motions south of 30°S from satellite altimetry. *Tectonophysics*, **253**, 167–208.

Spudich, P. & Orcutt, J. (1980) A new look at the seismic velocity structure of the oceanic crust. *Rep. Geophys. Space Phys.* **18**, 627–45.

Stampfli, G.M. & Borel, G.D. (2002) A plate tectonic model for the Paleozoic and Mesozoic constrained by dynamic plate boundaries and restored synthetic oceanic isochrons. *Earth planet. Sci. Lett.* **196**, 17–33.

Stanley, S.M. & Hardie, L.A. (1999) Hypercalcification: paleontology links plate tectonics and geochemistry to sedimentology. *GSA Today* **9**, 1–7.

Stauder, W. (1968) Mechanism of the Rat Island earthquake sequence of February, 1965, with reference to island arcs and sea-floor spreading. *J. geophys. Res.* **73**, 3847–58.

Steckler, M.S. (1985) Uplift and extension in the Gulf of Suez: indications of induced mantle convection. *Nature* **317**, 135–9.

Stein, C.A. & Stein, S. (1992) A model for the global variation in oceanic depth and heat flow with lithospheric age. *Nature* **359**, 123–9.

Stein, S. & Stein, C.A. (1996) Thermo-mechanical evolution of oceanic lithosphere: implications for the subduction process and deep earthquakes. *In* Bebout, G.E. *et al.* (eds) *Susduction Top to Bottom.* Geophys. Monogr. Ser. **96**, 1–17. American Geophysical Union, Washington, DC.

Stein, S. & Wysession, M. (2003) *An Introduction to Seismology, Earthquakes, and Earth Structure.* Blackwell Publishing, Oxford.

Steinberger, B. & O'Connell, R.J. (2000) Effects of mantle flow on Hotspot motion. *In* Richard, M., Gordon, R.G. & van der Hilst, R.O. (eds) *The History and Dynamics of Global Plate Motions.* Geophys. Monogr. Ser. **121**, pp. 377–398. American Geophysical Union, Washington, DC.

Stern, R.A., Syme, E.C. & Lucas, S.B. (1995) Geochemistry of 1.9 Ga MORB- and OIB-like basalts from the Amisk collage, Flin

Flon Belt, Canada, Evidence for an intra-oceanic origin. *Geochim. cosmochim. Acta* **59**, 3131–54.

Stern, R.J. (2002) Subduction zones. *Rev. Geophys.* **40**, RG4003, 1-38, doi:10.1029/2001RG000108.

Stern, R.J., Fouch, M.J. & Klemperer, S.L. (2003) An overview of the Izu–Bonin–Mariana Subduction Factory. *In* Eiler, J. (ed.) *Inside the Subduction Factory. Geophys. Monogr. Ser.* **138**, pp. 175–222. American Geophysical Union, Washington, DC.

Stern, T. *et al.* (2000) Teleseismic P-wave delays and modes of shortening the mantle lithosphere beneath South Island, New Zealand. *J. geophys. Res.* **105**, 21,615–31.

Stern, T. *et al.* (2001) Low seismic wave-speeds and enhanced fluid pressure beneath the Southern Alps of New Zealand. *Geology* **29**, 679–82.

Stern, T., Okaya, D. & Scherwath, M. (2002) Structure and strength of a continental transform from onshore–offshore seismic profiling of the South Island, New Zealand. *Earth Planets Space* **54**, 1011–19.

Stern, T.A. (1987) Asymmetric back-arc spreading, heat flux and structure associated with the central volcanic region of New Zealand. *Earth planet. Sci. Lett.* **85**, 265–76.

Stewart, B.M. & DePaolo, D.J. (1996) Isotopic studies of processes in mafic magma chambers: III. The Muskox Intrusion, Northwest Territories, Canada. *In* Basu, A.S.(ed.) *Earth Processes, Reading the Isotopic Code. Geophys. Monogr. Ser.* **95**, pp. 277–92 American Geophysical Union, Washington, DC.

Stewart, J.A. (1990) *Drifting Continents and Colliding Paradigms: perspectives on the geoscience revolution.* Indiana University Press, Bloomington.

Stock, J. & Molnar, P. (1988) Uncertainties and implications of the Late Cretaceous and Tertiary position of North America relative to the Farallon, Kula, and Pacific plates. *Tectonics* **7**, 1339–84.

Stockli, D.F. *et al.* (2001) Miocene unroofing of the Canyon Range during extension along the Sevier Desert detachment, west-central Utah. *Tectonics* **20**, 289–307.

Stolar, D.B., Willett, S.D. & Roe, G.H. (2006) Climate and tectonic forcing of a critical orogen. *In* Willett, S.D. *et al.* (eds) *Tectonics, Climate, and Landscape Evolution. Geol. Soc. Am. Sp. Paper* **398**, 241–50.

Stolper, E. & Newman, S. (1994) The role of water in the petrogenesis of Mariana Trough magmas. *Earth planet. Sci. Lett.* **121**, 293–325.

Storey, B.C. (1993) The changing face of late Precambrian and early Palaeozoic reconstructions. *J. geol. Soc. Lond.* **150**, 665–8.

Storey, B.C. (1995) The role of mantle plumes in continental breakup: case histories from Gondwanaland. *Nature* **377**, 301–8.

Stow, D.A.V. & Lovell, J.P.B. (1979) Contourites: their recognition in modern and ancient sediments. *Earth Sci. Rev.* **14**, 251–91.

Suda, Y. (2004) Crustal anatexis and evolution of granitoid magma in Permian intra-oceanic island arc, the Asago body of the Yakuno ophiolite, Southwest Japan. *J. Mineral. Petrol. Sci.* **99**, 339–56.

Sun, S.S., Nesbitt, R.W. & Sharashin, A.Y. (1979) Geochemical characteristics of mid-ocean ridge basalts. *Earth planet. Sci. Lett.* **44**, 119–38.

Sutherland, R., Berryman, K. & Norris, R. (2006) Quaternary slip rate and geomorphology of the Alpine fault: implications for kinematics and seismic hazard in southwest New Zealand. *Bull. geol. Soc. Am.* **118**, 464–74.

Sykes, L.R. (1966) The seismicity and deep structure of island arcs. *J. geophys. Res.* **71**, 2981–3006.

Sykes, L.R. (1967) Mechanism of earthquakes and nature of faulting on the mid-oceanic ridges. *J. geophys. Res.* **72**, 2131–53.

Sylvester, A.G. (1988) Strike-slip faults. *Bull. geol. Soc. Am.* **100**, 1666–703.

Tackley, P.J. (2000) Mantle convection and plate tectonics: toward an integrated physical and chemical theory. *Science* **288**, 2002–7.

Tackley, P.J. *et al.* (1993) Effects of an endothermic phase transition at 670 km depth in a spherical model of convection in the Earth's mantle. *Nature* **361**, 699–704.

Taira, A. (2001) Tectonic evolution of the Japanese Island Arc System. *Annu. Rev. Earth Planet. Sci.* **29**, 109–34.

Takagi, H. (1986) Implications of mylonitic microstructures for the geotectonic evolution of the Median Tectonic Line, central Japan. *J. struct. Geol.* **8**, 3–14.

Talwani, M. & Watts, A.B. (1974) Gravity anomalies seaward of deep-sea trenches and their tectonic implications. *Geophys. J. Roy. astr. Soc.* **36**, 57–90.

Talwani, M., Le Pichon, X. & Ewing, M. (1965) Crustal structure of the mid-ocean ridges 2. Computed model from gravity and seismic reduction data. *J. geophys. Res.* **70**, 341–52.

Tapley, B.D., Schutz, B.F. & Eanes, R.J. (1985) Station coordinates, baselines, and Earth rotation from LAGEOS laser ranging: 1976–1984. *J. geophys. Res.* **90**, 9235–48.

Tapponnier, P. & Molnar, P. (1976) Slip-line field theory and large-scale continental tectonics. *Nature* **264**, 319–24.

Tapponnier, P. *et al.* (1982) Propagating extrusion tectonics in Asia: new insights from simple experiments with plasticene. *Geology* **10**, 611–16.

Tapponnier, P. *et al.* (2001) Oblique stepwise rise and growth of the Tibet plateau. *Science.* **294**, 1671–7.

Tarduno, J.A. & Cottrell, R.D. (1997) Paleomagnetic evidence for motion of the Hawaiian hotspot during formation of the Emperor seamounts. *Earth planet. Sci. Lett.* **153**, 171–80.

Tarling, D.H. (ed.) (1981) *Economic Geology and Geotectonics.* Blackwell Scientific Publications. Oxford.

Tarling, D.H. (1983) *Palaeomagnetism.* Chapman & Hall, London.

Tarling, D.H. & Runcorn, S.K. (eds) (1973) *Implications of Continental Drift to the Earth Sciences*, vols 1 & 2. Academic Press, London.

Tarling, D.H. & Tarling, M.P. (1971) *Continental Drift. A study of the Earth's moving surface.* Bell, London.

Taylor, B. & Huchon, P. (2002) Active continental extension in the western Woodlark Basin: a synthesis of Leg 180 results. *In* Huchon, P., Taylor, B. & Klaus, A. (eds) *Proc. ODP, Sci. Results*, **180**, 1–36.[online].

Taylor, B. & Martínez, F. (2003) Back-arc basin basalt systematics. *Earth planet. Sci. Lett.* **210**, 481–97.

Taylor, B., Crook, K. & Sinton, J. (1994) Extensional transform zones and oblique spreading centers. *J. geophys. Res.* **99**, 19 707–18.

Taylor, B. *et al.* (1995) Continental rifting and initial sea-floor spreading in the Woodlark Basin. *Nature* **374**, 534–7.

Taylor, B., Goodliffe, A.M. & Martínez, F. (1999) How continents break up: insights from Papua New Guinea. *J. geophys. Res.* **104**, 7497–512.

Taylor, F.B. (1910) Bearing of the Tertiary mountain belt on the origin of the Earth's plan. *Bull. geol. Soc. Am.* **21**, 179–226.

Taylor, R.T. & Scott, M.M. (1985) *The Continental Crust: its composition and evolution.* Blackwell Scientific Publications, Oxford.

Tebbens, S.F. *et al.* (1997) The Chile Ridge: a tectonic framework. *J. geophys. Res.* **102**, 12,035–59.

Tegner, C. *et al.* (1998) ^{40}Ar–^{39}Ar geochronology of Tertiary mafic intrusions along the East Greenland rifted margin: relation to flood basalts and the Iceland hotspot track. *Earth planet. Sci. Lett.* **156**, 75–88.

ten Brink, U.S. *et al.* (1993) Structure of the Dead Sea pull-apart basin from gravity analysis. *J. geophys. Res.* **98**, 21 887–94.

Tepper, J.H. *et al.* (1993) Petrology of the Chilliwack batholith, North Cascades, Washington: generation of calc-alkaline granitoids by melting of mafic lower crust with variable water fugacity. *Contrib. Mineral. Petrol.* **113**, 333–51.

Tessema, A. & Antoine, L.A.G. (2004) Processing and interpretation of the gravity field of the East African Rift: implication for crustal extension. *Tectonophysics* **394**, 87–110.

Teyssier, C. & Tikoff, B. (1998) Strike-slip partitioned transpression of the San Andreas fault system: a lithospheric-scale approach. *In* Holdsworth, R.E., Strachan, R.A. & Dewey, J.F. (eds) *Continental Transpressional and Transtensional Tectonics. Spec. Pub. geol. Soc. Lond.* **135**, 143–58.

Thatcher, W. (1979) Systematic inversion of geodetic data in central California. *J. geophys. Res.* **84**, 2283–95.

Thatcher, W. (2003) GPS constraints on the kinematics of continental deformation. *Int. Geol. Rev.* **45**, 191–212.

Thomas, C., Livermore, R. & Pollitz, F. (2003) Motion of the Scotia Sea plates. *Geophys. J. Int.* **155**, 789–804.

Thompson, G. & Melson, W.G. (1972) The petrology of oceanic crust across fracture zones in the Atlantic Ocean: evidence of a new kind of sea-floor spreading. *J. Geol.* **80**, 526–38.

Thorkelson, D.J., *et al.* (2001) Early Proterozoic magmatism in Yukon, Canada: constraints on the evolution of northwestern Laurentia. *Can. J. Earth Sci.* **38**, 1479–94.

Thorpe, R.S. (ed.) (1982) *Andesites.* Wiley, London.

Thurber, C.H. & Aki, K. (1987) Three-dimensional seismic imaging. *Ann. Rev. Earth Planet. Sci.* **15**, 115–39.

Thybo, H., Ross, A.R. & Egorkin, A.V. (2003) Explosion seismic reflections from the Earth's core. *Earth planet. Sci. Lett.* **216**, 693–702.

Tiberi, C. *et al.* (2005) Inverse models of gravity data from the Red Sea–Aden–East African rifts triple junction zone. *Geophys. J. Int.* **163**, 775–87.

Tilmann, F. *et al.* (2003) Seismic imaging of the downwelling Indian lithosphere beneath central Tibet. *Science* **300**, 1424–7.

Tippett, J.M. &. Kamp, P.J.J. (1993) Fission track analysis of the late Cenozoic vertical kinematics of continental Pacific Crust, South Island, New Zealand. *J. geophys. Res.* **98**, 16,119–48.

Tissot, B. (1979) Effects on prolific petroleum source rocks and major coal deposits caused by sea-level changes. *Nature* **277**, 463–5.

Titus, S.J., DeMets, C. & Tikoff, B. (2005) New slip rate estimates for the creeping segment of the San Andreas fault, California. *Geology* **33**, 205–8.

Tohver, E. *et al.* (2002) Paleogeography of the Amazon Craton at 1.2 Ga; early Grenvillian collision with the Llano segment of Laurentia. *Earth planet. Sci. Lett.* **199**, 185–200.

Tolstoy, M. *et al.* (2001) Seismic character of volcanic activity at the ultraslow-spreading Gakkel Ridge. *Geology* **29**, 1139–42.

Tomlinson, K.Y. & Condie, K.C. (2001) Archean mantle plumes: evidence from greenstone belt geochemistry. *In* Ernst, R.E. & Buchan, K.L. (eds) *Mantle Plumes: their identification through time. Geol. Soc. Amer. Sp. Paper* **352**, 341–57.

Toomey, D.R. *et al.* (1990) The three-dimensional seismic velocity structure of the East Pacific Rise near latitude 9°30′N. *Nature* **347**, 639–45.

Torsvik, T.H. (2003) The Rodinia jigsaw puzzle. *Science* **300**, 1379–81.

Torsvik, T.H. & Van der Voo, R. (2002) Refining Gondwana and Pangea Paleogeography: estimates of Phanerozoic non-dipole (octupole) fields. *Geophys. J. Int.* **151**, 771–94.

Townend, J. & Zoback, M.D. (2004) Regional tectonic stress near the San Andreas fault in central and southern California. *Geophys. Res. Lett.* **31**, L15S11, doi:10.1029/2003GL018918.

Trendall, A.F. *et al.* (2004) SHRIMP zircon ages constraining the depositional history of the Hamersley Group, Western Australia. *Australian J. Earth Sci.* **51**, 621–44.

Tsikalas, F., Eldholm, O. & Faleide, J.I. (2005) Crustal structure of the Lofoten–Vesterålen continental margin, off Norway. *Tectonophysics* **404**, 151–74.

Tsuji, T. *et al.* (2006) Modern and ancient seismogenic out-of-sequence thrusts in the Nankai accretionary prism: comparison of laboratory-derived physical properties and seismic reflection data. *Geophys. Res. Lett.* **33**, L18309, doi:10.1029/2006GL027025.

Tuisku, P. & Huhma, H. (1998) Eclogite from the SW-marginal zone of the Lapland Granulite belt: evidence from the 1.90–1.88 Ga subduction zone. *In* Hanski, E. & Vuollo, J. (eds) *International Ophiolite Symposium and Field Excursion: generation and emplacement of ophiolites through time.* Geol. Surv. Finland, 61pp.

Tullis, J. (2002) Deformation of granitic rocks; experimental studies and natural examples. *Rev. Mineral. Geochem.* **51**, 51–95.

Tulloch, A.J. & Kimbrough, D.L. (2003) Paired plutonic belts in convergent margins and the development of high Sr/Y magmatism: Peninsular Ranges batholith of Baja–California and Median batholith of New Zealand. *In* Johnson, S.E. *et al.* (eds) *Tectonic evolution of Northwestern Mexico and the Southwestern USA. Geol. Soc. of Am. Sp. Paper* **374**, 275–95.

Turcotte, D.L. & Oxburgh, E.R (1978) Intra-plate volcanism. *Phil. Trans. Roy. Soc. Lond. A* **288**, 561–79.

Turcotte, D.L. & Schubert, G. (2002) *Geodynamics,* 2nd edn Cambridge University Press. Cambridge.

Turcotte, D.L., McAdoo, D.C. & Caldwell, J.G. (1978) An elastic-perfectly plastic analysis of the bending of the lithosphere at a trench. *Tectonophysics* **47**, 193–205.

Turner, J.P. & Williams, G.A. (2004) Sedimentary basin inversion and intra-plate shortening. *Earth Sci. Rev.* **65**, 277–304.

Turner, S. & Hawkesworth, C. (1997) Constraints on flux rates and mantle dynamics beneath island arcs from Tonga–Kermadec lava geochemistry. *Nature* **389**, 568–73.

Turner, S., Evans, P. & Hawkesworth, C. (2001) Ultrafast source-to-surface movement of melt at island arcs from $^{226}Ra–^{230}Th$ systematics. *Science* **292**, 1363–6.

Twiss, R.J. & Moores, E.M. (1992) *Structural Geology*. W.H. Freeman, New York.

Ujiie, K., Hisamitsu, T. & Taira, A. (2003) Deformation and fluid pressure variation during initiation and evolution of the plate boundary décollement zone in the Nankai accretionary prism. *J. geophys. Res.* **108**, 2398, doi:10.1029/2002JB002314.

Umhoefer, P.J. (1987) Northward translation of Baja British Columbia along the Late Cretaceous to Paleocene margin of western North America. *Tectonics* **6**, 377–94.

Umhoefer, P.J. (2000) Where are the missing faults in translated terranes? *Tectonophysics* **326**, 23–35.

Underwood, M.B. *et al.* (2003) Sedimentary and tectonic evolution of a trench–slope basin in the Nankai subduction zone of southwest Japan. *J. Sediment. Res.* **73**, 589–602.

Unruh, J.R. & Sawyer, T.L. (1997) Assessment of blind seismogenic sources, Livermore Valley, eastern San Francisco Bay region. *USGS National Earthquake Hazards Reduction Program*, pp. 1–88. Final Technical Report No. 1434-95-G-2611, US Geol. Survey, Reston,Virginia.

Unsworth, M. & Bedrosian, P.A. (2004) Electrical resistivity structure at the SAFOD site from magnetotelluric exploration. *J. geophys. Res.* **31**, L12S05, doi:10.1029/2003GL019405.

Unsworth, M.J. *et al.* (2005) Crustal rheology of the Himalaya and Southern Tibet inferred from magnetotelluric data. *Nature.* **438**, 78–81.

Upcott, N. *et al.* (1996) Along-axis segmentation and isostasy in the Western Rift, East Africa. *J. geophys. Res.* **101**, 3247–68.

Uyeda, S. & Kanamori, H. (1979) Back-arc opening and the mode of subduction. *J. geophys. Res.* **84**, 1049–61.

Uyeda, S. & Miyashiro, A. (1974) Plate tectonics and the Japanese Island: a synthesis. *Bull. geol. Soc. Am.* **85**, 1159–70.

Vacquier, V. (1965) Transcurrent faulting in the ocean floor. *Phil. Trans. Roy. Soc. Lond. A* **258**, 77–81.

Valentine, J.W. & Moores, E.M. (1970) Plate tectonic regulation of faunal diversity and sea level: a model. *Nature* **228**, 657–9.

Valentine, J.W. & Moores, E.M. (1972) Global tectonics and the fossil record. *J. Geol.* **80**, 167–84.

Valley, J.W. *et al.* (2006) Comment on "Heterogeneous Hadean Hafnium: evidence of Continental Crust at 4.4 to 4.5 Ga." *Science* **312**, 1139a.

Van Avendonk, H.J.A. *et al.* (2004) Continental crust under compression: a seismic refraction study of South Island Geophysical Transect I, South Island, New Zealand. *J. geophys. Res.* **109**, B06302, doi:10.1029/2003JB002790.

Van Couvering, J.A. *et al.* (1976) The terminal Miocene event. *Marine Micropaleontology* **1**, 263–86.

van der Beek, P.A. (1997) Flank uplift and topography at the central Baikal Rift (SE Siberia): a test of kinematic models for continental extension. *Tectonics* **16**, 122–36.

van der Beek, P.A. & Cloetingh, S. (1992) Lithospheric flexure and the tectonic evolution of the Betic Cordilleras (SE Spain). *Tectonophysics* **203**, 325–44.

van der Beek, P.A. *et al.* (1998) Denudation history of the Malawi and Rukwa Rift flanks from apatite fission track thermochronology. *J. Afr. Earth Sci.* **26**, 363–86.

van der Hilst, R. (1995) Complex morphology of subducted lithosphere in the mantle beneath the Tonga Trench. *Nature* **374**, 154–7.

van der Hilst, R.D., Widiyantoro, S. & Engdahl, E.R. (1997) Evidence for deep mantle circulation from global tomography. *Nature* **386**, 578–84.

van der Velden, A.J., van Staal, C.R. & Cook, F.A. (2004) Crustal structure, fossil subduction, and the tectonic evolution of the Newfoundland Appalachians: evidence from a reprocessed seismic reflection survey. *Bull. geol. Soc. Am.* **116**, 1485–98.

van der Velden, A.J. *et al.* (2006) Reflections of the Neoarchean: a global perspective. *In* Benn, K., Mareschal, J.C. & Condie, K.C. (eds) *Archean Geodynamics and Evironments. Geophys. Monogr. Ser.* **164**, pp. 255–65. American Geophysical Union, Washington, DC.

van Keken, P.E., Hauri, E.H. & Ballentine, C.J. (2002) Mantle mixing: the generation, preservation, and destruction of chemical heterogeneity. *Annu. Rev. Earth Planet. Sci.* **30**, 493–525.

Van Kranendonk, M.J. & Collins, W.J. (1998) Timing and tectonic significance of Late Archean, sinistral strike-slip deformation in the Central Pilbara Structural Corridor, Pilbara Craton, Western Australia. *Precambrian Res.* **88**, 207–32.

Van Kranendonk, M.J. *et al.* (2002) Geology and tectonic evolution of the Archean North Pilbara Terrain, Pilbara Craton, Western Australia. *Econ. Geol.* **97**, 695–732.

Van Kranendonk, M.J. *et al.* (2004) Critical tests of vertical vs. horizontal tectonic models for the Archean East Pilbara Granite–Greenstone Terrane, Pilbara Craton, Western Australia. *Precambrian Res.* **131**, 173–211.

Van Kranendonk, M.J. *et al.* (2007) Secular tectonic evolution of Archean continental crust: interplay between horizontal and vertical processes in the formation of the Pilbara Craton, Australia. *Terra Nova* **19**, 1–38.

Van Orman, J. *et al.* (1995) Distribution of shortening between the Indian and Australian plates in the central Indian Ocean. *Earth planet. Sci. Lett.* **133**, 35–46.

van Thienen, P., Vlaar, N.J. & van den Berg, A.P. (2005) Assessment of the cooling capacity of plate tectonics and flood volcanism in the evolution of Earth, Mars and Venus. *Phys. Earth planet. Int.* **150**, 287–315.

van Wijk, J.W. & Cloetingh, S.A.P.L. (2002) Basin migration caused by slow lithospheric extension. *Earth planet. Sci. Lett.* **198**, 275–88.

Vandamme, D. & Courtillot, V. (1990) Paleomagnetism of Leg 115 basement rocks and latitudinal evolution of the Reunion hotspot. *Proc. Ocean Drill. Program Sci. Results* **115**, 111–17.

Varsek, J.L. *et al.* (1993) Lithoprobe crustal reflection survey of the southern Canadian Cordillera 2: Coast Mountains transect. *Tectonics* **12**, 334–60.

Vaughan, A.P.M., Leat, P.T. & Pankhurst, R.J. (2005) Terrane processes at the margins of Gondwana: introduction. *In* Vaughan, A.P.M., Leat, P.T. & Pankhurst, R.J. (eds) *Terrane Processes at the Margins of Gondwana. Spec. Pub. geol. Soc. Lond.* **246**, 1–21.

Vening Meinesz, F.A. (1951) A third arc in many island arc areas. *Koninkl. Nederlandsch. Akad. Wetensch. Proc. Ser. B.* **54**, 432–42.

Venkataraman, A., Nyblade, A.A. & Ritsema, J. (2004) Upper mantle Q and thermal structure beneath Tanzania, East Africa from teleseismic P wave spectra. *Geophys. Res. Lett.* **31**, L15611, doi:10.1029/2004GL020351.

Vera, E.E. *et al.* (1990) The structure of 0 to 0.2 m.y. old oceanic crust at 9°N on the East Pacific Rise from expanded spread profiles. *J. geophys. Res.* **95**, 15 529–56.

Vermeersen, L.L.A. & Vlaar, N.J. (1993) Changes in the Earth's rotation by tectonic movements. *Geophys. Res. Lett.* **20**, 81–4.

Viljoen, M.J. & Viljoen, R.P. (1969) The geology and geochemistry of the lower ultramafic unit of the Onverwacht Group and a proposed new class of igneous rocks. *Spec. Publ. Geol. Soc. S. Afr.* **2**, 55–85.

Vine, F.J. (1966) Spreading of the ocean floor: new evidence. *Science* **154**, 1405–15.

Vine, F.J. (1977) The continental drift debate. *Nature* **266**, 19–22.

Vine, F.J. & Matthews, D.H. (1963) Magnetic anomalies over oceanic ridges. *Nature* **199**, 947–9.

Vogt, P.R. Ostenso, N.A. & Johnson, G.L. (1970) Magnetic and bathymetric data bearing on sea floor spreading north of Iceland. *J. geophys. Res.* **75**, 903–20.

von Huene, R. & Scholl, D.W. (1991) Observations at convergent margins concerning sediment subduction, subduction erosion, and the growth of continental crust. *Rev. Geophys.* **29**, 279–316.

von Huene, R. *et al.* (1998) Mass and fluid flux during accretion at the Alaskan margin. *Bull. geol. Soc. Am.* **110**, 468–82.

von Huene, R., Ranero, C.R. & Vannucchi, P. (2004) Generic model of subduction erosion. *Geology* **32**, 913–16.

von Raumer, J.F., Stampfli, G.D. & Bussy, F. (2003) Gondwana-derived microcontinents – the constituents of the Variscan and Alpine collisional orogens. *Tectonophysics* **365**, 7–22.

Wagner, D.L., Bortugno, E.J. & McJunkin, R.D. (compilers) (1990) Geologic map of the San Francisco–San Jose quadrangle. *California Division of Mines and Geology Regional Map Series*, Map No. 5A (Geology).

Wakabayashi J. & Dilek, Y. (2000) Spatial and temporal relationships between ophiolites and their metamorphic soles: a test of models of forearc ophiolite genesis. *In* Dilek, Y. *et al.* (eds) *Ophiolites and Oceanic Crust: new insights from field studies and the ocean drilling program. Geol. Soc. Am. Sp. Paper* **349**, 53–64.

Wakabayashi, J., Hengesh, J.V. & Sawyer, T.L. (2004) Four-dimensional transform fault processes: progressive evolution of step-overs and bends. *Tectonophysics* **392**, 279–301.

Walcott, R.I. (1970) Flexural rigidity, thickness, and viscosity of the lithosphere. *J. geophys. Res.* **75**, 3941–53.

Walcott, R.I. (1998) Modes of oblique compression: late Cenozoic tectonics of the South Island of New Zealand. *Rev. Geophys.* **36**, 1–26.

Waldhauser, F. & Ellsworth, W.L. (2002) Fault structure and mechanics of the Hayward Fault, California, from double-difference earthquake locations. *J. geophys. Res.* **107**, 1–15.

Walker, R.J. *et al.* (1989) Os, Sr, Nd, and Pb isotope systematics of southern African peridotite xenoliths: implications for the chemical evolution of subcontinental mantle, *Geochim. cosmochim. Acta* **53**, 1583–95.

Wallace, L.M. *et al.* (2004) GPS and seismological constraints on active tectonics and arc–continent collision in Papua New Guinea: implications for mechanics of microplate rotations in a plate boundary zone. *J. geophys. Res.* **109**, B05404, doi:10.1029/2003JB002481.

Wang, E. *et al.* (1998) Late Cenozoic Xianshuihe–Xiaojiang, Red River, and Dali fault systems of southwestern Sichuan and central Yunnan, China. *Geol. Soc. Am. Sp. Paper* **327**, 108pp.

Wang, Q. *et al.* (2001) Present-day crustal deformation in China constrained by global positioning system measurements. *Science* **294**, 574–7.

Wang, R. *et al.* (2004) The 2003 Bam (SE Iran) earthquake; precise source parameters from satellite radar interferometry. *Geophys. J. Int.* **159**, 917–22.

Wannamaker, P.E. *et al.* (2002) Fluid generation and pathways beneath an active compressional orogen, the New Zealand Southern Alps, inferred from magnetotelluric data. *J. geophys. Res.* **107**, 2117, doi:10.1029/2001JB000186.

Ward, M.A. (1963) On detecting changes in the Earth's radius. *Geophys. J. Roy. astr. Soc.* **8**, 217–25.

Ward, P.D. *et al.* (1997) Measurements of the Cretaceous paleolatitude of Vancouver Island: consistent with the Baja–British Columbia hypothesis. *Science* **277**, 1642–5.

Wareham, C.D., Millar, I.L. & Vaughan, A.P.M. (1997) The generation of sodic granite magmas, western Palmer Land, Antarctic Peninsula. *Contrib. Mineral. Petrol.* **128**, 81–96.

Watt J.P. & Shankland, T.J. (1975) Uniformity of mantle composition. *Geology* **3**, 91–4.

Watts, A.B. (2001) *Isostasy and Flexure of the Lithosphere.* Cambridge University Press, Cambridge.

Watts, A.B. & Burov, E. (2003) Lithospheric strength and its relationship to the elastic and seismogenic layer thickness. *Earth planet. Sci. Lett.* **213**, 113–31.

Watts, A.B. & Ryan, W.B.F. (1976) Flexure of the lithosphere and continental margin basins. *Tectonophysics* **36**, 25–44.

Watts, A.B., Cochran, J.R. & Selzer, G. (1975) Gravity anomalies and flexure of the lithosphere: a three dimensional study of the Great Meteor seamount, northeast Atlantic. *J. geophys. Res.* **80**, 1391–8.

Watts, A.B., Bodine, J.H. & Steckler, M.S. (1980) Observations of flexure and the state of stress in the oceanic lithosphere. *J. geophys. Res.* **85**, 6369–76.

Watts, A.B. *et al.* (1995) Lithospheric flexure and bending of the Central Andes. *Earth planet. Sci. Lett.* **134**, 9–24.

Wdowinski, S. (1992) Dynamically supported trench topography. *J. Geophys. Res.* **97**, 17,651–6.

Wdowinski, S. & Axen, G.J. (1992) Isostatic rebound due to tectonic denudation: a viscous flow model of a layered lithosphere. *Tectonics* **11**, 303–15.

Weeraratne, D.S. *et al.* (2003) Evidence for an upper mantle plume beneath the Tanzanian craton from Rayleigh wave tomography. *J. geophys Res.* **108**, 2427, doi:10.1029/2002JB002273.

Weertman, J. (1978) Creep laws for the mantle of the Earth. *Phil. Trans. Roy. Soc. Lond. A* **288**, 9–26.

Wegener, A. (1929) *The Origin of Continents and Oceans.*[English translation (1966), of the 4th German edition, by J. Biram.]Dover Publishing, New York, by arrangement with Vieweg, Braunschwieg.

Weissel, J.K. (1981) Magnetic lineations in marginal basins of the western Pacific. *Phil. Trans. Roy. Soc. Lond. A* **300**, 223–47.

Weissel, J.K. & Karner, G. (1989) Flexural uplift of rift flanks due to tectonic denudation of the lithosphere during extension. *J. geophys. Res.* **94**, 13,919–50.

Weissel, J.K. & Watts, A.B. (1979) Tectonic evolution of the Coral Sea basin. *J. geophys. Res.* **84**, 4572–82.

Wellman, H.W. (1953) Data for the study of Recent and late Pleistocene faulting in the South Island of New Zealand. *N.Z. J. Sci. Technol.* **B34**, 270–88.

Wellman, P. (2000) Upper crust of the Pilbara Craton, Australia; 3D geometry of a granite-greenstone terrain. *Precambrian Res.* **104**, 175–86.

Wernicke, B. (1981) Low angle normal faults in the Basin and Range Province – nappe tectonics in an extending orogen. *Nature* **291**, 645–8.

Wernicke, B. (1985) Uniform-sense simple shear of the continental lithosphere. *Can. J. Earth. Sci.* **22**, 108–25.

Wernicke, B. (1992) Cenozoic extensional tectonics of the United States Cordillera. *In* Burchfiel, B.C., Lipman, P.W. & Zoback, M.L. (eds) *The Cordilleran Orogen: conterminous US. Geology of North America* **G3**, pp. 552–81. Geological Society of America, Boulder, CO.

Wernicke, B. & Axen, G.J. (1988) On the role of isostasy in the evolution of normal fault systems. *Geology* **16**, 848–51.

Wernicke, B. & Snow, J.K. (1998) Cenozoic tectonism in the central Basin and Range: motion of the Sierran–Great Valley Block. *Int. geol. Rev.* **40**, 403–10.

Wesnousky, S.G. *et al.* (1999) Uplift and convergence along the Himalayan frontal thrust of India. *Tectonics*, **18**, 967–76.

Westbrook, G.K., Mascle, A. & Biju-Duval, B. (1984) Geophysics and structure of the Lesser Antilles forearc. *In* Biju–Duval, B. & Moore, J.C. (eds) *Init. Repts. DSDP 78A*, pp. 23–38. US Government Printing Office, Washington, DC.

White, G.W. (1980) Permian–Triassic continental reconstruction of the Gulf of Mexico–Caribbean area. *Nature* **283**, 823–6.

White, R.S. *et al.* (1984) Anomalous seismic crustal structure of oceanic fracture zones. *Geophysics* **79**, 779–98.

Whitman, D., Isacks, B.L. & Kay, S.M. (1996) Lithospheric structure and along-strike segmentation of the central Andean plateau: topography, tectonics and timing. *Tectonophysics* **259**, 29–40.

Whitmarsh, R.B. (1975) Axial intrusion zone beneath the median valley of the Mid-Atlantic ridge at 37°N detected by explosion seismology. *Geophys. J. Roy. astr. Soc.* **42**, 189–215.

Whitmarsh, R.B., Manatschal, G. & Minshull, T.A. (2001) Evolution of magma – poor continental margins from rifting to seafloor spreading. *Nature* **413**, 150–4.

Wiebe, R.A. (1992) Proterozoic anorthosite complexes. *In* Condie, K.C. (ed) *Proterozoic Crustal Evolution*, pp. 215–61. Elsevier, Amsterdam.

Wiens D.A. & Smith, G.P. (2003) Seismological constraints on structure and flow patterns within the mantle wedge. *In* Eiler, J. (ed.) *Inside the Subduction Factory. Geophys. Monogr. Ser.* **138**, pp. 59–81. American Geophysical Union, Washington, DC.

Wiens, D.A., McGuire, J.J. & Shore, P.J. (1993) Evidence for transformational faulting from a deep double seismic zone in Tonga. *Nature* **364**, 790–3.

Wijns, C. *et al.* (2005) Mode of crustal extension determined by rheological layering. *Earth planet. Sci. Lett.* **236**, 120–34.

Wilde, S.A. *et al.* (2001) Evidence from detrital zircons for the existence of continental crust and oceans on the Earth 4.4 Gyr ago. *Nature* **409**, 175–8.

Willems, H. *et al.* (1996) Stratigraphy of the Upper Cretaceous and Lower Tertiary strata in the Tethyan Himalayas of Tibet (Tingri area, China). *Geol. Rundsch.* **85**, 723–54.

Willett, S.D. (1992) Dynamic and kinematic growth and change of a Coulomb wedge. *In* McClay, K. (ed.) *Thrust Tectonics*, pp. 19–31. Chapman and Hall, New York.

Willett, S.D. (1999) Orogeny and orography: the effects of erosion on the structure of mountain belts. *J. geophys. Res.* **104**, 28,957–82.

Willett, S.D. & Beaumont, C. (1994) Subduction of Asian lithospheric mantle beneath Tibet inferred from models of continental collision. *Nature* **369**, 642–45.

Williams, C.F., Grubb, F.V. & Galanis Jr, S.P. (2004) Heat flow in the SAFOD pilot hole and implications for the strength of the San Andreas Fault. *J. geophys. Res.* **31**, L15S14, doi:10.1029/2003GL019352.

Williams, D.L. *et al.* (1974) The Galapagos spreading center: lithospheric cooling and hydrothermal circulation. *Geophys. J. Roy. astr. Soc.* **38**, 587–608.

Williams, H. *et al.* (1991) Anatomy of North America: thematic geologic portrayals of the continent. *Tectonophysics* **187**, 117–34.

Williams, Q. & Garnero, E.J. (1996) Seismic evidence for partial melt at the base of Earth's mantle. *Science* **273**, 1528–30.

Williams, T.B., Kelsey, H.M. & Freymueller, J.T. (2006) GPS-derived strain in northwestern California: termination of the San Andreas fault system and convergence of the Sierra Nevada–Great Valley block contribute to southern Cascadia forearc contraction. *Tectonophysics* **413**, 171–84.

Wilson, D.S. (1993) Confirmation of the astronomical calibration of the magnetic polarity timescale from seafloor spreading rates. *Nature* **364**, 788–90.

Wilson, D.S., Hey, R.N. & Nishimura, C. (1984) Propagation as a mechanism of reorientation of the Juan de Fuca Ridge. *J. geophys. Res.* **89**, 9215–25.

Wilson, J.T. (1963) Evidence from islands on the spreading of ocean floors. *Nature* **197**, 536–8.

Wilson, J.T. (1965) A new class of faults and their bearing on continental drift. *Nature* **207**, 343–7.

Windley, B.F. (1981) Precambrian rocks in the light of the plate-tectonic concept. *In* Kröner, A. (ed.) *Precambrian Plate Tectonics*, pp. 1–20. Elsevier, Amsterdam.

Windley, B.F. (1984) *The Evolving Continents*, 2nd edn Wiley, London.

Wingate, M.T.D., Pisarevsky, S.A. & Evans, D.A.D. (2002) Rodinia connections between Australia and Laurentia: no SWEAT, no AUSWUS? *Terra Nova* **14**, 121–8.

Winter, J.D. (2001) *An Introduction to Igneous and Metamorphic Petrology*. Prentice-Hall, New Jersey.

Wittlinger. G. *et al.* (2004) Teleseismic imaging of subducting lithosphere and Moho offsets beneath western Tibet. *Earth planet. Sci. Lett.* **221**, 117–30.

Wolfenden, E. *et al.* (2004) Evolution of the northern Main Ethiopian rift: birth of a triple junction. *Earth planet. Sci. Lett.* **224**, 213–28.

Wolfenden, E. *et al.* (2005) Evolution of a volcanic rifted margin: southern Red Sea, Ethiopia. *Bull. geol. Soc. Am.* **117**, 846–64.

Woodcock, N.H. & Fischer, M. (1986) Strike-slip duplexes. *J. struct. Geol.* **8**, 725–35.

Woodcock, N.H. & Rickards, B. (2003) Transpressive duplex and flower structure: Dent Fault System, NW England. *J. struct. Geol.* **25**, 1981–92.

Woodhouse, J.H. & Dziewonski, A.M. (1984) Mapping the upper mantle: three dimensional modelling of Earth structure by inversion of seismic waveforms. *J. geophys. Res.* **89**, 5953–86.

Wyllie, P.J. (1981) Plate tectonics and magma genesis. *Geol. Rundsch.* **70**, 128–53.

Wyllie, P.J. (1988) Magma genesis, plate tectonics and chemical differentiation of the Earth. *Rev. Geophys.* **26**, 370–404.

Wyman D.A. & Kerrich, R. (2002) Formation of Archean continental lithospheric roots: the role of mantle plumes. *Geology* **30**, 543–6.

Xie, J. *et al.* (2004) Lateral variations of crustal seismic attenuation along the INDEPTH profiles in Tibet from $Lg\ Q$ inversion. *J. geophys. Res.* **109**, B10308, doi:10.1029/2004JB002988.

Yáñez, G. & Cembrano, J. (2004) Role of viscous plate coupling in the late Tertiary Andean tectonics. *J. geophys. Res.* **109**, B02407, doi:10.1029/2003JB002494.

Yamazaki D. & Karato, S.-I. (2001) Some mineral physics constraints on the rheology and geothermal structure of Earth's lower mantle. *American Mineralogist* **86**, 385–91.

Yin, A. & Harrison, T.M. (2000) Geologic evolution of the Himalayan Tibetan Orogen. *Annu. Rev. Earth planet. Sci.* **28**, 211–80.

Yogodzinski, G.M., Lees, J.M. Churikova, T.G. *et al.* (2001) Geochemical evidence for the melting of subducting oceanic lithosphere at plate edges. *Nature* **409**, 500–4.

Young, G.M. (1992) Late Proterozoic stratigraphy and the Canada–Australia connection. *Geology* **20**, 215–18.

Yuan, X. *et al.* (2000) Subduction and collision processes in the central Andes constrained by converted seismic phases. *Nature* **408**, 958–61.

Yuan, X., Sobolev, S.V. & Kind, R. (2002) Moho topography in the central Andes and its geodynamic implications. *Earth planet. Sci. Lett.* **199**, 389–402.

Zachos, J. *et al.* (2001) Trends, rhythms and aberrations in global climate 65 Ma to present. *Science* **292**, 686–93.

Zandt, G., Myers, S.C. & Wallace, T.C. (1995) Crust and mantle structure across the Basin and Range–Colorado Plateau boundary at 37°N latitude and implications for Cenozoic extensional mechanism. *J. geophys. Res.* **100**, 10 529–48.

Zatman, S. (2000) On steady rate coupling between an elastic upper crust and a viscous interior. *Geophys. Res. Lett.* **27**, 2421–4.

Zegers, T.E. & van Keken, P.E. (2001) Middle Archean continent formation by crustal delamination. *Geology* **29**, 1083–6.

Zhao, D. *et al.* (1997) Depth extent of the Lau Back-arc spreading center and its relation to subduction processes. *Science* **278**, 254–7.

Zhao, G. *et al.* (2001) High-pressure granulites (retrograded eclogites) from the Hengshan Complex, North China Craton; petrology and tectonic implications. *J. Petrol.* **42**, 1141–70.

Zhao, G. *et al.* (2002) Review of global 2.1–1.8 Ga orogens: implications for a pre-Rodinia supercontinent. *Earth Sci. Rev.* **59**, 125–62.

Zhao, W., Nelson, K.D. & Project INDEPTH team (1993) Deep seismic reflection evidence for continental underthrusting beneath southern Tibet. *Nature* **366**, 557–9.

Zho, W. *et al.* (2001) Crustal structure of central Tibet as derived from the project INDEPTH wide-angle seismic data. *Geophys. J. Int.* **145**, 486–98.

Zhong, S. & Gurnis, M. (1993) Dynamic feedback between a continent-like raft and thermal convection. *J. geophys. Res.* **98**, 12 219–32.

Zhu, L. (2000) Crustal structure across the San Andreas Fault, southern California from teleseismic converted waves. *Earth planet. Sci. Lett.* **179**, 183–90.

Zhu, L. *et al.* (2006) Crustal thickness variations in the Aegean region and implications for the extension of continental crust. *J. geophys. Res.* **111**, B01301, doi:10.1029/2005JB003770.

Ziegler, P.A. (1993) Plate-moving mechanisms: their relative importance. *J. geol. Soc. Lond.* **150**, 927–40.

Zoback, M.D. (2000) Strength of the San Andreas. *Nature* **405**, 31–32.

Zoback, M.D. *et al.* (1987) New evidence on the state of stress of the San Andreas fault system. *Science* **238**, 1105–11.

Zoback, M.L. (1992) First- and second-order pattern of stress in the lithosphere: the World Stress Map Project. *J. geophys. Res.* **97**, 11 703–28.

Index

Page numbers in *italics* refer to figures; those in **bold** to tables